J. N. REDDY
Department of Mechanical Engineering
Texas A&M University
College Station, Texas, 77843-3123, USA

An Introduction to Nonlinear Finite Element Analysis

with applications to heat transfer, fluid mechanics, and solid mechanics

Second Edition

OXFORD
UNIVERSITY PRESS

OXFORD
UNIVERSITY PRESS

Great Clarendon Street, Oxford, OX2 6DP,
United Kingdom

Oxford University Press is a department of the University of Oxford.
It furthers the University's objective of excellence in research, scholarship,
and education by publishing worldwide. Oxford is a registered trade mark of
Oxford University Press in the UK and in certain other countries

© Oxford University Press 2015

The moral rights of the author have been asserted

First Edition published in 2004
Second Edition published in 2015

Impression: 1

Published in the United States of America by Oxford University Press
198 Madison Avenue, New York, NY 10016, United States of America

British Library Cataloguing in Publication Data

Data available

Library of Congress Control Number: 2014935440

ISBN 978–0–19–964175–8

Printed in Great Britain by
Clays Ltd, St Ives plc

To my beloved teacher and mentor
Professor John Tinsley Oden

Preface to the Second Edition

The development of realistic mathematical models that govern the response of systems or processes is intimately connected to the ability to translate them into meaningful discrete models that enable us to systematically evaluate various parameters of the systems and processes. Mathematical model development and numerical simulations are aids to designers, who are seeking to maximize the reliability of products and minimize the cost of production, distribution, and repairs. Mathematical models are developed using laws of physics and assumptions concerning a system's behavior. The most important step in arriving at a design that is both reliably functional and cost-effective is the construction of a suitable mathematical model of the system behavior and its translation into a powerful numerical simulation tool. While a select number of courses on continuum mechanics, material science, and dynamical systems, among others, provides engineers with the background to formulate a suitable mathematical model, courses on numerical methods prepare engineers and scientists to evaluate mathematical models and test material models in the context of the functionality and design constraints placed on the system. In cases where physical experiments are prohibitively expensive, numerical simulations are the only alternative, especially when the phenomena is governed by nonlinear differential equations, in evaluating various design options. It is in this context a course on nonlinear finite element analysis proves to be very useful.

Most books on nonlinear finite element analysis tend to be abstract in the presentation of details of the finite element formulations, derivation of element equations, and their solution by iterative methods. Such books serve as reference books but not as textbooks. The present textbook is unique (i.e. there is no parallel to this book in its class) since it actually helps the readers with details of finite element model development and implementation. In particular, it provides illustrative examples and problem sets that enable readers to test their understanding of the subject matter and utilize the tools developed in the formulation and finite element analysis of engineering problems.

The second edition of *An Introduction to Nonlinear Finite Element Analysis* has the same objective as the first edition, namely, to facilitate an easy and thorough understanding of the details that are involved in the theoretical formulation, finite element model development, and solutions of nonlinear problems. *The book offers easy-to-understand treatment of the subject of nonlinear finite element analysis, which includes element development from mathematical models and numerical evaluation of the underlying physics.* The new edition is extensively reorganized and contains substantially large amount of new material. In particular, Chapter 1 in the second edition contains a section on applied functional analysis; Chapter 2 on nonlinear continuum mechanics is entirely new; Chapters 3 through 8 in the new edition correspond to Chapter 2 through 8 of the first edition but with additional explanations, examples, and exercise problems (material on time dependent problems from Chapter 8 of the first edition is absorbed into Chapters 6 through 10 of the new edition); Chapter 9 is extensively revised and it contains up to date developments in the large deformation analysis of isotropic, composite, and functionally graded shells; Chapter 10 of the first edition on material nonlinearity and coupled problems is reorganized in the second edition by moving the material on solid mechanics to Chapter 12 in the new edition, and material on coupled problems to Chapter 10 on weak-form Galerkin finite element models of viscous incompressible fluids; finally, Chapter 11 in the second edition is entirely new and devoted to least-squares finite element models of viscous incompressible fluids. Chapter 12 of the second edition (available only online) contains material on one-dimensional formulations of nonlinear elasticity, plasticity, and viscoelasticity. In general, all of the chapters of the second edition contain additional explanations, detailed example problems, and additional exercise problems. Although all of the programming segments are in Fortran, the logic used in these Fortran programs is transparent and can be used in Matlab or C^{++} versions of the same. Thus the new edition more than replaces the first edition, and it is hoped that it is acquired by the library of every institution of higher learning as well as serious finite element analysts.

The book may be used as a textbook for an advanced course (after a first course) on the finite element method or the first course on nonlinear finite element analysis. A solutions manual has also been prepared for the book. The solution manual is available from the publisher only to instructors who adopt the book as a textbook for a course.

Since the publication of the first edition, many users of the book communicated their comments and compliments as well as errors they found, for which the author thanks them. All of the errors known to the author have been corrected in the current edition. The author is grateful to the following professional colleagues for their friendship, encouragement, and constructive comments on the book:

Hasan Akay, Purdue University at Indianapolis
Marcilio Alves, University of São Paulo, Brazil
Marco Amabili, McGill University, Canada
Ted Belytschko, Northwestern University
K. Chandrashekara, Missouri University of Science and Technology
A. Ecer, Purdue University at Indianapolis
Antonio Ferreira, University of Porto, Portugal
Antonio Grimaldi, University of Rome II, Italy
R. Krishna Kumar, Indian Institute of Technology, Madras
H. S. Kushwaha, Bhabha Atomic Research Centre, India
A. V. Krishna Murty, Indian Institute of Science, Bangalore
K. Y. Lam, Nanyang Technological University, Singapore

K. M. Liew, City University of Hong Kong
C. W. Lim, City University of Hong Kong
Franco Maceri, University of Rome II, Italy
C. S. Manohar, Indian Institute of Science, Bangalore
Antonio Miravete, Zaragoza University, Spain
Alan Needleman, Brown University
J. T. Oden, University of Texas at Austin
P. C. Pandey, Indian Institute of Science, Bangalore
Glaucio Paulino, University of Illinois at Urbana-Champaign
A. Rajagopal, Indian Institute of Technology, Hyderabad
Ekkehard Ramm, University of Stuttgart, Germany
Jani Romanoff, Aalto University, Finland
Samit Roy, University of Alabama, Tuscaloosa
Siva Prasad, Indian Institute of Technology, Madras
Elio Sacco, University of Cassino, Italy
Rüdger Schmidt, University of Aachen, Germany
E. C. N. Silva, University of São Paulo, Brazil
Fanis Strouboulis, Texas A&M University
Karan Surana, University of Kansas
Liqun Tang, South China University of Technology
Vinu Unnikrishnan, University of Alabama, Tuscaloosa
C. M. Wang, National University of Singapore
John Whitcomb, Texas A&M University
Y. B. Yang, National Taiwan University

Drafts of the manuscript of this book prior to its publication were read by the author's doctoral students, who have made suggestions for improvements. In particular, the author wishes to thank the following former and current students (listed in alphabetical order): Roman Arciniega, Ronald Averill, Ever Barbero, K. Chandrashekhara, Feifei Cheng, Stephen Engelstad, Eugénio Garcão, Miguel Gutierrez Rivera, Paul Heyliger, Filis Kokkinos, C. F. Liao, Goy Teck Lim, Ravisankar Mayavaram, John Mitchell, Filipa Moleiro, Felix Palmerio, Gregory Payette, Jan Pontaza, Vivek Prabhakar, Grama Praveen, N. S. Putcha, Rakesh Ranjan, Mahender Reddy, Govind Rengarajan, Donald Robbins, Jr., Samit Roy, Vinu Unnikrishnan, Ginu Unnikrihnan, Yetzirah Urthaler, Venkat Vallala; Archana Arbind, Parisa Khodabakhshi, Jinseok Kim, Wooram Kim, Helnaz Soltani, and Mohammad Torki. The author also expresses his sincere thanks to Mr. Sonke Adlung (Senior Editor, Engineering) and Ms. Victoria Mortimer (Senior Production editor) at Oxford University Press for their encouragement and help in producing this book. The author requests readers to send their comments and corrections to *jnreddy@exchange.tamu.edu.*

J. N. Reddy
College Station, Texas

Preface to the First Edition

The objective of this book is to present the theory and computer implementation of the finite element method as applied to simple nonlinear problems of heat transfer and similar field problems, fluid mechanics, and solid mechanics. Both geometric as well as material nonlinearities are considered, and static and transient (i.e. time-dependent) responses are studied. The guiding principle in writing the book was to make the presentation suitable for (a) adoption as a text book for a first course on nonlinear finite element analysis (or for a second course following an introductory course on the finite element method) and (b) for use by engineers and scientists from various disciplines for self-study and practice.

There exist a number of books on nonlinear finite elements. Most of these books contain a good coverage of the topics of structural mechanics, and few address topics of fluid dynamics and heat transfer. While these books serve as good references to engineers or scientists who are already familiar with the subject but wish to learn advanced topics or latest developments, they are not suitable as textbooks for a first course or for self study on nonlinear finite element analysis.

The motivation and encouragement that led to the writing of the present book have come from the users of the author's book, *An Introduction to the Finite Element Method* (McGraw-Hill, 1984; Second Edition, 1993; third edition scheduled for 2004), who have found the approach presented there to be most suitable for any one – irrespective of their scientific background – interested in learning the method, and also from the fact that there does not exist a book that is suitable as a textbook for a first course on nonlinear finite element analysis. The author has taught a course on nonlinear finite element analysis many times during the last twenty years, and the present book is an outcome of the lecture notes developed during this period. The same approach as that used in the aforementioned book, namely, the *differential equation approach*, is adopted in the present book to introduce the theory, formulation, and computer implementation of the finite element method as applied to nonlinear problems of science and engineering.

Beginning with a model (i.e. typical) second-order, nonlinear differential equation in one dimension, the book takes the reader through increasingly complex problems of nonlinear beam bending, nonlinear field problems in two dimensions, nonlinear plate bending, nonlinear formulations of solid continua, flows of viscous incompressible fluids in two dimensions (i.e. Navier–Stokes equations), time-approximation schemes, continuum formulations of shells, and material nonlinear problems of solid mechanics.

As stated earlier, the book is suitable as a textbook for a first course on nonlinear finite elements in civil, aerospace, mechanical, and mechanics departments as well as in applied sciences. It can be used as a reference by engineers and scientists working in industry, government laboratories, and academia. Introductory courses on the finite element method, continuum mechanics, and numerical analysis should prove to be helpful.

The author has benefited in writing the book by the encouragement and support of many colleagues around the world who have used his book, *An Introduction to the Finite Element Method*, and students who have challenged him to explain and implement complicated concepts and formulations in simple ways. While it is not possible to name all of them, the author expresses his sincere appreciation. The author expresses his deep sense of gratitude to his teacher and mentor, Professor J. T. Oden (University of Texas at Austin), without whose advice and support it would not have been possible for the author to modestly contribute to the field of applied mechanics in general and theory and application of the finite element method in particular, through his teaching, research, and writings.

J. N. Reddy
College Station, Texas

Contents

About the Author

J. N. Reddy is a University Distinguished Professor, Regents Professor, and the Holder of Oscar S. Wyatt Endowed Chair in the Department of Mechanical Engineering at Texas A&M University. Prior to the current position, he was the Clifton C. Garvin Professor in the Department of Engineering Science and Mechanics at Virginia Tech, and Associate Professor in the School of Mechanical, Aerospace and Nuclear Engineering at the University of Oklahoma.

Dr. Reddy is internationally known for his contributions to theoretical and applied mechanics and computational mechanics. He is the author of more than 500 journal papers and 18 textbooks with multiple editions. Professor Reddy is the recipient of numerous awards including the *Worcester Reed Warner Medal*, the *Charles Russ Richards Memorial Award*, and the *Honorary Member* award from the American Society of Mechanical Engineers, the *Nathan M. Newmark Medal* and the *Raymond D. Mindlin Medal* from the American Society of Civil Engineers, the *Distinguished Research Award* and the *Excellence in the Field of Composites Award* from the American Society of Composites, the *Computational Solid Mechanics Award* from the U.S. Association of Computational Mechanics, the *Computational Mechanics Award* from the Japanese Society of Mechanical Engineers, the *IACM Award* from the International Association of Computational Mechanics, and the *Archie Higdon Distinguished Educator Award* from the American Society of Engineering Education. Dr. Reddy received honorary degrees (Honoris Causa) from the Technical University of Lisbon (Portugal), and from Odlar Yurdu University (Azerbaijan). He is a Fellow of the American Society of Mechanical Engineers, the American Institute of Aeronautics and Astronautics, the American Society of Civil Engineers , the American Academy of Mechanics, the American Society of Composites, the U.S. Association of Computational Mechanics, the International Association of Computational Mechanics, and the Aeronautical Society of India.

Professor Reddy is the Editor-in-Chief of *Mechanics of Advanced Materials and Structures* and *International Journal of Computational Methods in Engineering Science and Mechanics*, and co-Editor of *International Journal of Structural Stability and Dynamics*; he also serves on the editorial boards of more than two dozen other journals, including *International Journal for Numerical Methods in Engineering, Computer Methods in Applied Mechanics and Engineering*, and *International Journal of Non-Linear Mechanics*.

Dr. Reddy is a selective researcher in engineering around the world who is recognized by *ISI Highly Cited Researchers* with over 15,000 citations (without self-citations over 14,000) with h-index of over 58 as per Web of Science; as per Google Scholar, the current number of citations exceed 39,000, and the h-index is 77. A more complete resume with links to journal papers can be found at **http://www.tamu.edu/acml/**.

List of Symbols

The symbols that are used throughout the book for various quantities are defined in the following list but the list is not exhaustive. In some cases, the same symbol has different meaning in different parts of the book, as it would be clear in the context.

Arabic alphabetical symbols

$\mathbf{a}$	Acceleration vector, $\frac{D\mathbf{v}}{Dt}$
$B(\cdot,\cdot)$	Bilinear form
$\mathbf{B}$	Left Cauchy–Green deformation tensor (or Finger tensor), $\mathbf{B} = \mathbf{F} \cdot \mathbf{F}^{\mathrm{T}}$; magnetic flux density vector
$\tilde{\mathbf{B}}$	Cauchy strain tensor, $\tilde{\mathbf{B}} = \mathbf{F}^{-\mathrm{T}} \cdot \mathbf{F}^{-1}$; $\tilde{\mathbf{B}}^{-1} = \mathbf{B}$
c	Specific heat, moisture concentration
c_v, c_p	Specific heat at constant volume and pressure
$\mathbf{c}$	Couple vector
$\mathbf{C}$	Right Cauchy–Green deformation tensor, $\mathbf{C} = \mathbf{F}^{\mathrm{T}} \cdot \mathbf{F}$; fourth-order elasticity tensor [see Eq. (2.4.3)] with coefficients C_{ij} or C_{ijkl}
$\mathbf{d}$	Symmetric part of the velocity gradient tensor, $\mathbf{l} = (\boldsymbol{\nabla}\mathbf{v})^{\mathrm{T}}$, that is $\mathbf{d} = \frac{1}{2}\left[(\boldsymbol{\nabla}\mathbf{v})^{\mathrm{T}} + \boldsymbol{\nabla}\mathbf{v}\right]$; electric flux vector; mass diffusivity tensor
$\mathcal{D}$	Internal dissipation
$d\mathbf{a}$	Area element (vector) in spatial description
$d\mathbf{A}$	Area element (vector) in material description
$d\mathbf{x}$	Line element (vector) in current configuration
$d\mathbf{X}$	Line element (vector) in reference configuration
$D/Dt, d/dt$	Material time derivative
ds	Surface element in current configuration
dS	Surface element in reference configuration
e_c	Internal energy per unit mass
$\mathbf{e}$	Almansi strain tensor, $\mathbf{e} = \frac{1}{2}\left(\mathbf{I} - \mathbf{F}^{-\mathrm{T}} \cdot \mathbf{F}^{-1}\right)$
$\mathbf{e}_i$	A basis vector in the x_i-direction
$\mathbf{e}_{ijk}$	Components of alternating or permutation tensor, $\mathcal{E}$
$\hat{\mathbf{e}}$	A unit vector
$\hat{\mathbf{e}}_A$	A unit basis vector in the direction of vector $\mathbf{A}$
E, E_1, E_2	Young's modulus (modulus of elasticity)
$\mathbf{E}$	Green–Lagrange strain tensor, $\mathbf{E} = \frac{1}{2}\left(\mathbf{F}^{\mathrm{T}} \cdot \mathbf{F} - \mathbf{I}\right)$ with components E_{ij}
$\hat{\mathbf{E}}_i$	Unit base vector along the X_i material coordinate direction
$\mathbf{f}$	Body force vector
f_x, f_y, f_z	Body force components in the x, y, and z directions
f	Load per unit length of a bar

$\mathbf{F}$	Deformation gradient, $\mathbf{F} = (\boldsymbol{\nabla}_0 \mathbf{x})^{\mathrm{T}}$; force vector with coefficient F_i
$\mathcal{F}$	Functional mapping
G	Shear modulus (modulus of rigidity)
$\mathbf{G}$	Geometric stiffness matrix with coefficients G_{ij} (beams and plates)
g	Acceleration due to gravity; function; internal heat generation per unit volume
$\mathbf{g}$	Gradient of temperature, $\mathbf{g} = \boldsymbol{\nabla}\theta$
h	Height of the beam; thickness; heat transfer coefficient; element length in 1-D
H	Total entropy; Unit step function
H_{net}	Net rate of heat transferred into the system
$\mathbf{H}$	Nonlinear deformation tensor; magnetic field intensity vector
I	Second moment of area of a beam cross section; functional
I_1, I_2, I_3	Principal invariants of stress tensor
$\mathbf{I}$	Unit second-order tensor
J	Determinant of the matrix of deformation gradient (Jacobian); polar second moment of area of a shaft cross section
J_i	Principal invariants of strain tensor $\mathbf{E}$ or rate of deformation tensor $\mathbf{D}$
$\mathbf{J}$	Current density vector; creep compliance; Jacobian matrix
k	Spring constant; thermal conductivity
$\mathbf{k}$	Thermal conductivity tensor
K	Kinetic energy
K_s	Shear correction factor in the Timoshenko beam theory
$\mathbf{K}$	Finite element coefficient matrix with coefficients K_{ij}
L	Length; Lagrangian function
L_i	Area coordinates; natural coordinates
$[L]$	Matrix of direction cosines, ℓ_{ij} [see Eq. (1.6.21)]
$\mathbf{l}$	Velocity gradient tensor, $\mathbf{l} = (\boldsymbol{\nabla}\mathbf{v})^{\mathrm{T}}$
m	A scalar memory function (or relaxation kernel)
$\mathbf{m}$	Couple traction vector
M	Bending moment in beam problems
m_0, m_1, m_2	Mass inertias used in beams, plates, and shells
N	Axial force in beam problems
n_i	ith component of the unit normal vector $\hat{\mathbf{n}}$
$\hat{\mathbf{n}}$	Unit normal vector in the current configuration
N_I	Ith component of the unit normal vector $\hat{\mathbf{N}}$
$\hat{\mathbf{N}}$	Unit normal vector in the reference configuration
p	Pressure (hydrostatic or thermodynamic)
P	Point load in beams; perimeter
$\mathbf{P}$	First Piola–Kirchhoff stress tensor; polarization vector
q	Distributed transverse load on a beam
q_0	Intensity of the distributed transverse load in beams

q_n	Heat flux normal to the boundary, $q_n = \nabla \cdot \hat{\mathbf{n}}$
$\mathbf{q}_0$	Heat flux vector in the reference configuration
$\mathbf{q}_i$	Force components
$\mathbf{q}$	Heat flux vector in the current configuration
Q	First moment of area; volume rate of flow
Q_h	Heat input
r	Radial coordinate in the cylindrical polar system
r_h	Internal heat generation per unit mass in the current configuration
r_0	Internal heat generation per unit mass in the reference configuration
R	Radial coordinate in the spherical coordinate system; universal gas constant
$\mathbf{R}$	Position vector in the spherical coordinate system; proper orthogonal tensor; residual vector
$\mathbf{S}$	A second-order tensor; second Piola–Kirchhoff stress tensor
S_{ij}	Elastic compliance coefficients
t	Time
$\mathbf{t}$	Stress vector; traction vector
T	Torque; temperature
$\mathbf{T}$	Tangent coefficient matrix with coefficients T_{ij}
$\mathbf{u}$	Displacement vector
u_1, u_2, u_3	Displacements in the x_1, x_2, and x_3 directions
U	Internal (or strain) energy
$\mathbf{U}$	Right Cauchy stretch tensor
u, v, w	Displacement components in the x, y, and z directions
u_x, u_y, u_z	Displacements in the x, y, and z directions
v	Velocity, $v = |\mathbf{v}|$
V	Shear force in beam problems; potential energy due to loads
V_f	Scalar potential
$\mathbf{v}$	Velocity vector in spatial coordinates, $\mathbf{v} = \frac{D\mathbf{x}}{Dt}$
$\mathbf{V}$	Velocity vector in material coordinates; left Cauchy stretch tensor
v_x, v_y, v_z	Velocity components in the x, y, and z directions
$\mathbf{w}$	Vorticity vector, $\mathbf{w} = \frac{1}{2}\nabla \times \mathbf{v}$
W_{net}	Net rate of power input
$\mathbf{W}$	Skew symmetric part of the velocity gradient tensor, $\mathbf{L} = (\nabla\mathbf{v})^{\text{T}}$; that is $\mathbf{W} = \frac{1}{2}\left[(\nabla\mathbf{v})^{\text{T}} - \nabla\mathbf{v}\right]$,
$\mathbf{x}$	Spatial coordinates
$\mathbf{X}$	Material coordinates
x_1, x_2, x_3	Rectangular Cartesian coordinates
x, y, z	Rectangular Cartesian coordinates
Y	Relaxation modulus
z	Transverse coordinate in the beam problem; axial coordinate in the torsion problem

Greek and parenthetical symbols

α	Angle; coefficient of thermal expansion; a parameter in time-approximation schemes
α_{ij}	Thermal coefficients of expansion
β	Acceleration parameter for convergence
β_{ij}	Material coefficients, $\beta_{ij} = C_{ijk\ell}\, \alpha_{k\ell}$
γ	Parameter in the Newmark scheme; penalty parameter
$\gamma_{yz}, \gamma_{xz}, \gamma_{xy}$	Shear strains in structural problems
Γ	Internal entropy production; boundary of Ω
δ	Variational operator used in Chapter 2; Dirac delta
δ_{ij}	Components of the unit tensor, $\mathbf{I}$ (Kronecker delta)
Δ	Change of (followed by another symbol)
ε	Infinitesimal strain tensor
$\tilde{\varepsilon}$	Symmetric part of the displacement gradient tensor, $(\boldsymbol{\nabla}\mathbf{u})^{\mathrm{T}}$; that is $\tilde{\varepsilon} = \frac{1}{2}\left[(\boldsymbol{\nabla}\mathbf{u})^{\mathrm{T}} + \boldsymbol{\nabla}\mathbf{u}\right]$
ϵ	Total stored energy per unit mass; convergence tolerance
ε_{ij}	Rectangular components of the infinitesimal strain tensor
ζ	Natural coordinate
η	Entropy density per unit mass; dashpot constant; natural coordinate
η_0	Viscosity coefficient
θ	Angular coordinate in the cylindrical and spherical coordinate systems; angle; twist per unit length; absolute temperature
κ_0, κ	Reference and current configurations
λ	Extension ratio; Lamé constant; eigenvalue
μ	Lamé constant; viscosity; principal value of strain
ν	Poisson's ratio; ν_{ij} Poisson's ratios
ξ	Natural coordinate
Π	Total potential energy functional
ρ	Density in the current configuration; charge density
ρ_0	Density in the reference configuration
σ	Boltzman constant
$\tilde{\sigma}$	Mean stress
$\boldsymbol{\sigma}$	Cauchy stress tensor
τ	Shear stress; time
$\boldsymbol{\tau}$	Viscous stress tensor
χ	Deformation mapping
φ_i	Approximation functions; Hermite interpolation functions
ϕ	A typical variable; angular coordinate in the spherical coordinate system; electric potential; relaxation function

ϕ_f Moisture source

Φ Viscous dissipation, $\Phi = \boldsymbol{\tau} : \mathbf{D}$; Gibb's potential;
        Airy stress function

ψ Warping function; stream function; creep function

ψ_i Finite element (Lagrange) interpolation functions

Ψ Helmholtz free energy density; Prandtl stress function

ω Angular velocity

$\boldsymbol{\omega}$ Infinitesimal rotation vector, $\boldsymbol{\omega} = \frac{1}{2} \boldsymbol{\nabla} \times \mathbf{u}$

Ω Domain of a problem

$\boldsymbol{\Omega}$ Skew symmetric part of the displacement gradient tensor,
        $(\boldsymbol{\nabla}\mathbf{u})^{\mathrm{T}}$; that is $\boldsymbol{\Omega} = \frac{1}{2}\left[(\boldsymbol{\nabla}\mathbf{u})^{\mathrm{T}} - \boldsymbol{\nabla}\mathbf{u}\right]$

$\boldsymbol{\nabla}$ Gradient operator with respect to $\mathbf{x}$

$\boldsymbol{\nabla}_0$ Gradient operator with respect to $\mathbf{X}$

$[\ ]$ Matrix associated with the enclosed quantity

$\{\ \}$ Column vector associated with the enclosed quantity

$|\ |$ Magnitude or determinant of the enclosed quantity

$\dot{(\ )}$ Time derivative of the enclosed quantity

$(\ )^*$ Enclosed quantity with superposed rigid-body motion

$(\ )'$ Deviatoric tensors associated with the enclosed tensor

---

Note:

Quotes by various people included in this book were found at various web sites, for example,

http://naturalscience.com/dsqhome.html/,
http://thinkexist.com/quotes/david_hilbert/, and
http://www.yalescientific.org/2010/10/from-the-editor-imagination-in-science/.

This author is inspired to include the quotes at various places in his book for their wit and wisdom; the author cannot vouch for their accuracy. *Train your mind to test every concept, definition, derivation, computation, train of reasoning, and claim to truth.*

General Introduction and Mathematical Preliminaries

Mathematics is the language with which God has written the universe. —— Galileo Galilei

1.1   General Comments

Engineers and scientists from applied sciences are involved in one or more of the following activities in studying engineering systems:

(1) Develop mathematical models of physical systems
(2) Carry out numerical simulations of the mathematical models
(3) Conduct experiments to determine and understand characteristics of the system
(4) Design the components of a system
(5) Manufacture the components and integrate them to build a system

All these activities are interdependent, and they are carried out consistent with goals of the study.

Manufacturing of a system or its components can take place only after the components are designed to meet the functionality and other requirements. On the other hand, design is an iterative process of selecting materials and configurations to meet the design requirements and cost-effectiveness. During each stage of the design, analysis is carried out for the selected configuration (i.e. geometry), materials, and loads. Analysis is deterministic and involves analytically determining the response of the system or its components with the help of a mathematical model (which may account for uncertainties in the data) and a numerical method. A mathematical model of a system or its components is a collection of relationships – algebraic, differential, and/or integral – among the quantities that describe the response. The present study is concerned with the first two tasks, namely, the development of mathematical models and numerical evaluation of the mathematical models using the finite element method. Additional discussion of these two topics is presented next.

J.N. Reddy, *An Introduction to Nonlinear Finite Element Analysis*, Second Edition. ©J.N. Reddy 2015. Published in 2015 by Oxford University Press.

1.2   Mathematical Models

One of the most important tasks engineers and scientists perform is to model natural phenomena. They develop conceptual and mathematical models to simulate physical events, whether they are aerospace, biological, chemical, geological, or mechanical. A *mathematical model* can be broadly defined as a set of analytical relationships among variables that express the essential features of a physical system or process. The mathematical models are described in terms of algebraic, differential, and/or integral equations relating various quantities of interest.

Mathematical models of physical phenomena are often based on fundamental scientific laws of physics, such as the principles of conservation of mass and balance of linear momentum, angular momentum, and energy, and also assumptions concerning the geometry, loads, boundary conditions, as well as the constitutive behavior. Mathematical models of biological and other phenomena may be based on observations and accepted theories. Keeping the scope of the present study in mind, the discussion is limited to engineering systems that are governed by principles of continuum mechanics.

The construction of a mathematical model is illustrated using the problem of finding the periodic motion of a simple pendulum. While this elementary example does not bring out all aspects of formulating a complex real-world problem, it sheds light on some basic steps involved in the development of a mathematical model.

Example 1.2.1 ───

Consider the problem of a simple pendulum. The system consists of a bob of mass m attached to one end of a rod of length l and the other end is pivoted to a fixed point O, as shown in Fig. 1.2.1(a). Assume that: (1) the bob as well as the rod connecting the bob to the fixed point O are rigid, (2) the mass of the rod is negligible relative to the mass of the bob that is constant, and (3) there is no friction at the pivot. Derive the differential equation governing the periodic motion of the simple pendulum and determine the analytical solution to the small-amplitude motion (i.e. linear solution) of the bob.

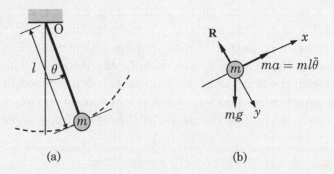

(a)   (b)

Fig. 1.2.1: (a) Simple pendulum. (b) Free-body-diagram showing forces on the bob.

Solution: The equation governing the motion of the bob can be derived, consistent with the stated assumptions, using the principle of balance of linear momentum, also known as Newton's second law of motion, which can be stated as *the vector sum of externally applied forces is equal to the time rate of change of the linear momentum of the body* (see Reddy [1]):

$$\mathbf{F} = m\mathbf{a} \tag{1}$$

where $\mathbf{F}$ is the vector sum of all forces acting on a system, m is the mass of the system (assumed to be independent of time t), and $\mathbf{a}$ is the acceleration of the system (i.e. $\mathbf{a} = \frac{d\mathbf{v}}{dt}$, where $\mathbf{v}$ is the velocity vector).

The x-component of Eq. (1) is of interest in the present problem. We have [see Fig. 1.2.1(b)]

$$F_x = -mg\,\sin\theta, \quad v_x = l\frac{d\theta}{dt}, \quad a_x = l\frac{d^2\theta}{dt^2} \tag{2}$$

where θ is the angular displacement, v_x is the velocity along x, g is the acceleration due to gravity, and t is the time. Then, the x-component of the equation of motion becomes

$$-mg\,\sin\theta = ml\frac{d^2\theta}{dt^2} \quad \text{or} \quad \frac{d^2\theta}{dt^2} + \frac{g}{l}\sin\theta = 0 \tag{3}$$

Thus, the problem at hand involves solving the nonlinear differential equation

$$\frac{d^2\theta}{dt^2} + \frac{g}{l}\sin\theta = 0, \quad 0 < t \leq T \tag{4}$$

subjected to the initial conditions (i.e. values of θ and its derivative at time $t = 0$)

$$\theta(0) = \theta_0, \quad \frac{d\theta}{dt}(0) = \frac{v_0}{l} \tag{5}$$

where θ_0 and v_0 are the initial values of angular displacement and velocity, respectively, and T is the final time. Mathematically, the problem is called an *initial-value problem* because it requires *initial* values (as opposed to *boundary* values) of θ and its derivative to solve the problem.

If the amplitude θ of the periodic motion is not small, the restoring moment is proportional to $\sin\theta$, and Eq. (3) represents a nonlinear equation. For small θ, $\sin\theta$ is approximately equal to the angle θ, and the motion is described by the linear equation

$$\frac{d^2\theta}{dt^2} + \lambda^2\theta = 0, \quad \lambda^2 = \frac{g}{l} \tag{6}$$

whose solution represents a simple harmonic motion.

The general analytical, which is exact, solution to the linear differential equation in Eq. (6) is given by

$$\theta(t) = c_1\sin\lambda t + c_2\cos\lambda t, \quad \lambda = \sqrt{\frac{g}{l}} \tag{7}$$

where c_1 and c_2 are constants to be determined using the initial conditions in Eq. (5). For the present case, they are

$$c_1 = \frac{v_0}{\lambda l}, \quad c_2 = \theta_0 \tag{8}$$

and the solution to the linear problem is

$$\theta(t) = \frac{v_0}{\lambda l}\sin\lambda t + \theta_0\cos\lambda t \tag{9}$$

For zero initial velocity and non-zero initial position θ_0, the solution becomes

$$\theta(t) = \theta_0\cos\lambda t \tag{10}$$

1.3   Numerical Simulations

Mathematical models of engineering systems are often characterized by very complex equations posed on geometrically complicated regions. Consequently, many of the mathematical models, until the advent of electronic computation, were drastically simplified in the interest of analytically solving them. Over the last three decades, however, the computer has made it possible, with the help of realistic mathematical models and numerical methods, to solve many practical problems of science and engineering. There now exists a large body of knowledge associated with the use of numerical methods and computers to analyze mathematical models of physical systems, and this body of knowledge is known as *computational mechanics*. Major established industries such as the automobile, aerospace, chemical, pharmaceutical, petroleum, electronics, and communications, as well as emerging industries such as nano and biotechnology, rely on computational mechanics-based capabilities to simulate complex systems for design and manufacture of high-technology products.

Numerical simulation of a process means that the solution of the governing equations (or mathematical model) of the process is obtained using a numerical method and a computer. While the derivation of the governing equations for most problems is not unduly difficult, their solution by exact methods of analysis is a formidable task. In such cases, numerical methods of analysis provide an alternative means of finding solutions. Numerical methods typically transform differential equations to algebraic equations that are to be solved using computers. For example, the mathematical formulation of the simple pendulum resulted in a nonlinear differential equation, Eq. (4) of Example 1.2.1, whose analytical solution cannot be obtained. Therefore, one must consider using a numerical method to solve it. Even linear problems may not admit exact solutions due to geometric and material complexities, but it is relatively easy to obtain approximate solutions using numerical methods. These ideas are illustrated below using the simple pendulum problem of Example 1.2.1. The finite difference method is used as a numerical method of solution.

Example 1.3.1 ――――――――――――――――――――――――――――

Determine the numerical solution to Eq. (4) of Example 1.2.1 governing a simple pendulum using a suitable finite difference method.

Solution: In the finite difference method, the derivatives are approximated by difference quotients (or the function is expanded in a Taylor series) that involve the unknown value of the solution at time t_{i+1} and the known value of the solution at time $t = t_i$. For example, consider the first-order equation

$$\frac{du}{dt} = f(t, u) \quad \text{with} \quad u(0) = u_0 \tag{1}$$

The derivative at $t = t_i$ is approximated using the *forward difference scheme*

$$\left(\frac{du}{dt}\right)\Bigg|_{t=t_i} \approx \frac{u(t_{i+1}) - u(t_i)}{t_{i+1} - t_i} \tag{2}$$

so that the discrete form of Eq. (1) is

$$u_{i+1} = u_i + \Delta t_{i+1} \ f(t_i, u_i) \tag{3}$$

$$u_i = u(t_i), \qquad \Delta t_{i+1} = t_{i+1} - t_i \tag{4}$$

Equation (3) can be solved repeatedly for u_{i+1}, starting from the known value u_0 of $u(t)$ at $t = t_0 = 0$. The repeated solution of Eq. (3) produces the values of u at times $t_1 = \Delta t_1$, $t_2 = \Delta t_1 + \Delta t_2$, ..., $t_n = \sum_{i=1}^{n} \Delta t_i = T$. This scheme is also known as *Euler's explicit method* or the *first-order Runge–Kutta method*. Note that the ordinary differential equation in Eq. (1) is converted to an algebraic equation, Eq. (3), which is then evaluated at different times to construct the discrete time history of $u(t)$, as shown in Fig. 1.3.1.

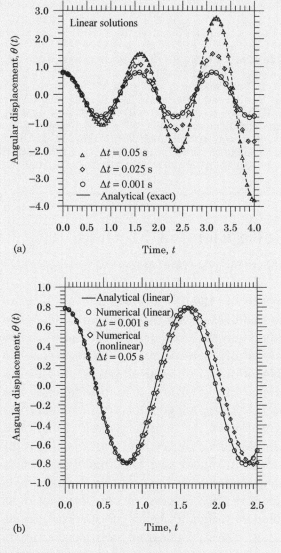

Fig. 1.3.1: Analytical and numerical solutions of the simple pendulum. (a) Comparison of the linear solutions predicted by different time steps. (b) Comparison of linear and nonlinear solutions.

Euler's explicit method can be applied to the nonlinear second-order differential equation in Eq. (4) of Example 1.2.1 after rewriting it as a pair of first-order differential equations:

$$\frac{d\theta}{dt} = \frac{v}{l} \equiv f_1(v), \qquad \frac{dv}{dt} = -g \sin\theta \equiv f_2(\theta) \tag{5}$$

which are *coupled* (i.e. one cannot be solved without the other). Applying the Euler's scheme of Eq. (3) to the two equations at hand, one obtains

$$\theta_{i+1} = \theta_i + \Delta t \, f_1(v_i); \quad v_{i+1} = v_i + \Delta t \, f_2(\theta_i) \tag{6}$$

where $\Delta t = \Delta t_1 = \Delta t_2 = \ldots = \Delta t_n$ (i.e. uniform time step is used). The expressions in Eq. (6) are repeatedly evaluated for θ_{i+1} and v_{i+1} using the known solution (θ_i, v_i) from the previous time step. At time $t = t_0 = 0$, the known initial values (θ_0, v_0) are used. Thus, one needs a computer and a computer language like Fortran (77 or 90) to write a program to compute numbers.

The numerical solutions of Eq. (6) for three different time steps, $\Delta t = 0.05$ s, $\Delta t = 0.025$ s, and $\Delta t = 0.001$ s along with the exact linear solution in Eq. (10) of Example 1.2.1 (with $\theta_0 = \pi/4$, $v_0 = 0$ m/s, $l = 2$ m, and $g = 9.81$ m/s^2) are presented in Fig. 1.3.1(a). The smaller the time step the more accurate the solution is. This is because the approximation of the derivative in Eq. (2) tends to the exact derivative with $\Delta t \to 0$. In fact, the numerical solution predicted by the Euler's scheme for $\Delta t = 0.05$ s is unstable; that is, error grows with time and the solution may go unbounded with time. The numerical solutions of the nonlinear problem are compared for the time step $\Delta t = 0.05$ s with the linear solution in Fig. 1.3.1(b). The nonlinear solution differs from the linear solution slightly and has longer period of oscillation.

1.4   The Finite Element Method

As illustrated in the previous section, numerical methods are extremely powerful tools for engineering analysis. With the advent of computers, there has been a tremendous explosion in the development and use of numerical methods in engineering analysis and design. Of these, the finite difference methods and the finite element method and their variants are the most commonly used methods today in the analysis of practical engineering problems. In finite difference methods, derivatives of various order are approximated using truncated Taylor's series approximations. The traditional finite difference methods suffer from two major drawbacks: (1) applying gradient-type boundary conditions requires additional approximation of the boundary data; (2) finite difference formulas are traditionally developed for rectangular grids, making it difficult to use them for irregular domains. Advances have been made in recent years to overcome these drawbacks but the remedies are problem-dependent.

The finite element method is based on the idea that every physical system is composed of different parts and, hence, its solution may also be represented in parts. In addition, the solution over each part is represented as a linear combination of undetermined parameters and known functions of position and possibly time, like in the traditional variational methods (e.g. the Ritz and Galerkin methods [2, 3]). The parts can differ from each other in shape, material properties, as well as physical behavior. Even when the system is of one geometric shape and made of one material, it is simpler to represent its solution in a piecewise manner, with certain continuity conditions between piecewise solutions. In recent years, generalizations of the finite element method have emerged (e.g. the *generalized finite element method* and element-free or *meshless methods*) but they are not considered in this book. Interested readers may consult the advanced titles [4–12] listed in the References at the end of this book.

The traditional finite element method is endowed with three basic features [2]. First, a domain of the system is represented as a collection of geometrically simple subdomains, called *finite elements*. Second, over each finite element, the unknown variable(s) are approximated by a linear combination of algebraic polynomials and undetermined parameters, and algebraic relations among the parameters are obtained by satisfying the governing equation(s), in a weighted-integral sense, over each element. The undetermined parameters represent the values of the unknown variables at a finite number of preselected points, called *nodes*, in the element. Third, the algebraic relations from all elements are put together (or "assembled") using continuity and balance considerations.

There are several reasons why an engineer or scientist should study the finite element method. They are outlined in the following:

(1) The finite element method is a powerful numerical procedure devised for the analysis of practical engineering problems. It is capable of handling geometrically complicated domains, all physically meaningful boundary conditions, nonlinearities, and coupled phenomena that are common in practical problems. The knowledge of how the method works greatly enhances the analysis skill and provides a greater understanding of the problem being solved.

(2) Commercial software packages or *canned* computer programs based on the finite element method are often used in industrial, research, and academic institutions for the solution of a variety of engineering and scientific problems. The intelligent use of these programs and a correct interpretation of the output are often predicated on knowledge of the basic theory underlying the method.

(3) It is not uncommon to find mathematical models derived in personal research and development that cannot be evaluated using canned programs. In such cases, an understanding of the finite element method and knowledge of computer programming can help design "user specified" programs to evaluate the mathematical models or parts thereof.

The basic ideas underlying the finite element method are presented in Chapter 3 using a model linear differential equation in a single variable in one and two dimensions. Readers who are familiar with these steps as applied to linear differential equations may skip Chapter 3 and go straight to Chapter 4.

1.5 Nonlinear Analysis

1.5.1 Introduction

Recall from the simple pendulum problem of Examples 1.2.1 and 1.3.1, that nonlinearity naturally arises in a rigorous mathematical formulation of most physical problems. Based on assumptions of smallness of certain quantities of the formulation, the problem may be reduced to a linear problem. Linear solutions may be obtained with considerable ease and less computational cost when compared to nonlinear solutions. Further, linear solutions due to various boundary conditions and "load" cases may be scaled and superimposed. In many instances, assumptions of linearity lead to reasonable idealization of the behavior of the system. However, in some cases assumption of linearity may result in an unrealistic approximation of the response or inefficient use of the materials used. The type of analysis, linear or nonlinear, depends on the goal of the analysis and errors in the system's response that may be tolerated. In some cases, nonlinear analysis is the only option left for the analyst as well as the designer (e.g. high-speed flows of inviscid fluids around solid bodies).

Nonlinear analysis is a necessity, for example, in (a) designing high-performance and efficient components of certain industries (e.g. aerospace, defense, and nuclear), (b) assessing functionality (e.g. residual strength and stiffness of structural elements) of existing systems that exhibit some types of damage and failure, (c) establishing causes of system failure, (d) simulating true material behavior of processes, and (e) research to gain a realistic understanding of physical phenomena.

The following features of nonlinear analysis should be noted:

- The principle of superposition does not hold.
- Analysis can be carried out for one *load* case at a time.
- The history (or sequence) of loading influences the response.
- The initial state of the system (e.g. prestress) may be important.

1.5.2 Classification of Nonlinearities

There are two common sources of nonlinearity: (1) geometry and (2) material. The geometric nonlinearity arises purely from geometric consideration (e.g. nonlinear strain-displacement relations), and the material nonlinearity is due to nonlinear constitutive behavior of the material of the system. A third type of nonlinearity may arise due to changing initial or boundary conditions. Various types of nonlinearities will be discussed through simple examples.

The simple pendulum problem of Example 1.2.1 is an example of geometric nonlinearity because the amplitude of angular motion is the deciding factor to treat it as the linear ($\sin\theta \approx \theta$) or nonlinear problem. Another example of the geometric nonlinearity is provided by a rigid link supported by a linear elastic torsional spring at one end and subjected to a vertical point load at the other end, as shown in Fig. 1.5.1 and discussed by Hinton [8]. Moment equilibrium (i.e. sum the moments about the hinge) and the linear moment-rotation relationship give

$$M - Fl \cos\theta = 0, \quad M = k_T\theta \quad \text{or} \quad \frac{Fl}{k_T} = \frac{\theta}{\cos\theta}, \quad \theta < \frac{\pi}{2} \qquad (1.5.1)$$

where k_T is the torsional spring constant, l is the length of the link, and M is the moment experienced by the torsional spring due to the angular displacement θ. Clearly, the force–rotation relationship is nonlinear for arbitrary θ. If θ is restricted to small values (say, $\theta < 10°$), $\cos\theta \approx 1$ and the governing equation, Eq. (1.5.1), is reduced to a linear equation

$$\frac{Fl}{k_T} = \theta \qquad (1.5.2)$$

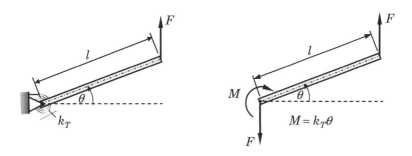

Fig. 1.5.1: Rigid link-linear torsional spring cantilever.

Plots of Fl/k_T versus θ are shown in Fig. 1.5.2 for the linear and nonlinear cases [using Eqs. (1.5.2) and (1.5.1), respectively]. Clearly, the nonlinearity in the present case is due to the large values of θ (i.e. due to the geometry of motion) and the nonlinear deflection is less than the linear deflection as the load is increased. Such a nonlinearity is known as the *hardening* type.

The material nonlinearity may be introduced into the problem if the moment–rotation relationship is nonlinear

$$M = k_T(\theta)\theta, \quad \text{say} \quad k_T = k_0 - k_1\theta \qquad (1.5.3)$$

where k_0 and k_1 are material parameters that are determined through tests. Note that the particular form used implies that the material becomes weaker as it deforms. Such nonlinearity is known as the *softening* type. If the relation in

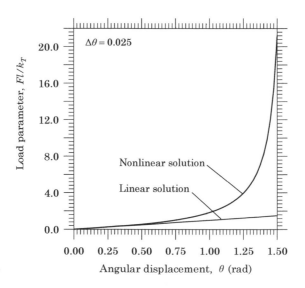

Fig. 1.5.2: Geometric linear and nonlinear response of a rigid link-linear torsional spring cantilever.

Eq. (1.5.3) is used in Eq. (1.5.1), a nonlinear equation that contains both geometric and material nonlinearities is obtained:

$$\frac{Fl}{k_0} = (1 - \gamma\theta)\frac{\theta}{\cos\theta}, \quad \gamma = \frac{k_1}{k_0} < 1 \tag{1.5.4}$$

If the expression for k_T from Eq. (1.5.3) is used in Eq. (1.5.2), the resulting equation is nonlinear, only due to the fact that the material behavior is a function of the angular displacement,

$$\frac{Fl}{k_0} = (1 - \gamma\theta)\theta, \quad \gamma = \frac{k_1}{k_0} \tag{1.5.5}$$

Figure 1.5.3 contains plots of the load factor Fl/k_0 versus rotation θ for the linear, geometrically nonlinear, materially nonlinear, and combined materially and geometrically nonlinear cases. Note that the material nonlinearity in the present case is due to reduction in the torsional spring stiffness, and therefore the materially nonlinear deflections are greater than the linear deflections as the load is increased. In the present case, the geometric nonlinearity dominates if both nonlinearities are included.

An example of another type of nonlinearity is provided by the axial deformation of an isotropic, homogeneous, linear elastic bar with constrained end displacement, as shown in Fig. 1.5.4 and discussed by Hinton [8]. The bar is of length $2L$, uniform cross-sectional area of A, and loaded with an axial force P at its midpoint, B. The right end of the rod is constrained so that it can at most have an axial displacement of u_0, which is assumed to be very small compared to the length L so that the induced axial strain is small.

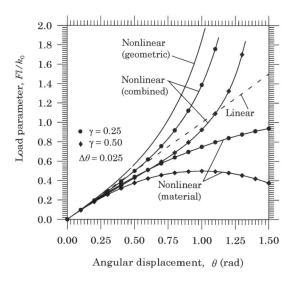

Fig. 1.5.3: Linear and nonlinear response of a rigid link-torsional spring cantilever.

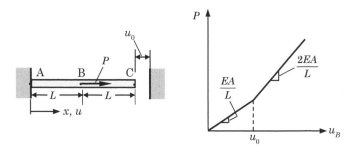

Fig. 1.5.4: Axial deformation of a rod with constrained end displacement.

The governing equation and boundary condition at point A are

$$\frac{dN}{dx} = 0 \quad \text{or} \quad EA\frac{d^2u}{dx^2} = 0, \ 0 < x < L; \quad u(0) = 0 \tag{1.5.6}$$

where N is the effective axial force, $N = EA(du/dx)$, and u is the axial displacement. The displacement at point B can be determined using the boundary condition at point C, which is dependent on whether the displacement of point C equals u_0. Thus, the solution is

$$N(L) = P \quad \text{or} \quad \left(EA\frac{du}{dx}\right)\bigg|_{x=L} = P, \quad \text{if} \quad u(L) = u(2L) < u_0 \tag{1.5.7}$$

$$u(2L) = u_0, \quad \text{if} \quad u(L) \geq u_0 \tag{1.5.8}$$

In the former case [i.e. when $u(L) = u(2L) < u_0$], the displacement is given by

$$u(x) = \frac{Px}{EA} \ (0 \le x \le L) \quad \text{and} \quad u(L) = \frac{PL}{EA} \quad \text{if} \quad u(2L) < u_0 \qquad (1.5.9)$$

Note that the axial displacements of points B and C are identical provided that the force P does not exceed the value $P = (EA/L)u_0$.

In the latter case (when the point B is already displaced axially by u_0), the additional displacement of point B is obtained by solving the governing equation in the two intervals and using the continuity conditions at $x = L + u_0$ and boundary condition at $x = 2L + u_0$

$$u(x) = \frac{Px}{EA} \frac{L}{2L + u_0}, \quad 0 \le x \le L + u_0 \qquad (1.5.10)$$

Assuming that u_0 is very small compared to L, the above expression can be approximated as

$$u(x) = \frac{Px}{2EA}, \quad 0 \le x \le L \qquad (1.5.11)$$

and

$$u_B = u(L) = \frac{PL}{2EA} \qquad (1.5.12)$$

Thus the force-displacement relationship of the bar is bilinear. Such problems are called *contact* or *nonlinear boundary* problems.

1.6   Review of Vectors and Tensors

1.6.1   Preliminary Comments

The quantities encountered in analytical description of physical phenomena may be classified into two groups according to the information needed to specify them completely: scalars and nonscalars. The scalars are given by a single number. Nonscalars have not only a magnitude specified, but also additional information, such as direction(s). Nonscalars that obey certain rules (such as the parallelogram law of addition) are called *vectors* or *tensors* of a certain order. Not all nonscalar quantities are vectors (e.g. a finite rotation is not a vector). In this section, a brief review of algebra and calculus of vectors and tensors and matrix algebra is presented. For more detailed account of these topics, the reader is referred to Chapter 2 of Reddy [1] and Reddy and Rasmussen [13].

Scalars are denoted with lightface letters like a or A, and a boldface letter **A** is used to denote a vector or tensor of any order; the precise meaning of such quantities will be known in the context. A *physical vector* is one that "has magnitude and direction and satisfies the parallelogram law of addition," and it is often shown as a directed line segment with an arrow head at the end of the line. The length of the line represents the magnitude of the vector and the arrow indicates the direction. The mathematical definition of an abstract vector, which can be a function satisfying certain rules of addition and multiplication by a scalar, is given in Section 1.7.

1.6.2 Definition of a Physical Vector

Physical vectors are entities that obey the following rules of vector addition and multiplication by a scalar.

1.6.2.1 Vector addition

Let $\mathbf{A}$, $\mathbf{B}$, and $\mathbf{C}$ be any vectors. Then there exists a vector $\mathbf{A} + \mathbf{B}$, called sum of $\mathbf{A}$ and $\mathbf{B}$, with the properties

$$
\begin{aligned}
&(1) \quad \mathbf{A} + \mathbf{B} = \mathbf{B} + \mathbf{A} \quad \text{(commutative)},\\
&(2) \quad (\mathbf{A} + \mathbf{B}) + \mathbf{C} = \mathbf{A} + (\mathbf{B} + \mathbf{C}) \quad \text{(associative)},\\
&(3) \quad \text{there exists a unique vector, } \mathbf{0}, \text{ independent of } \mathbf{A} \text{ such that}\\
&\qquad \mathbf{A} + \mathbf{0} = \mathbf{A} \quad \text{(existence of the zero vector)}, \qquad (1.6.1)\\
&(4) \quad \text{to every vector } \mathbf{A} \text{ there exists a unique vector } -\mathbf{A}\\
&\qquad \text{(that depends on } \mathbf{A}) \text{ such that}\\
&\qquad \mathbf{A} + (-\mathbf{A}) = \mathbf{0} \quad \text{(existence of the negative vector)}.
\end{aligned}
$$

The negative vector $-\mathbf{A}$ has the same magnitude as $\mathbf{A}$, but has the opposite *sense*. Subtraction of vectors is carried out along the same lines. To form the difference $\mathbf{A} - \mathbf{B}$, one may write $\mathbf{A} + (-\mathbf{B})$ and subtraction is reduced to the operation of addition.

1.6.2.2 Multiplication of a vector by a scalar

Let $\mathbf{A}$ and $\mathbf{B}$ be vectors and α and β be real numbers (scalars). To every vector $\mathbf{A}$ and every real number α, there corresponds a unique vector $\alpha\mathbf{A}$ such that

$$
\begin{aligned}
&(1) \quad \alpha(\beta\mathbf{A}) = (\alpha\beta)\mathbf{A} \text{ (scalar multiplication is associative)},\\
&(2) \quad (\alpha + \beta)\mathbf{A} = \alpha\mathbf{A} + \beta\mathbf{A} \text{ (scalar addition is distributive)},\\
&(3) \quad \alpha(\mathbf{A} + \mathbf{B}) = \alpha\mathbf{A} + \alpha\mathbf{B} \text{ (vector addition is distributive)},\\
&(4) \quad 1 \cdot \mathbf{A} = \mathbf{A} \cdot 1 = \mathbf{A}, \quad 0 \cdot \mathbf{A} = \mathbf{0}.
\end{aligned}
\qquad (1.6.2)
$$

The magnitude of the vector $\mathbf{A}$ is denoted by $|\mathbf{A}|$, $\|\mathbf{A}\|$, or A. The actual computation of $|\mathbf{A}|$ will be presented shortly. A vector of unit length is called a *unit vector*. The unit vector along $\mathbf{A}$ may be defined as $\hat{\mathbf{e}}_A = \mathbf{A}/A$, and vector $\mathbf{A}$ may now be expressed as

$$
\mathbf{A} = A\,\hat{\mathbf{e}}_A
$$

Thus, any vector may be represented as a product of its magnitude and a unit vector along the vector. A unit vector is used to designate direction; it does not have any physical dimensions. However, $|\mathbf{A}|$ has the physical dimensions. A "hat" (caret) above the boldface letter, $\hat{\mathbf{e}}$, is used to signify that it is a vector of unit magnitude. A vector of zero magnitude is called a *zero vector* or a *null vector*, and denoted by boldface zero, $\mathbf{0}$. Note that a lightface zero, 0, is a scalar and boldface zero, $\mathbf{0}$, is the zero vector. Also, a zero vector has no direction associated with it. Two vectors $\mathbf{A}$ and $\mathbf{B}$ are equal if their magnitudes are equal, $|\mathbf{A}| = |\mathbf{B}|$, *and* if their directions are equal.

1.6.3 Scalar and Vector Products

In engineering analysis, products of one vector with another vector are encountered. There are two such products involving vectors, one encountered in writing the work done and another in computing the moment due to a force.

1.6.3.1 Scalar product (or "dot" product)

Work done by a force (vector) $\mathbf{F}$ is the product of the magnitude of the force F and the magnitude of the displacement (vector) $\mathbf{u}$ in the direction of the force, that is the projection of the displacement vector in the direction of the force, $d = u \cos\theta$, where θ denotes the angle between the two vectors, $\mathbf{F}$ and $\mathbf{u}$. The product is denoted by $\mathbf{F} \cdot \mathbf{u}$ and it has the meaning

$$\mathbf{F} \cdot \mathbf{u} = |\mathbf{F}|\,|\mathbf{u}| \cos\theta = F\,u\,\cos\theta \qquad (1.6.3)$$

and it is known as the *scalar product* (or "dot" product) to indicate the fact that the result is a scalar. If the scalar product of two non-zero vectors is zero, $\mathbf{A} \cdot \mathbf{B} = 0$, then vector $\mathbf{A}$ is perpendicular to vector $\mathbf{B}$, and the vectors $\mathbf{A}$ and $\mathbf{B}$ are said to be *orthogonal* to each other. In view of the definition in Eq. (1.6.3) of the scalar product of vectors, the magnitude of a vector $\mathbf{A}$ can be defined as the square root of $\sqrt{\mathbf{A} \cdot \mathbf{A}}$ because

$$\mathbf{A} \cdot \mathbf{A} = |\mathbf{A}|\,|\mathbf{A}| \cos 0 = |\mathbf{A}|^2 \qquad (1.6.4)$$

1.6.3.2 Vector product

Next consider the concept of the moment of a force about a point. The magnitude of the moment about a point O of a force $\mathbf{F}$ acting at a point P is defined to be the magnitude of the force F multiplied by the perpendicular distance ℓ from the point O to the line of action of the force, $F\ell = F r \sin\theta$, where r is the magnitude of the vector $\mathbf{r}$ from the point O to point P, as shown in Fig. 1.6.1(a). A direction can be assigned to the moment. Drawing the vectors $\mathbf{F}$ and $\mathbf{r}$ from the common point O, we note that the rotation due to $\mathbf{F}$ tends to bring vector $\mathbf{r}$ into vector $\mathbf{F}$, as can be seen from Fig. 1.6.1(b). The axis of

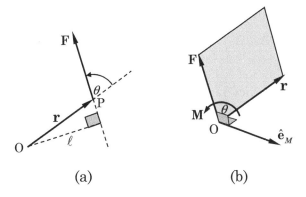

Fig. 1.6.1: (a) Moment of a force about a point. (b) Sign convention for moment of a force.

rotation is perpendicular to the plane formed by $\mathbf{F}$ and $\mathbf{r}$. Along this axis of rotation, a preferred direction is set up as that in which a right-handed screw would advance when turned in the direction of rotation due to the moment, as shown in Fig. 1.6.1(b). Along this axis of rotation, a unit vector $\hat{\mathbf{e}}_M$ is drawn and assume that it represents the direction of the moment $\mathbf{M}$

$$\mathbf{M} = Fr \, \sin\theta \, \hat{\mathbf{e}}_M \equiv \mathbf{r} \times \mathbf{F}$$

According to this expression, $\mathbf{M}$ may be looked upon as resulting from a special operation between the two vectors $\mathbf{F}$ and $\mathbf{r}$. Since the result of such a product is a vector, it is called the *vector product* or "cross product". Thus, the vector product of two vectors $\mathbf{A}$ and $\mathbf{B}$ is defined as

$$\mathbf{A} \times \mathbf{B} = |\mathbf{A}|\,|\mathbf{B}|\,\sin\theta\,\hat{\mathbf{e}}_n \qquad (1.6.5)$$

where $\hat{\mathbf{e}}_n$ denotes the unit vector normal to the plane of vectors $\mathbf{A}$ and $\mathbf{B}$. The vector product is skew-symmetric, that is $\mathbf{A} \times \mathbf{B} = -\mathbf{B} \times \mathbf{A}$.

1.6.3.3 Plane area as a vector

The magnitude of the vector $\mathbf{F} \times \mathbf{r}$ is equal to the area of the parallelogram formed by the vectors $\mathbf{F}$ and $\mathbf{r}$, shown as shaded in Fig. 1.6.1(b). In fact, the vector $\mathbf{C} = \mathbf{A} \times \mathbf{B}$ may be considered to represent *both* the magnitude and the direction of the product of $\mathbf{A}$ and $\mathbf{B}$. Thus, a plane area may be looked upon as possessing a direction in addition to a magnitude; the directional character arising out of the need to specify an orientation of the plane in space. It is customary to denote the direction of a plane area A by means of a unit vector $\hat{\mathbf{n}}$ drawn normal to that plane. The direction of the normal $\hat{\mathbf{n}}$ is taken by convention as that in which a right-handed screw advances as it is rotated according to the sense of travel along the boundary curve or contour. Thus, area can be expressed as a vector $\mathbf{A} = A\,\hat{\mathbf{n}}$.

1.6.3.4 Linear independence of vectors

A set of n vectors $\{\mathbf{A}_1, \mathbf{A}_2, \ldots, \mathbf{A}_n\}$ is said to be *linearly dependent* if a set of n numbers $c_1, c_2, \ldots, c_n$ can be found such that

$$c_1 \mathbf{A}_1 + c_2 \mathbf{A}_2 + \cdots + c_n \mathbf{A}_n = \mathbf{0}$$

where $c_1, c_2, \ldots, c_n$ cannot all be zero. If this expression cannot be satisfied, the vectors are said to be *linearly independent*. An orthonormal set of nonzero vectors is linearly independent.

1.6.3.5 Components of a vector

In a three-dimensional domain $\Omega \subset \Re^3$, a set of three linearly independent vectors is chosen as a basis. A basis is called *orthonormal* if they are mutually orthogonal and have unit magnitudes. To distinguish the basis $(\mathbf{e}_1, \mathbf{e}_2, \mathbf{e}_3)$ that is not orthonormal from one that is orthonormal, the orthonormal basis is denoted by $(\hat{\mathbf{e}}_1, \hat{\mathbf{e}}_2, \hat{\mathbf{e}}_3)$. The orthonormal basis vectors $(\hat{\mathbf{e}}_1, \hat{\mathbf{e}}_2, \hat{\mathbf{e}}_3)$ satisfy

$$\begin{aligned} \hat{\mathbf{e}}_1 \cdot \hat{\mathbf{e}}_2 = 0, \quad \hat{\mathbf{e}}_2 \cdot \hat{\mathbf{e}}_3 = 0, \quad \hat{\mathbf{e}}_3 \cdot \hat{\mathbf{e}}_1 = 0 \\ \hat{\mathbf{e}}_1 \cdot \hat{\mathbf{e}}_1 = 1, \quad \hat{\mathbf{e}}_2 \cdot \hat{\mathbf{e}}_2 = 1, \quad \hat{\mathbf{e}}_3 \cdot \hat{\mathbf{e}}_3 = 1 \end{aligned} \tag{1.6.6}$$

When the basis vectors are constant, that is fixed lengths and directions, the basis is called *Cartesian*. A general Cartesian system is oblique. When the basis vectors are unit and orthogonal, the basis system is called *rectangular Cartesian*. The familiar rectangular Cartesian coordinate system $(x, y, z) = (x_1, x_2, x_3)$ is shown in Fig. 1.6.2.

  Any vector in three-dimensional space can be represented as a linear combination of the basis vectors $(\hat{\mathbf{e}}_1, \hat{\mathbf{e}}_2, \hat{\mathbf{e}}_3)$

$$\mathbf{A} = A_1 \hat{\mathbf{e}}_1 + A_2 \hat{\mathbf{e}}_2 + A_3 \hat{\mathbf{e}}_3. \tag{1.6.7}$$

The vectors $A_1 \hat{\mathbf{e}}_1$, $A_2 \hat{\mathbf{e}}_2$, and $A_3 \hat{\mathbf{e}}_3$ are called the *vector components* of $\mathbf{A}$, and A_1, A_2, and A_3 are called the *scalar components* of $\mathbf{A}$ with respect to the basis

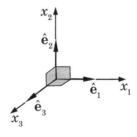

Fig. 1.6.2: Rectangular Cartesian coordinates.

$(\hat{\mathbf{e}}_1, \hat{\mathbf{e}}_2, \hat{\mathbf{e}}_3)$, as indicated in Fig. 1.6.2. Note that

$$\mathbf{A} \cdot \hat{\mathbf{e}}_i = A_i \quad \text{and therefore} \quad \mathbf{A} = \sum_{i=1}^{3} \left(\mathbf{A} \cdot \hat{\mathbf{e}}_i \right) \hat{\mathbf{e}}_i \tag{1.6.8}$$

1.6.4 Summation Convention and Kronecker Delta and Permutation Symbol

1.6.4.1 Summation convention

A typical term in Eq. (1.6.7) is of the form $A_i \mathbf{e}_i$, where i takes the values of 1, 2, and 3. Therefore, the expression can be abbreviated as

$$\mathbf{A} = \sum_{i=1}^{3} A_i \hat{\mathbf{e}}_i \quad \text{or} \quad \mathbf{A} = \sum_{j=1}^{3} A_j \hat{\mathbf{e}}_j$$

The summation index i or j is arbitrary as long as the same index is used for both A and $\hat{\mathbf{e}}$. The expression can be further shortened by omitting the summation sign and having the understanding that a repeated index means summation over all values of that index. Thus, the three-term expression for $\mathbf{A}$ can be simply written as

$$\mathbf{A} = A_i \hat{\mathbf{e}}_i = A_j \hat{\mathbf{e}}_j \tag{1.6.9}$$

This notation is called the *summation convention*. The index that is repeated is called a *dummy index*, as it can be replaced with any other index, as long as it is not already used in the expression. For example, an arbitrary vector $\mathbf{A}$ can be expressed in terms of its components as

$$\mathbf{A} = A_i \hat{\mathbf{e}}_i = \left(\mathbf{A} \cdot \hat{\mathbf{e}}_i \right) \hat{\mathbf{e}}_i = \left(\mathbf{A} \cdot \hat{\mathbf{e}}_j \right) \hat{\mathbf{e}}_j$$

In an expression of the form $A_i B_j C_j$, j is the dummy index, while i, which is not repeated, is known as the *free index*.

1.6.4.2 Kronecker delta symbol

The scalar product of vectors $\mathbf{A}$ and $\mathbf{B}$ in an orthogonal Cartesian system can be expressed in terms of their components as

$$\mathbf{A} \cdot \mathbf{B} = \left(A_i \hat{\mathbf{e}}_i \right) \cdot \left(B_j \hat{\mathbf{e}}_j \right) = A_i B_j \left(\hat{\mathbf{e}}_i \cdot \hat{\mathbf{e}}_j \right)$$

In view of the orthogonal relationships among the basis vectors $(\hat{\mathbf{e}}_1, \hat{\mathbf{e}}_2, \hat{\mathbf{e}}_3)$

$$\hat{\mathbf{e}}_i \cdot \hat{\mathbf{e}}_j = 1, \quad \text{whenever } i \text{ and } j \text{ have the same value,}$$
$$\hat{\mathbf{e}}_i \cdot \hat{\mathbf{e}}_j = 0, \quad \text{whenever } i \text{ and } j \text{ do not have the same value}$$

the scalar product of two vectors can be written as

$$\mathbf{A} \cdot \mathbf{B} = A_i B_j \, \delta_{ij}$$

where δ_{ij} denotes the *Kronecker delta* symbol

$$\delta_{ij} = \begin{cases} 1, & \text{if } i = j \text{ for fixed values of } i \text{ and } j \\ 0, & \text{if } i \neq j \text{ for fixed values of } i \text{ and } j \end{cases} \qquad (1.6.10)$$

The Kronecker delta constitutes the rectangular Cartesian components of the unit second-order tensor $\mathbf{I} = \delta_{ij}\, \hat{\mathbf{e}}_i\, \hat{\mathbf{e}}_j$, as will be shown in the section on tensors. The scalar product of two vectors can now be expressed in terms of their components as

$$\mathbf{A} \cdot \mathbf{B} = A_i B_j \, \delta_{ij} = A_i B_i = A_j B_j \qquad (1.6.11)$$

1.6.4.3   The permutation symbol

Consider the triple product

$$\mathbf{A} \cdot \mathbf{B} \times \mathbf{C} = A_i B_j C_k \, \hat{\mathbf{e}}_i \cdot \hat{\mathbf{e}}_j \times \hat{\mathbf{e}}_k$$

In view of the following vector products of the basis vectors:

$$\begin{aligned}
&\hat{\mathbf{e}}_1 \times \hat{\mathbf{e}}_1 = \mathbf{0}, \quad &&\hat{\mathbf{e}}_1 \times \hat{\mathbf{e}}_2 = \hat{\mathbf{e}}_3, \quad &&\hat{\mathbf{e}}_1 \times \hat{\mathbf{e}}_3 = -\hat{\mathbf{e}}_2 \\
&\hat{\mathbf{e}}_2 \times \hat{\mathbf{e}}_1 = -\hat{\mathbf{e}}_3, \quad &&\hat{\mathbf{e}}_2 \times \hat{\mathbf{e}}_2 = \mathbf{0}, \quad &&\hat{\mathbf{e}}_2 \times \hat{\mathbf{e}}_3 = \hat{\mathbf{e}}_1 \\
&\hat{\mathbf{e}}_3 \times \hat{\mathbf{e}}_1 = \hat{\mathbf{e}}_2, \quad &&\hat{\mathbf{e}}_3 \times \hat{\mathbf{e}}_2 = -\hat{\mathbf{e}}_1, \quad &&\hat{\mathbf{e}}_3 \times \hat{\mathbf{e}}_3 = \mathbf{0}
\end{aligned} \qquad (1.6.12)$$

the triple scalar product can be expressed as

$$\mathbf{A} \cdot \mathbf{B} \times \mathbf{C} = A_i B_j C_k \, e_{ijk}$$

where

$$e_{ijk} = \hat{\mathbf{e}}_i \cdot \hat{\mathbf{e}}_j \times \hat{\mathbf{e}}_k \quad \text{or} \quad \hat{\mathbf{e}}_i \times \hat{\mathbf{e}}_j = e_{ijk}\, \hat{\mathbf{e}}_k \qquad (1.6.13)$$

The symbol e_{ijk} is called the *alternating symbol* or *permutation symbol*, defined by

$$e_{ijk} = \begin{cases} 1, & \text{if } i, j, k \text{ are in cyclic order} \\ & \quad \text{and not repeated } (i \neq j \neq k) \\ -1, & \text{if } i, j, k \text{ are not in cyclic order} \\ & \quad \text{and not repeated } (i \neq j \neq k) \\ 0, & \text{if any of } i, j, k \text{ are repeated} \end{cases} \qquad (1.6.14)$$

By definition, the subscripts of the permutation symbol can be permuted without changing its value; an interchange of any two subscripts will change the sign (hence, interchange of two subscripts twice keeps the value unchanged): $e_{ijk} = e_{kij} = e_{jki}$, $e_{ijk} = -e_{jik} = e_{jki} = -e_{kji}$.

The rectangular Cartesian coordinates of a typical point $\mathbf{x}$ in Ω are denoted by (x_1, x_2, x_3) and $\mathbf{x}$ is expressed as

$$\mathbf{x} = x_1 \hat{\mathbf{e}}_1 + x_2 \hat{\mathbf{e}}_2 + x_3 \hat{\mathbf{e}}_3 = x_i \, \hat{\mathbf{e}}_i \qquad (1.6.15)$$

1.6.5 Tensors and their Matrix Representation

1.6.5.1 Concept of a second-order tensor

As is well known, the specification of stress, which is force per unit area, not only depends on the magnitude as well as the direction of the force but also on the orientation of the area on which the force acts. Thus, specification of stress at a point requires two vectors, one perpendicular to the plane on which the force is acting and the other in the direction of the force. Such quantities are termed *dyads* or *second-order tensors*. Because of its utilization in physical applications, a dyad is defined as two vectors standing side by side and acting as a unit, $\mathbf{S} = \mathbf{AB}$. The transpose of a dyad is obtained by the interchange of the two vectors in the dyad. Similarly, one can define a triad (or a third-order tensor) as an entity with three vectors standing next to each other, and so on.

Scalar products and cross products of vectors can be extended to tensors of any order. For example, the dot product of a second-order tensor $\mathbf{S} = \mathbf{AB}$ with a vector $\mathbf{V}$ is

$$\mathbf{S} \cdot \mathbf{V} = \mathbf{A}(\mathbf{B} \cdot \mathbf{V}) \quad \text{and} \quad \mathbf{V} \cdot \mathbf{S} = (\mathbf{V} \cdot \mathbf{A})\mathbf{B}$$

Thus, the dot product between a second-order tensor and a vector results in a vector. In the first case the tensor acts as a *prefactor* and in the second case as a *postfactor*. The two operations in general produce different vectors. The expressions in the above equation can also be written in alternative form using the definition of the transpose of a second-order tensor as

$$\mathbf{V} \cdot \mathbf{S} = \mathbf{S}^{\mathrm{T}} \cdot \mathbf{V}, \quad \mathbf{S} \cdot \mathbf{V} = \mathbf{V} \cdot \mathbf{S}^{\mathrm{T}}$$

In general, one can show that the transpose of the product of tensors (of any order) follows the rule

$$(\mathbf{R} \cdot \mathbf{S})^{\mathrm{T}} = \mathbf{S}^{\mathrm{T}} \cdot \mathbf{R}^{\mathrm{T}}, \quad (\mathbf{R} \cdot \mathbf{S} \cdot \mathbf{T})^{\mathrm{T}} = \mathbf{T}^{\mathrm{T}} \cdot \mathbf{S}^{\mathrm{T}} \cdot \mathbf{R}^{\mathrm{T}}$$

Much like a vector, a tensor can be expressed in component form with respect to a basis. For example, a second-order tensor $\mathbf{S}$ can be represented with respect to the rectangular Cartesian basis $(\hat{\mathbf{e}}_1, \hat{\mathbf{e}}_2, \hat{\mathbf{e}}_3)$ as

$$\begin{aligned}
\mathbf{S} = {} & S_{11}\hat{\mathbf{e}}_1\hat{\mathbf{e}}_1 + S_{12}\hat{\mathbf{e}}_1\hat{\mathbf{e}}_2 + S_{13}\hat{\mathbf{e}}_1\hat{\mathbf{e}}_3 \\
& + S_{21}\hat{\mathbf{e}}_2\hat{\mathbf{e}}_1 + S_{22}\hat{\mathbf{e}}_2\hat{\mathbf{e}}_2 + S_{23}\hat{\mathbf{e}}_2\hat{\mathbf{e}}_3 \\
& + S_{31}\hat{\mathbf{e}}_3\hat{\mathbf{e}}_1 + S_{32}\hat{\mathbf{e}}_3\hat{\mathbf{e}}_2 + S_{33}\hat{\mathbf{e}}_3\hat{\mathbf{e}}_3
\end{aligned}$$

or in the index notation

$$\mathbf{S} = S_{ij}\,\hat{\mathbf{e}}_i\hat{\mathbf{e}}_j \tag{1.6.16}$$

where S_{ij} are termed the components of the second-order tensor $\mathbf{S}$ with respect to the basis $(\hat{\mathbf{e}}_1, \hat{\mathbf{e}}_2, \hat{\mathbf{e}}_3)$.

It is also common to display the components of **S** in an array

$$[S] = \begin{bmatrix} S_{11} & S_{12} & S_{13} \\ S_{21} & S_{22} & S_{23} \\ S_{31} & S_{32} & S_{33} \end{bmatrix} \tag{1.6.17}$$

This rectangular array $[S]$ of scalars S_{ij} is called a *matrix*, and the quantities $S_{11}, S_{12}, \ldots, S_{33}$ are called the *elements* of matrix $[S]$. A matrix $[S]$ is designated sometimes by $[S_{ij}]$. The element in the ith row and jth column of a matrix $[S]$ is denoted by S_{ij}. When $S_{ij} = S_{ji}$ for all i and j, the matrix $[S]$ and the tensor **S** are said to be *symmetric*, and express $[S]^{\mathrm{T}} = [S]$ and $\mathbf{S} = \mathbf{S}^{\mathrm{T}}$.

If the matrix has only one row or one column, only a single subscript is used to designate its elements. For example,

$$\{X\} = \begin{Bmatrix} x_1 \\ x_2 \\ x_3 \end{Bmatrix} \quad \text{and} \quad \{Y\} = \{y_1 \ y_2 \ y_3\} \tag{1.6.18}$$

denote a column matrix and a row matrix, respectively. Row and column matrices can be used to represent the components of a vector.

1.6.5.2 Transformation laws for vectors and tensors

In much of the present study, only Cartesian bases are used. An orthonormal Cartesian basis is denoted by

$$\{\hat{\mathbf{e}}_x, \hat{\mathbf{e}}_y, \hat{\mathbf{e}}_z\} \quad \text{or} \quad \{\hat{\mathbf{e}}_1, \hat{\mathbf{e}}_2, \hat{\mathbf{e}}_3\}$$

The Cartesian coordinates are denoted by (x, y, z) or (x_1, x_2, x_3). The familiar rectangular Cartesian coordinate system is shown in Fig. 1.6.3(a). Only right-handed coordinate systems are used in this study.

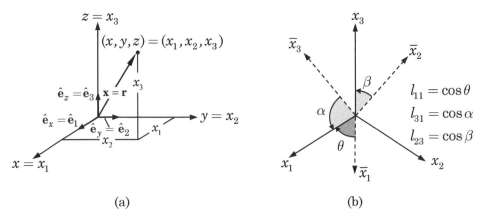

(a)   (b)

Fig. 1.6.3: (a) A rectangular Cartesian coordinate system. (b) Barred and unbarred coordinate systems.

A position vector $\mathbf{x} = \mathbf{r}$ to an arbitrary point $(x, y, z) = (x_1, x_2, x_3)$ is expressed as

$$\mathbf{x} = x\hat{\mathbf{e}}_x + y\hat{\mathbf{e}}_y + z\hat{\mathbf{e}}_z = x_1\hat{\mathbf{e}}_1 + x_2\hat{\mathbf{e}}_2 + x_3\hat{\mathbf{e}}_3$$

or, in summation notation,

$$\mathbf{x} = x_j\hat{\mathbf{e}}_j, \quad \mathbf{x} \cdot \mathbf{x} = r^2 = x_i x_i \tag{1.6.19}$$

Next, the relationship between the components of two different orthonormal coordinate systems, say, unbarred and barred [see Fig. 1.6.3(b)] is established. Consider the unbarred coordinate basis

$$(\hat{\mathbf{e}}_1, \hat{\mathbf{e}}_2, \hat{\mathbf{e}}_3) \quad \text{(unbarred system)}$$

and the barred coordinate basis

$$(\hat{\bar{\mathbf{e}}}_1, \hat{\bar{\mathbf{e}}}_2, \hat{\bar{\mathbf{e}}}_3) \quad \text{(barred system)}$$

The basis vectors are related by the following relations

$$\hat{\bar{\mathbf{e}}}_i = \ell_{ij}\hat{\mathbf{e}}_j, \quad \{\hat{\bar{\mathbf{e}}}\} = [L]\{\hat{\mathbf{e}}\} \tag{1.6.20}$$

where

$$\ell_{ij} = \hat{\bar{\mathbf{e}}}_i \cdot \hat{\mathbf{e}}_j \tag{1.6.21}$$

Note that the first subscript of ℓ_{ij} comes from the barred coordinate system and the second subscript from the unbarred system. Obviously, ℓ_{ij} is not symmetric (i.e. $\ell_{ij} \neq \ell_{ji}$); but the direction cosines have the property

$$\ell_{im}\ell_{jm} = \delta_{ij} \quad \text{or} \quad [L][L]^{\mathrm{T}} = [I] \tag{1.6.22}$$

Equation (1.6.20) gives the relationship between the base vectors of the barred and unbarred coordinate systems. The relationship also holds between the components $(\bar{A}_1, \bar{A}_2, \bar{A}_3)$ and (A_1, A_2, A_3):

$$\{\bar{A}\} = [L]\{A\} \tag{1.6.23}$$

and it is called the *transformation rule* between the barred and unbarred components of a vector in the two orthogonal coordinate systems. Similarly, a second-order tensor components in the two coordinate systems are related by

$$\bar{S}_{ij} = \ell_{im}\ell_{jn}S_{mn} \quad [\bar{S}] = [L][S][L]^{\mathrm{T}} \tag{1.6.24}$$

Mathematically speaking, $\mathbf{S}$ is said to be a second-order tensor if its components transform according to Eq. (1.6.24).

1.6.6   Calculus of Vectors and Tensors

Let us denote a scalar field by $\phi = \phi(\mathbf{x})$, $\mathbf{x}$ being the position vector in a rectangular Cartesian system (x_1, x_2, x_3). Let us now denote a differential element with $d\mathbf{x}$ and its magnitude by $ds \equiv |d\mathbf{x}|$. Then $\hat{\mathbf{e}} = d\mathbf{x}/ds$ is a unit vector in the direction of $d\mathbf{x}$, and it may be expressed as (by the use of the chain rule)

$$\left(\frac{d\phi}{ds}\right)_{\hat{\mathbf{e}}} = \frac{d\mathbf{x}}{ds} \cdot \frac{\partial\phi}{\partial\mathbf{x}} = \hat{\mathbf{e}} \cdot \left(\hat{\mathbf{e}}_1\frac{\partial\phi}{\partial x_1} + \hat{\mathbf{e}}_2\frac{\partial\phi}{\partial x_2} + \hat{\mathbf{e}}_3\frac{\partial\phi}{\partial x_3}\right)$$

The derivative $(d\phi/ds)_{\hat{\mathbf{e}}}$ is called the *directional derivative* of ϕ. Thus, it is the *rate of change* of ϕ with respect to distance and it depends on the direction $\hat{\mathbf{e}}$ in which the distance is taken. The vector $\partial\phi/\partial\mathbf{x}$ is called the *gradient vector* and is denoted by $\mathrm{grad}(\phi) = \nabla\phi$:

$$\nabla\phi = \mathrm{grad}\,\phi = \hat{\mathbf{e}}_1\frac{\partial\phi}{\partial x_1} + \hat{\mathbf{e}}_2\frac{\partial\phi}{\partial x_2} + \hat{\mathbf{e}}_3\frac{\partial\phi}{\partial x_3} \qquad (1.6.25)$$

The expression $\mathrm{grad}\,\phi$ is interpreted as some operator operating on function ϕ, that is $\mathrm{grad}\,\phi \equiv \nabla\phi$. This operator is denoted by

$$\nabla \equiv \hat{\mathbf{e}}_1\frac{\partial}{\partial x_1} + \hat{\mathbf{e}}_2\frac{\partial}{\partial x_2} + \hat{\mathbf{e}}_3\frac{\partial}{\partial x_3} = \hat{\mathbf{e}}_x\frac{\partial}{\partial x} + \hat{\mathbf{e}}_y\frac{\partial}{\partial y} + \hat{\mathbf{e}}_z\frac{\partial}{\partial z} \qquad (1.6.26)$$

and is called the *del operator*. The del operator is a *vector differential* operator. The gradient of a scalar function is a vector and the gradient of a vector function is a second-order tensor.

If a surface is given by $\phi(\mathbf{x}) = c$, the unit normal to the surface at point $\mathbf{x}$ is determined by

$$\hat{\mathbf{n}} = \pm\frac{\mathrm{grad}\,\phi}{|\mathrm{grad}\,\phi|} \qquad (1.6.27)$$

The plus or minus sign appears because the direction of $\hat{\mathbf{n}}$ may point in either direction away from the surface. If the surface is closed, the usual convention is to take $\hat{\mathbf{n}}$ pointing outward.

Two types of gradients are used in continuum mechanics: forward and backward gradients. The forward gradient is the usual gradient and backward gradient is the transpose of the forward gradient operator. To see the difference between the two types of gradients, consider a vector function $\mathbf{A} = A_i(\mathbf{x})\hat{\mathbf{e}}_i$. The forward and backward gradients of $\mathbf{A}$ are

$$\overrightarrow{\nabla}\mathbf{A} = \nabla\mathbf{A} = \hat{\mathbf{e}}_j\frac{\partial}{\partial x_j}\left(A_i\hat{\mathbf{e}}_i\right) = \frac{\partial A_i}{\partial x_j}\hat{\mathbf{e}}_j\hat{\mathbf{e}}_i = A_{i,j}\hat{\mathbf{e}}_j\hat{\mathbf{e}}_i \qquad (1.6.28)$$

$$\overleftarrow{\nabla}\mathbf{A} = \left(\nabla\mathbf{A}\right)^{\mathrm{T}} = \frac{\partial A_i}{\partial x_j}\left(\hat{\mathbf{e}}_j\hat{\mathbf{e}}_i\right)^{\mathrm{T}} = A_{i,j}\hat{\mathbf{e}}_i\hat{\mathbf{e}}_j \qquad (1.6.29)$$

where $A_{i,j} = \partial A_i/\partial x_j$.

Operations involving the del operator and second- and higher-order tensors follow the same rules as the first-order tensors (i.e. vectors), except that the order of the base vectors is kept intact (i.e. not switched from the order in which they appear). First, consider the gradient of a vector **A**

$$\text{grad } \mathbf{A} = \mathbf{\nabla}\mathbf{A} = \hat{\mathbf{e}}_i \frac{\partial}{\partial x_i}(A_j \hat{\mathbf{e}}_j) = \frac{\partial A_j}{\partial x_i} \hat{\mathbf{e}}_i \hat{\mathbf{e}}_j$$

which is a second-order tensor **S**

$$\mathbf{\nabla}\mathbf{A} \equiv \mathbf{S} = S_{ij}\,\hat{\mathbf{e}}_i\hat{\mathbf{e}}_j, \quad S_{ij} = \frac{\partial A_j}{\partial x_i} = A_{j,i} \tag{1.6.30}$$

Of course, in general, $\mathbf{S} = \mathbf{\nabla}\mathbf{A}$ is not a symmetric tensor. The gradient $\mathbf{\nabla}\mathbf{A}$ and its transpose can be expressed as the sum of symmetric and skew-symmetric tensors

$$\mathbf{\nabla}\mathbf{A} = \frac{1}{2}\left(\frac{\partial A_j}{\partial x_i} + \frac{\partial A_i}{\partial x_j}\right)\hat{\mathbf{e}}_i\hat{\mathbf{e}}_j - \frac{1}{2}\left(\frac{\partial A_i}{\partial x_j} - \frac{\partial A_j}{\partial x_i}\right)\hat{\mathbf{e}}_i\hat{\mathbf{e}}_j$$

$$= \frac{1}{2}\left[(\mathbf{\nabla}\mathbf{A}) + (\mathbf{\nabla}\mathbf{A})^{\mathrm{T}}\right] - \frac{1}{2}\left[(\mathbf{\nabla}\mathbf{A})^{\mathrm{T}} - (\mathbf{\nabla}\mathbf{A})\right] = \mathbf{V} - \mathbf{W} \tag{1.6.31}$$

$$\left(\mathbf{\nabla}\mathbf{A}\right)^{\mathrm{T}} = \frac{1}{2}\left(\frac{\partial A_j}{\partial x_i} + \frac{\partial A_i}{\partial x_j}\right)\hat{\mathbf{e}}_j\hat{\mathbf{e}}_i + \frac{1}{2}\left(\frac{\partial A_j}{\partial x_i} - \frac{\partial A_i}{\partial x_j}\right)\hat{\mathbf{e}}_j\hat{\mathbf{e}}_i$$

$$= \frac{1}{2}\left[(\mathbf{\nabla}\mathbf{A})^{\mathrm{T}} + (\mathbf{\nabla}\mathbf{A})\right] + \frac{1}{2}\left[(\mathbf{\nabla}\mathbf{A})^{\mathrm{T}} - (\mathbf{\nabla}\mathbf{A})\right] = \mathbf{V} + \mathbf{W} \tag{1.6.32}$$

where **V** and **W** are the symmetric and skew-symmetric second-order tensors, respectively,

$$\mathbf{V} = \frac{1}{2}\left[(\mathbf{\nabla}\mathbf{A})^{\mathrm{T}} + (\mathbf{\nabla}\mathbf{A})\right] = \mathbf{V}^{\mathrm{T}}; \qquad V_{ij} = \frac{1}{2}\left(\frac{\partial A_j}{\partial x_i} + \frac{\partial A_i}{\partial x_j}\right)$$

$$\mathbf{W} = \frac{1}{2}\left[(\mathbf{\nabla}\mathbf{A})^{\mathrm{T}} - (\mathbf{\nabla}\mathbf{A})\right] = -\mathbf{W}^{\mathrm{T}}; \quad W_{ij} = \frac{1}{2}\left(\frac{\partial A_i}{\partial x_j} - \frac{\partial A_j}{\partial x_i}\right) \tag{1.6.33}$$

The divergence of a second-order tensor **S** is a vector **A**

$$\mathbf{\nabla}\cdot\mathbf{S} = \hat{\mathbf{e}}_i\frac{\partial}{\partial x_i}\cdot(S_{jk}\hat{\mathbf{e}}_j\hat{\mathbf{e}}_k) = \frac{\partial S_{jk}}{\partial x_i}(\hat{\mathbf{e}}_i\cdot\hat{\mathbf{e}}_j)\hat{\mathbf{e}}_k = \frac{\partial S_{jk}}{\partial x_j}\hat{\mathbf{e}}_k$$

$$\equiv \mathbf{A} = A_k\,\hat{\mathbf{e}}_k, \quad A_k = \frac{\partial S_{jk}}{\partial x_j} = S_{jk,j} \tag{1.6.34}$$

The *curl* of a second-order tensor **S**, denoted by $\mathbf{\nabla}\times\mathbf{S}$, is a second-order tensor

$$\mathbf{\nabla}\times\mathbf{S} = \hat{\mathbf{e}}_i\frac{\partial}{\partial x_i}\times(S_{jl}\hat{\mathbf{e}}_j\hat{\mathbf{e}}_l) = \frac{\partial S_{jl}}{\partial x_i}(\hat{\mathbf{e}}_i\times\hat{\mathbf{e}}_j)\hat{\mathbf{e}}_l = \frac{\partial S_{jl}}{\partial x_i}e_{ijk}\,\hat{\mathbf{e}}_k\,\hat{\mathbf{e}}_l$$

$$\equiv \mathbf{T} = T_{kl}\,\hat{\mathbf{e}}_k\,\hat{\mathbf{e}}_l, \quad T_{kl} = e_{ijk}\frac{\partial S_{jl}}{\partial x_i}$$

Thus, the gradient of a tensor $\mathbf{S}$ of any order increases its order by one, the divergence of $\mathbf{S}$ decreases its order by one, and the curl of $\mathbf{S}$ keeps the order unchanged.

Two types of "double-dot products" between two second-order tensors are useful in the sequel. The horizontal double-dot product and the vertical double-dot product between a dyad $(\mathbf{AB})$ and another dyad $(\mathbf{CD})$ (i.e. $\mathbf{A}$, $\mathbf{B}$, $\mathbf{C}$, and $\mathbf{D}$ are vectors) are defined as the scalars

$$(\mathbf{AB})\cdot\cdot(\mathbf{CD}) \equiv (\mathbf{B}\cdot\mathbf{C})(\mathbf{A}\cdot\mathbf{D}); \quad (\mathbf{AB}):(\mathbf{CD}) \equiv (\mathbf{A}\cdot\mathbf{C})(\mathbf{B}\cdot\mathbf{D})$$

The double-dot products, by this definition, are commutative. The two double-dot products between two dyads $\mathbf{S}$ and $\mathbf{T}$ in a rectangular Cartesian system are

$$\mathbf{S}\cdot\cdot\mathbf{T} = S_{ij}T_{mn}(\hat{\mathbf{e}}_j \cdot \hat{\mathbf{e}}_m)(\hat{\mathbf{e}}_i \cdot \hat{\mathbf{e}}_n) = S_{ij}T_{mn}\delta_{jm}\delta_{in} = S_{ij}\,T_{ji} \quad (1.6.35)$$

$$\mathbf{S}:\mathbf{T} = S_{ij}T_{mn}(\hat{\mathbf{e}}_i \cdot \hat{\mathbf{e}}_m)(\hat{\mathbf{e}}_j \cdot \hat{\mathbf{e}}_n) = S_{ij}T_{mn}\delta_{im}\delta_{jn} = S_{ij}\,T_{ij} \quad (1.6.36)$$

The trace of a second-order tensor (i.e. the sum of the diagonal terms of the matrix representing the tensor) is defined to be the double-dot product of the tensor with the unit tensor

$$\operatorname{tr}\mathbf{S} = \mathbf{S}:\mathbf{I} = \mathbf{S}\cdot\cdot\mathbf{I} \qquad (1.6.37)$$

Certain combinations of the components of a tensor remain the same in all coordinate systems; that is they are invariant under coordinate transformations. Such quantities (involving sums and products of the components of a tensor) are termed *invariants*. For example, the determinant of a tensor is the same in all coordinate systems. Similarly, the trace of the matrix representing a tensor is an *invariant*. Among many invariants of a tensor, the following three invariants, called *principal invariants*, are identified because of their role in finding eigenvalues of the tensor (readers should be aware of the fact that the definition of the second and third principal invariants may differ from other books)

$$I_1 = \operatorname{tr}\mathbf{S}, \quad I_2 = \tfrac{1}{2}\left[(\operatorname{tr}\mathbf{S})^2 - \operatorname{tr}\left(\mathbf{S}^2\right)\right], \quad I_3 = |\mathbf{S}| \qquad (1.6.38)$$

In terms of the rectangular Cartesian components of $\mathbf{S}$, the three principal invariants have the form

$$I_1 = S_{ii}, \quad I_2 = \tfrac{1}{2}\left(S_{ii}S_{jj} - S_{ij}S_{ji}\right), \quad I_3 = |\mathbf{S}| \qquad (1.6.39)$$

In the general scheme that is developed so far, scalars are the *zeroth-order tensors*, vectors are the *first-order tensors*, and dyads are the *second-order tensors*. The order of a tensor can be determined by counting the number of basis vectors in the representation of a tensor. In order to qualify as a tensor its components must obey the transformation law (in rectangular Cartesian component form)

$$\bar{T}_{ijkl\cdots} = \ell_{ip}\ell_{jq}\ell_{kr}\ell_{ls} \cdots T_{pqrs\cdots} \qquad (1.6.40)$$

In view of the matrix representation of a second-order tensor, many of the definitions and properties introduced for matrices can be extended to second-order tensors, $\mathbf{S}$. They are summarized here.

(1) $\mathbf{S}$ is *symmetric* if and only if $\mathbf{S} = \mathbf{S}^{\mathrm{T}}$ ($S_{ij} = S_{ji}$).

(2) $\mathbf{S}$ is *skew-symmetric* if and only if $\mathbf{S} = -\mathbf{S}^{\mathrm{T}}$ ($S_{ij} = -S_{ji}$ for $i \neq j$ and $S_{(i)(i)} = 0$ for any fixed i).

(3) $\mathbf{S}$ can be represented as a sum of symmetric and skew-symmetric parts:

$$\mathbf{S} = \tfrac{1}{2}\left(\mathbf{S} + \mathbf{S}^{\mathrm{T}}\right) + \tfrac{1}{2}\left(\mathbf{S} - \mathbf{S}^{\mathrm{T}}\right) \equiv \mathbf{S}^{\mathrm{sym}} + \mathbf{S}^{\mathrm{skew}} \tag{1.6.41}$$

(4) If $\mathbf{S}$ is symmetric, $\mathbf{W}$ is skew-symmetric, and $\mathbf{T}$ is an arbitrary tensor, then

$$\mathbf{S} : \mathbf{W} = \mathbf{W} : \mathbf{S} = 0, \quad \mathbf{S} : \mathbf{T} = \mathbf{S} : \mathbf{T}^{\mathrm{sym}},$$

$$\mathbf{W} : \mathbf{T} = -\mathbf{W} : \mathbf{T}^{\mathrm{T}} = \mathbf{W} : \mathbf{T}^{\mathrm{skew}}$$

Also, if $\mathbf{S} : \mathbf{T} = 0$ for any tensor $\mathbf{S}$, then $\mathbf{T} = \mathbf{0}$. If $\mathbf{S} : \mathbf{T} = 0$ for any symmetric tensor $\mathbf{S}$, then $\mathbf{T}$ is skew-symmetric. The converse also holds: If $\mathbf{S} : \mathbf{W} = 0$ for any skew-symmetric tensor $\mathbf{W}$, then $\mathbf{S}$ is symmetric.

(5) The inverse $\mathbf{T}$ of any second-order tensor $\mathbf{S}$, denoted $\mathbf{T} = \mathbf{S}^{-1}$, is defined to be $\mathbf{S}^{-1} \cdot \mathbf{S} = \mathbf{S} \cdot \mathbf{S}^{-1} = \mathbf{I}$. The notation $(\mathbf{S}^{-1})^{\mathrm{T}} = (\mathbf{S}^{\mathrm{T}})^{-1} = \mathbf{S}^{-\mathrm{T}}$ is used, and note that $(\mathbf{S} \cdot \mathbf{T})^{-1} = \mathbf{T}^{-1} \cdot \mathbf{S}^{-1}$ and $(\mathbf{S} \cdot \mathbf{T})^{-\mathrm{T}} = \mathbf{S}^{-\mathrm{T}} \cdot \mathbf{T}^{-\mathrm{T}}$.

(6) A necessary and sufficient condition for a second-order tensor $\mathbf{Q}$ to be *orthogonal* is $\mathbf{Q} \cdot \mathbf{Q}^{\mathrm{T}} = \mathbf{I}$. The determinant of an orthogonal matrix is $|\mathbf{Q}| = \pm 1$. When $|\mathbf{Q}| = +1$, $\mathbf{Q}$ is called a *proper orthogonal matrix*. It can be shown that an orthogonal tensor $\mathbf{Q}$ preserves the inner product in the sense that $(\mathbf{Q} \cdot \mathbf{u}) \cdot (\mathbf{Q} \cdot \mathbf{v}) = \mathbf{u} \cdot \mathbf{v}$. In the case of physical vectors, this amounts to preserving the lengths of the vectors $\mathbf{u}$ and $\mathbf{v}$ as well as the angle between them.

(7) A second-order tensor $\mathbf{S}$ is said to be *positive-definite* if and only if $\mathbf{u} \cdot \mathbf{S} \cdot \mathbf{u} > 0$ for all nonzero vectors $\mathbf{u}$.

The gradient and divergence of a tensor $\mathbf{T}$ in other coordinate systems can be readily obtained by writing ∇ and $\mathbf{T}$ in that coordinate system and accounting for the derivatives of the basis vectors. One of the commonly used orthogonal curvilinear coordinate system is the *cylindrical* coordinate system (see Fig. 1.6.4), and Table 1.6.1 contains a summary of the basic information for the coordinate system. The transformation equations between the orthogonal rectangular Cartesian system (x, y, z) and the cylindrical coordinate system (r, θ, z) are obvious from Table 1.6.1.

Table 1.6.1: Base vectors and del and Laplace operators in the cylindrical coordinate system.

$x = r\cos\theta,\ y = r\sin\theta,\ z = z,\ \mathbf{R} = r\hat{\mathbf{e}}_r + z\hat{\mathbf{e}}_z$

$\mathbf{A} = A_r\hat{\mathbf{e}}_r + A_\theta\hat{\mathbf{e}}_\theta + A_z\hat{\mathbf{e}}_z$ (typical vector)

$\hat{\mathbf{e}}_r = \cos\theta\ \hat{\mathbf{e}}_x + \sin\theta\ \hat{\mathbf{e}}_y$

$\hat{\mathbf{e}}_\theta = -\sin\theta\ \hat{\mathbf{e}}_x + \cos\theta\ \hat{\mathbf{e}}_y,\ \ \hat{\mathbf{e}}_z = \hat{\mathbf{e}}_z$

Nonzero derivatives of the base vectors are

$\frac{\partial\hat{\mathbf{e}}_r}{\partial\theta} = -\sin\theta\ \hat{\mathbf{e}}_x + \cos\theta\ \hat{\mathbf{e}}_y = \hat{\mathbf{e}}_\theta$

$\frac{\partial\hat{\mathbf{e}}_\theta}{\partial\theta} = -\cos\theta\ \hat{\mathbf{e}}_x - \sin\theta\ \hat{\mathbf{e}}_y = -\hat{\mathbf{e}}_r$

$\boldsymbol{\nabla} = \hat{\mathbf{e}}_r\frac{\partial}{\partial r} + \frac{1}{r}\hat{\mathbf{e}}_\theta\frac{\partial}{\partial\theta} + \hat{\mathbf{e}}_z\frac{\partial}{\partial z}$

$\nabla^2 = \frac{1}{r}\left[\frac{\partial}{\partial r}\left(r\frac{\partial}{\partial r}\right) + \frac{1}{r}\frac{\partial^2}{\partial\theta^2} + r\frac{\partial^2}{\partial z^2}\right]$

$\boldsymbol{\nabla}\mathbf{A} = \hat{\mathbf{e}}_r\hat{\mathbf{e}}_r\frac{\partial A_r}{\partial r} + \hat{\mathbf{e}}_r\hat{\mathbf{e}}_\theta\frac{\partial A_\theta}{\partial r} + \hat{\mathbf{e}}_\theta\hat{\mathbf{e}}_r\frac{1}{r}\left(\frac{\partial A_r}{\partial\theta} - A_\theta\right) + \hat{\mathbf{e}}_r\hat{\mathbf{e}}_z\frac{\partial A_z}{\partial r} + \hat{\mathbf{e}}_z\hat{\mathbf{e}}_r\frac{\partial A_r}{\partial z}$

$\qquad + \hat{\mathbf{e}}_\theta\hat{\mathbf{e}}_\theta\frac{1}{r}\left(A_r + \frac{\partial A_\theta}{\partial\theta}\right) + \frac{1}{r}\hat{\mathbf{e}}_\theta\hat{\mathbf{e}}_z\frac{\partial A_z}{\partial\theta} + \hat{\mathbf{e}}_z\hat{\mathbf{e}}_\theta\frac{\partial A_\theta}{\partial z} + \hat{\mathbf{e}}_z\hat{\mathbf{e}}_z\frac{\partial A_z}{\partial z}$

$\boldsymbol{\nabla}\cdot\mathbf{A} = \frac{1}{r}\left[\frac{\partial(rA_r)}{\partial r} + \frac{\partial A_\theta}{\partial\theta} + r\frac{\partial A_z}{\partial z}\right]$

$\boldsymbol{\nabla}\times\mathbf{A} = \left(\frac{1}{r}\frac{\partial A_z}{\partial\theta} - \frac{\partial A_\theta}{\partial z}\right)\hat{\mathbf{e}}_r + \left(\frac{\partial A_r}{\partial z} - \frac{\partial A_z}{\partial r}\right)\hat{\mathbf{e}}_\theta + \frac{1}{r}\left[\frac{\partial(rA_\theta)}{\partial r} - \frac{\partial A_r}{\partial\theta}\right]\hat{\mathbf{e}}_z$

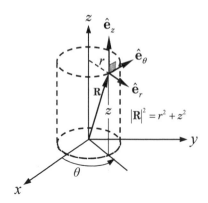

Fig. 1.6.4: Cylindrical coordinate system.

In closing this section, the gradient, divergence, and curl theorems are expressed in a single statement for future reference. Let Ω denote a bounded domain in $\Re^3$ with closed boundary Γ. Then the following integral identity holds for a tensor $\mathbf{T}$ of any order:

$$\int_\Omega \boldsymbol{\nabla} * \mathbf{T}\, d\mathbf{x} = \oint_\Gamma \hat{\mathbf{n}} * \mathbf{T}\, ds \qquad (1.6.42)$$

where a circle around a boundary integral denotes closed boundary integral and the operation * denotes gradient, divergence, or curl operation.

1.7 Concepts from Functional Analysis

1.7.1 Introduction

In the study of solutions to various physical problems that are unrelated, one may find that they share certain common mathematical structure. For example, the approximate solution of problems by numerical methods ultimately leads to the solution of a set of linear equations. Then it is necessary to know under what conditions these equations, without regard to a specific problem, have a solution (existence) and it is the only solution (uniqueness). That is, instead of studying a particular kind of problem, the particular problem is studied in the context of a general problem. A systematic study of the mathematical and computational properties of a given *class* of problems leads to a formal treatment that is more abstract than the treatment of a specific problem. This in turn leads to deeper understanding and, more importantly, provides insights into the essential – physical, mathematical, and computational – features of the problem. A good example of this kind is provided by the notion of an abstract vector, which includes as a special case the geometric vector studied in Section 1.6. The notion of "direction" is generalized to include a "property" that the vector should possess, among other things, in order to qualify as a vector. The collection of all vectors which share a common property, such as satisfying a given differential equation, and satisfying certain rules of addition and scalar multiplication, is termed a *vector space*. An *abstract vector space* contains points that could be numbers, points in ordinary Euclidean space of functions, and so on. The study of those properties of a completely arbitrary collection of points in a space which hold independently of their particular nature, whether they be numbers, points, or functions, is the subject of *functional analysis.*

In functional analysis the fundamental concepts and methods from elementary analysis, algebra, and geometry are studied from a unified point of view and for more general objects. For example, the definitions of a functional, its extremum, and the conditions for the existence of an extremum in the calculus of variations are entirely analogous to the definitions of a function, its extremum, and the conditions for the existence of an extremum in the differential calculus. Another example of analogy between functional analysis and geometry is given by the development of functions with respect to elements of an orthogonal system, which resemble orthogonal systems of vectors in the Euclidean space. Thus, functional analysis is an indispensable tool in approximation theories and numerical analysis. Therefore, a brief exposure to the subject will make the reader comfortable when abstract variational statements are encountered in the context of finite element formulations of physical problems. To keep the scope of this review within reasonable limits, only the concepts useful in the present study from linear algebra and analysis are included here. For more details the reader may consult any one of the many books on functional analysis [14–16].

The following notation is used in presenting concepts from linear algebra:

$$
\begin{aligned}
\subset \quad & \text{means "a subset of"} \\
\not\subset \quad & \text{means "not a subset of"} \\
\in \quad & \text{means "an element of"} \\
\notin \quad & \text{means "not an element of"} \\
\ni \quad & \text{means "such that"} \\
\forall \quad & \text{means "for all"} \\
\Re = & \{x : x \text{ is a real number,} \ -\infty < x < \infty\} \\
[a, b] = & \{x : x \text{ is real,} \ a \leq x \leq b\} \\
[a, b) = & \{x : x \text{ is real,} \ a \leq x < b\} \\
(a, b] = & \{x : x \text{ is real,} \ a < x \leq b\} \\
(a, b) = & \{x : x \text{ is real,} \ a < x < b\}
\end{aligned}
$$

The set $[a, b]$ is called a *closed interval* and (a, b) is called an *open interval*.

1.7.2 Linear Vector Spaces

In Section 1.6, a geometric or physical vector is considered. In this section, more general, abstract objects than physical vectors, which are also called vectors, are studied. A collection of *vectors, $u, v, w, \ldots$,* is called a *linear vector space V* over the real number field $\Re$ if the following rules of vector addition and scalar multiplication of a vector are satisfied by the elements of the vector space.

1.7.2.1 Vector addition

To every pair of vectors $u, v \in V$, there corresponds a unique vector $u + v \in V$, called the *sum* of u and v, with the following properties:

(1) $u + v = v + u$ (commutative)
(2) $(u + v) + w = u + (v + w)$ (associative)
(3) there exists a unique vector, $0 \in V$, independent of u such that
$u + 0 = u$ for every $u \in V$ (existence of an identity or zero element)
(4) to every u there exists a unique vector, $-u$
(that depends on u) such that $u + (-u) = 0$ for every $u \in V$
(existence of the additive inverse element)

All properties listed above for an abstract vector are a generalization of the properties of a physical vector that we discussed in Section 1.6.2. We note that the notion of a vector in abstract analysis does not require the vector to have a magnitude. However, in nearly all cases of engineering analysis the vector is endowed with a magnitude, in which case the vector is said to belong to a *normed vector space*, which is discussed in Section 1.7.3. Thus, physical vectors of Section 1.6 belong to a special normed vector space.

1.7.2.2   Scalar multiplication

To every vector $u \in V$ and every real number $\alpha \in \Re$ there corresponds a unique vector $\alpha u \in V$ such that the following properties hold:

(1) $\alpha(\beta u) = (\alpha\beta)u$ (associative)
(2) $(\alpha + \beta)u = \alpha u + \beta u$ (distributive with respect to the scalar addition)
(3) $\alpha(u + v) = \alpha u + \alpha v$ (distributive with respect to the vector addition)
(4) $1 \cdot u = u \cdot 1; \quad 0 \cdot u = 0$

1.7.2.3   Linear subspaces

A linear subspace S of a given linear vector space V over $\Re$ is a nonempty subset of V which is (i.e. S) itself a linear vector space with respect to the operations of addition and scalar multiplication defined over V. To determine whether a given subset of a linear vector space qualifies as a subspace, one must first check if it is a linear vector space by itself. In other words, a subset S of a linear space V is a subspace if and only if $\alpha u + \beta v \in S$ for all $u, v \in S$ and scalars $\alpha, \beta \in \Re$ (closure property), and the identity and inverse elements are the same as those defined in V.

1.7.2.4   Linear dependence and independence of vectors

Let V be a linear vector space. If an element $u \in V$ can be expressed as $u = \sum \alpha_i u_i$, $\alpha_i \in \Re$ and $u_i \in V$, then u is said to be a linear combination of u_i's. An expression of the form $\sum \alpha_i u_i = 0$ is called a *linear relation* among the u_i's. A relation with all $\alpha_i = 0$ is called a *trivial relation*, and a relation with at least one coefficient nonzero is called a *nontrivial relation*.

A set of vectors $\{u_i\}$ is said to be *linearly dependent* if there exist real numbers $\{\alpha_i\}$, not all zero, such that $\sum \alpha_i u_i = 0$, that is there exists a nontrivial relation among them. Otherwise the set is said to be *linearly independent*.

The following observations are simple consequences of the definitions:

(1) If $(u_1, u_2, \ldots, u_n)$ are dependent, at least one vector is a linear combination of the others.

(2) If the zero vector belongs to a set of vectors, the set is linearly dependent (since for $\alpha \in \Re$ and $0 \in V$, $\alpha 0 = 0$).

(3) A set consisting of exactly one nonzero vector is linearly independent (since for $\alpha \in \Re$ and $u \in V$, $\alpha u = 0$ implies $\alpha = 0$).

(4) If $(u_1, u_2, \ldots, u_n)$ is a linearly dependent set, so is any set that includes $(u_1, u_2, \ldots, u_n)$. A set of vectors is linearly dependent if some subset of it is linearly dependent.

(5) If $(u_1, u_2, \ldots, u_n)$ is a linearly independent set, any subset of it is also linearly independent.

1.7.3   Normed Vector Spaces

Let V be a linear vector space over the real number field $\Re$. The notation $||\cdot||$ is used to denote the norm of real-valued functions $u(\mathbf{x})$, $\mathbf{x} \in \Omega \subset \Re^3$. The integral

$$\int_\Omega u(\mathbf{x}) \, d\mathbf{x}$$

denotes the Lebesgue integral[1] of u. For $1 \leq p \leq \infty$, let

$$||u||_{L_p(\Omega)} \equiv ||u||_p = \left[\int_\Omega |u(\mathbf{x})|^p \, d\mathbf{x} \right]^{1/p} < \infty \tag{1.7.1}$$

and for $p = \infty$ set

$$||u||_{L_\infty(\Omega)} \equiv ||u||_\infty = \sup \left\{ |u(\mathbf{x})| : \mathbf{x} \in \Omega \right\} \tag{1.7.2}$$

For $1 \leq p \leq \infty$, the *Lebesgue spaces* are defined by

$$L_p(\Omega) = \{u : ||u||_p < \infty\} \tag{1.7.3}$$

The following inequalities are useful in establishing the continuity of linear and bilinear forms.

1.7.3.1   Hölder inequality

For $1 \leq p, q \leq \infty$ such that $1/p + 1/q = 1$ and if $f \in L_p(\Omega)$ and $g \in L_q(\Omega)$, then $fg \in L_1(\Omega)$ and

$$||fg||_1 = \int_\Omega |fg| d\mathbf{x} \leq \left(\int_\Omega |f|^p d\mathbf{x} \right)^{1/p} \left(\int_\Omega |g|^q d\mathbf{x} \right)^{1/q} \tag{1.7.4}$$

For $p = q = 2$, the above inequality is known as the *Cauchy–Schwartz inequality*.

1.7.3.2   Minkowski inequality

For $1 \leq p \leq \infty$ and $f, g \in L_p(\Omega)$, the following inequality holds:

$$||f + g||_p \leq ||f||_p + ||g||_p \tag{1.7.5}$$

A norm $||\cdot||$ can be used to define a notion of distance, or *natural metric*

$$d(u, v) \equiv ||u - v|| \quad \text{for} \quad u, v \in V \tag{1.7.6}$$

---

[1]Formal definitions of Riemann and Lebesgue integrable functions are quite involved and require considerable mathematical machinery [12]. In simple terms, the Riemann integral is one in which the integrand is continuous and integrable in the usual sense, whereas piecewise continuous functions may have finite jump discontinuities, but they are locally Lebesgue integrable.

A linear vector space endowed with the topology induced by this metric is called a *normed vector space*. A linear subspace S of a normed vector space V is a linear subspace equipped with the norm of V. A subspace is itself a vector space.

A vector space can have more than one norm defined in it. Two norms $\| \cdot \|_1$ and $\| \cdot \|_2$ on the normed linear space V are said to be *equivalent* if there exist positive numbers c_1 and c_2, independent of $u \in V$, such that the following double inequality holds:

$$c_1 \|u\|_1 \leq \|u\|_2 \leq c_2 \|u\|_1 \tag{1.7.7}$$

A normed space V is called *complete* if every Cauchy sequence $\{u_j\}$ of elements of V has a limit $u \in V$. For a normed vector space, a Cauchy sequence is one such that

$$\|u_j - u_k\| \to 0 \quad \text{as} \quad j, k \to \infty$$

and completeness means that

$$\|u - u_j\| \to 0 \quad \text{as} \quad j \to \infty$$

A normed vector space which is complete in its natural metric is called a *Banach space*. A linear subspace of a Banach space is itself a Banach space if and only if the subspace is closed (hence complete). Every finite-dimensional linear subspace of a normed vector space (that is not necessarily finite dimensional) is complete. The n-dimensional Euclidean space $\Re^n$ is a Banach space with respect to the *Euclidean norm*

$$||\mathbf{x}|| \equiv \sqrt{\sum_{i=1}^{n} x_i^2} \tag{1.7.8}$$

The space $C[0, 1]$ of real-valued continuous functions $f(x)$ defined on the closed interval $[0, 1]$, equipped with the sup-norm defined in Eq. (1.7.2), is also a Banach space. It is a linear vector space with respect to the vector addition and scalar multiplication defined as

$$(f + g)(x) = f(x) + g(x), \quad (\alpha f)(x) = \alpha f(x), \quad \alpha \in \Re$$

Further, it is complete with respect to the sup-norm defined in Eq. (1.7.2):

$$||f||_\infty \equiv \max |f(x)|$$

Let $C^m(\Omega)$ denote the set of all real-valued functions with m continuous derivatives defined in $\Omega \in \Re^3$, and let $C^\infty(\Omega)$ denote the set of infinitely differentiable continuous functions. The norm on $C^m(\Omega)$, called the *Sobolev norm*, is defined by

$$||u||_{m,p} = \left[\int_\Omega \sum_{|\alpha| \leq m} |D^\alpha u(\mathbf{x})|^p d\mathbf{x} \right]^{1/p} \tag{1.7.9}$$

for $1 \leq p \leq \infty$ and for all $u \in C^m(\Omega)$. In Eq. (1.7.9), $\boldsymbol{\alpha}$ denotes an n-tuple of integers

$$\boldsymbol{\alpha} = (\alpha_1, \alpha_2, \cdots, \alpha_n), \quad |\boldsymbol{\alpha}| = \sum_i^n \alpha_i, \quad \alpha_i \geq 0$$

$$D^\alpha = \frac{\partial^{|\alpha|}}{\partial x_1^{\alpha_1} \partial x_2^{\alpha_2} \cdots \partial x_n^{\alpha_n}}$$

(1.7.10)

For $m = 1, n = 2$, and $1 \leq p < \infty$ $[\boldsymbol{\alpha} = (\alpha_1, \alpha_2), \alpha_1, \alpha_2 = 0, 1]$, the norm is given by

$$\|u\|_{1,p} = \left\{ \int_{\Omega \subset \mathfrak{R}^2} \left[|u|^p + \left| \frac{\partial u}{\partial x} \right|^p + \left| \frac{\partial u}{\partial y} \right|^p \right] dx dy \right\}^{1/p}$$

(1.7.11)

The space $C^m(\Omega)$ is not complete with respect to the Sobolev norm $\| \cdot \|_{m,p}$. The completion of $C^m(\Omega)$ with respect to the norm $\| \cdot \|_{m,p}$ is called the *Sobolev space of order* (m, p), denoted by $W^{m,p}(\Omega)$. The completion of $C(\Omega)$ is the $L_2(\Omega)$ space. Hence the Sobolev space is a Banach space. Of course, the Lebesque space $L_p(\Omega)$ is a special case of the Sobolev space $W^{m,p}$ for $m = 0$, and of $L_2(\Omega)$ for $p = 2$, with the norms defined in Eq. (1.7.9).

If U and V are each normed linear spaces, a norm can be defined on the product space $U \times V \equiv \mathbf{W}$. First, addition of vectors and scalar multiplication of numbers from $U \times V$ must be defined so that $\mathbf{W} = U \times V$ can qualify as a linear vector space. Let $\mathbf{w}_1 = (u_1, v_1)$ and $\mathbf{w}_2 = (u_2, v_2)$ be two vectors in $\mathbf{W}$ with $u_1, u_2 \in U$ and $v_1, v_2 \in V$. The addition of two vectors is defined as

$$\mathbf{w}_1 + \mathbf{w}_2 = (u_1, v_1) + (u_2, v_2) = (u_1 + u_2, v_1 + v_2)$$

and multiplication by a scalar α as

$$\alpha \mathbf{w} = \alpha(u, v) = (\alpha u, \alpha v)$$

One can verify that $\mathbf{W} = U \times V$ is a linear vector space with respect to these operations.

1.7.4   Inner Product Spaces

The notion of a norm provides us with a way of measuring the length of a vector, or the difference between two vectors of a linear vector space. In order to have a means of measuring the angle between two vectors, the notion of an inner product (analogous to the scalar product or dot product of ordinary physical vectors) must be introduced.

An *inner product* on a linear vector space V over $\mathfrak{R}$ is a real-valued function that associates with each pair of vectors $u, v \in V$ a scalar, denoted $(u, v) \in \mathfrak{R}$, which satisfies the following axioms:

(1) $(u, v) = (v, u)$ (symmetry)
(2a) $(\alpha u, v) = \alpha(u, v)$ (homogeneous)   (1.7.12)
(2b) $(u_1 + u_2, v) = (u_1, v) + (u_2, v)$ (additive)
(3) $(u, v) \geq 0$ and $(u, u) = 0$ if and only if $u = 0$ (positive-definite)

for every $u, u_1, u_2, v \in V$ and $\alpha \in \Re$. A linear vector space on which an inner product can be defined is called an *inner product space*. A linear subspace S of an inner product space V is a subspace with the inner product of V.

Note that the square root of the inner product of a vector with itself satisfies the axioms of a norm. Consequently, one can associate with every inner product a norm by defining

$$||u|| = \sqrt{(u, u)} \tag{1.7.13}$$

The norm thus obtained is called the *natural norm induced by the inner product*. Since a norm can be associated with each inner product, every inner product space is also a normed vector space. It should be obvious to the reader that the converse does not hold in general.

1.7.4.1   Orthogonality of vectors

The concept of orthogonality from the ordinary physical vectors can be generalized to the elements of linear vector spaces with the aid of an inner product. Two vectors u and v in an inner product space are said to be *orthogonal* if

$$(u, v) = 0 \tag{1.7.14}$$

A set of vectors $\{u_1, u_2, \ldots\}$ is called an *orthogonal set* if each pair of the set is orthogonal: $(u_i, u_j) = 0$ for $i \neq j$. The set $\{u_n\}$ is called *orthonormal* if

$$(u_i, u_j) = \delta_{ij} \tag{1.7.15}$$

As already stated earlier, an orthonormal set of nonzero vectors is linearly independent.

The following lemma, known as the *fundamental lemma of variational calculus*, plays an important role in the variational theory.

Lemma 1.7.1:   Let V be an inner product space. If $(u, v) = 0$ for all $v \in V$, then $u = 0$.
Proof: Since $(u, v) = 0$ for all v, it must also hold for $v = u$. Then $(u, u) = 0$ implies that $u = 0$.

1.7.4.2   Cauchy–Schwartz inequality

Recall that the Hölder inequality for $p = q = 2$ is known as the Cauchy–Schwartz inequality. The inequality is given here in terms of a norm. Let u and

v be any elements of an inner product space V. Then the Cauchy–Schwartz inequality can be stated as

$$|(u, v)| \leq \sqrt{(u, u)(v, v)} = ||u|| \, ||v|| \qquad (1.7.16)$$

Let V be the set of ordinary (physical) vectors in $\Re^3$. A typical element $\mathbf{a} \in V$ has the form $\mathbf{a} = (a_1, a_2, a_3)$. It is an inner product space with respect to the inner product,

$$(\mathbf{a}, \mathbf{b}) \equiv \mathbf{a} \cdot \mathbf{b} = a_1 b_1 + a_2 b_2 + a_3 b_3$$
$$|\mathbf{a}| = \sqrt{a_1^2 + a_2^2 + a_3^2}$$

The angle θ between two vectors $\mathbf{a}$ and $\mathbf{b}$ is given by

$$\cos \theta = \frac{\mathbf{a} \cdot \mathbf{b}}{|\mathbf{a}||\mathbf{b}|} = \frac{a_1 b_1 + a_2 b_2 + a_3 b_3}{\sqrt{a_1^2 + a_2^2 + a_3^2} \cdot \sqrt{b_1^2 + b_2^2 + b_3^2}}$$

If two vectors $\mathbf{a}$ and $\mathbf{b}$ are orthogonal in V, then

$$\mathbf{a} \cdot \mathbf{b} = 0, \quad \cos \theta = 0 \quad \text{or} \quad \theta = \pi/2$$

This means vectors $\mathbf{a}$ and $\mathbf{b}$ are perpendicular to each other. The projection of the vector $\mathbf{a}$ on the line generated by $\mathbf{b}$ is given by

$$P\mathbf{a} = \frac{(\mathbf{a} \cdot \mathbf{b})\mathbf{b}}{|\mathbf{b}|^2} = (\mathbf{a} \cdot \hat{\mathbf{e}}_b)\hat{\mathbf{e}}_b = |\mathbf{a}| \cos \theta \, \hat{\mathbf{e}}_b, \quad \hat{\mathbf{e}}_b = \frac{\mathbf{b}}{|\mathbf{b}|}$$

Note that the projection is a vector.

Let V be the two-dimensional linear subspace of $C[0, 1]$ spanned by $\{\sin 2\pi x, \cos 2\pi x\}$. Define the inner product on $C[0, 1]$ (hence on V) by

$$(u, v) = \int_0^1 u(x)v(x)dx$$

The inner product of $u = a_1 \sin 2\pi x + a_2 \cos 2\pi x$ and $v = b_1 \sin 2\pi x + b_2 \cos 2\pi x$ is the real number

$$(u, v) = \frac{1}{2}(a_1 b_1 + a_2 b_2)$$

An element $u = a_1 \sin 2\pi x + a_2 \cos 2\pi x$ is orthogonal to $v \in V$ if and only if v is of the form

$$v = -a_2 \sin 2\pi x + a_1 \cos 2\pi x$$

The "angle" between u and v is given by

$$\cos \theta = \frac{(u, v)}{||u|| \, ||v||} = \frac{(a_1 b_1 + a_2 b_2)}{\sqrt{a_1^2 + a_2^2}\sqrt{b_1^2 + b_2^2}}$$

The projection of the vector u on the line generated by v is

$$Pu = (a_1 b_1 + a_2 b_2)(b_1 \sin 2\pi x + b_2 \cos 2\pi x)(b_1^2 + b_2^2)^{-1}$$

The Schwartz inequality takes the form

$$|(a_1 b_1 + a_2 b_2)| \le \sqrt{a_1^2 + a_2^2} \cdot \sqrt{b_1^2 + b_2^2}$$

From previous discussions it is clear that the geometry of normed and inner product vector spaces is much like the familiar two- and three-dimensional Euclidean geometry. The geometry of Hilbert spaces is even closer to the Euclidean geometry. For example, if two vectors u and v of an inner product space V are orthogonal, then the Pythagorean theorem holds even in function spaces:

$$||(u + v)||^2 = (u + v, u + v) = (u, u) + 2(u, v) + (v, v) = ||u||^2 + ||v||^2$$

1.7.4.3 Hilbert spaces

A complete (in its natural metric) inner product space is called a *Hilbert space*. Every inner product space (hence a normed space) has a completion.

Euclidean Space, $\Re^n$. The n-dimensional Euclidean space is a Hilbert space. The inner product in $\Re^n$ is defined by

$$(\mathbf{x}, \mathbf{y}) = \sum_{i=1}^{n} x_i y_i, \quad \mathbf{x}, \mathbf{y} \in \Re^n \tag{1.7.17}$$

The space $\Re^n$ is complete. To see this, let $\mathbf{x}^k = (x_1^k, x_2^k, \ldots, x_n^k)$ be a Cauchy sequence in $\Re^n$. Then for any $\epsilon > 0$ there exists an N such that

$$d(\mathbf{x}^m, \mathbf{x}^p) \equiv (\mathbf{x}^m - \mathbf{x}^p, \mathbf{x}^m - \mathbf{x}^p) = \left[\sum_{i=1}^{n} (x_i^{(m)} - x_i^{(p)})^2 \right]^{1/2} \le \epsilon$$

whenever $m, p > N$. This implies that

$$|x_1^{(m)} - x_1^{(p)}| \le \epsilon, |x_2^{(m)} - x_2^{(p)}| \le \epsilon, \ldots, |x_n^{(m)} - x_n^{(p)}| \le \epsilon, \text{ for } m, p > N$$

Thus, each of the sequences $x_n^{(k)}$ must converge, since $\Re$ is complete, as $k \to \infty$. Let $\lim_{k \to \infty} x_i^{(k)} = x_i$; then $\lim_{k \to \infty} \mathbf{x}^k = (x_1, x_2, \ldots, x_n)$, and the space $\Re^n$ is complete. Hence, it is a Hilbert space. The Schwartz inequality becomes

$$\left[\sum_{k=1}^{n} x_k y_k \right] \le \left[\sum_{k=1}^{n} x_k^2 \right]^{1/2} \left[\sum_{k=1}^{n} y_k^2 \right]^{1/2}$$

The completeness of $\Re^n$ implies the completeness of all finite-dimensional vector spaces.

$H^m(\Omega)$-inner product. Consider the Sobolev space $W^{m,2}(\Omega) \equiv H^m(\Omega)$ with $\Omega \subset \Re^3$. For $u, v \in H^m(\Omega)$ define

$$(u, v)_m = \int_\Omega \sum_{|\alpha| \leq m} D^\alpha u \cdot D^\alpha v \, d\mathbf{x} \tag{1.7.18}$$

This defines an inner product on $H^m(\Omega)$. Once again, the $L_2(\Omega)$ inner product is obtained as a special case of the $H^m(\Omega)$ inner product by setting $m = 0$:

$$(u, v)_0 = \int_\Omega u \cdot v \, d\mathbf{x} \tag{1.7.19}$$

For example, let $\Omega \subset \Re^2$ ($n = 2$) and $m = 1$ in Eq. (1.7.18). Then the $H^1(\Omega)$ inner product becomes

$$(u, v)_1 = \int_\Omega \left(uv + \frac{\partial u}{\partial x}\frac{\partial v}{\partial x} + \frac{\partial u}{\partial y}\frac{\partial v}{\partial y} \right) dx dy \tag{1.7.20}$$

Similarly,

$$
\begin{aligned}
(u, v)_2 = (u, v)_1 &+ \int_\Omega \left(\frac{\partial^2 u}{\partial x \partial y}\frac{\partial^2 v}{\partial x \partial y} + \frac{\partial^2 u}{\partial x^2}\frac{\partial^2 v}{\partial x^2} + \frac{\partial^2 u}{\partial y^2}\frac{\partial^2 v}{\partial y^2} \right) dx\, dy \\
= \int_\Omega \Big(uv &+ \frac{\partial u}{\partial x}\frac{\partial v}{\partial x} + \frac{\partial u}{\partial y}\frac{\partial v}{\partial y} + \frac{\partial^2 u}{\partial x \partial y}\frac{\partial^2 v}{\partial x \partial y} + \frac{\partial^2 u}{\partial x^2}\frac{\partial^2 v}{\partial x^2} + \frac{\partial^2 u}{\partial y^2}\frac{\partial^2 v}{\partial y^2} \Big) dx\, dy
\end{aligned}
\tag{1.7.21}
$$

The norm induced by the inner product gives back the Sobolev norm $|| \cdot ||_{m,2}$. However, there is no inner product associated with the Sobolev norm $|| \cdot ||_{m,p}$ (i.e. when $p \neq 2$).

Hilbert spaces, $H^m(\Omega)$. Since the Sobolev space $W^{m,p}(\Omega)$ is a Banach space, $W^{m,2}(\Omega)$ is a Hilbert space with the inner product defined in Eq. (1.7.18). It should be noted once again that the Hilbert space $H^m(\Omega)$ is a special case of the Sobolev space $W^{m,p}(\Omega)$. For $p \neq 2$, $W^{m,p}(\Omega)$ is not a Hilbert space. The Hilbert space $H_0^m(\Omega)$ is a linear subspace of functions from $H^m(\Omega)$ that vanish, along with their derivatives up to order $m - 1$, on the boundary of Ω.

1.7.5   Linear Transformations

The notion of a function or transformation from one set into another can be extended to vector spaces. A transformation T from a linear vector space U into another linear vector space V (both vector spaces are defined on the same field of scalars) is a correspondence which assigns to each element $u \in U$ a unique element $v = Tu \in V$. The terms *transformation, mapping,* and *operator* are used interchangeably and the transformation is expressed as $T : U \to V$.

     A transformation $T : U \to V$, where U and V are vector spaces that have the same scalar field, is said to be *linear* if

(1) $T(\alpha u) = \alpha T(u)$, for all $u \in U$, $\alpha \in \Re$ (homogeneous)
(2) $T(u_1 + u_2) = T(u_1) + T(u_2)$, for all $u_1, u_2 \in U$ (additive) \hfill (1.7.22)

Otherwise, it is said to be a *nonlinear transformation*. Conditions 1 and 2 can be combined into one: a transformation $T : U \to V$ is said to be linear if

$$T(\alpha u_1 + \beta u_2) = \alpha T(u_1) + \beta T(u_2) \tag{1.7.23}$$

for all vectors $u_1, u_2 \in U$ and scalars $\alpha, \beta \in \Re$. A transformation $T : U \to V$ is said to be a *continuous linear transformation* if it is linear and continuous. The operator T is said to be bounded if there exists a real number $M \geq 0$ such that

$$||Tu||_V \leq M||u||_U, \qquad \text{for all } u \in U \tag{1.7.24}$$

where $|| \cdot ||_U$ is the norm in U and $|| \cdot ||_V$ is the norm in V. A linear transformation $T : U \to V$ is continuous if and only if it is bounded. In finite-dimensional normed vector spaces, *all* linear transformations are continuous (hence bounded).

If $T_1 : U \to V$ and $T_2 : V \to W$ are linear transformations, then their composition $T_2 T_1$ is also linear:

$$\begin{aligned}
(T_2 T_1)(\alpha u_1 + \beta u_2) &= T_2(\alpha T_1(u_1) + \beta T_1(u_2)) \\
&= \alpha T_2(T_1(u_1)) + \beta T_2(T_1(u_2)) \\
&= \alpha (T_2 T_1)(u_1) + \beta (T_2 T_1)(u_2)
\end{aligned}$$

The *norm* $||T||$ of a bounded linear transformation $T : U \to V$, where U and V are normed spaces, is defined by

$$||T|| = \inf \{M : ||Tu||_V \leq M||u||_U \qquad \text{for all } u \in U\} \tag{1.7.25}$$

Other equivalent operator norms are

$$\begin{aligned}
||T|| &= \sup \{||Tu||_V : ||u||_U \leq 1\} \\
||T|| &= \sup \left\{ \frac{||Tu||_V}{||u||_U} : ||u||_U \neq 0 \right\}
\end{aligned} \tag{1.7.26}$$

Physically, the number $||Tu||/||u||$ can be viewed as the *amplification* of T at the point u, and $||T||$ as the maximum amplification of a system.

1.7.6 Linear Functionals, Bilinear Forms, and Quadratic Forms

Linear transformations that map linear vector spaces or products of linear vector spaces into the real numbers $\Re$ are of considerable interest in the study of variational formulation of operator equations. Such transformations are called *functionals*.

1.7.6.1   Linear functional

Let U be a linear vector space over the real number field $\Re$. A linear transformation $\ell : U \to \Re$ is called a *linear form* or *linear functional*. Since linear functionals are a special case of linear transformations, the concepts and results given in Section 1.7.5 still hold for linear functionals. The set of all linear functionals on a linear vector space is itself a vector space, called the *dual* or *conjugate space* of U.

1.7.6.2   Bilinear forms

Let U and V be two vector spaces with the same field of scalars. The mapping $B : U \times V \to \Re$ of pairs (u, v), $u \in U$, $v \in V$, into the field of scalars $\Re$ such that

$$B(\alpha u_1 + \beta u_2, \mu v_1 + \lambda v_2) = \alpha\mu B(u_1, v_1) + \alpha\lambda B(u_1, v_2) + \beta\mu B(u_2, v_1)$$
$$+ \beta\lambda B(u_2, v_2) \tag{1.7.27}$$

is called a *bilinear form*. A simple example of the bilinear form is the inner product, in which $U = V$ and $B(u, v) = (u, v)$.

A bilinear form $B(\cdot, \cdot) : U \times U \to \Re$ is said to be *symmetric* if $B(u, v) = B(v, u)$ for all $u, v \in U$. If $B(u, u) = 0$ and $B(u, v) = -B(v, u)$ for all $u, v \in U$, the bilinear form is said to be *skew-symmetric*. Every bilinear form can be represented uniquely as a sum of a symmetric $B_s(\cdot, \cdot)$ and a skew-symmetric $B_{ss}(\cdot, \cdot)$ bilinear form,

$$B(u, v) = \frac{1}{2}[B(u, v) + B(v, u)] + \frac{1}{2}[B(u, v) - B(v, u)] = B_s(u, v) + B_{ss}(u, v) \tag{1.7.28}$$

1.7.6.3   Quadratic forms

Let U be a vector space and $B(\cdot, \cdot)$ be a bilinear form on $U \times U$. A *quadratic form* is a functional Q on U defined by setting

$$Q(u) = B(u, u) \tag{1.7.29}$$

Note that if $B(\cdot, \cdot)$ is represented as the sum of a symmetric and a skew-symmetric bilinear form. Hence,

$$Q(u) = B(u, u) = B_s(u, u) + B_{ss}(u, u) = B_s(u, u) \tag{1.7.30}$$

That is $Q(u)$ is completely determined by the symmetric part of the bilinear form. Therefore, two different bilinear forms with the same symmetric part must generate the same quadratic form.

Linear functionals in $\Re^n$ are of the form,

$$f(\mathbf{x}) = \sum_{i=1}^{n} f_i x_i$$

where x_i are components of the vector $\mathbf{x}$ in $\Re^n$, and f_i are any numbers which can be thought of as components of a vector $\mathbf{f}$. Clearly, the scalar product of ordinary vectors is a special case of the above statement. Indeed, every bounded linear functional can be written as a scalar product,

$$f(\mathbf{x}) = (\mathbf{f}, \mathbf{x})$$

Let $V = L_2(0, 1)$ and define a linear functional on V by

$$\ell(v) = \int_0^1 f(x)v(x)\, dx$$

where f is an arbitrary function. The functional ℓ is linear because it is homogeneous,

$$\ell(\alpha v) = \int_0^1 f(x)\alpha v(x)\, dx = \alpha \int_0^1 f(x)v(x)\, dx = \alpha \ell(v)$$

and additive,

$$\ell(v_1 + v_2) = \int_0^1 f(x)(v_1 + v_2)dx = \ell(v_1) + \ell(v_2)$$

The functional is also continuous. If $||v_n - v|| \to 0$, then

$$|\ell(v_n) - \ell(v)| = |\ell(v_n - v)| \le ||f||\, ||v_n - v||$$

Since $||f||$ is independent of v and v_n, it follows that ℓ is continuous.

Consider the bilinear form defined by

$$B(u, v) = \int_{\Omega \subset \Re^3} (c_1\,\mathrm{grad}\, u \cdot \mathrm{grad}\, v + c_0 uv)d\mathbf{x}, \quad c_1, c_0 = \text{constants}$$

on $H_0^1(\Omega) \times H_0^1(\Omega)$, where Ω is a bounded domain in $\Re^3$, and $H_0^1(\Omega)$ is the Hilbert space. It is easy to verify that $B(\cdot, \cdot)$ is linear in its arguments. The bilinear form $B(\cdot, \cdot)$ is also continuous. Indeed, by appealing to the Cauchy–Schwartz inequality, one has

$$|B(u, v)| \le M \left[\int_\Omega (|u|^2 + |\,\mathrm{grad}\, u|^2)d\mathbf{x}\right]^{1/2} \left[\int_\Omega (|v|^2 + |\,\mathrm{grad}\, v|^2)d\mathbf{x}\right]^{1/2}$$

$$= M||u||_1 ||v||_1, \quad M = \max(c_1, c_0)$$

where $|| \cdot ||_1$ denotes the $H^1(\Omega)$-norm in Eq. (1.7.20). Further, the associated quadratic form is bounded below (or positive-definite):

$$Q(u) = B(u, u) = \int_\Omega (c_1 |\text{ grad } u|^2 + c_0 |u|^2) d\mathbf{x} \geq \mu ||u||_1^2$$

where $\mu = \min(c_1, c_0)$. The continuity and positive-definite properties play a crucial role in the existence and uniqueness of solutions to functional equations.

The functional analysis concepts introduced in this section will be used only where it is appropriate in this book. In particular, importance will be given to physical and computational aspects over functional analysis framework.

1.8   The Big Picture

Engineering design is the process of altering dimensions, shapes, and materials to find the best (optimum) configuration of the system to carry out certain specific function. Analysis is an aid to design, and it involves (1) mathematical model development, (2) data acquisition by measurements, (3) numerical simulation, and (4) evaluation of the results in light of known information and corrections to the mathematical model. The mathematical model is developed using laws of physics and assumptions concerning the process behavior. The data include the actual system parameters such as the geometry, material properties, loading, and boundary conditions. The material properties are determined through laboratory experiments. The mathematical model, in most practical cases, does not admit an analytical solution due to the geometric complexity and/or nonlinearities. Nonlinearities in a mathematical model arise from changing geometry or material behavior. Thus, it is necessary to employ numerical methods to compute an approximate solution to the mathematical model. The finite element method has emerged as a powerful computational procedure that can be used to analyze a variety of practical engineering problems.

A typical finite element analysis exercise begins with the actual physical system or part thereof to be analyzed. First, a set of objectives for the analysis is identified. If the analysis objective is to help develop a preliminary design of the system, the analysis can be very simple. On the other hand, if the analysis objective is to verify and certify the final design of a system, the analysis must be the most sophisticated one. Thus, the objective will dictate the type of idealization of the system to be adopted; for example, should it be modeled as a two-dimensional or three-dimensional problem, analyzed as a linear or nonlinear problem and what type of nonlinearities to be considered, what type constitutive model be used, how the loads and boundary conditions of the actual system are idealized, what coupling effects, if any, to be considered, and so on.

Once the system idealization has been completed (i.e. mathematical model is in place), one must decide on type of numerical approximation (and software to be used). This involves (1) choice of unknowns, which in turn dictates the type of finite element model, (2) type of elements, (3) type of the mesh (i.e. density of elements), and if nonlinear analysis is to be carried out, select (4) magnitude of *load increments*, (5) type of iterative method of solution, (6) error criterion, (7) error tolerance, and (8) maximum allowable number of iterations for the termination of the program. Of course, engineering background in the topic area coupled with experience with the software being used are always helpful in creating a suitable computational model of the problem at hand.

The final step in creating a computational model is verification of the code and validation of the mathematical model [17–21]. *Verification* is the process of determining if the computational model is an accurate discrete analog of the mathematical model. Thus, if the round-off errors introduced due to finite arithmetic in a computer are negligible, the computational model should give the exact solution of the mathematical model. Thus verification involves comparing the numerical results with known exact solutions of benchmark problems. A *benchmark problem* is one that has a number of features to test a computer program for correctness, including the following:

(1) has a set of standard reference results (analytical and numerical) that can be used to compare the accuracy and efficiency of computation;

(2) provides a list of pitfalls and difficulties;

(3) has a means to assess the adequacy of theoretical formulations and numerical algorithms;

(4) allows assessing robustness of the program (i.e. the ability of the program to handle ill-defined or sensitive conditions); and

(5) illustrates the use of a computer program.

On the other hand, *validation* is the process of determining the degree to which the mathematical model (hence the computer code that is verified) represents the physical reality of the system from the perspective of the intended uses of the model. For example, a mathematical model based on linear elasticity is adequate for determining linear elastic solutions of a solid but inadequate for determining its nonlinear response. The validation exercise allows one to modify the mathematical model to include the missing elements of the mathematical model that make the computed response come closer to the physical response. In fact, a mathematical model can never be validated because one does not know all the physics of the problem; it can only be invalidated. It is always a good idea, when developing a new computer program (also called a *code*), to undertake the verification exercise. Validation is a must when studying new problems and/or simultaneously accounting for multiple components of the physics involved (also called *multiphysics* problems). Additional concepts related to verification and validation of simulation models can be found in [17, 19, 20].

1.9 Summary

In this chapter a motivation for the present study is provided and the meaning of mathematical models and numerical simulations is discussed. The main features of the finite element method and nonlinear analysis are also discussed, and some mathematical preliminaries that are useful in the coming chapters are presented. In particular, a review of ordinary vectors and tensors, index notation, and linear vector spaces is presented.

Readers familiar with the material on vectors and tensors and functional analysis may browse through the material and skip the details of the respective sections. However, Chapter 2, which deals with nonlinear continuum mechanics, makes use of the material on vectors and tensors. Therefore readers should, at least, get familiar with the notation introduced here.

Problems

VECTORS AND TENSORS

1.1 Establish the following identities:

$$(a) \quad e_{ijk} = \begin{vmatrix} \delta_{i1} & \delta_{i2} & \delta_{i3} \\ \delta_{j1} & \delta_{j2} & \delta_{j3} \\ \delta_{k1} & \delta_{k2} & \delta_{k3} \end{vmatrix}. \qquad (b) \quad e_{ijk}e_{pqr} = \begin{vmatrix} \delta_{ip} & \delta_{iq} & \delta_{ir} \\ \delta_{jp} & \delta_{jq} & \delta_{jr} \\ \delta_{kp} & \delta_{kq} & \delta_{kr} \end{vmatrix}.$$

$$(c) \quad e_{ijk}e_{mnk} = \delta_{im}\delta_{jn} - \delta_{in}\delta_{jm}. \qquad (d) \quad e_{ijk}\,e_{ijk} = 6.$$

1.2 Determine the transformation matrix relating the orthonormal basis vectors $(\hat{e}_1, \hat{e}_2, \hat{e}_3)$ and $(\hat{e}_1', \hat{e}_2', \hat{e}_3')$, when $\hat{e}_i'$ are given by

(a) $\hat{e}_1'$ is along the vector $\hat{e}_1 - \hat{e}_2 + \hat{e}_3$ and $\hat{e}_2'$ is perpendicular to the plane $2x_1 + 3x_2 + x_3 - 5 = 0$.

(b) $\hat{e}_1'$ is along the line segment connecting point $(1, -1, 3)$ to $(2, -2, 4)$ and $\hat{e}_3' = (-\hat{e}_1 + \hat{e}_2 + 2\hat{e}_3)/\sqrt{6}$.

1.3 The displacement vector at a point referred to the basis $(\hat{e}_1, \hat{e}_2, \hat{e}_3)$ is $\mathbf{u} = 2\hat{e}_1 + 2\hat{e}_2 - 4\hat{e}_3$. Determine $\bar{u}_i$ with respect to the basis $(\hat{\bar{e}}_1, \hat{\bar{e}}_2, \hat{\bar{e}}_3)$, where $\hat{\bar{e}}_1 = (2\hat{e}_1 + 2\hat{e}_2 + \hat{e}_3)/3$ and $\hat{\bar{e}}_2 = (\hat{e}_1 - \hat{e}_2)/\sqrt{2}$.

1.4 Determine the rotation transformation matrix such that the new base vector $\hat{\bar{e}}_1$ is along $\hat{e}_1 - \hat{e}_2 + \hat{e}_3$, and $\hat{\bar{e}}_2$ is along the normal to the plane $2x_1 + 3x_2 + x_3 = 5$. If $\mathbf{S}$ is the tensor whose components in the unbarred system are given by $s_{11} = 1$, $s_{12} = s_{21} = 0$, $s_{13} = s_{31} = -1$, $s_{22} = 3$, $s_{23} = s_{32} = -2$, and $s_{33} = 0$, find the components in the barred coordinates.

1.5 Let $\mathbf{r}$ denote a position vector $\mathbf{r} = \mathbf{x} = x_i\hat{e}_i$ $(r^2 = x_ix_i)$ and $\mathbf{A}$ be an arbitrary constant vector. Use index notation to show that:

(a) $\nabla^2(r^n) = n(n+1)r^{n-2}$. \qquad (b) $\nabla(\mathbf{r} \cdot \mathbf{A}) = \mathbf{A}$.

(c) $\nabla \cdot (\mathbf{r} \times \mathbf{A}) = 0$. \qquad (d) $\nabla \times (\mathbf{r} \times \mathbf{A}) = -2\mathbf{A}$.

(e) $\nabla \cdot (r\mathbf{A}) = \dfrac{1}{r}(\mathbf{r} \cdot \mathbf{A})$. \qquad (f) $\nabla \times (r\mathbf{A}) = \dfrac{1}{r}(\mathbf{r} \times \mathbf{A})$.

1.6 Use the index notation to rewrite the vector expression as a sum or difference of two vector expressions, when $\mathbf{A}$ is a vector-valued function:

$$\nabla \times (\nabla \times \mathbf{A})$$

1.7 Establish the following identities for a second-order tensor $\mathbf{S}$:

(a) $|\mathbf{S}| = e_{ijk}\, s_{1i}\, s_{2j}\, s_{3k}$.

(b) $|\mathbf{S}| = \frac{1}{6} s_{ir}\, s_{js}\, s_{kt}\, e_{rst}\, e_{ijk}$.

(c) $e_{rst}|\mathbf{S}| = e_{ijk}\, s_{ir}\, s_{js}\, s_{kt}$.

(d) $\begin{vmatrix} s_{im} & s_{in} & s_{ip} \\ s_{jm} & s_{jn} & s_{jp} \\ s_{km} & s_{kn} & s_{kp} \end{vmatrix} = e_{ijk}\, e_{mnp}\, |\mathbf{S}|$.

1.8 Express the components of the *symmetric part* of the gradient of a vector $\mathbf{A}$

$$\mathbf{V} = \frac{1}{2}\left[(\boldsymbol{\nabla}\mathbf{A})^{\mathrm{T}} + (\boldsymbol{\nabla}\mathbf{A}) \right]$$

in terms of the components of vector $\mathbf{A}$ and definition of the del operator $\boldsymbol{\nabla}$ in the cylindrical coordinate system (see Table 1.6.1). In particular, show that

$$V_{rr} = \frac{\partial A_r}{\partial r}, \quad V_{r\theta} = \frac{1}{2}\left(\frac{1}{r}\frac{\partial A_r}{\partial \theta} + \frac{\partial A_\theta}{\partial r} - \frac{A_\theta}{r} \right), \quad V_{rz} = \frac{1}{2}\left(\frac{\partial A_r}{\partial z} + \frac{\partial A_z}{\partial r} \right)$$

$$V_{\theta\theta} = \frac{A_r}{r} + \frac{1}{r}\frac{\partial A_\theta}{\partial \theta}, \quad V_{z\theta} = \frac{1}{2}\left(\frac{\partial A_\theta}{\partial z} + \frac{1}{r}\frac{\partial A_z}{\partial \theta} \right), \quad V_{zz} = \frac{\partial A_z}{\partial z}$$

1.9 Express the components of the *skew-symmetric part* of the gradient of a vector $\mathbf{A}$

$$\mathbf{W} = \frac{1}{2}\left[(\boldsymbol{\nabla}\mathbf{A})^{\mathrm{T}} - (\boldsymbol{\nabla}\mathbf{A}) \right]$$

in terms of the components of vector $\mathbf{A}$ and definition of the del operator $\boldsymbol{\nabla}$ in the cylindrical coordinate system. In particular, show that

$$W_{r\theta} = \frac{1}{2}\left(\frac{1}{r}\frac{\partial A_r}{\partial \theta} - \frac{A_\theta}{r} - \frac{\partial A_\theta}{\partial r} \right) = -W_{\theta r}$$

$$W_{rz} = \frac{1}{2}\left(\frac{\partial A_r}{\partial z} - \frac{\partial A_z}{\partial r} \right) = -W_{zr}$$

$$W_{z\theta} = \frac{1}{2}\left(\frac{1}{r}\frac{\partial A_z}{\partial \theta} - \frac{\partial A_\theta}{\partial z} \right) = -W_{\theta z}$$

1.10 For an arbitrary second-order tensor $\mathbf{S}$ show that $\boldsymbol{\nabla}\cdot\mathbf{S}$ in the cylindrical coordinate system is given by

$$\boldsymbol{\nabla}\cdot\mathbf{S} = \left[\frac{\partial S_{rr}}{\partial r} + \frac{1}{r}\frac{\partial S_{\theta r}}{\partial \theta} + \frac{\partial S_{zr}}{\partial z} + \frac{1}{r}\left(S_{rr} - S_{\theta\theta} \right) \right]\hat{\mathbf{e}}_r$$

$$+ \left[\frac{\partial S_{r\theta}}{\partial r} + \frac{1}{r}\frac{\partial S_{\theta\theta}}{\partial \theta} + \frac{\partial S_{z\theta}}{\partial z} + \frac{1}{r}\left(S_{r\theta} + S_{\theta r} \right) \right]\hat{\mathbf{e}}_\theta$$

$$+ \left[\frac{\partial S_{rz}}{\partial r} + \frac{1}{r}\frac{\partial S_{\theta z}}{\partial \theta} + \frac{\partial S_{zz}}{\partial z} + \frac{1}{r} S_{rz} \right]\hat{\mathbf{e}}_z$$

1.11 For an arbitrary second-order tensor $\mathbf{S}$ show that $\boldsymbol{\nabla}\cdot\mathbf{S}$ in the spherical coordinate system is given by

$$\boldsymbol{\nabla}\cdot\mathbf{S} = \left\{ \frac{\partial S_{RR}}{\partial R} + \frac{1}{R}\frac{\partial S_{\phi R}}{\partial \phi} + \frac{1}{R\sin\phi}\frac{\partial S_{\theta R}}{\partial \theta} + \frac{1}{R}\left[2S_{RR} - S_{\phi\phi} - S_{\theta\theta} + S_{\phi R}\cot\phi \right] \right\}\hat{\mathbf{e}}_R$$

$$+ \left\{ \frac{\partial S_{R\phi}}{\partial R} + \frac{1}{R}\frac{\partial S_{\phi\phi}}{\partial \phi} + \frac{1}{R\sin\phi}\frac{\partial S_{\theta\phi}}{\partial \theta} + \frac{1}{R}\left[\left(S_{\phi\phi} - S_{\theta\theta} \right)\cot\phi + S_{\phi R} + 2S_{R\phi} \right] \right\}\hat{\mathbf{e}}_\phi$$

$$+ \left\{ \frac{\partial S_{R\theta}}{\partial R} + \frac{1}{R}\frac{\partial S_{\phi\theta}}{\partial \phi} + \frac{1}{R\sin\phi}\frac{\partial S_{\theta\theta}}{\partial \theta} + \frac{1}{R}\left[\left(S_{\phi\theta} + S_{\theta\phi} \right)\cot\phi + 2S_{R\theta} + S_{\theta R} \right] \right\}\hat{\mathbf{e}}_\theta.$$

where

$$\nabla = \hat{e}_R \frac{\partial}{\partial R} + \frac{1}{R} \hat{e}_\phi \frac{\partial}{\partial \phi} + \frac{1}{R \sin \phi} \hat{e}_\theta \frac{\partial}{\partial \theta}$$

$$\frac{\partial \hat{e}_R}{\partial \phi} = \hat{e}_\phi, \quad \frac{\partial \hat{e}_R}{\partial \theta} = \sin \phi \, \hat{e}_\theta, \quad \frac{\partial \hat{e}_\phi}{\partial \phi} = -\hat{e}_R, \quad \frac{\partial \hat{e}_\phi}{\partial \theta} = \cos \phi \, \hat{e}_\theta, \quad \frac{\partial \hat{e}_\theta}{\partial \theta} = -\sin \phi \, \hat{e}_R - \cos \phi \, \hat{e}_\phi$$

All other derivatives of the base vectors are zero.

1.12 For an arbitrary second-order tensor $\mathbf{S}$ show that $\nabla \times \mathbf{S}$ in the cylindrical coordinate system is given by

$$\begin{aligned}
\nabla \times \mathbf{S} = &\hat{e}_r \hat{e}_r \left(\frac{1}{r} \frac{\partial S_{zr}}{\partial \theta} - \frac{\partial S_{\theta r}}{\partial z} - \frac{1}{r} S_{z\theta} \right) + \hat{e}_\theta \hat{e}_\theta \left(\frac{\partial S_{r\theta}}{\partial z} - \frac{\partial S_{z\theta}}{\partial r} \right) \\
&+ \hat{e}_z \hat{e}_z \left(\frac{1}{r} S_{\theta z} - \frac{1}{r} \frac{\partial S_{rz}}{\partial \theta} + \frac{\partial S_{\theta z}}{\partial r} \right) + \hat{e}_r \hat{e}_\theta \left(\frac{1}{r} \frac{\partial S_{z\theta}}{\partial \theta} - \frac{\partial S_{\theta\theta}}{\partial z} + \frac{1}{r} S_{zr} \right) \\
&+ \hat{e}_\theta \hat{e}_r \left(\frac{\partial S_{rr}}{\partial z} - \frac{\partial S_{zr}}{\partial r} \right) + \hat{e}_r \hat{e}_z \left(\frac{1}{r} \frac{\partial S_{zz}}{\partial \theta} - \frac{\partial S_{\theta z}}{\partial z} \right) \\
&+ \hat{e}_z \hat{e}_r \left[\frac{\partial S_{\theta r}}{\partial r} - \frac{1}{r} \frac{\partial S_{rr}}{\partial \theta} + \frac{1}{r} (S_{r\theta} + S_{\theta r}) \right] + \hat{e}_\theta \hat{e}_z \left(\frac{\partial S_{rz}}{\partial z} - \frac{\partial S_{zz}}{\partial r} \right) \\
&+ \hat{e}_z \hat{e}_\theta \left[\frac{\partial S_{\theta\theta}}{\partial r} + \frac{1}{r} (S_{\theta\theta} - S_{rr}) - \frac{1}{r} \frac{\partial S_{r\theta}}{\partial \theta} \right]
\end{aligned}$$

FUNCTIONAL ANALYSIS

1.13 Let $\mathcal{P}$ be the space of all polynomials with real coefficients. Determine which of the following subsets of $\mathcal{P}$ are subspaces:

(a) $S_1 = \{p(x) : \; p(1) = 0\}$

(b) $S_2 = \{p(x) : \; \text{degree of } p(x) = 3\}$

(c) $S_3 = \{p(x) : \; \text{degree of } p(x) \leq 3\}$

(d) $S_4 = \{p(x) : \; \text{constant term is zero}\}$

1.14 Determine which of the following sets in $\Re^3$ are linearly independent:

(a) $\{(-1, 1, 0), (-1, 1, 1), (-2, -1, 1), (1, 1, 1)\}$

(b) $\{(1, 0, 0), (1, 1, 0), (1, 1, 1)\}$

(c) $\{(1, 1, 1), (1, 2, 3), (2, -1, 1)\}$

1.15 Determine which of the following sets span $\Re^3$:

(a) $\{(2, 1, 0), (1, -1, 2), (0, 3, -4)\}$

(b) $\{(1, 1, 1), (-2, -1, 2), (-1, 1, 1), (-1, 1, 0)\}$

(c) $\{(-1, 1, 0), (-1, 1, 1), (-2, -1, 1), (1, 1, 1)\}$

(d) $\{(1, 0, 0), (1, 2, 0), (1, 2, 3)\}$

1.16 Compute the L_2 norm and the sup-norm of the following functions in the interval indicated:

(a) $u(x) = \sin \pi x - x$, on $0 \leq x \leq 1$

(b) $u(x) = x^{1/3}$, on $0 \leq x \leq 1$

(c) $u(x) = \cos \pi x + 2x - 1$, on $0 \leq x \leq 1$

(d) $u(x) = \sqrt{1 + x^2}$, on $0 \leq x \leq 1$

1.17 Prove the following relations in a real inner product space:

    (a) Parallelogram law: $||u + v||^2 + ||u - v||^2 = 2(||u||^2 + ||v||^2)$

    (b) $(u, v) = \frac{1}{4}[||u + v||^2 - ||u - v||^2]$

    (c) $|\,||u|| - ||v||\,| \leq ||u - v||$

1.18 If two vectors u and v of an inner product space V are orthogonal, show that the Pythagorean theorem holds even in function spaces: $||(u + v)||^2 = ||u||^2 + ||v||^2$.

1.19 Compute the inner product of the following pairs of functions on the interval indicated. Use the L_2 inner product and the $H^1(0, 1)$ inner product.

    (a) $u = (1 + x), v = 3x^2 - 1$, on $-1 \leq x \leq 1$

    (b) $u = \sin \pi x, v = \cos \pi x$, on $0 \leq x \leq 1$

    (c) $u = \sin \pi x, v = a + bx + cx^2$, on $0 \leq x \leq 1$

    (d) $u = \sin \pi x \sin \pi y, v = (1 - x^2 - y^2)$, on $0 \leq x, y \leq 1$

1.20 Determine a vector in the space $\mathcal{P}$ of polynomials of degree 2 such that the vector is orthogonal to the polynomials $p_1 = 1 + x - 2x^2$ and $p_2 = -2 + 4x + x^2$ in $L_2(0, 1)$.

1.21 Determine the constants a and b such that $w(x) = a + bx + 3x^2$ is orthogonal in $L_2(0, 1)$ to both $u(x) = 2 + 3x^2 - x$ and $v(x) = \frac{1}{3} + 3x - 5x^2$.

1.22 Let T_1 and T_2 be the linear operators defined on $\Re^3$ as $T_1(\mathbf{x}) = (0, x_1 - x_3, x_2 + x_3)$ and $T_2(\mathbf{x}) = (x_2 - x_3, 2x_1 - x_2, x_1)$. Determine (a) $T_1 + T_2$, (b) $T_1 T_2$, and (c) $T_2 T_1$.

1.23 Determine which of the following operators represent bilinear forms:

    (a) $B : \Re^2 \times \Re^2 \to \Re, \quad B(\mathbf{x}, \mathbf{y}) = (x_1 + y_1)^2 - (x_1 - y_1)^2$

    (b) $B : \Re^2 \times \Re^2 \to \Re, \quad B(\mathbf{x}, \mathbf{y}) = x_1 y_2 - x_2 y_1$

Elements of Nonlinear Continuum Mechanics

Although to penetrate into the intimate mysteries of nature and thence to learn the true causes of phenomena is not allowed to us, nevertheless it can happen that a certain fictive hypothesis may suffice for explaining many phenomena. —— Leonhard Euler

2.1 Introduction

The study of matter at molecular or atomistic levels is very useful for understanding a variety of phenomena. However, studies at these scales are not useful to solve common engineering problems. The understanding gained at the molecular level needs to be taken to the macroscopic scale (i.e. the scale visible to an unaided eye) to be able to study its behavior. At the macroscopic scale, it is assumed that the length scales of interest are large compared to the length scales of the discrete molecular structure. In other words, the system is treated as a *continuum* in which all physical quantities such as the density, displacements, velocities, and stresses vary continuously so that their spatial derivatives exist and are continuous. The continuum assumption allows us to shrink an arbitrary volume of material to a point, in much the same way as we take the limit in defining a derivative, so that we can define quantities of interest at a point.[1] A mathematical study of mechanics of such an idealized continuum is called *continuum mechanics*.

The present chapter deals with a review of continuum mechanics concepts as applied to an idealized matter, whether a solid or fluid, subjected to external forces. In particular, we discuss kinematics of deformation, conservation of mass, balance of momenta and energy, and constitutive relations, and the derivation of field equations resulting from the conservation and balance principles valid over a continuum. Most of the material presented here is taken from the author's text book on continuum mechanics [1]; the readers may also consult the books by Holzapfel [22], Bonet and Wood [23], and Malvern [24].

---

[1]The continuum assumption is not valid when one considers discontinuity in density and velocity (e.g. shock waves). In such cases, certain jump conditions are employed to deal with the discontinuities. We do not consider such situations in this book.

J.N. Reddy, *An Introduction to Nonlinear Finite Element Analysis*, Second Edition. ©J.N. Reddy 2015. Published in 2015 by Oxford University Press.

2.2   Description of Motion

2.2.1   Configurations of a Continuous Medium

Consider a body $\mathcal{B}$ of known geometry, constitution, and loading in a three-dimensional space $\Re^3$; $\mathcal{B}$ may be viewed as a set of material particles, each particle representing a large collection of molecules, having a continuous distribution of matter in space and time. For a given geometry and set of forces or stimuli, the body $\mathcal{B}$ will undergo macroscopic change in geometry, which is known as *deformation*. Geometric changes are accompanied by stresses induced in the body. If the applied loads are time-dependent, the geometry of the body $\mathcal{B}$ will change continuously with time. If the forces are applied slowly so that the deformation is only dependent on the loads, the body will occupy a continuous sequence of configurations. By the term *configuration* we mean the simultaneous positions occupied in space $\Re^3$ by all material points of the continuum $\mathcal{B}$ at an instant of time, and we denote it by Ω.

Suppose that the continuum initially, say at time $t = 0$, occupies a configuration Ω_0, in which a particle X occupies the position $\mathbf{X}$, referred to a rectangular Cartesian system (X_1, X_2, X_3). Note that X (lightface letter) is the name of the particle that occupies the location $\mathbf{X}$ (boldface letter) in configuration Ω_0, and therefore (X_1, X_2, X_3) are called the *material coordinates*. After the application of the loads, the continuum changes its geometric shape and thus assumes a new configuration Ω, called the *current* or *deformed configuration*. The particle X now occupies the position $\mathbf{x}$ in the deformed configuration Ω, as shown in Fig. 2.2.1. The mapping $\boldsymbol{\chi} : \Omega_0 \to \Omega$ is called the *deformation mapping* of the body $\mathcal{B}$ from Ω_0 to Ω. The deformation mapping $\boldsymbol{\chi}(\mathbf{X})$ takes the position vector $\mathbf{X}$ from the reference configuration Ω_0 and places the same point in the deformed configuration Ω as $\mathbf{x} = \boldsymbol{\chi}(\mathbf{X})$.

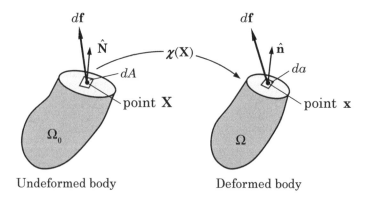

Fig. 2.2.1: Configurations and mapping of a continuous body.

A frame of reference is chosen to describe the deformation. The same reference frame for reference and current configurations is chosen, with the origins of the basis vectors $\hat{\mathbf{E}}_i$ and $\hat{\mathbf{e}}_i$ coinciding. The components X_i and x_i of vectors $\mathbf{X} = X_i \hat{\mathbf{E}}_i$ and $\mathbf{x} = x_i \hat{\mathbf{e}}_i$ are taken along the coordinates used.

2.2.2 Material and Spatial Descriptions

The mathematical description of the deformation of a continuous body follows one of the two approaches: (1) the material description and (2) the spatial description. The material description is also known as the *Lagrangian description,* and the spatial description is known as the *Eulerian description.*

In the material description, the motion of the body is referred to a reference configuration Ω_R, which is often chosen to be the undeformed configuration, $\Omega_R = \Omega_0$. Thus, in the Lagrangian description, the current coordinates ($\mathbf{x} \in \Omega$) are expressed in terms of the reference coordinates ($\mathbf{X} \in \Omega_0$):

$$\mathbf{x} = \chi(\mathbf{X}, t), \quad \chi(\mathbf{X}, 0) = \mathbf{X} \tag{2.2.1}$$

and the variation of a typical variable ϕ over the body is described with respect to the material coordinates $\mathbf{X}$ and time t:

$$\phi = \phi(\mathbf{X}, t) \tag{2.2.2}$$

For a fixed value of $\mathbf{X} \in \Omega_0$, $\phi(\mathbf{X}, t)$ gives the value of ϕ at time t associated with the fixed material point X whose position in the reference configuration is $\mathbf{X}$, as shown in Fig. 2.2.2. Thus, a change in time t implies that the *same* material particle X, occupying position $\mathbf{X}$ in Ω_0, has a different value ϕ. Thus, the attention is focused on material particles of the continuum.

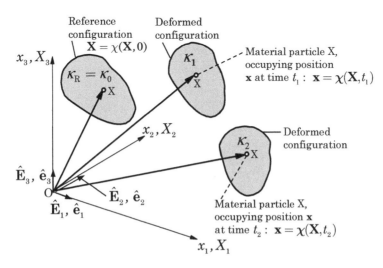

Fig. 2.2.2: Reference configuration and deformed configurations at two different times in *material description.*

In the spatial description, the motion is referred to the current configuration Ω occupied by the body $\mathcal{B}$, and ϕ is described with respect to the current position ($\mathbf{x} \in \Omega$) in space, currently occupied by material particle X:

$$\phi = \phi(\mathbf{x}, t), \quad \mathbf{X} = \mathbf{X}(\mathbf{x}, t) \tag{2.2.3}$$

The coordinates $\mathbf{x}$ are termed the *spatial coordinates*. For a fixed value of $\mathbf{x} \in \Omega$, $\phi(\mathbf{x}, t)$ gives the value of ϕ associated with a fixed point $\mathbf{x}$ in space, which will be the value of ϕ associated with different material particles at different times, because different material particles occupy the position $\mathbf{x} \in \Omega$ at different times, as shown in Fig. 2.2.3. Thus, a change in time t implies that a different value ϕ is observed at the *same* spatial location $\mathbf{x} \in \Omega$, now probably occupied by a different material particle X. Thus, in the spatial description attention is focused on a spatial position $\mathbf{x} \in \Omega$.

When ϕ is known in the material description, $\phi = \phi(\mathbf{X}, t)$, its *material time derivative*, denoted by d/dt, is simply the partial derivative with respect to time while holding $\mathbf{X}$ fixed:

$$\frac{d}{dt}[\phi(\mathbf{X}, t)] = \frac{\partial}{\partial t}[\phi(\mathbf{X}, t)]\Big|_{\mathbf{X} \text{ fixed}} = \frac{\partial \phi}{\partial t} \tag{2.2.4}$$

However, when ϕ is known in the spatial description, $\phi = \phi(\mathbf{x}, t)$, its material time derivative, by chain rule of differentiation, is

$$\frac{d}{dt}[\phi(\mathbf{x}, t)] = \frac{\partial}{\partial t}[\phi(\mathbf{x}, t)] + \frac{\partial}{\partial x_i}[\phi(\mathbf{x}, t)]\frac{dx_i}{dt}$$

$$= \frac{\partial \phi}{\partial t} + v_i \frac{\partial \phi}{\partial x_i} = \frac{\partial \phi}{\partial t} + \mathbf{v} \cdot \boldsymbol{\nabla}\phi \tag{2.2.5}$$

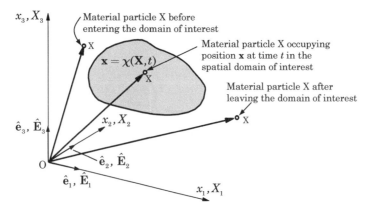

Fig. 2.2.3: Material points within and outside the spatial domain of interest in *spatial description*.

where $\mathbf{v}$ is the velocity vector

$$\mathbf{v} = \frac{d\mathbf{x}}{dt} \equiv \dot{\mathbf{x}} \tag{2.2.6}$$

Thus, the material time derivative of a spatial field is the operator

$$\frac{d}{dt} = \frac{\partial}{\partial t} + \mathbf{v} \cdot \nabla \tag{2.2.7}$$

For example, the acceleration of a particle X occupying position $\mathbf{x}$ is given by

$$\mathbf{a}(\mathbf{x}, t) = \frac{d\mathbf{v}(\mathbf{x}, t)}{dt} = \frac{\partial \mathbf{v}}{\partial t} + \mathbf{v} \cdot \nabla \mathbf{v}; \quad a_i = \frac{\partial v_i}{\partial t} + v_j \frac{\partial v_i}{\partial x_j} \tag{2.2.8}$$

Example 2.2.1 (see Reddy [1]) illustrates the determination of the inverse of a given mapping and computation of the material time derivative of a given function in material and spatial descriptions.

Example 2.2.1

Suppose that the motion of a continuous medium $\mathcal{B}$ is described by the mapping $\chi : \Omega_0 \to \Omega$,

$$\chi(\mathbf{X}, t) = \mathbf{x} = (X_1 + \alpha t X_2)\hat{\mathbf{e}}_1 + (X_2 - \alpha t X_1)\hat{\mathbf{e}}_2 + X_3 \hat{\mathbf{e}}_3$$

where α is a constant, and suppose that the temperature T in the continuum in the spatial description is given by

$$T(\mathbf{x}, t) = x_1 + t x_2$$

Determine (a) the inverse of the mapping, (b) the velocity components, and (c) the time derivatives of T in the two descriptions.

Solution: A known deformation mapping $\chi(\mathbf{X}, t)$ relates the material coordinates (X_1, X_2, X_3) to the spatial coordinates (x_1, x_2, x_3) of a particle X. In the present case, we have

$$x_1 = X_1 + \alpha t X_2, \quad x_2 = X_2 - \alpha t X_1, \quad x_3 = X_3 \quad \text{or} \quad \begin{Bmatrix} x_1 \\ x_2 \\ x_3 \end{Bmatrix} = \begin{bmatrix} 1 & \alpha t & 0 \\ -\alpha t & 1 & 0 \\ 0 & 0 & 1 \end{bmatrix} \begin{Bmatrix} X_1 \\ X_2 \\ X_3 \end{Bmatrix} \tag{1}$$

Clearly, the relationships between x_i and X_i are linear (i.e. the mapping is linear). Therefore, it maps polygons into polygons. In particular, a unit square is mapped into a square that is rotated in clockwise direction, as shown in Fig. 2.2.4. Note that, in general, the deformed square is not a unit square as the side now has a length of $1/\cos\theta$, where $\theta = \tan^{-1}(\alpha t)$. The reference configuration and deformed configurations at four different times, $t = 1, 2, 3$, and 4, for a value of $\alpha = 0.25$, are shown in Fig. 2.2.5.

(a) The inverse mapping can be determined, when possible, by expressing (x_1, x_2, x_3) in terms of (X_1, X_2, X_3). In the present case, it is possible to invert the relations in Eq. (1) and obtain

$$\begin{Bmatrix} X_1 \\ X_2 \\ X_3 \end{Bmatrix} = \frac{1}{(1 + \alpha^2 t^2)} \begin{bmatrix} 1 & -\alpha t & 0 \\ \alpha t & 1 & 0 \\ 0 & 0 & 1 + \alpha^2 t^2 \end{bmatrix} \begin{Bmatrix} x_1 \\ x_2 \\ x_3 \end{Bmatrix} \tag{2}$$

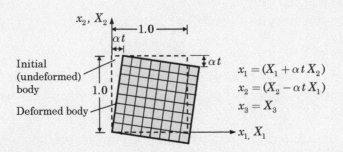

Fig. 2.2.4: A sketch of the mapping χ as applied to a unit square.

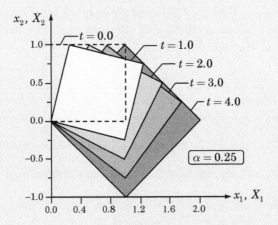

Fig. 2.2.5: Deformed configurations of the unit square at four different times.

Therefore, we can write the inverse mapping as $\chi^{-1} : \Omega \to \Omega_0$ as

$$\chi^{-1}(\mathbf{x},t) = \left(\frac{x_1 - \alpha\,t\,x_2}{1 + \alpha^2\,t^2}\right)\hat{\mathbf{E}}_1 + \left(\frac{x_2 + \alpha\,t\,x_1}{1 + \alpha^2\,t^2}\right)\hat{\mathbf{E}}_2 + x_3\,\hat{\mathbf{E}}_3 \tag{3}$$

(b) The velocity vector is given by $\mathbf{v} = v_1\hat{\mathbf{e}}_1 + v_2\hat{\mathbf{e}}_2$, with

$$v_1 = \frac{dx_1}{dt} = \alpha\,X_2, \quad v_2 = \frac{dx_2}{dt} = -\alpha\,X_1 \tag{4}$$

(c) The time rate of change of temperature of a material particle in $\mathcal{B}$ is simply

$$\frac{d}{dt}[T(\mathbf{X},t)] = \frac{\partial}{\partial t}[T(\mathbf{X},t)]\bigg|_{\mathbf{X}\text{ fixed}} = -2\alpha\,t\,X_1 + (1+\alpha)X_2 \tag{5}$$

On the other hand, the time rate of change of temperature at point $\mathbf{x}$, which is now occupied by particle X, is

$$\frac{d}{dt}[T(\mathbf{x},t)] = \frac{\partial T}{\partial t} + v_i\frac{\partial T}{\partial x_i} = x_2 + v_1\cdot 1 + v_2\cdot t = -2\alpha\,t\,X_1 + (1+\alpha)X_2 \tag{6}$$

The material description is commonly used to study the stress and deformation of solid bodies, as one is interested in the body irrespective of what spatial location it occupies. On the other hand, spatial description is adopted in studying fluid motions, where one is interested in the conditions of flow (for example, density, temperature, pressure, and so on) at a fixed spatial location, rather than in the material particles that happen to occupy instantly the fixed spatial location.

2.2.3 Displacement Field

The phrase deformation of a continuum refers to relative displacements and changes in the geometry experienced by the continuum $\mathcal{B}$ under the influence of a force system. The displacement of the particle X is given, as can be seen from Fig. 2.2.6, by

$$\mathbf{u} = \mathbf{x} - \mathbf{X} \tag{2.2.9}$$

In the Lagrangian description, the displacements are expressed in terms of the material coordinates X_i

$$\mathbf{u}(\mathbf{X}, t) = \mathbf{x}(\mathbf{X}, t) - \mathbf{X} \tag{2.2.10}$$

If the displacement of every particle in the body $\mathcal{B}$ is known, we can construct the current configuration Ω from the reference configuration Ω_0, $\chi(\mathbf{X}) = \mathbf{X} + \mathbf{u}(\mathbf{X})$. On the other hand, in the Eulerian description the displacements are expressed in terms of the spatial coordinates x_i

$$\mathbf{u}(\mathbf{x}, t) = \mathbf{x} - \mathbf{X}(\mathbf{x}, t) \tag{2.2.11}$$

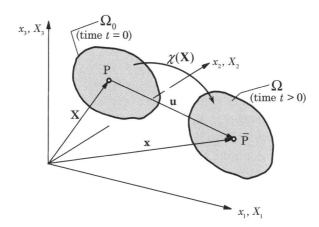

Fig. 2.2.6: Material points P and $\bar{\text{P}}$ in the undeformed configuration Ω_0 and deformed configuration Ω, respectively.

A *rigid-body motion* is one in which all material particles of the continuum $\mathcal{B}$ undergo the same linear and angular displacements. On the other hand, a deformable body is one in which the material particles can move relative to each other. Then the deformation of a continuum can be determined only by considering the change of distance between any two arbitrary but infinitesimally close points of the continuum.

2.3   Analysis of Deformation

2.3.1   Deformation Gradient

One of the key quantities in deformation analysis is the *deformation gradient,* which gives the relationship of a material line $d\mathbf{X}$ before deformation to the line $d\mathbf{x}$ (consisting of the same material as $d\mathbf{X}$) after deformation. It is defined as

$$d\mathbf{x} = \mathbf{F} \cdot d\mathbf{X} = d\mathbf{X} \cdot \mathbf{F}^{\mathrm{T}}, \quad \mathbf{F} \equiv \overleftarrow{\boldsymbol{\nabla}}_0 \mathbf{x} = (\boldsymbol{\nabla}_0 \mathbf{x})^{\mathrm{T}} = \left(\frac{\partial \mathbf{x}}{\partial \mathbf{X}} \right)^{\mathrm{T}} \tag{2.3.1}$$

where $\boldsymbol{\nabla}_0$ is the gradient operator with respect to $\mathbf{X}$. The definition of $\mathbf{F}$ given in Eq. (2.3.1) may appear as the transpose of that defined in other books but these books use backward gradient operator without explicitly stating. We note that $\mathbf{F} \equiv (\boldsymbol{\nabla}_0 \mathbf{x})^{\mathrm{T}} = F_{iJ}\,\hat{\mathbf{e}}_i\,\hat{\mathbf{E}}_J$, whereas $\boldsymbol{\nabla}_0 \mathbf{x} = \hat{\mathbf{E}}_J \frac{\partial x_i}{\partial X_J}\,\hat{\mathbf{e}}_i = F_{iJ}\,\hat{\mathbf{E}}_J\,\hat{\mathbf{e}}_i = \mathbf{F}^{\mathrm{T}}$, and the definition of F_{iJ} used here and in other books is the same.

The inverse relations are given by

$$d\mathbf{X} = \mathbf{F}^{-1} \cdot d\mathbf{x} = d\mathbf{x} \cdot \mathbf{F}^{-\mathrm{T}}, \quad \text{where} \quad \mathbf{F}^{-1} = \left(\frac{\partial \mathbf{X}}{\partial \mathbf{x}} \right)^{\mathrm{T}} \equiv (\boldsymbol{\nabla} \mathbf{X})^{\mathrm{T}} \tag{2.3.2}$$

and $\boldsymbol{\nabla}$ is the gradient operator with respect to $\mathbf{x}$. In indicial notation, the matrix forms of $\mathbf{F}$ and $\mathbf{F}^{-1}$ are

$$[F] = \begin{bmatrix} \frac{\partial x_1}{\partial X_1} & \frac{\partial x_1}{\partial X_2} & \frac{\partial x_1}{\partial X_3} \\ \frac{\partial x_2}{\partial X_1} & \frac{\partial x_2}{\partial X_2} & \frac{\partial x_2}{\partial X_3} \\ \frac{\partial x_3}{\partial X_1} & \frac{\partial x_3}{\partial X_2} & \frac{\partial x_3}{\partial X_3} \end{bmatrix}, \quad [F]^{-1} = \begin{bmatrix} \frac{\partial X_1}{\partial x_1} & \frac{\partial X_1}{\partial x_2} & \frac{\partial X_1}{\partial x_3} \\ \frac{\partial X_2}{\partial x_1} & \frac{\partial X_2}{\partial x_2} & \frac{\partial X_2}{\partial x_3} \\ \frac{\partial X_3}{\partial x_1} & \frac{\partial X_3}{\partial x_2} & \frac{\partial X_3}{\partial x_3} \end{bmatrix} \tag{2.3.3}$$

In Eqs. (2.3.2) and (2.3.3), the lower case indices refer to the current (spatial) Cartesian coordinates, whereas upper case indices refer to the reference (material) Cartesian coordinates. The determinant of $\mathbf{F}$ is called the *Jacobian of the motion,* and it is denoted by $J = \det \mathbf{F}$. The equation $\mathbf{F} \cdot d\mathbf{X} = 0$ for $d\mathbf{X} \neq 0$ implies that a material line in the reference configuration is reduced to zero by the deformation. Since this is physically not realistic, we conclude that $\mathbf{F} \cdot d\mathbf{X} \neq 0$ for $d\mathbf{X} \neq 0$. That is, $\mathbf{F}$ is a non-singular tensor, $J \neq 0$. Hence, $\mathbf{F}$ has an inverse $\mathbf{F}^{-1}$. The deformation gradient can be expressed in terms of the displacement vector as

$$\mathbf{F} = (\boldsymbol{\nabla}_0 \mathbf{x})^{\mathrm{T}} = (\boldsymbol{\nabla}_0 \mathbf{u} + \mathbf{I})^{\mathrm{T}} \quad \text{or} \quad \mathbf{F}^{-1} = (\boldsymbol{\nabla} \mathbf{X})^{\mathrm{T}} = (\mathbf{I} - \boldsymbol{\nabla} \mathbf{u})^{\mathrm{T}} \tag{2.3.4}$$

2.3.2 Volume and Surface Elements in the Material and Spatial Descriptions

Here we discuss how deformation mapping affects surface areas and volumes of a continuum. First, we consider three non-coplanar line elements $d\mathbf{X}^{(1)}$, $d\mathbf{X}^{(2)}$, and $d\mathbf{X}^{(3)}$ forming the edges of a parallelepiped at a point with position vector $\mathbf{X}$ in the reference configuration Ω_0 of body $\mathcal{B}$ so that

$$d\mathbf{x}^{(i)} = \mathbf{F} \cdot d\mathbf{X}^{(i)}, \quad i = 1, 2, 3 \tag{2.3.5}$$

Note that the vectors $d\mathbf{x}^{(i)}$ are not necessarily parallel to or have the same length as the vectors $d\mathbf{X}^{(i)}$ due to shearing and stretching of the parallelepiped. We assume that the triad $(d\mathbf{X}^{(1)}, d\mathbf{X}^{(2)}, d\mathbf{X}^{(3)})$ is positively oriented in the sense that the triple scalar product $d\mathbf{X}^{(1)} \cdot d\mathbf{X}^{(2)} \times d\mathbf{X}^{(3)} > 0$. The volume of the parallelepiped is

$$dV = d\mathbf{X}^{(1)} \cdot d\mathbf{X}^{(2)} \times d\mathbf{X}^{(3)} = \left(\hat{\mathbf{N}}_1 \cdot \hat{\mathbf{N}}_2 \times \hat{\mathbf{N}}_3 \right) dX^{(1)} dX^{(2)} dX^{(3)}$$
$$= dX^{(1)} dX^{(2)} dX^{(3)} \tag{2.3.6}$$

where $\hat{\mathbf{N}}_i$ denote the unit vector along $d\mathbf{X}^{(i)}$. The corresponding volume in the current deformed configuration Ω is given by

$$dv = d\mathbf{x}^{(1)} \cdot d\mathbf{x}^{(2)} \times d\mathbf{x}^{(3)}$$
$$= \left(\mathbf{F} \cdot \hat{\mathbf{N}}_1 \right) \cdot \left(\mathbf{F} \cdot \hat{\mathbf{N}}_2 \right) \times \left(\mathbf{F} \cdot \hat{\mathbf{N}}_3 \right) dX^{(1)} dX^{(2)} dX^{(3)}$$
$$= \det \mathbf{F}\ dX^{(1)} dX^{(2)} dX^{(3)} = J\ dV \tag{2.3.7}$$

We assume that the volume elements are positive so that the relative orientation of the line elements is preserved under the deformation, that is, $J > 0$. Thus, J has the physical meaning of being the local ratio of current volume to reference volume of a material volume element.

Next, we consider an infinitesimal vector element of material surface $d\mathbf{A}$ in a neighborhood of the point $\mathbf{X}$ in the undeformed configuration. The surface vector can be expressed as $d\mathbf{A} = dA\,\hat{\mathbf{N}}$, where $\hat{\mathbf{N}}$ is the positive unit normal to the surface in the reference configuration. Suppose that $d\mathbf{A}$ becomes $d\mathbf{a}$ in the deformed body, where $d\mathbf{a} = da\,\hat{\mathbf{n}}$, $\hat{\mathbf{n}}$ being the positive unit normal to the surface in the deformed configuration. The areas of the parallelograms in the two configurations are related by (see Problem 2.7)

$$d\mathbf{a} = J\mathbf{F}^{-\mathrm{T}} \cdot d\mathbf{A} \quad \text{or} \quad \hat{\mathbf{n}}\,da = J\mathbf{F}^{-\mathrm{T}} \cdot \hat{\mathbf{N}}\,dA \tag{2.3.8}$$

which is known as the *Nanson's formula*.

The relations in Eqs. (2.3.7) and (2.3.8) are useful in expressing volume and area integrals defined in the current configuration in terms of the integrals over the volume and surface elements in the reference configuration.

2.4   Strain Measures

2.4.1   Deformation Tensors

The geometric changes that a continuous medium experiences can be measured in a number of ways. Here, we discuss a general measure of deformation of a continuous medium, independent of both translation and rotation.

Consider two material particles P and Q in the neighborhood of each other, separated by distance $d\mathbf{X}$ in the reference configuration, which occupy positions $\bar{\mathrm{P}}$ and $\bar{\mathrm{Q}}$, respectively, in the current configuration, and they are separated by distance $d\mathbf{x}$ (see Fig. 2.4.1). We wish to determine the change in the length $d\mathbf{X}$ of the line segment PQ as the body deforms and the material points move to the new locations $\bar{\mathrm{P}}$ and $\bar{\mathrm{Q}}$, respectively.

The squares of the lengths of the line segments PQ and $\bar{\mathrm{P}}\bar{\mathrm{Q}}$ are

$$(dS)^2 = d\mathbf{X} \cdot d\mathbf{X} \tag{2.4.1}$$

$$(ds)^2 = d\mathbf{x} \cdot d\mathbf{x} = d\mathbf{X} \cdot (\mathbf{F}^{\mathrm{T}} \cdot \mathbf{F}) \cdot d\mathbf{X} \equiv d\mathbf{X} \cdot \mathbf{C} \cdot d\mathbf{X} \tag{2.4.2}$$

where $\mathbf{C}$ is called the *right Cauchy–Green deformation tensor*

$$\mathbf{C} = \mathbf{F}^{\mathrm{T}} \cdot \mathbf{F} \tag{2.4.3}$$

By definition, $\mathbf{C}$ is a symmetric second-order tensor. The inverse of $\mathbf{C}$ is denoted by $\mathbf{B}$ and it is called *Piola deformation tensor*

$$\mathbf{B} = \mathbf{C}^{-1} = \mathbf{F}^{-1} \cdot \mathbf{F}^{-\mathrm{T}} \tag{2.4.4}$$

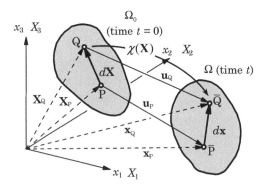

Fig. 2.4.1: Points P and Q separated by a distance $d\mathbf{X}$ in configuration Ω_0 take up positions $\bar{\mathrm{P}}$ and $\bar{\mathrm{Q}}$, respectively, in configuration Ω, where they are separated by distance $d\mathbf{x}$.

2.4.2   The Green–Lagrange Strain Tensor

The change in the squared lengths that occurs as a body deforms from the reference to the current configuration can be expressed relative to the original length as

$$(ds)^2 - (dS)^2 = 2\, d\mathbf{X} \cdot \mathbf{E} \cdot d\mathbf{X} \tag{2.4.5}$$

where $\mathbf{E}$ is called the *Green–St. Venant (Lagrangian) strain tensor* or simply the *Green strain tensor*[2]. The Green strain tensor can be expressed, in view of Eqs. (2.4.1)–(2.4.3), as

$$\mathbf{E} = \tfrac{1}{2}\left(\mathbf{F}^{\mathrm{T}} \cdot \mathbf{F} - \mathbf{I}\right) = \tfrac{1}{2}\left(\mathbf{C} - \mathbf{I}\right) \tag{2.4.6}$$

$$= \tfrac{1}{2}\left[\boldsymbol{\nabla}_0 \mathbf{u} + (\boldsymbol{\nabla}_0 \mathbf{u})^{\mathrm{T}} + (\boldsymbol{\nabla}_0 \mathbf{u}) \cdot (\boldsymbol{\nabla}_0 \mathbf{u})^{\mathrm{T}}\right] \tag{2.4.7}$$

By definition, the Green strain tensor is a symmetric second-order tensor. Also, the change in the squared lengths is zero if and only if $\mathbf{E} = \mathbf{0}$.

The vector form of the Green strain tensor in Eq. (2.4.7) allows us to express it in terms of its components in any coordinate system. In particular, in rectangular Cartesian coordinate system (X_1, X_2, X_3), the components of $\mathbf{E}$ and $\mathbf{C}$ are given by

$$
\begin{aligned}
E_{IJ} &= \frac{1}{2}\left(\frac{\partial u_I}{\partial X_J} + \frac{\partial u_J}{\partial X_I} + \frac{\partial u_K}{\partial X_I}\frac{\partial u_K}{\partial X_J}\right) \\
C_{IJ} &= \frac{\partial x_k}{\partial X_I}\frac{\partial x_k}{\partial X_J}
\end{aligned}
\tag{2.4.8}
$$

where sum on repeated indices is implied.

In expanded notation, the components of $\mathbf{E}$ are given by

$$
\begin{aligned}
E_{11} &= \frac{\partial u_1}{\partial X_1} + \frac{1}{2}\left[\left(\frac{\partial u_1}{\partial X_1}\right)^2 + \left(\frac{\partial u_2}{\partial X_1}\right)^2 + \left(\frac{\partial u_3}{\partial X_1}\right)^2\right] \\[4pt]
E_{22} &= \frac{\partial u_2}{\partial X_2} + \frac{1}{2}\left[\left(\frac{\partial u_1}{\partial X_2}\right)^2 + \left(\frac{\partial u_2}{\partial X_2}\right)^2 + \left(\frac{\partial u_3}{\partial X_2}\right)^2\right] \\[4pt]
E_{33} &= \frac{\partial u_3}{\partial X_3} + \frac{1}{2}\left[\left(\frac{\partial u_1}{\partial X_3}\right)^2 + \left(\frac{\partial u_2}{\partial X_3}\right)^2 + \left(\frac{\partial u_3}{\partial X_3}\right)^2\right] \\[4pt]
E_{12} &= \frac{1}{2}\left(\frac{\partial u_1}{\partial X_2} + \frac{\partial u_2}{\partial X_1} + \frac{\partial u_1}{\partial X_1}\frac{\partial u_1}{\partial X_2} + \frac{\partial u_2}{\partial X_1}\frac{\partial u_2}{\partial X_2} + \frac{\partial u_3}{\partial X_1}\frac{\partial u_3}{\partial X_2}\right) \\[4pt]
E_{13} &= \frac{1}{2}\left(\frac{\partial u_1}{\partial X_3} + \frac{\partial u_3}{\partial X_1} + \frac{\partial u_1}{\partial X_1}\frac{\partial u_1}{\partial X_3} + \frac{\partial u_2}{\partial X_1}\frac{\partial u_2}{\partial X_3} + \frac{\partial u_3}{\partial X_1}\frac{\partial u_3}{\partial X_3}\right) \\[4pt]
E_{23} &= \frac{1}{2}\left(\frac{\partial u_2}{\partial X_3} + \frac{\partial u_3}{\partial X_2} + \frac{\partial u_1}{\partial X_2}\frac{\partial u_1}{\partial X_3} + \frac{\partial u_2}{\partial X_2}\frac{\partial u_2}{\partial X_3} + \frac{\partial u_3}{\partial X_2}\frac{\partial u_3}{\partial X_3}\right)
\end{aligned}
\tag{2.4.9}
$$

Note that both $\mathbf{C}$ and $\mathbf{E}$ are material strain tensors.

---

[2]The reader should distinguish between the symbol $\mathbf{E}$ used for the Lagrangian strain tensor and $\mathbf{E}_i$ used for the basis vectors in the reference configuration.

2.4.3   The Cauchy and Euler Strain Tensors

Returning to the strain measures, the change in the squared lengths that occurs as the body deforms from the initial to the current configuration can be expressed relative to the current length. First, we express dS in terms of $d\mathbf{x}$ as

$$(dS)^2 = d\mathbf{X} \cdot d\mathbf{X} = d\mathbf{x} \cdot (\mathbf{F}^{-\mathrm{T}} \cdot \mathbf{F}^{-1}) \cdot d\mathbf{x} \equiv d\mathbf{x} \cdot \mathbf{b}^{-1} \cdot d\mathbf{x} \qquad (2.4.10)$$

where $\mathbf{b}$ is called the *left Cauchy–Green strain tensor*

$$\mathbf{b} = \mathbf{F} \cdot \mathbf{F}^{\mathrm{T}} \qquad (2.4.11)$$

The tensor $\mathbf{b}$ is sometimes referred to as the *Finger deformation tensor*, which is a symmetric second-order tensor in spatial coordinates.

Now the change in the squares of the lengths can be expressed in the current coordinates (i.e. spatial description) as

$$(ds)^2 - (dS)^2 = 2\, d\mathbf{x} \cdot \mathbf{e} \cdot d\mathbf{x} \qquad (2.4.12)$$

where $\mathbf{e}$, called the *Euler–Almansi strain tensor* or the *Euler strain tensor*, is

$$\mathbf{e} = \tfrac{1}{2} \left(\mathbf{I} - \mathbf{F}^{-\mathrm{T}} \cdot \mathbf{F}^{-1} \right) = \tfrac{1}{2} \left(\mathbf{I} - \mathbf{b}^{-1} \right) \qquad (2.4.13)$$

$$= \tfrac{1}{2} \left[\boldsymbol{\nabla}\mathbf{u} + (\boldsymbol{\nabla}\mathbf{u})^{\mathrm{T}} - (\boldsymbol{\nabla}\mathbf{u}) \cdot (\boldsymbol{\nabla}\mathbf{u})^{\mathrm{T}} \right] \qquad (2.4.14)$$

The rectangular Cartesian components of $\mathbf{e}$ are given by

$$e_{ij} = \tfrac{1}{2} \left(\delta_{ij} - \frac{\partial X_K}{\partial x_i} \frac{\partial X_K}{\partial x_j} \right) = \tfrac{1}{2} \left(\frac{\partial u_i}{\partial x_j} + \frac{\partial u_j}{\partial x_i} - \frac{\partial u_k}{\partial x_i} \frac{\partial u_k}{\partial x_j} \right) \qquad (2.4.15)$$

Note that $\mathbf{b}$ and $\mathbf{e}$ are spatial strain tensors.

The transformations between spatial and material fields are often called *push-forward* operations and *pull-back* operations. For example, a push-forward operation transforms a field expressed in a reference configuration Ω_0 to the current configuration Ω. A pull-back is an inverse operation, where a field in the current configuration Ω is transformed to the reference configuration Ω_0. As a specific example, consider the push-forward operation between $\mathbf{e}$ and $\mathbf{E}$:

$$\mathbf{e} = \tfrac{1}{2} \left(\mathbf{I} - \mathbf{F}^{-\mathrm{T}} \cdot \mathbf{F}^{-1} \right) = \tfrac{1}{2}\mathbf{F}^{-\mathrm{T}} \cdot \mathbf{F}^{\mathrm{T}} \cdot \left(\mathbf{I} - \mathbf{F}^{-\mathrm{T}} \cdot \mathbf{F}^{-1} \right) \cdot \mathbf{F} \cdot \mathbf{F}^{-1}$$

$$= \tfrac{1}{2}\mathbf{F}^{-\mathrm{T}} \cdot \left(\mathbf{F}^{\mathrm{T}} \cdot \mathbf{F} - \mathbf{I} \right) \cdot \mathbf{F}^{-1} = \mathbf{F}^{-\mathrm{T}} \cdot \mathbf{E} \cdot \mathbf{F}^{-1} \qquad (2.4.16)$$

The pull-back operation gives

$$\mathbf{E} = \mathbf{F}^{\mathrm{T}} \cdot \mathbf{e} \cdot \mathbf{F} \qquad (2.4.17)$$

2.4.4 Infinitesimal Strain Tensor and Rotation Tensor

2.4.4.1 Infinitesimal strain tensor

When all displacements gradients are small, that is, $|\nabla_0 \mathbf{u}| \ll 1$, we can neglect the nonlinear terms in the definition of the Green–Lagrange strain tensor $\mathbf{E}$ defined in Eq. (2.4.7). In order to derive the infinitesimal strain tensor from $\mathbf{E}$, we must linearize $\mathbf{E}$ by using a measure of smallness. We introduce the nonnegative function

$$\epsilon(t) = \|\nabla_0 \mathbf{u}\|_\infty = \sup_{\mathbf{X} \in \Omega} |\nabla_0 \mathbf{u}|$$

where "sup" stands for supremum or the least upper bound of the set of all absolute values of $\nabla_0 \mathbf{u}$ defined for all $\mathbf{X} \in \Omega$.

If $\mathbf{E}$ is of the order $O(\epsilon)$ in $\nabla_0 \mathbf{u}$, then we mean

$$\frac{\partial u_I}{\partial X_J} = O(\epsilon) \quad \text{as} \quad \epsilon \to 0$$

If terms of the order $O(\epsilon^2)$ as $\epsilon \to 0$ can be omitted in $\mathbf{E}$, then E_{IJ} can be approximated as

$$E_{IJ} \approx \tfrac{1}{2} \left(\frac{\partial u_I}{\partial X_J} + \frac{\partial u_J}{\partial X_I} \right) = O(\epsilon) \text{ as } \epsilon \to 0$$

Thus, it is immaterial whether the partial derivative of the displacement field $\mathbf{u}$ is taken with respect to x_j or X_j so that $\frac{\partial u_i}{\partial x_j} = \frac{\partial u_i}{\partial X_j}$; that is, $|\nabla \mathbf{u}| \approx |\nabla_0 \mathbf{u}| = O(\epsilon)$. In other words, in the case of infinitesimal strains, no distinction is made between the material coordinates $\mathbf{X}$ and the spatial coordinates $\mathbf{x}$, and it is not necessary to distinguish between the Green–Lagrange strain tensor $\mathbf{E}$ and the Eulerian strain tensor $\mathbf{e}$. The *infinitesimal strain tensor* is denoted by $\boldsymbol{\varepsilon}$, and it is defined as

$$\mathbf{E} \approx \boldsymbol{\varepsilon} = \tfrac{1}{2} \left[\nabla_0 \mathbf{u} + (\nabla_0 \mathbf{u})^{\mathrm{T}} \right] \tag{2.4.18}$$

The rectangular Cartesian components of the infinitesimal strain tensor are given by

$$\varepsilon_{ij} = \tfrac{1}{2} \left(\frac{\partial u_i}{\partial X_j} + \frac{\partial u_j}{\partial X_i} \right) \tag{2.4.19}$$

or, in expanded form:

$$\begin{aligned}
\varepsilon_{11} &= \frac{\partial u_1}{\partial X_1}, & \varepsilon_{22} &= \frac{\partial u_2}{\partial X_2}, & \varepsilon_{12} &= \tfrac{1}{2} \left(\frac{\partial u_1}{\partial X_2} + \frac{\partial u_2}{\partial X_1} \right) \\
\varepsilon_{33} &= \frac{\partial u_3}{\partial X_3}, & \varepsilon_{13} &= \tfrac{1}{2} \left(\frac{\partial u_1}{\partial X_3} + \frac{\partial u_3}{\partial X_1} \right), & \varepsilon_{23} &= \tfrac{1}{2} \left(\frac{\partial u_2}{\partial X_3} + \frac{\partial u_3}{\partial X_2} \right)
\end{aligned} \tag{2.4.20}$$

The strain components ε_{11}, ε_{22}, and ε_{33} are the infinitesimal normal strains and ε_{12}, ε_{13}, and ε_{23} are the infinitesimal shear strains. The shear strains $2\varepsilon_{12} = \gamma_{12}$, $2\varepsilon_{13} = \gamma_{13}$, and $2\varepsilon_{23} = \gamma_{23}$ are called the *engineering shear strains*.

2.4.4.2   Infinitesimal rotation tensor

The displacement gradient tensor $\nabla_0 \mathbf{u}$ can be expressed as the sum of a symmetric tensor and a skew-symmetric tensor

$$\mathbf{L} \equiv (\nabla_0 \mathbf{u})^{\mathrm{T}} = \tfrac{1}{2}\left[(\nabla_0 \mathbf{u})^{\mathrm{T}} + \nabla_0 \mathbf{u}\right] + \tfrac{1}{2}\left[(\nabla_0 \mathbf{u})^{\mathrm{T}} - \nabla_0 \mathbf{u}\right] \equiv \tilde{\boldsymbol{\varepsilon}} + \boldsymbol{\Omega} \quad (2.4.21)$$

where the symmetric part is similar to the infinitesimal strain tensor, and the skew-symmetric part is known as the *infinitesimal rotation tensor*

$$\boldsymbol{\Omega} = \tfrac{1}{2}\left[(\nabla_0 \mathbf{u})^{\mathrm{T}} - \nabla_0 \mathbf{u}\right] \qquad (2.4.22)$$

From the definition in Eq. (2.4.22), it follows that $\boldsymbol{\Omega}$ is a skew-symmetric tensor, that is, $\boldsymbol{\Omega}^{\mathrm{T}} = -\boldsymbol{\Omega}$. In Cartesian component form we have

$$\Omega_{ij} = \tfrac{1}{2}\left(\frac{\partial u_i}{\partial X_j} - \frac{\partial u_j}{\partial X_i}\right), \qquad \Omega_{ij} = -\Omega_{ji} \qquad (2.4.23)$$

Thus, there are only three independent components of $\boldsymbol{\Omega}$:

$$[\Omega] = \begin{bmatrix} 0 & \Omega_{12} & \Omega_{13} \\ -\Omega_{12} & 0 & \Omega_{23} \\ -\Omega_{13} & -\Omega_{23} & 0 \end{bmatrix} \qquad (2.4.24)$$

Since $\boldsymbol{\Omega}$ has only three independent components, the three components can be used to define the components of a vector $\boldsymbol{\omega}$,

$$\boldsymbol{\Omega} = -\mathcal{E} \cdot \boldsymbol{\omega} \qquad \text{or} \qquad \boldsymbol{\omega} = -\tfrac{1}{2}\mathcal{E} : \boldsymbol{\Omega}$$
$$\Omega_{ij} = -e_{ijk}\omega_k \qquad \text{or} \qquad \omega_i = -\tfrac{1}{2}e_{ijk}\Omega_{jk} \qquad (2.4.25)$$

where $\mathcal{E}$ is the permutation (alternating) tensor, $\mathcal{E} = e_{ijk}\hat{\mathbf{e}}_i\hat{\mathbf{e}}_j\hat{\mathbf{e}}_k$. In view of Eqs. (2.4.22) and (2.4.25), it follows that

$$\omega_i = \tfrac{1}{2}e_{ijk}\frac{\partial u_k}{\partial X_j} \qquad \text{or} \qquad \boldsymbol{\omega} = \tfrac{1}{2}\nabla_0 \times \mathbf{u} \qquad (2.4.26)$$

2.4.5   Time Derivatives of the Deformation Tensors

It is useful in the sequel to know how the tensor fields introduced in the previous sections change with time when the motion $\mathbf{x} = \boldsymbol{\chi}(\mathbf{X}, t)$ is known. First, we write the derivative of the spatial velocity field with respect to the spatial coordinates. The *spatial velocity gradient* $\mathbf{l}$ is the derivative of the spatial velocity field $\mathbf{v}(\mathbf{x}, t)$ with respect to the spatial coordinate $\mathbf{x}$

$$\mathbf{l} = \left(\frac{\partial \mathbf{v}(\mathbf{x}, t)}{\partial \mathbf{x}}\right)^{\mathrm{T}} = (\nabla \mathbf{v}(\mathbf{x}, t))^{\mathrm{T}} \qquad \text{or} \qquad \ell_{ij} = \frac{\partial v_i}{\partial x_j} \qquad (2.4.27)$$

We note that l is an unsymmetric second-order tensor, which can be expressed as the sum of symmetric and skew-symmetric tensors

$$l = \frac{1}{2} \left[\boldsymbol{\nabla}\mathbf{v} + (\boldsymbol{\nabla}\mathbf{v})^\mathrm{T} \right] + \frac{1}{2} \left[(\boldsymbol{\nabla}\mathbf{v})^\mathrm{T} - (\boldsymbol{\nabla}\mathbf{v}) \right] \equiv \mathbf{d} + \mathbf{w} \qquad (2.4.28)$$

The symmetric part of the spatial velocity gradient is denoted by $\mathbf{d}$, called the *rate of deformation tensor*, and the skew-symmetric part is denoted by $\mathbf{w}$, called the *vorticity tensor* or the *spin tensor*:

$$\mathbf{d} = \frac{1}{2} \left[\boldsymbol{\nabla}\mathbf{v} + (\boldsymbol{\nabla}\mathbf{v})^\mathrm{T} \right] = \mathbf{d}^\mathrm{T}, \quad \mathbf{w} = \frac{1}{2} \left[(\boldsymbol{\nabla}\mathbf{v})^\mathrm{T} - \boldsymbol{\nabla}\mathbf{v} \right] = -\mathbf{w}^\mathrm{T} \qquad (2.4.29)$$

or in index notation

$$d_{ij} = \tfrac{1}{2} \left(\frac{\partial v_i}{\partial x_j} + \frac{\partial v_j}{\partial x_i} \right) = d_{ji}, \quad w_{ij} = \tfrac{1}{2} \left(\frac{\partial v_i}{\partial x_j} - \frac{\partial v_j}{\partial x_i} \right) = -w_{ji} \qquad (2.4.30)$$

The material time derivative of the deformation gradient $\mathbf{F} = \mathbf{F}(\mathbf{X}, t)$ is given by

$$\frac{d\mathbf{F}(\mathbf{X}, t)}{dt} = \dot{\mathbf{F}} = \frac{\partial}{\partial t} (\boldsymbol{\nabla}_0 \mathbf{x})^\mathrm{T} = \left[\boldsymbol{\nabla}_0 \left(\frac{\partial \mathbf{x}}{\partial t} \right) \right]^\mathrm{T} = (\boldsymbol{\nabla}_0 \mathbf{V})^\mathrm{T} \qquad (2.4.31)$$

where $\mathbf{V}$ denotes the material velocity vector. Thus, $\dot{\mathbf{F}}^\mathrm{T}$ is equal to $\boldsymbol{\nabla}_0 \mathbf{V}$, called the *material velocity gradient* tensor.

We can express the spatial velocity gradient l in terms of the material velocity gradient $\dot{\mathbf{F}} = (\boldsymbol{\nabla}_0 \mathbf{V})^\mathrm{T}$ as

$$l = (\boldsymbol{\nabla}\mathbf{v})^\mathrm{T} = \left(\frac{\partial \mathbf{X}}{\partial \mathbf{x}} \cdot \frac{\partial \dot{\mathbf{x}}}{\partial \mathbf{X}} \right)^\mathrm{T} = \left(\mathbf{F}^{-\mathrm{T}} \cdot \dot{\mathbf{F}}^\mathrm{T} \right)^\mathrm{T} = \dot{\mathbf{F}} \cdot \mathbf{F}^{-1} \qquad (2.4.32)$$

or

$$\dot{\mathbf{F}} = l \cdot \mathbf{F} \qquad (2.4.33)$$

The material time derivative of the Green–Lagrange strain tensor $\mathbf{E}$ ($d\mathbf{E}/dt = \partial\mathbf{E}/\partial t = \dot{\mathbf{E}}$) is

$$\dot{\mathbf{E}} = \tfrac{1}{2} \left(\dot{\mathbf{F}}^\mathrm{T} \cdot \mathbf{F} + \mathbf{F}^\mathrm{T} \cdot \dot{\mathbf{F}} \right) = \tfrac{1}{2} \left(\mathbf{F}^\mathrm{T} \cdot l^\mathrm{T} \cdot \mathbf{F} + \mathbf{F}^\mathrm{T} \cdot l \cdot \mathbf{F} \right)$$

$$= \mathbf{F}^\mathrm{T} \cdot \left[\tfrac{1}{2} \left(l^\mathrm{T} + l \right) \right] \cdot \mathbf{F} = \mathbf{F}^\mathrm{T} \cdot \mathbf{d} \cdot \mathbf{F} \qquad (2.4.34)$$

The material time derivative of $\mathbf{E}$ is called *material strain rate tensor*, which is a pull-back of rate of deformation tensor $\mathbf{d}$. One can show that the material time derivative of the Euler–Almansi strain tensor $\mathbf{e}$ is given by

$$\dot{\mathbf{e}} = \mathbf{d} - l^\mathrm{T} \cdot \mathbf{e} - \mathbf{e} \cdot l \qquad (2.4.35)$$

2.5   Measures of Stress

2.5.1   Stress Vector

First we introduce the true stress, i.e. stress in the deformed configuration Ω that is measured per unit area of the deformed configuration Ω. The surface force acting on a small element of area in a continuous medium depends not only on the magnitude of the area but also upon the orientation of the area. It is customary to denote the direction of a plane area by means of a unit vector drawn normal to that plane. The direction of the normal is taken by convention as that in which a right-handed screw advances as it is rotated according to the sense of travel along the boundary curve or contour.

Let the unit normal vector be denoted by $\hat{\mathbf{n}}$, and let $d\mathbf{f}(\hat{\mathbf{n}})$ be the force on a small area $\hat{\mathbf{n}}da$ located at the position $\mathbf{x}$. Then the *stress vector* is defined, shown graphically in Fig. 2.5.1, as

$$\mathbf{t}(\hat{\mathbf{n}}) = \lim_{\Delta a \to 0} \frac{\Delta \mathbf{f}(\hat{\mathbf{n}})}{\Delta a} \tag{2.5.1}$$

We see that the stress vector is a point function of the unit normal $\hat{\mathbf{n}}$ which denotes the orientation of the surface Δa. The component of $\mathbf{t}$ that is in the direction of $\hat{\mathbf{n}}$, $(\mathbf{t} \cdot \hat{\mathbf{n}})\hat{\mathbf{n}}$, is called the *normal stress* vector. The component of $\mathbf{t}$ that is normal to $\hat{\mathbf{n}}$, but lies in the plane of $\mathbf{t}$ and $\hat{\mathbf{n}}$, is called the *shear* stress. Because of Newton's third law for action and reaction, we see that

$$\mathbf{t}(-\hat{\mathbf{n}}) = -\mathbf{t}(\hat{\mathbf{n}})$$

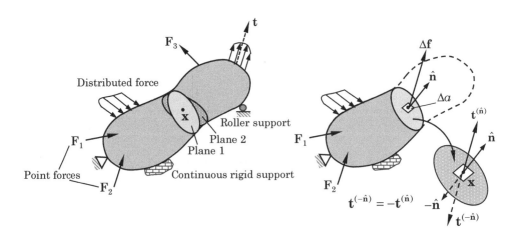

Fig. 2.5.1: Stress vector at a point on a plane normal to $\hat{\mathbf{n}}$.

2.5.2   Cauchy's Formula and Stress Tensor

To establish the relationship between $\mathbf{t}$ and $\hat{\mathbf{n}}$ we set up an infinitesimal tetrahedron in Cartesian coordinates, as shown in Fig. 2.5.2. If $-\mathbf{t}_1, -\mathbf{t}_2, -\mathbf{t}_3$, and $\mathbf{t}$ denote the stress vectors in the outward directions on the faces of the infinitesimal tetrahedron whose areas are Δa_1, Δa_2, Δa_3, and Δa, respectively, we have by Newton's second law for the mass inside the tetrahedron,

$$\mathbf{t}\,\Delta a - \mathbf{t}_1\,\Delta a_1 - \mathbf{t}_2\,\Delta a_2 - \mathbf{t}_3\,\Delta a_3 + \rho\,\Delta v\,\mathbf{f} = \rho\,\Delta v\,\mathbf{a} \qquad (2.5.2)$$

where Δv is the volume of the tetrahedron, ρ the density, $\mathbf{f}$ the body force per unit mass, and $\mathbf{a}$ the acceleration. Since the total vector area of a closed surface is zero, we have

$$\Delta a\,\hat{\mathbf{n}} - \Delta a_1\,\hat{\mathbf{e}}_1 - \Delta a_2\,\hat{\mathbf{e}}_2 - \Delta a_3\,\hat{\mathbf{e}}_3 = \mathbf{0} \qquad (2.5.3)$$

It follows that

$$\Delta a_i = (\hat{\mathbf{n}} \cdot \hat{\mathbf{e}}_i)\Delta a = (\hat{\mathbf{e}}_i \cdot \hat{\mathbf{n}})\Delta a \quad \text{for} \quad i = 1, 2, 3 \qquad (2.5.4)$$

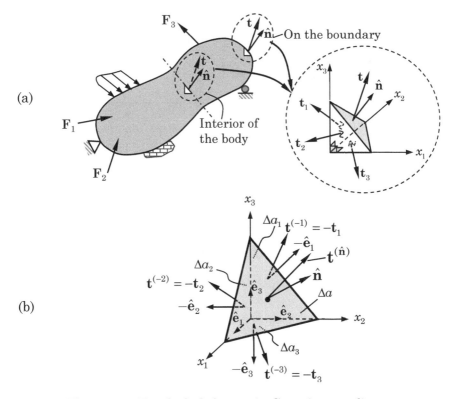

Fig. 2.5.2: Tetrahedral element in Cartesian coordinates.

The volume of the element Δv can be expressed as

$$\Delta v = \frac{\Delta h}{3} \Delta a \tag{2.5.5}$$

where Δh is the perpendicular distance from the origin to the slant face.

Substitution of Eqs. (2.5.3) and (2.5.5) into Eq. (2.5.2) and dividing throughout by Δa, yields

$$\mathbf{t} = (\hat{\mathbf{n}} \cdot \hat{\mathbf{e}}_i)\mathbf{t}_i + \rho\frac{\Delta h}{3}(\mathbf{a} - \mathbf{f}) = \mathbf{t}_i(\hat{\mathbf{e}}_i \cdot \hat{\mathbf{n}}) + \rho\frac{\Delta h}{3}(\mathbf{a} - \mathbf{f}) \tag{2.5.6}$$

where the summation convention is used. In the limit when the tetrahedron shrinks to a point, $\Delta h \to 0$, we are left with

$$\mathbf{t} = (\hat{\mathbf{n}} \cdot \hat{\mathbf{e}}_i)\mathbf{t}_i = \mathbf{t}_i(\hat{\mathbf{e}}_i \cdot \hat{\mathbf{n}}) \tag{2.5.7}$$

It is now convenient to display the above equation as

$$\mathbf{t} = \hat{\mathbf{n}} \cdot (\hat{\mathbf{e}}_i \mathbf{t}_i) \quad \text{or} \quad \mathbf{t} = (\mathbf{t}_i \hat{\mathbf{e}}_i) \cdot \hat{\mathbf{n}} \tag{2.5.8}$$

We *define* the expression $\mathbf{t}_i \hat{\mathbf{e}}_i$ as the *stress dyadic*[3] or *stress tensor* $\boldsymbol{\sigma}$:

$$\boldsymbol{\sigma} \equiv \mathbf{t}_1\,\hat{\mathbf{e}}_1 + \mathbf{t}_2\,\hat{\mathbf{e}}_2 + \mathbf{t}_3\,\hat{\mathbf{e}}_3 = \mathbf{t}_i\,\hat{\mathbf{e}}_i \tag{2.5.9}$$

Thus, from Eqs. (2.5.8) and (2.5.9), we have

$$\mathbf{t}(\hat{\mathbf{n}}) = \boldsymbol{\sigma} \cdot \hat{\mathbf{n}} = \hat{\mathbf{n}} \cdot \boldsymbol{\sigma}^{\mathrm{T}} \tag{2.5.10}$$

Equation (2.5.10) is known as the *Cauchy stress formula*, and $\boldsymbol{\sigma}$ is termed the *Cauchy stress tensor*. Thus, the Cauchy stress tensor $\boldsymbol{\sigma}$ is defined to be the *current force* per unit *deformed area*, $d\mathbf{f} = \mathbf{t}\,da = \boldsymbol{\sigma} \cdot d\mathbf{a}$, where Cauchy's formula, $\mathbf{t} = \hat{\mathbf{n}} \cdot \boldsymbol{\sigma}^{\mathrm{T}} = \boldsymbol{\sigma} \cdot \hat{\mathbf{n}}$, is used.

It is useful to resolve the stress vectors $\mathbf{t}_1, \mathbf{t}_2,$ and $\mathbf{t}_3$ into their orthogonal components in a rectangular Cartesian system

$$\mathbf{t}_j = \hat{\mathbf{e}}_1\,\sigma_{1j} + \hat{\mathbf{e}}_2\,\sigma_{2j} + \hat{\mathbf{e}}_3\,\sigma_{3j} = \hat{\mathbf{e}}_i\,\sigma_{ij} \tag{2.5.11}$$

for $j = 1, 2, 3$. Hence, the stress tensor can be expressed in the Cartesian component form as

$$\boldsymbol{\sigma} = \mathbf{t}_j\,\hat{\mathbf{e}}_j = \hat{\mathbf{e}}_i\,\sigma_{ij}\,\hat{\mathbf{e}}_j = \sigma_{ij}\,\hat{\mathbf{e}}_i\,\hat{\mathbf{e}}_j \tag{2.5.12}$$

The component σ_{ij} represents the stress in the x_i coordinate direction on a plane perpendicular to the x_j coordinate, as shown in Fig. 2.5.3.

---

[3]One can also define the stress tensor as $\boldsymbol{\sigma} = \hat{\mathbf{e}}_i\,\mathbf{t}_i$, which is the transpose of the stress tensor defined in Eq. (2.5.9). Both definitions can be found in the literature, and when the stress tensor is symmetric the difference vanishes.

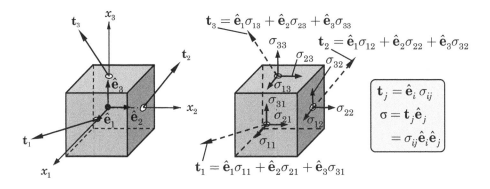

Fig. 2.5.3: Display of stress components in Cartesian rectangular coordinates.

2.5.3 Piola–Kirchhoff Stress Tensors

The Cauchy stress tensor is the most natural and physical measure of the state of stress at a point in the deformed configuration and measured per unit area of the deformed configuration. It is the quantity most commonly used in spatial description of problems in fluid mechanics. The equations of motion or equilibrium of a material body in the Lagrange description must be derived for the deformed configuration of the body at time t. However, since the geometry of the deformed configuration is not known, the equations must be written in terms of the known reference configuration. In doing so we introduce various measures of stress. They emerge in a natural way as we transform volumes and areas from the deformed configuration to undeformed (or reference) configuration. These measures are purely mathematical but facilitate analysis. These are discussed next.

2.5.3.1 First Piola–Kirchhoff stress tensor

Consider a continuum $\mathcal{B}$ subjected to a deformation mapping χ that results in the deformed configuration Ω, as shown in Fig. 2.5.4. Let the force vector on an elemental area da with normal $\hat{\mathbf{n}}$ in the deformed configuration be $d\mathbf{f}$. Suppose that the area element in the undeformed configuration that corresponds to da is dA. The force $d\mathbf{f}$ can be expressed in terms of a stress vector $\mathbf{t}$ times the deformed area da is

$$d\mathbf{f} = \mathbf{t}^{(\mathbf{n})}\, da \qquad (2.5.13)$$

We define a stress vector $\mathbf{T}^{(\mathbf{N})}$ over the area element dA with normal $\mathbf{N}$ in the undeformed configuration such that it results in the same total force

$$d\mathbf{f} = \mathbf{t}^{(\mathbf{n})}\, da = \mathbf{T}^{(\mathbf{N})} dA \qquad (2.5.14)$$

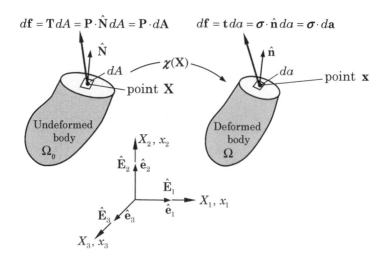

Fig. 2.5.4: Definition of the first Piola–Kirchhoff stress tensor.

The vector $\mathbf{T}^{(\mathbf{N})}$ is known as the *pseudo stress vector* associated with the first Piola–Kirchhoff stress tensor. The stress vector $\mathbf{T}^{(\mathbf{N})}$ is measured per unit undeformed area, while the stress vector $\mathbf{t}^{(\mathbf{n})}$ is measured per deformed area. Clearly, both stress vectors have the same direction but different magnitudes owing to the different areas.

From Cauchy's formula, we have $\mathbf{t}^{(\mathbf{n})} = \boldsymbol{\sigma} \cdot \hat{\mathbf{n}}$, where $\boldsymbol{\sigma}$ is the Cauchy stress tensor. In a similar fashion, we introduce a stress tensor $\mathbf{P}$, called the *first Piola–Kirchhoff stress tensor*, such that $\mathbf{T}^{(\mathbf{N})} = \mathbf{P} \cdot \hat{\mathbf{N}}$. Then using Eq. (2.5.14) we can write

$$\boldsymbol{\sigma} \cdot \hat{\mathbf{n}}\, da = \mathbf{P} \cdot \hat{\mathbf{N}}\, dA \quad \text{or} \quad \boldsymbol{\sigma} \cdot d\mathbf{a} = \mathbf{P} \cdot d\mathbf{A} \qquad (2.5.15)$$

where

$$d\mathbf{a} = da\,\hat{\mathbf{n}}, \quad d\mathbf{A} = dA\,\hat{\mathbf{N}} \qquad (2.5.16)$$

The first Piola–Kirchhoff stress tensor, also referred to as the *nominal stress* or *Lagrangian stress* tensor, is the *current force* per unit *undeformed area*. The tensor Cartesian component representation of $\mathbf{P}$ is given by

$$\mathbf{P} = P_{iI}\,\hat{\mathbf{e}}_i\,\hat{\mathbf{E}}_I \qquad (2.5.17)$$

Clearly, the first Piola–Kirchhoff stress tensor is a mixed tensor.

Nanson's formula in Eq. (2.3.8) can be used to relate the first Piola–Kirchhoff stress tensor to the Cauchy stress tensor. From Eqs. (2.3.8) and (2.5.15), we obtain

$$\mathbf{P} \cdot d\mathbf{A} = \boldsymbol{\sigma} \cdot d\mathbf{a} = J\boldsymbol{\sigma} \cdot \mathbf{F}^{-\mathrm{T}} \cdot d\mathbf{A} \qquad (2.5.18)$$

or

$$\mathbf{P} = J\boldsymbol{\sigma} \cdot \mathbf{F}^{-\mathrm{T}} \qquad (2.5.19)$$

where J is the Jacobian. In general, the first Piola–Kirchhoff stress tensor $\mathbf{P}$ is unsymmetric even when the Cauchy stress tensor σ is symmetric.

2.5.3.2 Second Piola–Kirchhoff stress tensor

The *second Piola–Kirchhoff stress tensor* $\mathbf{S}$, which is used in the study of large deformation analysis, is introduced as the stress tensor associated with the force $d\mathcal{F}$ in the undeformed elemental area $d\mathbf{A}$ that corresponds to the force $d\mathbf{f}$ on the deformed elemental area $d\mathbf{a}$

$$d\mathcal{F} = \mathbf{S} \cdot d\mathbf{A} \tag{2.5.20}$$

Thus, the second Piola–Kirchhoff stress tensor gives the *transformed current force* per unit *undeformed area*.

Similar to the relationship between $d\mathbf{x}$ and $d\mathbf{X}$, $d\mathbf{X} = \mathbf{F}^{-1} \cdot d\mathbf{x}$, the force $d\mathbf{f}$ on the deformed elemental area $d\mathbf{a}$ is related to the force $d\mathcal{F}$ on the undeformed elemental area $d\mathbf{A}$

$$d\mathcal{F} = \mathbf{F}^{-1} \cdot d\mathbf{f} = \mathbf{F}^{-1} \cdot (\mathbf{P} \cdot d\mathbf{A}) = \mathbf{S} \cdot d\mathbf{A} \tag{2.5.21}$$

Hence, the second Piola–Kirchhoff stress tensor is related to the first Piola–Kirchhoff stress tensor and Cauchy stress tensor according to the equations

$$\mathbf{S} = \mathbf{F}^{-1} \cdot \mathbf{P} = J\mathbf{F}^{-1} \cdot \sigma \cdot \mathbf{F}^{-T} \tag{2.5.22}$$

Clearly, $\mathbf{S}$ is symmetric whenever σ is symmetric. Cartesian component representation of $\mathbf{S}$ is

$$\mathbf{S} = S_{IJ}\hat{\mathbf{E}}_I\hat{\mathbf{E}}_J \tag{2.5.23}$$

We can introduce the pseudo stress vector $\tilde{\mathbf{T}}$ associated with the second Piola–Kirchhoff stress tensor by

$$d\mathcal{F} = \tilde{\mathbf{T}}\, dA = \mathbf{S} \cdot \hat{\mathbf{N}}\, dA = \mathbf{S} \cdot d\mathbf{A} \tag{2.5.24}$$

2.6 Material Frame Indifference

2.6.1 The Basic Idea

In the analytical description of physical events, the following two requirements must be followed:

1. Invariance of the equations with respect to stationary coordinate frames of reference.
2. Invariance of the equations with respect to frames of reference that move in arbitrary relative motion.

The first requirement is readily met by expressing the equations in vector/tensor form, which is invariant. The assertion that an equation is in "invariant form" refers to the vector form that is independent of the choice of a coordinate system. The coordinate systems used in the present study were assumed to be relatively at rest. The second requirement is that the invariance property holds for reference frames (or observers) moving arbitrarily with respect to each other. This requirement is dictated by the need for forces to be the same as measured by all observers irrespective of their relative motions. The concept of frames of reference should not be confused with that of coordinate systems, as they are not the same at all. A given observer is free to choose any coordinate system as may be convenient to observe or analyze the system response. Invariance with respect to changes of observer is termed *material frame indifference* or *material objectivity*. In general, *if a quantity is objective, then it is independent of an observer*.

A detailed discussion of frame indifference is outside the scope of the present study, and only a brief discussion is presented here. Let $\mathcal{F}$ denote a reference frame with origin at O in which $\mathbf{x}$ is the current position of a particular particle at time t. Let $\mathcal{F}^*$ be another reference frame with origin at O* with time denoted with t^*. Let ϕ be a scalar field when described in the frame $\mathcal{F}$ and ϕ^* is the same scalar field described with respect to the frame $\mathcal{F}^*$, and let $(\mathbf{u}, \mathbf{u}^*)$ and $(\mathbf{S}, \mathbf{S}^*)$ be the vector and tensor fields in the two frames. Scalar, vector, and tensor fields are called *frame indifferent* or *objective* if they transform according to the following equations:

1. Events	$\mathbf{x}^* = \mathbf{c}(t) + \mathbf{Q}(t) \cdot \mathbf{x}, \quad t^* = t - a$
2. Scalar field	$\phi^*(\mathbf{x}^*, t^*) = \phi(\mathbf{x}, t)$
3. Displacement vector	$\mathbf{u}^*(\mathbf{x}^*, t^*) = \mathbf{Q}(t) \cdot \mathbf{u}(\mathbf{x}, t)$ (2.6.1)
4. General second-order tensors	$\mathbf{S}^*(\mathbf{x}^*, t^*) = \mathbf{Q}(t) \cdot \mathbf{S}(\mathbf{x}, t) \cdot \mathbf{Q}^{\mathrm{T}}(t)$
5. Two-point second-order tensors	$\mathbf{F}^*(\mathbf{x}^*, t^*) = \mathbf{Q}(t) \cdot \mathbf{F}(\mathbf{x}, t)$

where $\mathbf{Q}(t)$ is a proper orthogonal tensor that rotates frame $\mathcal{F}^*$ into frame $\mathcal{F}$, $\mathbf{c}(t)$ is a vector from O to O* that only depends on time t, and a is a constant. For example, $\mathbf{x}$ and $\mathbf{x}^*$ refer to the same motion, but mathematically $\mathbf{x}^*$ is the motion obtained from $\mathbf{x}$ by superposition of a rigid rotation and translation. The mapping $\mathbf{x}^* = \mathbf{c}(t) + \mathbf{Q}(t) \cdot \mathbf{x}$ may be interpreted as one that takes $(\mathbf{x}, t)$ to $(\mathbf{x}^*, t^*)$ as a change of observer from O to O*, so that the event which is observed at place $\mathbf{x}$ at time t by observer O is the *same* event as that observed at place $\mathbf{x}^*$ at time t^* by observer O*, where $t^* = t - a$, and a is a constant. Thus, a change of observer merely changes the *description* of an event. In short, the objectivity insures that the direction(s) and magnitude are independent of the coordinate frame used to describe them.

2.6.2   Objectivity of Strains and Strain Rates

Under the following general rigid body mapping

$$\mathbf{x}^*(\mathbf{X}, t^*) = \mathbf{c}(t) + \mathbf{Q}(t) \cdot \mathbf{x}, \quad t^* = t - a \tag{2.6.2}$$

between observer O and observer O* (if the reference configuration is independent of the observer) the right Cauchy–Green deformation tensor $\mathbf{C}$ and the Green–Lagrange strain tensor $\mathbf{E}$ change as follows:

$$\mathbf{C}^* = (\mathbf{F}^*)^{\mathrm{T}} \cdot \mathbf{F}^* = \left(\mathbf{F}^{\mathrm{T}} \cdot \mathbf{Q}^{\mathrm{T}}\right) \cdot \left(\mathbf{Q} \cdot \mathbf{F}\right) = \mathbf{F}^{\mathrm{T}} \cdot \mathbf{F} = \mathbf{C} \tag{2.6.3}$$

where the property $\mathbf{Q}^{\mathrm{T}} \cdot \mathbf{Q} = \mathbf{I}$ of an orthogonal matrix $\mathbf{Q}$ is used. Hence, by definition, the Green–Lagrange strain tensor $\mathbf{E}$ and the right Cauchy–Green deformation tensor $\mathbf{C}$, being defined with respect to the reference configuration, are unaffected by the superposed rigid-body motion:

$$\mathbf{E} = \mathbf{E}^*, \qquad \mathbf{C} = \mathbf{C}^* \tag{2.6.4}$$

However, the velocities and accelerations of a material point are affected by the superposed rigid-body motion. For example, consider velocity after imposing the rigid-body motion (note that $dt/dt^* = 1$)

$$\mathbf{v}^*(\mathbf{x}^*, t^*) = \frac{d\mathbf{x}^*}{dt^*} = \frac{d}{dt^*}(\mathbf{c}(t) + \mathbf{Q}(t) \cdot \mathbf{x}) = \dot{\mathbf{c}}(t) + \dot{\mathbf{Q}}(t) \cdot \mathbf{x} + \mathbf{Q}(t) \cdot \mathbf{v} \tag{2.6.5}$$

which shows that $\mathbf{v}^*$ and $\mathbf{v}$ are not the same, but one can be calculated from the other when $\mathbf{c}$ and $\mathbf{Q}$ are known for the superposed rigid-body motion.

In addition, the symmetric part of $\mathbf{l}$ (or $\mathbf{l}^{\mathrm{T}}$), namely, $\mathbf{d}$ is also objective in the sense

$$\mathbf{d}^* = \mathbf{Q} \cdot \mathbf{d} \cdot \mathbf{Q}^{\mathrm{T}} \tag{2.6.6}$$

We have noted that the two observers' view of the velocity and acceleration of a given motion are different, even though the rate of change at fixed $\mathbf{X}$ is the same in each case. Thus, velocity and acceleration vectors are *not* objective.

2.6.3   Objectivity of Stress Tensors

2.6.3.1   Cauchy stress tensor

The Cauchy stress tensor is objective if we can show that $\boldsymbol{\sigma}^* = \mathbf{Q} \cdot \boldsymbol{\sigma} \cdot \mathbf{Q}^{\mathrm{T}}$ [see Eq. (2.6.1) for the definition of the objectivity of various order tensors]. We begin with the relations

$$\mathbf{t} = \boldsymbol{\sigma} \cdot \mathbf{n}, \quad \mathbf{t}^* = \boldsymbol{\sigma}^* \cdot \mathbf{n}^*; \qquad \mathbf{t}^* = \mathbf{Q} \cdot \mathbf{t}, \quad \mathbf{n}^* = \mathbf{Q} \cdot \mathbf{n} \tag{2.6.7}$$

Then

$$\mathbf{t}^* = \boldsymbol{\sigma}^* \cdot \mathbf{n}^* = \boldsymbol{\sigma}^* \cdot (\mathbf{Q} \cdot \mathbf{n}),$$

$$\mathbf{t}^* = \mathbf{Q} \cdot \mathbf{t} = \mathbf{Q} \cdot \boldsymbol{\sigma} \cdot \mathbf{n} \tag{2.6.8}$$

Then, we have

$$\boldsymbol{\sigma}^* \cdot \mathbf{Q} = \mathbf{Q} \cdot \boldsymbol{\sigma}$$

from which it follows that

$$\boldsymbol{\sigma}^* = \mathbf{Q} \cdot \boldsymbol{\sigma} \cdot \mathbf{Q}^{\mathrm{T}} \tag{2.6.9}$$

Thus, the Cauchy stress tensor is objective.

2.6.3.2   First Piola–Kirchhoff stress tensor

Since the first Piola–Kirchhoff stress tensor $\mathbf{P}$ is a two-point tensor, it transforms like the other two-point tensor $\mathbf{F}$. To establish this, we begin with the relation between $\mathbf{P}$ and $\boldsymbol{\sigma}$ after superposed rigid-body motion and make use of the relations $\mathbf{F}^* = \mathbf{Q} \cdot \mathbf{F}$ and $J^* = J$,

$$\begin{aligned}
\mathbf{P}^* &= J^* \boldsymbol{\sigma}^* \cdot (\mathbf{F}^*)^{-\mathrm{T}} = J(\mathbf{Q} \cdot \boldsymbol{\sigma} \cdot \mathbf{Q}^{\mathrm{T}}) \cdot (\mathbf{Q} \cdot \mathbf{F})^{-\mathrm{T}} \\
&= J\mathbf{Q} \cdot \boldsymbol{\sigma} \cdot (\mathbf{Q}^{\mathrm{T}} \cdot \mathbf{Q}^{-\mathrm{T}}) \cdot \mathbf{F}^{-\mathrm{T}} = J\mathbf{Q} \cdot \boldsymbol{\sigma} \cdot \mathbf{F}^{-\mathrm{T}} = \mathbf{Q} \cdot \mathbf{P} \tag{2.6.10}
\end{aligned}$$

Thus $\mathbf{P}$, being a two-point tensor, transforms like vector under superposed rigid-body motion, and hence is objective.

2.6.3.3   Second Piola–Kirchhoff stress tensor

The second Piola–Kirchhoff stress tensor $\mathbf{S}$ is the stress tensor of choice in the study of solid mechanics. Since it is defined with respect to the reference configuration, rigid-body motion should not alter it. Using the relations $\mathbf{F}^* = \mathbf{Q} \cdot \mathbf{F}$ and $\boldsymbol{\sigma}^* = \mathbf{Q} \cdot \boldsymbol{\sigma} \cdot \mathbf{Q}^{\mathrm{T}}$, we obtain

$$\begin{aligned}
\mathbf{S}^* &= J^* (\mathbf{F}^*)^{-1} \cdot \boldsymbol{\sigma}^* \cdot (\mathbf{F}^*)^{-\mathrm{T}} = J(\mathbf{F}^{-1} \cdot \mathbf{Q}^{-1}) \cdot (\mathbf{Q} \cdot \boldsymbol{\sigma} \cdot \mathbf{Q}^{\mathrm{T}}) \cdot (\mathbf{Q}^{-\mathrm{T}} \cdot \mathbf{F}^{-\mathrm{T}}) \\
&= J\mathbf{F}^{-1} \cdot \boldsymbol{\sigma} \cdot \mathbf{F}^{-\mathrm{T}} = \mathbf{S} \tag{2.6.11}
\end{aligned}$$

Thus, $\mathbf{S}$ is not affected by the superposed rigid-body motion and, therefore, it is objective.

2.7   Equations of Continuum Mechanics

2.7.1   Introduction

This section is devoted to a review of the conservation principles and balance laws of physics in analytical terms. The laws of physics that we consider are: (1) the principle of conservation of mass, (2) the balance of linear momentum, (3) the balance of angular momentum, and (4) the balance of energy. The equations resulting from these principles are outlined here. For additional details, reader may consult the continuum mechanics text by the author [1].

2.7.2 Conservation of Mass

The principle of conservation of mass can be stated as: *the total mass of any part of the body does not change in any motion.* The mathematical form of this principle is different in different descriptions of motion. The equation resulting from the principle of conservation of mass is also known as the *continuity equation.*

2.7.2.1 Spatial form of the continuity equation

First we derive the equation resulting from the principle of conservation of mass in the spatial description. Let each element of mass in the medium move with the velocity $\mathbf{v}(\mathbf{x}, t)$ and consider a spatial region Ω such that the bounding surface Γ is attached to a fixed set of material elements. Then each point of this surface moves itself with the material velocity, and the region Ω thus contains a fixed mass since no mass crosses the boundary surface Γ. The time rate of change of an integral of a function $\phi(\mathbf{x}, t)$ over this material region is the sum of the integral of the instantaneous change of ϕ with time and the rate of outflow of ϕ through the surface:

$$\frac{d}{dt} \int_\Omega \phi(\mathbf{x}, t) \, dv = \int_\Omega \frac{\partial \phi}{\partial t} \, dv + \oint_\Gamma \phi \mathbf{v} \cdot \hat{\mathbf{n}} \, ds \qquad (2.7.1)$$

which holds for a material region, that is, a region of fixed total mass.

Let $\rho(\mathbf{x}, t)$ denote the mass density of a continuous region Ω with closed boundary Γ. Then the principle of conservation of mass for a fixed *material* region requires that

$$0 = \frac{d}{dt} \int_\Omega \rho \, dv \equiv \int_\Omega \frac{\partial \rho}{\partial t} \, dv + \oint_\Gamma \rho \mathbf{v} \cdot \hat{\mathbf{n}} \, ds \qquad (2.7.2)$$

where Eq. (2.7.1) with $\phi = \rho$ is used in arriving at the last statement. Converting the surface integral in Eq. (2.7.2) to a volume integral by means of the divergence theorem, we obtain

$$\int_\Omega \left[\frac{\partial \rho}{\partial t} + \operatorname{div}(\rho \mathbf{v}) \right] dv = 0 \qquad (2.7.3)$$

Since this integral vanishes, for a continuous medium, for any arbitrary region Ω, we deduce that this can be true only if the integrand itself vanishes identically, giving the following local (i.e. point-wise) form:

$$\frac{\partial \rho}{\partial t} + \operatorname{div}(\rho \mathbf{v}) = 0 \qquad (2.7.4)$$

This equation, called the *continuity equation*, expresses local conservation of mass at any point in a continuous medium.

2.7.2.2   Material form of the continuity equation

Next, we derive the continuity equation in the material description. Consider a material body $\mathcal{B}$ that occupies configuration Ω_0 with density $\rho_0(\mathbf{X}, t)$. The same material body occupies the configuration Ω under the action of external forces, and it has a density $\rho(\mathbf{x}, t)$. Then the principle of conservation of mass gives

$$\int_{\Omega_0} \rho_0 \, dV = \int_{\Omega} \rho \, dv \tag{2.7.5}$$

Using the relation (2.3.7) between the volume elements of the body in the undeformed and deformed configurations, $dv = J \, dV$, where J is the determinant of the deformation gradient $\mathbf{F}$, we arrive at

$$\int_{\Omega_0} (\rho_0 - J\rho) \, dV = 0 \tag{2.7.6}$$

This is the *global form* of the *continuity equation* in the material description. Since the material volume Ω_0 we selected is arbitrarily small, as we shrink the volume to a point, we obtain the *local form* of the continuity equation

$$\rho_0 = J\rho \tag{2.7.7}$$

2.7.3   Reynolds Transport Theorem

The material derivative operator d/dt corresponds to changes with respect to a fixed mass, that is, $\rho \, dv$ is constant with respect to this operator. Therefore, from Eq. (2.7.1) it follows that, for $\phi = \rho Q(\mathbf{x}, t)$, the result

$$\frac{d}{dt} \int_{\Omega} \rho Q \, dv = \frac{\partial}{\partial t} \int_{\Omega} \rho Q \, dv + \oint_{\Gamma} \rho Q \, \mathbf{v} \cdot \hat{\mathbf{n}} \, ds \tag{2.7.8}$$

or

$$\frac{d}{dt} \int_{\Omega} \rho Q \, dv = \int_{\Omega} \left[\rho \frac{\partial Q}{\partial t} + Q \frac{\partial \rho}{\partial t} + \boldsymbol{\nabla} \cdot (\rho Q \mathbf{v}) \right] dv$$

$$= \int_{\Omega} \left[\rho \left(\frac{\partial Q}{\partial t} + \mathbf{v} \cdot \boldsymbol{\nabla} Q \right) + Q \left(\frac{\partial \rho}{\partial t} + \boldsymbol{\nabla} \cdot (\rho \mathbf{v}) \right) \right] dv \tag{2.7.9}$$

and using the continuity equation, Eq. (2.7.3), and the definition of the material time derivative, we arrive at the result

$$\frac{d}{dt} \int_{\Omega} \rho Q \, dv = \int_{\Omega} \rho \frac{dQ}{dt} \, dv \tag{2.7.10}$$

Equation (2.7.10) is known as the *Reynolds transport theorem*.

2.7.4 Balance of Linear Momentum

The principle of balance of linear momentum (i.e. mass times velocity) applied to a given mass of a medium $\mathcal{B}$, instantaneously occupying a region Ω with bounding surface Γ, can be stated as: *the time rate of change of linear momentum be equal to sum of forces acting on the body.* The principle results in equations of motion governing a continuum.

2.7.4.1 Spatial form of the equations of motion

Let $\mathbf{f}$ be the body force per unit volume and $\mathbf{t}$ be the surface force per unit area. Consider an elemental volume dv inside Ω. The body force of the elemental volume dv is equal to $dv\,\mathbf{f}$ and the surface force on an elemental surface ds is $\mathbf{t}\,ds$. Then the principle of balance of linear momentum requires that

$$\frac{d}{dt}\int_{\Omega}\rho\mathbf{v}\,dv = \oint_{\Gamma}\mathbf{t}\,ds + \int_{\Omega}\mathbf{f}\,dv = \int_{\Omega}\left(\boldsymbol{\nabla}\cdot\boldsymbol{\sigma}^{\mathrm{T}} + \mathbf{f}\right)dv \qquad (2.7.11)$$

where $\mathbf{v}$ is the velocity vector, and the stress vector $\mathbf{t}$ is expressed in terms of the stress tensor $\boldsymbol{\sigma}$ by Cauchy's formula in Eq. (2.5.10) and the divergence theorem is used in arriving at the final expression. Using the Reynolds Transport Theorem, Eq. (2.7.10), we arrive at

$$0 = \int_{\Omega}\left(\boldsymbol{\nabla}\cdot\boldsymbol{\sigma}^{\mathrm{T}} + \mathbf{f} - \rho\frac{d\mathbf{v}}{dt}\right)dv \qquad (2.7.12)$$

which is the global form of the equation of motion. The local form is given by

$$\boldsymbol{\nabla}\cdot\boldsymbol{\sigma}^{\mathrm{T}} + \mathbf{f} = \rho\frac{d\mathbf{v}}{dt} \qquad (2.7.13)$$

or

$$\boldsymbol{\nabla}\cdot\boldsymbol{\sigma}^{\mathrm{T}} + \mathbf{f} = \rho\left(\frac{\partial\mathbf{v}}{\partial t} + \mathbf{v}\cdot\boldsymbol{\nabla}\mathbf{v}\right) \qquad (2.7.14)$$

In Cartesian rectangular system, we have

$$\frac{\partial\sigma_{ij}}{\partial x_j} + f_i = \rho\left(\frac{\partial v_i}{\partial t} + v_j\frac{\partial v_i}{\partial x_j}\right) \qquad (2.7.15)$$

Equations (2.7.13)–(2.7.15) are valid in the current configuration, Ω.

2.7.4.2 Material form of the equations of motion

To express the spatial form of the equation of motion, Eq. (2.7.13), in material description, we rewrite Eq. (2.7.12) in the reference configuration using appropriate transformations. The following relations hold:

$$dv = J\,dV, \quad J\rho\frac{d\mathbf{v}}{dt} = \rho_0\frac{\partial\mathbf{v}}{\partial t}, \quad \int_{\Omega}\boldsymbol{\nabla}\cdot\boldsymbol{\sigma}^{\mathrm{T}}\,dv = \int_{\Omega_0}\boldsymbol{\nabla}_0\cdot\mathbf{P}^{\mathrm{T}}\,dV \qquad (2.7.16)$$

Substitution of the above relations into Eq. (2.7.12), we arrive at

$$0 = \int_{\Omega_0} \left(\boldsymbol{\nabla}_0 \cdot \mathbf{P}^{\mathrm{T}} + \mathbf{f}^0 - \rho_0 \frac{\partial \mathbf{V}}{\partial t} \right) dV$$

where

$$\mathbf{f}^0 = J \, \mathbf{f} \qquad (2.7.17)$$

and $\mathbf{V}(\mathbf{X}, t) = \mathbf{v}(\mathbf{x}, t)$ is the material velocity field. Thus, the local form of the equation of motion in terms of the first Piola–Kirchhoff stress tensor $\mathbf{P}$ in the material description is given by

$$\boldsymbol{\nabla}_0 \cdot \mathbf{P}^{\mathrm{T}} + \mathbf{f}^0 = \rho_0 \frac{\partial \mathbf{V}}{\partial t}, \quad \mathbf{X} \in \Omega_0 \qquad (2.7.18)$$

The local form of the equation of motion in terms of the second Piola–Kirchhoff stress tensor $\mathbf{S}$ in the material description is given by $[\mathbf{P}^{\mathrm{T}} = (\mathbf{F} \cdot \mathbf{S})^{\mathrm{T}} = \mathbf{S}^{\mathrm{T}} \cdot \mathbf{F}^{\mathrm{T}}]$

$$\boldsymbol{\nabla}_0 \cdot (\mathbf{S}^{\mathrm{T}} \cdot \mathbf{F}^{\mathrm{T}}) + \mathbf{f}^0 = \rho_0 \frac{\partial \mathbf{V}}{\partial t}, \quad \mathbf{X} \in \Omega_0 \qquad (2.7.19)$$

2.7.5   Balance of Angular Momentum

The principle of balance of angular momentum can be stated as: *the time rate of change of the total moment of momentum for a continuum is equal to vector sum of the moments of external forces acting on the continuum.* The principle yields, in the absence of body couples, symmetry of the Cauchy stress tensor. A continuum said to have no body couples (that is volume-dependent couples $\mathbf{M}$) if $\lim_{\Delta V \to 0} \Delta \mathbf{M} / \Delta V = \mathbf{0}$ holds.

The principle of balance of angular momentum can be expressed analytically as

$$\oint_{\Gamma} \mathbf{x} \times \mathbf{t} \, ds + \int_{\Omega} \mathbf{x} \times \mathbf{f} \, dv = \frac{d}{dt} \int_{\Omega} \mathbf{x} \times \rho \mathbf{v} \, dv \qquad (2.7.20)$$

which yields, in view of the equations of motion, Eq. (2.7.13), the symmetry of the Cauchy stress tensor

$$\boldsymbol{\sigma} = \boldsymbol{\sigma}^{\mathrm{T}} \quad \text{or} \quad \sigma_{ij} = \sigma_{ji} \qquad (2.7.21)$$

2.7.6   Thermodynamic Principles

There are two laws of thermodynamics. The first law of thermodynamics can be stated as: *the time rate of the total energy is equal to the sum of the rate of work done by the external forces and the change of heat content per unit mass.* The second law of thermodynamics provides a restriction on the inter-convertibility of energies (e.g. thermal to mechanical). In this section, a simple form of the energy equation is derived and the second law is only briefly discussed. For additional details, the reader may consult [1].

2.7.6.1 Energy equation in the spatial description

The first law of thermodynamics for a system occupying the domain (control volume) Ω can be expressed as

$$\frac{d}{dt} \int_\Omega \rho \epsilon \, dv = W_{\text{net}} + H_{\text{net}} \tag{2.7.22}$$

where ϵ is the total energy stored per unit mass, W_{net} is the net rate of work transferred into the system, and H_{net} is the net rate of heat transfer into the system. The total stored energy per unit mass ϵ consists of the internal energy (sum of all microscopic forms of energy) per unit mass e_c and the kinetic energy per unit mass $v^2/2$

$$\epsilon = e_c + \tfrac{1}{2}(\mathbf{v} \cdot \mathbf{v}) \tag{2.7.23}$$

The power input, in the *nonpolar* case (i.e. for a continuum without body couples), consists of the rate of work done by external surface tractions $\mathbf{t}$ per unit area and body forces $\mathbf{f}$ per unit volume of the region Ω bounded by Γ:

$$
\begin{aligned}
W_{\text{net}} &= \oint_\Gamma \mathbf{t} \cdot \mathbf{v} \, ds + \int_\Omega \mathbf{f} \cdot \mathbf{v} \, dv \\
&= \int_\Omega [(\nabla \cdot \boldsymbol{\sigma} + \mathbf{f}) \cdot \mathbf{v} + \boldsymbol{\sigma} : \nabla \mathbf{v}] \, dv \\
&= \int_\Omega \left(\rho \frac{d\mathbf{v}}{dt} \cdot \mathbf{v} + \sigma : \nabla \mathbf{v} \right) dv
\end{aligned}
\tag{2.7.24}
$$

where : denotes the "double-dot product" defined in Eq. (1.6.36). In arriving at the second line and the third line of Eq. (2.7.24), we have used Cauchy's formula, Eq. (2.5.10), the divergence theorem in Eq. (1.6.42), and the equation of motion (2.7.14). Because of the symmetry of the stress tensor $\boldsymbol{\sigma}$, only the symmetric part of $\boldsymbol{\sigma} : \nabla \mathbf{v}$, namely, $\boldsymbol{\sigma} : \mathbf{d}$ is nonzero. Hence, we can write Eq. (2.7.24) as

$$
\begin{aligned}
W_{\text{net}} &= \tfrac{1}{2} \int_\Omega \rho \frac{d}{dt} (\mathbf{v} \cdot \mathbf{v}) \, dv + \int_\Omega \boldsymbol{\sigma} : \mathbf{d} \, dv \\
&= \tfrac{1}{2} \frac{d}{dt} \int_\Omega \rho \, \mathbf{v} \cdot \mathbf{v} \, dv + \int_\Omega \boldsymbol{\sigma} : \mathbf{d} \, dv
\end{aligned}
\tag{2.7.25}
$$

where $\mathbf{d}$ is the symmetric part of the velocity gradient tensor $\nabla \mathbf{v}$ [see Eq. (2.4.29)]

$$\mathbf{d} = \tfrac{1}{2} \left[\nabla \mathbf{v} + (\nabla \mathbf{v})^{\mathsf{T}} \right]$$

and the Reynolds transport theorem (2.7.10) is used to write the final expression.

The rate of heat input consists of conduction through the surface s and heat generation inside the region Ω (possibly from a radiation field or transmission of electric current). Let $\mathbf{q}$ be the heat flux vector and g be the internal heat generation per unit volume. Then the heat inflow across the surface element ds is $-\mathbf{q} \cdot \hat{\mathbf{n}} \, ds$, and internal heat generation in volume element dv is $g \, dv$. Hence, the total heat input is

$$H_{\text{net}} = -\oint_\Gamma \mathbf{q} \cdot \hat{\mathbf{n}} \, ds + \int_\Omega g \, dv = \int_\Omega (-\nabla \cdot \mathbf{q} + g) \, dv \tag{2.7.26}$$

Substituting the expressions for ϵ, W_{net}, and H_{net} from Eqs. (2.7.23)–(2.7.26) into Eq. (2.7.22), we obtain

$$\frac{d}{dt} \int_\Omega \rho \left(\tfrac{1}{2} \mathbf{v} \cdot \mathbf{v} + e_c \right) dv = \tfrac{1}{2} \frac{d}{dt} \int_\Omega \rho \mathbf{v} \cdot \mathbf{v} \, dv + \int_\Omega (\boldsymbol{\sigma} : \mathbf{d} - \nabla \cdot \mathbf{q} + g) \, dv$$

or

$$0 = \int_\Omega \left(\rho \frac{de_c}{dt} - \boldsymbol{\sigma} : \mathbf{d} + \nabla \cdot \mathbf{q} - g \right) dv \tag{2.7.27}$$

which is the global form of the energy equation. The local form of the energy equation is given by

$$\rho \frac{de_c}{dt} = \boldsymbol{\sigma} : \mathbf{d} - \nabla \cdot \mathbf{q} + g \tag{2.7.28}$$

which is known as the *thermodynamic form* of the energy equation for a continuum. The term $\boldsymbol{\sigma} : \mathbf{d}$ is known as the *stress power*.

The heat flux vector $\mathbf{q}$ is related to the gradient of temperature by the Fourier heat conduction equation (a constitutive relation for conduction of heat)

$$\mathbf{q} = -\mathbf{k} \cdot \nabla T \tag{2.7.29}$$

where $\mathbf{k}$ is the second-order conductivity tensor and T denotes the temperature.

2.7.6.2   Energy equation in the material description

To derive the energy equation in the material description, we express all quantities in the material coordinates (see Reddy [1] for details):

$$\hat{\mathbf{n}} \, ds = J \mathbf{F}^{-\text{T}} \cdot \hat{\mathbf{N}} \, dS = J \hat{\mathbf{N}} \cdot \mathbf{F}^{-1} \, dS, \quad \rho \, dv = \rho_0 \, dV \tag{2.7.30}$$

and

$$\int_\Omega \rho \epsilon \, dv = \int_{\Omega_0} \rho_0 \left(\tfrac{1}{2} \mathbf{V} \cdot \mathbf{V} + e_c \right) dV \tag{2.7.31}$$

$$W_{\text{net}} = \int_{\Omega_0} \left[(\nabla_0 \cdot \mathbf{P}^\text{T} + \mathbf{f}^0) \cdot \mathbf{V} + \mathbf{P}^\text{T} : \nabla_0 \mathbf{V} \right] dV$$

$$= \int_{\Omega_0} \left[\tfrac{1}{2} \rho_0 \frac{\partial}{\partial t} (\mathbf{V} \cdot \mathbf{V}) + \mathbf{P}^\text{T} : \nabla_0 \mathbf{V} \right] dV \tag{2.7.32}$$

$$H_{\text{net}} = -\oint_\Gamma \hat{\mathbf{n}} \cdot \mathbf{q} \, ds + \int_\Omega g \, dv = -\oint_{\Gamma_0} \hat{\mathbf{N}} \cdot \mathbf{q}_0 \, dS + \int_{\Omega_0} g_0 \, dV$$

$$= \int_{\Omega_0} \left[-\boldsymbol{\nabla}_0 \cdot \mathbf{q}_0 + g_0 \right] dV \tag{2.7.33}$$

where Eqs. (2.7.16) and (2.7.18) are used to write the final expressions. Substitution of expressions from Eqs. (2.7.32)–(2.7.33) into Eq. (2.7.22), we obtain the following local form of the energy equation in the material description:

$$\rho_0 \frac{\partial e_c}{\partial t} = \mathbf{P}^{\mathrm{T}} : \boldsymbol{\nabla}_0 \mathbf{V} - \boldsymbol{\nabla}_0 \cdot \mathbf{q}_0 + g_0 \tag{2.7.34}$$

In terms of the second Piola–Kirchhoff stress tensor $\mathbf{S}$, we have

$$\rho_0 \frac{\partial e_c}{\partial t} = (\mathbf{S} \cdot \mathbf{F}^{\mathrm{T}}) : \boldsymbol{\nabla}_0 \mathbf{V} - \boldsymbol{\nabla}_0 \cdot \mathbf{q}_0 + g_0 \tag{2.7.35}$$

2.7.6.3 Entropy inequality

The concept of *entropy* is a difficult one (at least for the author) to explain in simple terms. It is generally considered as a measure of the tendency of the atoms toward a disorder. Crystals with atoms closely bound in a highly ordered array have less entropy. The second law of thermodynamics says that *the internal entropy production is always positive*,[4] and it is known as the *entropy inequality principle* or *the Clausius–Duhem inequality*.

To derive the Clausius–Duhem inequality, we introduce the entropy density per unit mass, η, so that the total entropy is

$$\mathcal{E} = \int_{\Omega} \rho \eta \, dv \tag{2.7.36}$$

We introduce θ as the absolute temperature whose greatest lower bound is zero. We also recall that in an admissible deformation, the deformation gradient $\mathbf{F}$ should be nonsingular.[5] The *entropy production* is then defined as

$$\begin{aligned}
\Gamma &= \frac{d\mathcal{E}}{dt} - \left[-\oint_{\Gamma} \frac{1}{\theta} \mathbf{q} \cdot \hat{\mathbf{n}} \, ds + \int_{\Omega} \frac{g}{\theta} \, dv \right] \\
&= \int_{\Omega} \left[\rho \frac{d\eta}{dt} + \boldsymbol{\nabla} \cdot \left(\frac{\mathbf{q}}{\theta} \right) - \frac{g}{\theta} \right] dv
\end{aligned} \tag{2.7.37}$$

The second law of thermodynamics requires $\Gamma \geq 0$:

$$\int_{\Omega} \rho \frac{d\eta}{dt} \, dv \geq \int_{\Omega} \left[\frac{g}{\theta} - \boldsymbol{\nabla} \cdot \left(\frac{\mathbf{q}}{\theta} \right) \right] dv \tag{2.7.38}$$

which is the global form of the Clausius–Duhem inequality. The local form of the *entropy inequality* is

---

[4]That is, things can only get more disorderly than orderly from their current state.
[5]Thus, each thermodynamic process must satisfy the conditions $\theta \geq 0$ and $J \neq 0$.

$$\rho \frac{d\eta}{dt} \geq \frac{g}{\theta} - \boldsymbol{\nabla} \cdot \left(\frac{\mathbf{q}}{\theta}\right) \quad \text{or} \quad \rho\theta\frac{d\eta}{dt} - g + \boldsymbol{\nabla} \cdot \mathbf{q} - \frac{1}{\theta}\mathbf{q} \cdot \boldsymbol{\nabla}\theta \geq 0 \qquad (2.7.39)$$

The quantity $\mathbf{q}/\theta$ is known as the *entropy flux*.

The sum of internal energy (e_c) and irreversible heat energy $(-\theta\eta)$ is known as *Helmhotz free energy density*

$$\Psi = e_c - \theta\eta \qquad (2.7.40)$$

Substituting Eq. (2.7.40) into Eq. (2.7.28), we obtain

$$\rho \frac{d\Psi}{dt} = \boldsymbol{\sigma} : \mathbf{d} - \rho \frac{d\theta}{dt}\eta - \mathcal{D} \qquad (2.7.41)$$

where $\mathbf{d}$ is the symmetric part of the velocity gradient tensor and $\mathcal{D}$ is the *internal dissipation*

$$\mathcal{D} = \rho\,\theta\frac{d\eta}{dt} + \boldsymbol{\nabla} \cdot \mathbf{q} - g \qquad (2.7.42)$$

In view of Eq. (2.7.39) we can write

$$\mathcal{D} - \frac{1}{\theta}\mathbf{q} \cdot \boldsymbol{\nabla}\theta \geq 0 \qquad (2.7.43)$$

We have $\mathcal{D} > 0$ for an irreversible process and $\mathcal{D} = 0$ for a reversible process.

2.8   Constitutive Equations for Elastic Solids

2.8.1   Introduction

Constitutive equations are those relations that connect the kinematic variables to kinetic variables. Materials for which the constitutive behavior is only a function of the current state of deformation are known as *elastic*. Special cases of these materials are the Hookean solids. A study of these "theoretical" materials is important because these materials provide good mathematical models for the behavior of "real" materials. There exist other materials, for example, polymers and elastomers, whose constitutive relations cannot be adequately described by those of a Hookean solid.

Constitutive equations are often postulated directly from experimental observations. While experiments are necessary in the determination of various parameters (e.g. elastic constants) appearing in the constitutive equations, the formulation of the constitutive equations for a given material is very difficult in nonlinear continuum mechanics, and their development is guided by certain physical requirements [1], including invariance of the equations and material frame indifference discussed in Section 2.6.

2.8.2 Restrictions Placed by the Entropy Inequality

All constitutive equations must be consistent with the thermodynamic principles; in particular, they must satisfy the conditions resulting from the entropy inequality. The stress tensor $\boldsymbol{\sigma}$, the Helmholtz free energy density Ψ, specific entropy η, and heat flux vector $\mathbf{q}$ must be the dependent variables in the constitutive models for a homogeneous and isotropic material. Suppose that the free energy potential Ψ is a function of $\mathbf{F}$, θ, and $\mathbf{g}$. The entropy inequality, from Eq. (2.7.41), is

$$-\rho\dot{\Psi} + \boldsymbol{\sigma} : \mathbf{l} - \rho\dot{\theta}\eta - \frac{1}{\theta}\mathbf{q}\cdot\mathbf{g} \geq 0 \tag{2.8.1}$$

where $\mathbf{g} = \boldsymbol{\nabla}\theta$, and $\mathbf{l}$ is the velocity gradient tensor, $\mathbf{l} = (\boldsymbol{\nabla}\mathbf{v})^{\mathrm{T}} = \mathbf{d} + \mathbf{w}$, and $\mathbf{d}$ is the symmetric part and $\mathbf{w}$ is the skew-symmetric part of $\mathbf{l}$. We can write

$$\dot{\Psi} = \frac{\partial\Psi}{\partial\mathbf{F}} : \dot{\mathbf{F}}^{\mathrm{T}} + \frac{\partial\Psi}{\partial\theta}\,\dot{\theta} + \frac{\partial\Psi}{\partial\mathbf{g}}\cdot\dot{\mathbf{g}} \tag{2.8.2}$$

Using Eq. (2.4.33) in Eq. (2.8.2) and the result into Eq. (2.8.1), we obtain

$$\left(\boldsymbol{\sigma} - \rho\frac{\partial\Psi}{\partial\mathbf{F}}\cdot\mathbf{F}^{\mathrm{T}}\right) : \mathbf{l}^{\mathrm{T}} - \rho\left(\eta + \frac{\partial\Psi}{\partial\theta}\right)\dot{\theta} - \frac{\partial\Psi}{\partial\mathbf{g}}\cdot\dot{\mathbf{g}} - \frac{1}{\theta}\mathbf{q}\cdot\mathbf{g} \geq 0 \tag{2.8.3}$$

Since $\mathbf{l}$, $\dot{\theta}$, and $\dot{\mathbf{g}}$ are linearly independent of each other, it follows that

$$\boldsymbol{\sigma} - \rho\frac{\partial\Psi}{\partial\mathbf{F}}\cdot\mathbf{F}^{\mathrm{T}} = \mathbf{0} \tag{2.8.4}$$

$$\eta + \frac{\partial\Psi}{\partial\theta} = 0 \tag{2.8.5}$$

$$-\frac{\partial\Psi}{\partial\mathbf{g}} = 0 \tag{2.8.6}$$

$$-\mathbf{q}\cdot\mathbf{g} \geq 0 \tag{2.8.7}$$

Equation (2.8.6) implies that Ψ is not a function of the temperature gradient $\mathbf{g}$. Also, Eq. (2.8.7) implies that

$$\mathbf{q}\cdot\mathbf{g} \leq 0 \tag{2.8.8}$$

Therefore, $\mathbf{q}$ is proportional to the negative of the gradient of the temperature $\mathbf{g} = \boldsymbol{\nabla}\theta$, as can be seen from the Fourier heat conduction law in Eq. (2.7.29). Equation (2.8.5) implies that η can be determined from Ψ and, hence, cannot be a dependent variable in the constitutive model. Furthermore, we conclude from Eq. (2.8.4), because Ψ is only a function of $\mathbf{F}$ and θ, that $\boldsymbol{\sigma}$ can only depend on $\mathbf{F}$ and θ.

2.8.3   Elastic Materials and the Generalized Hooke's Law

A material is called *Cauchy-elastic* or *elastic* if the stress field at time t depends only on the state of deformation and temperature at that time, and not on the history of these variables. A *hyperelastic material*, also known as the *Green-elastic material*, is one for which there exists a *Helmholtz free-energy potential* Ψ (measured per unit volume), whose derivative with respect to a strain gives corresponding stress and whose derivative with respect to temperature gives heat flux vector. When Ψ is solely a function of $\mathbf{F}$, $\mathbf{C}$, or some strain tensor, it is called the *strain energy density function*, and denoted by U_0 (measured per unit mass). For example, if $U_0 = U_0(\mathbf{F})$, we have [see Eq. (2.8.4)]

$$\mathbf{P} = \rho_0 \frac{\partial U_0(\mathbf{F})}{\partial \mathbf{F}} \quad \left(P_{iJ} = \rho_0 \frac{\partial U_0}{\partial F_{iJ}} \right) \tag{2.8.9}$$

$$\boldsymbol{\sigma} = \frac{1}{J} \mathbf{P} \cdot \mathbf{F}^{\mathrm{T}} = \rho \frac{\partial U_0(\mathbf{F})}{\partial \mathbf{F}} \cdot \mathbf{F}^{\mathrm{T}} \tag{2.8.10}$$

$$\mathbf{S} = \mathbf{F}^{-1} \cdot \mathbf{P} = \rho_0 \mathbf{F}^{-1} \cdot \frac{\partial U_0(\mathbf{F})}{\partial \mathbf{F}} \tag{2.8.11}$$

The linear relations between the stress and strain components for the case of infinitesimal deformation (i.e. $|\boldsymbol{\nabla}\mathbf{u}| = O(\epsilon) << 1$) of linear elastic solids are known as the *generalized Hooke's law*. Here, we do not distinguish between various measures of stress and strain, and use $\mathbf{S} \approx \boldsymbol{\sigma}$ for the stress tensor and $\mathbf{E} \approx \boldsymbol{\varepsilon}$ for the strain tensor in the material description used in solid mechanics. For such materials, the Helmholtz free energy density Ψ is the same as the strain energy density U_0, and it is more meaningful to assume that the strain energy density is a function of the strain, $\boldsymbol{\varepsilon}$, rather than the deformation gradient, although one may also assume that U_0 is a function of the strain invariants.

The constitutive relations to be discussed here for stress tensor $\boldsymbol{\sigma}$ do not include creep at constant stress and stress relaxation at constant strain. Thus, the material coefficients that specify the constitutive relationship between the stress and strain components are assumed to be constant during the deformation. We account for the thermal expansion of the material, which can produce strains or stresses as large as those produced by the applied mechanical forces.

The hyperelastic constitutive equations for linearized elasticity can be derived using

$$\boldsymbol{\sigma} = \rho_0 \frac{\partial U_0(\boldsymbol{\varepsilon})}{\partial \boldsymbol{\varepsilon}} \quad \left(\sigma_{ij} = \rho_0 \frac{\partial U_0}{\partial \varepsilon_{ij}} \right) \tag{2.8.12}$$

In developing a mathematical model of the constitutive behavior of a hyperelastic material, $\rho_0 U_0$ is expanded in Taylor's series about $\boldsymbol{\varepsilon} = 0$ as

$$\rho_0 U_0 = C_0 + C_{ij}\,\varepsilon_{ij} + \frac{1}{2!}C_{ijk\ell}\,\varepsilon_{ij}\,\varepsilon_{k\ell} + \frac{1}{3!}C_{ijk\ell mn}\,\varepsilon_{ij}\,\varepsilon_{k\ell}\,\varepsilon_{mn} + \dots \tag{2.8.13}$$

where C_0, C_{ij}, and so on are material stiffnesses. For nonlinear elastic materials, Ψ is a cubic and higher-order function of the strains. For linear elastic materials

Ψ is a quadratic function of strains and we can write, assuming that the body is free of stress prior to the load application,

$$\sigma = \mathbf{C} : \varepsilon \quad (\sigma_{ij} = C_{ijkl}\, \varepsilon_{kl}) \tag{2.8.14}$$

where $\mathbf{C}$ is a fourth-order elasticity tensor. The coefficients C_{ijkl} are called elastic *stiffness* coefficients. In general, there are $81(= 3^4)$ scalar components of the fourth-order tensor $\mathbf{C}$. However, the components C_{ijkl} satisfy the following symmetry conditions by virtue of the symmetry of stress and strain tensor components:

$$C_{ijkl} = C_{klij}, \quad C_{ijkl} = C_{jikl}, \quad C_{ijlk} = C_{ijkl}, \quad C_{ijkl} = C_{jilk} \tag{2.8.15}$$

Thus, the number of independent coefficients in C_{ijkl} is reduced to 21.

For an orthotropic material, we have the following strain–stress relations [1] (introducing single-subscript notation, $\{\varepsilon\} = [S]\{\sigma\}$):

$$
\begin{Bmatrix}
\varepsilon_{11} \equiv \varepsilon_1 \\
\varepsilon_{22} \equiv \varepsilon_2 \\
\varepsilon_{33} \equiv \varepsilon_3 \\
2\varepsilon_{23} \equiv \varepsilon_4 \\
2\varepsilon_{13} \equiv \varepsilon_5 \\
2\varepsilon_{12} \equiv \varepsilon_6
\end{Bmatrix}
=
\begin{bmatrix}
\frac{1}{E_1} & -\frac{\nu_{21}}{E_2} & -\frac{\nu_{31}}{E_3} & 0 & 0 & 0 \\
-\frac{\nu_{12}}{E_1} & \frac{1}{E_2} & -\frac{\nu_{32}}{E_3} & 0 & 0 & 0 \\
-\frac{\nu_{13}}{E_1} & -\frac{\nu_{23}}{E_2} & \frac{1}{E_3} & 0 & 0 & 0 \\
0 & 0 & 0 & \frac{1}{G_{23}} & 0 & 0 \\
0 & 0 & 0 & 0 & \frac{1}{G_{13}} & 0 \\
0 & 0 & 0 & 0 & 0 & \frac{1}{G_{12}}
\end{bmatrix}
\begin{Bmatrix}
\sigma_{11} \equiv \sigma_1 \\
\sigma_{22} \equiv \sigma_2 \\
\sigma_{33} \equiv \sigma_3 \\
\sigma_{23} \equiv \sigma_4 \\
\sigma_{13} \equiv \sigma_5 \\
\sigma_{12} \equiv \sigma_6
\end{Bmatrix}
\tag{2.8.16}
$$

where E_1, E_2, E_3 are Young's moduli in 1, 2, and 3 material directions, respectively, ν_{ij} is Poisson's ratio, defined as the ratio of transverse strain in the jth direction to the axial strain in the ith direction when stressed in the i-direction, and G_{23}, G_{13}, G_{12} are the shear moduli in the 2-3, 1-3, and 1-2 planes, respectively. Since $[S]$ is the inverse of $[C]$ and the $[C]$ is symmetric, then $[S]$ is also a symmetric matrix. This in turn implies that the following reciprocal relations hold:

$$\frac{\nu_{ij}}{E_i} = \frac{\nu_{ji}}{E_j} \tag{2.8.17}$$

for $i, j = 1, 2, 3$. The nine independent material coefficients for an orthotropic material are

$$E_1, \; E_2, \; E_3, \; G_{23}, \; G_{13}, \; G_{12}, \; \nu_{12}, \; \nu_{13}, \; \nu_{23} \tag{2.8.18}$$

Isotropic materials are those for which the material properties are independent of the direction; that is, there exist an infinite number of material symmetry planes. An isotropic fourth-order tensor, whose components are symmetric with respect to the first two and the last two indices, can be expressed as

$$C_{ijkl} = \lambda \delta_{ij}\delta_{kl} + 2\mu\delta_{ik}\delta_{jl} \tag{2.8.19}$$

where μ and λ are called the *Lamé constants*. Then, Eq. (2.8.14) takes the simple form

$$\sigma = 2\mu\,\varepsilon + \lambda\,\mathrm{tr}(\varepsilon)\mathbf{I} \quad (\sigma_{ij} = 2\mu\,\varepsilon_{ij} + \lambda\,\varepsilon_{kk}\,\delta_{ij}) \qquad (2.8.20)$$

where $\mathrm{tr}(\cdot)$ denotes the *trace* (sum of the diagonal elements) of the enclosed tensor. Thus, only two material parameters, μ and λ, are needed to characterize the mechanical response of an isotropic material. The Lamé constants μ and λ are related to E and ν by

$$\mu = \frac{E}{2(1+\nu)}, \quad \lambda = \frac{\nu E}{(1+\nu)(1-2\nu)}, \quad 2\mu + \lambda = \frac{(1-\nu)E}{(1+\nu)(1-2\nu)} \qquad (2.8.21)$$

The stress–strain relations in Eq. (2.8.20) can be expressed in terms of E and ν as

$$\sigma = \frac{E}{1+\nu}\varepsilon + \frac{\nu E}{(1+\nu)(1-2\nu)}\,\mathrm{tr}(\varepsilon)\,\mathbf{I} \quad \left[\sigma_{ij} = \frac{E}{1+\nu}\varepsilon_{ij} + \frac{\nu E}{(1+\nu)(1-2\nu)}\,\varepsilon_{kk}\,\delta_{ij}\right] \qquad (2.8.22)$$

and the inverse relations are

$$\varepsilon = \left(\frac{1+\nu}{E}\right)\sigma - \frac{\nu}{E}\,\mathrm{tr}(\sigma)\,\mathbf{I} \quad \left[\varepsilon_{ij} = \left(\frac{1+\nu}{E}\right)\sigma_{ij} - \frac{\nu}{E}\,\sigma_{kk}\,\delta_{ij}\right] \qquad (2.8.23)$$

The strain energy functional $\rho_0 U_0$ for an isotropic material takes the form

$$\rho_0 U_0(\varepsilon) = \frac{\lambda}{2}(\mathrm{tr}\,\varepsilon)^2 + \mu\,\mathrm{tr}(\varepsilon\cdot\varepsilon) \quad \left[\rho_0 U_0(\varepsilon_{ij}) = \mu\varepsilon_{ij}\varepsilon_{ij} + \tfrac{1}{2}\lambda(\varepsilon_{kk})^2\right] \qquad (2.8.24)$$

Note that the strain energy density U_0 is positive-definite, that is,

$$U_0(\varepsilon) > 0 \text{ whenever } \varepsilon \neq \mathbf{0}, \text{ and } U_0(\varepsilon) = 0 \text{ only when } \varepsilon = \mathbf{0} \qquad (2.8.25)$$

For the isotropic case, the stress–strain relations (2.8.22) for an isotropic material can be expressed in matrix form as

$$
\begin{Bmatrix} \sigma_1 \\ \sigma_2 \\ \sigma_3 \\ \sigma_4 \\ \sigma_5 \\ \sigma_6 \end{Bmatrix}
=
\frac{E}{(1+\nu)(1-2\nu)}
\begin{bmatrix}
1-\nu & \nu & \nu & 0 & 0 & 0 \\
\nu & 1-\nu & \nu & 0 & 0 & 0 \\
\nu & \nu & 1-\nu & 0 & 0 & 0 \\
0 & 0 & 0 & \frac{1-2\nu}{2} & 0 & 0 \\
0 & 0 & 0 & 0 & \frac{1-2\nu}{2} & 0 \\
0 & 0 & 0 & 0 & 0 & \frac{1-2\nu}{2}
\end{bmatrix}
\begin{Bmatrix} \varepsilon_1 \\ \varepsilon_2 \\ \varepsilon_3 \\ \varepsilon_4 \\ \varepsilon_5 \\ \varepsilon_6 \end{Bmatrix}
$$

$$\qquad (2.8.26)$$

If the only nonzero normal stress component is $\sigma_{11} = \sigma$ and the only nonzero shear component is $\sigma_{12} = \tau$, and if we denote $\varepsilon_{11} = \varepsilon$ and $2\varepsilon_{12} = \gamma$, then Eq. (2.8.23) gives the following uniaxial strain–stress relations:

$$\varepsilon = \frac{1}{E}\sigma \;\rightarrow\; \sigma = E\,\varepsilon; \quad \gamma = \frac{2(1+\nu)}{E}\tau \;\rightarrow\; \tau = G\gamma \qquad (2.8.27)$$

2.9 Energy Principles of Solid Mechanics

2.9.1 Virtual Displacements and Virtual Work

A mechanical system can take many possible geometric configurations consistent with the geometric constraints placed on the system. Of all the possible configurations, only one corresponds to the equilibrium configuration of the system under the prescribed set of forces. It is this configuration that satisfies Newton's second law of motion of the system. The set of configurations that satisfy the geometric constraints, but not necessarily Newton's second law, is called the *set of admissible configurations*. These configurations are restricted to a neighborhood of the true configuration so that they are obtained from infinitesimal variations of the true configuration. During such variations, the geometric constraints of the system are not violated and all the forces are fixed at their actual values. When a mechanical system experiences such variations in its configuration, it is said to undergo *virtual displacements* from its actual configuration. The difference between the displacements of two neighboring configurations is denoted by $\delta \mathbf{u}$, and it is called the *first variation* of the displacement field $\mathbf{u}$; $\delta \mathbf{u}$ is arbitrary and infinitesimal. These displacements need not have any relationship to the actual displacements that might occur due to a change in the applied loads. The displacements are called *virtual* because they are *imagined* to take place while the actual loads are acting at their fixed values [3].

When a force $\mathbf{F}$ acts on a material point and moves it through a displacement $\mathbf{u}$, the work done by the force is defined by the projection of the force in the direction of the displacement times the magnitude of the displacement, that is, the work done is $\mathbf{F} \cdot \mathbf{u}$. When $\mathbf{F}$ and $\mathbf{u}$ are vector functions of position in a three-dimensional domain, the work done is obtained by the volume integral of the product $\mathbf{F} \cdot \mathbf{u}$:

$$W = \int_\Omega \mathbf{F} \cdot \mathbf{u} \, dv$$

The work done by the actual forces $\mathbf{F}$ moving through virtual displacements $\delta \mathbf{u}$ is called the *virtual work*, and it is given by

$$\delta W = \int_\Omega \mathbf{F} \cdot \delta \mathbf{u} \, dv \tag{2.9.1}$$

where dv denotes a volume element in the region occupied by Ω.

2.9.2 First Variation or Gâteaux Derivative

The virtual displacement $\delta \mathbf{u}$ is totally independent of the actual displacement $\mathbf{u}$ and it is taken at a *fixed* instant of time. It may be expressed in terms of spatial coordinates or material coordinates

$$\delta \mathbf{u}(\mathbf{x}) = \delta \mathbf{u}(\chi(\mathbf{X})) = \delta \mathbf{U}(\mathbf{X}) \tag{2.9.2}$$

For simplicity, we use $\delta\mathbf{U}(\mathbf{X}) = \delta\mathbf{u}(\mathbf{X})$ for the virtual displacement in the material coordinate system.

Let $\varphi = \varphi(\mathbf{u}(\mathbf{X}))$ be a smooth (i.e. differentiable) function of the vector $\mathbf{u}$ in the material description. The directional derivative of $\varphi(\mathbf{u})$ at any fixed $\mathbf{u}$ in the direction of $\delta\mathbf{u}$, denoted as $D_{\delta\mathbf{u}}\varphi(\mathbf{u})$ and called the *Gâteaux derivative*, is defined as

$$\delta\varphi(\mathbf{u}, \delta\mathbf{u}) = D_{\delta\mathbf{u}}\varphi(\mathbf{u}) \equiv \frac{d}{d\epsilon}\varphi(\mathbf{u} + \epsilon\,\delta\mathbf{u})\Big|_{\epsilon=0} \tag{2.9.3}$$

and say that $\delta\varphi(\mathbf{u}, \delta\mathbf{u})$ is the first variation of the function $\varphi(\mathbf{u})$ in the direction of the virtual displacement field $\delta\mathbf{u}$. Thus, δ *is a differential operator with respect to the dependent variables.*

The delta symbol "δ" used in conjunction with virtual displacements can be interpreted as an operator, called the *variational operator*, and it is linear. Indeed, the laws of variation of sums, products, ratios, powers, and so forth are completely analogous to the corresponding laws of differentiation. In addition, the variational operator can be interchanged with differential and integral operators (commutativity):

$$\delta(\boldsymbol{\nabla}\mathbf{u}) = \left[\frac{d}{d\epsilon}\boldsymbol{\nabla}(\mathbf{u} + \epsilon\,\delta\mathbf{u})\right]_{\epsilon=0} = \boldsymbol{\nabla}(\delta\mathbf{u}) \tag{2.9.4}$$

$$\delta\left(\int_{\Omega}\mathbf{u}\,dv\right) = \left[\frac{d}{d\epsilon}\int_{\Omega}(\mathbf{u} + \epsilon\,\delta\mathbf{u})\,dv\right]_{\epsilon=0} = \int_{\Omega}\delta\mathbf{u}\,dv \tag{2.9.5}$$

The variational operator proves to be very useful in constructing virtual work statements and deriving governing equations from virtual work principles, as will be shown shortly. The following identities can be established (see Problem 2.20):

$$\boldsymbol{\nabla}\delta\mathbf{u} = \mathbf{F}^{-\mathrm{T}}\cdot\boldsymbol{\nabla}_0\,\delta\mathbf{u}; \quad \boldsymbol{\nabla}\delta\mathbf{v} = \mathbf{F}^{-\mathrm{T}}\cdot\boldsymbol{\nabla}_0\,\delta\mathbf{v} \tag{2.9.6}$$

$$\delta\mathbf{F} = (\boldsymbol{\nabla}_0\,\delta\mathbf{u})^{\mathrm{T}}; \quad \delta\dot{\mathbf{F}} = (\boldsymbol{\nabla}_0\,\delta\mathbf{v})^{\mathrm{T}}; \quad \delta\mathbf{F}^{-1} = -(\boldsymbol{\nabla}\,\delta\mathbf{u})^{\mathrm{T}} \tag{2.9.7}$$

$$\delta\boldsymbol{\varepsilon} = \tfrac{1}{2}\left[(\boldsymbol{\nabla}_0\,\delta\mathbf{u})^{\mathrm{T}} + \boldsymbol{\nabla}_0\,\delta\mathbf{u}\right] \tag{2.9.8}$$

$$\delta\mathbf{E} = \tfrac{1}{2}\mathbf{F}^{\mathrm{T}}\cdot\left[(\boldsymbol{\nabla}\,\delta\mathbf{u})^{\mathrm{T}} + \boldsymbol{\nabla}\,\delta\mathbf{u}\right]\cdot\mathbf{F} \approx \mathbf{F}^{\mathrm{T}}\cdot\delta\boldsymbol{\varepsilon}\cdot\mathbf{F} \tag{2.9.9}$$

$$\delta\mathbf{d} = \tfrac{1}{2}\left[(\boldsymbol{\nabla}\,\delta\mathbf{v})^{\mathrm{T}} + \boldsymbol{\nabla}\,\delta\mathbf{v}\right] = \delta\dot{\mathbf{F}}\cdot\mathbf{F}^{-1} \tag{2.9.10}$$

where $\mathbf{F}$ is the deformation gradient, $\mathbf{E}$ is the Green–Lagrange strain tensor, and $\boldsymbol{\varepsilon}$ is the infinitesimal strain tensor [see Eqs. (2.3.1)–(2.3.4), (2.4.6), (2.4.7), and (2.4.18) for various definitions].

2.9.3   The Principle of Virtual Displacements

Consider a rigid body acted upon by a set of applied forces $\mathbf{F}_1, \mathbf{F}_2, \ldots, \mathbf{F}_n$, and suppose that the points of application of these forces are subjected to the virtual displacements $\delta\mathbf{u}_1, \delta\mathbf{u}_2, \ldots, \delta\mathbf{u}_n$, respectively. The virtual displacement

$\delta \mathbf{u}_i$ has no relation to $\delta \mathbf{u}_j$, for $i \neq j$. The external virtual work done by the virtual displacements is

$$\delta W_E \equiv -[\mathbf{F}_1 \cdot \delta \mathbf{u}_1 + \mathbf{F}_2 \cdot \delta \mathbf{u}_2 + \cdots + \mathbf{F}_n \cdot \delta \mathbf{u}_n] = -\mathbf{F}_i \cdot \delta \mathbf{u}_i \qquad (2.9.11)$$

where the sum on repeated indices (over the range of 1 to n) is implied. The minus sign is used to indicate that the work is done *on* the body as opposed to work stored *in* the body. Since the body is assumed to be rigid, no internal forces are generated. In addition, the virtual displacements $\delta \mathbf{u}_1, \delta \mathbf{u}_2, \ldots, \delta \mathbf{u}_n$ should all be the same, say $\delta \mathbf{u}$, for a rigid body. Thus, we have

$$\delta W_E = -\mathbf{F}_i \cdot \delta \mathbf{u}_i = -\left(\sum_{i=1}^{n} \mathbf{F}_i\right) \cdot \delta \mathbf{u} \qquad (2.9.12)$$

But by Newton's second law, the vector sum of the forces acting on a body in equilibrium is zero. Thus, for a rigid body in equilibrium, the total virtual work done due to virtual displacements is zero. This statement is known as *the principle of virtual displacements*.

The principle of virtual displacements also holds for a deformable continuum for which the work done by actual internal forces in moving through the virtual internal displacements, δW_I, is not zero. To establish this, we begin with the virtual work done per unit time by body force $\mathbf{f}$ and surface traction $\mathbf{t}$:

$$\delta W_E = -\left(\int_{\Omega} \mathbf{f} \cdot \delta \mathbf{v} \, dv + \int_{\Gamma_\sigma} \mathbf{t} \cdot \delta \mathbf{v} \, ds\right) \qquad (2.9.13)$$

where $\mathbf{v}$ is the velocity vector and Γ_σ is the portion of the boundary Γ of Ω on which tractions are specified. On the remaining portion, $\Gamma_u = \Gamma - \Gamma_\sigma$, displacements are specified.[6] Virtual displacements (or velocities) are arbitrary, continuous functions except that they satisfy the homogeneous form of specified geometric boundary conditions (i.e. they must belong to the *set of admissible variations*). Thus, $\delta \mathbf{v}$ is necessarily zero on Γ_u. Therefore, we can write Eq. (2.9.13) as (notice that the boundary integral is now a closed integral)

$$\delta W_E \equiv -\left(\int_{\Omega} \mathbf{f} \cdot \delta \mathbf{v} \, dv + \oint_{\Gamma} \mathbf{t} \cdot \delta \mathbf{v} \, ds\right) \qquad (2.9.14)$$

Using Cauchy's formula in Eq. (2.5.10) and the divergence theorem, we can write[7]

$$\int_{\Omega} \mathbf{f} \cdot \delta \mathbf{v} \, dv + \oint_{\Gamma} \mathbf{t} \cdot \delta \mathbf{v} \, ds = \int_{\Omega} \mathbf{f} \cdot \delta \mathbf{v} \, dv + \oint_{\Gamma} \hat{\mathbf{n}} \cdot \boldsymbol{\sigma} \cdot \delta \mathbf{v} \, ds$$

$$= \int_{\Omega} \mathbf{f} \cdot \delta \mathbf{v} \, dv + \int_{\Omega} \boldsymbol{\nabla} \cdot (\boldsymbol{\sigma} \cdot \delta \mathbf{v}) \, dv$$

$$= \int_{\Omega} [(\boldsymbol{\nabla} \cdot \boldsymbol{\sigma} + \mathbf{f}) \cdot \delta \mathbf{v} + \boldsymbol{\sigma} : \boldsymbol{\nabla}(\delta \mathbf{v})] \, dv \qquad (2.9.15)$$

---

[6]The boundary portions Γ_u and Γ_σ are disjoint (i.e. do not overlap), and their sum is the total boundary Γ.

[7]Unless stated otherwise, hereafter we assume that the Cauchy stress tensor is symmetric.

Using the equation of equilibrium

$$\boldsymbol{\nabla} \cdot \boldsymbol{\sigma} + \mathbf{f} = \mathbf{0} \qquad (2.9.16)$$

we obtain

$$\int_{\Omega} \boldsymbol{\sigma} : \boldsymbol{\nabla}(\delta \mathbf{v}) \, dv - \left(\int_{\Omega} \mathbf{f} \cdot \delta \mathbf{v} \, dv + \oint_{\Gamma} \mathbf{t} \cdot \delta \mathbf{v} \, ds \right) = 0 \qquad (2.9.17)$$

We shall call the first expression in Eq. (2.9.17) *the internal virtual work* per unit time stored in the body, δW_I (symmetry of $\boldsymbol{\sigma}$ is assumed)

$$\delta W_I \equiv \int_{\Omega} \boldsymbol{\sigma} : \boldsymbol{\nabla}(\delta \mathbf{v}) \, dv = \int_{\Omega} \boldsymbol{\sigma} : \delta \mathbf{l}^{\mathrm{T}} \, dv = \int_{\Omega} \boldsymbol{\sigma} : \delta \mathbf{d} \, dv \qquad (2.9.18)$$

The last expression is arrived at using the following identity [see Eq. (2.4.28)]:

$$\boldsymbol{\sigma} : \delta \mathbf{l}^{\mathrm{T}} = \boldsymbol{\sigma} : (\delta \mathbf{d} - \delta \mathbf{w}) = \boldsymbol{\sigma} : \delta \mathbf{d} \qquad (2.9.19)$$

where $\boldsymbol{\sigma} : \delta \mathbf{w}$ is zero because $\mathbf{w}$ is a skew-symmetric tensor.

The principle of virtual displacements can now be stated as follows: *If a continuous system is in equilibrium, the virtual work of all actual forces in moving through a virtual displacement is zero*:

$$\delta W \equiv \delta W_I + \delta W_E = 0 \qquad (2.9.20)$$

In deriving Eq. (2.9.20) no assumption is made concerning the material constitutive behavior. Therefore, the principle of virtual work is independent of any constitutive law and applies to both elastic (linear and nonlinear) and inelastic continua. Since δW_I is the virtual work (or virtual energy) per unit current volume and time produced by the product, $\boldsymbol{\sigma} : \mathbf{d}$, the pair $(\boldsymbol{\sigma}, \mathbf{d})$ is said to be *energetically conjugate* with respect to the current deformed volume.

For equilibrium problems of linearized elasticity, the principle of virtual work, Eq. (2.9.20), is valid with δW_E and δW_I defined in terms of the virtual displacement $\delta \mathbf{u}$

$$\delta W_E = - \left[\int_{\Omega} \mathbf{f} \cdot \delta \mathbf{u} \, dv + \int_{\Gamma_\sigma} \hat{\mathbf{t}} \cdot \delta \mathbf{u} \, ds \right] \qquad (2.9.21)$$

$$\delta W_I = \int_{\Omega} \boldsymbol{\sigma} : \delta \boldsymbol{\varepsilon} \, dv \qquad (2.9.22)$$

where $\boldsymbol{\sigma}$ is the Cauchy stress tensor and $\boldsymbol{\varepsilon}$ is the infinitesimal strain tensor

$$\boldsymbol{\varepsilon} = \tfrac{1}{2} \left[(\boldsymbol{\nabla} \mathbf{u})^{\mathrm{T}} + \boldsymbol{\nabla} \mathbf{u} \right] \qquad (2.9.23)$$

Then the principle of virtual displacements for configuration Ω in equilibrium can be expressed as

$$\int_{\Omega} \boldsymbol{\sigma} : \delta \boldsymbol{\varepsilon} \, dv - \int_{\Omega} \mathbf{f} \cdot \delta \mathbf{u} \, dv - \int_{\Gamma_\sigma} \hat{\mathbf{t}} \cdot \delta \mathbf{u} \, ds = 0 \qquad (2.9.24)$$

Writing in terms of the Cartesian rectangular components, Eq. (2.9.24) takes the form (sum on repeated subscripts is implied)

$$\int_\Omega (\sigma_{ij}\delta\varepsilon_{ij} - f_i\delta u_i)\, dv - \int_{\Gamma_\sigma} \hat{t}_i\delta u_i\, ds = 0 \qquad (2.9.25)$$

We can show that the statement in Eq. (2.9.24) is equivalent to the following equilibrium equations of the 3-D elasticity in Ω and the traction boundary conditions on Γ_σ (see Problem 2.22):

$$\sigma_{ji,j} + \rho f_i = 0 \qquad (\boldsymbol{\nabla}\cdot\boldsymbol{\sigma} + \mathbf{f} = \mathbf{0}) \ \text{ in } \Omega \qquad (2.9.26)$$

$$\sigma_{ji}n_j - \hat{t}_i = 0 \qquad (\hat{\mathbf{n}}\cdot\boldsymbol{\sigma} = \hat{\mathbf{t}}) \ \text{ on } \Gamma_\sigma \qquad (2.9.27)$$

Equations (2.9.26) and (2.9.27) are called the *Euler equations* associated with the principle of virtual displacements for a body in spatial description. The boundary conditions in Eq. (2.9.27) are known as the *natural boundary conditions*.

The integrals in Eq. (2.9.17) can be transformed to the initial volume and area using the relation $dv = J\, dV$ (see Reddy [1])

$$\int_{\Omega_0} J\boldsymbol{\sigma} : \boldsymbol{\nabla}(\delta\mathbf{V})\, dV - \left(\int_{\Omega_0} \mathbf{f}^0 \cdot \delta\mathbf{V}\, dV + \oint_{\Gamma_0} \mathbf{t}^0 \cdot \delta\mathbf{V}\, dS \right) = 0 \qquad (2.9.28)$$

where $\mathbf{f}^0 = J\mathbf{f}$ is the body force vector measured per unit undeformed volume and $\mathbf{t}^0 = \mathbf{t}(ds/dS)$ is the traction vector measured per unit undeformed surface. The quantity $J\boldsymbol{\sigma}$ is called the *Kirchhoff stress tensor*.

Next, consider the internal virtual work done per unit time from Eq. (2.9.18):

$$\delta W_I = \int_{\Omega_0} J\boldsymbol{\sigma} : (\boldsymbol{\nabla}\,\delta\mathbf{V})\, dV = \int_{\Omega_0} J\boldsymbol{\sigma} : \delta\mathbf{d}\, dV$$

$$= \int_{\Omega_0} J\boldsymbol{\sigma} : (\delta\dot{\mathbf{F}} \cdot \mathbf{F}^{-1})\, dV = \int_{\Omega_0} (J\boldsymbol{\sigma} \cdot \mathbf{F}^{-T}) : \delta\dot{\mathbf{F}}\, dV \qquad (2.9.29)$$

where Eq. (2.9.10) and the symmetry of $\boldsymbol{\sigma}$ are used in arriving at the last statement. The expression $J\boldsymbol{\sigma}\cdot\mathbf{F}^{-T}$, by Eq. (2.5.19), is equal to the first Piola–Kirchhoff stress tensor, $\mathbf{P}$. Thus, the pair $(\mathbf{P}, \dot{\mathbf{F}})$ is energetically conjugate with respect to the undeformed volume. The principle of virtual displacements in terms of $\mathbf{P}$ is

$$\int_{\Omega_0} \mathbf{P} : \delta\dot{\mathbf{F}}\, dV - \left(\int_{\Omega_0} \mathbf{f}^0 \cdot \delta\mathbf{V}\, dV + \oint_{\Gamma_0} \mathbf{t}^0 \cdot \delta\mathbf{V}\, dS \right) = 0 \qquad (2.9.30)$$

Finally, using the identity [see Eq. (2.4.34)][8]

---

[8]Based on the *Lie derivative* of a tensor; **d** is the push forward of $\dot{\mathbf{E}}$, whereas **e** is the push forward of **E**.

$$\delta \dot{\mathbf{E}} = \mathbf{F}^{\mathrm{T}} \cdot \delta \mathbf{d} \cdot \mathbf{F} \quad \text{or} \quad \delta \mathbf{d} = \mathbf{F}^{-\mathrm{T}} \cdot \delta \dot{\mathbf{E}} \cdot \mathbf{F}^{-1} \tag{2.9.31}$$

in the first line of Eq. (2.9.29), we obtain

$$\delta W_I = \int_{\Omega_0} J \, \boldsymbol{\sigma} : \delta \mathbf{d} \, dV = \int_{\Omega_0} J \, \boldsymbol{\sigma} : (\mathbf{F}^{-\mathrm{T}} \cdot \delta \dot{\mathbf{E}} \cdot \mathbf{F}^{-1}) \, dV$$

$$= \int_{\Omega_0} (J \, \mathbf{F}^{-1} \cdot \boldsymbol{\sigma} \cdot \mathbf{F}^{-\mathrm{T}}) : \delta \dot{\mathbf{E}} \, dV = \int_{\Omega_0} \mathbf{S} : \delta \dot{\mathbf{E}} \, dV \tag{2.9.32}$$

where $\mathbf{S}$ is the second Piola–Kirchhoff stress tensor [see Eq. (2.5.22)]. Thus, the pair $(\mathbf{S}, \dot{\mathbf{E}})$ is energetically conjugate with respect to the undeformed volume. The principle of virtual displacements in terms of the second Piola–Kirchhoff stress tensor $\mathbf{S}$ is

$$\int_{\Omega_0} \mathbf{S} : \delta \dot{\mathbf{E}} \, dV - \left(\int_{\Omega_0} \mathbf{f}^0 \cdot \delta \mathbf{V} \, dV + \oint_{\Gamma_0} \mathbf{t}^0 \cdot \delta \mathbf{V} \, dS \right) = 0 \tag{2.9.33}$$

For equilibrium problems, the principle of virtual displacements takes the form (work done unit time)

$$\int_{\Omega_0} \mathbf{S} : \delta \mathbf{E} \, dV - \left(\int_{\Omega_0} \mathbf{f}^0 \cdot \delta \mathbf{u} \, dV + \oint_{\Gamma_0} \mathbf{t}^0 \cdot \delta \mathbf{u} \, dS \right) = 0 \tag{2.9.34}$$

The principles of virtual work are valid only for bounded continuum of a fixed material. The concept of virtual work is not meaningful for a continuum that instantly occupies a fixed region because the principle of conservation of mass must also be satisfied by the virtual velocity field. In this study, we develop integral statements, called *weak forms*, of governing equations of a continuum, which form the basis of finite element models. For structural problems, the weak forms are the same as the statements resulting from the application of the principle of virtual displacements.

2.10   Summary

In this chapter, the two descriptions of motion, namely, the spatial (Eulerian) and material (Lagrangian) descriptions of motion are discussed, and the deformation gradient, Cauchy–Green deformation tensors, and various measures of strain and stress were introduced. The strain tensors discussed include the Green–Lagrange strain tensor, Cauchy strain tensor, and the Euler strain tensor. The stress measures introduced include the Cauchy stress tensor and first and second Piola–Kirchhoff stress tensors. Equations governing a continuum, resulting from the conservation of mass and energy and balance of momenta, are derived, and constitutive relations for an elastic body are presented. The principle of conservation of angular momentum yields, in the absence of body couples, in the symmetry of Cauchy stress tensor. The concept of virtual work and first variation are introduced and principles of virtual displacements in the spatial and material descriptions are presented. The governing equations are derived in invariant (i.e. vector and tensor) form so that they can be expressed in any chosen coordinate system (e.g. rectangular, cylindrical, spherical, or even a curvilinear system). The continuum mechanics concepts presented and equations derived in this chapter are useful in the remaining chapters of this book.

The main equations of the chapter are summarized here for ready reference. Generally, capital letters are used for quantities in the material description and lowercase letters are used for those in the spatial description.

- Coordinates, mapping, and gradient operator:

$$
\begin{aligned}
\text{Material coordinates:} \quad & \mathbf{X} = (X_1, X_2, X_3) \\
\text{Spatial coordinates:} \quad & \mathbf{x} = (x_1, x_2, x_3) \\
\text{Motion mapping:} \quad & \mathbf{x} = \chi(\mathbf{X}, t), \quad \mathbf{X} = \chi^{-1}(\mathbf{x}, t) \\
\text{Material and spatial gradients:} \quad & \mathbf{\nabla}_0 = \partial/\partial\mathbf{X}, \quad \mathbf{\nabla} = \partial/\partial\mathbf{x}
\end{aligned}
\tag{2.10.1}
$$

- Displacement vectors:

$$
\begin{aligned}
\mathbf{u}(\mathbf{x}, t) &= \mathbf{x} - \mathbf{X}(\mathbf{x}, t) = \mathbf{x} - \chi^{-1}(\mathbf{x}, t) \\
\mathbf{U}(\mathbf{X}, t) &= \mathbf{x}(\mathbf{X}, t) - \mathbf{X} = \chi(\mathbf{X}, t) - \mathbf{X} \\
\mathbf{U}(\mathbf{X}, t) &= \mathbf{U}(\chi^{-1}(\mathbf{x}, t)) = \mathbf{u}(\mathbf{x}, t)
\end{aligned}
\tag{2.10.2}
$$

- Velocity vectors:

$$
\begin{aligned}
\mathbf{V}(\mathbf{x}, t) &= \frac{d\mathbf{x}}{dt}\Big|_{\mathbf{X}=\text{fixed}} = \left(\frac{\partial\chi(\mathbf{X}, t)}{\partial t}\right)\Big|_{\mathbf{X}=\text{fixed}} = \left(\frac{\partial\mathbf{U}(\mathbf{X}, t)}{\partial t}\right)\Big|_{\mathbf{X}=\text{fixed}} \\
\mathbf{v}(\mathbf{x}, t) &= \frac{d\mathbf{x}}{dt}\Big|_{\mathbf{X}=\text{fixed}} = \left(\frac{\partial\chi(\mathbf{X}, t)}{\partial t}\right)\Big|_{\mathbf{X}=\chi^{-1}(\mathbf{x}, t)} \\
\mathbf{V}(\mathbf{X}, t) &= \mathbf{V}[\chi^{-1}(\mathbf{x}, t)] = \mathbf{v}(\mathbf{x}, t)
\end{aligned}
\tag{2.10.3}
$$

- Acceleration vectors:

$$
\begin{aligned}
\mathbf{A}(\mathbf{x}, t) &= \frac{d\mathbf{V}}{dt}\Big|_{\mathbf{X}=\text{fixed}} = \left(\frac{\partial\mathbf{V}(\mathbf{X}, t)}{\partial t}\right)\Big|_{\mathbf{X}=\text{fixed}} \\
\mathbf{a}(\mathbf{x}, t) &= \frac{d\mathbf{v}}{dt} = \frac{\partial\mathbf{v}(\mathbf{x}, t)}{\partial t} + \mathbf{v} \cdot \mathbf{\nabla}\mathbf{v} \\
\mathbf{A}(\mathbf{X}, t) &= \mathbf{A}[\chi^{-1}(\mathbf{x}, t)] = \mathbf{a}(\mathbf{x}, t)
\end{aligned}
\tag{2.10.4}
$$

- Deformation gradient:

$$
\mathbf{F} = \overleftarrow{\mathbf{\nabla}}_0\mathbf{x} = (\mathbf{\nabla}_0\mathbf{x})^{\mathrm{T}}
\tag{2.10.5}
$$

- Green–Lagrange strain tensor:

$$
\mathbf{E} = \tfrac{1}{2}\left(\mathbf{F}^{\mathrm{T}} \cdot \mathbf{F} - \mathbf{I}\right) = \tfrac{1}{2}\left[\mathbf{\nabla}_0\mathbf{u} + (\mathbf{\nabla}_0\mathbf{u})^{\mathrm{T}} + (\mathbf{\nabla}_0\mathbf{u}) \cdot (\mathbf{\nabla}_0\mathbf{u})^{\mathrm{T}}\right]
\tag{2.10.6}
$$

- Euler–Almansi strain tensor:

$$e = \tfrac{1}{2}\left(\mathbf{I} - \mathbf{F}^{-\mathrm{T}} \cdot \mathbf{F}^{-1}\right) = \tfrac{1}{2}\left[\boldsymbol{\nabla}\mathbf{u} + (\boldsymbol{\nabla}\mathbf{u})^{\mathrm{T}} - (\boldsymbol{\nabla}\mathbf{u}) \cdot (\boldsymbol{\nabla}\mathbf{u})^{\mathrm{T}}\right] \qquad (2.10.7)$$

- Velocity gradient tensor and its symmetric and skew-symmetric parts:

$$\mathbf{l} = \left(\frac{\partial \mathbf{v}(\mathbf{x}, t)}{\partial \mathbf{x}}\right)^{\mathrm{T}} = (\boldsymbol{\nabla}\mathbf{v}(\mathbf{x}, t))^{\mathrm{T}} = \mathbf{d} + \mathbf{w} \qquad (2.10.8)$$

$$\mathbf{d} = \tfrac{1}{2}\left[\boldsymbol{\nabla}\mathbf{v} + (\boldsymbol{\nabla}\mathbf{v})^{\mathrm{T}}\right] = \mathbf{d}^{\mathrm{T}}, \quad \mathbf{w} = \tfrac{1}{2}\left[(\boldsymbol{\nabla}\mathbf{v})^{\mathrm{T}} - \boldsymbol{\nabla}\mathbf{v}\right] = -\mathbf{w}^{\mathrm{T}} \qquad (2.10.9)$$

- Nanson's formula:

$$\hat{\mathbf{n}}\, da = J\mathbf{F}^{-\mathrm{T}} \cdot \hat{\mathbf{N}}\, dA \qquad (2.10.10)$$

- Relationships:

$$e = \mathbf{F}^{-\mathrm{T}} \cdot \mathbf{E} \cdot \mathbf{F}^{-1}, \quad \mathbf{E} = \mathbf{F}^{\mathrm{T}} \cdot e \cdot \mathbf{F} \qquad (2.10.11)$$

$$\dot{\mathbf{F}} \equiv \frac{d\mathbf{F}}{dt} = (\boldsymbol{\nabla}_0 \mathbf{v})^{\mathrm{T}}, \quad \dot{\mathbf{F}} = \mathbf{l} \cdot \mathbf{F}, \quad \dot{\mathbf{E}} \equiv \frac{d\mathbf{E}}{dt} = \mathbf{F}^{\mathrm{T}} \cdot \mathbf{d} \cdot \mathbf{F} \qquad (2.10.12)$$

$$\dot{e} = \mathbf{d} - \mathbf{l}^{\mathrm{T}} \cdot e - e \cdot \mathbf{l} \qquad (2.10.13)$$

- Stress measures:
  $\boldsymbol{\sigma}$ = Cauchy stress tensor; $\mathbf{P}$ = first Piola–Kirchhoff stress tensor;
  $\mathbf{S}$ = second Piola–Kirchhoff stress tensor

$$\mathbf{P} = J\boldsymbol{\sigma} \cdot \mathbf{F}^{-\mathrm{T}}, \quad \mathbf{S} = \mathbf{F}^{-1} \cdot \mathbf{P} = J\mathbf{F}^{-1} \cdot \boldsymbol{\sigma} \cdot \mathbf{F}^{-\mathrm{T}} \qquad (2.10.14)$$

- Conservation of mass in spatial and material descriptions:

$$\frac{\partial \rho}{\partial t} + \boldsymbol{\nabla} \cdot (\rho \mathbf{v}) = 0 \qquad (2.10.15)$$

$$\rho_0 = J\rho \qquad (2.10.16)$$

- Equations of motion in spatial and material descriptions:

$$\boldsymbol{\nabla} \cdot \boldsymbol{\sigma}^{\mathrm{T}} + \mathbf{f} = \rho\left(\frac{\partial \mathbf{v}}{\partial t} + \mathbf{v} \cdot \boldsymbol{\nabla}\mathbf{v}\right) \qquad (2.10.17)$$

$$\boldsymbol{\nabla}_0 \cdot \left(\mathbf{S}^{\mathrm{T}} \cdot \mathbf{F}^{\mathrm{T}}\right) + \mathbf{f}^0 = \rho_0 \frac{\partial^2 \mathbf{u}}{\partial t^2} \qquad (2.10.18)$$

- Energy equation in spatial description:
  e_c = internal energy per unit mass; $\mathbf{q}$ = heat flux vector;
  g = internal heat generation per unit volume;
  $\mathbf{k}$ = conductivity tensor, and T is the temperature:

$$\rho\frac{de_c}{dt} = \boldsymbol{\sigma} : \mathbf{d} - \boldsymbol{\nabla} \cdot \mathbf{q} + g \qquad (2.10.19)$$

$$\mathbf{q} = -\mathbf{k} \cdot \boldsymbol{\nabla}T \qquad (2.10.20)$$

- Principle of virtual power in spatial and material descriptions:

$$0 = \int_{\Omega} \boldsymbol{\sigma} : \delta \mathbf{d} \, dv - \left(\int_{\Omega} \mathbf{f} \cdot \delta \mathbf{v} \, dv + \oint_{\Gamma} \mathbf{t} \cdot \delta \mathbf{v} \, ds \right) \tag{2.10.21}$$

$$0 = \int_{\Omega_0} \boldsymbol{S} : \delta \dot{\mathbf{E}} \, dV - \left(\int_{\Omega_0} \mathbf{f}^0 \cdot \delta \mathbf{V} \, dV + \oint_{\Gamma_0} \mathbf{t}^0 \cdot \delta \mathbf{V} \, dS \right) \tag{2.10.22}$$

Problems

More problems can be found in the textbook by the author [1].

KINEMATICS AND KINETICS

2.1 Given the motion
$$\mathbf{x} = (1 + t)\mathbf{X}$$
determine the velocity and acceleration fields of the motion.

2.2 Show that in the spatial description the acceleration components in the cylindrical coordinates are
$$a_r = \frac{\partial v_r}{\partial t} + v_r \frac{\partial v_r}{\partial r} + \frac{v_\theta}{r} \frac{\partial v_r}{\partial \theta} + v_z \frac{\partial v_r}{\partial z} - \frac{v_\theta^2}{r}$$
$$a_\theta = \frac{\partial v_\theta}{\partial t} + v_r \frac{\partial v_\theta}{\partial r} + \frac{v_\theta}{r} \frac{\partial v_\theta}{\partial \theta} + v_z \frac{\partial v_\theta}{\partial z} + \frac{v_r v_\theta}{r}$$
$$a_z = \frac{\partial v_z}{\partial t} + v_r \frac{\partial v_z}{\partial r} + \frac{v_\theta}{r} \frac{\partial v_z}{\partial \theta} + v_z \frac{\partial v_z}{\partial z}$$

2.3 The motion of a body is described by the mapping
$$\boldsymbol{\chi}(\mathbf{X}) = (X_1 + t^2 X_2)\,\hat{\mathbf{e}}_1 + (X_2 + t^2 X_1)\,\hat{\mathbf{e}}_2 + X_3\,\mathbf{e}_3$$
where t denotes time. Determine

(a) the components of the deformation gradient $\mathbf{F}$,

(b) the components of the displacement, velocity, and acceleration vectors, and

(c) the position (X_1, X_2, X_3) of the particle in undeformed configuration that occupies the position $(x_1, x_2, x_3) = (9, 6, 1)$ at time $t = 2\,\text{s}$ in the deformed configuration.

2.4 Suppose that the motion of a continuous medium is given by
$$x_1 = X_1 \cos \alpha t + X_2 \sin \alpha t$$
$$x_2 = -X_1 \sin \alpha t + X_2 \cos \alpha t$$
$$x_3 = (1 + ct)X_3$$
where α and c are constants. Determine the components of

(a) the displacement vector in the material description,

(b) the displacement vector in the spatial description, and

(c) the Green–Lagrange and Euler–Almansi strain tensors.

2.5 If the deformation mapping of a body is given by
$$\boldsymbol{\chi}(\mathbf{X}) = (X_1 + k_1 X_2)\,\hat{\mathbf{e}}_1 + (X_2 + k_2 X_1)\,\hat{\mathbf{e}}_2 + X_3\,\hat{\mathbf{e}}_3$$
where k_1 and k_2 are constants, determine

(a) the displacement components in the material description,

(b) the displacement components in the spatial description, and

(c) the components of the Green–Lagrange and Euler–Almansi strain tensors.

2.6 Consider the uniform deformation of a square of side 2 units initially centered at $\mathbf{X} = (0,0)$. The deformation is given by the mapping

$$x_1 = \tfrac{1}{4}(18 + 4X_1 + 6X_2), \quad x_2 = \tfrac{1}{4}(14 + 6X_2)$$

(a) Sketch the deformed configuration of the body.

(b) Compute the components of the deformation gradient $\mathbf{F}$ and its inverse (display them in matrix form).

(c) Compute the components of the right and left Cauchy–Green deformation tensors (display them in matrix form).

(d) Compute the Green–Lagrange and Euler–Almansi strain tensor components (display them in matrix form).

2.7 *Nanson's formula* Let the differential area in the reference configuration be dA. Then

$$\hat{\mathbf{N}} dA = d\mathbf{X}^{(1)} \times d\mathbf{X}^{(2)} \quad \text{or} \quad N_I dA = e_{IJK} dX_J^{(1)} dX_K^{(2)}$$

where $d\mathbf{X}^{(1)}$ and $d\mathbf{X}^{(2)}$ are two nonparallel differential vectors in the reference configuration. The mapping from the undeformed configuration to the deformed configuration maps $d\mathbf{X}^{(1)}$ and $d\mathbf{X}^{(2)}$ into $d\mathbf{x}^{(1)}$ and $d\mathbf{x}^{(2)}$, respectively. Then $\hat{\mathbf{n}} da = d\mathbf{x}^{(1)} \times d\mathbf{x}^{(2)}$. Show that

$$\hat{\mathbf{n}}\, da = J \mathbf{F}^{-\mathrm{T}} \cdot \hat{\mathbf{N}}\, dA$$

2.8 Show that the components of the Green–Lagrange strain tensor in cylindrical coordinate system are given by

$$E_{rr} = \frac{\partial u_r}{\partial r} + \frac{1}{2}\left[\left(\frac{\partial u_r}{\partial r}\right)^2 + \left(\frac{\partial u_\theta}{\partial r}\right)^2 + \left(\frac{\partial u_z}{\partial r}\right)^2\right]$$

$$E_{r\theta} = \frac{1}{2}\left(\frac{1}{r}\frac{\partial u_r}{\partial \theta} + \frac{\partial u_\theta}{\partial r} - \frac{u_\theta}{r} + \frac{1}{r}\frac{\partial u_r}{\partial r}\frac{\partial u_r}{\partial \theta} + \frac{1}{r}\frac{\partial u_\theta}{\partial r}\frac{\partial u_\theta}{\partial \theta}\right.$$
$$\left. + \frac{1}{r}\frac{\partial u_z}{\partial r}\frac{\partial u_z}{\partial \theta} + \frac{u_r}{r}\frac{\partial u_\theta}{\partial r} - \frac{u_\theta}{r}\frac{\partial u_r}{\partial r}\right)$$

$$E_{rz} = \frac{1}{2}\left(\frac{\partial u_r}{\partial z} + \frac{\partial u_z}{\partial r} + \frac{\partial u_r}{\partial r}\frac{\partial u_r}{\partial z} + \frac{\partial u_\theta}{\partial r}\frac{\partial u_\theta}{\partial z} + \frac{\partial u_z}{\partial r}\frac{\partial u_z}{\partial z}\right)$$

$$E_{\theta\theta} = \frac{u_r}{r} + \frac{1}{r}\frac{\partial u_\theta}{\partial \theta} + \frac{1}{2}\left[\left(\frac{1}{r}\frac{\partial u_r}{\partial \theta}\right)^2 + \left(\frac{1}{r}\frac{\partial u_\theta}{\partial \theta}\right)^2 + \left(\frac{1}{r}\frac{\partial u_z}{\partial \theta}\right)^2\right.$$
$$\left. - \frac{2}{r^2}u_\theta\frac{\partial u_r}{\partial \theta} + \frac{2}{r^2}u_r\frac{\partial u_\theta}{\partial \theta} + \left(\frac{u_\theta}{r}\right)^2 + \left(\frac{u_r}{r}\right)^2\right]$$

$$E_{\theta z} = \frac{1}{2}\left(\frac{\partial u_\theta}{\partial z} + \frac{1}{r}\frac{\partial u_z}{\partial \theta} + \frac{1}{r}\frac{\partial u_r}{\partial \theta}\frac{\partial u_r}{\partial z} + \frac{1}{r}\frac{\partial u_\theta}{\partial \theta}\frac{\partial u_\theta}{\partial z}\right.$$
$$\left. + \frac{1}{r}\frac{\partial u_z}{\partial \theta}\frac{\partial u_z}{\partial z} - \frac{u_\theta}{r}\frac{\partial u_r}{\partial z} + \frac{u_r}{r}\frac{\partial u_\theta}{\partial z}\right)$$

$$E_{zz} = \frac{\partial u_z}{\partial z} + \frac{1}{2}\left[\left(\frac{\partial u_r}{\partial z}\right)^2 + \left(\frac{\partial u_\theta}{\partial z}\right)^2 + \left(\frac{\partial u_z}{\partial z}\right)^2\right]$$

2.9 Show that

$$\frac{d}{dt}\left[(ds)^2\right] = 2d\mathbf{x} \cdot \mathbf{d} \cdot d\mathbf{x}$$

2.10 Verify that

$$\dot{\mathbf{v}} = \frac{\partial \mathbf{v}}{\partial t} + \tfrac{1}{2}\,\mathrm{grad}\,(\mathbf{v}\cdot\mathbf{v}) + 2\mathbf{w}\cdot\mathbf{v}$$

$$= \frac{\partial \mathbf{v}}{\partial t} + \tfrac{1}{2}\,\mathrm{grad}\,(\mathbf{v}\cdot\mathbf{v}) + 2\omega \times \mathbf{v}$$

where $\mathbf{w}$ is the spin tensor and ω is the vorticity vector.

2.11 Show that the components of the spin tensor $\mathbf{w}$ in cylindrical coordinate system are

$$w_{\theta r} = \frac{1}{2}\left(\frac{1}{r}\frac{\partial v_r}{\partial \theta} - \frac{v_\theta}{r} - \frac{\partial v_\theta}{\partial r}\right) = -w_{r\theta}$$

$$w_{zr} = \frac{1}{2}\left(\frac{\partial v_r}{\partial z} - \frac{\partial v_z}{\partial r}\right) = -w_{rz}$$

$$w_{\theta z} = \frac{1}{2}\left(\frac{1}{r}\frac{\partial v_z}{\partial \theta} - \frac{\partial v_\theta}{\partial z}\right) = -w_{z\theta}$$

2.12 Establish the identity in Eq. (2.4.35): $\dot{\mathbf{e}} = \mathbf{d} - \mathbf{l}^{\mathrm{T}}\cdot\mathbf{e} - \mathbf{e}\cdot\mathbf{l}$.

2.13 Show that

$$\boldsymbol{\nabla}_0 \cdot \mathbf{P}^{\mathrm{T}} = J\,\boldsymbol{\nabla}\cdot\boldsymbol{\sigma}^{\mathrm{T}}, \quad \boldsymbol{\nabla}_0 = \frac{\partial}{\partial \mathbf{X}}, \quad \boldsymbol{\nabla} = \frac{\partial}{\partial \mathbf{x}}$$

where $\mathbf{P}$ is the first Piola–Kirchhoff stress tensor and σ is the Cauchy stress tensor.

CONSERVATION OF MASS AND BALANCE PRINCIPLES

2.14 Let an arbitrary region in a continuous medium be denoted by Ω and the bounding closed surface of this region be continuous and denoted by Γ. Let each point on the bounding surface move with the velocity $\mathbf{v}$. It can be shown that the time derivative of the volume integral over some continuous function $Q(\mathbf{x}, t)$ is given by

$$\frac{d}{dt}\int_\Omega Q(\mathbf{x}, t)\,dv \equiv \int_\Omega \frac{\partial Q}{\partial t}\,dv + \oint_\Gamma Q\mathbf{v}\cdot\hat{\mathbf{n}}\,ds \qquad (1)$$

This expression for the differentiation of a volume integral with variable limits is sometimes known as the three-dimensional *Leibniz rule*. The material derivative operator d/dt corresponds to changes with respect to a fixed mass, that is $\rho\,dv$ is constant with respect to this operator. Show formally by means of Leibniz's rule, the divergence theorem, and conservation of mass that

$$\frac{d}{dt}\int_\Omega \rho\phi\,dv \equiv \int_\Omega \rho\frac{d\phi}{dt}\,dv \qquad (2)$$

2.15 Use the statement in Eq. (2.7.20) to establish the symmetry of the Cauchy stress tensor, Eq. (2.7.21).

2.16 Establish the symmetry of the Cauchy stress tensor, Eq. (2.7.21), using the statement

$$\sigma_{ij} = \frac{\partial \Psi}{\partial \varepsilon_{ij}} \qquad (1)$$

2.17 Let e_c denote the thermodynamic internal energy per unit mass of a material. Then the principle of balance of total energy of a material region can be expressed as

$$\frac{d}{dt}\int_\Omega \rho\left(e_c + \frac{v^2}{2}\right)dv = \oint_\Gamma \hat{\mathbf{n}}\cdot\boldsymbol{\sigma}^{\mathrm{T}}\cdot\mathbf{v}\,ds + \int_\Omega \mathbf{f}\cdot\mathbf{v}\,dv - \oint_\Gamma \mathbf{q}\cdot\hat{\mathbf{n}}\,ds \qquad (1)$$

The first two terms on the right-hand side describe the rate of work done on the material region by the surface stresses and the body forces. The third integral describes the net *outflow* of heat from the region, causing a decrease of energy inside the region. The heat-flux vector $\mathbf{q}$ describes the magnitude and direction of the flow of heat energy per unit time and per unit area. By suitable operations, obtain the differential form of the energy equation

$$\rho \frac{d}{dt} \left(e_c + \frac{v^2}{2} \right) = \boldsymbol{\nabla} \cdot \left(\boldsymbol{\sigma}^{\mathrm{T}} \cdot \mathbf{v} \right) + \mathbf{f} \cdot \mathbf{v} - \boldsymbol{\nabla} \cdot \mathbf{q} \tag{2}$$

and subsequently the relation

$$\rho \frac{de_c}{dt} = \boldsymbol{\nabla} \cdot \left(\boldsymbol{\sigma}^{\mathrm{T}} \cdot \mathbf{v} \right) - \boldsymbol{\nabla} \cdot \boldsymbol{\sigma}^{\mathrm{T}} \cdot \mathbf{v} - \boldsymbol{\nabla} \cdot \mathbf{q} \tag{3}$$

2.18 Specialize the equations of motion in Eq. (2.7.14) and the continuity equation in Eq. (2.7.4) for two-dimensional flows of viscous, *incompressible* (i.e. ρ is a constant), isotropic fluids in terms of the velocity components (v_1, v_2) and pressure P. Assume that the total stress tensor $\boldsymbol{\sigma}$ can be expressed as

$$\boldsymbol{\sigma} = 2\mu\mathbf{d} - P\mathbf{I}, \qquad\qquad \sigma_{ij} = 2\mu d_{ij} - P\delta_{ij}$$

$$\mathbf{d} = \frac{1}{2} \left[(\boldsymbol{\nabla}\mathbf{v})^{\mathrm{T}} + \boldsymbol{\nabla}\mathbf{v} \right], \quad d_{ij} = \frac{1}{2} \left(\frac{\partial v_i}{\partial x_j} + \frac{\partial v_j}{\partial x_i} \right) \tag{1}$$

where μ is the viscosity of the fluid.

2.19 Use the Fourier heat conduction law in Eq. (2.7.29) to express the energy equation in Eq. (2.7.28) for incompressible fluids in terms of velocity components v_1 and v_2 and temperature T in rectangular Cartesian coordinates (x_1, x_2) for an isotropic ($\mathbf{k} = k\mathbf{I}$) two-dimensional medium. Neglect stress power and assume that the internal energy e_c is equal to $c_p T$, where c_p is the specific heat at constant pressure.

VARIATIONAL PRINCIPLES

2.20 Establish the identities in Eqs. (2.9.6)–(2.9.10).

2.21 Show that $\delta J = J (\boldsymbol{\nabla} \cdot \delta \mathbf{u})$.

2.22 Show that the principle of virtual displacements, Eq. (2.9.24) or Eq. (2.9.25) yields Eqs. (2.9.26) and (2.9.27) as the Euler equations.

2.23 Establish the equivalence

$$\int_{\Omega} \boldsymbol{\sigma} : \delta \mathbf{e} \, dv = \int_{\Omega_0} \mathbf{S} : \delta \mathbf{E} \, dV$$

so that the principle of virtual displacements for the static case takes the form

$$\int_{\Omega_0} \mathbf{S} : \delta \mathbf{E} \, dV - \left(\int_{\Omega_0} \mathbf{f}^0 \cdot \delta \mathbf{u} \, dV + \oint_{\Gamma_0} \mathbf{t}^0 \cdot \delta \mathbf{u} \, dS \right) = 0$$

Use the identity $\delta \mathbf{e} = \mathbf{F}^{-\mathrm{T}} \cdot \delta \mathbf{E} \cdot \mathbf{F}^{-1}$.

2.24 Derive the Euler equations in terms of the second Piola–Kirchhoff stress tensor $\mathbf{S}$ using the principle of virtual displacements given in Problem 2.23.

2.25 A special case of the principle of virtual displacements that deals with linear as well as nonlinear elastic bodies is known as the *principle of minimum total potential energy*. For elastic bodies (in the absence of temperature variations), there exists a *strain energy density* function Ψ (measured per unit volume) such that

$$\sigma_{ij} = \frac{\partial \Psi}{\partial \varepsilon_{ij}} \tag{1}$$

Equation (1) represents the constitutive equation of a hyperelastic material. The strain energy density Ψ is a single-valued function of strains at a point, and is assumed to be positive-definite. The statement of the principle of virtual displacements can be expressed in terms of the strain energy density Ψ:

$$\int_\Omega \frac{\partial \Psi}{\partial \varepsilon_{ij}} \, \delta \varepsilon_{ij} \, dv - \left[\int_\Omega f_i \, \delta u_i \, dv + \int_{\Gamma_\sigma} t_i \, \delta u_i \, ds \right] = 0 \tag{2}$$

The first integral in Eq. (2) is equal to [see Eq. (2.8.12)]

$$\int_\Omega \sigma_{ij} \, \delta \varepsilon_{ij} \, dv = \int_\Omega \delta \Psi \, dv = \delta \Psi \tag{3}$$

where Ψ is the strain energy density

$$U = \int_\Omega \Psi \, dv = \int_\Omega \left(\int_0^{\varepsilon_{ij}} \sigma_{ij} \, d\varepsilon_{ij} \right) dv \tag{4}$$

For linear elastic material, the strain energy U can be expressed as

$$U = \frac{1}{2} \int_\Omega \sigma_{ij} \, \varepsilon_{ij} \, dv \tag{5}$$

The expression in the square brackets of Eq. (2) can be identified as $-\delta W_E$. We suppose that there exists a potential $V = W_E$ whose first variation is

$$\delta V = \delta W_E = - \left[\int_\Omega f_i \, \delta u_i \, dv + \int_{\Gamma_\sigma} t_i \, \delta u_i \, ds \right] \tag{6}$$

Then the principle of virtual work takes the form

$$\delta U + \delta V = \delta(U + V) \equiv \delta \Pi = 0 \tag{7}$$

The sum $U + V = \Pi$ is called the *total potential energy* of the elastic body. The statement in Eq. (7) is known as the *principle of minimum total potential energy*. Derive the Euler equations for an isotropic Hookean solid using the principle of minimum total potential energy, Eq. (7). Assume small strains (i.e. linearized elasticity).

2.26 Hamilton's principle is a generalization of the principle of virtual displacements to dynamics. The dynamics version of the principle of virtual displacements, i.e. Hamilton's principle, is based on the assumption that a dynamical system is characterized by two energy functions: a *kinetic energy K* and a *potential energy W*. Hamilton used the principle of D'Alembert, which asserts that any law of statics becomes transformed into a law of kinetics if the forces acting on the system are augmented by inertial forces, to extend the principle of virtual displacements to kinetics. Hamilton's principle states that *of all possible paths that a material particle could travel from its position at time t_1 to its position at time t_2, its actual path will be one for which the integral*

$$\int_{t_1}^{t_2} (K - W) \, dt \tag{1}$$

is an extremum. The difference between the kinetic and potential energies is called the *Lagrangian* function

$$L = K - W = K - (W_I + W_E) \tag{2}$$

Using D'Alembert's principle, the virtual work done by external forces in the dynamic case can be expressed as

$$\delta W_E = - \left[\int_\Omega \left(\mathbf{f} - \rho \frac{d\mathbf{v}}{dt} \right) \cdot \delta \mathbf{v} \, dv + \int_{\Gamma_\sigma} \hat{\mathbf{t}} \cdot \delta \mathbf{v} \, ds \right] \tag{3}$$

where **v** is the velocity vector of a material particle in the body. Then, Hamilton's principle can be expressed as

$$0 = \int_{t_1}^{t_2} (\delta W_I + \delta W_E) \, dt \tag{4}$$

Use Hamilton's principle to derive the equations of motion in Eq. (2.7.13) as the *Euler–Lagrange equations*. See Reddy [3, 13] for more details and examples of applications of Hamilton's principle.

The Finite Element Method: A Review

Once the whole is divided, the parts need names.
There are already enough names.
One must know when to stop.

—— Lao Tsu (and *Taoism*)

3.1 Introduction

The main ideas of the finite element method were briefly visited in Chapter 1. To summarize, the finite element method has the following basic features:

1. Divide the whole domain of the system into parts, called *finite elements* (because the size of each part remains finite, no matter how small the part is).

2. Over each representative finite element, develop algebraic relations among pairs of dual variables, called the *primary* and *secondary variables* (e.g. relations between pairs of dual variables; e.g. "forces" and "displacements", "heats" and "temperatures", and so on).

3. Assemble the elements (i.e. put together the algebraic relations of all elements), using certain physical and mathematical requirements, to obtain relations between the primary and secondary variables of the whole system.

The meaning of the phrases "primary" and "secondary" variables will be made clear in the coming sections. The first feature, namely, dividing the whole into parts, of the finite element method is shared by no other method. Not all geometric shapes can be used as "elements"; only those shapes that allow unique derivation of their approximation functions qualify as elements.

In the present chapter, we review the basic steps involved in the finite element model development as applied to one- and two-dimensional problems described by typical linear second-order differential equations involving a single dependent unknown. The main objective is to familiarize the reader with the terminology and specific steps involved in the finite element model development and applications. Readers who are familiar with the author's approach, as presented in the textbook by Reddy [2], to finite element model development from a set of differential equations may skip this chapter.

J.N. Reddy, *An Introduction to Nonlinear Finite Element Analysis*, Second Edition. ©J.N. Reddy 2015. Published in 2015 by Oxford University Press.

3.2   One-Dimensional Problems

3.2.1   Governing Differential Equation

Consider the differential equation

$$-\frac{d}{dx}\left(a\frac{du}{dx}\right) + cu = f \qquad \text{for} \qquad 0 < x < L \tag{3.2.1}$$

where $a = a(x)$, $c = c(x)$, and $f = f(x)$ are called the *data* (i.e. known quantities) of the problem, and $u(x)$ is the solution to be determined. The data depends on the material properties, geometry, and "loads." Equation (3.2.1) represents the mathematical model of a number of problems of engineering and applied science, and it must be solved subject to appropriate boundary conditions. Table 3.2.1 contains a list of field problems, by no means exhaustive, in which Eq. (3.2.1) arises.

3.2.2   Finite Element Approximation

The domain $\Omega = (0, L)$ of the problem consists of all points between $x = 0$ and $x = L$. The points $x = 0$ and $x = L$ are the boundary points of the total domain. In the finite element method, the domain $(0, L)$ is divided into a set of intervals, $\Omega^e = (x_a^e, x_b^e)$, where x_a^e and x_b^e denote the coordinates of the end points of the finite element Ω^e with respect to the coordinate x, as shown in Fig. 3.2.1. A typical interval Ω^e is called a *finite element*, and it is of length $h_e = x_b^e - x_a^e$.

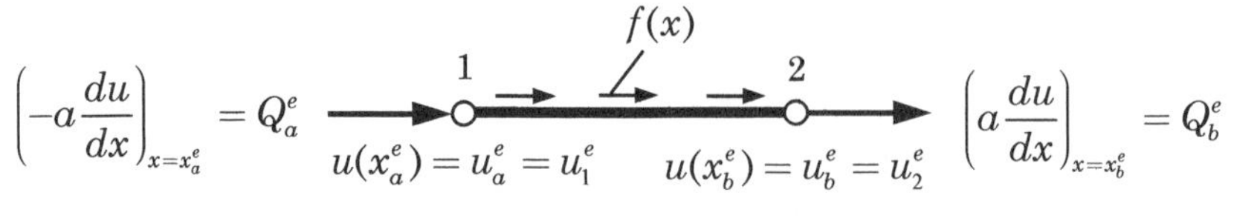

$$\left(-a\frac{du}{dx}\right)_{x=x_a^e} = Q_a^e \qquad \qquad \left(a\frac{du}{dx}\right)_{x=x_b^e} = Q_b^e$$

$$u(x_a^e) = u_a^e = u_1^e \qquad u(x_b^e) = u_b^e = u_2^e$$

Fig. 3.2.1: A typical finite element in one dimension.

In the finite element method, we seek an approximate solution to Eq. (3.2.1) over each finite element Ω^e; a typical finite element is shown in Fig. 3.2.1. The finite element approximation $u_h^e(x)$ of actual solution $u(x)$ over the element Ω^e is sought in the form

$$u(x) \approx u_h^e(x) = c_1^e\varphi_1^e(x) + c_2^e\varphi_2^e(x) + \ldots + c_n^e\varphi_n^e(x) = \sum_{j=1}^{n} c_j^e\varphi_j^e(x) \tag{3.2.2}$$

where $\varphi_j^e(x)$ are functions to be selected and c_j^e are constants to be determined such that Eq. (3.2.2) satisfies the differential equation in Eq. (3.2.1) and

Table 3.2.1: List of fields in which the model equation in Eq. (3.2.1) arises, with meaning of various parameters and variables; see the bottom of the table for the meaning of the parameters*.

Field of study	Primary variable u	Coefficient a	Coefficient c	Source term f	Secondary variable Q
Heat transfer	Temperature $T - T_\infty$	Thermal conductance kA	Surface convection $p\beta$	Heat generation f	Heat Q
Flow through porous medium	Fluid head ϕ	Permeability μ	$--$ 0	Infiltration f	Point source Q
Flow through pipes	Pressure P	Pipe resistance $1/R$	$--$ 0	$--$ 0	Point source Q
Flow of viscous fluids	Velocity v_z	Viscosity μ	$--$ 0	Pressure gradient $-dP/dx$	Shear stress σ_{xz}
Elastic cables	Displacement u	Tension T	$--$ 0	Transverse force f	Point force P
Elastic bars	Displacement u	Axial stiffness EA	$--$ 0	Axial force f	Point load P
Torsion of bars	Angle of twist θ	Shear stiffness GJ	$--$ 0	$--$ 0	Torque T
Electro-statics	Electrical potential ϕ	Dielectric constant ϵ	$--$ 0	Charge density ρ	Electric flux E

* k = thermal conductance; β = convective film conductance; p = perimeter; P = pressure or force; T_∞ = ambient temperature of the surrounding fluid medium; $R = 128\mu h/(\pi d^4)$ with μ being the viscosity; h, the length and d the diameter of the pipe; E = Young's modulus; A = area of cross-section; J = polar moment of inertia.

appropriate end conditions over the element. Since there are n unknown parameters, we need n linearly independent algebraic relations to determine them. Substituting the approximate solution in Eq. (3.2.2) into the left-hand side of Eq. (3.2.1), we obtain an expression that, in general, will not be equal to the right-hand side of the equation, $f(x)$. The difference between the two sides of the equation is called the *residual* (i.e. error in the differential equation)

$$-\frac{d}{dx}\left(a\frac{du_h^e}{dx}\right) + cu_h^e(x) - f(x) \equiv R^e(x, c_1^e, c_2^e, \ldots, c_n^e) \neq 0 \qquad (3.2.3)$$

We wish to determine c_j^e $(j = 1, 2, \ldots, n)$ such that the residual R^e is zero, in some meaningful sense, over the element.

One way of making the residual zero over the element domain is to set the weighted-integral of the residual to zero:

$$\int_{x_a^e}^{x_b^e} w_i^e(x) R^e(x, c_1^e, c_2^e, \ldots, c_n^e) \, dx = 0, \qquad i = 1, 2, \ldots, n \qquad (3.2.4)$$

where w_i^e $(i = 1, 2, \ldots, n)$ are weight functions. Equation (3.2.4) provides a set of n algebraic relations among the parameters c_j^e $(j = 1, 2, \ldots, n)$. The set $\{w_i^e(x)\}_{i=1}^n$ must be linearly independent so that the algebraic equations in Eq. (3.2.4) are also linearly independent and invertible. Note that if $w^e = 1$, we will have only one relation among n unknowns $c_1^e, c_2^e, \ldots, c_n^e$.

There are several choices of w_i^e that may be used. In the present study, we take $w_i^e(x)$ to be the same as the approximation functions $\varphi_i^e(x)$. This particular choice is adopted in the *Galerkin method*. Different choice of the weight functions $\{w_i^e\}_{i=1}^n$ will result in a different set of algebraic relations or a different *finite element model* of the same differential equation. In particular, the algebraic relations resulting from the weighted-residual statement with $w_i^e = \varphi_i^e$ is called the *Galerkin finite element model*. We note that R^e contains the same order derivatives of the dependent unknown $u(x)$ as in the differential equation, Eq. (3.2.1), and hence requires at least quadratic representation of $u_h^e(x)$. To reduce or weaken the differentiability of φ_i^e, we mathematically distribute differentiation equally (when the given differential equation is of even order) between w_i^e and u_h^e, and the resulting integral form is known as the *weak form*, as will be discussed in more detail shortly. The resulting finite element model will be termed *weak form Galerkin finite element model*.

In solid and structural mechanics, the governing equations are often derived from the principle of virtual displacements or its special case, *the principle of minimum total potential energy*. These principles are based on energy concepts and constitute the basis of classical variational methods like the Ritz method (see Reddy [1, 2] for additional details), and most of the finite element models in the literature are based on an element-wise application of the Ritz method. The statement of the principle of virtual displacements is an integral statement, which is the same as the *weak form* derived directly from the governing differential equation. However, the derivation of the weak form of a differential equation is independent of any variational principle and thus has a general setting independent of the field of study. All commercial finite element programs are based on weak forms of the differential equations arising in various physical fields. Therefore, more importance is given in this study to the weak form development of governing equations while making connection to the principle of virtual displacements wherever appropriate.

3.2.3 Derivation of the Weak Form

The starting point to construct the weak form of Eq. (3.2.1) is the weighted-residual statement in Eq. (3.2.4). The next step is to weaken the differentiability required of $u_h^e(x)$ [and hence φ_j of Eq. (3.2.2)]. This requires trading half of the differentiation from u_h^e to w_i^e such that both u_h^e and w_i^e are differentiated equally, once each in the present case. The resulting integral form is termed the *weak form* of Eq. (3.2.1) because it allows approximation functions with less or weaker continuity than that required in the integral statement in Eq. (3.2.1). A three-step procedure of constructing the weak form of Eq. (3.2.1) is presented next.

Step 1. The first step is to write the weighted-residual statement as in Eq. (3.2.4)

$$0 = \int_{x_a^e}^{x_b^e} w_i^e \left[-\frac{d}{dx}\left(a\frac{du_h^e}{dx} \right) + cu_h^e - f \right] dx \qquad (3.2.5)$$

Step 2. The second step is to trade differentiation from u_h^e to w_i^e, using integration by parts. We obtain

$$0 = \int_{x_a^e}^{x_b^e} \left(a\frac{dw_i^e}{dx}\frac{du_h^e}{dx} + cw_i^e u_h^e - w_i^e f \right) dx - \left[w_i^e \cdot a\frac{du_h^e}{dx} \right]_{x_a^e}^{x_b^e} \qquad (3.2.6)$$

Consider the boundary term appearing in Eq. (3.2.6), namely, the expression

$$\left[w_i^e \cdot a\frac{du}{dx} \right]_{x_a^e}^{x_b^e}$$

The coefficient of the weight function w_i^e in the boundary expression, $a(du/dx)$, is called the *secondary variable*, and its specification constitutes the *natural* or *Neumann boundary condition*. The *primary variable* is the dependent unknown of the differential equation, u, *in the same form as the weight function* w_i^e in the boundary expression. The specification of a primary variable on the boundary constitutes the *essential* or *Dirichlet boundary condition*. For the model equation at hand, the primary and secondary variables are

$$\text{Primary variable:}\quad u\ ;\qquad \text{Secondary variable:}\quad n_x\left(a\frac{du}{dx} \right) \equiv Q(x)\ (3.2.7)$$

where $n_x = -1$ at the left end and $n_x = 1$ at the right end of the element.

Step 3. In writing the final weak form, we denote the secondary variables at the ends of the element as

$$Q_a^e = Q(x_a^e) = -\left(a\frac{du}{dx}\right)_{x_a^e}, \qquad Q_b^e = Q(x_b^e) = \left(a\frac{du}{dx}\right)_{x_b^e} \qquad (3.2.8)$$

The primary and secondary variables at the nodes are shown on the typical element in Fig. 3.2.1. It can be viewed as the *free-body diagram* of a typical but arbitrary portion of a bar, with Q_a^e and Q_b^e denoting the axial forces; Q_a^e is a compressive force while Q_b^e is a tensile force (algebraically, both are in the positive x direction, as shown in Fig. 3.2.1). For heat conduction problems, Q_a^e denotes the heat input at the left end and Q_b^e the heat output from the right end of the element. Although Q replaced $a(du/dx)$, it is *not* considered as function of u; it is a variable that is dual to u.

With the notation in Eq. (3.2.8), the final expression for the weak form is given by

$$0 = \int_{x_a^e}^{x_b^e} \left(a\frac{dw_i^e}{dx}\frac{du_h^e}{dx} + cw_i^e u_h^e - w_i^e f\right) dx - w_i^e(x_a^e)Q_a^e - w_i^e(x_b^e)Q_b^e \qquad (3.2.9)$$

This completes the three-step procedure of constructing the weak form.

Some remarks on the weak form derivation are in order.

1. The integration-by-parts used in Step 2 has two implications: (i) reduces the degree of finite element approximation used to represent u_h^e; and (ii) introduces secondary variables Q_a^e and Q_b^e that are physically meaningful in the sense that they can be specified at a point whenever the primary variable u_h^e is not specified there. If a secondary variable resulting from the integration by parts is not a physical quantity, then integration by parts should not be carried out even if it results in the weakening of the differentiability of u_h^e.

2. The weak form in Eq. (3.2.9) contains two types of expressions: those containing both w_i^e and u_h^e, and those containing only w_i^e. The expression containing both w_i^e and u_h^e is called a *bilinear functional* when it is linear in w_i^e as well as u_h^e (see Section 1.7.6):

$$B(w_i^e, u_h^e) \equiv \int_{x_a^e}^{x_b^e} \left(a\frac{dw_i^e}{dx}\frac{du_h^e}{dx} + cw_i^e u_h^e\right) dx \qquad (3.2.10)$$

When a and/or c is a function of u, $B(w_i^e, u_h^e)$ is (always) linear in w_i^e but nonlinear in u_h^e. The expression containing only w_i^e is called a *linear functional* because it is (always) linear in w_i^e

$$\ell(w_i^e) = \int_{x_a^e}^{x_b^e} w_i^e f \, dx + w_i^e(x_a)Q_a^e + w_i^e(x_b)Q_b^e \qquad (3.2.11)$$

3. In view of the definitions in Eqs. (3.2.10) and (3.2.11), the weak form in Eq. (3.2.9) can now be expressed as

$$B(w_i^e, u_h^e) = \ell(w_i^e)$$

which is called the *variational problem* associated with Eq. (3.2.1). In the functional analysis terminology, the variational problem can be stated as one of finding $u^e \in \mathcal{U}$ such that

$$B(w_i^e, u_h^e) = \ell(w_i^e) \tag{3.2.12}$$

holds for all $w_i^e \in \mathcal{U}$, where $\mathcal{U} \subset H^1(\Omega^e)$, the Hilbert space of order 1

$$H^1(\Omega^e) = \left\{ u : u, \frac{du}{dx} \in L_2(x_a, x_b), \ x_a < x < x_b \right\} \tag{3.2.13}$$

$$L_2(\Omega^e) = \left\{ u : \int_{x_a^e}^{x_b^e} |u| \, dx < \infty \right\} \tag{3.2.14}$$

where $\Omega^e = (x_a^e, x_b^e)$. As will be seen later, the bilinear form $B(w_i^e, u_h^e)$ results in the left-hand side $\mathbf{K}^e \mathbf{u}^e$ and the linear form $\ell(w_i^e)$ leads to the right-hand side column vector $\mathbf{F}^e$ of the finite element equations, $\mathbf{K}^e \mathbf{u}^e = \mathbf{F}^e$.

4. Those who have a background in solid and structural mechanics will appreciate the fact that the weak form in Eq. (3.2.9) or the variational problem in Eq. (3.2.12) is nothing but the statement of the principle of the minimum total potential energy $\Pi(u_h^e)$ applied to a *bar element* (see Reddy [2, 3]): $\delta\Pi = B(\delta u_h^e, u_h^e) - \ell(\delta u_h^e) = 0$, where δ denotes the variational symbol and $\Pi(u_h^e)$ is the quadratic functional

$$\Pi(u_h^e) = \frac{1}{2} B(u_h^e, u_h^e) - \ell(u_h^e) \tag{3.2.15}$$

$$= \int_{x_a^e}^{x_b^e} \left[\frac{a}{2} \left(\frac{du_i^e}{dx} \right)^2 + \frac{c}{2} (u_h^e)^2 - u_h^e f \right] dx - u_h^e(x_a^e) Q_a^e - u_h^e(x_b^e) Q_b^e \tag{3.2.16}$$

Equation (3.2.15) holds only when $B(w_i^e, u_h^e)$ is bilinear and symmetric in u_h^e and $w_i^e = \delta u_h^e$,

$$B(w_i^e, u_h^e) = B(u_h^e, w_i^e) \tag{3.2.17}$$

and $\ell(w_i^e)$ is linear in w_i^e. The expression $\frac{1}{2} B(u_h^e, u_h^e)$ represents the elastic strain energy stored in the bar finite element and $\ell(u_h^e)$ represents the work done by the applied distributed force $f(x)$ and point loads Q_a^e and Q_b^e.

3.2.4 Approximation Functions

Recall that the weak form over an element is equivalent to the differential equation, and it contains the end conditions on the "forces" Q_a^e and Q_b^e. Therefore, the approximate solution $u_h^e(x)$ should be selected such that the differentiability (or continuity) conditions implied by the weak form are met and the end conditions on the primary variables $u(x_i) = u_i^e$ are satisfied. Since the weak form contains the first-order derivative of u_h^e, any function with a non-zero first derivative would be a candidate for u_h^e, i.e. $u_h^e \in H^1(x_a, x_b)$. Thus, the finite element approximation u_h^e of $u(x)$ can be an interpolation, that is, must be equal to u_a^e at x_a and u_b^e at x_b. Thus, a linear polynomial (see Fig. 3.2.2)

$$u_h^e(x) = c_1^e + c_2^e x \tag{3.2.18}$$

is admissible if one can select c_1^e and c_2^e such that

$$u_h^e(x_a^e) = c_1^e + c_2^e x_a^e = u_a^e, \qquad u_h^e(x_b^e) = c_1^e + c_2^e x_b^e = u_b^e$$

or

$$\begin{bmatrix} 1 & x_a^e \\ 1 & x_b^e \end{bmatrix} \begin{Bmatrix} c_1^e \\ c_2^e \end{Bmatrix} = \begin{Bmatrix} u_a^e \\ u_b^e \end{Bmatrix} \rightarrow c_1^e = \frac{u_a^e x_b^e - u_b^e x_a^e}{x_b^e - x_a^e}, \quad c_2^e = \frac{u_b^e - u_a^e}{x_b^e - x_a^e} \tag{3.2.19}$$

Substitution of Eq. (3.2.19) for c_i^e into Eq. (3.2.18) yields

$$u_h^e(x) = \psi_1^e(x)\, u_1^e + \psi_2^e(x)\, u_2^e = \sum_{j=1}^{2} \psi_j^e(x)\, u_j^e \tag{3.2.20}$$

where (noting that $h_e = x_b^e - x_a^e$)

$$\psi_1^e(x) = \frac{x_b^e - x}{h_e}, \qquad \psi_2^e(x) = \frac{x - x_a^e}{h_e} \tag{3.2.21}$$

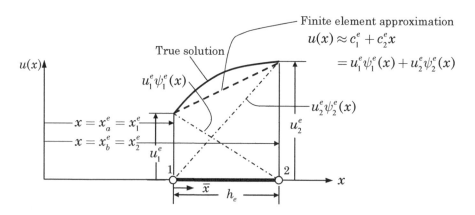

Fig. 3.2.2: Linear approximation over a finite element.

are the *linear Lagrange interpolation functions*, and

$$u_1^e = u_a^e, \qquad u_2^e = u_b^e \tag{3.2.22}$$

are the nodal values of $u_h^e(x)$ at $x = x_a^e$ and $x = x_b^e$, respectively. An element with linear approximation is called a *linear element*. In this case, $\mathcal{U}_h \subset H^1(\Omega^e)$ is the two-dimensional[1] subspace spanned by the set $\{\psi_1^e \; \psi_2^e\}$. Note that $\psi_i^e(x)$ satisfies the *interpolation property*

$$\psi_i^e(x_j^e) = \delta_{ij} = \begin{cases} 1, & \text{if } i = j \\ 0, & \text{if } i \neq j \end{cases} \tag{3.2.23}$$

where $x_1^e = x_a^e$ and $x_2^e = x_b^e$ (see Fig. 3.2.2). In addition, the Lagrange interpolation functions satisfy the property known as the *partition of unity*:

$$\sum_{j=1}^{n} \psi_j^e(x) = 1 \tag{3.2.24}$$

If we wish to approximate $u(x)$ with a quadratic polynomial, we write

$$u_h^e(x) = c_1^e + c_2^e x + c_3^e x^2 \tag{3.2.25}$$

Since there are three parameters c_1^e, c_2^e, and c_3^e, we must identify one more nodal point, in addition to the two end points in the element, to express all three c's in terms of the values of u_h^e at the three nodes. On the other hand, one may use the nodal values u_a^e and u_b^e (so that they can be used to join adjacent elements) and parameter c_3^e as the *nodeless* unknown of the approximation. Here we discuss the conventional element where all three c_i^e are expressed in terms of the three nodal values of $u^e(x)$. Identifying the third node at the center of the element, as indicated in Fig. 3.2.3(a), with nodal locations

$$x_1^e = x_a^e, \qquad x_2^e = x_a^e + \frac{h_e}{2}, \qquad x_3^e = x_a^e + h_e = x_b^e \tag{3.2.26}$$

we obtain

$$u_h^e(x) = \psi_1^e(x)u_1^e + \psi_2^e(x)u_2^e + \psi_3^e(x)u_3^e = \sum_{j=1}^{3} \psi_j^e(x)u_j^e \tag{3.2.27}$$

where $\psi_i^e(x)$ are the quadratic Lagrange interpolation functions [see Fig. 3.2.3(b)], which can be expressed in terms of the element coordinate $\bar{x} = x - x_a^e$ as

$$\psi_1^e(\bar{x}) = \left(1 - \frac{\bar{x}}{h_e}\right)\left(1 - \frac{2\bar{x}}{h_e}\right), \quad \psi_2^e(\bar{x}) = 4\frac{\bar{x}}{h_e}\left(1 - \frac{\bar{x}}{h_e}\right), \quad \psi_3^e(\bar{x}) = -\frac{\bar{x}}{h_e}\left(1 - \frac{2\bar{x}}{h_e}\right)$$
$$\tag{3.2.28}$$

---

[1]The dimension of the function space $\mathcal{U}_h$ should not be confused with the dimension of the geometric domain $\Omega^e = (x_a, x_b)$, which is 1. Equation (3.2.20) can be viewed as a representation of a typical function $u_h^e \in \mathcal{U}_h$ with respect to its basis $\{1 \; x\}$ or $\{\psi_1^e \; \psi_2^e\}$.

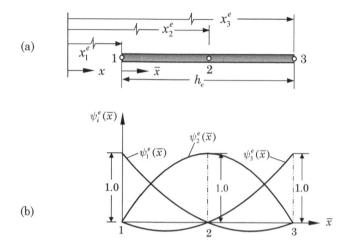

Fig. 3.2.3: Quadratic (a) finite element and (b) Lagrange interpolation functions.

An element with quadratic approximation of the field variable $u_h^e(x)$ is called a *quadratic element*.

Higher-order Lagrange interpolations of $u(x)$ can be developed along similar lines. Thus, an $(n-1)$st degree Lagrange interpolation of $u(x)$ can be written as

$$u_h^e(x) = \psi_1^e(x)u_1^e + \psi_2^e(x)u_2^e + \ldots + \psi_n^e(x)u_n^e = \sum_{j=1}^{n} \psi_j^e(x)u_j^e \qquad (3.2.29)$$

where the Lagrange interpolation functions of degree $n-1$ are given by

$$\psi_j^e(x) = \prod_{i=1,i\neq j}^{n} \left(\frac{x - x_i^e}{x_j^e - x_i^e} \right) \qquad (3.2.30)$$

The element contains n nodes. In general, the nodes are not equally-spaced.

The finite element solution $u_h^e(x)$ must fulfill certain requirements in order that it be convergent to the actual solution $u(x)$ as the number of elements (h refinement) or the degree of the polynomials (p refinement) is increased. These requirements are:

1. The approximate solution should be continuous and differentiable as required by the weak form.

2. It should be a complete polynomial, that is, the polynomial includes all lower-order terms, including the constant term, up to the highest order term used.

3. It should be an interpolant of the primary variables at the nodes of the finite element; it should at least interpolate the solution at the end points.

The reason for the first requirement is obvious; it ensures that every term of the governing equation has a non-zero contribution to the coefficient matrix. The second requirement is necessary in order to capture all possible states – say, constant, linear and so on – of the actual solution. For example, if a linear polynomial without the constant term is used to represent the temperature distribution in a one-dimensional system, the approximate solution can never be able to represent a uniform state of temperature field in the element. The third requirement is necessary in order to enforce continuity of the primary variables across the elements, i.e. at the end points, where the element is connected to other elements.

3.2.5 Finite Element Model

Substitution of Eq. (3.2.29) into Eq. (3.2.9) gives the necessary algebraic equations among the nodal values u_i^e and Q_i^e of the element, which will be termed the *finite element model*. In order to formulate the finite element model (i.e. derive a set of algebraic relations among u_i^e and Q_i^e) based on the weak form in Eq. (3.2.9), it is not necessary to decide the degree of approximation of $u_h^e(x)$ *a priori*. The model can be developed using an arbitrary degree of interpolation. For $n > 2$, the weak form in Eq. (3.2.9) must be modified to include non-zero secondary variables, if any, at interior nodes:

$$0 = \int_{x_a}^{x_b} \left(a \frac{dw_i^e}{dx} \frac{du_h^e}{dx} + c w_i^e u_h^e \right) dx - \int_{x_a}^{x_b} w_i^e f \, dx - \sum_{j=1}^n w_i^e(x_j^e) Q_j^e \quad (3.2.31)$$

where x_i^e is the global coordinate of the ith node of element $\Omega^e = (x_a, x_b)$. If nodes 1 and n denote the end points of the element, then $Q_1^e = Q_a^e$ and $Q_n^e = Q_b^e$ represent the *unknown* point sources, and all other Q_i^e ($i = 2, 3, \ldots, n-1$) are the externally applied (hence, known) point sources at nodes $2, 3, \ldots, n-1$, respectively.

Substituting Eq. (3.2.29) for u_h^e and $w_1^e = \psi_1^e$, $w_2^e = \psi_2^e, \ldots, w_i^e = \psi_i^e, \ldots,$ $w_n^e = \psi_n^e$ into the weak form Eq. (3.2.31), we obtain n algebraic equations. This choice of replacing w_i^e with ψ_i^e amounts to using the Galerkin method; however, the original Galerkin method was based on the weighted-integral statement of the residual – not on the weak form. In fact, it amounts to using the Ritz method [1, 2]. The element equations are numbered according to the choice of the weight function w_i^e. In other words, the ith algebraic equation is the one that is obtained from Eq. (3.2.31) by replacing w_i^e with ψ_i^e,

$$0 = \int_{x_a}^{x_b} \left[a \frac{d\psi_i^e}{dx} \left(\sum_{j=1}^n u_j^e \frac{d\psi_j^e}{dx} \right) + c\psi_i^e \left(\sum_{j=1}^n u_j^e \psi_j^e \right) - \psi_i^e f \right] dx - \sum_{j=1}^n \psi_i^e(x_j^e) Q_j^e$$

$$= \sum_{j=1}^n \left[\int_{x_a}^{x_b} \left(a \frac{d\psi_i^e}{dx} \frac{d\psi_j^e}{dx} + c\psi_i^e \psi_j^e \right) dx \right] u_j^e - \int_{x_a}^{x_b} \psi_i^e f \, dx - Q_i^e$$

$$0 = \sum_{j=1}^{n} K_{ij}^e u_j^e - f_i^e - Q_i^e \tag{3.2.32}$$

for $i = 1, 2, \ldots, n$, where

$$K_{ij}^e = \int_{x_a}^{x_b} \left(a \frac{d\psi_i^e}{dx} \frac{d\psi_j^e}{dx} + c\psi_i^e \psi_j^e \right) dx = B(\psi_i^e, \psi_j^e) \tag{3.2.33}$$

$$f_i^e = \int_{x_a}^{x_b} f \psi_i^e \, dx = \ell(\psi_i^e) \tag{3.2.34}$$

Note that the interpolation property in Eq. (3.2.23) is used to write (the summation convention is not invoked here)

$$\sum_{j=1}^{n} \psi_i^e(x_j^e) Q_j^e = \sum_{j=1}^{n} \delta_{ij} Q_j^e = Q_i^e \tag{3.2.35}$$

where δ_{ij} is the Kronecker delta symbol defined in Eq. (3.2.23). In matrix notation, these algebraic equations can be written as

$$\mathbf{K}^e \mathbf{u}^e = \mathbf{f}^e + \mathbf{Q}^e \equiv \mathbf{F}^e \tag{3.2.36}$$

The matrix $\mathbf{K}^e$, which is symmetric in the present case, is called the *coefficient matrix* and the column vector $\mathbf{f}^e$ is the *source vector*, and it is a nodal representation of the distributed source $f(x)$. In structural mechanics applications, $\mathbf{K}^e$ is known as the *stiffness matrix* and $\mathbf{f}^e$ as the *force vector*. The algebraic system of equations in Eq. (3.2.36) is called the weak-form Galerkin or Ritz finite element model of Eq. (3.2.1), because the equations are obtained using the weak-form Galerkin/Ritz method.

Equation (3.2.36) consists of n equations among $n + 2$ unknowns, namely, n primary nodal values $(u_1^e, u_2^e, \ldots, u_n^e)$ and 2 secondary nodal values (Q_1^e, Q_n^e); $(u_1^e, u_2^e, \ldots, u_n^e)$ are also called the element *primary nodal degrees of freedom*. Due to the fact that there are more unknowns than the number of equations, the element equations cannot be solved without assembling the elements. The assembly of elements (i.e. putting the elements together) is carried out by imposing the *continuity of the primary variable* and *equilibrium of secondary variables* at nodes common to different elements. Upon assembly and imposition of boundary conditions, we shall obtain exactly the same number of algebraic equations as the total number of unknown primary and secondary nodal degrees of freedom.

The coefficient matrix $\mathbf{K}^e$, and source vector $\mathbf{f}^e$ can be evaluated for a given interpolation and data $(a, c, \text{and } f)$. When a, c, and f are functions of x, it may be necessary to evaluate $\mathbf{K}^e$ and $\mathbf{f}^e$ using numerical integration. We will discuss the numerical integration concepts in the sequel. Here we give the exact element equations for linear and quadratic elements for the case in which the data $(a, c, \text{ and } f)$ is element-wise constant. Suppose that a_e, c_e, and f_e denote the element-wise constant values of $a(x)$, $c(x)$, and $f(x)$, respectively. Then the element (matrix) equations are obtained by analytically evaluating the integrals appearing in K_{ij}^e and f_i^e. The results are summarized here for elements with linear and quadratic approximations.

Linear element (i.e. element with linear approximation)

$$\psi_1^e(\bar{x}) = 1 - \frac{\bar{x}}{h_e}, \qquad \psi_2^e(\bar{x}) = \frac{\bar{x}}{h_e} \tag{3.2.37}$$

$$\left(\frac{a_e}{h_e} \begin{bmatrix} 1 & -1 \\ -1 & 1 \end{bmatrix} + \frac{c_e h_e}{6} \begin{bmatrix} 2 & 1 \\ 1 & 2 \end{bmatrix} \right) \begin{Bmatrix} u_1^e \\ u_2^e \end{Bmatrix} = \frac{f_e h_e}{2} \begin{Bmatrix} 1 \\ 1 \end{Bmatrix} + \begin{Bmatrix} Q_1^e \\ Q_2^e \end{Bmatrix} \tag{3.2.38}$$

Quadratic element (i.e. element with quadratic approximation)

$$\psi_1^e(\bar{x}) = \left(1 - \frac{2\bar{x}}{h_e}\right)\left(1 - \frac{\bar{x}}{h_e}\right), \ \psi_2^e(\bar{x}) = \frac{4\bar{x}}{h_e}\left(1 - \frac{\bar{x}}{h_e}\right), \ \psi_3^e(\bar{x}) = -\frac{\bar{x}}{h_e}\left(1 - \frac{2\bar{x}}{h_e}\right) \tag{3.2.39}$$

$$\left(\frac{a_e}{3h_e} \begin{bmatrix} 7 & -8 & 1 \\ -8 & 16 & -8 \\ 1 & -8 & 7 \end{bmatrix} + \frac{c_e h_e}{30} \begin{bmatrix} 4 & 2 & -1 \\ 2 & 16 & 2 \\ -1 & 2 & 4 \end{bmatrix} \right) \begin{Bmatrix} u_1^e \\ u_2^e \\ u_3^e \end{Bmatrix} = \frac{f_e h_e}{6} \begin{Bmatrix} 1 \\ 4 \\ 1 \end{Bmatrix} + \begin{Bmatrix} Q_1^e \\ Q_2^e \\ Q_3^e \end{Bmatrix} \tag{3.2.40}$$

We note that the contribution of a uniform source to the nodes in a quadratic element is non-uniform, that is, $f_i^e \neq f_e h_e/3$; also, the source vector of a quadratic element of length h_e is *not* equivalent to that of two linear elements of length $h_e/2$ each; it is the source vector that is consistent with the coefficient matrix of the quadratic element. We also note that there are more unknowns than the number of equations, independent of the degree of interpolation. Therefore, the element equations cannot be solved without properly assembling finite element equations associated with all elements in the mesh. When only one element is used, there will be only n unknowns because two known values are provided by the boundary conditions. Next we consider an example to illustrate the ideas presented here.

Example 3.2.1

Consider a homogeneous, isotropic bar of length L (m), cross-sectional area A (m^2), and conductivity k [W/(m·°C)], as shown in Fig. 3.2.4(a). The bar is maintained at a temperature of T_0 (°C) at the left end, insulated throughout the length so that there is no heat loss from the surface, and the right end is exposed to air at an ambient temperature of T_{inf}(°C) with a uniform internal heat generation of g_0 (W/m). The heat transfer coefficient associated with fin material and the air is β [W/(m^2 · °C)]. Use two different cases of uniform mesh (i.e. elements of equal size), (a) four linear elements and (b) two quadratic elements, in the domain to determine the temperature distribution along the length of the bar and heat at the left end.

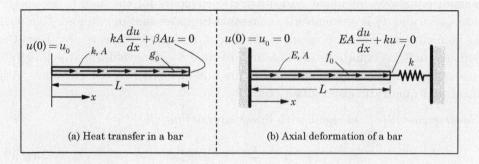

Fig. 3.2.4: (a) Heat transfer in a bar. (b) Axial deformation of a bar.

Solution: The governing differential equation and boundary conditions of the problem are

$$-\frac{d}{dx}\left(kA\frac{dT}{dx}\right) = g_0, \quad 0 < x < L \tag{1}$$

$$T(0) = T_0, \quad \left[kA\frac{dT}{dx} + \beta A\left(T - T_\infty\right)\right]_{x=L} = 0 \tag{2}$$

where the origin of the coordinate system is taken at the left end of the rod.

Let $u = T - T_\infty$. Then Eqs. (1) and (2) take the form

$$-\frac{d}{dx}\left(a\frac{du}{dx}\right) = g_0, \ 0 < x < L; \quad u(0) = u_0, \quad \left[a\frac{du}{dx} + c_s u\right]_{x=L} = 0 \tag{3}$$

where

$$a = kA, \quad u_0 = T_0 - T_\infty, \quad c_s = \beta A \tag{4}$$

The differential equation and boundary conditions in Eq. (3) are also valid for the axial deformation of a bar fixed at the left end, connected to a linear elastic spring at the right end, and subjected to a uniformly distributed axial body force f_0 [see Fig. 3.2.4(b)]. One only need to note the following correspondence between the two problems:

$$a = EA, \quad u_0 = 0, \quad c_s = \text{spring constant}, \quad g_0 = f_0 \tag{5}$$

with u being the axial displacement. In other words, the temperature distribution obtained for heat transfer problem can be interpreted as the axial displacement of a bar with $k = E$, $u_0 = T(0) - T_\infty = 0$, $g_0 = f_0$, and the spring constant equal to $c_s = \beta A$. Thus, physically two different problems share the same mathematical structure, facilitating the analysis of one problem but yielding solutions for both problems. Here we analyze Eq. (3) using the heat transfer problem.

(a) Mesh of Linear Elements

Element equations. The finite element equations in Eq. (3.2.38) for a linear element, with $a_e = a = $ constant, $c_e = 0$, and $f_e = g_0$, are valid for the problem at hand. For this case, all elements have the same coefficient matrix. The element equations are

$$\frac{a}{h}\begin{bmatrix} 1 & -1 \\ -1 & 1 \end{bmatrix}\begin{Bmatrix} u_1^e \\ u_2^e \end{Bmatrix} = \frac{g_0 h}{2}\begin{Bmatrix} 1 \\ 1 \end{Bmatrix} + \begin{Bmatrix} Q_1^e \\ Q_2^e \end{Bmatrix} \tag{6}$$

Assembled equations. Since there are no external heat inputs at the global nodes common to the elements, we require that

$$Q_2^1 + Q_1^2 = 0, \quad Q_2^2 + Q_1^3 = 0, \quad Q_2^3 + Q_1^4 = 0$$

Therefore, the assembled system of equations is given by

$$\frac{a}{h}\begin{bmatrix} 1 & -1 & 0 & 0 & 0 \\ -1 & 2 & -1 & 0 & 0 \\ 0 & -1 & 2 & -1 & 0 \\ 0 & 0 & -1 & 2 & -1 \\ 0 & 0 & 0 & -1 & 1 \end{bmatrix}\begin{Bmatrix} U_1 \\ U_2 \\ U_3 \\ U_4 \\ U_5 \end{Bmatrix} = \frac{g_o h}{2}\begin{Bmatrix} 1 \\ 2 \\ 2 \\ 2 \\ 1 \end{Bmatrix} + \begin{Bmatrix} Q_1^1 \\ Q_2^1 + Q_1^2 = 0 \\ Q_2^2 + Q_1^3 = 0 \\ Q_2^3 + Q_1^4 = 0 \\ Q_2^4 \end{Bmatrix} \tag{7}$$

where U_I denotes the temperature at the Ith global node, $I = 1, 2, \ldots, 5$. The assembled equations presented in Eq. (7) for a mesh of four linear elements that are connected in series are valid for any set of boundary conditions.

Imposition of boundary conditions. The boundary conditions in Eq. (3) can be expressed in terms of the global nodal variables as follows:

$$u(0) = u_0 \rightarrow U_1 = u_0; \quad \left[a\frac{du}{dx}\right]_{x=L} + c_s u(L) = 0 \rightarrow Q_2^4 + c_s U_5 = 0 \tag{8}$$

Then the assembled system of equations become

$$\frac{a}{h}\begin{bmatrix} 1 & -1 & 0 & 0 & 0 \\ -1 & 2 & -1 & 0 & 0 \\ 0 & -1 & 2 & -1 & 0 \\ 0 & 0 & -1 & 2 & -1 \\ 0 & 0 & 0 & -1 & 1 \end{bmatrix}\begin{Bmatrix} u_0 \\ U_2 \\ U_3 \\ U_4 \\ U_5 \end{Bmatrix} = \frac{g_o h}{2}\begin{Bmatrix} 1 \\ 2 \\ 2 \\ 2 \\ 1 \end{Bmatrix} + \begin{Bmatrix} Q_1^1 \\ 0 \\ 0 \\ 0 \\ -c_s U_5 \end{Bmatrix} \tag{9}$$

Condensed equations. The assembled equations can be grouped into two kinds, one in terms of the unknown primary global nodal variables and the other in terms of the unknown secondary variables. They are called condensed equations. Since only one element of each pair of primary and secondary variables, (U_I, F_I), at a node are known, we will always have only N unknowns, where N is the total number of global nodes. In the present case, the unknown primary global variables are (U_2, U_3, U_4, U_5); hence, the condensed equations for unknown primary global variables are

$$\frac{a}{h}\begin{bmatrix} 2 & -1 & 0 & 0 \\ -1 & 2 & -1 & 0 \\ 0 & -1 & 2 & -1 \\ 0 & 0 & -1 & 1+\alpha \end{bmatrix}\begin{Bmatrix} U_2 \\ U_3 \\ U_4 \\ U_5 \end{Bmatrix} = \frac{g_o h}{2}\begin{Bmatrix} 2 \\ 2 \\ 2 \\ 1 \end{Bmatrix} + \begin{Bmatrix} \frac{a}{h}u_0 \\ 0 \\ 0 \\ 0 \end{Bmatrix} \tag{10}$$

where $\alpha = c_s h/a$. The algebraic equations in Eq. (10) can be readily solved, since the coefficient matrix is invertible, for U_2, U_3, U_4, and U_5. The unknown heat at nodes 1 is given by

$$Q_1^1 = \frac{a}{h}(u_0 - U_2) - \frac{g_o h}{2} \tag{11}$$

(b) Mesh of Quadratic Elements

Element equations. For the uniform mesh of two quadratic elements (i.e. $h_1 = h_2 = h = L/2$), the element equations are

$$\frac{a}{3h}\begin{bmatrix} 7 & -8 & 1 \\ -8 & 16 & -8 \\ 1 & -8 & 7 \end{bmatrix}\begin{Bmatrix} u_1^e \\ u_2^e \\ u_3^e \end{Bmatrix} = \frac{g_o h}{6}\begin{Bmatrix} 1 \\ 4 \\ 1 \end{Bmatrix} + \begin{Bmatrix} Q_1^e \\ Q_2^e \\ Q_3^e \end{Bmatrix} \tag{12}$$

Assembled equations. The assembled equations are

$$\frac{a}{3h}\begin{bmatrix} 7 & -8 & 1 & 0 & 0 \\ -8 & 16 & -8 & 0 & 0 \\ 1 & -8 & 7+7 & -8 & 1 \\ 0 & 0 & -8 & 16 & -8 \\ 0 & 0 & 1 & -8 & 7 \end{bmatrix}\begin{Bmatrix} U_1 \\ U_2 \\ U_3 \\ U_4 \\ U_5 \end{Bmatrix} = \frac{g_0 h}{6}\begin{Bmatrix} 1 \\ 4 \\ 2 \\ 4 \\ 1 \end{Bmatrix} + \begin{Bmatrix} Q_1^1 \\ Q_2^1 \\ Q_3^1 + Q_1^2 \\ Q_2^2 \\ Q_3^2 \end{Bmatrix} \tag{13}$$

The balance of heats at the interface of the two elements (i.e. at global node 3) requires $Q_3^1 + Q_1^2 = 0$.

Imposition of boundary conditions. As in the case of linear element mesh, the boundary conditions given in Eq. (8) are also valid for the mesh of quadratic elements (because both meshes have the same number of total global nodes). Since there are no heat input at nodes 2, 3, and 4, we require $Q_2^1 = 0$, $Q_3^1 + Q_1^2 = 0$, and $Q_2^2 = -c_s U_5$. The assembled equations become

$$\frac{a}{3h}\begin{bmatrix} 7 & -8 & 1 & 0 & 0 \\ -8 & 16 & -8 & 0 & 0 \\ 1 & -8 & 7+7 & -8 & 1 \\ 0 & 0 & -8 & 16 & -8 \\ 0 & 0 & 1 & -8 & 7+\mu \end{bmatrix}\begin{Bmatrix} u_0 \\ U_2 \\ U_3 \\ U_4 \\ U_5 \end{Bmatrix} = \frac{g_0 h}{6}\begin{Bmatrix} 1 \\ 4 \\ 2 \\ 4 \\ 1 \end{Bmatrix} + \begin{Bmatrix} Q_1^1 \\ 0 \\ 0 \\ 0 \\ 0 \end{Bmatrix} \tag{14}$$

where $\mu = 3c_s h/a$.

Condensed equations. The set of condensed equations for the unknown primary variables is

$$\frac{a}{3h}\begin{bmatrix} 16 & -8 & 0 & 0 \\ -8 & 14 & -8 & 1 \\ 0 & -8 & 16 & -8 \\ 0 & 1 & -8 & 7+\mu \end{bmatrix}\begin{Bmatrix} U_2 \\ U_3 \\ U_4 \\ U_5 \end{Bmatrix} = \frac{g_0 h}{6}\begin{Bmatrix} 4 \\ 2 \\ 4 \\ 1 \end{Bmatrix} + \begin{Bmatrix} \frac{8a}{3h}u_0 \\ -\frac{a}{3h}u_0 \\ 0 \\ 0 \end{Bmatrix} \tag{15}$$

The heat at node 1 is give by

$$Q_1^1 = \frac{7a}{3h}u_0 - \frac{8a}{3h}U_2 + \frac{a}{3h}U_3 - \frac{g_0 h}{6} \tag{16}$$

(c) Numerical results

Problem data. We use the following data (material of the bar is assumed to be copper),

$$k = 385 \text{ W/(m}\cdot^\circ\text{ C)}, \quad \beta = 25 \text{ W/(m}^2\cdot^\circ\text{ C)}, \quad T_0 = T_\infty = 20^\circ\text{C}$$
$$L = 0.1 \text{ m}, \quad A = 5 \times 10^{-6}\text{m}^2, \quad g_0 = 10.0 \text{ W/m} \tag{17}$$

Then various parameters used in the problem have the values

$$\frac{a}{h} = \frac{kA}{h} = \frac{385 \times 5 \times 10^{-6}}{0.025} = 0.077, \quad c_s = \beta A = 0.125 \times 10^{-3}, \quad u_0 = 0$$

Numerical results for the linear element mesh. The condensed equations for the unknown nodal temperatures become

$$10^{-2}\begin{bmatrix} 15.4 & -7.7 & 0 & 0 \\ -7.7 & 15.4 & -7.7 & 0 \\ 0 & -7.7 & 15.4 & -7.7 \\ 0 & 0 & -7.7 & 7.7 \end{bmatrix}\begin{Bmatrix} U_2 \\ U_3 \\ U_4 \\ U_5 \end{Bmatrix} = \begin{Bmatrix} 0.25 \\ 0.25 \\ 0.25 \\ 0.125 \end{Bmatrix} \tag{18}$$

The solution of these equations (obtained with the help of an equation solver in a computer) is (note that $U_1 = 0.0°$ C)

$$U_2 = 11.322° \text{ C}, \quad U_3 = 19.397° \text{ C}, \quad U_4 = 24.225° \text{ C}, \quad U_5 = 25.806° \text{ C} \tag{19}$$

The heat input at node 1 from Eq. (11) is

$$Q_1^1 = \frac{a}{h}(U_1 - U_2) - \frac{g_0 h}{2} = 0.077(0 - 11.322) - 0.125 = -0.9968 \text{ W} \tag{20}$$

On the other hand, from the definition

$$(Q_1^1)_{def} = -a\left(\frac{du}{dx}\right)_{x=0} \approx -a\left(\frac{du_h}{dx}\right)_{x=0} = -a\left(U_1\frac{d\psi_1}{dx} + U_2\frac{d\psi_2}{dx}\right)_{x=0}$$
$$= -\frac{a}{h}(U_2 - U_1) = -0.872 \text{ W} \tag{21}$$

where ψ_i are the linear functions defined in Eq. (3.2.21). The value of heat computed from the definition differs from the heat computed from Eq. (11), and the difference is 0.125. The difference is exactly equal to $g_0 h/2$. As the mesh is refined in the neighborhood of $x = 0$, the post-computed heat loss from the definition converges to that computed from Eq. (11), which is exact.

Numerical results for the quadratic element mesh. For the mesh of two quadratic elements, the system of the condensed equations for the unknown temperatures is

$$10^{-2}\begin{bmatrix} 20.533 & -10.267 & 0 & 0 \\ -10.267 & 17.967 & -10.267 & 12.883 \\ 0 & -10.267 & 20.533 & -10.267 \\ 0 & 12.883 & -10.267 & 8.996 \end{bmatrix}\begin{Bmatrix} U_2 \\ U_3 \\ U_4 \\ U_5 \end{Bmatrix} = 10^{-2}\begin{Bmatrix} 33.333 \\ 16.667 \\ 33.333 \\ 8.333 \end{Bmatrix} \tag{22}$$

whose solution is

$$U_2 = 11.322° \text{ C}, \quad U_3 = 19.397° \text{ C}, \quad U_4 = 24.225° \text{ C}, \quad U_5 = 25.806° \text{ C}$$

The solution is the same as that predicted by the mesh of four linear elements. The heat input at node 1 from Eq. (16) is

$$Q_1^1 = \frac{a}{3h}(7U_1 - 8U_2 + U_3) - \frac{g_0 h}{6} = -0.9968 \text{ W}$$

and from the definition we have

$$(Q_1^1)_{def} = -\frac{a}{h}\left(\frac{du}{dx}\right)_{x=0} \approx -\frac{a}{h}\left(\frac{du_h}{dx}\right)_{x=0} = -a\left(U_1\frac{d\psi_1}{dx} + U_2\frac{d\psi_2}{dx} + U_3\frac{d\psi_3}{dx}\right)_{x=0}$$
$$= -\frac{a}{h}(-3U_1 + 4U_2 - U_3) = -0.9968 \text{ W}$$

where ψ_i are the quadratic functions defined in Eq. (3.2.28). Thus, for the mesh of quadratic elements, the value of heat at node 1 obtained with Eq. (16) and the definition is the same.

The exact solution of the problem described by Eq. (3) is

$$u(x) = -\frac{g_0 x^2}{2a} + c_1 x + c_2, \quad 0 < x < L$$

$$c_1 = \left[g_0 L\left(1 + 0.5\frac{c_s L}{a}\right) - c_s u_0\right](a + c_s L)^{-1}, \quad c_2 = u_0$$

Evaluating the exact solution at the nodes, we obtain the same values as the finite element solution at the nodes. The exact value of heat at node 1 is -0.9968.

   To interpret the results for the axial deformation of a bar, the conductivity k should be replaced by modulus E (which is several orders of magnitude larger than k); since the displacement is inversely proportional to the modulus, the values of u would be approximately (not exactly because of the nonzero spring constant) of the order ku/E.

3.2.6   Natural Coordinates

From Section 3.2.4, it is clear that the approximation functions ψ_i^e in the finite element method are defined over an element and they do not extend to the neighboring elements. Therefore, they can be derived in terms of a coordinate, called a *local coordinate*, that is fixed in the element. The coordinate $\bar{x}$ used in Fig. 3.2.3 is an example of a local coordinate. This facilitates easier evaluation (especially, when numerical integration is used) of the element coefficients K_{ij}^e and f_i^e. In fact, in two or three dimensions, it is not practical to derive the approximation functions using the global coordinates. A *natural coordinate* is a local coordinate that is dimensionless. Here we consider two different natural coordinate systems in one dimension.

   The first natural coordinate is $\bar{x}/h$, with the origin at the left end of the element. A pair of functions that measure the distance of an arbitrary point P in the element relative to the two ends of the element are defined by [see Fig. 3.2.5(a)]:

$$L_1(\bar{x}) = 1 - \frac{\bar{x}}{h}, \quad L_2(\bar{x}) = 1 - L_1 = \frac{\bar{x}}{h} \qquad (3.2.41)$$

The element label is omitted for brevity. We note that L_1 and L_2 are the same functions as the linear approximation functions defined in Eq. (3.2.37)

$$\psi_1(\bar{x}) = L_1(\bar{x}), \quad \psi_2(\bar{x}) = L_2(\bar{x}) \qquad (3.2.42)$$

which satisfy the "Kronecker delta property" in Eq. (3.2.23) and "partition of unity" in Eq. (3.2.24). Similarly, the quadratic functions in Eq. (3.2.39) can be expressed in terms of L_1 and L_2 as

$$\psi_1(\bar{x}) = (2L_1 - 1)L_1, \quad \psi_2(\bar{x}) = 4L_1L_2, \quad \psi_3(\bar{x}) = (2L_2 - 1)L_2 \qquad (3.2.43)$$

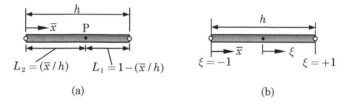

(a)   (b)

Fig. 3.2.5: (a) Natural coordinates L_1 and L_2. (b) Natural coordinate ξ.

In view of the relation $\bar{x} = x - x_a$ and $\bar{x} = \bar{x}(L_1, L_2)$, we have

$$\frac{d}{dx} = \frac{d}{d\bar{x}} = \frac{dL_1}{d\bar{x}}\frac{\partial}{\partial L_1} + \frac{dL_2}{d\bar{x}}\frac{\partial}{\partial L_2} = \frac{1}{h}\left(\frac{\partial}{\partial L_2} - \frac{\partial}{\partial L_1}\right) \qquad (3.2.44)$$

and

$$\frac{dL_i}{dx} = \frac{1}{h}\left(\frac{\partial L_i}{\partial L_2} - \frac{\partial L_i}{\partial L_1}\right) = \frac{1}{h}(\delta_{2i} - \delta_{1i}) \quad (i = 1, 2) \qquad (3.2.45)$$

The integral of the products of various powers of L_i can be evaluated using the formula

$$\int_0^h L_1^m L_2^n \, d\bar{x} = \frac{m!\,n!}{(m+n+1)!}h \qquad (3.2.46)$$

where $0! = 1$, $!$ is the factorial symbol.

The second natural coordinate, denoted as ξ, is the one that is used more commonly with Gauss quadrature (i.e. numerical integration; see Section 3.6) rule. The origin of ξ is taken to be at the center of the element, with $\xi = -1$ and $\xi = +1$ being the left and right ends, respectively, of the element [see Fig. 3.2.5(b)]. The relationships between x and ξ and between $\bar{x}$ and ξ are given by

$$x = x_a + \frac{h}{2}(1 + \xi), \quad \frac{\bar{x}}{h} = \frac{1}{2}(1 + \xi), \quad \xi = \frac{2}{h}(x - x_a) - 1 = \frac{2\bar{x}}{h} - 1 \qquad (3.2.47)$$

Then the linear and quadratic interpolation functions in Eqs. (3.2.37) and (3.2.39) can be expressed in terms of ξ as

Linear: $\qquad\qquad \psi_1 = \frac{1}{2}(1 - \xi), \quad \psi_2 = \frac{1}{2}(1 + \xi)$

$$\qquad\qquad\qquad\qquad\qquad\qquad\qquad\qquad\qquad\qquad\qquad (3.2.48)$$

Quadratic: $\quad \psi_1 = -\frac{1}{2}\xi(1 - \xi), \quad \psi_2 = (1 - \xi^2), \quad \psi_3 = \frac{1}{2}\xi(1 + \xi)$

The relationship between the derivative with respect to x and the derivative with respect to ξ is given by

$$\frac{d}{dx} = \frac{2}{h}\frac{d}{d\xi}, \quad \frac{d\psi_i}{dx} = \frac{2}{h}\frac{d\psi_i}{d\xi} \qquad (3.2.49)$$

Integrals over the element domain $\Omega^e = (x_a, x_b)$ can be expressed as the one on interval $(-1, 1)$ as

$$\int_{x_a}^{x_b} \psi_i(x) \, dx = \int_0^h \psi_i(\bar{x}) \, d\bar{x} = \frac{h}{2}\int_{-1}^1 \psi_i(\xi) \, d\xi$$

$$\int_{x_a}^{x_b} \frac{d\psi_i}{dx} \, dx = \int_0^h \frac{d\psi_i}{d\bar{x}} \, d\bar{x} = \int_{-1}^1 \frac{d\psi_i}{d\xi} \, d\xi \qquad (3.2.50)$$

$$\int_{x_a}^{x_b} \frac{d\psi_i}{dx}\frac{d\psi_j}{dx} \, dx = \frac{2}{h}\int_{-1}^1 \frac{d\psi_i}{d\xi}\frac{d\psi_j}{d\xi} \, d\xi$$

3.3 Two-Dimensional Problems

3.3.1 Governing Differential Equation

Consider the problem of finding $u(x, y)$ such that the following partial differential equation is satisfied

$$-\left[\frac{\partial}{\partial x}\left(a_{xx}\frac{\partial u}{\partial x}\right) + \frac{\partial}{\partial y}\left(a_{yy}\frac{\partial u}{\partial y}\right)\right] = f(x, y) \quad \text{in } \Omega \qquad (3.3.1)$$

where Ω is a two-dimensional domain with boundary Γ, as shown in Fig. 3.3.1(a). Here $a_{xx}(x, y)$ and $a_{yy}(x, y)$ are material coefficients in the x and y directions, respectively, and $f(x, y)$ is the known source function. For example, in a heat transfer problem, u denotes temperature T, $a_{xx} = k_{xx}$ and $a_{yy} = k_{yy}$ denote the conductivities of an orthotropic medium, and f is the internal heat generation. For an isotropic medium, we set $k_{xx} = k_{yy} = k$. Similarly, for a ground water flow problem u denotes the water head (i.e. velocity potential), a_{xx} and a_{yy} are the permeabilities in the x and y directions, respectively, and $f(x, y)$ is the distributed water source. Equation (3.3.1) also arises in other fields of science and engineering, and some of them are listed in Table 3.3.1.

Equation (3.3.1) must be solved in conjunction with specified boundary conditions of the problem. The following two types of boundary conditions are assumed:

$$u = \hat{u}(s) \quad \text{on} \quad \Gamma_u \qquad (3.3.2)$$

$$\left(a_{xx}\frac{\partial u}{\partial x}n_x + a_{yy}\frac{\partial u}{\partial y}n_y\right) + q_c = \hat{q}_n(s) \quad \text{on} \quad \Gamma_q \qquad (3.3.3)$$

where Γ_u and Γ_q are disjoint portions of the boundary Γ such that $\Gamma = \Gamma_u \cup \Gamma_q$, q_c refers to the convective component of flux (e.g. in heat transfer problems)

$$q_c = h_c(u - u_c) \qquad (3.3.4)$$

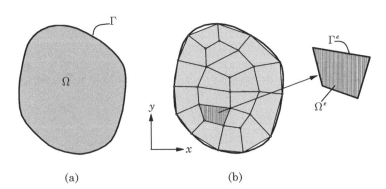

(a)   (b)

Fig. 3.3.1: (a) A two-dimensional domain Ω with its boundary Γ. (b) Finite element discretization of $\bar{\Omega} = \Omega \cup \Gamma$ and a typical finite element Ω^e with its boundary Γ^e.

Table 3.3.1: List of fields in which the model equation (3.3.1) arises, with meaning of various parameters and variables*.

Field of study	Primary variable u	Data a_{xx}	Data a_{yy}	Data f	Secondary variable q_n
Heat transfer	Temperature T	Thermal conductance k_{xx}	Thermal conductance k_{yy}	Heat generation f	Heat flux q_n
Flow through porous medium	Fluid head ϕ	Permeability μ_{xx}	Permeability μ_{yy}	Infiltration f	Flux q_n
Torsion of cylindrical members	Warping function ϕ	1	1	0	$\frac{\partial \phi}{\partial n}$
Torsion of cylindrical members	Stress function ψ	1	1	$2G\theta$	$\frac{\partial \psi}{\partial n}$
Deflection of membranes	Displacement u	Tension a_{11}	Tension a_{22}	Transverse force f	$\frac{\partial u}{\partial n}$
Flows of inviscid flows	Velocity potential ϕ	1	1	0	$\frac{\partial \phi}{\partial n}$
Flows of inviscid flows	Stream function ψ	1	1	0	$\frac{\partial \psi}{\partial n}$
Electro-statics	Electrical potential ϕ	Dielectric constant ε	Dielectric constant ε	Charge density ρ	Electric flux $\frac{\partial \psi}{\partial n}$

* k = thermal conductance; G = shear modulus; θ = angle of twist.

(n_x, n_y) are the direction cosines of the unit normal $\hat{\mathbf{n}}$ on the boundary Γ

$$n_x = \cos(x, \hat{\mathbf{n}}), \text{ cosine of the angle between +ve } x\text{-axis and } \hat{\mathbf{n}}$$
$$n_y = \cos(y, \hat{\mathbf{n}}), \text{ cosine of the angle between +ve } y\text{-axis and } \hat{\mathbf{n}}$$

(3.3.5)

and s denotes the coordinate along the boundary. In Eq. (3.3.4), h_c denotes the convective heat transfer coefficient and u_c is the ambient temperature. Similar interpretation of the boundary condition in Eq. (3.3.3) is possible in other fields.

3.3.2   Finite Element Approximation

In the finite element method, the domain $\bar{\Omega} = \Omega \cup \Gamma$ is divided into a set of sub-domains $\bar{\Omega}^e = \Omega^e \cup \Gamma^e$, called finite elements [see Fig. 3.3.1(b)]. Any geometric shape qualifies as an element provided that the approximation functions ψ_i^e can be derived uniquely for the shape. We shall discuss simple geometric shapes and orders of approximation shortly. To keep the formulation steps very general, that is, not confine the formulation to a specific geometric shape, we have denoted the domain of a typical element by Ω^e and its boundary by Γ^e. The element $\bar{\Omega}^e$ can be a triangle or quadrilateral in shape, and the degree of interpolation over it can be linear, quadratic, and higher. The non-overlapping sum of all elements $\bar{\Omega}^e$ is denoted by Ω^h, and it is called the *finite element mesh* of the domain Ω. In general, Ω^h may not equal $\bar{\Omega}$ when the boundary Γ is curved. Of course, for polygonal domains, the finite element mesh exactly represents the actual domain.

Suppose that the dependent unknown u is approximated over a typical finite element $\bar{\Omega}^e$ by $u_h^e(\mathbf{x}) \in \mathcal{U}_h$

$$u(\mathbf{x}) \approx u_h^e(\mathbf{x}) = \sum_{j=1}^{n} u_j^e \psi_j^e(\mathbf{x}), \quad \mathbf{x} = (x, y) \in \bar{\Omega}^e \qquad (3.3.6)$$

where u_j^e denote the values of the function $u_h^e(\mathbf{x})$ at a selected number of points (i.e. element nodes) in the element $\bar{\Omega}^e$, and ψ_j^e are the Lagrange interpolation functions associated with the element. Here the symbol '$\in$' means "an element of" or "a point in". As we shall see shortly, the interpolation functions depend not only on the number of nodes in the element, but also on the shape of the element. The shape of the element must be such that its geometry is uniquely defined by a set of nodes and the approximated function varies among nodes according to the adopted approximation. A triangle is the simplest two-dimensional geometric shape in two dimensions because it is uniquely defined by three points ($n = 3$) in a plane and the approximated function varies between any two points according to $u_h^e(\mathbf{x}) = c_1^e + c_2^e x + c_3^e y$. A triangle with three nodes per side (a total of six nodes in the element) will uniquely define the geometry while representing the quadratic variation $u_h^e(\mathbf{x}) = c_1^e + c_2^e x + c_3^e y + c_4^e xy + c_5^e x^2 + c_6^e y^2$ uniquely along any of the three segments. On the other hand, a pentagon requires five points to define its geometry uniquely, but there is no five-parameter polynomial in (x, y) that uniquely defines the linear variation along any segment of the pentagon.

3.3.3   Weak Formulation

The n nodal values u_j^e in Eq. (3.3.6) must be determined such that the approximate solution $u_h^e(\mathbf{x})$ satisfies Eq. (3.3.1) and interpolates the nodal values of $u_h^e(\mathbf{x})$ over the element Ω^e. As described in Sections 3.2.2 and 3.2.3, we seek to satisfy the governing differential equation in a weak-form sense, with the weight functions being the same as the approximation functions. The resulting finite element model i.e. set of algebraic equations) will be called the *weak-form Galerkin finite element model* or the *Ritz finite element model*.

We use the three-step procedure described in Section 3.2.3 to develop the weak form of Eq. (3.3.1) that accounts for the form of the natural boundary condition in Eq. (3.3.3) over the typical element Ω^e.

Step 1. The first step is to take all non-zero expressions in Eq. (3.3.1) to one side of the equality, multiply the resulting equation with a weight function w_i^e from a set of linearly independent functions $\{w_i^e\}_{i=1}^n$, and integrate the equation over the element domain Ω^e:

$$0 = \int_{\Omega^e} w_i^e \left[-\frac{\partial}{\partial x}\left(a_{xx}\frac{\partial u_h^e}{\partial x}\right) - \frac{\partial}{\partial y}\left(a_{yy}\frac{\partial u_h^e}{\partial y}\right) - f(x,y) \right] dA \qquad (3.3.7)$$

where $dA = dx\,dy$. The expression in the square brackets of the above equation represents the residual of the approximation of the differential equation (3.3.1) because $u_h^e(\mathbf{x})$ is only an approximation of $u(\mathbf{x})$. For n independent choices of w_i^e, we obtain a set of n linearly independent algebraic equations, called *weighted-residual finite element model*.

Step 2. In the second step, we distribute the differentiation among u_h^e and w_i^e equally so that both u_h^e and w_i^e are required to be differentiable only once with respect to x and y. To achieve this we use the component form of the gradient (or divergence) theorem [see Eq. (1.6.42)]

$$\int_{\Omega^e} \frac{\partial}{\partial x}\left(w_i^e F_1\right) dA = \oint_{\Gamma^e} (w_i^e F_1) n_x\, ds$$
$$\int_{\Omega^e} \frac{\partial}{\partial y}\left(w_i^e F_2\right) dA = \oint_{\Gamma^e} (w_i^e F_2) n_y\, ds \qquad (3.3.8)$$

where n_x and n_y are the components of the unit normal vector $\hat{\mathbf{n}}$

$$\hat{\mathbf{n}} = n_x\hat{\mathbf{e}}_x + n_y\hat{\mathbf{e}}_y = \cos\alpha\ \hat{\mathbf{e}}_x + \sin\alpha\ \hat{\mathbf{e}}_y \qquad (3.3.9)$$

on the boundary Γ^e, and ds is the arc length of an infinitesimal line element along the boundary. With

$$F_1 = a_{xx}\frac{\partial u_h^e}{\partial x}, \quad F_2 = a_{yy}\frac{\partial u_h^e}{\partial y}$$

and identities

$$-w_i^e\frac{\partial F_1}{\partial x} = -\frac{\partial}{\partial x}\left(w_i^e F_1\right) + F_1\frac{\partial w_i^e}{\partial x}, \quad -w_i^e\frac{\partial F_2}{\partial y} = -\frac{\partial}{\partial y}\left(w_i^e F_2\right) + F_2\frac{\partial w_i^e}{\partial y}$$

we obtain

$$0 = \int_{\Omega^e} \left(a_{xx}\frac{\partial w_i^e}{\partial x}\frac{\partial u_h^e}{\partial x} + a_{yy}\frac{\partial w_i^e}{\partial y}\frac{\partial u_h^e}{\partial y} - w_i^e f \right) dA$$

$$-\oint_{\Gamma^e} w_i^e \left(a_{xx} \frac{\partial u_h^e}{\partial x} n_x + a_{yy} \frac{\partial u_h^e}{\partial y} n_y \right) ds \qquad (3.3.10)$$

From an inspection of the boundary term in Eq. (3.3.10), we note that u_h^e is the primary variable, and specification of u_h^e constitutes the essential boundary condition. The coefficient of the weight function in the boundary expression, namely

$$q_n = a_{xx} \frac{\partial u_h^e}{\partial x} n_x + a_{yy} \frac{\partial u_h^e}{\partial y} n_y \qquad (3.3.11)$$

is the secondary variable. Its specification constitutes the natural boundary condition. By definition q_n is positive outward from the surface as one travels counterclockwise along the boundary Γ^e. This is the reason why the element node numbers are counted in the counterclockwise direction and the evaluation of boundary integrals is carried out in the counterclockwise sense. The secondary variable q_n denotes the outward flux normal to the boundary of the element because the flux vector $\mathbf{q}$ is given by

$$\mathbf{q} = q_x \hat{\mathbf{e}}_x + q_y \hat{\mathbf{e}}_y, \quad q_x = a_{xx} \frac{\partial u_h^e}{\partial x}, \quad q_y = a_{yy} \frac{\partial u_h^e}{\partial y} \qquad (3.3.12)$$

and thus the outward flux normal to the boundary is defined by

$$q_n \equiv \hat{\mathbf{n}} \cdot \mathbf{q} = a_{xx} \frac{\partial u_h^e}{\partial x} n_x + a_{yy} \frac{\partial u_h^e}{\partial y} n_y \qquad (3.3.13)$$

Step 3. The third and last step of the formulation is to use the boundary condition from Eqs. (3.3.3) and (3.3.4) in Eq. (3.3.10) and write it as

$$0 = \int_{\Omega^e} \left(a_{xx} \frac{\partial w_i^e}{\partial x} \frac{\partial u_h^e}{\partial x} + a_{yy} \frac{\partial w_i^e}{\partial y} \frac{\partial u_h^e}{\partial y} - w_i^e f \right) dA - \oint_{\Gamma^e} w_i^e \left[\hat{q}_n - h_c(u_h^e - u_c) \right] ds$$

$$= \int_{\Omega^e} \left(a_{xx} \frac{\partial w_i^e}{\partial x} \frac{\partial u_h^e}{\partial x} + a_{yy} \frac{\partial w_i^e}{\partial y} \frac{\partial u_h^e}{\partial y} - w_i^e f \right) dA + \oint_{\Gamma^e} h_c w_i^e u_h^e \, ds$$

$$- \oint_{\Gamma^e} w_i^e (\hat{q}_n + h_c u_c) ds$$

or

$$B(w_i^e, u_h^e) = \ell(w_i^e) \qquad (3.3.14)$$

where $B(w_i^e, u_h^e)$ and $\ell(w_i^e)$ are defined by

$$B(w_i^e, u_h^e) = \int_{\Omega^e} \left(a_{xx} \frac{\partial w_i^e}{\partial x} \frac{\partial u_h^e}{\partial x} + a_{yy} \frac{\partial w_i^e}{\partial y} \frac{\partial u_h^e}{\partial y} \right) dA + \oint_{\Gamma^e} h_c w_i^e u_h^e \, ds$$

$$\ell(w_i^e) = \int_{\Omega^e} w_i^e f \, dA + \oint_{\Gamma^e} w_i^e (\hat{q}_n + h_c u_c) \, ds \qquad (3.3.15)$$

The following remarks on the weak form are in order:

1. The variational problem associated with Eqs. (3.3.1)–(3.3.3) over a finite element Ω^e is to find $u_h^e \in \mathcal{U}_h \subset H^1(\Omega^e)$ such that

$$B(w_i^e, u_h^e) = \ell(w_i^e) \tag{3.3.16}$$

holds for all $w_i^e \in \mathcal{U}_h$. In the finite element method, the subspace $\mathcal{U}_h \subset H^1(\Omega^e)$ is a finite-dimensional subspace spanned by set of polynomial basis functions, as will be discussed shortly.

2. Note that $B(w_i^e, u_h^e)$ is bilinear and symmetric in its arguments w_i^e and u_h^e, and $\ell(w_i^e)$ is linear in w_i^e. Therefore, it is possible to construct the associated quadratic functional from the formula (see Reddy [2, 3])

$$I^e(u_h^e) = \frac{1}{2} B(u_h^e, u_h^e) - \ell(u_h^e) \tag{3.3.17}$$

which is of considerable interest in the sense that the minimum of the functional $I(u_h^e)$ (often dictated by some principle of mechanics) is equivalent to solving the variational problem in Eq. (3.3.14). Although it is possible to construct the weak form of any second- and higher-order differential equations (linear or not), it is not always possible to construct a functional whose first variation is equivalent to the weak form [i.e. $\delta I^e = B(\delta u_h^e, u_h^e) - \ell(\delta u_h^e) = 0$]. However, when the variational problem does not correspond to the necessary condition that a functional be a minimum, Eq. (3.3.14) merely provides a means for computing the solution u_h^e, and it does not necessarily guaranty that the error in the approximation is a minimum.

3.3.4 Finite Element Model

The weak form in Eq. (3.3.14) requires that u_h^e be at least linear in both x and y so that there are no terms in Eq. (3.3.14) that become identically zero. Suppose that u_h^e is represented over a typical finite element Ω^e by expression of the form in Eq. (3.3.6). Substituting the finite element approximation, Eq. (3.3.6), into the weak form in Eq. (3.3.14), we obtain

$$0 = \sum_{j=1}^{n} \left\{ \int_{\Omega^e} \left[\frac{\partial w_i^e}{\partial x} \left(a_{xx} \frac{\partial \psi_j^e}{\partial x} \right) + \frac{\partial w_i^e}{\partial y} \left(a_{yy} \frac{\partial \psi_j^e}{\partial y} \right) \right] dA + \oint_{\Gamma^e} h_c w_i^e \psi_j^e \, ds \right\} u_j^e$$

$$- \int_{\Omega^e} w_i^e f \, d\mathbf{x} - \oint_{\Gamma^e} w_i^e (\hat{q}_n + h_c u_c) \, ds \tag{3.3.18}$$

This equation must hold for any weight function w_i^e from the set $\{w_i^e\}_{i=1}^{n}$. In particular it must hold for the case $w_i^e = \psi_i^e$. This particular choice of weight functions is a natural one when the weight function is viewed as a virtual variation of the dependent unknown [i.e. $w_i^e(\mathbf{x}) = \delta u_h^e(\mathbf{x}) = \sum_{i=1}^{n} \delta u_i^e \, \psi_i^e(\mathbf{x})$].

For each choice of w_i^e we obtain an algebraic relation among $(u_1^e, u_2^e, \ldots, u_n^e)$. The ith algebraic equation is obtained by substituting $w_i^e = \psi_i^e$ into Eq. (3.3.14):

$$\sum_{j=1}^{n} K_{ij}^e u_j^e = f_i^e + q_i^e \quad \text{or} \quad \mathbf{K}^e \mathbf{u}^e = \mathbf{f}^e + \mathbf{q}^e \equiv \mathbf{F}^e \qquad (3.3.19)$$

where the coefficients $K_{ij}^e = K_{ji}^e$ (i.e. $\mathbf{K}^e$ is symmetric), f_i^e, and q_i^e are defined by

$$K_{ij}^e = \int_{\Omega^e} \left(a_{xx} \frac{\partial \psi_i^e}{\partial x} \frac{\partial \psi_j^e}{\partial x} + a_{yy} \frac{\partial \psi_i^e}{\partial y} \frac{\partial \psi_j^e}{\partial y} \right) dA + \oint_{\Gamma^e} h_c \psi_i^e \psi_j^e \, ds$$

$$f_i^e = \int_{\Omega^e} f \, \psi_i^e \, dA, \quad q_i^e = \oint_{\Gamma^e} \left(\hat{q}_n + h_c u_c \right) \psi_i^e \, ds \qquad (3.3.20)$$

3.3.5 Approximation Functions: Element Library

As before, we seek polynomial approximations of $u(\mathbf{x})$. The finite element approximation $u_h^e(\mathbf{x})$ of $u(x, y)$ over an element Ω^e must satisfy the following conditions in order for the approximate solution to converge to the true solution:

1. $u_h^e(\mathbf{x})$ must be continuous as required in the weak form of the problem; that is, all terms in the weak form are represented as non-zero values.

2. The polynomials used to represent $u_h^e(\mathbf{x})$ must be complete and contain both x and y of the same order (i.e. equi-presence of x and y). This means all terms, beginning with a constant term up to the highest order desired, in both x and y, should be included in the expression of $u_h^e(\mathbf{x})$.

3. All terms in the polynomial should be linearly independent.

As already discussed, the number of linearly independent terms in the representation of u_h^e dictates the shape and the number of nodes in the element. It turns out that only triangular and quadrilateral shapes meet the requirements stated above. Here we review the interpolation functions of linear and quadratic triangular and rectangular elements.

3.3.5.1 Linear triangular element

An examination of the statement (3.3.18) and the finite element matrices in Eq. (3.3.20) shows that ψ_i^e $(i = 1, 2, \ldots, n)$ should be, at least, a linear function of x and y. The lowest-order linear polynomial that meets the requirements is

$$u_h^e(\mathbf{x}) = c_1^e + c_2^e \, x + c_3^e \, y \qquad (3.3.21)$$

The polynomial is complete because constant and linear terms in x and y are included, both x and y are represented equally, and the set $\{1, x, y\}$ is linearly independent and, hence, forms a basis for the vector space $\mathcal{U}_h$. In this case, $\mathcal{U}_h$ is a three-dimensional (complete) subspace of $H^1(\Omega^e)$. To write the three parameters (c_1^e, c_2^e, c_3^e) in terms of the values of u_h^e at certain points of the domain Ω^e, three points, called nodes, must be identified such that they uniquely define the geometry Ω^e of the element and allow the imposition of interelement continuity of the variable $u_h^e(\mathbf{x})$, as dictated by the weak form. Obviously, the geometric shape defined by three points in a two-dimensional domain is a triangle. Thus, the polynomial in Eq. (3.3.21) is associated with a triangular element, with the vertices of the triangle being identified as the nodes (see Fig. 3.3.2).

The linear interpolation functions for an arbitrary three-node triangle are

$$\psi_i^e(\mathbf{x}) = \frac{1}{2A_e}(\alpha_i^e + \beta_i^e x + \gamma_i^e y), \qquad (i = 1, 2, 3) \tag{3.3.22}$$

where A_e is the area of the triangle, and α_i^e, β_i^e, and γ_i^e are constants known in terms of the nodal coordinates (x_i, y_i), that is, dependent only on the geometry of the element

$$\alpha_i^e = x_j y_k - x_k y_j; \qquad \beta_i^e = y_j - y_k; \qquad \gamma_i^e = -(x_j - x_k) \tag{3.3.23}$$

for $i \neq j \neq k$, and i, j, and k permute in a natural order (see Reddy [2] for the derivation of ψ_i^e). Note that (x, y) are the *global coordinates* used in the governing equation, Eq. (3.3.1), over the domain Ω. The interpolation functions ψ_i^e $(i = 1, 2, \ldots, n)$ satisfy the following interpolation properties:

$$\text{(i)} \quad \psi_i^e(x_j, y_j) = \delta_{ij}, \quad (i, j = 1, 2, 3); \qquad \text{(ii)} \quad \sum_{i=1}^{3} \psi_i^e(x, y) = 1 \tag{3.3.24}$$

and they are the linear Lagrange interpolation functions associated with a triangle. The use of linear approximation of the actual function $u(x, y)$, which is possibly a surface, results in a planar shape, as shown in Fig. 3.3.3.

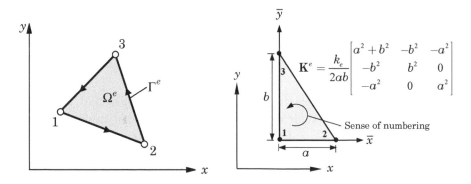

Fig. 3.3.2: The linear triangular finite element.

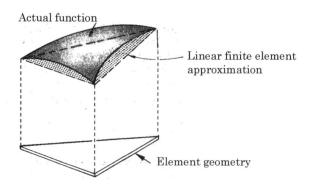

Fig. 3.3.3: The linear triangular finite element.

The integrals in the definition of the coefficients K_{ij}^e and f_i^e can be evaluated for given data: a_{xx}, a_{yy}, and f. For example, for element-wise constant values of the data, that is, $a_{xx} = a_{xx}^e$, $a_{yy} = a_{yy}^e$, and $f = f_e$, we have

$$K_{ij}^e = \frac{1}{4A_e}(a_{xx}^e \beta_i^e \beta_j^e + a_{yy}^e \gamma_i^e \gamma_j^e); \quad f_i^e = \frac{f_e A_e}{3} \tag{3.3.25}$$

where A_e is the area of the triangular element, and β_i^e and γ_i^e are known in terms of the global nodal coordinates of the element nodes, as given in Eq. (3.3.23). For a right-angled triangular element with base a and height b, and node 1 at the right angle (nodes are numbered counterclockwise as shown in Fig. 3.3.2), $\mathbf{K}^e$ takes the form

$$\mathbf{K}^e = \frac{a_{xx}^e}{2} \begin{bmatrix} \alpha^e & -\alpha^e & 0 \\ -\alpha^e & \alpha^e & 0 \\ 0 & 0 & 0 \end{bmatrix} + \frac{a_{yy}^e}{2} \begin{bmatrix} \beta^e & 0 & -\beta^e \\ 0 & 0 & 0 \\ -\beta^e & 0 & \beta^e \end{bmatrix} \tag{3.3.26}$$

where $\alpha^e = b/a$ and $\beta^e = a/b$. Of course, for cases in which the conductivities are functions of (x, y), numerical integration can be used to evaluate the coefficients (see Section 3.6.4). We note that the first part of the coefficient matrix contains an aspect ratio b/a and the second part contains the reciprocal of the aspect ratio. When a is very large compared to b, the second part contains large numbers compared to the first part, for the same orders of magnitudes of a_{xx} and a_{yy}; hence, one must maintain an aspect ratio that does not unduly affect the physics.

We note that q_i^e is not evaluated for an element connected on all sides to other elements because their contributions to the global nodes is canceled by similar contributions from the neighboring elements. For elements with nodes on the boundary of the computational domain, $q_n^e(s)$ is either known or determined in the post-computation. When q_n^e is known, the evaluation of boundary integral

$$q_i^e = \oint_{\Gamma^e} \hat{q}_n \psi_i^e(s) \, ds \tag{3.3.27}$$

involves the evaluation of line integrals. *It is necessary to compute such integrals only when* Γ^e, *or a portion of it, coincides with the boundary* Γ_q *of the total domain* Ω *on which the flux* q_n *is specified.* On portions of Γ^e that are in the interior of the domain Ω, q_n^e on side (i, j) of element Ω^e cancels with q_n^f on side (p, q) of element Ω^f when sides (i, j) of element Ω^e and (p, q) of element Ω^f are the same (i.e. at the interface of elements Ω^e and Ω^f). This can be viewed as the balance of the internal flux. When Γ^e falls on the boundary Γ_u of the domain Ω, q_n^e is not known there and can be determined in the post-computation. Note that the primary variable u is specified on Γ_u. For additional details, see Reddy [2].

3.3.5.2 Linear rectangular element

The next polynomial that meets the requirements is

$$u_h^e(\mathbf{x}) = c_1^e + c_2^e x + c_3^e y + c_4^e xy \qquad (3.3.28)$$

which contains four linearly independent terms, $\{1\, x\, y\, xy\}$, which span the four-dimensional vector space $\mathcal{U}_h \subset H^1(\Omega^e)$. In order to represent Eq. (3.3.28) in terms of the values of $u_h^e(x, y)$, an element (i.e. geometric shape) with four points, with linear variation along any two points in the element, must be identified. It is a rectangle with nodes at the four corners of the rectangle (see Fig. 3.3.4). When the element is a quadrilateral in shape, we use coordinate transformations to represent the integrals posed on the quadrilateral element equivalent to those over a square geometry and then use numerical integration to evaluate the integrals over the quadrilateral domain.

For a linear rectangular element (some times called *bilinear element* because it is linear in both x and y), we have

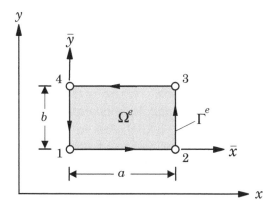

Fig. 3.3.4: The linear rectangular finite element.

$$u_h^e(\bar{\mathbf{x}}) = \sum_{i=1}^{4} u_i^e \psi_i^e(\bar{x}, \bar{y}) \tag{3.3.29}$$

where ψ_i^e are the Lagrange interpolation functions expressed in terms of the element coordinates $(\bar{x}, \bar{y})$

$$\psi_1^e = \left(1 - \frac{\bar{x}}{a}\right)\left(1 - \frac{\bar{y}}{b}\right), \qquad \psi_2^e = \frac{\bar{x}}{a}\left(1 - \frac{\bar{y}}{b}\right)$$
$$\psi_3^e = \frac{\bar{x}}{a}\frac{\bar{y}}{b}, \qquad \psi_4^e = \left(1 - \frac{\bar{x}}{a}\right)\frac{\bar{y}}{b} \tag{3.3.30}$$

and $(\bar{x}, \bar{y})$ denote the local coordinates with the origin located at node 1 of the element, and (a, b) denote the horizontal and vertical dimensions of the rectangle, as shown in Fig. 3.3.4.

The integrals in the definition of K_{ij}^e and f_i^e can be easily evaluated over a rectangular element of sides a and b. For example, for element-wise constant values of the data, that is, $a_{xx} = a_{xx}^e$, $a_{yy} = a_{yy}^e$, and $f = f_e$, we have (see Reddy [1, p. 313; p. 387]) the following results:

$$\mathbf{K}^e = a_{xx}^e \mathbf{S}^{11} + a_{yy}^e \mathbf{S}^{22}, \qquad f_i^e = \frac{f_e a b}{4} \tag{3.3.31}$$

where

$$\mathbf{S}^{11} = \frac{1}{6}\begin{bmatrix} 2\alpha & -2\alpha & -\alpha & \alpha \\ -2\alpha & 2\alpha & \alpha & -\alpha \\ -\alpha & \alpha & 2\alpha & -2\alpha \\ \alpha & -\alpha & -2\alpha & 2\alpha \end{bmatrix}, \quad \mathbf{S}^{22} = \frac{1}{6}\begin{bmatrix} 2\beta & \beta & -\beta & -2\beta \\ \beta & 2\beta & -2\beta & -\beta \\ -\beta & -2\beta & 2\beta & \beta \\ -2\beta & -\beta & \beta & 2\beta \end{bmatrix} \tag{3.3.32}$$

and $\alpha = b/a$ and $\beta = a/b$. Once again, we note that the element aspect ratio should be kept at a reasonable value so that the coefficient matrix is not ill-conditioned.

3.3.5.3 Higher-order triangular elements

Higher-order triangular elements (i.e. triangular elements with interpolation functions of higher degree) can be systematically developed with the help of the so-called *area coordinates*. For triangular elements, it is possible to construct three natural coordinates L_i ($i = 1, 2, 3$), like in the case of one-dimensional elements, which vary in a direction normal to the sides directly opposite to each node, as shown in Fig. 3.3.5(a). The coordinates are defined as

$$\psi_i = L_i = \frac{A_i}{A}, \quad A = \text{area of the triangle} \tag{3.3.33}$$

where A_i is the area of the triangle formed by nodes j and k and an arbitrary point P in the element, and A is the total area of the element. For example, A_1

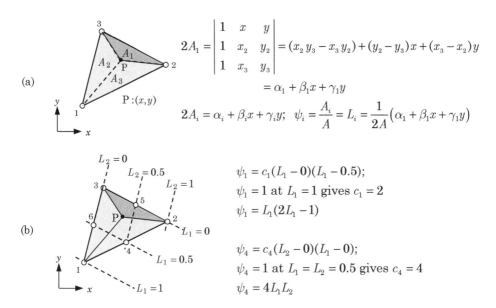

(a)

$$2A_1 = \begin{vmatrix} 1 & x & y \\ 1 & x_2 & y_2 \\ 1 & x_3 & y_3 \end{vmatrix} = (x_2\,y_3 - x_3\,y_2) + (y_2 - y_3)x + (x_3 - x_2)y$$

$$= \alpha_1 + \beta_1 x + \gamma_1 y$$

$$2A_i = \alpha_i + \beta_i x + \gamma_i y; \quad \psi_i = \frac{A_i}{A} = L_i = \frac{1}{2A}(\alpha_i + \beta_i x + \gamma_i y)$$

(b)

$$\psi_1 = c_1(L_1 - 0)(L_1 - 0.5);$$
$$\psi_1 = 1 \text{ at } L_1 = 1 \text{ gives } c_1 = 2$$
$$\psi_1 = L_1(2L_1 - 1)$$

$$\psi_4 = c_4(L_2 - 0)(L_1 - 0);$$
$$\psi_4 = 1 \text{ at } L_1 = L_2 = 0.5 \text{ gives } c_4 = 4$$
$$\psi_4 = 4L_1 L_2$$

Fig. 3.3.5: (a) Definition of area coordinates L_i for triangular elements. (b) Determination of ψ_i in terms of L_i for the quadratic triangular element.

is the area of the shaded triangle which is formed by nodes 2 and 3 and point P. Therefore, A_1 is given by [see Fig. 3.3.5(a)]

$$A_1 = \frac{1}{2}(\alpha_1 + \beta_1 x + \gamma_1 y); \quad \psi_1 = L_1 = \frac{A_1}{A} = \frac{1}{2A}(\alpha_1 + \beta_1 x + \gamma_1 y) \quad (3.3.34)$$

Similarly, $2A_2 = \alpha_2 + \beta_2 x + \gamma_2 y$ and $2A_3 = \alpha_3 + \beta_3 x + \gamma_3 y$. Thus, for the linear triangular element, we have

$$\psi_i = L_i \quad (3.3.35)$$

The area coordinates (L_1, L_2, L_3) can be used to construct interpolation functions for higher-order triangular elements. For example, in the case of quadratic triangular element (with midside nodes) ψ_1 is given by [see Fig. 3.3.5(b)]

$$\psi_1 = c_1(L_1 - 0)(L_1 - 0.5)$$

where c_1 is determined such that $\psi_1 = 1$ when $L_1 = 1$, giving $c_1 = 2$. Hence,

$$\psi_1 = L_1(2L_1 - 1)$$

The explicit forms of the interpolation functions for the linear and quadratic elements are

$$\mathbf{\Psi}^e = \begin{Bmatrix} L_1 \\ L_2 \\ L_3 \end{Bmatrix}; \quad \mathbf{\Psi}^e = \begin{Bmatrix} L_1(2L_1 - 1) \\ L_2(2L_2 - 1) \\ L_3(2L_3 - 1) \\ 4L_1 L_2 \\ 4L_2 L_3 \\ 4L_3 L_1 \end{Bmatrix} \quad (3.3.36)$$

Note that the order of the interpolation functions in the above arrays corresponds to the node numbers shown in Figs. 3.3.6(a) and (b).

Integrals of the products of the area coordinates L_i over an element can be evaluated using the following formula:

$$\int_A L_1^p \, L_2^q \, L_3^r \, dA = \frac{p! \, q! \, r!}{(p+q+r+1)!} \, A \tag{3.3.37}$$

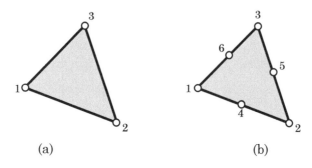

Fig. 3.3.6: (a) Linear and (b) quadratic triangular elements.

3.3.5.4 Higher-order rectangular elements

The linear and quadratic Lagrange interpolation functions associated with rectangular family of master finite elements can be obtained from the tensor product of the corresponding one-dimensional Lagrange interpolation functions in Eq. (3.2.48). We use a local coordinate system (ξ, η) such that $-1 \le (\xi, \eta) \le 1$ to derive the approximation functions for the (master) rectangular elements.

The linear interpolation functions in Eq. (3.2.48) can be expressed in terms of the local coordinates (ξ, η) for the two coordinate directions, and taking the product of the column vector with the row vector

$$\frac{1}{2} \left\{ \begin{array}{c} (1 - \xi) \\ (1 + \xi) \end{array} \right\} \frac{1}{2} \left\{ (1 - \eta) \quad (1 + \eta) \right\}$$

gives four functions, which are placed in the same order as the node numbers shown in Fig. 3.3.7(a):

$$\mathbf{\Psi}^e = \frac{1}{4} \left\{ \begin{array}{c} (1 - \xi)(1 - \eta) \\ (1 + \xi)(1 - \eta) \\ (1 + \xi)(1 + \eta) \\ (1 - \xi)(1 + \eta) \end{array} \right\} \tag{3.3.38}$$

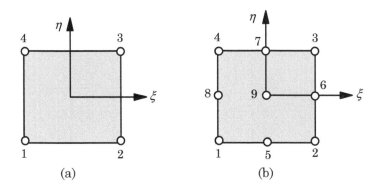

Fig. 3.3.7: (a) Linear and (b) quadratic rectangular elements.

Similarly, the product of the quadratic functions from Eq. (3.2.48) for the two coordinate directions

$$\frac{1}{4}\left\{\begin{array}{c} -\xi(1-\xi) \\ 2(1-\xi^2) \\ \xi(1+\xi) \end{array}\right\}\left\{-\eta(1-\eta) \quad 2(1-\eta^2) \quad \eta(1+\eta)\right\}$$

yields the interpolation functions for the quadratic element [see Fig. 3.3.7(b) for the node numbers; the functions are in an equivalent modified form]:

$$\boldsymbol{\Psi}^e = \frac{1}{4}\left\{\begin{array}{c} (1-\xi)(1-\eta)(-\xi-\eta-1)+(1-\xi^2)(1-\eta^2) \\ (1+\xi)(1-\eta)(\xi-\eta-1)+(1-\xi^2)(1-\eta^2) \\ (1+\xi)(1+\eta)(\xi+\eta-1)+(1-\xi^2)(1-\eta^2) \\ (1-\xi)(1+\eta)(-\xi+\eta-1)+(1-\xi^2)(1-\eta^2) \\ 2[(1-\xi^2)(1-\eta)-(1-\xi^2)(1-\eta^2)] \\ 2[(1+\xi)(1-\eta^2)-(1-\xi^2)(1-\eta^2)] \\ 2[(1-\xi^2)(1+\eta)-(1-\xi^2)(1-\eta^2)] \\ 2[(1-\xi)(1-\eta^2)-(1-\xi^2)(1-\eta^2)] \\ 4(1-\xi^2)(1-\eta^2) \end{array}\right\} \tag{3.3.39}$$

The *serendipity elements* are those elements that have no interior nodes. These elements have fewer nodes compared to the same order complete Lagrange elements. The approximation functions ψ_i^e of the serendipity elements are not complete (in the sense that they are not complete polynomials), and they cannot be obtained using tensor products of one-dimensional Lagrange interpolation functions. Instead, an alternative procedure must be employed. The lowest-order serendipity element in two dimensions is the eight-node quadratic element shown in Fig. 3.3.8. The approximation functions of the eight-node quadratic element, compared to the nine-quadratic element, do not contain the bi-quadratic term $\xi^2\eta^2$ and, therefore, they are incomplete quadratic polynomials. However, ψ_i^e of the serendipity elements do satisfy the Kronecker-delta property [i.e. property (i) in Eq. (3.3.24)]. The interpolation functions for the eight-node quadratic serendipity element are given by

$$\boldsymbol{\Psi}^e = \frac{1}{4} \left\{ \begin{array}{c} (1-\xi)(1-\eta)(-\xi-\eta-1) \\ (1+\xi)(1-\eta)(\xi-\eta-1) \\ (1+\xi)(1+\eta)(\xi+\eta-1) \\ (1-\xi)(1+\eta)(-\xi+\eta-1) \\ 2(1-\xi^2)(1-\eta) \\ 2(1+\xi)(1-\eta^2) \\ 2(1-\xi^2)(1+\eta) \\ 2(1-\xi)(1-\eta^2) \end{array} \right\} \tag{3.3.40}$$

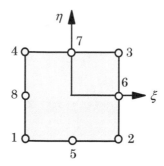

Fig. 3.3.8: Quadratic rectangular serendipity element.

3.3.6 Assembly of Elements

The assembly of finite elements to obtain the equations of the entire domain is based on the following two rules:

1. Continuity of the primary variable $u_h^e(\mathbf{x})$ (e.g. displacements, temperature, etc.)

2. Equilibrium of secondary variables q_i^e (e.g. forces, heats, etc.)

We illustrate the assembly procedure by considering a finite element mesh consisting of a triangular element and a quadrilateral element, as shown in Fig. 3.3.9.

The nodes of the finite element mesh are called *global nodes*. From the mesh shown in Fig. 3.3.9, it is clear that the following correspondence between global and element nodes exists: nodes 1, 2, and 3 of element 1 correspond to global nodes 1, 2, and 3, respectively. Nodes 1, 2, 3, and 4 of element 2 correspond to global nodes 2, 4, 5, and 3, respectively. Hence, the correspondence between the local and global nodal values of the primary variable u is as follows:

$$u_1^1 = u_1, \quad u_2^1 = u_1^2 = u_2, \quad u_3^1 = u_4^2 = u_3, \quad u_2^2 = u_4, \quad u_3^2 = u_5 \tag{3.3.41}$$

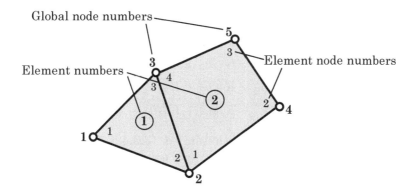

Fig. 3.3.9: Global-local correspondence of nodes for element assembly.

where the superscripts refer to the element numbers and subscripts to element node numbers. This amounts to imposing the continuity of the primary variables at the nodes common to elements 1 and 2. Note that the continuity of the primary variables at the interelement nodes guarantees the continuity of the primary variable along the entire interelement boundary, that is $u^1(s) = u^2(s)$, where s denotes a coordinate along the interface of the two elements.

Next, we consider the equilibrium of secondary variables at the interelement boundaries. At the interface between any two elements, the flux from the two elements should be equal in magnitude and opposite in sign. For the two elements shown in Fig. 3.3.9, the interface is along the side connecting global nodes 2 and 3. Hence, the internal flux q_n^1 on side 2–3 of element 1 should balance, in integral sense, the flux q_n^2 on side 4–1 of element 2 (recall the sign convention on q_n^e):

$$\int_{h_{23}^1} q_n^1 \psi_2^1 \, ds = - \int_{h_{14}^2} q_n^2 \psi_1^2 \, ds \quad \text{or} \quad \int_{h_{23}^1} q_n^1 \psi_3^1 \, ds = - \int_{h_{14}^2} q_n^2 \psi_4^2 \, ds \quad (3.3.42)$$

where h_{pq}^e denotes length of the side connecting nodes p and q of element Ω^e.

Now we are ready to assemble the element equations for the two-element mesh. Let K_{ij}^1 $(i, j = 1, 2, 3)$ denote the coefficient matrix corresponding to the triangular element and let K_{ij}^2 $(i, j = 1, 2, 3, 4)$ denote the coefficient matrix corresponding to the quadrilateral element. The element equations of the two elements are written separately first. The element equations of the triangular element are of the form

$$K_{11}^1 u_1^1 + K_{12}^1 u_2^1 + K_{13}^1 u_3^1 = f_1^1 + q_1^1$$
$$K_{21}^1 u_1^1 + K_{22}^1 u_2^1 + K_{23}^1 u_3^1 = f_2^1 + q_2^1 \qquad (3.3.43)$$
$$K_{31}^1 u_1^1 + K_{32}^1 u_2^1 + K_{33}^1 u_3^1 = f_3^1 + q_3^1$$

Similarly, for the rectangular element we have

$$
\begin{aligned}
K_{11}^2 u_1^2 + K_{12}^2 u_2^2 + K_{13}^2 u_3^2 + K_{14}^2 u_4^2 &= f_1^2 + q_1^2 \\
K_{21}^2 u_1^2 + K_{22}^2 u_2^2 + K_{23}^2 u_3^2 + K_{24}^2 u_4^2 &= f_2^2 + q_2^2 \\
K_{31}^2 u_1^2 + K_{32}^2 u_2^2 + K_{33}^2 u_3^2 + K_{34}^2 u_4^2 &= f_3^2 + q_3^2 \\
K_{41}^2 u_1^2 + K_{42}^2 u_2^2 + K_{43}^2 u_3^2 + K_{44}^2 u_4^2 &= f_4^2 + q_4^2
\end{aligned}
\tag{3.3.44}
$$

In order to impose the balance condition in Eq. (3.3.42), it is necessary to add the second equation of element 1 to the first of element 2, and also add the third equation of element 1 to the fourth equation of element 2:

$$
\begin{aligned}
(K_{21}^1 u_1^1 + K_{22}^1 u_2^1 + K_{23}^1 u_3^1) + (K_{11}^2 u_1^2 + K_{12}^2 u_2^2 + K_{13}^2 u_3^2 + K_{14}^2 u_4^2) \\
= (f_2^1 + q_2^1) + (f_1^2 + q_1^2) \\
(K_{31}^1 u_1^1 + K_{32}^1 u_2^1 + K_{33}^1 u_3^1) + (K_{41}^2 u_1^2 + K_{42}^2 u_2^2 + K_{43}^2 u_3^2 + K_{44}^2 u_4^2) \\
= (f_3^1 + q_3^1) + (f_4^2 + q_4^2)
\end{aligned}
$$

Using the local-global nodal variable correspondence in Eq. (3.3.41), we can rewrite the above equations as

$$
\begin{aligned}
K_{21}^1 u_1 + (K_{22}^1 + K_{11}^2) u_2 + (K_{23}^1 + K_{14}^2) u_3 + K_{12}^2 u_4 + K_{13}^2 u_5 \\
= f_2^1 + f_1^2 + (q_2^1 + q_1^2) \\
K_{31}^1 u_1 + (K_{32}^1 + K_{41}^2) u_2 + (K_{33}^1 + K_{44}^2) u_3 + K_{42}^2 u_4 + K_{43}^2 u_5 \\
= f_3^1 + f_4^2 + (q_3^1 + q_4^2)
\end{aligned}
\tag{3.3.45}
$$

Now we can impose the conditions in Eq. (3.3.42) by setting appropriate portions of the expressions in parentheses on the right-hand side of Eq. (3.3.45) to zero. This is done by means of the connectivity relations, that is, the correspondence of the local node number to the global node number. The ideas presented in Sections 3.3.4–3.3.6 are illustrated using Example 3.3.1.

Example 3.3.1 ———————————————————————————

Consider steady-state heat conduction in an isotropic rectangular region of dimensions $3a \times 2a$, as shown in Fig. 3.3.10(a). The origin of the x and y coordinates is taken at the lower left corner such that x is parallel to the side $3a$ and y is parallel to the side $2a$. Boundaries $x = 0$ and $y = 0$ are insulated, boundary $x = 3a$ is maintained at zero temperature, and boundary $y = 2a$ is maintained at temperature $T = T_0 \cos(\pi x / 6a)$. Determine the temperature distribution using (1) mesh of linear triangular elements shown in Fig. 3.3.10(b) and (2) mesh of linear rectangular elements shown in Fig. 3.3.10(c).

Solution: The governing equation is a special case of the model equation (3.3.1) with zero internal heat generation $f = 0$ and coefficients $a_{xx} = a_{yy} = k$. Thus, Eq. (3.3.1) takes the form

$$
-k \left(\frac{\partial^2 T}{\partial x^2} + \frac{\partial^2 T}{\partial y^2} \right) = 0
\tag{1}
$$

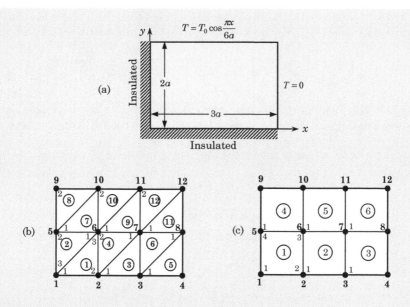

Fig. 3.3.10: Finite element analysis of a heat conduction problem over a rectangular domain: (a) domain; (b) mesh of linear triangular elements; and (c) mesh of linear rectangular elements.

The exact solution of Eq. (1) for the boundary conditions shown in Fig. 3.3.10(a) is

$$T(x,y) = T_0 \frac{\cosh\left(\pi y/6a\right)\cos\left(\pi x/6a\right)}{\cosh(\pi/3)} \tag{2}$$

The finite element model of Eq. (1) is given by

$$\mathbf{K}^e\mathbf{T}^e = \mathbf{q}^e \tag{3}$$

where T_i^e is the temperature at node i of element Ω_e, and

$$K_{ij}^e = \int_{\Omega_e} k\left(\frac{\partial\psi_i}{\partial x}\frac{\partial\psi_j}{\partial x} + \frac{\partial\psi_i}{\partial y}\frac{\partial\psi_j}{\partial y}\right) dA, \quad q_i^e = \oint_{\Gamma_e} q_n\psi_i\, ds \tag{4}$$

Suppose that we use a 3×2 mesh (i.e. 3 subdivisions along the x-axis and 2 subdivisions along the y-axis) of linear triangular elements and then with a 3×2 mesh of linear rectangular elements, as shown in Figs. 3.3.10(b) and 3.3.10(c). Both meshes have the same number of global nodes, namely 12, but differing numbers of elements.

Mesh of Triangular Elements. The global node numbers, element numbers, and element node numbers used are shown in Fig. 3.3.10(b). Of course, the global node numbering and element numbering is arbitrary (does not have to follow any particular pattern) although the global node numbering dictates the size of the half bandwidth of the assembled equations, which in turn affects the computational time of Gauss elimination methods used in the solution of algebraic equations in a computer. The element node numbering scheme should be the one that is used in the development of element interpolation functions. By a suitable numbering of the element nodes, all similar elements can be made to have identical element coefficient matrices. Such considerations are important only when hand calculations are carried out.

For a typical element of the mesh of triangles in Fig. 3.3.10(b), the element coefficient matrix is given by [see Eq. (3.3.26) and Fig. 3.3.2 for the element matrix and element geometry],

$$\mathbf{K}^e = \frac{k}{2} \begin{bmatrix} 1 & -1 & 0 \\ -1 & 2 & -1 \\ 0 & -1 & 1 \end{bmatrix}, \quad \mathbf{f}^e = \mathbf{0} \tag{5}$$

where k is the conductivity of the medium. Note that the element matrix is independent of the size of the element, as long as the element is a right-angle triangle with its base equal to its height.

The assembly of the elements follows the logic discussed earlier. For example, we have

$$K_{11} = K_{11}^1 + K_{33}^2 = \frac{k}{2}(1+1), \quad K_{12} = K_{12}^1 = \frac{k}{2}(-1), \quad K_{13} = 0$$

$$K_{15} = K_{32}^2 = \frac{k}{2}(-1), \quad K_{16} = K_{13}^1 + K_{31}^2 = 0+0, \quad \text{etc.} \tag{6}$$

$$F_1 = q_1^1 + q_3^2, \quad F_6 = q_3^1 + q_1^2 + q_2^4 + q_2^7 + q_1^9 + q_3^{10}, \quad \text{etc.}$$

In view of the boundary conditions, we have

$$U_4 = U_8 = U_{12} = 0, \quad U_9 = T_0, \quad U_{10} = \frac{\sqrt{3}}{2}T_0, \quad U_{11} = \frac{T_0}{2}$$

$$F_1 = F_2 = F_3 = F_5 = 0 \tag{7}$$

and the balance of internal heat flow requires that

$$F_6 = F_7 = 0 \tag{8}$$

Thus, the unknown primary variables and secondary variables of the problem are:

$$U_1, \quad U_2, \quad U_3, \quad U_5, \quad U_6, \quad U_7; \quad F_4, \quad F_8, \quad F_9, \quad F_{10}, \quad F_{11}, \quad F_{12} \tag{9}$$

We write the six finite element equations for the six unknown primary variables. These equations come from rows 1, 2, 3, 5, 6, and 7 (corresponding to the same global nodes):

$$K_{11}U_1 + K_{12}U_2 + \cdots + K_{1(12)}U_{12} = F_1 = (q_1^1 + q_3^2) = 0$$
$$K_{21}U_1 + K_{22}U_2 + \cdots + K_{2(12)}U_{12} = F_2 = (q_2^1 + q_1^3 + q_3^4) = 0$$
$$\vdots \tag{10}$$
$$K_{71}U_1 + K_{72}U_2 + \cdots + K_{7(12)}U_{12} = F_7 = (q_3^3 + q_1^4 + q_2^6 + q_2^9 + q_1^{11} + q_3^{12}) = 0$$

Using the boundary conditions and the values of K_{IJ}, we obtain

$$k(U_1 - \frac{1}{2}U_2 - \frac{1}{2}U_5) = 0$$

$$k(-\frac{1}{2}U_1 + 2U_2 - \frac{1}{2}U_3 - U_6) = 0$$

$$k(-\frac{1}{2}U_2 + 2U_3 - U_7) = 0$$

$$k(-\frac{1}{2}U_1 + 2U_5 - U_6 - \frac{1}{2}U_9) = 0 \quad (U_9 = T_0) \tag{11}$$

$$k(-U_2 - U_5 + 4U_6 - U_7 - U_{10}) = 0 \quad (U_{10} = \frac{\sqrt{3}}{2}T_0)$$

$$k(-U_3 - U_6 + 4U_7 - U_{11}) = 0 \quad (U_{11} = \frac{1}{2}T_0)$$

or, in the matrix form

$$
\frac{k}{2}
\begin{bmatrix}
2 & -1 & 0 & -1 & 0 & 0 \\
-1 & 4 & -1 & 0 & -2 & 0 \\
0 & -1 & 4 & 0 & 0 & -2 \\
-1 & 0 & 0 & 4 & -2 & 0 \\
0 & -2 & 0 & -2 & 8 & -2 \\
0 & 0 & -2 & 0 & -2 & 8
\end{bmatrix}
\begin{Bmatrix}
U_1 \\ U_2 \\ U_3 \\ U_5 \\ U_6 \\ U_7
\end{Bmatrix}
=
\frac{k}{2}
\begin{Bmatrix}
0 \\ 0 \\ 0 \\ T_0 \\ \sqrt{3}T_0 \\ T_0
\end{Bmatrix}
\tag{12}
$$

The solution of these equations is (in °C)

$$
U_1 = 0.6362\,T_0, \quad U_2 = 0.5510\,T_0, \quad U_3 = 0.3181\,T_0
$$
$$
U_5 = 0.7214\,T_0, \quad U_6 = 0.6248\,T_0, \quad U_7 = 0.3607\,T_0
\tag{13}
$$

Evaluating the exact solution in Eq. (2) at the nodes, we have (in °C)

$$
T_1 = 0.6249\,T_0, \quad T_2 = 0.5412\,T_0, \quad T_3 = 0.3124\,T_0
$$
$$
T_5 = 0.7125\,T_0, \quad T_6 = 0.6171\,T_0, \quad T_7 = 0.3563\,T_0
\tag{14}
$$

The heat at node 4, for example, can be computed from the 4th finite element equation

$$
F_4 = Q_2^5 = K_{41}U_1 + K_{42}U_2 + K_{43}U_3 + K_{44}U_4 + K_{45}U_5
$$
$$
+ K_{46}U_6 + K_{47}U_7 + K_{48}U_8 + \ldots
\tag{15}
$$

Noting that $K_{41} = K_{42} = K_{45} = \cdots = K_{4(12)} = 0$ and $U_4 = U_8 = 0$, we obtain

$$
Q_2^5 = -\frac{1}{2}kU_3 = -0.1591kT_0 \quad \text{(in W)}
\tag{16}
$$

Mesh of Rectangular Elements. For a 3×2 mesh of linear rectangular elements [see Fig. 3.3.10(c)], the element coefficient matrix is given by Eq. (3.3.31)

$$
\mathbf{K}^e = \frac{k}{6}
\begin{bmatrix}
4 & -1 & -2 & -1 \\
-1 & 4 & -1 & -2 \\
-2 & -1 & 4 & -1 \\
-1 & -2 & -1 & 4
\end{bmatrix},
\quad \mathbf{f}^e = \mathbf{0}
\tag{17}
$$

The present mesh of rectangular elements is node-wise equivalent to the triangular element mesh considered in Fig. 3.3.10(b). Hence the boundary conditions in Eqs. (7) and (8) are valid for the present case. The assembled global coefficients are

$$
K_{11} = K_{11}^1, \quad K_{12} = K_{12}^1, \quad K_{15} = K_{14}^1, \quad K_{16} = K_{13}^1, \quad K_{22} = K_{22}^1 + K_{11}^2
$$
$$
K_{23} = K_{12}^2, \quad K_{25} = K_{24}^1, \quad K_{26} = K_{23}^1 + K_{14}^2, \quad K_{27} = K_{13}^1, \quad \text{etc.}
\tag{18}
$$
$$
F_1 = q_1^1, \quad F_2 = q_2^1 + q_1^2, \quad F_3 = q_2^2 + q_1^3, \quad F_4 = q_2^3, \quad \text{etc.}
$$

The equations for the unknown temperatures (i.e. condensed equations for the unknown primary variables) are

$$
\frac{k}{6}
\begin{bmatrix}
4 & -1 & 0 & -1 & -2 & 0 \\
-1 & 8 & -1 & -2 & -2 & -2 \\
0 & -1 & 8 & 0 & -2 & -2 \\
-1 & -2 & 0 & 8 & -2 & 0 \\
-2 & -2 & -2 & -2 & 16 & -2 \\
0 & -2 & -2 & 0 & -2 & 16
\end{bmatrix}
\begin{Bmatrix}
U_1 \\ U_2 \\ U_3 \\ U_5 \\ U_6 \\ U_7
\end{Bmatrix}
=
\frac{k}{6}
\begin{Bmatrix}
0 \\ 0 \\ 0 \\ T_0 + \sqrt{3}T_0 \\ 2T_0 + \sqrt{3}T_0 + T_0 \\ \sqrt{3}T_0 + T_0
\end{Bmatrix}
\tag{19}
$$

The solution of these equations is

$$U_1 = 0.6128\,T_0, \quad U_2 = 0.5307\,T_0, \quad U_3 = 0.3064\,T_0$$
$$U_5 = 0.7030\,T_0, \quad U_6 = 0.6088\,T_0, \quad U_7 = 0.3515\,T_0 \tag{20}$$

The value of the heat at node 4 is given by

$$q_2^3 = K_{43}U_3 + K_{47}U_7 = -\frac{k}{6}U_3 - \frac{2k}{6}U_7 = -0.1682\,kT_0 \ (\mathrm{W}) \tag{21}$$

Table 3.3.2 contains a comparison of the finite element solutions with the analytical solution in Eq. (2) for two different meshes of linear triangular and rectangular elements. We note that the results obtained using the 3×2 mesh of rectangular elements is not as accurate as that obtained with the 3×2 mesh of triangular elements. This is due to the fact that there are only half as many elements in the former case when compared to the latter.

Table 3.3.2: Comparison of the nodal temperatures $T(x,y)/T_0$, obtained using various finite element meshes with the analytical solution in Eq. (2).

		FEM Solution				Analytical solution
		Triangles		Rectangles		
x	y	3×2	6×4	3×2	6×4	
0.0	0.0	0.6362	0.6278	0.6128	0.6219	0.6249
0.5	0.0	– – –	0.6064	– – –	0.6007	0.6036
1.0	0.0	0.5510	0.5437	0.5307	0.5386	0.5412
1.5	0.0	– – –	0.4439	– – –	0.4398	0.4419
2.0	0.0	0.3181	0.3139	0.3064	0.3110	0.3124
2.5	0.0	– – –	0.1625	– – –	0.1610	0.1617
0.0	1.0	0.7214	0.7148	0.7030	0.7102	0.7125
0.5	1.0	– – –	0.6904	– – –	0.6860	0.6882
1.0	1.0	0.6248	0.6190	0.6088	0.6150	0.6171
1.5	1.0	– – –	0.5054	– – –	0.5022	0.5038
2.0	1.0	0.3607	0.3574	0.3515	0.3551	0.3563
2.5	1.0	– – –	0.1850	– – –	0.1838	0.1844

See Fig. 3.3.10 for the geometry and meshes.

3.4 Axisymmetric Problems

3.4.1 Introduction

Consider the second-order differential equation in vector form

$$-\boldsymbol{\nabla} \cdot (\mathbf{a} \cdot \boldsymbol{\nabla} u) = f(\mathbf{x}) \tag{3.4.1}$$

where $\boldsymbol{\nabla}$ is the gradient operator, $\mathbf{a}$ is a known second-order tensor, u is the field variable to be determined, and f is a known function. An example of Eq. (3.4.1) is provided by heat conduction equation, where u is the temperature, $\mathbf{a}$ is the conductivity tensor, and f is the internal heat generation.

The equations governing physical processes in cylindrical geometries are described analytically in terms of cylindrical coordinates (r, θ, z). For a cylindrically orthotropic material (i.e. when the material axes of an orthotropic material coincide with the r, θ, and z axes), Eq. (3.4.1) takes the form

$$-\frac{1}{r}\frac{\partial}{\partial r}\left(ra_{rr}\frac{\partial u}{\partial r}\right) - \frac{1}{r^2}\frac{\partial}{\partial \theta}\left(a_{\theta\theta}\frac{\partial u}{\partial \theta}\right) - \frac{\partial}{\partial z}\left(a_{zz}\frac{\partial u}{\partial z}\right) = f(r, \theta, z) \qquad (3.4.2)$$

When the geometry, loading, and boundary conditions are independent of the circumferential direction (i.e. θ-coordinate direction), the problem is said to be axisymmetric and the governing equation becomes two-dimensional in terms of r and z:

$$-\frac{1}{r}\frac{\partial}{\partial r}\left(ra_{rr}\frac{\partial u}{\partial r}\right) - \frac{\partial}{\partial z}\left(a_{zz}\frac{\partial u}{\partial z}\right) = f(r, z) \qquad (3.4.3)$$

In addition, if the problem geometry and data are independent of z, for example, when the cylinder is very long, the equation is a function of only the radial coordinate r:

$$-\frac{1}{r}\frac{d}{dr}\left(ra_{rr}\frac{du}{dr}\right) = f(r) \quad \text{for} \quad a < r < b \qquad (3.4.4)$$

In this section we develop the finite element models of one-dimensional and two-dimensional axisymmetric problems described by Eqs. (3.4.3) and (3.4.4), which may arise in various different fields, and the coefficients a_{rr}, $a_{\theta\theta}$, and a_{zz} may be functions of u and its derivatives. In each case, we begin with the development of the weak form, where the volume element dv is replaced by $dv = r\,dr\,d\theta\,dz$.

3.4.2 One-Dimensional Problems

In developing the weak form of Eq. (3.4.4), we replace u with its approximation u_h, multiply the resulting residual with a weight function $w_i(r)$, and integrate over the element volume $v^e = (r_a, r_b) \times (0, 2\pi) \times (0, 1)$ of the cylinder of unit length $(a \equiv a_{rr})$

$$0 = \int_{v^e} w_i\left[-\frac{1}{r}\frac{d}{dr}\left(ra\frac{du_h}{dr}\right) - f\right]dv = 2\pi\int_{r_a}^{r_b} w_i\left[-\frac{1}{r}\frac{d}{dr}\left(ra\frac{du_h}{dr}\right) - f\right]r\,dr \qquad (3.4.5)$$

where (r_a, r_b) is the domain of a typical element along the radial direction. Next, we carry out the remaining two steps of the weak formulation:

$$0 = 2\pi\int_{r_a}^{r_b}\left(a\frac{dw_i}{dr}\frac{du_h}{dr} - w_i f\right)r\,dr - 2\pi\left[w_i\,ra\frac{du_h}{dr}\right]_{r_a}^{r_b}$$

$$0 = 2\pi\int_{r_a}^{r_b}\left(a\frac{dw_i}{dr}\frac{du_h}{dr} - w_i f\right)r\,dr - w_i(r_a)Q_1^e - w_i(r_b)Q_2^e \qquad (3.4.6)$$

where

$$Q_1^e \equiv -2\pi \left(ra\frac{du_h}{dr} \right) \bigg|_{r_a}, \quad Q_2^e \equiv 2\pi \left(ra\frac{du_h}{dr} \right) \bigg|_{r_b} \qquad (3.4.7)$$

The finite element model is obtained by substituting the approximation

$$u(r) \approx \sum_{j=1}^{n} u_j^e \psi_j^e(r) \qquad (3.4.8)$$

where $\psi_j^e(r)$ are the Lagrange interpolation functions; the only change is that the axial coordinate x is replaced by the radial r. For example, the linear interpolation functions are of the form ($h_e = r_b - r_a$)

$$\psi_1^e(r) = \frac{r_b - r}{h_e}, \quad \psi_2^e(r) = \frac{r - r_a}{h_e} \qquad (3.4.9)$$

or, in terms of the local coordinate $\bar{r}$, we have

$$\psi_1^e(\bar{r}) = 1 - \frac{\bar{r}}{h_e}, \quad \psi_2^e(\bar{r}) = \frac{\bar{r}}{h_e} \qquad (3.4.10)$$

As before, we replace w_i with ψ_i^e to obtain the ith algebraic equation of the Ritz (or weak-form Galerkin) finite element model

$$\sum_{j=1}^{n} K_{ij}^e u_j^e = f_i^e + Q_i^e \quad \text{or} \quad \mathbf{K}^e \mathbf{u}^e = \mathbf{f}^e + \mathbf{Q}^e \qquad (3.4.11)$$

where

$$K_{ij}^e = 2\pi \int_{r_a}^{r_b} a\frac{d\psi_i^e}{dr}\frac{d\psi_j^e}{dr} \, r dr, \quad f_i^e = 2\pi \int_{r_a}^{r_b} \psi_i^e f \, r \, dr \qquad (3.4.12)$$

3.4.3 Two-Dimensional Problems

Following the three-step procedure, we obtain the weak form (the constant multiplier 2π is omitted)

$$0 = \int_{\Omega_e} \left(a_{rr}\frac{\partial w_i}{\partial r}\frac{\partial u_h}{\partial r} + a_{zz}\frac{\partial w_i}{\partial z}\frac{\partial u_h}{\partial z} - w_i f \right) r \, dr \, dz - \oint_{\Gamma_e} w_i q_n \, ds \qquad (3.4.13)$$

where w_i is the ith weight function and q_n is the flux normal to the boundary

$$q_n = r \left(a_{rr}\frac{\partial u_h}{\partial r} n_r + a_{zz}\frac{\partial u_h}{\partial z} n_z \right) \qquad (3.4.14)$$

Let us assume that $u(r, z)$ is approximated by the finite element interpolation u_h^e over the element Ω_e

$$u \approx u_h^e(r, z) = \sum_{j=1}^{n} u_j^e \psi_j^e(r, z) \qquad (3.4.15)$$

where interpolation functions $\psi_j^e(r, z)$ are the same as those developed for linear triangular and rectangular elements, with $x = r$ and $y = z$. Substitution of Eq. (3.4.15) for u_h and ψ_i^e for w_i into the weak form gives the ith equation of the finite element model

$$0 = \sum_{j=1}^{n} \left[\int_{\Omega_e} \left(a_{rr} \frac{\partial \psi_i^e}{\partial r} \frac{\partial \psi_j^e}{\partial r} + a_{zz} \frac{\partial \psi_i^e}{\partial z} \frac{\partial \psi_j^e}{\partial z} \right) r dr dz \right] u_j^e$$

$$- \int_{\Omega_e} \psi_i^e f r \, dr \, dz - \oint_{\Gamma_e} \psi_i^e q_n \, ds \tag{3.4.16}$$

or

$$0 = \sum_{j=1}^{n} K_{ij}^e u_j^e - f_i^e - Q_i^e \tag{3.4.17}$$

where

$$K_{ij}^e = \int_{\Omega_e} \left(a_{rr} \frac{\partial \psi_i^e}{\partial r} \frac{\partial \psi_j^e}{\partial r} + a_{zz} \frac{\partial \psi_i^e}{\partial z} \frac{\partial \psi_j^e}{\partial z} \right) r \, dr \, dz$$

$$f_i^e = \int_{\Omega_e} \psi_i^e f r \, dr \, dz, \quad Q_i^e = \oint_{\Gamma_e} \psi_i^e q_n \, ds \tag{3.4.18}$$

3.5 The Least-Squares Method

3.5.1 Background

The success of weak-form Galerkin finite element models in the field of solid and structural mechanics is intimately connected with the notion of global minimization of quadratic functionals, which often represent the total energy of the system. When weak forms of a set of partial differential equations can be obtained equivalently through the minimization of a quadratic functional, the finite element solution becomes an orthogonal projection of the exact solution onto the space of approximations associated with the finite element discretization. The resulting numerical solution represents the best possible approximation of the exact solution in the space of approximation as measured with respect to the norm induced by the functional, called the *energy norm*. When the energy norm can be shown to be equivalent to a norm in an appropriate Hilbert space, such as the H^1, optimal convergence rates of the finite element solution can be established. Such a setting, often referred to as a *variational* setting, is ideal for finite element approximation.

The finite element models based on the weak-form Galerkin approach often depart from the ideal variational setting. For example, the weak-form formulation of the Stokes equations yields a constrained variational problem, whose discrete solution must satisfy restrictive compatibility conditions on the approximations used for velocity components and pressure. The weak-form Galerkin finite element model of the Navier–Stokes equations, on the other hand, is completely divorced from any minimization principles and further inherits the inf-sup condition [25–27] of the Stokes problem. Numerical solutions are far from optimal as characterized by the need for severe mesh refinement in order to suppress spurious oscillations of the solution. A considerable amount of research in recent years has been devoted to modifications of the weak form Galerkin approach in the hope of obtaining a more favorable setting for the numerical solution. These include the meshless and particle methods [28–37] and least-squares finite element formulations [38–45], including the k-version finite element method advanced by Surana, Reddy, and their colleagues [40–42]. Stabilized finite element formulations based on the penalty, SUPG and Galerkin least-squares methods have been proposed and extensively researched [46–48] (also see [10, 49, 50] for additional references). These schemes have yet to gain wide acceptance, due in part to the associated time- and mesh-dependent ad hoc parameters that must be fine tuned in each formulation.

Finite element models based on least-squares formulations offer an attractive alternative to the more popular weak-form Galerkin approach. In the case of linear problems, the least-squares formulation always admits a symmetric positive-definite coefficient matrix, regardless of whether or not such symmetry is manifest in the governing partial differential equations. As a result, robust iterative solution algorithms such as the preconditioned conjugate gradient method can be employed in the solution process, and only half of the coefficient matrix need be stored in memory. This is not the case when the weak-form Galerkin scheme is applied to nonself-adjoint systems of equations. In the context of fluid mechanics problems, the least-squares formulation does not suffer from the restrictive inf-sup condition. Approximations of the velocity and pressure can therefore be sought using the same bases of interpolation.

Least-squares variational formulations allow us to define a convex least-squares functional constructed in terms of the sum of the squares of the norms of the residuals of approximation of a set of differential equations. It is also possible to define a norm associated with the least-squares functional. If it can be shown that the norm induced by the least-squares functional is equivalent to a H^m norm, optimal convergence rates can be established for the least-squares finite element approximations. Under such conditions, the least-squares finite element model constitutes an ideal *variational* setting, regardless of whether or not such a setting is achieved by the associated weak-form Galerkin formulation. Unfortunately, norm equivalence (or H^1-coercivity of the least squares functional) cannot always be established for a given least-squares functional. However, unlike weak-form Galerkin formulations, such a departure from the ideal variational setting does not typically result in disastrous consequences for least-squares finite element models. In particular, Pontaza and Reddy [43, 44] and Prabhakar and Reddy [45] demonstrated numerically that hp least-squares finite element models are capable of yielding highly accurate results even when the least-squares functional cannot be shown to be H^1-coercive *a priori*.

Least-squares formulations are certainly not without their own shortcomings. Most problems in engineering physics possess, at the very minimum, second-order spatial differential operators. Least-squares formulations of such equations typically require higher-order approximations of solution variables. This significantly affects the conditioning of the coefficient matrix and the number of iterations needed for convergence of an iterative solution scheme used to solve linear equations. To circumvent these challenges, the governing equations are often recast as a set of first-order equations through the introduction of auxiliary variables. This not only allows for a better conditioning of the matrices, but also leads to a larger global system of equations. It can be argued that such a formulation is at least somewhat useful, since the auxiliary variables often represent important physical quantities of interest (e.g. heat flux and components of the stress tensor). The objective of this section is to discuss the basic idea of the least-squares formulation. A more detailed discussion will be presented in Chapter 11.

3.5.2 The Basic Idea

The least-squares method is based on the idea of minimizing the error functional, which is the integral of the squares of the errors introduced in the approximation of a differential equation and associated boundary conditions, with respect to the parameters (i.e. nodal values) of the approximation. To illustrate the idea, we consider a problem described by linear operator equations involving a single variable u defined over a domain Ω

$$A(u) - f = 0, \quad \text{in} \ \ \Omega \quad \text{and} \quad B(u) - g = 0 \ \ \text{on} \ \ \Gamma \tag{3.5.1}$$

where A and B are linear differential operators and f and g are the data.

Suppose that the domain Ω is decomposed into a set of finite elements Ω^e and the dependent variable u is approximated in Ω^e by expression of the form

$$u(\mathbf{x}) \approx u_h^e(\mathbf{x}) = \sum_{j=1}^{n} \Delta_j^e \varphi_j^e(\mathbf{x}) \tag{3.5.2}$$

where Δ_j are unknown parameters, related to the values of u and possibly its derivatives, and $\varphi_j(\mathbf{x})$ are suitably selected approximation functions. Then

$$\mathcal{R}_1^e \equiv A(u_h^e) - f = \sum_{j=1}^{n} \Delta_j^e A(\varphi_j^e(\mathbf{x})) - f \neq 0, \quad \text{in} \ \ \Omega^e$$

$$\mathcal{R}_2^e \equiv B(u_h^e) - g = \sum_{j=1}^{n} \Delta_j^e B(\varphi_j^e(\mathbf{x})) - g \neq 0, \quad \text{on} \ \ \Gamma^e$$

are the *residuals* in the differential equation and the boundary conditions due to the approximation of u by u_h^e.

To cast the problem in a suitable functional analysis setting, recall from Section 1.7 the standard notation used for the Sobolev (or Hilbert) space of order $m \geq 0$, $H^m(\Omega^e)$ and $H^m(\Gamma^e)$ of functions defined in the domain Ω^e and its boundary Γ^e. When $m = 0$, we use the notation $L_2(\Omega^e) = H^0(\Omega^e)$ and $L_2(\Gamma^e) = H^0(\Gamma^e)$. Therefore, one can construct the least-squares functional in terms of the sum of the squares of the L_2 norms of the residuals

$$\mathcal{J}^e(u_h^e; f, g) = \frac{1}{2} \int_{\Omega^e} [\mathcal{R}_1^e]^2 \, d\mathbf{x} + \frac{1}{2} \int_{\Gamma^e} [\mathcal{R}_2^e]^2 \, ds$$

$$= \frac{1}{2} \left(\|A(u_h^e) - f\|_{\Omega^e,0}^2 + \|B(u_h^e) - g\|_{\Gamma^e,0}^2 \right) \qquad (3.5.3)$$

The least-squares (minimum) principle associated with the least-squares functional $\mathcal{J}^e$ can be stated as follows: find $u_h^e \in \mathcal{U}_h \subset H(\Omega^e)$ such that

$$\mathcal{J}(u_h^e; f, g) \leqslant \mathcal{J}(w_h^e; f, g) \quad \text{for all } w_h^e \in \mathcal{U}_h \qquad (3.5.4)$$

where $\mathcal{U}_h$ is an appropriate vector space, such as $H^1(\Omega^e)$, spanned by the set $\{\varphi_i^e\}_{i=1}^n$. The necessary condition for the minimum of $\mathcal{J}(u_h^e)$ is that its first variation be zero

$$\delta\mathcal{J}(u_h^e, \delta u_h^e; f, g) = (A(\delta u_h^e), A(u_h^e) - f)_{\Omega^e,0} + (B(\delta u_h^e), B(u_h^e) - g)_{\Gamma^e,0} \quad (3.5.5)$$

where $\delta u_h^e \in \mathcal{U}_h$ is an admissible variation of u_h^e, $(\cdot, \cdot)_{\Omega^e,0}$ and $(\cdot, \cdot)_{\Gamma^e,0}$ are the L_2-inner products, and $\|\cdot\|_{\Omega^e,0}$ and $\|\cdot\|_{\Gamma^e,0}$ are the associated norms. Now the variational problem associated with Eq. (3.5.5) can be stated as: find $u_h^e \in \mathcal{U}_h$ such that

$$\mathcal{B}(w_h^e, u_h^e) = \mathcal{F}(w_h^e) \quad \text{for all } w_h^e \in \mathcal{U}_h \qquad (3.5.6)$$

where the bilinear form $\mathcal{B}$ and linear form $\mathcal{F}$ are given as

$$\mathcal{B}(w_h^e, u_h^e) = (A(w_h^e), A(u_h^e))_{\Omega^e,0} + (B(w_h^e), B(u_h^e))_{\Gamma^e,0}$$
$$\mathcal{F}(w_h^e) = (A(w_h^e), f)_{\Omega^e,0} + (B(w_h^e), g)_{\Gamma^e,0} \qquad (3.5.7)$$

We can define a norm $\|\cdot\|_A$ associated with the least-squares functional as

$$\|u\|_A = \sqrt{\mathcal{J}(u; 0, 0)} \qquad (3.5.8)$$

We note that, when A and B are linear, the bilinear form $\mathcal{B}(\cdot, \cdot)$ is symmetric regardless of whether or not operator A is self-adjoint. When the differential operator A is nonlinear, this is no longer the case.

Substitution of the finite element approximation, Eq. (3.5.2) into the variational statement in Eq. (3.5.6), we obtain the finite element model

$$\mathbf{K}^e \mathbf{\Delta}^e = \mathbf{F}^e \qquad (3.5.9)$$

where

$$K_{ij}^e = \left(A(\varphi_i^e), A(\varphi_j^e)\right)_{\Omega^e,0} + \left(B(\varphi_i^e), B(\varphi_j^e)\right)_{\Gamma^e,0}$$
$$= \int_{\Omega^e} A(\varphi_i^e)\, A(\varphi_j^e)\; d\mathbf{x} + \int_{\Gamma^e} B(\varphi_i^e)\, B(\varphi_j^e)\; ds$$
$$F_i^e = \left(A(\varphi_i^e), f\right)_{\Omega^e,0} + \left(B(\varphi_i^e), g\right)_{\Gamma^e,0}$$
$$= \int_{\Omega^e} A(\varphi_i^e)\, f\; d\mathbf{x} + \int_{\Gamma^e} B(\varphi_i^e)\, g\; ds$$

(3.5.10)

We note that the approximations φ_i^e should be selected such that $A(\varphi_i^e)$ exists and is square-integrable, and $B(u_h^e)$ is continuous across the interelement boundary. Thus, $\varphi_i^e(\mathbf{x})$ must belong to the Hermite class of functions when A is a second- or higher-order differential operator.

3.6 Numerical Integration

3.6.1 Preliminary Comments

An accurate representation of irregular domains (i.e. domains with curved boundaries) can be accomplished by the use of refined meshes and/or irregularly shaped curvilinear elements. For example, a non-rectangular region cannot be represented using rectangular elements; however, it can be represented by quadrilateral elements. Since the interpolation functions are easily derivable for a rectangular element and it is easier to evaluate integrals over rectangular geometries, we transform the finite element integral statements defined over quadrilaterals to a rectangle. The transformation results in complicated expressions for the integrands in terms of the coordinates used for the rectangular element. Therefore, numerical integration is used to evaluate such complicated integrals. The numerical integration schemes, such as the Gauss–Legendre numerical integration scheme, require the integral to be evaluated on a specific domain or with respect to a specific coordinate system.

3.6.2 Coordinate Transformations

Gauss quadrature requires the integral to be expressed over a square region $\hat{\Omega}$ of dimension 2×2 with respect to the coordinate system, (ξ, η) such that $-1 \leq (\xi, \eta) \leq 1$. The transformation of the geometry and the variable coefficients of the differential equation from the problem coordinates (x, y) to the local coordinates (ξ, η) results in algebraically complex expressions, and they preclude analytical (i.e. exact) evaluation of the integrals. Thus, the transformation of a given integral expression defined over an element Ω^e to one on the domain $\hat{\Omega}$ facilitates the numerical integration. Each element Ω^e of the finite element mesh Ω_h is transformed to $\hat{\Omega}$, only for the purpose of numerically

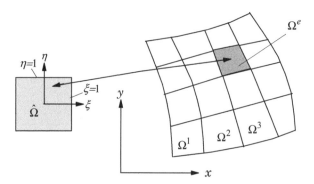

Fig. 3.6.1: Transformation of quadrilateral elements to the master rectangular element for numerical evaluation of integrals.

evaluating the integrals (see Fig. 3.6.1). The element $\hat{\Omega}$ is called a *master element*. For example, every quadrilateral element can be transformed to a square element with a side of length 2 and $-1 \leq (\xi, \eta) \leq 1$ that facilitates the use of Gauss–Legendre quadrature to evaluate integrals defined over the quadrilateral element.

The transformation between a typical element Ω^e in the mesh and the master element $\hat{\Omega}$ [or equivalently, between (x, y) and (ξ, η)] is accomplished by a coordinate transformation of the form

$$x = \sum_{j=1}^{m} x_j^e \phi_j^e(\xi, \eta) , \qquad y = \sum_{j=1}^{m} y_j^e \phi_j^e(\xi, \eta) \tag{3.6.1}$$

where ϕ_j^e denotes the finite element interpolation functions of the master element $\hat{\Omega}$. The coordinates in the master element are chosen to be the natural coordinates (ξ, η) such that $-1 \leq (\xi, \eta) \leq 1$. This choice is dictated by the limits of integration in the Gauss quadrature rule used to evaluate the integrals.

The transformation in Eq. (3.6.1) maps, for example, the line $\xi = 1$ in $\hat{\Omega}$ to the line defined parametrically by $x = x(1, \eta)$ and $y = y(1, \eta)$ in the xy-plane. In other words, the master element $\hat{\Omega}$ is transformed, under the linear transformation, into a quadrilateral element (i.e. a four-sided element whose sides are not parallel) in the xy-plane. Conversely, every quadrilateral element of a mesh can be transformed to the same four-node square (master) element $\hat{\Omega}$ in the (ξ, η)-plane.

In general, the dependent variable(s) of the problem are approximated by expressions of the form

$$u(\mathbf{x}) \approx u_h^e(\mathbf{x}) = \sum_{j=1}^{n} u_j^e \psi_j^e(\mathbf{x}) \tag{3.6.2}$$

The interpolation functions ψ_j^e used for the approximation of the dependent variable, in general, are different from ϕ_j^e used in the approximation of the geometry. Depending on the relative degree of approximations used for the geometry [see Eq. (3.6.1)] and the dependent variable(s) [see Eq. (3.6.2)], the finite element formulations are classified into three categories:

1. *Superparametric* $(m > n)$. The approximation used for the geometry is higher order than that used for the dependent variable.
2. *Isoparametric* $(m = n)$. Equal degrees of approximation are used for both geometry and dependent variables.
3. *Subparametric* $(m < n)$. Higher-order approximation of the dependent variable is used.

It should be noted that the transformation of a quadrilateral element of a mesh to the master element $\hat{\Omega}$ is solely for the purpose of numerically evaluating the integrals (see Fig. 3.6.1). *No transformation of the physical domain or elements is involved in the finite element analysis.* The resulting algebraic equations of the finite element formulation are always in terms of the nodal values of the physical domain. Different elements of the finite element mesh can be generated from the same master element by assigning appropriate global coordinates to each of the elements. Master elements of a different order define different transformations and hence different collections of finite elements within the mesh. For example, a quadratic rectangular master element can be used to generate a mesh of quadratic curvilinear quadrilateral elements. The transformations of a master element should be such that no spurious gaps exist between elements, and no element overlaps occur. For example, consider the element coefficients

$$K_{ij}^e = \int_{\Omega^e} \left[a_{xx}(\mathbf{x}) \frac{\partial \psi_i^e}{\partial x} \frac{\partial \psi_j^e}{\partial x} + a_{yy}(\mathbf{x}) \frac{\partial \psi_i^e}{\partial y} \frac{\partial \psi_j^e}{\partial y} \right] dA \qquad (3.6.3)$$

The integrand (i.e. the expression in the square brackets under the integral) is a function of the global coordinates x and y. We must rewrite it in terms of ξ and η using the transformation in Eq. (3.6.1). Note that the integrand contains not only functions but also derivatives with respect to the global coordinates (x, y). Therefore, we must relate $\left(\frac{\partial \psi_i^e}{\partial x}, \frac{\partial \psi_i^e}{\partial y} \right)$ to $\left(\frac{\partial \psi_i^e}{\partial \xi}, \frac{\partial \psi_i^e}{\partial \eta} \right)$ using the transformation in Eq. (3.6.1).

The functions $\psi_i^e(x, y)$ can be expressed in terms of the local coordinates (ξ, η) by means of the transformation in Eq. (3.6.1). Hence, by the chain rule of partial differentiation, we have

$$\frac{\partial \psi_i^e}{\partial \xi} = \frac{\partial \psi_i^e}{\partial x} \frac{\partial x}{\partial \xi} + \frac{\partial \psi_i^e}{\partial y} \frac{\partial y}{\partial \xi}; \qquad \frac{\partial \psi_i^e}{\partial \eta} = \frac{\partial \psi_i^e}{\partial x} \frac{\partial x}{\partial \eta} + \frac{\partial \psi_i^e}{\partial y} \frac{\partial y}{\partial \eta}$$

or, in matrix notation

$$
\left\{ \begin{array}{c} \frac{\partial \psi_i^e}{\partial \xi} \\ \frac{\partial \psi_i^e}{\partial \eta} \end{array} \right\} = \left[\begin{array}{cc} \frac{\partial x}{\partial \xi} & \frac{\partial y}{\partial \xi} \\ \frac{\partial x}{\partial \eta} & \frac{\partial y}{\partial \eta} \end{array} \right] \left\{ \begin{array}{c} \frac{\partial \psi_i^e}{\partial x} \\ \frac{\partial \psi_i^e}{\partial y} \end{array} \right\} \tag{3.6.4}
$$

which gives the relation between the derivatives of ψ_i^e with respect to the global and local coordinates. The matrix in Eq. (3.6.4) is called the *Jacobian matrix* of the transformation in Eq. (3.6.1):

$$
\mathbf{J}^e = \left[\begin{array}{cc} \frac{\partial x}{\partial \xi} & \frac{\partial y}{\partial \xi} \\ \frac{\partial x}{\partial \eta} & \frac{\partial y}{\partial \eta} \end{array} \right] \tag{3.6.5}
$$

Note from the expression given for K_{ij}^e in Eq. (3.6.3) that we must relate $(\frac{\partial \psi_i^e}{\partial x},$ $\frac{\partial \psi_i^e}{\partial y})$ to $(\frac{\partial \psi_i^e}{\partial \xi}, \frac{\partial \psi_i^e}{\partial \eta})$ whereas Eq. (3.6.4) provides the inverse relations. Therefore, Eq. (3.6.4) must be inverted. We have

$$
\left\{ \begin{array}{c} \frac{\partial \psi_i^e}{\partial x} \\ \frac{\partial \psi_i^e}{\partial y} \end{array} \right\} = (\mathbf{J}^e)^{-1} \left\{ \begin{array}{c} \frac{\partial \psi_i^e}{\partial \xi} \\ \frac{\partial \psi_i^e}{\partial \eta} \end{array} \right\} \tag{3.6.6}
$$

This requires that the Jacobian matrix $\mathbf{J}^e$ be non-singular (i.e. $J_e = \det \mathbf{J}^e \neq 0$). In fact, to transform a right-handed coordinate system into a right-handed coordinate system, we require $J_e > 0$.

Using the transformation in Eq. (3.6.1), we can write

$$
\frac{\partial x}{\partial \xi} = \sum_{j=1}^{m} x_j^e \frac{\partial \phi_j^e}{\partial \xi} , \quad \frac{\partial y}{\partial \xi} = \sum_{j=1}^{m} y_j^e \frac{\partial \phi_j^e}{\partial \xi}
$$

$$
\frac{\partial x}{\partial \eta} = \sum_{j=1}^{m} x_j^e \frac{\partial \phi_j^e}{\partial \eta} , \quad \frac{\partial y}{\partial \eta} = \sum_{j=1}^{m} y_j^e \frac{\partial \phi_j^e}{\partial \eta} \tag{3.6.7}
$$

and by means of Eq. (3.6.5) one can compute the Jacobian matrix and then its inverse. Thus, given the global coordinates (x_j^e, y_j^e) of node j of element Ω^e and the interpolation functions ϕ_j^e used for geometry approximation, the Jacobian matrix can be evaluated using Eq. (3.6.5). A necessary and sufficient condition for $(\mathbf{J}^e)^{-1}$ to exist is that the determinant J_e, called the Jacobian, be non-zero at every point (ξ, η) in $\hat{\Omega}$:

$$
J_e \equiv \det \mathbf{J}^e = \frac{\partial x}{\partial \xi} \frac{\partial y}{\partial \eta} - \frac{\partial x}{\partial \eta} \frac{\partial y}{\partial \xi} \neq 0. \tag{3.6.8}
$$

From Eq. (3.6.8) it is clear that the functions $\xi(x, y)$ and $\eta(x, y)$ must be continuous, differentiable, and invertible. Moreover, the transformation should be algebraically simple so that the Jacobian matrix can be easily evaluated.

Transformations of the form in Eq. (3.6.1) satisfy these requirements and the requirement that no spurious gaps between elements or overlapping of elements occur.

Returning to numerical evaluation of integrals, we have from Eq. (3.6.6),

$$
\left\{ \begin{array}{c} \frac{\partial \psi_i^e}{\partial x} \\ \frac{\partial \psi_i^e}{\partial y} \end{array} \right\} = (\mathbf{J}^e)^{-1} \left\{ \begin{array}{c} \frac{\partial \psi_i^e}{\partial \xi} \\ \frac{\partial \psi_i^e}{\partial \eta} \end{array} \right\} \equiv \mathbf{J}^* \left\{ \begin{array}{c} \frac{\partial \psi_i^e}{\partial \xi} \\ \frac{\partial \psi_i^e}{\partial \eta} \end{array} \right\} \tag{3.6.9}
$$

where J_{ij}^* is the element in position (i, j) of the inverse of the Jacobian matrix $\mathbf{J}^e$. The element area $dA = dx\, dy$ in element Ω^e is transformed to $d\xi d\eta$ in the master element $\hat{\Omega}$ with the relation

$$
dA = J_e\, d\xi\, d\eta \tag{3.6.10}
$$

Equations (3.6.7)–(3.6.10) provide the necessary relations to transform integral expressions on any element Ω^e to the master element $\hat{\Omega}^e$. For instance, consider the integral expression in Eq. (3.6.3), where a_{xx} and a_{yy} are functions of x and y. Suppose that the finite element Ω^e can be generated by the master element $\hat{\Omega}^e$. Under transformation, Eq. (3.6.1), we can write

$$
K_{ij}^e = \int_{\Omega^e} \left[a_{xx}(\mathbf{x}) \frac{\partial \psi_i^e}{\partial x} \frac{\partial \psi_j^e}{\partial x} + a_{yy}(\mathbf{x}) \frac{\partial \psi_i^e}{\partial y} \frac{\partial \psi_j^e}{\partial y} \right] dA \equiv \int_{\hat{\Omega}^e} F_{ij}^e(\xi, \eta)\, d\xi\, d\eta \tag{3.6.11}
$$

where

$$
\begin{aligned}
F_{ij}^e = \Big[& a_{xx}(\xi, \eta) \left(J_{11}^* \frac{\partial \psi_i^e}{\partial \xi} + J_{12}^* \frac{\partial \psi_i^e}{\partial \eta} \right) \left(J_{11}^* \frac{\partial \psi_j^e}{\partial \xi} + J_{12}^* \frac{\partial \psi_j^e}{\partial \eta} \right) \\
& + a_{yy}(\xi, \eta) \left(J_{21}^* \frac{\partial \psi_i^e}{\partial \xi} + J_{22}^* \frac{\partial \psi_i^e}{\partial \eta} \right) \left(J_{21}^* \frac{\partial \psi_j^e}{\partial \xi} + J_{22}^* \frac{\partial \psi_j^e}{\partial \eta} \right) \Big] J_e \tag{3.6.12}
\end{aligned}
$$

The discussion presented above is valid for master elements of both rectangular and triangular geometry. In the present study, the geometry will be approximated using the same degree of Lagrange interpolation functions as those used for the field variable(s). When all or some of the variables are approximated using Hermite interpolation functions, linear approximation of the geometry is used; thus, isoparametric or subparametric formulations are adopted.

3.6.3 Integration Over a Master Rectangular Element

Integrals defined over a rectangular master element $\hat{\Omega}_R^e = [-1, 1] \times [-1, 1]$ can be numerically evaluated using the Gauss–Legendre quadrature formulas

$$
\int_{\hat{\Omega}_R^e} F(\xi, \eta)\, d\xi\, d\eta = \int_{-1}^{1} \int_{-1}^{1} F(\xi, \eta)\, d\xi\, d\eta \approx \sum_{I=1}^{M} \sum_{J=1}^{N} F(\xi_I, \eta_J)\, W_I W_J \tag{3.6.13}
$$

Table 3.6.1: Gauss quadrature points and weights.
$$\int_{-1}^{1} F(\xi) \, d\xi = \sum_{I=1}^{N} F(\xi_I) \, W_I$$

N	Points, ξ_I	Weights, W_I
1	0.0000000000	2.0000000000
2	$\pm$ 0.5773502692	1.0000000000
3	0.0000000000	0.8888888889
	$\pm$ 0.7745966692	0.5555555555
4	$\pm$ 0.3399810435	0.6521451548
	$\pm$ 0.8611363116	0.3478548451
5	0.0000000000	0.5688888889
	$\pm$ 0.5384693101	0.4786286705
	$\pm$ 0.9061798459	0.2369268850
6	$\pm$ 0.2386191861	0.4679139346
	$\pm$ 0.6612093865	0.3607615730
	$\pm$ 0.9324695142	0.1713244924

where M and N denote the number of Gauss quadrature points, (ξ_I, η_J) denote the Gauss point coordinates, and W_I and W_J denote the corresponding Gauss weights as shown in Table 3.6.1.

The selection of the number of Gauss points is based on the formula $N = \text{int}[(p+1)/2]$, where p is the polynomial degree to which the integrand is approximated, and "int" means the nearest integer greater than or equal to $[(p+1)/2]$. In most cases, the interpolation functions are of the same degree in both ξ and η, and therefore one has $M = N$. When the integrand is of a different degree in ξ and η, we use $max(M, N)$. The minimum allowable quadrature rule is one that yields the area or volume of the element exactly. The maximum degree of the polynomial refers to the degree of the highest polynomial in ξ or η that is present in the integrands of the element matrices of the type in Eq. (3.6.3). Note that the polynomial degree of the coefficients a_{xx} and a_{yy} as well as $J_e(\xi, \eta)$ should be accounted for in determining the total polynomial degree of the integrand. The polynomial degree of J_e in ξ and η depends on $\phi_i^e(\xi, \eta)$ as well as the geometric shape of element Ω^e.

3.6.4   Integration Over a Master Triangular Element

In the preceding section, we discussed numerical integration on quadrilateral elements which can be used to represent very general geometries as well as field variables in a variety of problems. It is possible to distort a quadrilateral element to obtain a required triangular element by moving the position of the corner nodes, and the fourth corner in the quadrilateral is merged with one of the neighboring nodes (see Bathe [4] for the details). Here we discuss the transformations from a master triangular element to an arbitrary triangular element.

We choose the unit right isosceles triangle as the master element, $\hat{\Omega}_T^e$. An arbitrary triangular element Ω^e can be generated from the master triangular element $\hat{\Omega}_T^e$ by transformation of the form in Eq. (3.6.1). The derivatives of the area coordinate L_i^e [see Eq. (3.3.33)] with respect to the global coordinates can be computed from Eq. (3.6.6), which take the form

$$\left\{\begin{array}{c} \frac{\partial \psi_i^e}{\partial x} \\ \frac{\partial \psi_i^e}{\partial y} \end{array}\right\} = (\mathbf{J}^e)^{-1} \left\{\begin{array}{c} \frac{\partial \psi_i^e}{\partial L_1} \\ \frac{\partial \psi_i^e}{\partial L_2} \end{array}\right\}, \qquad \mathbf{J}^e = \left[\begin{array}{cc} \frac{\partial x}{\partial L_1} & \frac{\partial y}{\partial L_1} \\ \frac{\partial x}{\partial L_2} & \frac{\partial y}{\partial L_2} \end{array}\right] \qquad (3.6.14)$$

Note that only L_1 and L_2 are treated as linearly independent coordinates because $L_3 = 1 - L_1 - L_2$. After transformation, integrals on $\hat{\Omega}_T^e$ have the form

$$\int_{\Omega^e} G(x, y)\, dx\, dy = \int_{\hat{\Omega}_T^e} G(L_1, L_2, L_3)\, dL_1\, dL_2 \qquad (3.6.15)$$

which can be approximated by the quadrature formula

$$\int_{\hat{\Omega}_T^e} G(L_1, L_2, L_3)\, dL_1\, dL_2 \approx \sum_{I=1}^{N} G(\mathbf{S}_I) W_I \qquad (3.6.16)$$

where W_I and $\mathbf{S}_I$ denote the weights and integration points of the quadrature rule. Table 3.6.2 contains the location of integration points and weights for one-, three-, and four-point quadrature rules over a triangular element.

3.7 Computer Implementation

3.7.1 General Comments

In this section, computer implementation of finite element calculations is presented to illustrate the ease with which theoretical ideas can be transformed into practice. The material presented here is based on a more detailed account presented by Reddy [2] and Reddy and Gartling [10]. We begin with some general comments on a typical finite element program.

A typical computer program consists of three basic parts, as shown in the flow chart presented in Fig. 3.7.1. They are (1) Preprocessor, (2) Processor, and (3) Postprocessor. In the preprocessor part of a program, the input data of the problem are read in and/or generated. This includes the geometry of the domain, analysis option (e.g. static, eigenvalue, or transient analysis), the data of the problem (e.g. definition of the coefficients appearing in the differential equation), boundary conditions, finite element analysis information (e.g. element type, number of elements, geometric information required to generate the finite element mesh and element connectivity), and indicators for various

Table 3.6.2: Quadrature weights and points for triangular elements.

Number of integration points	Degree of polynomial Order of the residual	Geometric locations	S_1	S_2	S_3	W_I
1	1 $O(h^2)$	a	1/3	1/3	1/3	1
3	2 $O(h^3)$	a	1/2	0	1/2	1/3
		b	1/2	1/2	0	1/3
		c	0	1/2	1/2	1/3
4	3 $O(h^4)$	a	1/3	1/3	1/3	$-27/48$
		b	0.6	0.2	0.2	25/48
		c	0.2	0.6	0.2	25/48
		d	0.2	0.2	0.6	25/48

postprocessing options (e.g. printing of calculated values, quantities to be post-computed, and so on). In the post-processor part of the program, the primary variables are computed by interpolation at points of the domain other than the nodes, secondary variables that are derivable from the solution are also computed, and the output data are processed in a desired format for printout and/or graphs. The preprocessor and post-processor computer modules may contain a few Fortran statements to read and print pertinent information, simple subroutines (e.g. subroutines to generate mesh and compute the gradient of the solution), or complex programs linked to other units via disk and tape files.

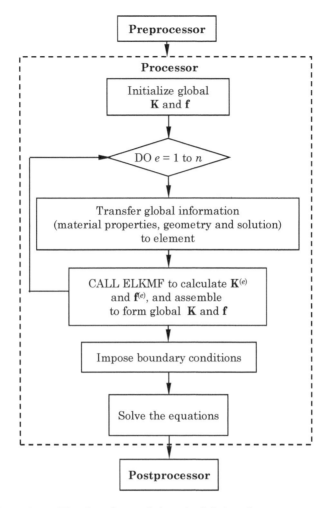

Fig. 3.7.1: The flowchart of a typical finite element program.

The processor module, where typically large amounts of computing time are spent, may consist of several subroutines, each having a special purpose, such as generation of element matrices, assembly of element equations, imposition of boundary conditions, and solution of equations. The degree of sophistication and the complexity of a finite element program depend on the general class of problems being programmed, the generality of the data in the equation, and the intended user of the program. Interested readers may consult Chapters 7 and 13 of the linear finite element book by Reddy [2], where details of the computer implementation and use of computer programs FEM1D and FEM2D are presented. Appendix 1 contains Fortran subroutines for the solution of systems of symmetric and unsymmetric equations.

3.7.2   One-Dimensional Problems

Here we discuss the main ideas behind the calculation of element coefficient matrices, $\mathbf{K}^e$ and $\mathbf{f}^e$, for the model problem discussed in Sections 3.2 and 3.3. Recall that the coefficients of $\mathbf{K}^e$ and $\mathbf{f}^e$ involve integrals of the form

$$K_{ij}^e = \int_{x_a}^{x_b} \left[a(x) \frac{d\psi_i^e}{dx} \frac{d\psi_j^e}{dx} + c(x)\psi_i^e \psi_j^e \right] dx, \quad f_i^e = \int_{x_a}^{x_b} f(x)\psi_i^e(x)\, dx \quad (3.7.1)$$

To evaluate these integrals using the Gauss quadrature, we first transform the integrals posed over interval (x_a, x_b) to the ones posed on interval $(-1, 1)$ by the use of the coordinate transformation [see Eq. (3.6.1)]

$$x = \sum_{j=1}^{m} x_j^e \phi_j^e(\xi) \quad (3.7.2)$$

In Eq. (3.7.2), x_j^e denotes the global coordinate of node j of element $\Omega^e = (x_a, x_b)$, and ϕ_j^e is the approximation function used to approximate the geometry. For example, if we use linear interpolation functions to represent the geometry of the element, we have $x_1^e = x_a$, $x_2^e = x_b$, $\phi_1^e = 0.5(1-\xi)$, $\phi_2^e = 0.5(1+\xi)$, and Eq. (3.7.2) becomes

$$x = x_a \phi_1^e(\xi) + x_b \phi_2^e(\xi) = \frac{x_a + x_b}{2} + \frac{x_b - x_a}{2}\xi = x_a + \frac{h_e}{2}\xi \quad (3.7.3)$$

and the Jacobian of transformation is a constant

$$J_e \equiv \frac{dx}{d\xi} = \frac{h_e}{2} \quad (3.7.4)$$

The derivatives of $\psi_i^e(x)$ with respect to the global coordinate x is given by

$$\frac{d\psi_i^e}{dx} = \frac{d\psi_i^e}{d\xi} \frac{d\xi}{dx} = \frac{d\psi_i^e}{d\xi} J_e^{-1}$$

Then the integrals in Eq. (3.7.1) become

$$K_{ij}^e = \int_{x_a}^{x_b} \left[a(x) \frac{d\psi_i^e}{dx} \frac{d\psi_j^e}{dx} + c(x)\psi_i^e \psi_j^e \right] dx$$

$$= \int_{-1}^{1} \left[\hat{a}(\xi) \left(\frac{d\psi_i^e}{d\xi} J_e^{-1} \right) \left(\frac{d\psi_j^e}{d\xi} J_e^{-1} \right) + \hat{c}(\xi)\psi_i^e \psi_j^e \right] J_e\, d\xi \quad (3.7.5)$$

$$f_i^e = \int_{x_a}^{x_b} f(x)\psi_i^e(x)\, dx = \int_{-1}^{1} \hat{f}(\xi)\psi_i^e(\xi)\, J_e(\xi)\, d\xi$$

where $\hat{a}(\xi) = a(x(\xi))$ and so on. Each of the integral expressions above can be evaluated using the Gauss quadrature

$$\int_{-1}^{1} F(\xi) J_e\, d\xi = \sum_{I=1}^{NGP} F(\xi_I)\, J_e(\xi_I)\, W_I \quad (3.7.6)$$

where NGP is the number of Gauss points, and ξ_I is the Ith Gauss point and W_I is the Ith Gauss weight from the Gauss rule with NGP points.

To implement the above development into a computer subroutine for an arbitrary degree of ψ_i^e, we must first create a subroutine of all interpolation functions and their derivatives with respect to ξ that we intend to use in our analysis. For the present discussion, we limit them to the linear and quadratic Lagrange family of functions. The functions and their derivatives are given below.

Linear

$$\psi_1(\xi) = \frac{1}{2}(1-\xi), \quad \frac{d\psi_1}{d\xi} = -0.5; \quad \psi_2(\xi) = \frac{1}{2}(1+\xi), \quad \frac{d\psi_2}{d\xi} = 0.5 \quad (3.7.7)$$

Quadratic

$$\psi_1(\xi) = -\frac{1}{2}\xi(1-\xi), \quad \psi_2(\xi) = \left(1-\xi^2\right), \quad \psi_3(\xi) = \frac{1}{2}\xi(1+\xi)$$

$$\frac{d\psi_1}{d\xi} = 0.5(2\xi-1), \quad \frac{d\psi_2}{d\xi} = -2\xi, \quad \frac{d\psi_3}{d\xi} = 0.5(1+2\xi)$$

$$(3.7.8)$$

Approximation functions ψ_i and their derivatives with respect to the global coordinate x at point ξ are evaluated in subroutine **INTERPLN1D**, which is presented in Box 3.7.1.

The following variable names are used in the subroutine **INTERPLN1D**:

$$\text{SFL}(i) = \psi_i^e, \quad \text{DSFL}(i) = \frac{d\psi_i^e}{d\xi}, \quad \text{GDSFL}(i) = \frac{d\psi_i^e}{dx} \quad (3.7.9)$$

The notation should be transparent to the reader: SFL = shape functions of the Lagrange family; DSFL = derivative of SFL with respect to ξ; and GDSFL = the global derivative (i.e. derivative with respect to x) of SFL. All of them are $n \times 1$ arrays, where n is the number of nodes per element, NPE. The notation GJ is used for the Jacobian J^e and XI for the normalized coordinate ξ.

Next, we implement the steps to evaluate the matrix coefficients K_{ij}^e and f_i^e ($i = 1$ to NPE) into a subroutine, **ELMATRCS1D** (element coefficient matrix). To this end, we assume the following polynomial representation of the data, $a(x)$, $c(x)$, and $f(x)$ for the entire domain of the problem:

$$a(x) = a_0 + a_1 x, \quad c(x) = c_0 + c_1 x, \quad f(x) = f_0 + f_1 x$$
$$\text{AX} = \text{AX0} + \text{AX1} \cdot x, \quad \text{CX} = \text{CX0} + \text{CX1} \cdot x, \quad \text{FX} = \text{FX0} + \text{FX1} \cdot x$$

The coefficients $a_0 = \text{AX0}$, $a_1 = \text{AX1}$, etc. of the polynomials must be read in the preprocessor and transferred to the subroutine **ELMATRCS1D** through a COMMON block (or the argument list). For this choice of data, it is sufficient to use NGP = 2 for the linear element and NGP = 3 for the quadratic element to exactly evaluate the integrals appearing in the definition of $\mathbf{K}^e$ and $\mathbf{f}^e$.

Box 3.7.1: Listing of subroutine **INTERPLN1D**.

```
      SUBROUTINE INTERPLN1D(ELX,GJ,IEL,MODEL,NPE,XI)
C
C     Subroutine to compute 1-D Lagrange (L) linear and quadratic,
C     and Hermite (H) cubic interpolation functions and their global
C     derivatives. Here SF stands for "shape functions".
C
C     SFL(I)    Interpolation functions of the Lagrange type
C     GDSFL(I)  Global derivative (with respect to x) of SFL(I)
C     DSFL(I)   Local derivative (with respect to xi) of SFL(I)
C     SFH(I)    Interpolation functions of the Hermite type
C     GDSFH(I)  Global derivative (with respect to x) of SFH(I)
C     DSFH(I)   Local derivative (with respect to xi) of SFH(I)
C     EL        Length of the line element
C     XI        Normalized local coordinate xi such that -1 < xi < 1
C
      IMPLICIT REAL*8(A-H,O-Z)
      COMMON/SHP/SFH(4),GDSFH(4),GDDSFH(4),SFL(4),GDSFL(4)
      DIMENSION DSFH(4),DSFL(4),DDSFH(4),ELX(3)
C
      EL=ELX(NPE)-ELX(1)
C
C     LAGRANGE LINEAR interpolation functions and their local derivatives
C
      IF(IEL.EQ.1)THEN
         SFL(1) =0.5*(1.0-XI)
         SFL(2) =0.5*(1+XI)
         DSFL(1)=-0.5
         DSFL(2)=0.5
      ENDIF
C
C     LAGRANGE QUADRATIC interpolation functions and their derivatives
C
      IF(IEL.EQ.2)THEN
         SFL(1) =0.5*(XI-1.0)*XI
         SFL(2) =1.0-XI*XI
         SFL(3) =0.5*(XI+1.0)*XI
         DSFL(1)=XI-0.5
         DSFL(2)=-2.0*XI
         DSFL(3)=XI+0.5
      ENDIF
C
C     Compute the Jacobian (GJ) and the GLOBAL DERIVATIVES
C     of the interpolation functions
C
      GJ=0.0
      DO 10 I=1,NPE
10    GJ=GJ+DSFL(I)*ELX(I)
C
      DO 20 I=1,NPE
20    GDSFL(I)=DSFL(I)/GJ
```

Box 3.7.1: Listing of subroutine **INTERPLN1D** (continued).

```
C
C    HERMITE CUBIC interpolation functions   [see Eq. (5.2.28)]
C
      IF(MODEL.EQ.2)THEN
         SFH(1)=0.25*(2.0-3.0*XI+XI**3)
         SFH(2)=-EL*(1.0-XI)*(1.0-XI*XI)/8.0
         SFH(3)=0.25*(2.0+3.0*XI-XI**3)
         SFH(4)=+EL*(1.0+XI)*(1.0-XI*XI)/8.0
         DSFH(1)    = -0.75*(1.0-XI*XI)
         DSFH(2)    = +EL*(1.0+2.0*XI-3.0*XI*XI)/8.0
         DSFH(3)    = 0.75*(1.0-XI*XI)
         DSFH(4)    = +EL*(1.0-2.0*XI-3.0*XI*XI)/8.0
         DDSFH(1)   = 1.5*XI
         DDSFH(2)   = +0.25*EL*(1.0-3.0*XI)
         DDSFH(3)   = -1.5*XI
         DDSFH(4)   = -0.25*(1.0+3.0*XI)*EL
         DO 30 I=1,4
            GDSFH(I) = DSFH(I)/GJ
30          GDDSFH(I)= DDSFH(I)/GJ/GJ
      ENDIF
      RETURN
      END
```

Since the evaluation of the matrix coefficients involves summation on the number of Gauss points, the arrays used for $\mathbf{K}^e$ and $\mathbf{f}^e$ must be initialized outside the do-loop on the number of Gauss points, NGP. The following notation is used:

$$\text{ELK}(i,j) = K_{ij}^e, \quad \text{ELF}(i) = f_i^e, \quad \text{ELX}(i) = x_i^e, \quad i = 1, 2, \ldots, n = \text{NPE}$$

$$\text{GAUSPT}(I, J) = \xi_I, \ I\text{th Gauss point of the } J\text{-point Gauss rule}$$

$$\text{GAUSWT}(I, J) = W_I, \ I\text{th Gauss weight of the } J\text{-point Gauss rule}$$

The Gauss points are arranged in a two-dimensional matrix array form so that the Jth column corresponds to the Jth-order Gauss rule. The same notation is used for the Gauss weights. Inside the do-loop on $I = 1$ to NGP, subroutine INTRPLN1D is called to compute SFL(i) and GDSFL(i) at the Ith Gauss point, and then all necessary parameters and coefficients are computed. The integrals required in the definition of K_{ij}^e are denoted as follows:

$$\text{S00}(i,j) \equiv S_{ij}^{00} = \int_{x_a}^{x_b} \psi_i^e \, \psi_j^e \, dx, \quad \text{S01}(i,j) \equiv S_{ij}^{01} = \int_{x_a}^{x_b} \psi_i^e \frac{d\psi_j^e}{dx} \, dx$$

$$\text{S10}(i,j) \equiv S_{ij}^{10} = \int_{x_a}^{x_b} \frac{d\psi_i^e}{dx} \psi_j^e \, dx, \quad \text{S11}(i,j) \equiv S_{ij}^{11} = \int_{x_a}^{x_b} \frac{d\psi_i^e}{dx} \frac{d\psi_j^e}{dx} \, dx$$

$$(3.7.10)$$

Box 3.7.2 contains a listing of Subroutine ELMATRCS1D.

Box 3.7.2: Listing of subroutine **ELMATRCS1D**.

```
      SUBROUTINE ELMATRCS1D(IEL,NDF,NGP,NPE)
C     ---------------------------------------------------------------------
C     IEL............... Element type (1, Linear; 2, Quadratic)
C     H................. Element length
C     X, XI............. Global and local coordinates, x and xi
C     [GAUSPT],[GAUSWT]. Matrices of Gauss points and Gauss weights
C     [ELK],{ELF} ...... Element coefficient matrix [K] source vector {F}
C     {ELX}............. Vector of the global coordinates of element nodes
C     ---------------------------------------------------------------------
      IMPLICIT REAL*8(A-H,O-Z)
      COMMON/STF1/ELK(9,9),ELF(9),ELX(4)
      COMMON/STF2/AX0,AX1,CX0,CX1,FX0,FX1
      COMMON/SHP/SFL(4),GDSFL(4)
      DIMENSION GAUSPT(5,5),GAUSWT(5,5)
      DATA GAUSPT/5*0.0D0,-0.57735027D0,0.57735027D0,3*0.0D0,-0.77459667D0,
     *  .0D0,0.77459667D0,2*0.0D0,-0.86113631D0,-0.33998104D0,0.33998104D0,
     *  .86113631D0,.0D0,-0.906180D0,-0.538469D0,.0D0,.538469D0,.906180D0/
      DATA GAUSWT/2.0D0,4*0.0D0,2*1.0D0,3*0.0D0,0.55555555D0,0.88888888D0,
     *  0.55555555D0,2*0.0D0,0.34785485D0,2*0.65214515D0,0.34785485D0,0.0D0,
     *  0.236927D0,0.478629D0,0.568889D0,0.478629D0,0.236927D0/
C
C     Initialize the arrays
      DO 10 J=1,NPE
      ELF(J) = 0.0
         DO 10 I=1,NPE
10    ELK(I,J)=0.0
C
C DO-LOOP on number of Gauss points begins here
C
      DO 50 NI=1,NGP
         XI=GAUSPT(NI,NGP)
         CALL INTERPLN1D(ELX,GJ,IEL,NPE,XI)
         CNST=GJ*GAUSWT(NI,NGP)
         X=ELX(1)+0.5*(1.0+XI)*H
         AX=AX0+AX1*X
         CX=CX0+CX1*X
         FX=FX0+FX1*X
C
C  Calculate element coefficients for Model 1 equation (see Chs. 3-5)
C
         DO 40 I=1,NPE
            ELF(I)=ELF(I)+FX*SFL(I)*CNST
            DO 40 J=1,NPE
               S00=SFL(I)*SFL(J)*CNST
               S11=GDSFL(I)*GDSFL(J)*CNST
               ELK(I,J)=ELK(I,J)+AX*S11+CX*S00
40          CONTINUE
50       CONTINUE
      RETURN
      END
```

3.7.3   Two-Dimensional Problems

The ideas presented in Section 3.7.2 for one-dimensional problems extend in a straightforward way to two-dimensional problems. The main differences are: (a) numerical integration in two dimensions; (b) computation of the Jacobian matrix $\mathbf{J}^e$ defined in Eq. (3.6.5); and (c) computation of global derivatives of the interpolation functions using Eq. (3.6.9).

Subroutine **INTERPLN2D** is used to evaluate the interpolation functions and its derivatives for four-node, eight-node, and nine-node quadrilateral elements, and the element coefficient matrices $\mathbf{K}^e$ and $\mathbf{f}^e$ are generated in the subroutine **ELMATRCS2D**. Once again, we note that the boundary integrals q_i^e are not programmed because the sum of q_i^e at an interior global node is set to zero to enforce 'equilibrium' of secondary variables, and one needs to specify the value at nodes where the sum is known [through specified secondary variables; see Eq. (3.3.27) for additional discussion]. As stated earlier, in the present study, the geometry is approximated using the lowest-order Lagrange interpolation functions, that is, isoparametric or subparametric formulations are used.

The number of Gauss points [NGP $= (p+1)/2$] used in each coordinate direction to evaluate K_{ij}^e is determined using the polynomial degree p of the integrand in that coordinate. For example, the integral

$$\int_{\Omega^e} c(\mathbf{x})\,\psi_i^e(\mathbf{x})\,\psi_j^e(\mathbf{x})\,dA = \int_{-1}^{1}\int_{-1}^{1} c(\xi,\eta)\,\psi_i^e(\xi,\eta)\,\psi_j^e(\xi,\eta)J_e(\xi,\eta)\,d\xi d\eta$$

can be evaluated using the $M \times N$ Gauss rule (M Gauss points along the ξ direction and N points along the η direction), where the values of M and N depend on the degree p in ξ and q in η of the integrand F_{ij}^e

$$M = \left[\frac{p+1}{2}\right], \quad N = \left[\frac{q+1}{2}\right], \quad F_{ij}^e(\xi,\eta) = c(\xi,\eta)\,\psi_i^e(\xi,\eta)\,\psi_j^e(\xi,\eta)J_e(\xi,\eta)$$

Thus, the combined polynomial variation of c, ψ_i^e, and J_e in ξ and η dictates the values of M and N.

In general, the Jacobian J_e is a function of ξ and η. When the element is rectangular in shape, the Jacobian is a constant, independent of ψ_i^e. If the element is irregular but with straight sides, then J_e is a linear function of ξ and η whenever ψ_i^e is linear and it is a cubic function of ξ and η whenever ψ_i^e is quadratic. When the element has quadratic curved sides, the Jacobian is a fourth-degree polynomial of ξ and η. For example, if c is a constant and ψ_i^e are linear, then F_{ij}^e is cubic in both ξ and η, and therefore NGP $= M = N = 2$. Similarly, if c is a constant, ψ_i^e are quadratic, and the element is straight-sided, then F_{ij}^e is fifth degree in both ξ and η and therefore NGP $= M = N = 3$; if the element has curved boundaries, then NGP $= M = N = 5$.

The Lagrange interpolation functions given in Eqs. (3.3.38)–(3.3.40) for rectangular elements can be expressed succinctly as follows:

Linear element $(i = 1, 2, 3, 4)$

$$\psi_i = \tfrac{1}{4}(1 + \xi_i\,\xi)(1 + \eta_i\,\eta), \quad \frac{\partial \psi_i}{\partial \xi} = \tfrac{1}{4}\xi_i(1 + \eta_i\,\eta), \quad \frac{\partial \psi_i}{\partial \eta} = \tfrac{1}{4}(1 + \xi_i\,\xi)\eta_i \quad (3.7.11)$$

Nine-node quadratic element (see Fig. 3.3.7 for the node numbers)
- for corner nodes $(i = 1, 2, 3, 4)$

$$\psi_i = \tfrac{1}{4}(\xi^2 + \xi_i\,\xi)(\eta^2 + \eta_i\,\eta)$$
$$\frac{\partial \psi_i}{\partial \xi} = \tfrac{1}{4}(2\xi + \xi_i)(\eta^2 + \eta_i\,\eta) \qquad (3.7.12)$$
$$\frac{\partial \psi_i}{\partial \eta} = \tfrac{1}{4}(\xi^2 + \xi_i\,\xi)(2\eta + \eta_i)$$

- for midside nodes $(i = 5, 6, 7, 8)$

$$\psi_i = \tfrac{1}{2}\left[\xi_i\xi(1 + \xi\,\xi_i)(1 - \eta^2) + \eta_i\eta(1 + \eta\eta_i)(1 - \xi^2)\right]$$
$$\frac{\partial \psi_i}{\partial \xi} = \tfrac{1}{2}\xi_i(1 + 2\xi_i\xi)(1 - \eta^2) - \eta_i\eta(1 + \eta\,\eta_i)\xi \qquad (3.7.13)$$
$$\frac{\partial \psi_i}{\partial \eta} = -\xi_i\xi(1 + \xi\,\xi_i)\eta + \tfrac{1}{2}\eta_i(1 + 2\eta_i\eta)(1 - \xi^2)$$
$$\psi_9 = (1 - \xi^2)(1 - \eta^2)$$
$$\frac{\partial \psi_9}{\partial \xi} = -2\xi(1 - \eta^2), \quad \frac{\partial \psi_9}{\partial \eta} = -2(1 - \xi^2)\eta \qquad (3.7.14)$$

Eight-node quadratic element (see Fig. 3.3.8 for the node numbers)
- for corner nodes $(i = 1, 2, 3, 4)$

$$\psi_i = \tfrac{1}{4}(1 + \xi_i\,\xi)(1 + \eta_i\,\eta)(\xi_i\,\xi + \eta_i\,\eta - 1)$$
$$\frac{\partial \psi_i}{\partial \xi} = \tfrac{1}{4}\xi_i\left[(1 + \eta_i\,\eta)(\xi_i\,\xi + \eta_i\,\eta - 1) + (1 + \xi_i\,\xi)(1 + \eta_i\,\eta)\right] \quad (3.7.15)$$
$$\frac{\partial \psi_i}{\partial \eta} = \tfrac{1}{4}\eta_i\left[(1 + \xi_i\,\xi)(\xi_i\,\xi + \eta_i\,\eta - 1) + (1 + \xi_i\,\xi)(1 + \eta_i\,\eta)\right]$$

- for midside nodes $(i = 5, 6, 7, 8)$

$$\psi_i = \tfrac{1}{2}\xi_i^2(1 + \xi_i\,\xi)(1 - \eta^2) + \tfrac{1}{2}\eta_i^2(1 + \eta_i\,\eta)(1 - \xi^2)$$
$$\frac{\partial \psi_i}{\partial \xi} = \tfrac{1}{2}\xi_i^3(1 - \eta^2) - \eta_i^2\,\xi(1 + \eta_i\,\eta) \qquad (3.7.16)$$
$$\frac{\partial \psi_i}{\partial \eta} = -\xi_i^2\,\eta(1 + \xi_i\,\xi) + \tfrac{1}{2}\eta_i^3(1 - \xi^2)$$

where (ξ_i, η_i) denote the coordinates of the ith node of the element $\hat{\Omega}$. Note that we have (see Fig. 3.3.7 for the node numbers)

$$
\begin{aligned}
(\xi_1, \eta_1) &= (-1, -1), & (\xi_2, \eta_2) &= (1, -1), & (\xi_3, \eta_3) &= (1, 1) \\
(\xi_4, \eta_4) &= (-1, 1), & (\xi_5, \eta_5) &= (0, -1), & (\xi_6, \eta_6) &= (1, 0) \\
(\xi_7, \eta_7) &= (0, 1), & (\xi_8, \eta_8) &= (-1, 0), & (\xi_9, \eta_9) &= (0, 0)
\end{aligned}
\tag{3.7.17}
$$

Fortran programs of **INTERPLN2D** and **ELMATRCS2D** are presented in Box 3.7.3 and Box 3.7.4, respectively. The following variables and definitions, in addition to those defined earlier, are used in the subroutines; other variables are introduced in the programs to define the required expressions (e.g. array NP, ξ_0, η_0, etc):

NDF = number of primary degrees of freedom per node (here NDF=1)

NGP = number of Gauss points used to evaluate K_{ij}^e and f_i^e

$$
\mathrm{SF}(i) = \psi_i^e, \quad \mathrm{DSFL}(1, i) = \frac{\partial \psi_i^e}{\partial \xi}, \quad \mathrm{DSFL}(2, i) = \frac{\partial \psi_i^e}{\partial \eta}, \quad \mathrm{GDSFL}(1, i) = \frac{\partial \psi_i^e}{\partial x}
$$

$$
\mathrm{GDSFL}(2, i) = \frac{\partial \psi_i^e}{\partial y}, \quad \mathrm{ELXY}(i, 1) = x_i^e, \quad \mathrm{ELXY}(i, 2) = y_i^e, \quad \mathrm{DET} = J^e
$$

$$
\mathrm{GJ} = \mathbf{J}^e, \quad \mathrm{GJINV} = (\mathbf{J}^e)^{-1}, \quad \mathrm{XNODE}(i, 1) = \xi_i, \quad \mathrm{XNODE}(i, 2) = \eta_i
$$

where (ξ_i, η_i) are the coordinates of the ith node of the master element

$$
S_{ij}^{00} = \int_{\Omega^e} \psi_i^e \psi_j^e \, dxdy, \qquad S_{ij}^{10} = \int_{\Omega^e} \frac{\partial \psi_i^e}{\partial x} \psi_j^e \, dxdy, \qquad S_{ij}^{01} = \int_{\Omega^e} \psi_i^e \frac{\partial \psi_j^e}{\partial x} \, dxdy
$$

$$
S_{ij}^{11} = \int_{\Omega^e} \frac{\partial \psi_i^e}{\partial x} \frac{\partial \psi_j^e}{\partial x} \, dxdy, \quad S_{ij}^{20} = \int_{\Omega^e} \frac{\partial \psi_i^e}{\partial y} \psi_j^e \, dxdy, \quad S_{ij}^{02} = \int_{\Omega^e} \psi_i^e \frac{\partial \psi_j^e}{\partial y} \, dxdy
$$

$$
S_{ij}^{22} = \int_{\Omega^e} \frac{\partial \psi_i^e}{\partial y} \frac{\partial \psi_j^e}{\partial y} \, dxdy, \quad S_{ij}^{\alpha\beta} = \int_{\Omega^e} \frac{\partial \psi_i^e}{\partial x_\alpha} \frac{\partial \psi_j^e}{\partial x_\beta} \, dxdy
\tag{3.7.18}
$$

where $S_{ij}^{\alpha\beta}$ denotes the integral of the product of first derivative of ψ_i^e with respect to x_α and the first derivative of ψ_j^e with respect to x_β (with $x_1 = x$ and $x_2 = y$). The matrix coefficients defined in Eq. (3.7.18) can be used conveniently to define almost any element matrix. For example, the element matrix in Eq. (3.6.3) can be expressed as [see Eq. (3.3.31)]

$$
\mathrm{ELK}(i, j) = \mathrm{AXX} * \mathrm{S11}(i, j) + \mathrm{AYY} * \mathrm{S22}(i, j)
$$

The computer implementation of the finite element models developed for axisymmetric problems in Section 3.4 follows the ideas discussed here. In fact, by an appropriate definition of the coefficients a, c, a_{xx}, a_{yy}, and f, one can use the programs developed in this section for axisymmetric problems as well.

Box 3.7.3: Listing of subroutine **INTERPLN2D**.

```
      SUBROUTINE INTERPLN2D(NPE,XI,ETA,DET,ELXY)
C
C     The subroutine computes SF and GDSF for 2-D rectangular elements
C
      IMPLICIT REAL*8 (A-H,O-Z)
      DIMENSION ELXY(9,2),XNODE(9,2),NP(9),DSFL(2,9),GJ(2,2),GJINV(2,2)
      COMMON/SHP/SFL(9),GDSFL(2,9)
      COMMON/IO/IN,IT
      DATA XNODE/-1.0D0, 2*1.0D0, -1.0D0, 0.0D0, 1.0D0, 0.0D0, -1.0D0,
     *        0.0D0, 2*-1.0D0, 2*1.0D0, -1.0D0, 0.0D0, 1.0D0, 2*0.0D0/
      DATA NP/1,2,3,4,5,7,6,8,9/
      FNC(A,B) = A*B
      IF(NPE.EQ.4) THEN
C   LINEAR Lagrange interpolation functions for FOUR-NODE element
         DO 10 I = 1, NPE
         XP  = XNODE(I,1)
         YP  = XNODE(I,2)
         XI0 = 1.0+XI*XP
         ETA0=1.0+ETA*YP
         SFL(I)    = 0.25*FNC(XI0,ETA0)
         DSFL(1,I)= 0.25*FNC(XP,ETA0)
  10     DSFL(2,I)= 0.25*FNC(YP,XI0)
      ELSE
         IF(NPE.EQ.8) THEN
C   QUADRATIC Lagrange interpolation functions for EIGHT-NODE element
            DO 20 I = 1, NPE
            NI   = NP(I)
            XP   = XNODE(NI,1)
            YP   = XNODE(NI,2)
            XI0  = 1.0+XI*XP
            ETA0 = 1.0+ETA*YP
            XI1  = 1.0-XI*XI
            ETA1 = 1.0-ETA*ETA
            IF(I.LE.4) THEN
               SFL(NI)    = 0.25*FNC(XI0,ETA0)*(XI*XP+ETA*YP-1.0)
               DSFL(1,NI) = 0.25*FNC(ETA0,XP)*(2.0*XI*XP+ETA*YP)
               DSFL(2,NI) = 0.25*FNC(XI0,YP)*(2.0*ETA*YP+XI*XP)
            ELSE
               IF(I.LE.6) THEN
                  SFL(NI)    = 0.5*FNC(XI1,ETA0)
                  DSFL(1,NI) = -FNC(XI,ETA0)
                  DSFL(2,NI) = 0.5*FNC(YP,XI1)
               ELSE
                  SFL(NI)    = 0.5*FNC(ETA1,XI0)
                  DSFL(1,NI) = 0.5*FNC(XP,ETA1)
                  DSFL(2,NI) = -FNC(ETA,XI0)
               ENDIF
            ENDIF
  20        CONTINUE
      ELSE
C     QUADRATIC Lagrange interpolation functions for NINE-NODE element
```

Box 3.7.3: Listing of subroutine **INTERPLN2D** (continued).

```
              DO 30 I=1,NPE
              NI   = NP(I)
              XP   = XNODE(NI,1)
              YP   = XNODE(NI,2)
              XIO  = 1.0+XI*XP
              ETAO = 1.0+ETA*YP
              XI1  = 1.0-XI*XI
              ETA1 = 1.0-ETA*ETA
              XI2  = XP*XI
              ETA2 = YP*ETA
              IF(I .LE. 4) THEN
                  SFL(NI)    = 0.25*FNC(XIO,ETAO)*XI2*ETA2
                  DSFL(1,NI)= 0.25*XP*FNC(ETA2,ETAO)*(1.0+2.0*XI2)
                  DSFL(2,NI)= 0.25*YP*FNC(XI2,XIO)*(1.0+2.0*ETA2)
              ELSE
                  IF(I .LE. 6) THEN
                     SFL(NI)    = 0.5*FNC(XI1,ETAO)*ETA2
                     DSFL(1,NI) = -XI*FNC(ETA2,ETAO)
                     DSFL(2,NI) = 0.5*FNC(XI1,YP)*(1.0+2.0*ETA2)
                  ELSE
                     IF(I .LE. 8) THEN
                        SFL(NI)    = 0.5*FNC(ETA1,XIO)*XI2
                        DSFL(2,NI) = -ETA*FNC(XI2,XIO)
                        DSFL(1,NI) = 0.5*FNC(ETA1,XP)*(1.0+2.0*XI2)
                     ELSE
                        SFL(NI)    = FNC(XI1,ETA1)
                        DSFL(1,NI) = -2.0*XI*ETA1
                        DSFL(2,NI) = -2.0*ETA*XI1
                     ENDIF
                  ENDIF
              ENDIF
30            CONTINUE
          ENDIF
      ENDIF
      DO 40 I = 1,2
      DO 40 J = 1,2
      GJ(I,J) = 0.0
      DO 40 K = 1,NPE
40   GJ(I,J) = GJ(I,J) + DSFL(I,K)*ELXY(K,J)
      DET = GJ(1,1)*GJ(2,2)-GJ(1,2)*GJ(2,1)
      GJINV(1,1) = GJ(2,2)/DET
      GJINV(2,2) = GJ(1,1)/DET
      GJINV(1,2) = -GJ(1,2)/DET
      GJINV(2,1) = -GJ(2,1)/DET
      DO 50 I  = 1,2
      DO 50 J  = 1,NPE
      GDSFL(I,J) = 0.0
      DO 50 K  = 1, 2
50   GDSFL(I,J) = GDSFL(I,J) + GJINV(I,K)*DSFL(K,J)
      RETURN
      END
```

Box 3.7.4: Listing of subroutine **ELMATRCS2D**.

```
      SUBROUTINE ELMATRCS2D(NGP,NPE)
C
C  Element calculations for linear and quadratic quadrilateral elements
      IMPLICIT REAL*8(A-H,O-Z)
      COMMON/STF/ELF(18),ELK(18,18),ELXY(9,2),ELU(18)
      COMMON/PST/A10,A1X,A1Y,A20,A2X,A2Y,A00,A0X,A0Y,F0,FX,FY
      COMMON/SHP/SFL(9),GDSFL(2,9)
      DIMENSION  GAUSPT(5,5),GAUSWT(5,5)
      DATA GAUSPT/5*0.0D0, -0.57735027D0, 0.57735027D0, 3*0.0D0,
     2  -0.77459667D0, 0.0D0, 0.77459667D0, 2*0.0D0, -0.86113631D0,
     3  -0.33998104D0, 0.33998104D0, 0.86113631D0, 0.0D0, -0.90617984D0,
     4  -0.53846931D0,0.0D0,0.53846931D0,0.90617984D0/
C
      DATA GAUSWT/2.0D0, 4*0.0D0, 2*1.0D0, 3*0.0D0, 0.55555555D0,
     2   0.88888888D0, 0.55555555D0, 2*0.0D0, 0.34785485D0,
     3   2*0.65214515D0, 0.34785485D0, 0.0D0, 0.23692688D0,
     4   0.47862867D0, 0.56888888D0, 0.47862867D0, 0.23692688D0/
C
C Initialize the element arrays
      DO 100 I = 1,NPE
      ELF(I)   = 0.0
      DO 100 J = 1,NPE
100   ELK(I,J) = 0.0
C
C Do-loops on Gauss quadrature begin here
      DO 200 NI = 1,NGP
      DO 200 NJ = 1,NGP
      XI  = GAUSPT(NI,NGP)
      ETA = GAUSPT(NJ,NGP)
      CALL INTERPLN2D (NPE,XI,ETA,DET,ELXY)
      CNST = DET*GAUSWT(NI,NGP)*GAUSWT(NJ,NGP)
      X=0.0
      Y=0.0
      DO 120 I=1,NPE
         X=X+ELXY(I,1)*SFL(I)
120      Y=Y+ELXY(I,2)*SFL(I)
      FXY=F0+FX*X+FY*Y
      A00=A00+A0X*X+A0Y*Y
      A11=A10+A1X*X+A1Y*Y
      A22=A20+A2X*X+A2Y*Y
      DO 180 I=1,NPE
         ELF(I) = ELF(I)+FXY*SFL(I)*CNST
         DO 180 J=1,NPE
         S00=SFL(I)*SFL(J)*CNST
         S11=GDSFL(1,I)*GDSFL(1,J)*CNST
         S22=GDSFL(2,I)*GDSFL(2,J)*CNST
         ELK(I,J) = ELK(I,J) + A00*S00 + A11*S11 + A22*S22
180      CONTINUE
200   CONTINUE
      RETURN
      END
```

3.8 Summary

The present chapter was devoted to a study of: (1) the weak-form Galerkin (or Ritz) finite element models of one- and two-dimensional problems involving generalized Poisson's equation, (2) a derivation of interpolation functions for basic one- and two-dimensional elements, (3) description of the least-squares finite element model of a linear operator equation, (4) numerical evaluation of integrals, (5) computer implementation ideas, and (6) finite element models of axisymmetric problems. An understanding of the topics presented in this chapter is a necessary prerequisite for the subsequent chapters of this book.

The advantages of the weak-form finite element models of second-order boundary-value problems are: (1) identification of the primary and secondary variables – a duality that exists in all phenomena of nature, (2) weakening the differentiability required of approximation functions used for the primary variables and leading to C^0–approximations, and hence simpler finite elements, and (3) possible symmetry of the finite element coefficient matrix for most problems of engineering physics. The weak-form formulations are the most natural in solid mechanics problems because the weak forms are the same as those resulting from the principle of virtual displacements. In general, the derivation of the weak forms from the governing differential equations is subject to the requirement that the resulting secondary variables be physically meaningful in that they can be specified on the portions of the boundary where the corresponding primary variables are unknown.

A disadvantage of the C^0–approximation is that the derivatives of the variables being approximated are discontinuous across interelement boundaries. Another disadvantage of the weak-form formulations for problems without a physical principle (that give rise to the weak forms) is that they only offer a means for computing the solution, without the idea of minimization of the error in the approximation. Of course, one may interpret the weak forms as those equivalent to forcing the weighted-integral of the error in the differential equation to zero. Such statements are "weaker" than minimizing the error, as in the least-squares method. On the other hand, least-squares formulations involve higher-order approximations (the same as in the original differential equation) and/or lead to more unknowns in the resulting finite element model.

The strong point of methods that use global approximations (i.e. no subdivision of the domain into elements, like in the traditional Ritz and Galerkin methods) is that they do not introduce discontinuities in the derivatives of the solution between elements. At the moment there are no clear consensus on the overall effectiveness and robustness of such computational procedures to be competitive with or displace the current weak-form finite element technology. It is possible that a hybrid technology that utilizes different computational procedures for different problems or different parts of the same problem may emerge in the years ahead.

Problems

The readers may find additional examples and problems in the author's textbook [2] to test and extend their understanding of the finite element method as applied to linear problems of engineering and applied science.

FINITE ELEMENT MODELS OF ONE-DIMENSIONAL PROBLEMS

3.1 Consider the differential equation

$$-\frac{d}{dx}\left(a\frac{du}{dx}\right) + \frac{d^2}{dx^2}\left(b\frac{d^2u}{dx^2}\right) + cu = f$$

where a, b, c, and f are known functions of position x. (a) Develop the weak form over a typical element $\Omega^e = (x_a, x_b)$ such that the bilinear form is symmetric, (b) identify the bilinear and linear forms and construct the quadratic functional, (c) discuss the type of finite element approximation of u that may be used, and (d) develop the finite element model of the equation.

3.2 Derive the Lagrange cubic interpolation functions for a four-node (one-dimensional) element (with equally spaced nodes) using the alternative procedure based on interpolation properties listed in Eqs. (3.2.23) and (3.2.24). Use the local coordinate $\bar{x}$ for simplicity.

3.3 Derive the finite element model of the differential equation

$$-\frac{d}{dx}\left(a(x)\frac{du}{dx}\right) = f(x) \quad \text{for} \quad 0 < x < L \tag{1}$$

over an element while accounting for the mixed boundary condition of the following form at both ends of the element

$$\left[n_x\left(a(x)\frac{du}{dx}\right) + \beta(u - u_c)\right] = P \tag{2}$$

where $n_x = -1$ at $x = x_a$ and $n_x = 1$ at $x = x_b$.

3.4 Rewrite Eq. (3.2.1) as a pair of first-order equations

$$-\frac{dv}{dx} + cu - f = 0, \quad v - a\frac{du}{dx} = 0 \tag{1}$$

and develop the weak-form Galerkin finite element model of the pair.

FINITE ELEMENT MODELS OF TWO-DIMENSIONAL PROBLEMS

3.5 Evaluate the coefficients K_{ij}^e and F_i^e of Eq. (3.3.20) for a linear triangular element when a_{xx}, a_{yy}, f, h_c and u_c are constants.

3.6 Repeat Problem 3.5 for a linear rectangular element.

3.7 (*Plane Elasticity*) The governing equations of plane (i.e. two-dimensional) elasticity problems are summarized below.

Equilibrium of forces

$$\frac{\partial \sigma_{xx}}{\partial x} + \frac{\partial \sigma_{xy}}{\partial y} + f_x = 0, \tag{1}$$

$$\frac{\partial \sigma_{xy}}{\partial x} + \frac{\partial \sigma_{yy}}{\partial y} + f_y = 0, \tag{2}$$

where $(\sigma_{xx}, \sigma_{yy}, \sigma_{xy})$ are the stress components and f_x and f_y are the components of the body force vector (measured per unit volume) along the x- and y-directions, respectively.

Stress-displacement (or constitutive) relations

$$\left\{ \begin{array}{c} \sigma_{xx} \\ \sigma_{yy} \\ \sigma_{xy} \end{array} \right\} = \left[\begin{array}{ccc} c_{11} & c_{12} & 0 \\ c_{12} & c_{22} & 0 \\ 0 & 0 & c_{66} \end{array} \right] \left\{ \begin{array}{c} \frac{\partial u_x}{\partial x} \\ \frac{\partial u_y}{\partial y} \\ \frac{\partial u_x}{\partial y} + \frac{\partial u_y}{\partial x} \end{array} \right\} \tag{3}$$

where $c_{ij}(c_{ji} = c_{ij})$ are the elasticity (material) constants for an orthotropic medium with the material principal directions (x_1, x_2, x_3) coinciding with the coordinate axes (x, y, z) used to describe the problem and (u_x, u_y) are the displacements. The c_{ij} can be expressed in terms of the engineering constants $(E_1, E_2, \nu_{12}, G_{12})$ for an orthotropic material. For *plane stress* problems the elastic constants are given by

$$c_{11} = \frac{E_1}{(1 - \nu_{12}\nu_{21})}, \quad c_{22} = \frac{E_2}{(1 - \nu_{12}\nu_{21})}, \quad c_{12} = \nu_{12}c_{22} = \nu_{21}c_{11}, \quad c_{66} = G_{12} \tag{4}$$

There are **four** independent material constants for plane stress case: E_1, E_2, ν_{12}, and G_{12}. For isotropic case we have

$$c_{11} = c_{22} = \frac{E}{(1 - \nu^2)}, \quad c_{12} = \frac{\nu E}{(1 - \nu^2)}, \quad c_{66} = G \tag{5}$$

For *plane strain* problems they are given by

$$c_{11} = \frac{(1 - \nu_{23}\nu_{32})E_1}{(1 - \nu_{23}\nu_{32} - \nu_{13}\nu_{31} - \nu_{12}\nu_{21} - 2\nu_{12}\nu_{23}\nu_{31})},$$

$$c_{12} = \frac{(\nu_{12} + \nu_{13}\nu_{32})E_2}{(1 - \nu_{23}\nu_{32} - \nu_{13}\nu_{31} - \nu_{12}\nu_{21} - 2\nu_{12}\nu_{23}\nu_{31})}$$

$$= \frac{(\nu_{21} + \nu_{23}\nu_{31})E_1}{(1 - \nu_{23}\nu_{32} - \nu_{13}\nu_{31} - \nu_{12}\nu_{21} - 2\nu_{12}\nu_{23}\nu_{31})}, \tag{6}$$

$$c_{22} = \frac{(1 - \nu_{13}\nu_{31})E_2}{(1 - \nu_{23}\nu_{32} - \nu_{13}\nu_{31} - \nu_{12}\nu_{21} - 2\nu_{12}\nu_{23}\nu_{31})}$$

Thus, there are **seven** independent material constants for the plane strain case: E_1, E_2, E_3 (to determine ν_{31} and ν_{32}) ν_{12}, ν_{13}, ν_{23}, and G_{12}. For isotropic case, the constitutive equations reduce to

$$c_{11} = c_{22} = \frac{(1 - \nu)E}{(1 + \nu)(1 - 2\nu)}, \quad c_{12} = \frac{\nu E}{(1 + \nu)(1 - 2\nu)}, \quad c_{66} = G \tag{7}$$

Boundary conditions

$$\sigma_{xx}n_x + \sigma_{xy}n_y = \hat{t}_x, \quad \sigma_{xy}n_x + \sigma_{yy}n_y = \hat{t}_y \tag{8}$$

where (n_x, n_y) denote the components (or direction cosines) of the unit normal vector on the boundary Γ; $\hat{t}_x$ and $\hat{t}_y$ denote the components of the specified traction vector, and $\hat{u}_x$ and $\hat{u}_y$ are the components of specified displacement vector. Only one element of each pair, (u_x, t_x) and (u_y, t_y), may be specified at a boundary point. Eliminate the stresses from Eqs. (1) and (2) by substituting the stress–displacement relations (3). Develop weak-form Galerkin finite element model of the resulting equations.

3.8 Consider the stress equilibrium equations, Eqs. (1) and (2), and inverse of the stress-displacement relations, Eq. (3), in Problem 3.7 to formulate the mixed finite element model.

FINITE ELEMENT MODELS OF AXISYMMETRIC PROBEMS

3.9 Give the explicit forms of the coefficients K_{ij}^e and f_i^e in Eq. (3.4.12) for element-wise constant values of $a = a_e$ and $f = f_e$ for linear and quadratic elements.

3.10 The principle of minimum total potential energy for axisymmetric bending of polar orthotropic plates according to the first-order shear deformation theory requires $\delta\Pi(w, \phi_r) = 0$, where

$$\delta\Pi = \int_b^a \left[\left(D_{11}\frac{d\phi_r}{dr} + D_{12}\frac{\phi_r}{r} \right) \frac{d\delta\phi_r}{dr} + \frac{1}{r}\left(D_{12}\frac{d\phi_r}{dr} + D_{22}\frac{\phi_r}{r} \right) \delta\phi_r \right.$$
$$\left. + A_{55}\left(\phi_r + \frac{dw}{dr} \right)\left(\delta\phi_r + \frac{d\delta w}{dr} \right) - q\,\delta w \right] r\,dr \tag{1}$$

where b is the inner radius and a the outer radius. Derive the displacement finite element model of the equations. In particular, show that the finite element model is of the form

$$\begin{bmatrix} [K^{11}] & [K^{12}] \\ [K^{12}]^T & [K^{22}] \end{bmatrix} \begin{Bmatrix} \{w\} \\ \{\phi_r\} \end{Bmatrix} = \begin{Bmatrix} \{F^1\} \\ \{F^2\} \end{Bmatrix} \tag{2}$$

and define the matrix coefficients.

3.11 The following differential equation arises in connection with the axisymmetric deformation (i.e. only the radial displacement u_r is nonzero) of a thick-walled cylinder of inside radius a and outside radius b, made of an isotropic material (with Lamé constants, μ and λ), pressurized at $r = a$ and/or at $r = b$, and rotating about its own axis at a uniform angular velocity ω:

$$-\frac{1}{r}\frac{d}{dr}(r\sigma_{rr}) + \frac{1}{r}\sigma_{\theta\theta} - \rho\omega^2 r = 0, \quad a < r < b \tag{1}$$

where σ_{rr} and $\sigma_{\theta\theta}$ are radial and circumferential stresses, respectively, and they are related to the radial and circumferential components of the linear strains

$$\varepsilon_{rr} = \frac{du_r}{dr}, \quad \varepsilon_{\theta\theta} = \frac{u_r}{r} \tag{2}$$

by

$$\sigma_{rr} = (2\mu + \lambda)\varepsilon_{rr} + \lambda\varepsilon_{\theta\theta}, \quad \sigma_{\theta\theta} = (2\mu + \lambda)\varepsilon_{\theta\theta} + \lambda\varepsilon_{rr} \tag{3}$$

Develop the displacement finite element model of Eq. (1).

3.12 Consider the partial differential equation governing heat transfer in an axisymmetric geometry

$$-\frac{1}{r}\frac{\partial}{\partial r}\left(rk_{rr}\frac{\partial T}{\partial r} \right) - \frac{\partial}{\partial z}\left(k_{zz}\frac{\partial T}{\partial z} \right) = f(r, z) \tag{1}$$

where (k_{rr}, k_{zz}) and f are the conductivities and internal heat generation per unit volume, respectively. In developing the weak form, we integrate over the elemental volume of the axisymmetric geometry: $r\,dr\,d\theta\,dz$. Develop the weak form and associated finite element model over an element.

LEAST-SQUARES FINITE ELEMENT MODELS

3.13 Rewrite Eq. (3.2.1) as a pair of first-order equations

$$-\frac{dv}{dx} + cu - f = 0, \quad v - a\frac{du}{dx} = 0 \tag{1}$$

and develop the least-squares finite element model of the pair.

3.14 Consider the residual R^e of Eq. (3.2.3). In the least-squares method, we minimize the least-squares functional

$$\delta \int_{x_a}^{x_b} [R^e]^2 \, dx = 0 \quad \text{or} \quad \frac{\partial}{\partial c_i} \int_{x_a}^{x_b} [R^e(x, c_1, c_2, \cdots, c_n)]^2 \, dx = 0, \quad i = 1, 2, \cdots, n$$

(a) Identify the weight function w_i^e if the least-squares method is to be deduced from Eq. (3.2.4), (b) develop the least-squares finite element model, and (c) discuss the type of finite element approximation u_h^e (i.e. the nodal variables and approximation functions) that may be used.

3.15 Consider the equations

$$-\frac{dV}{dx} + kw - q = 0, \quad \theta + \frac{dw}{dx} = 0, \quad \frac{M}{EI} - \frac{d\theta}{dx} = 0, \quad -V + \frac{dM}{dx} = 0 \quad (1)$$

This set of equations arise in bending of beams according to the Euler–Bernoulli beam theory, where the first equation represents equilibrium of transverse forces on the beam, the second equation defines the slope θ in terms of the transverse deflection w, the third equation is a constitutive equation (moment-slope relation), and the fourth equation represents equilibrium of moments; here M is the bending moment, V the shear force, q the transversely distributed load, k the modulus of elastic foundation, and $EI > 0$ is the product of Young's modulus and second moment of area.
Eliminating V and θ from the above set yields the pair

$$-\frac{d^2 M}{dx^2} + kw - q = 0, \quad \frac{M}{EI} + \frac{d^2 w}{dx^2} = 0 \quad (2)$$

Develop the weak-form Galerkin finite element model of the pair and discuss the nature of approximation of the variables.

3.16 Develop a least-squares finite element model of the pair of equations in Eq. (2) of Problem 3.15.

3.17 Consider the following system of equations that arise in connection with the Timoshenko beam theory of straight beams:

$$-\frac{dV}{dx} + kw - q = 0, \quad \frac{V}{a} - \left(\phi + \frac{dw}{dx}\right) = 0, \quad (1,2)$$

$$\frac{M}{b} - \frac{d\phi}{dx} = 0, \quad -V + \frac{dM}{dx} = 0 \quad (3,4)$$

where the first equation represents equilibrium of transverse forces on the beam, the second equation defines the shear force V in terms of the rotation ϕ, the third equation is a constitutive equation between the bending moment M and rotation ϕ, and the fourth equation represents equilibrium of moments; here q is the transversely distributed load, k is the modulus of elastic foundation, $a > 0$ is the shear stiffness ($a = GAK_s$, G is the shear modulus, A is the area of cross section, and K_s is the shear correction factor), and $b > 0$ is the product of Young's modulus E and second moment of area I ($b = EI$). Eliminating the shear force V and bending moment M from the above equations, we arrive at

$$-\frac{d}{dx}\left[a\left(\phi + \frac{dw}{dx}\right)\right] + kw - q = 0, \quad (5)$$

$$-\frac{d}{dx}\left(b\frac{d\phi}{dx}\right) + a\left(\phi + \frac{dw}{dx}\right) = 0, \quad (6)$$

Develop the weak-form Galerkin finite element model of the equations.

3.18 Develop the least-squares finite element model of Eqs. (5) and (6) of Problem 3.17.

3.19 Develop the least-squares finite element model of Navier's equations of linearized plane elasticity:

$$\frac{\partial}{\partial x}\left(c_{11}\frac{\partial u_x^h}{\partial x} + c_{12}\frac{\partial u_y^h}{\partial y}\right) + \frac{\partial}{\partial y}\left[c_{66}\left(\frac{\partial u_x^h}{\partial y} + \frac{\partial u_y^h}{\partial x}\right)\right] + f_x = 0$$

$$\frac{\partial}{\partial x}\left[c_{66}\left(\frac{\partial u_x^h}{\partial y} + \frac{\partial u_y^h}{\partial x}\right)\right] + \frac{\partial}{\partial y}\left(c_{12}\frac{\partial u_x^h}{\partial x} + c_{22}\frac{\partial u_y^h}{\partial y}\right) + f_y = 0$$

3.20 Develop the mixed least-squares finite element model of Eqs. (1)–(4) of Problem 3.17.

Finite Element Analysis (problem-solving)

3.21 The following differential equation arises in connection with heat transfer in an insulated rod:

$$-\frac{d}{dx}\left(k\frac{dT}{dx}\right) = q \quad \text{for} \quad 0 < x < L$$

$$T(0) = T_0, \quad \left[k\frac{dT}{dx} + \beta(T - T_\infty) + \hat{q}\right]\Bigg|_{x=L} = 0$$

where T is the temperature, k the thermal conductivity, and q the heat generation. Take the following values for the data: $q = 0$, $\hat{q} = 0$, $L = 0.1$ m, $k = 0.01$ W m^{-1} $^\circ$C^{-1}, $\beta = 25$ W m$^{-2\circ}$C^{-1}, $T_0 = 50^\circ$C, and $T_\infty = 5^\circ$C. Solve the problem using two linear finite elements for temperature values at $x = \frac{1}{2}L$ and $x = L$, and heat at $x = 0$.
Answer: $T_2 = 27.59^\circ$C, $T_3 = 5.179^\circ$C, $Q_1^{(1)} = 4.482$ W m$^{-2} = -Q_2^{(2)}$.

3.22 An insulating wall is constructed of three homogeneous layers with conductivities k_1, k_2, and k_3 in intimate contact (see Fig. P3.22). Under steady-state conditions, the temperatures at the boundaries of the layers are characterized by the external surface temperatures T_1 and T_4 and the interface temperatures T_2 and T_3. Formulate the problem to determine the temperatures T_i ($i = 1, \ldots, 4$) when the ambient temperatures T_0 and T_5 and the (surface) film coefficients β_0 and β_5 are known for the following cases: *Case 1:* $T_0 = T_5 = 20^\circ$C and $\beta_0 = \beta_5 = 500$ W/m^2·C; *Case 2:* $T_0 = 120^\circ$C, $T_5 = 20^\circ$C, $\beta_0 = 500$ W/m^2·C, and $\beta_5 = 560$ W/m^2·C; *Case 3:* $T_1 = 100^\circ$C, $T_5 = 20^\circ$C, and $\beta_5 = 500$W/m^2·C. Assume that there is no internal heat generation and that the heat flow is one-dimensional ($\partial T/\partial y = 0$).

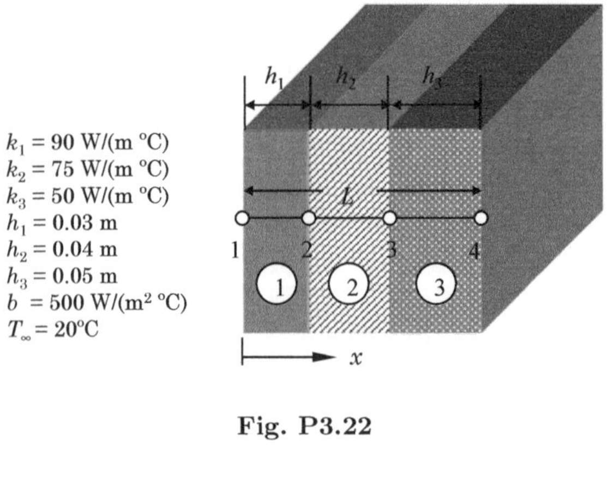

$k_1 = 90$ W/(m $^\circ$C)
$k_2 = 75$ W/(m $^\circ$C)
$k_3 = 50$ W/(m $^\circ$C)
$h_1 = 0.03$ m
$h_2 = 0.04$ m
$h_3 = 0.05$ m
$b = 500$ W/(m^2 $^\circ$C)
$T_\infty = 20^\circ$C

Fig. P3.22

3.23 Consider the steady laminar flow of a viscous fluid through a long circular cylindrical tube. The governing equation is

$$-\frac{1}{r}\frac{d}{dr}\left(r\mu\frac{dw}{dr}\right) = \frac{P_0 - P_L}{L} \equiv f_0, \quad 0 < r < R_0 \tag{1}$$

where w is the axial (i.e. z) component of velocity, μ is the viscosity, and f_0 is the gradient of pressure (which includes the combined effect of static pressure and gravitational force). The boundary conditions are

$$\left(r\frac{dw}{dr}\right)\bigg|_{r=0} = 0, \quad w(R_0) = 0 \tag{2}$$

Using (a) two linear elements and (b) one quadratic element, determine the velocity field and compare with the exact solution at the nodes:

$$w_e(r) = \frac{f_0 R_0^2}{4\mu}\left[1 - \left(\frac{r}{R_0}\right)^2\right] \tag{3}$$

3.24 In the problem of the flow of a viscous fluid through a circular cylinder (see Problem 3.23), assume that the fluid slips at the cylinder wall; that is, instead of assuming that $w = 0$ at $r = R_0$, use the boundary condition that

$$kw = -\mu\frac{dw}{dr} \quad \text{at} \quad r = R_0 \tag{1}$$

in which k is the "coefficient of sliding friction." Solve the problem with two linear elements.

3.25 The two members in Fig. P3.25 are fastened together and to rigid walls. If the members are stress free before they are loaded, what will be the stresses and deformations in each after two 50,000 lbs loads are applied? Use $E_s = 30 \times 10^6$ psi and $E_a = 10^7$ psi; the aluminum rod is of 2 in diameter and the steel rod is of 1.5 in diameter.

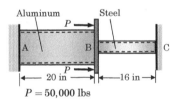

Fig. P3.25

3.26 Consider the steady-state heat transfer (or other phenomenon) in a square region shown in Fig. P3.26. The governing equation is given by

$$-\frac{\partial}{\partial x}\left(k\frac{\partial u}{\partial x}\right) - \frac{\partial}{\partial y}\left(k\frac{\partial u}{\partial y}\right) = f_0 \tag{1}$$

The boundary conditions for the problem are:

$$u(0,y) = y^2, \quad u(x,0) = x^2, \quad u(1,y) = 1 - y, \quad u(x,1) = 1 - x \tag{2}$$

Assuming $k = 1$ and $f_0 = 2$, determine the unknown nodal value of u using the finite element mesh shown in Fig. P3.26.

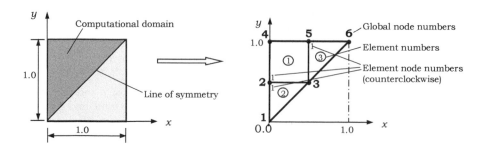

Fig. P3.26

3.27 Consider a single-degree freedom problem governed by the differential equation

$$-\nabla^2 u = 0 \tag{1}$$

over the domain shown in Fig. P3.27. Write the global finite element equation associated with global node 8 and identify all of the coefficients (other than unknown U_I) in the equation in terms of the element coefficients $K_{ij}^{(e)}$ (algebraic) and problem data only (i.e. you must give the numerical values of other contributions other than $K_{ij}^{(e)}$).

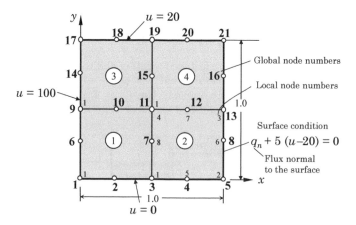

Fig. P3.27

3.28 Consider a single-degree freedom problem governed by the differential equation

$$-\frac{\partial}{\partial x}\left(k\frac{\partial u}{\partial x}\right) - \frac{\partial}{\partial y}\left(k\frac{\partial u}{\partial y}\right) = f_0 \tag{1}$$

over the domain shown in Fig. P3.28.

(a) Write the finite element equation associated with global node 1 in terms of element coefficients (include only the non-zero contributions).

(b) Compute the contribution of the flux q_0 to global nodes 1 and 4.

(c) Compute the contribution of the boundary condition $q_n + 5u = 0$ to global node 1 using the coefficients H_{ij}^e and P_i^e due to convection.

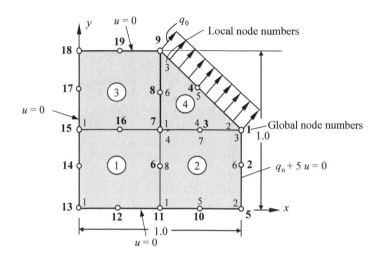

Fig. P3.28

3.29 The nodal values of a linear triangular element (see Fig. P3.29) in the finite-element analysis of the field problem $-\nabla^2 u = f_0$ are $u_1^e = U_{10} = 389.79$, $u_2^e = U_{12} = 337.19$, and $u_3^e = U_{11} = 395.08$. Find the (a) gradient of the solution

$$\nabla u \equiv \frac{\partial u}{\partial x}\hat{\mathbf{e}}_x + \frac{\partial u}{\partial y}\hat{\mathbf{e}}_y$$

in the element, and (b) the places where the $u = 392$ isoline intersects the boundary of the element.

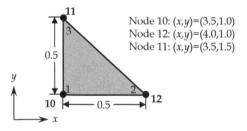

Node 10: $(x,y)=(3.5,1.0)$
Node 12: $(x,y)=(4.0,1.0)$
Node 11: $(x,y)=(3.5,1.5)$

Fig. P3.29

3.30 For the plane elasticity problem shown in Fig. P3.30, give

(a) the *known* nodal displacement and force (zero as well as nonzero) degrees of freedom and their specified values, and

(b) the global stiffness coefficients K_{22}, K_{19}, and $K_{5(13)}$ in terms of the element stiffness coefficients $K_{ij}^{(e)}$.

Assume that the element displacement degrees of freedom are placed in the order $\{u_1, v_1, u_2, v_2, \ldots, \}$ where u and v are the x and y components of the displacement vector $\mathbf{u}$, and global displacement vector is ordered as $\{U_1, V_1, U_2, V_2, \ldots, \}$.

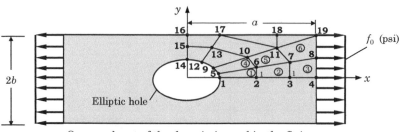

One quadrant of the domain is used in the finite
element analysis (isotropic body of thickness h).

Fig. P3.30

PROGRAMMING PROBLEMS

3.31 Define the array NOD(e,j) such that the eth row consists of the global node number
corresponding to the element node number j, where $e = 1, 2, \ldots,$ NEM, and $j = 1, 2, \ldots,$ NPE (NEM denotes the number of elements in the mesh and NPE is the
number of nodes per element). Assuming that the number of degrees of freedom per
node is NDF, put together a computer program to assemble element coefficient matrices
K^e_{ij} and f^e_i into a global coefficient matrices K_{IJ} and F_I.

3.32 Let us define the following arrays and variables:

Variable	Definition
NDF	Number of degrees of freedom per node
NSPV	Number of specified primary degrees of freedom (DoF), U
NSSV	Number of specified secondary DoF, $\hat{Q} \equiv n_x\, a\frac{du}{dx}$
NSMB	number of specified mixed boundary conditions of the type
	$$Q + \beta(U - U_{\text{ref}}) = \hat{Q} \qquad (1)$$
ISPV(I,J)	The array of the specified primary DoF (I $= 1, 2, \ldots,$ NSPV and J $= 1, 2, \ldots,$ NDF)
ISPV(I,1)	Global node number at which the primary variable is specified
ISPV(I,2)	Degree of freedom at the global node ISPV(I,1) that is specified (for the class of problems discussed in this chapter, NDF $= 1$ and therefore ISPV(I,2) $= 1$ for all cases)
VSPV(I)	Array of specified values of primary variable DoF in ISPV(I,J) [VSPV(I) should be in the same sequence as the specified DoF in array ISPV(I,1)]
ISSV(I,J)	Similar to ISPV(I,J), but for secondary variables
VSSV(I)	Similar to VSPV(I), but for secondary variables
ISMB(I,J)	Array of specified mixed boundary conditions in Eq. (1), I $= 1, 2, \ldots,$ NSMB, and has a meaning similar to arrays ISPV(I,J) and ISSV(I,J)
BETA0(I)	Array of specified values of β_0 in Eq. (1)
BETAU(I)	Array of specified values of β_u in Eq. (1)
UREF(I)	Array of specified values of U_{ref} in Eq. (1)

Write a subroutine to impose the specified boundary conditions on the primary and
secondary variables, as well as boundary conditions of the mixed kind.

3.33 Modify the subroutine **ELMATRCS1D** to evaluate the element coefficient matrix and force vector in Eq. (3.4.12) for axisymmetric problems.

3.34 Modify the subroutine **ELMATRCS2D** to evaluate the element coefficient matrix and force vector in Eq. (3.4.18) for axisymmetric problems.

4

One-Dimensional Problems Involving a Single Variable

It is really quite amazing by what margins competent but conservative scientists and engineers can miss the mark, when they start with the preconceived idea that what they are investigating is impossible. When this happens, the most well-informed men become blinded by their prejudices and are unable to see what lies directly ahead of them. —— Arthur C. Clarke

4.1 Model Differential Equation

We shall consider a model nonlinear differential equation involving a single un-known in one-dimensional problems to illustrate its finite element model development and discuss solution methods to solve the nonlinear algebraic equations. As we shall see, the weak form and finite element model development discussed in Section 3.2 for a linear model equation are also valid for a nonlinear equation.

We consider differential operator equations of the form $\mathcal{A}(u(x)) = f(x)$ in the interval $0 < x < L$, with boundary condition of the form $\mathcal{B}(u) = g$. If the operators $\mathcal{A}$ and $\mathcal{B}$ satisfy the conditions of the type $\mathcal{A}(\alpha u + \beta v) = \alpha\mathcal{A}(u) + \beta\mathcal{A}(v)$ for any real numbers α and β, we say that the problem is linear; if either $\mathcal{A}$ or $\mathcal{B}$ are not linear, the problem is said to be nonlinear (see Section 1.7.5 for a formal definition of a linear transformation or operator).

Consider the differential equation (see Table 3.2.1 for fields of study)

$$-\frac{d}{dx}\left[a(x,u)\frac{du}{dx}\right] + b(x,u)\frac{du}{dx} + c(x,u)u = f(x), \quad 0 < x < L \quad (4.1.1)$$

subjected to boundary conditions of the form

$$n_x\, a\frac{du}{dx} + \beta(x,u)\,(u - u_\infty) = \hat{Q}, \quad \text{or} \quad u = \hat{u} \quad (4.1.2)$$

at a boundary point. Here $u(x)$ denotes the dependent variable to be determined, a, b, and c are known functions of x and u (and possibly derivatives of u), f is a known function of x, $(u_\infty, \hat{u}, \beta, \hat{Q})$ are known (or specified) quantities, and n_x is the cosine of the angle between the positive x-axis and the outward normal to the edge at the node (note that $n_x = -1$ and $\beta = \beta_1$ at $x = x_a$ and $n_x = 1$ and $\beta = \beta_2$ at $x = x_b$).

J.N. Reddy, *An Introduction to Nonlinear Finite Element Analysis*, Second Edition. ©J.N. Reddy 2015. Published in 2015 by Oxford University Press.

The first boundary condition in Eq. (4.1.2) is called a *mixed* boundary condition or *Newton's* type boundary condition, and it is a statement of the balance of "forces" at the boundary point. In the case of axial deformation of bars, the mixed boundary condition (with $u_\infty = 0$) represents the balance of the internal force $n_x a(du/dx)$, with $n_x = 1$, the external applied force $\hat{Q}$, and the spring force $-\beta_2 u$, as illustrated in Fig. 4.1.1. Similarly, for one-dimensional heat flow, the mixed boundary condition is a statement of balance of energy due to the internal heat $a(du/dx)$, the externally applied heat $\hat{Q}$, and the heat due to convection $\beta_2(u - u_\infty)$. The second boundary condition in Eq. (4.1.2) is a special case of the first boundary condition (with $u_\infty = \hat{u}$) in the limit $\beta_2 \to \infty$.

Equation (4.1.1) and the first boundary condition in Eq. (4.1.2) can be expressed in operator form as

$$\mathcal{A}(u) = f \text{ in } 0 < x < L; \qquad \mathcal{B}(u) = g \text{ at } x = 0 \text{ or } L$$

where $\mathcal{A}$ and $\mathcal{B}$ are the differential operators

$$\mathcal{A} \equiv -\frac{d}{dx}\left(a\frac{d}{dx}\right) + b\frac{d}{dx} + c, \quad \mathcal{B} \equiv n_x a\frac{d}{dx} + \beta, \quad g \equiv \beta u_\infty + \hat{Q}$$

Clearly, operator $\mathcal{A}$ or $\mathcal{B}$ is not linear whenever any of the coefficients (a, b, c) are functions of u. The source of nonlinearity can be coefficients a, b, c and/or β which, in most engineering systems, include geometric and material parameters that may be functions of position x and the dependent variable u. For example, in the study of heat conduction in a rod with convective heat transfer through the surface, a is equal to the product $a = kA$, $b = 0$, and $c = P\beta$, where k denotes conductivity, β the convective heat transfer coefficient, A the cross-sectional area, and P the perimeter of the rod. Then the nonlinearity arises from the conductivity and heat transfer coefficients being functions of temperature u. Such nonlinearity is known as the *material nonlinearity*. We wish to solve this nonlinear differential equation using the finite element method.

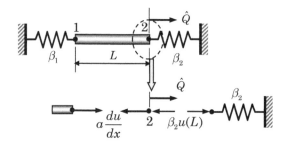

Fig. 4.1.1: Interpretation of the mixed boundary condition $a(du/dx) + \beta_2 u = \hat{Q}$.

4.2 Weak Formulation

Suppose that the domain $\Omega = (0, L)$ is divided into N line elements. A typical element from the collection of N elements is denoted as $\Omega^e = (x_a, x_b)$, where x_a and x_b denote the global coordinates of the end nodes of the line element. The weak form of Eq. (4.1.1) over the element can be developed as follows (see Section 3.2 for details):

$$
0 = \int_{x_a}^{x_b} \left(a \frac{dw_i^e}{dx} \frac{du_h^e}{dx} + bw_i^e \frac{du_h^e}{dx} + cw_i^e u_h^e - w_i^e f \right) dx - \left[w_i^e \left(a \frac{du_h^e}{dx} \right) \right]_{x_a}^{x_b}
$$

$$
= \int_{x_a}^{x_b} \left[a(x, u) \frac{dw_i^e}{dx} \frac{du_h^e}{dx} + b(x, u) w_i^e \frac{du_h^e}{dx} + c(x, u) w_i^e u_h^e - w_i^e f(x) \right] dx
$$

$$
- \left\{ Q_a^e - \beta_a \left[u_h^e(x_a) - u_\infty^a \right] \right\} w_i^e(x_a) - \left\{ Q_b^e - \beta_b \left[u_h^e(x_b) - u_\infty^b \right] \right\} w_i^e(x_b)
$$

$$
\text{(4.2.1)}
$$

where $w_i^e(x)$ is the ith weight function. The number of weight functions is equal to the number of unknowns in the approximation of u_h. The first line of Eq. (4.2.1) suggests that u is the primary variable and $Q = a(du/dx)$ is the secondary variable of the formulation. Using the mixed boundary condition in Eq. (4.1.2), we can express $a(du/dx)$ in terms of (Q_a^e, Q_b^e) and (u_∞^a, u_∞^b) as

$$
\begin{aligned}
- \left[a \frac{du_h^e}{dx} \right]_{x=x_a} &= Q_a^e - \beta_a \left[u_h^e(x_a) - u_\infty^a \right] \\
\left[a \frac{du_h^e}{dx} \right]_{x=x_b} &= Q_b^e - \beta_b \left[u_h^e(x_b) - u_\infty^b \right]
\end{aligned}
\tag{4.2.2}
$$

where (Q_a^e, Q_b^e) are the nodal values, (u_∞^a, u_∞^b) denote the values of the variable u_∞, and (β_a, β_b) denote certain physical parameters (e.g. film conductances) at the left and right ends of the element, respectively. When a node is in the interior of the element, the corresponding β and u_∞ are zero.

As discussed in Chapter 3, specifying a primary variable is termed the *essential* (or *geometric*) boundary condition and specifying a secondary variable is known as the *natural* (or *force*) boundary condition. The first boundary condition in Eq. (4.1.2) is of the mixed type since it includes both the primary and secondary variables, and it is nonlinear because of the dependence of β on u; the second boundary condition in Eq. (4.1.2) is of the essential type.

4.3 Finite Element Model

Suppose that the dependent unknown $u(x)$ is approximated over element Ω^e by the finite element approximation of the form

$$
u(x) \approx u_h^e(x) = \sum_{j=1}^{n} u_j^e \psi_j^e(x)
\tag{4.3.1}
$$

Substituting the approximation from Eq. (4.3.1) for u and $w_i^e = \psi_i^e$ (Galerkin's approach) into the weak form, Eq. (4.2.1), we obtain the following *weak-form Galerkin* or *Ritz finite element model*:

$$\mathbf{K}^e(\mathbf{u}^e)\,\mathbf{u}^e = \mathbf{F}^e \tag{4.3.2}$$

where

$$K_{ij}^e = \int_{x_a}^{x_b}\left[a(x, u_h^e)\frac{d\psi_i^e}{dx}\frac{d\psi_j^e}{dx} + b(x, u_h^e)\psi_i^e\frac{d\psi_j^e}{dx} + c(x, u_h^e)\psi_i^e\psi_j^e\right]dx$$
$$+\,\beta_a\psi_i^e(x_a)\psi_j^e(x_a) + \beta_b\psi_i^e(x_b)\psi_j^e(x_b) \tag{4.3.3}$$

$$F_i^e = \int_{x_a}^{x_b} f(x)\psi_i^e\,dx + \beta_a u_\infty^a\psi_i^e(x_a) + \beta_b u_\infty^b\psi_i^e(x_b) + Q_a\psi_i^e(x_a) + Q_b\psi_i^e(x_b)$$

Note that the coefficient matrix $\mathbf{K}^e$ is a function of the unknown nodal values u_j^e, and it is an unsymmetric matrix when $b \neq 0$; when $b = 0$, $\mathbf{K}^e$ is a symmetric matrix. The term involving c is symmetric, independent of whether it depends on u and/or du/dx. Therefore, it is advisable to include nonlinear terms of the type $u\,(du/dx)$ in a differential equation as the c-term in the equation by writing it as $u\,(du/dx) = cu$, with $c = du/dx$; otherwise, it will be unsymmetric and convergence of the solution may become a problem. The coefficients involving β in $\mathbf{K}^e$ and $\mathbf{F}^e$ should be included only in elements that have end nodes with the convection type boundary condition. Example 4.3.1 provides more insight into the make-up of the coefficient matrix $\mathbf{K}^e$.

Example 4.3.1 ——————————————————————————————————

Consider the problem described by Eqs. (4.1.1) and (4.1.2). Suppose that $a(x) = a_0 + a_u\,u(x)$ and $b = c = 0$, where a_0 and a_u are functions of x only. Determine the explicit form of the element matrices $\mathbf{K}^e$ and $\mathbf{f}^e$ using linear approximation of u and element-wise constant values a_0^e and a_u^e of a_0 and a_u, respectively.

Solution: For all elements except for the last one, the coefficients β_a and β_b are zero; for the last element we have $\beta_a = 0$ and $\beta_b \neq 0$ when the end $x = L$ is subjected to the convection boundary condition. Thus, we have for $e \neq N$, where N denotes the total number of elements in a mesh of elements connected end-to-end:

$$K_{ij}^e = \int_{x_a}^{x_b}\left[a_0^e + a_u^e\left(\sum_{k=1}^{n} u_k^e\psi_k^e\right)\right]\frac{d\psi_i^e}{dx}\frac{d\psi_j^e}{dx}\,dx$$
$$= \int_{x_a}^{x_b} a_0^e\frac{d\psi_i^e}{dx}\frac{d\psi_j^e}{dx}\,dx + \sum_{k=1}^{n} u_k^e\int_{x_a}^{x_b} a_u^e\psi_k^e\frac{d\psi_i^e}{dx}\frac{d\psi_j^e}{dx}\,dx \tag{1}$$

For the linear ($n = 2$) approximation of $u(x)$, we have [see Eq. (3.2.21)]

$$u_h^e(x) = u_1^e\psi_1^e(x) + u_2^e\psi_2^e(x) = u_1^e\left(\frac{x_b - x}{h_e}\right) + u_2^e\left(\frac{x - x_a}{h_e}\right) \tag{2}$$

and

$$K_{ij}^e = a_0^e\int_{x_a}^{x_b}\frac{d\psi_i^e}{dx}\frac{d\psi_j^e}{dx}\,dx + \sum_{k=1}^{2} a_u^e u_k^e\int_{x_a}^{x_b} \psi_k^e\frac{d\psi_i^e}{dx}\frac{d\psi_j^e}{dx}\,dx$$

$$
= (-1)^{i+j} \frac{1}{h_e^2} \left[a_0^e \int_{x_a}^{x_b} 1 \cdot dx + a_u^e \sum_{k=1}^{2} u_k^e \int_{x_a}^{x_b} \psi_k^e \, dx \right]
$$

$$
= (-1)^{i+j} \frac{1}{h_e^2} \left[a_0^e h_e + a_u^e \frac{h_e}{2} \left(\sum_{k=1}^{2} u_k^e \right) \right] = (-1)^{i+j} \frac{1}{h_e} \left[a_0^e + \frac{a_u^e}{2} \left(u_1^e + u_2^e \right) \right] \tag{3}
$$

or

$$
\mathbf{K}^e = \frac{a_0^e + 0.5 a_u^e (u_1^e + u_2^e)}{h_e} \begin{bmatrix} 1 & -1 \\ -1 & 1 \end{bmatrix}, \quad 1 < e < N \tag{4}
$$

The last (i.e. Nth) element coefficient matrix and source vector are given by

$$
\mathbf{K}^{(N)} = \frac{a_0^e + 0.5 a_u^e (u_1^e + u_2^e)}{h_e} \begin{bmatrix} 1 & -1 \\ -1 & 1 \end{bmatrix} + \begin{bmatrix} 0 & 0 \\ 0 & \beta_b \end{bmatrix}, \quad \mathbf{f}^{(N)} = \begin{Bmatrix} f_1^{(N)} \\ f_2^{(N)} \end{Bmatrix} + \begin{Bmatrix} 0 \\ \beta_b u_\infty^b \end{Bmatrix} \tag{5}
$$

The assembly of element equations follows the same procedure as in linear finite element analysis. If we denote the global nodal vector by $\mathbf{U}$, the assembled system of equations can be written as

$$
\mathbf{K}(\mathbf{U})\mathbf{U} = \mathbf{F}(\mathbf{U}) \tag{4.3.4}
$$

where $\mathbf{K}$ and $\mathbf{F}$ denote the global coefficient matrix and the right-hand side vector, respectively. In Example 4.3.1, $\mathbf{K}$ is a linear function of the nodal values U_i and it is a symmetric matrix. Consequently, the resulting finite element equations are nonlinear; in the present case, the algebraic equations are quadratic in U_i. The right-hand side vector $\mathbf{F}$ depends on $\mathbf{U}$ only when the boundary conditions are nonlinear. Example 4.3.2 illustrates these ideas.

Example 4.3.2

Consider the problem of Example 4.3.1 with the following specific boundary conditions:

$$
u(0) = u_0, \quad \left[a \frac{du}{dx} + \beta (u - u_\infty) \right]_{x=L} = \hat{Q} \tag{1}
$$

with

$$
\beta = \beta_0 + \beta_1 u \tag{2}
$$

Give the assembled finite element equations after the imposition of boundary conditions for a uniform mesh of two linear elements.

Solution: The assembled equations associated with a mesh of two linear elements of equal length h and with uniform data a_0 and a_u, but *without* the contribution due to the convection boundary condition (1) at $x = L$, are

$$
\frac{1}{h} \begin{bmatrix} a_0 + 0.5 a_u (U_1 + U_2) & -a_0 - 0.5 a_u (U_1 + U_2) & 0 \\ -a_0 - 0.5 a_u (U_1 + U_2) & 2a_0 + 0.5 a_u (U_1 + 2U_2 + U_3) & -a_0 - 0.5 a_u (U_2 + U_3) \\ 0 & -a_0 - 0.5 a_u (U_2 + U_3) & a_0 + 0.5 a_u (U_2 + U_3) \end{bmatrix} \begin{Bmatrix} U_1 \\ U_2 \\ U_3 \end{Bmatrix}
$$

$$
= \begin{Bmatrix} f_1^{(1)} \\ f_2^{(1)} + f_1^{(2)} \\ f_2^{(2)} \end{Bmatrix} + \begin{Bmatrix} Q_1^{(1)} \\ Q_2^{(1)} + Q_1^{(2)} \\ Q_2^{(2)} \end{Bmatrix} \tag{3}
$$

The boundary conditions for the present mesh of two linear elements imply

$$U_1 = u_0, \quad Q_2^{(2)} = \hat{Q} - (\beta_0 + \beta_1 U_3)(U_3 - u_\infty) \tag{4}$$

Hence, after imposing the boundary conditions and balance of the secondary variables $Q_2^{(1)} + Q_1^{(2)} = 0$, the assembled equations in (3) become

$$\left(\frac{a_0}{h} \begin{bmatrix} 1 & -1 & 0 \\ -1 & 2 & -1 \\ 0 & -1 & 1 \end{bmatrix} + \frac{a_u}{2h} \begin{bmatrix} (U_1 + U_2) & -(U_1 + U_2) & 0 \\ -(U_1 + U_2) & (U_1 + 2U_2 + U_3) & -(U_2 + U_3) \\ 0 & -(U_2 + U_3) & (U_2 + U_3) \end{bmatrix} \right) \begin{Bmatrix} u_0 \\ U_2 \\ U_3 \end{Bmatrix}$$

$$+ \begin{bmatrix} 0 & 0 & 0 \\ 0 & 0 & 0 \\ 0 & 0 & \beta_0 + \beta_1 U_3 \end{bmatrix} \begin{Bmatrix} u_0 \\ U_2 \\ U_3 \end{Bmatrix} = \begin{Bmatrix} f_1^{(1)} \\ f_2^{(1)} + f_1^{(2)} \\ f_2^{(2)} \end{Bmatrix} + \begin{Bmatrix} Q_1^{(1)} \\ 0 \\ \hat{Q} + (\beta_0 + \beta_1 U_3)u_\infty \end{Bmatrix} \tag{5}$$

Clearly, the assembled equations in Eq. (5) are the same as the system of equations obtained by assembling the element matrices in Eqs. (4) and (5) of Example 4.3.1. Note that in the present case, due to the nonlinearity in the boundary condition, $\mathbf{F}$ is a function of $\mathbf{U}$. One may move the term involving the unknown U_3 from the right-hand side vector to the left side of the equality in Eq. (5).

4.4 Solution of Nonlinear Algebraic Equations

4.4.1 General Comments

Numerical procedures used to solve nonlinear algebraic equations of the type in Eq. (4.3.4) are iterative in nature. All iterative methods of solution of nonlinear algebraic equations are based on linearization of the element coefficient matrix $\mathbf{K}^e$ (and the source vector $\mathbf{F}^e$ if it is dependent on $\mathbf{u}^e$). Then the element equations are assembled and boundary conditions are imposed before solving them by an iterative procedure. Two iterative procedures are outlined here for the problem at hand. Some general features of iterative methods used for nonlinear equations are discussed before getting into the details of each method (see Appendix 2 for additional discussion).

4.4.2 Direct Iteration Procedure

Suppose that we wish to solve the assembled system, Eq. (4.3.4), which already has the boundary conditions imposed on them. Since $\mathbf{K}(\mathbf{U})$ cannot be evaluated until $\mathbf{U}$ is known and the solution is not known until we solve the equations, one must resort to an iterative procedure in which we assume that solution at the $(r-1)$st iteration is known and seek solution at the rth iteration:

$$\mathbf{K}(\mathbf{U}^{(r-1)})\mathbf{U}^{(r)} = \mathbf{F}(\mathbf{U}^{(r-1)}) \tag{4.4.1}$$

This procedure of solving Eq. (4.4.1) iteratively is called the *direct iteration* technique, also known as the *Picard iteration* or the *method of successive sub-stitutions*.

At the beginning of the iteration process, that is, when $r = 1$, solution $\mathbf{U}^{(0)}$ must be assumed or "guessed" consistent with the problem boundary conditions. Using the solution from the $(r-1)$st iteration, we can compute the coefficient matrix $\mathbf{K}^{(r-1)} \equiv \mathbf{K}(\mathbf{U}^{(r-1)})$ and vector $\mathbf{F}^{(r-1)} \equiv \mathbf{F}(\mathbf{U}^{(r-1)})$. The solution at the rth iteration is determined by inverting Eq. (4.4.1):

$$\mathbf{U}^{(r)} = [\mathbf{K}^{(r-1)}]^{-1}\mathbf{F}^{(r-1)} \tag{4.4.2}$$

In writing Eq. (4.4.2), we have assumed that the coefficient matrix $\mathbf{K}^{(r-1)}$ is invertible after the imposition of boundary conditions. Thus, the initial *guess vector* $\mathbf{U}^0$ should be such that: (a) it satisfies the specified essential boundary conditions and (b) $\mathbf{K}^{(0)}$ is invertible.

Once the solution $\mathbf{U}^{(r)}$ is obtained, we check to see if the residual vector

$$\mathbf{R}^{(r)} \equiv \mathbf{K}^{(r)}\mathbf{U}^{(r)} - \mathbf{F}^{(r)} \tag{4.4.3}$$

is zero. The magnitude of this residual vector will be small enough if the solution has converged. In other words, we terminate the iteration if the magnitude of the residual vector, measured in a suitable norm, is less than some preselected tolerance ϵ. If the problem data are such that $\mathbf{KU}$ as well as $\mathbf{F}$ are very small, the norm of the residual vector may also be very small even when the solution $\mathbf{U}$ has not converged. Therefore, it is necessary to normalize the residual vector with respect to $\mathbf{F}$. Using the Euclidean norm[1], we can express the error criterion as

$$\sqrt{\frac{\mathbf{R}^{(r)} \cdot \mathbf{R}^{(r)}}{\mathbf{F}^{(r)} \cdot \mathbf{F}^{(r)}}} \leq \epsilon \tag{4.4.4}$$

Alternatively, one may check to see if the normalized difference between solution vectors from two consecutive iterations, measured with the Euclidean norm, is less than a preselected tolerance ϵ:

$$\sqrt{\frac{\Delta\mathbf{U} \cdot \Delta\mathbf{U}}{\mathbf{U}^{(r)} \cdot \mathbf{U}^{(r)}}} = \sqrt{\frac{\sum_{I=1}^{N} |U_I^{(r)} - U_I^{(r-1)}|^2}{\sum_{I=1}^{N} |U_I^{(r)}|^2}} \leq \epsilon \tag{4.4.5}$$

where $\Delta\mathbf{U} = \mathbf{U}^{(r)} - \mathbf{U}^{(r-1)}$. Thus, the iteration process is continued until the error criterion in Eq. (4.4.4) or Eq. (4.4.5) is satisfied or the number of iterations exceeds a preselected maximum number of iterations, so that the process does not go into a loop of indefinite number of iterations in the case of non-convergence.

Acceleration of convergence for some types of nonlinearities may be achieved by using a weighted-average of solutions from the last two iterations rather than the solution from the last iteration to evaluate the coefficient matrix:

---

[1]The *Euclidean norm* of a vector $\mathbf{r} = (r_1, r_2, \ldots, r_n)$, where r_i are real numbers, is $\|\mathbf{r}\| = (r_1^2 + r_2^2 + \cdots + r_n^2)^{1/2}$, and it is a measure of the length of the vector $\mathbf{r}$; see Eq. (1.7.8).

$$\mathbf{U}^{(r)} = [\mathbf{K}(\bar{\mathbf{U}})]^{-1}\mathbf{F}(\bar{\mathbf{U}}), \quad \bar{\mathbf{U}} \equiv \beta\mathbf{U}^{(r-2)} + (1 - \beta)\mathbf{U}^{(r-1)}, \quad 0 \leq \beta \leq 1 \quad (4.4.6)$$

where β is called the *acceleration parameter*. The value of β depends on the nature of nonlinearity and the type of problem. Often, one has to play with the value of β to obtain convergence. One should begin with a value of $\beta = 0$ and, if the solution does not converge, use a value of $\beta = 0.25$ and gradually increase but never set β equal to 1 (only an advice).

Figure 4.4.1 depicts the general idea of the direct iteration procedure for a single-degree-of-freedom system. Here K denotes the slope of the line joining the origin to the point $K(U)$ on the curve $F = K(U)U \equiv f(U)$. Note that $K(U)$ is *not* the slope of the tangent to the curve at U. The direct iteration converges if the nonlinearity is mild; otherwise, it diverges. The steps involved in the direct iteration solution scheme are described in Box 4.4.1. Unless stated otherwise, the error criterion based on the solution rather than the residual, that is, Eq. (4.4.5) is used in this book.

Box 4.4.1: Steps involved in the direct iteration scheme.

---

1. **Initial solution vector.** Assume an initial solution vector $\mathbf{U}^{(0)}$ such that it (a) satisfies the specified boundary conditions on $\mathbf{U}$ and (b) does not make $\mathbf{K}^e$ singular.

2. **Computation of K and F.** Use the latest known vector $\mathbf{U}^{(r-1)}$ ($\mathbf{U}^{(0)}$ during the first iteration) to evaluate $\mathbf{K}^e$ and $\mathbf{F}^e$, assemble them to obtain global $\mathbf{K}$ and $\mathbf{F}$, and apply the specified boundary conditions on the assembled system.

3. **Computation of $\mathbf{U}^{(r)}$.** Compute the solution at the rth iteration

$$\mathbf{U}^{(r)} = [\mathbf{K}(\mathbf{U}^{(r-1)})]^{-1}\mathbf{F}^{(r-1)}$$

4. **Convergence check.** Compute the residual

$$\mathbf{R}^{(r)} = \mathbf{K}(\mathbf{U}^{(r)})\mathbf{U}^{(r)} - \mathbf{F}^{(r)}$$

with the latest known solution and check if

$$\|\mathbf{R}^{(r)}\| \leq \epsilon\|\mathbf{F}^{(r)}\| \quad \text{or} \quad \|\mathbf{U}^{(r)} - \mathbf{U}^{(r-1)}\| \leq \epsilon\|\mathbf{U}^{(r)}\|$$

where $\|\cdot\|$ denotes the euclidean norm and ϵ is the convergence tolerance (read as an input). If the solution has converged, print the solution and move to the next "load" level or quit if it is the only or final load; otherwise, continue.

5. **Maximum iteration check.** Check if $r < itmax$, where $itmax$ is the maximum number of iterations allowed (read as an input). If *yes*, set $r \rightarrow r + 1$ and go to Step 2; if *no*, print a message that the iteration scheme did not converge and quit.

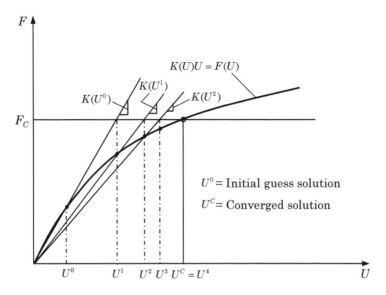

Fig. 4.4.1: Convergence of the direct iteration procedure (calculation of U for a specified source value F).

Example 4.4.1

Solve the nonlinear differential equation

$$-\frac{d}{dx}\left(u\frac{du}{dx}\right) = f_0, \quad 0 < x < 1 \tag{1}$$

subjected to the boundary conditions

$$\left[u\frac{du}{dx}\right]_{x=0} = \hat{Q}, \qquad u(1) = \hat{u} \tag{2}$$

using a uniform mesh of 2 linear finite elements. Use the data

$$\hat{Q} = 0, \qquad \hat{u} = \sqrt{2} = 1.4142, \qquad f_0 = -1, \qquad h = L/2 = 0.5 \tag{3}$$

Solution: Compared to the model equation in Eq. (4.1.1), we have $L = 1$, $a = a_0 + a_u u$, $b = c = 0$, and $f = f_0$ with $a_0 = 0$ and $a_u = 1$. For this case the element matrix is given by Eq. (4) of Example 4.3.1, and for a mesh of two linear elements with $h_1 = h_2 = h = 0.5$, the assembled equations are given by Eq. (3) of Example 4.3.2. After imposing the boundary conditions $Q_1^{(1)} = \hat{Q}$, $Q_2^{(1)} + Q_1^{(2)} = 0$, and $U_3 = \hat{u}$, Eq. (3) of Example 4.3.2 becomes ($a_0 = 0$ and $a_u = 1.0$)

$$\frac{1}{2h}\begin{bmatrix} (\bar{U}_1 + \bar{U}_2) & -(\bar{U}_1 + \bar{U}_2) & 0 \\ -(\bar{U}_1 + \bar{U}_2) & (\bar{U}_1 + 2\bar{U}_2 + \bar{U}_3) & -(\bar{U}_2 + \bar{U}_3) \\ 0 & -(\bar{U}_2 + \bar{U}_3) & (\bar{U}_2 + \bar{U}_3) \end{bmatrix}\begin{Bmatrix} U_1 \\ U_2 \\ \hat{u} \end{Bmatrix}^{(r)} = \begin{Bmatrix} f_1^{(1)} \\ f_2^{(1)} + f_1^{(2)} \\ f_2^{(2)} \end{Bmatrix} + \begin{Bmatrix} \hat{Q} \\ 0 \\ Q_2^{(2)} \end{Bmatrix} \tag{4}$$

where $\bar{U}_I = U_I^{(r-1)}$ $(I = 1, 2, 3)$ denote the nodal values from the previous iteration. Omitting the last equation (which can be used to compute $Q_2^{(2)}$), we obtain the following condensed set of linear equations:

$$\frac{1}{2h} \begin{bmatrix} (\bar{U}_1 + \bar{U}_2) & -(\bar{U}_1 + \bar{U}_2) \\ -(\bar{U}_1 + \bar{U}_2) & (\bar{U}_1 + 2\bar{U}_2 + \bar{U}_3) \end{bmatrix} \begin{Bmatrix} U_1 \\ U_2 \end{Bmatrix}^{(r)} = \begin{Bmatrix} f_1^{(1)} \\ f_2^{(1)} + f_1^{(2)} \end{Bmatrix} + \begin{Bmatrix} \hat{Q} \\ \frac{(\bar{U}_2 + \bar{U}_3)}{2h} \hat{u} \end{Bmatrix} \tag{5}$$

For the data given in Eq. (3), we select the initial guess vector, which must be consistent with specified essential boundary conditions (i.e. $U_3 = \hat{u} = \sqrt{2}$), to be

$$U_1^{(0)} = 1.0, \qquad U_2^{(0)} = 1.0, \qquad U_3^{(0)} = \sqrt{2} = 1.4142 \tag{6}$$

Then Eq. (5) becomes $(f_i^{(e)} = -h/2 = -0.25)$

$$\begin{bmatrix} 2 & -2 \\ -2 & 4.4142 \end{bmatrix} \begin{Bmatrix} U_1 \\ U_2 \end{Bmatrix}^{(1)} = -\begin{Bmatrix} 0.25 \\ 0.50 \end{Bmatrix} + \begin{Bmatrix} 0 \\ (2.4142)\sqrt{2} \end{Bmatrix} = \begin{Bmatrix} -0.2500 \\ 2.9142 \end{Bmatrix} \tag{7}$$

The solution at the end of the first iteration is

$$U_1^{(1)} = 0.9785, \qquad U_2^{(1)} = 1.1035, \qquad U_3^{(1)} = 1.4142 \tag{8}$$

For the second iteration, we use the solution from Eq. (8) to evaluate the coefficient matrix in Eq. (5) [i.e. $\beta = 0$ in Eq. (4.4.6)] and obtain

$$\begin{bmatrix} 2.0821 & -2.0821 \\ -2.0821 & 4.5998 \end{bmatrix} \begin{Bmatrix} U_1 \\ U_2 \end{Bmatrix}^{(2)} = -\begin{Bmatrix} 0.25 \\ 0.50 \end{Bmatrix} + \begin{Bmatrix} 0.0000 \\ 3.5607 \end{Bmatrix} \tag{9}$$

and the solution becomes

$$U_1^{(2)} = 0.9962, \qquad U_2^{(2)} = 1.1163, \qquad U_3^{(2)} = 1.4142 \tag{10}$$

The root-mean-square error in the solution (with respect to solution obtained in the first iteration) [see Eq. (4.4.5)] is 0.0106.

Table 4.4.1: Comparison of finite element solutions with the exact solution of the problem in Example 4.4.1 (*direct iteration*).

x	Iteration	2L	4L	1Q	2Q	Exact
0.00	1	0.9785	0.9517	0.9699	0.9415	
	2	0.9962	0.9903	0.9954	0.9881	
	3	0.9995	0.9988	0.9995	0.9986	
	4	0.9999	0.9999	0.9999	0.9999	1.0000
0.25	1	—	0.9830	—	0.9727	
	2	—	1.0226	—	1.0208	
	3	—	1.0299	—	1.0298	
	4	—	1.0307	—	1.0308	1.0308
0.50	1	1.1036	1.0767	1.1013	1.0665	
	2	1.1163	1.1136	1.1168	1.1127	
	3	1.1178	1.1176	1.1118	1.1176	
	4	1.1180	1.1180	1.1187	1.1180	1.1180
0.75	1	—	1.2330	—	1.2285	
	2	—	1.2489	—	1.2488	
	3	—	1.2499	—	1.2500	
	4	—	1.2500	—	1.2500	1.2500
1.00		1.4142	1.4142	1.4142	1.4142	1.4142

The exact solution to this problem, with the data given in Eq. (3), is (one can verify by direct substitution)

$$u(x) = \sqrt{1 + x^2}$$

The convergence tolerance is taken to be $\epsilon = 10^{-3}$, and the maximum number of iterations is prescribed to be ten. The finite element solutions obtained with the linear (L) and quadratic (Q) elements are compared with the exact solution in Table 4.4.1. The finite element solution converges to the exact solution in four iterations.

4.4.3 Newton's Iteration Procedure

In Newton's method, we expand the residual vector $\mathbf{R}^{(r)}$ of Eq. (4.4.3) in Taylor's series about the known solution $\mathbf{U}^{(r-1)}$:

$$\mathbf{R}^{(r)} = \mathbf{R}^{(r-1)} + \left(\frac{\partial \mathbf{R}}{\partial \mathbf{U}}\right)^{(r-1)} \cdot \Delta\mathbf{U} + \frac{1}{2!}\left[\left(\frac{\partial^2 \mathbf{R}}{\partial \mathbf{U}^2}\right)^{(r-1)} \cdot \Delta\mathbf{U}\right] \cdot \Delta\mathbf{U} + \cdots \quad (4.4.7)$$

where

$$\Delta\mathbf{U} = \mathbf{U}^{(r)} - \mathbf{U}^{(r-1)} \quad (4.4.8)$$

Omitting the terms of order 2 and higher in $\Delta\mathbf{U}$ and requiring that $\mathbf{R}^{(r)}$ be zero, we obtain

$$\left(\frac{\partial \mathbf{R}}{\partial \mathbf{U}}\right)^{(r-1)} \Delta\mathbf{U} = -\mathbf{R}^{(r-1)} \quad (4.4.9)$$

or

$$\mathbf{T}^{(r-1)} \Delta\mathbf{U} = -\mathbf{R}^{(r-1)} \quad \rightarrow \quad \mathbf{T}^{(r-1)} \mathbf{U}^{(r)} = -\mathbf{R}^{(r-1)} + \mathbf{T}^{(r-1)} \mathbf{U}^{(r-1)} \quad (4.4.10)$$

where $\mathbf{T}$ is called the *tangent matrix*, and it is defined as

$$\mathbf{T}^{(r-1)} \equiv \left(\frac{\partial \mathbf{R}}{\partial \mathbf{U}}\right)^{(r-1)} \quad (4.4.11)$$

The component definition of the tangent matrix is

$$T_{IJ} \equiv \frac{\partial R_I}{\partial U_J} = \frac{\partial}{\partial U_J}\left(\sum_{m=1}^{N} K_{Im}U_m - F_I\right)$$

$$= \sum_{m=1}^{N}\left(K_{Im}\frac{\partial U_m}{\partial U_J} + \frac{\partial K_{Im}}{\partial U_J}U_m\right) - \frac{\partial F_I}{\partial U_J}$$

$$= K_{IJ} + \sum_{m=1}^{N}\frac{\partial K_{Im}}{\partial U_J}U_m - \frac{\partial F_I}{\partial U_J}$$

for $I, J = 1, 2, \ldots, N$, where N is the total number of nodes in the mesh.

All of the ideas presented above also apply to the element equation in Eq. (4.3.2). Indeed, since element matrices are computed at the element level, we apply Newton's procedure at the element level, and then assemble element equations to form the global equations. The element equation for the Newton iteration procedure is

$$\mathbf{T}^e(\mathbf{u}^{(r-1)})\,\Delta\mathbf{u}^e = -\mathbf{R}^e(\mathbf{u}^{(r-1)}) \tag{4.4.12}$$

and the tangent coefficient matrix at the element level is

$$T_{ij}^e \equiv \frac{\partial R_i^e}{\partial u_j^e} = \frac{\partial}{\partial u_j^e}\left(\sum_{m=1}^{n} K_{im}^e u_m^e - F_i^e\right) = K_{ij}^e + \sum_{m=1}^{n}\frac{\partial K_{im}^e}{\partial u_j^e}u_m^e - \frac{\partial F_i^e}{\partial u_j^e} \tag{4.4.13}$$

When the vector $\mathbf{F}^e$ is independent of $\mathbf{u}^e$, we have $\partial F_i^e/\partial u_j^e = 0$.

The element equations in Eq. (4.4.12) are assembled and the boundary conditions are imposed on the increment $\Delta\mathbf{U}$ before solving the assembled system for $\Delta\mathbf{U} = -\mathbf{T}^{-1}\mathbf{R}$. The total solution at the rth iteration is then

$$\mathbf{U}^r = \mathbf{U}^{(r-1)} + \Delta\mathbf{U} \tag{4.4.14}$$

Figure 4.4.2 shows convergence of the Newton iteration procedure for a single degree-of-freedom system. Here $T(U)$ denotes the slope of the tangent to the curve $F = K(U)U$ at U. The Newton iteration converges for hardening as well as softening type nonlinearities. For the hardening type, convergence may be accelerated using the under-relaxation given in Eq. (4.4.6). The method may diverge for a saddle-point behavior. The steps involved in Newton's iteration solution approach are described in Box 4.4.2.

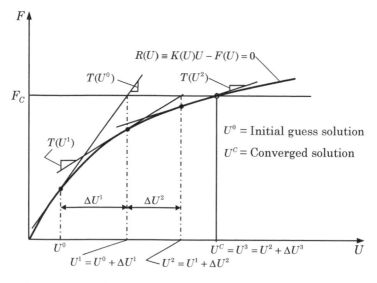

Fig. 4.4.2: Convergence of the Newton iteration procedure (calculation of ΔU for a specified residual value R).

Box 4.4.2: Steps involved in the Newton iteration scheme.

1. **Initial solution vector.** Assume an initial solution vector $\mathbf{U}^{(0)}$ such that: (a) it satisfies the specified boundary conditions on $\mathbf{U}$ and (b) it does not make $\mathbf{T}^e$ singular.

2. **Computation of T and R.** Use the latest known vector $\mathbf{U}^{(r-1)}$ ($\mathbf{U}^{(0)}$ during the first iteration) to: (a) evaluate $\mathbf{K}^e$, $\mathbf{F}^e$, $\mathbf{T}^e$, and $-\mathbf{R}^e = \mathbf{F}^e - \mathbf{K}^e\mathbf{U}^e$, (b) assemble $\mathbf{T}^e$ and $\mathbf{R}^e$ to obtain global $\mathbf{T}$ and $\mathbf{R}$, and (c) apply the specified *homogeneous* boundary conditions (since $\mathbf{U}^{(0)}$ already satisfies the actual boundary conditions) on the assembled system, $\mathbf{TU} = -\mathbf{R}$.

3. **Computation of $\mathbf{U}^{(r)}$.** Compute the solution increment at the rth iteration

$$\Delta\mathbf{U} = -[\mathbf{T}(\mathbf{U}^{(r-1)})]^{-1}\mathbf{R}^{(r-1)}$$

   and update the total solution

$$\mathbf{U}^{(r)} = \mathbf{U}^{(r-1)} + \Delta\mathbf{U}$$

4. **Convergence check.** Compute the residual

$$\mathbf{R}^{(r)} = \mathbf{K}(\mathbf{U}^{(r)})\mathbf{U}^{(r)} - \mathbf{F}^{(r)}$$

   and check if

$$\|\mathbf{R}^{(r)}\| \le \epsilon\|\mathbf{F}^{(r)}\| \quad \text{or} \quad \|\Delta\mathbf{U}\| \le \epsilon\|\mathbf{U}^{(r)}\|$$

   where ϵ is the convergence tolerance (read as an input). If the solution has converged, print the solution and move to the next "load" level or quit. Otherwise, continue.

5. **Maximum iteration check.** Check if $r < itmax$, where $itmax$ is the maximum number of iterations allowed (read as an input). If *yes*, set $r \to r+1$ and go to Step 2; if *no*, print a message that the iteration scheme did not converge and quit.

Remarks

Some remarks are in order on Newton's iteration scheme.

1. In the direct iteration method, the actual boundary conditions are applied to the assembled system $\mathbf{KU} = \mathbf{F}$ during each iteration, prior to solving them. In Newton's method, at the end of each iteration, we only compute the increment $\Delta\mathbf{U}$ to the known solution $\mathbf{U}^{(r-1)}$. Therefore, if $\mathbf{U}^{(r-1)}$ already satisfies the specified boundary conditions, then $\Delta\mathbf{U}$ must only satisfy the corresponding *homogeneous* boundary conditions so that $\mathbf{U}^{(r)} = \mathbf{U}^{(r-1)} + \Delta\mathbf{U}$ satisfies the specified boundary conditions. For example, if U_m is specified to be $U_m = \alpha_m$, and the initial guess is such that $U_m^{(0)} = \alpha_m$, then $\Delta U_m = 0$ so that $U_m^{(1)} = U_m^{(0)} + \Delta U_m = \alpha_m$. Alternatively, if the initial guess is such that $U_m^{(0)} = 0$, then $\Delta U_m = \alpha_m$ for the first iteration so that $U_m^{(1)} = U_m^{(0)} + \Delta U_m = \alpha_m$, and the boundary condition on ΔU_m during the second iteration onwards should be $\Delta U_m = 0$.

2. The symmetry of the coefficient matrices $\mathbf{K}$ and $\mathbf{T}$ depends on the original differential equation as well as weak form used to develop the finite element equations. Even when $\mathbf{K}$ is symmetric, $\mathbf{T}$ may not be symmetric, and vice versa; the symmetry of $\mathbf{T}$ depends on $\mathbf{K}$ as well as on the nature of nonlinearity in the governing differential equations.

3. The tangent matrix $\mathbf{T}$ does not have to be exact; an approximate $\mathbf{T}$ can also provide the solution but it may take more iterations. In any case, the residual vector will be computed using the definition $\mathbf{R} = \mathbf{KU} - \mathbf{F}$ and convergence is declared only when the residual vector $\mathbf{R}$ or the solution increment $\Delta\mathbf{U}$ is sufficiently small as indicated in Eqs. (4.4.4) and (4.4.5).

4. When the tangent matrix is updated only once in a certain pre-specified number of iterations (to save computational time) while updating the residual vector during each iteration, the procedure is known as the *modified Newton's method* or the *Newton–Raphson method*. Generally, the Newton–Raphson iteration solution method takes more iterations to converge than the full Newton's method, and it may even diverge for certain types of nonlinearity.

Example 4.4.2 ───────────────────────────────────

Consider the element coefficient matrix in Eq. (4.3.3). Assuming that

$$a(x,u) = a_0 + a_1 u + a_2 u^2 + a_3 \frac{du}{dx}, \quad b = 0$$

$$c(x,u) = c_0 + c_1 u + c_2 u^2 + c_3 \frac{du}{dx}, \quad \beta_a = 0, \quad \beta_b = \beta_0 + \beta_1 u \tag{1}$$

determine the tangent matrix coefficients.

Solution: We have

$$K_{ij}^e = \int_{x_a}^{x_b} \left[\left(a_0 + a_1 u_h + a_2 u_h^2 + a_3 \frac{du_h}{dx} \right) \frac{d\psi_i^e}{dx} \frac{d\psi_j^e}{dx} + \left(c_0 + c_1 u_h + c_2 u_h^2 + c_3 \frac{du_h}{dx} \right) \psi_i^e \psi_j^e \right] dx$$

$$+ \left[\beta_0 + \beta_1 u_h(x_b) \right] \psi_i^e(x_b) \psi_j^e(x_b) \tag{2}$$

$$F_i^e = \int_{x_a}^{x_b} f(x) \psi_i^e \, dx + \left[\beta_0 + \beta_1 u_h(x_b) \right] u_\infty^b \psi_i^e(x_b) + \psi_i^e(x_a) Q_a + \psi_i^e(x_b) Q_b$$

where $a_0, a_1, a_2, a_3, c_0, c_1, c_2$, and c_3 are functions of x, and β_0 and β_1 are constants. The element tangent matrix coefficients are computed from the definition

$$T_{ij}^e \equiv \frac{\partial R_i^e}{\partial u_j^e} = K_{ij}^e + \sum_{m=1}^{n} \frac{\partial K_{im}^e}{\partial u_j^e} u_m^e - \frac{\partial F_i^e}{\partial u_j^e} \tag{3}$$

where

$$\sum_{m=1}^{n} \frac{\partial K_{im}^e}{\partial u_j^e} u_m^e = \sum_{m=1}^{n} \frac{\partial}{\partial u_j^e} \left\{ \int_{x_a}^{x_b} \left[\left(a_0 + a_1 u_h + a_2 u_h^2 + a_3 \frac{du_h}{dx} \right) \frac{d\psi_i^e}{dx} \frac{d\psi_m^e}{dx} \right.\right.$$

$$\left.\left. + \left(c_0 + c_1 u_h + c_2 u_h^2 + c_3 \frac{du_h}{dx} \right) \psi_i^e \psi_m^e \right] dx + \left[\beta_0 + \beta_1 u_h(x_b) \right] \psi_i^e(x_b) \psi_m^e(x_b) \right\} u_m^e$$

$$= \sum_{m=1}^{n} u_m^e \left\{ \int_{x_a}^{x_b} \left[\left(a_1 \psi_j^e + 2a_2 u_h \psi_j^e + a_3 \frac{d\psi_j^e}{dx} \right) \frac{d\psi_i^e}{dx} \frac{d\psi_m^e}{dx} \right. \right.$$

$$\left. \left. + \left(c_1 \psi_j^e + 2c_2 u_h \psi_j^e + c_3 \frac{d\psi_j^e}{dx} \right) \psi_i^e \psi_m^e \right] dx + \beta_1 \psi_j^e(x_b) \psi_i^e(x_b) \psi_m^e(x_b) \right\}$$

$$= \int_{x_a}^{x_b} \left[\frac{d\psi_i^e}{dx} \left(a_1 \psi_j^e + 2a_2 u_h \psi_j^e + a_3 \frac{d\psi_j^e}{dx} \right) \frac{du_h}{dx} + \psi_i^e \left(c_1 \psi_j^e + 2c_2 u_h \psi_j^e + c_3 \frac{d\psi_j^e}{dx} \right) u_h \right] dx$$

$$+ \beta_1 u_h(x_b) \psi_i^e(x_b) \psi_j^e(x_b) \tag{4}$$

$$\frac{\partial F_i^e}{\partial u_j^e} = \frac{\partial}{\partial u_j^e} \left\{ \int_{x_a}^{x_b} f(x) \psi_i^e \, dx + [\beta_0 + \beta_1 u_h(x_b)] u_\infty^b \psi_i^e(x_b) + \psi_i^e(x_a) Q_a + \psi_i^e(x_b) Q_b \right\}$$

$$= \beta_1 u_\infty^b \psi_i^e(x_b) \psi_j^e(x_b) \tag{5}$$

where the identities

$$\frac{\partial u_h}{\partial u_j^e} = \psi_j^e, \quad \sum_{m=1}^{n} u_m^e \psi_m^e = u_h, \quad \sum_{m=1}^{n} u_m^e \frac{d\psi_m^e}{dx} = \frac{du_h}{dx}$$

are used in arriving at Eq. (4). Thus, the tangent matrix coefficients are given by

$$T_{ij}^e = K_{ij}^e + \int_{x_a}^{x_b} \left\{ \frac{d\psi_i^e}{dx} \left[(a_1 + 2a_2 u_h) \psi_j^e + a_3 \frac{d\psi_j^e}{dx} \right] \frac{du_h}{dx} \right.$$

$$\left. + \psi_i^e \left[(c_1 + 2c_2 u_h) \psi_j^e + c_3 \frac{d\psi_j^e}{dx} \right] u_h \right\} dx$$

$$+ \beta_1 \left[u_h(x_b) - u_\infty^b \right] \psi_i^e(x_b) \psi_j^e(x_b) \tag{6}$$

Since K_{ij}^e is symmetric, T_{ij}^e is symmetric if and only if $a_1 = a_2 = c_3 = 0$.

Example 4.4.3

Solve the nonlinear differential equation of Example 4.4.1

$$-\frac{d}{dx} \left(u \frac{du}{dx} \right) = f_0, \quad 0 < x < 1; \quad \left[u \frac{du}{dx} \right]_{x=0} = \hat{Q}, \quad u(1) = \hat{u} \tag{1}$$

using the Newton iteration technique.

Solution: First, we compute the tangent matrix for a typical element. We have

$$T_{ij}^e = K_{ij}^e + \sum_{m=1}^{n} \frac{\partial K_{im}^e}{\partial u_j^e} u_m^e = K_{ij}^e + \sum_{m=1}^{n} \frac{\partial}{\partial u_j^e} \left(\int_{x_a}^{x_b} u_h \frac{d\psi_i^e}{dx} \frac{d\psi_m^e}{dx} dx \right) u_m^e$$

$$= K_{ij}^e + \int_{x_a}^{x_b} \frac{\partial u_h}{\partial u_j^e} \frac{d\psi_i^e}{dx} \left(\sum_{m=1}^{n} u_m^e \frac{d\psi_m^e}{dx} \right) dx = K_{ij}^e + \int_{x_a}^{x_b} \frac{du_h}{dx} \frac{d\psi_i^e}{dx} \psi_j^e \, dx$$

$$\equiv K_{ij}^e + \hat{K}_{ij}^e \tag{2}$$

where

$$\hat{K}_{ij}^e = \int_{x_a}^{x_b} \frac{du_h}{dx} \frac{d\psi_i^e}{dx} \psi_j^e \, dx, \quad \sum_{m=1}^{n} u_m^e \frac{d\psi_m^e}{dx} = \frac{du_h}{dx}$$

are used in arriving at the final result. Clearly, $\mathbf{T}^e$ is not symmetric. For the choice of linear approximation functions ψ_i^e, numerical form of $\mathbf{K}^e$ is given by (with $a_0 = 0$ and $a_u = 1$) Eq. (4) of Example 4.3.1. The coefficients of matrix $\hat{\mathbf{K}}^e$ become

$$\hat{K}_{ij}^e = \int_{x_a}^{x_b} \frac{du_h}{dx} \frac{d\psi_i^e}{dx} \psi_j^e \, dx = \frac{u_2^e - u_1^e}{h_e} \int_{x_a}^{x_b} \frac{d\psi_i^e}{dx} \psi_j^e \, dx$$

or

$$\hat{K}^e = \frac{u_2^e - u_1^e}{2h_e} \begin{bmatrix} -1 & -1 \\ 1 & 1 \end{bmatrix} \tag{3}$$

Thus, the tangent matrix and the residual vector are

$$\mathbf{T}^e = \mathbf{K}^e + \hat{\mathbf{K}}^e = \frac{(\bar{u}_1^e + \bar{u}_2^e)}{2h_e} \begin{bmatrix} 1 & -1 \\ -1 & 1 \end{bmatrix} + \frac{(\bar{u}_2^e - \bar{u}_1^e)}{2h_e} \begin{bmatrix} -1 & -1 \\ 1 & 1 \end{bmatrix} = \frac{1}{h_e} \begin{bmatrix} \bar{u}_1^e & -\bar{u}_2^e \\ -\bar{u}_1^e & \bar{u}_2^e \end{bmatrix}$$

$$\mathbf{R}^e = \mathbf{K}^e \mathbf{u}^e - \mathbf{F}^e = \frac{(\bar{u}_1^e)^2 - (\bar{u}_2^e)^2}{2h_e} \begin{Bmatrix} 1 \\ -1 \end{Bmatrix} - \begin{Bmatrix} f_1^e + Q_1^e \\ f_2^e + Q_2^e \end{Bmatrix} \tag{4}$$

where $\bar{u}_i^e$ denote the nodal values known from the iteration before the current iteration. The element equations $\mathbf{T}^e \Delta \mathbf{u}^e = -\mathbf{R}^e$ are (hold for any iteration)

$$\frac{1}{h_1} \begin{bmatrix} \bar{u}_1^1 & -\bar{u}_2^1 \\ -\bar{u}_1^1 & \bar{u}_2^1 \end{bmatrix} \begin{Bmatrix} \Delta u_1^1 \\ \Delta u_2^1 \end{Bmatrix} = \frac{(\bar{u}_1^1)^2 - (\bar{u}_2^1)^2}{2h_1} \begin{Bmatrix} -1 \\ 1 \end{Bmatrix} + \begin{Bmatrix} f_1^1 + Q_1^1 \\ f_2^1 + Q_2^1 \end{Bmatrix}$$

$$\frac{1}{h_2} \begin{bmatrix} \bar{u}_1^2 & -\bar{u}_2^2 \\ -\bar{u}_1^2 & \bar{u}_2^2 \end{bmatrix} \begin{Bmatrix} \Delta u_1^2 \\ \Delta u_2^2 \end{Bmatrix} = \frac{(\bar{u}_1^2)^2 - (\bar{u}_2^2)^2}{2h_2} \begin{Bmatrix} -1 \\ 1 \end{Bmatrix} + \begin{Bmatrix} f_1^2 + Q_1^2 \\ f_2^2 + Q_2^2 \end{Bmatrix} \tag{5}$$

and the assembled system of equations for the case when $h_1 = h_2 = h$ is

$$\frac{1}{h} \begin{bmatrix} \bar{U}_1 & -\bar{U}_2 & 0 \\ -\bar{U}_1 & 2\bar{U}_2 & -\bar{U}_3 \\ 0 & -\bar{U}_2 & \bar{U}_3 \end{bmatrix} \begin{Bmatrix} \Delta U_1 \\ \Delta U_2 \\ \Delta U_3 \end{Bmatrix} = \frac{1}{2h} \begin{Bmatrix} \bar{U}_2^2 - \bar{U}_1^2 \\ \bar{U}_1^2 - 2\bar{U}_2^2 + \bar{U}_3^2 \\ \bar{U}_2^2 - \bar{U}_3^2 \end{Bmatrix} + \begin{Bmatrix} f_1^1 + Q_1^1 \\ (f_2^1 + Q_2^1) + (f_1^2 + Q_1^2) \\ f_2^2 + Q_2^2 \end{Bmatrix} \tag{6}$$

As in Example 4.4.1, we use the following data (for a uniform mesh of two linear elements, $h_1 = h_2 = L/2 = 0.5$)

$$\hat{Q} = 0, \qquad \hat{u} = \sqrt{2} = 1.4142, \qquad f_0 = -1 \ (f_i^{(e)} = 0.5 f_0 h = -0.25) \tag{7}$$

For the choice of initial guess vector, as in the direct iteration,

$$\bar{\mathbf{U}} \equiv \mathbf{U}^{(0)} = \begin{Bmatrix} 1.0 \\ 1.0 \\ \sqrt{2} \end{Bmatrix}, \quad \text{or} \quad \bar{\mathbf{u}}^1 = \begin{Bmatrix} 1.0 \\ 1.0 \end{Bmatrix} \quad \text{and} \quad \bar{\mathbf{u}}^2 = \begin{Bmatrix} 1.0 \\ \sqrt{2} \end{Bmatrix} \tag{8}$$

the assembled equations at the beginning of iteration $r = 1$ are

$$2 \begin{bmatrix} 1 & -1 & 0 \\ -1 & 2 & -\sqrt{2} \\ 0 & -1 & \sqrt{2} \end{bmatrix} \begin{Bmatrix} \Delta U_1 \\ \Delta U_2 \\ \Delta U_3 \end{Bmatrix} = \begin{Bmatrix} 0 \\ 1 \\ -1 \end{Bmatrix} - \begin{Bmatrix} 0.25 \\ 0.50 \\ 0.25 \end{Bmatrix} + \begin{Bmatrix} Q_1^1 \\ Q_2^1 + Q_1^2 \\ Q_2^2 \end{Bmatrix} \tag{10}$$

Using the boundary conditions and balance conditions

$$\Delta U_3 = 0.0, \quad Q_1^1 = 0, \quad Q_2^1 + Q_1^2 = 0 \tag{11}$$

we obtain the following condensed equations for the increments ΔU_1 and ΔU_2

$$\begin{bmatrix} 2 & -2 \\ -2 & 4 \end{bmatrix} \begin{Bmatrix} \Delta U_1 \\ \Delta U_2 \end{Bmatrix} = \begin{Bmatrix} -0.25 \\ 0.50 \end{Bmatrix} \tag{12}$$

The solution to these equations is $\Delta U_1 = 0.0$ and $\Delta U_2 = 0.125$, and the complete solution becomes

$$\mathbf{U}^{(1)} = \mathbf{U}^{(0)} + \Delta \mathbf{U} = \begin{Bmatrix} 1.0 \\ 1.0 \\ \sqrt{2} \end{Bmatrix} + \begin{Bmatrix} 0.0000 \\ 0.1250 \\ 0.0000 \end{Bmatrix} = \begin{Bmatrix} 1.0000 \\ 1.1250 \\ 1.4142 \end{Bmatrix} \tag{13}$$

For the second iteration, we have [from Eq. (6) with $\bar{\mathbf{U}} = \mathbf{U}^{(1)}$]

$$\begin{bmatrix} 2 & -2.25 \\ -2 & 4.50 \end{bmatrix} \begin{Bmatrix} \Delta U_1 \\ \Delta U_2 \end{Bmatrix} = \begin{Bmatrix} 0.265625 \\ 0.468750 \end{Bmatrix} - \begin{Bmatrix} 0.25 \\ 0.50 \end{Bmatrix} = \begin{Bmatrix} 0.015625 \\ -0.031250 \end{Bmatrix} \tag{14}$$

The solution to these equations is

$$\Delta U_1 = 0.0, \qquad \Delta U_2 = -0.007, \qquad \Delta U_3 = 0.0 \tag{15}$$

and the complete solution becomes

$$\mathbf{U}^{(2)} = \mathbf{U}^{(1)} + \Delta \mathbf{U} = \begin{Bmatrix} 1.0000 \\ 1.1250 \\ 1.4142 \end{Bmatrix} + \begin{Bmatrix} 0.000 \\ -0.007 \\ 0.000 \end{Bmatrix} = \begin{Bmatrix} 1.0000 \\ 1.1180 \\ 1.4142 \end{Bmatrix} \tag{16}$$

The Euclidean norm of the error in the solution (with respect to the solution obtained in the first iteration) [see Eq. (4.4.5)] is 0.0034.

Table 4.4.2 contains the numerical results obtained with the Newton iteration method. The Newton method gives converged solution in only three iterations as opposed to four iterations taken by the direct iteration method. For the mesh of one quadratic element, the error criterion is met while the solution did not actually coincide with the exact solution; the exact value may be obtained at the end of the fourth iteration.

Table 4.4.2: Comparison of finite element solutions with the exact solution of the problem in Example 4.4.3 (Newton's iteration method).

x	Iteration	2L	4L	1Q	2Q	Exact
0.00	1	1.0000	1.0000	1.0000	1.0000	
	2	1.0000	1.0000	1.0000	1.0000	
	3	1.0000	1.0000	1.0000	1.0000	1.0000
0.25	1	–	1.0312	–	1.0312	
	2	–	1.0308	–	1.0308	
	3	–	1.0308	–	1.0308	1.0308
0.50	1	1.1250	1.1250	1.1241	1.1250	
	2	1.1180	1.1180	1.1187	1.1180	
	3	1.1180	1.1180	1.1187	1.1180	1.1180
0.75	1	–	1.2812	–	1.2766	
	2	–	1.2504	–	1.2502	
	3	–	1.2500	–	1.2500	1.2500

4.5   Computer Implementation

4.5.1   Introduction

The nonlinear formulations and solution procedures described in the preceding sections can be implemented on a computer using a programming language. Fortran is chosen here mainly because of its transparent nature compared to C++ language and the author's own background. Of course, once the logic of implementation is understood, one may carry out the computations in any programming language. A flow chart of a typical nonlinear analysis program is shown in Fig. 4.5.1.

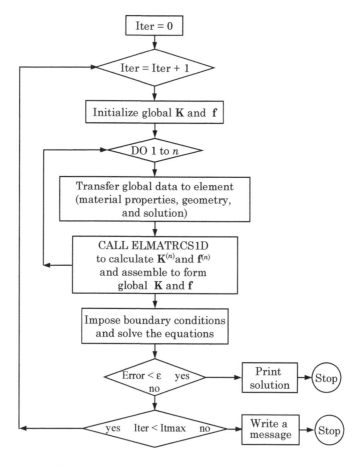

Fig. 4.5.1: A flow chart of the nonlinear finite element analysis program.

It is useful for the reader to familiarize with programs FEM1D and FEM2D described in the finite element book by Reddy [2] to fully understand the logic and variable names used here (Fortran source codes are available from the author). Fortran programs of banded symmetric and unsymmetric equation solvers are listed in Appendix 1.

4.5.2 Preprocessor Unit

The preprocessor has the objective of reading the problem data (e.g. coefficients of the differential equation) and analysis type, generating the mesh information (e.g. connectivity array, global coordinates of the nodes) and printing the data and analysis information.

1. Read or define the problem and analysis data

   (a) Define the coefficients $a(x, u)$, $b(x, u)$, and $c(x, u)$, and $f(x)$ that appear in the model differential equation, Eq. (4.1.1); for example, one may assume a, b, c, and f to depend on x, u, and du/dx as follows:

$$a(x, u) = a_{x0} + a_{x1}x + a_{u1}u + a_{u2}u^2 + a_{ux1}\left(\frac{du}{dx}\right) + a_{ux2}\left(\frac{du}{dx}\right)^2$$

$$b(x, u) = b_{x0} + b_{x1}x + b_{u1}u + b_{u2}u^2 + b_{ux1}\left(\frac{du}{dx}\right) + b_{ux2}\left(\frac{du}{dx}\right)^2$$

$$c(x, u) = c_{x0} + c_{x1}x + c_{u1}u + c_{u2}u^2 + c_{ux1}\left(\frac{du}{dx}\right) + c_{ux2}\left(\frac{du}{dx}\right)^2$$

$$f(x) = f_{x0} + f_{x1}x + f_{x2}x^2 \qquad (4.5.1)$$

   so that a large number of nonlinear problems described by the model equation in Eq. (4.1.1) may be solved with the same computer program. Thus, one must read all parameters appearing in the above expressions; the data is assumed to be the same for the whole domain: AX0, AX1, AU1, AU2, AUX1, AUX2; BX0, BX1, BU1, etc.

   (b) Read and generate geometry, material, and FE analysis data (the domain is assumed to be a straight line):

Variable	Definition
X0	Global coordinate of the first node of the domain
AL	Domain length
IEL	Degree of interpolation used (1: linear; 2: quadratic)
NEM	Number of elements in the mesh
NNM	Number of nodes in the mesh
NDF	Number of degrees of freedom per node (here, NDF=1)
NEQ	Number of equations in the mesh, NEQ=NNM*NDF
NLS	Number of "load" steps (= 1 for the present case)
NONLIN	Parameter for the type of analysis: = 0, linear analysis; = 1, nonlinear analysis using the direct iteration; > 1, nonlinear analysis using Newton's iteration
EPS	Allowable error tolerance ϵ for the convergence test
ITMAX	Maximum allowable number of iterations for convergence

   (c) Read data to specify the boundary and initial conditions. The variables names and their meaning is given here.

Variable	Definition
NSPV	Number of specified primary degrees of freedom (DoF), U
NSSV	Number of specified secondary DoF, $\hat{Q} \equiv n_x\, a\frac{du}{dx}$
NSMB	number of specified mixed boundary conditions of the type

$$Q + (\beta_0 + \beta_u U)(U - U_{\text{ref}}) = \hat{Q} \qquad (4.5.2)$$

Variable	Definition
ISPV(I,J)	The array of the specified primary DoF ($I = 1, 2, \ldots,$ NSPV and $J = 1, 2, \ldots,$ NDF)
ISPV(I,1)	Global node number at which the primary variable is specified
ISPV(I,2)	Degree of freedom at the global node ISPV(I,1) that is specified (for the class of problems discussed in this chapter, NDF=1 and therefore ISPV(I,2)=1 for all cases)
VSPV(I)	Array of specified values of primary variable DoF in ISPV(I,J) [VSPV(I) should be in the same sequence as the specified DoF in array ISPV(I,1)]
ISSV(I,J)	Similar to ISPV(I,J), but for secondary variables
VSSV(I)	Similar to VSPV(I), but for secondary variables
ISMB(I,J)	Array of specified mixed boundary conditions in Eq. (4.5.2), $I = 1, 2, \ldots,$ NSMB, and has a meaning similar to arrays ISPV(I,J) and ISSV(I,J)
BETA0(I)	Array of specified values of β_0 in Eq. (4.5.2)
BETAU(I)	Array of specified values of β_u in Eq. (4.5.2)
UREF(I)	Array of specified values of U_{ref} in Eq. (4.5.2)
GU0(I)	Initial (guess) solution vector ($I = 1, 2, \ldots,$ NEQ)

2. Generate a mesh of line elements; in particular, compute arrays GLX(I) and NOD(I,J).

Variable	Definition
GLX(I)	Global x coordinate of global node I
NOD(N,I)	Global node number corresponding to the Ith node of the Nth element in the mesh
NHBW	Half bandwidth of the assembled coefficient matrix (the maximum number of entries from the diagonal term in any row to the last nonzero term in the same row; both the diagonal entry and the last entry are counted in NHBW)
NBW	Twice the half bandwidth, NBW=2*NHBW

3. Write out the input data and necessary warning messages. Box 4.5.1 shows the necessary preprocessor Fortran statements.

4.5.3 Processor Unit

Here we discuss the calculation of element coefficients, assembly of element equations, imposition of various types of boundary conditions, and error check inside the iteration algorithm.

4.5.3.1 Calculation of element coefficients

The number of Gauss points (NGP) needed to evaluate the coefficients is determined by the highest polynomial degree p [$NGP = (p+1)/2$], appearing in the integrand of any integral being evaluated. Typically, the expression that dictates the highest degree is the integral involving the product $c(x, u)\psi_i^e\psi_j^e$.

Box 4.5.1: Fortran statements to read the input data and generate GLX(I) and NOD(I,J) arrays in a preprocessor.

```
        READ(IN,*)AX0,AX1,AU1,AU2,AUX1,AUX2
        READ(IN,*)BX0,BX1,BU1,BU2,BUX1,BUX2
        READ(IN,*)CX0,CX1,CU1,CU2,CUX1,CUX2
        READ(IN,*)FX0,FX1,FX2
        READ(IN,*)X0,AL
        READ(IN,*)IEL,NEM
        NDF = 1
        NPE = IEL+1
        NNM = IEL*NEM+1
        NEQ = NNM*NDF
        READ(IN,*)NSPV,NSSV,NSMB
        IF(NSPV.GT.0)THEN
            DO 10 NB=1,NSPV
10          READ(IN,*) (ISPV(NB,J),J=1,2),VSPV(NB)
        ENDIF
        IF(NSSV.GT.0)THEN
            DO 20 NB=1,NSSV
20          READ(IN,*) (ISSV(NB,J),J=1,2),VSSV(NB)
        ENDIF
        IF(NSMB.NE.0)THEN
            DO 30 I=1, NSMB
30          READ(IN,*) (ISMB(I,J),J=1,2),UREF(I),BETA0(I),BETAU(I)
        ENDIF
        READ(IN,*)NONLIN,NPRNT
        IF(NONLIN.GT.0)THEN
            READ(IN,*)NLS,ITMAX,EPS,GAMA
            READ(IN,*)(DP(I),I=1,NLS)
            READ(IN,*)(GP1(I),I=1,NNM)
        ENDIF
C
C   Generate GLX(I) and NOD(I,J) arrays and compute bandwidths
        DL=AL/NEM/IEL
        DO I=1,NNM
        GLX(I)=X0+DL*(I-1)
        ENDDO
        NHBW = (IEL+1)*NDF
        NBW  = 2*NHBW
        NN   = NPE*NDF
        DO I=1,NPE
        NOD(1,I)=I
        ENDDO
        DO 40 N=2,NEM
        DO 40 I=1,NPE
40  NOD(N,I)=NOD(N-1,I)+IEL
```

If we assume c to be a linear function of u, and if u is approximated using linear polynomials, the expression is a cubic ($p = 3$) polynomial. Hence, a two-point integration is needed to evaluate it; for quadratic interpolation of u, the expression is a fifth degree polynomial ($p = 5$), and a three-point Gauss quadrature is required.

The element coefficients K_{ij}^e of Eq. (4.3.3), for $\beta = 0$, can be evaluated as

```
ELK(I,J) = ELK(I,J) + (AX*GDSFL(I)*GDSFL(J) + BX*SFL(I)*GDSFL(J)
         +  CX*SFL(I)*SFL(J))*GJ*GAUSWT(NI,NGP)
```

which is calculated inside a Gauss quadrature loop on NI $= 1, 2, \ldots$, NGP and inner loops on I,J $= 1, 2, \ldots$, NPE. The Fortran subroutine, **ELMATRCS1D**, for the calculation of $\mathbf{K}^e$ and $\mathbf{f}^e$ for the model nonlinear problem is presented in Box 4.5.2. Subroutine **INTERPLN1D** that is called from **ELMATRCS1D**[2] was presented in Box 3.7.1. The variable NONLIN is introduced for the type of analysis (0: linear; 1: nonlinear with direct iteration, and; > 1: nonlinear with Newton's iteration).

Box 4.5.2: Subroutine **ELMATRCS1D** to generate $\mathbf{K}^e$, $\mathbf{f}^e$, $\mathbf{T}^e$, and $\mathbf{R}^e$.

```
      SUBROUTINE ELMATRCS1D(IEL,NPE,NONLIN,FO)
C     ------------------------------------------------------------------
C     SFL(I)     Shape (interpolation) functions of the Lagrange type
C     GDSFL(I)   Global derivative (with respect to x) of SFL(I)
C     GJ                          Jacobian of the transformation, J
C     GAUSPT(I,J) Ith Gauss point in the J-point Gauss rule
C     GAUSWT(I,J) Ith Gauss weight in the J-point Gauss rule
C     NGP        Number of Gauss points used to evaluate integrals
C     EL         Length of the element, h
C     XI         Normalized local coordinate such that -1 < x < 1
C     U          Primary variable (function of x)
C     DU         Global derivative of u: du/dx
C     ------------------------------------------------------------------
C
      IMPLICIT REAL*8(A-H,O-Z)
      DIMENSION GAUSPT(5,5),GAUSWT(5,5),TANG(3,3)
      COMMON/SHP/SFL(4),GDSFL(4)
      COMMON/STF/ELK(3,3),ELF(3),ELX(3),ELU(3)
     1            AX0,AX1,AU1,AU2,AUX1,AUX2,CX0,CX1,CU1,CU2,CUX1,CUX2,
     2            BX0,BX1,BU1,BU2,BUX1,BUX2,FX0,FX1,FX2
C
      DATA GAUSPT/5*0.0D0,-.57735027D0,.57735027D0,3*0.0D0,-.77459667D0,
     1    0.0D0,.77459667D0,2*0.0D0,-.86113631D0,-0.33998104D0,.33998104D0,
     2    0.86113631D0,.0D0,-.906180D0,-.538469D0,.0D0,.538469D0,.906180D0/
      DATA GAUSWT/2.0D0,4*0.0D0,2*1.0D0,3*0.0D0,.55555555D0,.88888888D0,
     1    0.55555555D0,2*0.0D0,.34785485D0,2*.65214515D0,.34785485D0,0.0D0,
     2    0.236927D0,0.478629D0,0.568889D0,0.478629D0,0.236927D0/
```

---

[2]Delete the variable MODEL from the argument list of **INTERPLN1D** in Box 3.7.1 and delete the Hermite interpolation functions from the subroutine.

Box 4.5.2: Subroutine **ELMATRCS1D** (continued).

```
        NGP=IEL+1
        EL=ELX(NPE)-ELX(1)
        DO 10 I=1,NPE
        ELF(I)=0.0
        DO 10 J=1,NPE
        IF(NONLIN.GT.1)THEN
             TANG(I,J)=0.0
        ENDIF
 10   ELK(I,J) =0.0
        DO 50 NI=1,NGP
        XI=GAUSPT(NI,NGP)
        CALL INTERPLN1D(ELX,GJ,IEL,NPE,XI)
        CNST=GJ*GAUSWT(NI,NGP)
        X=ELX(1)+0.5*(1.0+XI)*EL
        AX=AX0+AX1*X
        BX=BX0+BX1*X
        CX=CX0+CX1*X
        FX=FX0+FX1*X+FX2*X*X
        IF(NONLIN.GT.0)THEN
             U=0.0
             DU=0.0
             DO 20 I=1,NPE
             U =U+SFL(I)*ELU(I)
             DU=DU+GDSFL(I)*ELU(I)
 20          CONTINUE
 C
 C   Compute coefficients needed to define the nonlinear element matrix
             AX=AX+AU1*U+AUX1*DU+AU2*U*U+AUX2*DU*DU
             BX=BX+BU1*U+BUX1*DU+BU2*U*U+BUX2*DU*DU
             CX=CX+CU1*U+CUX1*DU+CU2*U*U+CUX2*DU*DU
 C
 C   Compute additional coefficients needed to define the tangent matrix
             IF(NONLIN.GT.1)THEN
                  AXT1=(AU1+2.0*AU2*U)*DU
                  AXT2=(AUX1+2.0*AUX2*DU)*DU
                  BXT1=(BU1+2.0*BU2*U)*DU
                  BXT2=(BUX1+2.0*BUX2*DU)*DU
                  CXT1=(CU1+2.0*CU2*U)*U
                  CXT2=(CUX1+2.0*CUX2*DU)*U
             ENDIF
        ENDIF
 C   Compute element matrices [K]=[ELK], [TANG], and {F}={ELF}
 C   [TANG] is the addition to [K] to obtain [T]; see Eq.~(4.4.13)
        DO 40 I=1,NPE
        ELF(I)=ELF(I)+F0*FX*SFL(I)*CNST
        DO 40 J=1,NPE
        S00=SFL(I)*SFL(J)*CNST
        S01=SFL(I)*GDSFL(J)*CNST
        S10=GDSFL(I)*SFL(J)*CNST
        S11=GDSFL(I)*GDSFL(J)*CNST
        ELK(I,J)=ELK(I,J)+AX*S11+BX*S01+CX*S00
```

Box 4.5.2: Subroutine **ELMATRCS1D** (continued).

```
         IF(NONLIN.GT.1)THEN
            TANG(I,J)=TANG(I,J)+AXT1*S10+AXT2*S11+BXT1*S00
     *                         +BXT2*S01+CXT1*S00+CXT2*S01
         ENDIF
   40 CONTINUE
   50 CONTINUE
   C
   C   Compute the residual vector {R}={ELF} and tangent matrix {T}=[ELK]
   C
         IF(NONLIN.GT.1)THEN
            DO 60 I=1,NPE
            DO 60 J=1,NPE
   60       ELF(I)=ELF(I)-ELK(I,J)*ELU(J)
            DO 80 I=1,NPE
            DO 80 J=1,NPE
   80       ELK(I,J)=ELK(I,J)+TANG(I,J)
         ENDIF
         RETURN
         END
```

4.5.3.2   Assembly of element coefficients

The assembly of element equations is carried out as soon as the element matrices $\mathbf{K}^e$ and $\mathbf{f}^e$ are computed, rather than waiting until element coefficients of all elements are computed. The latter requires storage of the element coefficients of each element of the mesh, and it is an unnecessary waste of computer storage. The assembly process can be performed in the same loop in which subroutine **ELCOEFN1D** is called to compute element matrices $\mathbf{K}^e$ and $\mathbf{f}^e$.

The local nature of the finite element interpolation functions (i.e. ψ_i^e is defined to be nonzero only over the element Ω^e) is responsible for the banded character of the assembled matrix. If two global nodes do not belong to the same element then the corresponding entry in the global matrix is zero:

$$K_{IJ} = 0, \text{ if global nodes } I \text{ and } J \text{ do not correspond to}$$
$$\text{any two nodes of the same element} \tag{4.5.3}$$

The banded nature of the finite element equations enables us to save storage and computing time by assembling the element equations into a banded form. When element matrices $\mathbf{K}^e$ are symmetric, the resulting global or assembled coefficient matrix $\mathbf{K}$ is also symmetric, with many zeros away from the main diagonal. Therefore, it is sufficient to store only the upper *half-band* of the assembled matrix. When $\mathbf{K}^e$ is unsymmetric, we can store the assembled matrix into a full banded form. General-purpose equation solvers are available for such banded systems of equations.

The half bandwidth of a matrix is defined as follows: let N_i be the number of matrix elements between the diagonal element and the last nonzero element in the ith row, after which all elements in that row are zero; the half-bandwidth is the maximum of $(N_i + 1) \times$ NDF (NDF is the number of degrees of freedom per node):

$$\begin{array}{c} \max \\ 1 \leq i \leq \text{NEQ} \end{array} [(N_i + 1) \times \text{NDF}]$$

where NEQ is the number of equations (or rows) in the matrix.

The half-bandwidth NHBW of the assembled (i.e. global) finite element matrix can be determined in the finite element program itself. The property in Eq. (4.5.3) enables us to determine the half-bandwidth NHBW of the assembled matrix as

$$\text{NHBW} = \begin{array}{c} \max \\ 1 \leq N \leq \text{NEM} \\ 1 \leq I, J, \leq \text{NPE} \end{array} \Big\{ \text{abs}[\text{NOD}(N,I) - \text{NOD}(N,J)] + 1 \Big\} \times \text{NDF} \quad (4.5.4)$$

For example, for one-dimensional problems with elements connected in series (i.e. connected in one line), the maximum difference between nodes of an element is equal to NPE-1. Hence, we have

$$\text{NHBW} = [(\text{NPE} - 1) + 1] \times \text{NDF} = \text{NPE} \times \text{NDF} \quad (4.5.5)$$

The logic behind assembling the element matrices $\mathbf{K}^e$ into the upper-banded form of the global coefficient matrix $\mathbf{K}$ is that the assembly be skipped whenever the column number is both less than the row number and greater than the half bandwidth plus the row number. Similarly, the full-banded form is obtained by skipping assembly whenever the column number is both less than the row number minus NHBW and greater than the row number plus NHBW. In the case of the upper-banded form, the main diagonal of the actual matrix $\mathbf{K}$ becomes the first column of the assembled banded matrix $\mathbf{K}_{FB}$, as shown in Fig. 4.5.2 (also see Box 4.5.3). For the full-banded form the actual matrix,

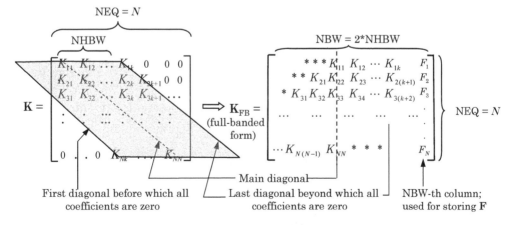

Fig. 4.5.2: Assembled coefficient matrix storage in (upper) half-banded and full-banded forms.

Box 4.5.3: Fortran statements for the assembly of element matrices into a full-banded form.

```
            DO 140 N=1,NEM
              DO 100 I=1,NPE
                NI=NOD(N,I)
                ELU(I)=GAMA*GP2(NI)+(1.0-GAMA)*GP1(NI)
100             ELX(I)=GLX(NI)
                CALL ELMATRCS1D(IEL,NPE,NONLIN,F0)
                IF(NPRNT.GT.0)THEN
                    IF(N.EQ.1 .AND. NCOUNT.EQ.1)THEN
                      WRITE(IT,450)
                      DO 110 I=1,NN
110                   WRITE(IT,350)(ELK(I,J),J=1,NN)
                      WRITE(IT,460)
                      WRITE(IT,350)(ELF(I),I=1,NN)
                    ENDIF
                ENDIF
                DO 130 I=1,NPE
                NR=(NOD(N,I)-1)*NDF
                DO 130 II=1,NDF
                NR=NR+1
                L=(I-1)*NDF+II
                GLK(NR,NBW)=GLK(NR,NBW)+ELF(L)
                DO 120 J=1,NPE
                NCL=(NOD(N,J)-1)*NDF
                DO 120 JJ=1,NDF
                M=(J-1)*NDF+JJ
                NC=NCL-NR+JJ+NHBW
                IF(NC.GT.0)THEN
                    GLK(NR,NC)=GLK(NR,NC)+ELK(L,M)
                ENDIF
120             CONTINUE
130         CONTINUE
140     CONTINUE
```

K becomes the NHBW-th column of the assembled banded matrix $\mathbf{K}_{FB}$. The diagonals parallel to the main diagonal take the position of respective columns in the banded matrix. Thus, the banded matrix $\mathbf{K}_{HB}$ has the dimension of NEQ $\times$ NHBW and $\mathbf{K}_{FB}$ has the dimension of NEQ $\times$ (NBW-1); by making $\mathbf{K}_{FB}$ an array of dimension NEQ $\times$ NBW, we can use the NBW-th column to store the assembled source vector **F**.

4.5.3.3 Imposition of boundary conditions

The imposition of boundary conditions on a banded system of non-symmetric equations is discussed next. The assembled coefficient matrix $\mathbf{K}_{FB}$ =[GLK] is stored in a full bandwidth (NBW) form, and it is of the order NEQ$\times$NBW, the last column GLK(I,NBW) being reserved for the source vector **F**. Note that GLK($\cdot$, NHBW) is the main diagonal of the actual matrix **K**.

Essential boundary conditions

Now suppose that U_I is specified to be $\hat{U}_I$. Then, the Ith equation of the system is replaced with the equation $U_I = \hat{U}_I$, which can be expressed as $K_{II}U_I = F_I$ (no sum on I) with $K_{II} = \text{GLK(I,NHBW)} = 1.0$ and $F_I = \text{GLK(I,NBW)} = \hat{U}_I$. This can be implemented as follows:

$$\text{GLK(I,J)} = 0.0, \quad \text{for} \quad J = 1, 2, \ldots, \text{NBW}$$
$$\text{GLK(I,NHBW)} = 1.0, \quad \text{GLK(I,NBW)} = \hat{U}_I \tag{4.5.6}$$

This is repeated NSPV times, NSPV being the number of specified primary degrees of freedom. Note that all other equations remain unchanged.

Natural boundary conditions

Next, suppose that Q_I is specified to be $\hat{Q}_I$. Then the force in the Ith equation is augmented with $\hat{Q}_I$ [this applies to the nonzero $\hat{Q}$ in the mixed boundary condition in Eq. (4.5.2)]:

$$\text{GLK(I,NBW)} = \text{GLK(I,NBW)} + \hat{Q}_I \tag{4.5.7}$$

This is repeated NSSV times, NSSV being the number of specified secondary variables.

Mixed boundary conditions

Finally, we consider mixed boundary conditions of the form in Eq. (4.5.2). Since $\hat{Q}$, if nonzero, is already accounted for in NSSV, Eq. (4.5.2) can be expressed in the homogeneous form

$$Q_I + \beta_I \left(U_I - U_{\text{ref}}\right) = 0 \quad \text{(no sum on } I) \tag{4.5.8}$$

where β_I is a function of u, say $\beta_I = \beta_I^0 + \beta_I^1 U_I$. Recall from Eq. (4.3.3) and Eq. (5) of Example 4.3.1 that the boundary condition in Eq. (4.5.8) adds the value β_I to the diagonal term $K_{II} = \text{GLK(I,NHBW)}$ and $\beta_I U_{\text{ref}}$ to $F_I = \text{GLK(I,NBW)}$. Thus, the boundary condition in Eq. (4.5.8) may be implemented as follows:

$$\text{GLK(I,NHBW)} = \text{GLK(I,NHBW)} + \left(\beta_I^0 + \beta_I^1 U_I\right)$$
$$\text{GLK(I,NBW)} = \text{GLK(I,NBW)} + \left(\beta_I^0 + \beta_I^1 U_I\right) U_{\text{ref}} \tag{4.5.9}$$

The above statement modifies the assembled direct stiffness matrix and source vector that are used in the direct iteration procedure. In the case of Newton's iteration procedure, the assembled tangent matrix $\mathbf{T}_{FB} = [\text{GLK}]$ is modified as follows [see Eqs. (5) and (6) of Example 4.4.2 for the derivation]:

$$\text{GLK(I,NHBW)} = \text{GLK(I,NHBW)} + \left(\beta_I^0 + \beta_I^1 U_I\right) + \beta_I^1 \left(U_I - U_{\text{ref}}\right)$$
$$= \text{GLK(I,NHBW)} + \left(\beta_I^0 + 2\beta_I^1 U_I\right) - \beta_I^1 U_{\text{ref}} \tag{4.5.10}$$

The residual vector $-\mathbf{R} = \{\text{GLK(I,NBW)}\}$ must be modified by adding the contributions due to the mixed boundary condition only, $-R_I = F_I - K_{II}U_I = \beta_I U_{\text{ref}} - \beta_I U_I$:

$$\text{GLK(I,NBW)} = \text{GLK(I,NBW)} + \left(\beta_I^0 + \beta_I^1 U_I\right)\left(U_{\text{ref}} - U_I\right) \qquad (4.5.11)$$

Box 4.5.4 contains a listing of the subroutine **BNDRYUNS1D** that imposes various types of boundary conditions discussed here on a banded unsymmetric system of equations. Fortran programs of symmetric and unsymmetric equation solvers are included in Appendix 1.

Box 4.5.4: Subroutine **BNDRYUNS1D** for the implementation of various types of boundary conditions.

```
      SUBROUTINE BNDRYUNS1D(NONLIN,MXNEQ,MXFBW,MXEBC,MXNBC,NDF,NHBW,
    1 GLK,GLU,NSPV,NSSV,NSMB,ISPV,ISSV,ISMB,VSPV,VSSV,BETAO,BETAU,UREF)
C
C     Implement boundary conditions on BANDED UNSYMMETRIC equations
C
      IMPLICIT REAL*8 (A-H,O-Z)
      DIMENSION ISPV(MXEBC,2),ISSV(MXNBC,2),ISMB(MXNBC,2),
    1          VSPV(MXEBC),VSSV(MXEBC),UREF(MXNBC),BETAO(MXNBC),
    2          BETAU(MXNBC),GLU(MXNEQ),GLK(MXNEQ,MXFBW)
      NBW=2*NHBW
C
C     Include specified PRIMARY degrees of freedom
C
      IF(NSPV.NE.0)THEN
          DO 120 NP=1,NSPV
          NB=(ISPV(NP,1)-1)*NDF+ISPV(NP,2)
          DO 110 J=1,NBW
  110     GLK(NB,J)=0.0D0
          GLK(NB,NHBW)=1.0D0
  120     GLK(NB,NBW)=VSPV(NP)
      ENDIF
C
C     Modify source vector to include nonzero SECONDARY VARIABLES
C
      IF(NSSV.NE.0)THEN
          DO 130 NS=1,NSSV
          NB=(ISSV(NS,1)-1)*NDF+ISSV(NS,2)
  130     GLK(NB,NBW)=GLK(NB,NBW)+VSSV(NS)
      ENDIF
C     Implement the specified MIXED BOUNDARY CONDITIONS
      IF(NSMB.NE.0)THEN
          DO 150 MB=1,NSMB
          NB=(ISMB(MB,1)-1)*NDF+ISMB(MB,2)
          IF(NONLIN.LE.1)THEN
              GLK(NB,NHBW)=GLK(NB,NHBW)+BETAO(MB)+BETAU(MB)*GLU(NB)
              GLK(NB,NBW)=GLK(NB,NBW)+(BETAO(MB)+BETAU(MB)*GLU(NB))*UREF(MB)
          ELSE
```

Box 4.5.4: Subroutine **BNDRYUNS1D** (continued).

```
            GLK(NB,NHBW)=GLK(NB,NHBW)+BETAO(MB)+2.0*BETAU(MB)*GLU(NB)
                                      -UREF(MB)*BETAU(MB)
            GLK(NB,NBW)=GLK(NB,NBW)+(BETAO(MB)+BETAU(MB)*GLU(NB))*UREF(MB)
     2                   -(BETAO(MB)+BETAU(MB)*GLU(NB))*GLU(NB)
          ENDIF
150       CONTINUE
        ENDIF
        RETURN
        END
```

Example 4.5.1

Consider heat transfer in an isotropic bar of length $L = 0.18$ m with left end maintained at a temperature of $500°$K and the right end at $300°$K. Suppose that there is no internal heat generation $(f = 0)$ and the surface of the bar is insulated so that there is no convection from the surface. The governing differential equation [the same as Eq. (4.1.1) with $a = k$, $b = c = f = 0$, and $u = T$, the temperature] and boundary conditions of the problem are

$$- \frac{d}{dx} \left[k(T) \frac{dT}{dx} \right] = 0, \quad 0 < x < L$$

$$T(0) = 500°\text{K}, \quad T(L) = 300°\text{K}$$

(1)

In addition, suppose that the conductivity $k(T)$ varies according to the relation

$$k(T) = k_0 (1 + k_1 T)$$

(2)

where k_0 is the thermal conductivity $[k_0 = 0.2$ W/(m $°$K)] and k_1 the temperature coefficient of thermal conductivity $[k_1 = 2 \times 10^{-3}$ ($°$K^{-1})]. Determine the finite element solution $T_h(x)$ using meshes of 4 and 8 linear elements and 2 and 4 quadratic elements. Use both direct and Newton's iteration schemes.

Solution: Table 4.5.1 contains linear and nonlinear finite element solutions $T_h(x)$ for different values of x. The results obtained with the direct iteration (DI) method and Newton's iteration (NI) method with $\epsilon = 10^{-3}$ are tabulated in Table 4.5.1. In both methods, the convergent solution was obtained for three iterations. Both methods and both meshes give the same solution for this specific problem.

Table 4.5.1: Finite element solutions of a nonlinear heat conduction problem.

	DI/NI	Direct iteration		Newton iteration	
x	Linear	8L	4Q	8L	4Q
0.0000	500.00	500.00	500.00	500.00	500.00
0.0225	475.00	477.24	477.24	477.24	477.24
0.0450	450.00	453.94	453.94	453.94	453.94
0.0675	425.00	430.06	430.06	430.05	430.05
0.0900	400.00	405.54	405.54	405.54	405.54
0.1125	375.00	380.35	380.35	380.34	380.34
0.1350	350.00	354.40	354.40	354.40	354.40
0.1575	325.00	327.65	327.65	327.65	327.65
0.1800	300.00	300.00	300.00	300.00	300.00

Example 4.5.2

Consider the nonlinear differential equation

$$-\frac{d^2u}{dx^2} + 2u^3 = 0, \quad 1 < x < 2 \tag{1}$$

subjected to the boundary conditions

$$u(1) = 1, \quad \left[\frac{du}{dx} + u^2\right]_{x=2} = 0 \tag{2}$$

Solve the problem using four and eight linear elements and two and four quadratic elements (i.e. four different meshes) and the direct and Newton iteration procedures with error tolerance of $\epsilon = 10^{-3}$ and maximum number of iterations of 15. Tabulate the finite element solution along with the exact solution $u(x) = 1/x$ at the nodes.

Solution: As compared to the model equation in Eq. (4.1.1), we have $a = 1$, $b = 0$, and $c = 2u^2$. The boundary condition at $x = 2$ is a mixed boundary condition with $\beta_0 = 0$ and $\beta_1 = 1$, and $u_\infty = 0$.

The coefficients of the element matrix are

$$K_{ij}^e = \int_{x_a}^{x_b} \left(\frac{d\psi_i^e}{dx}\frac{d\psi_j^e}{dx} + 2u^2\psi_i^e\psi_j^e\right)dx \tag{3}$$

The direct iteration scheme with a mesh of four linear elements and a guess vector of $(1.0, 0.5, \ldots, 0.5)$ takes 11 iterations to converge within a convergence tolerance of 10^{-3} [the guess vector of $(1.0, 1.0, \ldots, 1.0)$ may take even more iterations to converge]. The results obtained with the direct iteration for various meshes are presented in Table 4.5.2.

For Newton's scheme, we need to compute the coefficients of the tangent matrix

$$T_{ij}^e = K_{ij}^e + 4\sum_{k=1}^{n} u_k^e \int_{x_a}^{x_b} u\frac{\partial u}{\partial u_j^e}\psi_i^e\psi_k^e\,dx = \int_{x_a}^{x_b} \left(\frac{d\psi_i^e}{dx}\frac{d\psi_j^e}{dx} + 6u^2\psi_i^e\psi_j^e\right)dx \tag{4}$$

$$K_{MM} = K_{nn}^N + u(2), \quad Q_n^N = 0, \quad T_{MM} = T_{nn}^N + 1, \quad M = Np + 1 \tag{5}$$

Table 4.5.2: Comparison of finite element solutions with the exact solution of Example 4.5.2 (*direct iteration scheme*).

x	Iteration	4L	8L	2Q	4Q	Exact
1.25	1	0.8417	0.8495	0.8469	0.8516	
	2	0.7833	0.7823	0.7827	0.7821	
	10	0.7984	0.7995	0.7995	0.7998	
	11	0.7987	0.7998	0.7999	0.8001	0.8000
1.50	1	0.7220	0.7287	0.7289	0.7306	
	2	0.6394	0.6375	0.6383	0.6371	
	10	0.6648	0.6659	0.6664	0.6663	
	11	0.6654	0.6665	0.6670	0.6669	0.6666
1.75	1	0.6250	0.6308	0.6310	0.6325	
	2	0.5385	0.5362	0.5369	0.5356	
	10	0.5696	0.5706	0.5710	0.5709	
	11	0.5703	0.5714	0.5718	0.5717	0.5714
2.00	1	0.5475	0.5527	0.5529	0.5542	
	2	0.4648	0.4622	0.4628	0.4615	
	10	0.4984	0.4991	0.4995	0.4994	
	11	0.4992	0.5001	0.5004	0.5004	0.5000

where N denotes the total number of elements in the mesh, n is the number of nodes in the element, and $p = n - 1$ is the polynomial degree. Newton's iteration scheme only takes 3 iterations to converge in comparison to 11 iterations for the direct iteration scheme. A summary of the results obtained using Newton's iteration scheme is presented in Table 4.5.3.

Table 4.5.3: Comparison of finite element solutions with the exact solution of Example 4.5.2 (*Newton's iteration scheme*).

x	Iteration	4L	8L	2Q	4Q	Exact
1.25	1	0.8125	0.8186	0.8170	0.8205	
	2	0.7987	0.7998	0.7999	0.8002	
	3	0.7986	0.7997	0.7998	0.8000	0.8000
1.50	1	0.6814	0.6862	0.6870	0.6879	
	2	0.6653	0.6665	0.6670	0.6669	
	3	0.6652	0.6663	0.6668	0.6667	0.6666
1.75	1	0.5834	0.5871	0.5878	0.5885	
	2	0.5702	0.5713	0.5717	0.5717	
	3	0.5701	0.5711	0.5715	0.5714	0.5714
2.00	1	0.5092	0.5121	0.5126	0.5132	
	2	0.4990	0.4999	0.5003	0.5003	
	3	0.4989	0.4997	0.5001	0.5000	0.5000

Example 4.5.3

Consider the large-deformation analysis of a bar of length L, uniform cross-sectional area A, and made of an isotropic, linear elastic material (modulus E). Suppose that the bar is fixed at one end and subjected to axial load P at the other end. Determine the load-displacement curve, treating the bar as a one-dimensional problem.

Solution: We note that Eq. (1.5.6) is not valid to analyze this nonlinear bar problem. To analyze the problem using the finite element method, first we must identify the mathematical model of the problem. For this structural problem, it is appropriate to use the principle of virtual displacements [see Section 2.9 and Eq. (2.9.24)] to set up the weak form. For a typical element (x_a, x_b) we have

$$0 = \int_{V^e} \delta\varepsilon_{xx}^e \, \sigma_{xx}^e \, dA \, dx - \int_{x_a}^{x_b} \delta u \, f(x) \, dx - \delta u^e(x_a) P_a^e - \delta u^e(x_b) P_b^e \tag{1}$$

where $f(x)$ is the body force, P_a^e and P_b^e are the end forces of a bar element, and $\delta\varepsilon_{xx}^e$ is the virtual strain, which can be computed from the actual strain (the element label 'e' is omitted in the rest of the discussion)

$$\varepsilon_{xx} = \frac{du}{dx} + \frac{1}{2}\left(\frac{du}{dx}\right)^2, \quad \delta\varepsilon_{xx} = \frac{d\delta u}{dx} + \frac{du}{dx}\frac{d\delta u}{dx} \tag{2}$$

where $u(x)$ is the axial displacement. Then the principle of virtual displacement statement in Eq. (1) can be expressed as

$$0 = \int_{x_a}^{x_b} \left(\frac{d\delta u}{dx} + \frac{du}{dx}\frac{d\delta u}{dx}\right) N_{xx} \, dx - \delta u(x_a) P_a - \delta u(x_b) P_b \tag{3}$$

where

$$N_{xx} = \int_{A^e} \sigma_{xx} \, dA = EA\varepsilon_{xx} = EA\left[\frac{du}{dx} + \frac{1}{2}\left(\frac{du}{dx}\right)^2\right] \tag{4}$$

Equation (3) forms the basis of the finite element model. The governing differential equation of the problem can be determined from Eq. (3) as

$$-\frac{d}{dx}\left[\left(1 + \frac{du}{dx}\right)N_{xx}\right] - f = 0, \quad x_a < x < x_b \tag{5}$$

The boundary conditions involve specifying one element of each of the pairs $(u(x_a), P_a)$ and $(u(x_b), P_b)$, where

$$P_a = -\left[\left(1 + \frac{du}{dx}\right)N_{xx}\right]_{x_a}, \quad P_b = \left[\left(1 + \frac{du}{dx}\right)N_{xx}\right]_{x_b} \tag{6}$$

The finite element model $\mathbf{K}^e\mathbf{u}^e = \mathbf{F}^e$ is obtained by assuming approximation of the form

$$u(x) \approx \sum_{j=1}^{n} u_j^e \psi_j^e(x) \tag{7}$$

into the virtual work statement in Eq. (3) with

$$\begin{aligned}
K_{ij}^e &= \int_{x_a}^{x_b} E^e A^e\left(1 + \frac{du}{dx}\right)\left(1 + \frac{1}{2}\frac{du}{dx}\right)\frac{d\psi_i^e}{dx}\frac{d\psi_j^e}{dx}\, dx \\
F_i^e &= \int_{x_a}^{x_b} f\psi_i^e(x)\, dx + P_a^e\psi_i^e(x_a) + P_b^e\psi_i^e(x_b)
\end{aligned} \tag{8}$$

A comparison of the element coefficients in Eq. (8) with those in Eq. (4.3.3) indicates that the present model is a special case of Eq. (4.3.3) with $\beta_a = \beta_b = b = c = 0$, and

$$a(x, u) = E(x)A(x)\left(1 + \frac{du}{dx}\right)\left(1 + \frac{1}{2}\frac{du}{dx}\right) \tag{9}$$

Hence, the finite element program developed to analyze problems described by Eq. (4.1.1) can be used to solve the problem at hand. The tangent stiffness matrix in the present case is given by

$$T_{ij}^e = K_{ij}^e + \int_{x_a}^{x_b} EA\left(\frac{du}{dx}\right)\left(\frac{3}{2} + \frac{du}{dx}\right)\frac{d\psi_i^e}{dx}\frac{d\psi_j^e}{dx}\, dx \tag{10}$$

Note that, in the present problem, both $\mathbf{K}^e$ and $\mathbf{T}^e$ are symmetric matrices.

Figure 4.5.3 contains the nondimensional load $\bar{P} = P/AE$ versus the dimensionless displacement $\bar{u} = u(L)/L$ for a bar (with constant EA) fixed at one end and subjected to axial load P at the other end $(f = 0)$. The exact solution is

$$u(x) = kx, \quad \text{where } k \text{ is the real root of } \ k^3 + 3k^2 + 2k - 2\bar{P} = 0, \quad \bar{P} = \frac{P}{EA} > \frac{1}{27} \tag{11}$$

The finite element results were obtained using Newton's iteration method with eight linear elements (a mesh of four quadratic elements also gives the same results, and both coincide with the exact solution); a tolerance of $\epsilon = 10^{-3}$ was used with a load increment of $\Delta\bar{P} = 0.2$. Convergence was achieved for five and four iterations for the first two load steps, respectively, and only three iterations were taken during the remaining load steps. It should be noted that the direct iteration method takes many more iterations to converge, especially at higher loads. From the load-deflection curve, it is clear that the geometric nonlinearity is that of hardening type, that is, the bar becomes stiff as it deforms. Of course, we assumed that the modulus remains unchanged during the deformation, which most likely does not hold for large strains.

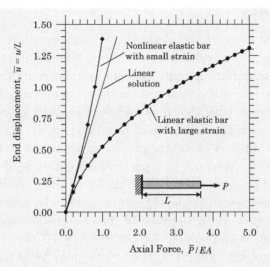

Fig. 4.5.3: Nondimensional load versus displacement curves for geometric and material nonlinear analyses of an elastic bar fixed at one end subjected to an axial load at the other end (all solutions were obtained with eight linear elements or four quadratic elements).

Next, consider the case of small strains but with material nonlinearity in the form

$$E = E_0(1 - \alpha\varepsilon_{xx}), \quad \varepsilon_{xx} = \frac{du}{dx} \tag{12}$$

where E_0 and α are material parameters known from experiments. In this case, the coefficient a is given by

$$a = EA = E_0A\left(1 - \alpha\frac{du}{dx}\right) \tag{13}$$

The tangent stiffness coefficients then become

$$T_{ij}^e = K_{ij}^e - \alpha\int_{x_a}^{x_b} E_0A\frac{du}{dx}\frac{d\psi_i^e}{dx}\frac{d\psi_j^e}{dx}\,dx \tag{14}$$

The exact solution is given by

$$u(x) = kx, \quad \text{where} \quad k = \frac{1}{\alpha}\left(0.5 - \sqrt{0.25 - \alpha\bar{P}}\right), \quad \bar{P} = \frac{P}{EA} < \frac{0.25}{\alpha} \tag{15}$$

Figure 4.5.3 shows the softening effect of the material nonlinearity (with $\alpha = 0.2$) because the material is assumed to degrade as the bar deforms (as implied by the negative sign in front of α). Not all material models show the softening effect; in fact, biological materials (e.g. soft tissues) are known to exhibit the stiffening effect. Thus, the geometric and material nonlinearities do not always have the opposite effect on the response.

A complete computer program can be put together with the help of the discussion presented in Chapters 3 and 4 and Fortran statements provided in various boxes to solve different problems based on the model equation in Eq. (4.1.1).

4.6 Summary

In this chapter, the weak-form Galerkin finite element model of a typical second-order nonlinear boundary-value problem in one dimension is presented, iterative methods of solving nonlinear finite element equations are discussed, and computer implementation of the nonlinear finite element analysis is outlined. The model equation used is such that it can account for both geometric and material nonlinearities through the definition of the data $a(x, u)$, $b(x, u)$, and $c(x, u)$.

The most difficult part of solving nonlinear equations is that they have multiple solutions and convergence to a particular solution is often dictated by the choice of the solution vector, $\mathbf{U}^{(0)}$. The choice of the iterative solution scheme also makes a difference in achieving convergence. In general, the coefficient matrix being inverted, $\mathbf{K}$ in the case of direct iteration and $\mathbf{T}$ in the case of Newton's iteration, is responsible for convergence. If the coefficient matrix is symmetric or diagonally dominant, the process may yield faster convergence.

Another comment relates to the coefficient $b(x, u)$ in the model equation, Eq. (4.1.1). Even when the problem is linear, the coefficient changes the character of the equation from elliptic to hyperbolic, and it is responsible for the unsymmetry of $\mathbf{K}$. In general, hyperbolic equations are difficult to solve (even for linear problems) numerically. Therefore, when there is a choice during linearization of the equation, we should pick one that yields a symmetric $\mathbf{K}$ for direct iteration scheme or a symmetric $\mathbf{T}$ for Newton's iteration scheme, because there is a greater chance of convergence with such choices. As an example, consider the equation

$$-\frac{d}{dx}\left(a\frac{du}{dx}\right) + \hat{b}u\frac{du}{dx} + \hat{c}u = f, \quad 0 < x < L$$

There are two choices of b and c that will produce the model equation: (1) $b = \hat{b}u$ and $c = \hat{c}$ and (2) $c = \hat{c} + \hat{b}(du/dx)$. The second choice may work better for the direct iteration scheme while either choice may be alright, depending on resulting $\mathbf{T}$, for Newton's iteration scheme.

In some cases, several terms of a given differential equation can be combined to make it a special case of the model equation, Eq. (4.1.1). An example is provided by the equation

$$-(c_0 + c_1 u)\frac{d^2 u}{dx^2} - \left(\frac{dc_0}{dx} + \frac{dc_1}{dx}u + c_1\frac{du}{dx}\right)\frac{du}{dx} = f(x)$$

where c_0 and c_1 are functions of x only. This equation is the same as

$$-\frac{d}{dx}\left[(c_0 + c_1 u)\frac{du}{dx}\right] = f(x)$$

and a special case of Eq. (4.1.1) with $a(x, u) = (c_0 + c_1 u)$, $b(x, u) = 0$ and $c(x, u) = 0$.

With a good understanding of the contents of this chapter, one should be able to develop a nonlinear finite element analysis program to solve boundary value problems described by Eq. (4.1.1) or its variation (see Problem 4.19). When faced with analyzing a physical problem, one must first determine if it can be modeled as a one-dimensional problem, and then see if its mathematical model is a special case of Eq. (4.1.1) or Eq. (1) of Problem 4.19.

Problems

See **Problem 4.22** for the development of a computer program to solve the problems of this chapter. Additional nonlinear differential equations can be found in [51] and [52].

4.1 Consider the second-order differential equation

$$-\frac{d}{dx}\left(\mu\frac{du}{dx}\right) = f(x), \qquad \mu = \mu_0\left(\frac{du}{dx}\right)^{n-1} \tag{1}$$

where $u(x)$ is the dependent unknown, $f(x)$ is a known function of position x, and μ is a function of the dependent variable, as given in Eq. (1). Write the finite element model and derive the tangent stiffness matrix coefficients.

4.2 Consider Eqs. (4.1.1) and (4.1.2) with $a(x, u) = a_0^e(x)\frac{du}{dx}$, $b = 0$, $c(x, u) = 0$, and $\beta = 0$. Compute the tangent matrix.

4.3 Consider the nonlinear differential equation

$$-(\sqrt{2} + u)\frac{d^2u}{dx^2} - \left(\frac{du}{dx}\right)^2 = 1, \quad 0 < x < 1 \tag{1}$$

$$\frac{du}{dx}(0) = 0, \qquad u(1) = 0 \tag{2}$$

Analyze the nonlinear problem using the finite element method with direct iteration procedure. Tabulate the nodal values of $u(x)$ for four and eight linear elements and two and four quadratic elements.

4.4 Formulate Problem 4.3 with Newton's iteration procedure, and compute the tangent coefficient matrix. Tabulate the nodal values of $u(x)$ for four and eight linear elements and two and four quadratic elements. Use an error tolerance of $\epsilon = 10^{-3}$.

4.5 Develop the finite element model of the nonlinear differential equation

$$-\frac{d^2u}{dx^2} - 2u\frac{du}{dx} = 0, \quad 0 < x < 1 \tag{1}$$

$$u(0) = 1, \qquad u(1) = 0.5 \tag{2}$$

and solve the problem using direct iteration procedure. Tabulate the nodal values of $u(x)$ for four and eight linear elements and two and four quadratic elements. The exact solution is given by $u(x) = 1/(1 + x)$.

4.6 Compute the tangent stiffness matrix associated with Problem 4.5, and solve it with Newton's iteration procedure. Tabulate the nodal values of $u(x)$ for four and eight linear elements and two and four quadratic elements (use an error tolerance of $\epsilon = 10^{-3}$).

4.7 Formulate the nonlinear differential equation in Problem 4.5 subject to the boundary conditions

$$\left[\frac{du}{dx}\right]_{x=0} = -1, \qquad \left[\frac{du}{dx} + u^2\right]_{x=1} = 0 \tag{1}$$

and solve it with (a) direct and (b) Newton's iteration procedures with a convergence tolerance of 10^{-4}. Tabulate the nodal values of $u(x)$ for four and eight linear elements and two and four quadratic elements, along with the exact solution $u(x) = 1/(1+x)$.

4.8 Formulate the nonlinear differential equation in Problem 4.5 subject to the boundary conditions

$$u(0) = 1, \qquad \left[\frac{du}{dx} + u^2\right]_{x=1} = 0 \tag{1}$$

and solve the problem using direct iteration procedure (use an error tolerance of $\epsilon = 10^{-3}$). Tabulate the nodal values of $u(x)$ for four and eight linear elements and two and four quadratic elements. The exact solution is given by $u(x) = 1/(1+x)$.

4.9 Compute the tangent stiffness matrix associated with Problem 4.8, and solve it with Newton's iteration procedure (use an error tolerance of $\epsilon = 10^{-4}$). Tabulate the nodal values of $u(x)$ for four and eight linear elements and two and four quadratic elements.

4.10 Develop the finite element model of the nonlinear differential equation in Problem 4.5 subject to the boundary conditions

$$\left[\frac{du}{dx} + 2u\right]_{x=0} = 1, \qquad \left[\frac{du}{dx} + u^2\right]_{x=1} = 0 \tag{1}$$

and solve the problem using direct iteration procedure (use an error tolerance of $\epsilon = 10^{-4}$). Tabulate the nodal values of $u(x)$ for four and eight linear elements and two and four quadratic elements. The exact solution is given by $u(x) = 1/(1+x)$.

4.11 Compute the tangent stiffness matrix associated with Problem 4.10, and solve it with Newton's iteration procedure (use an error tolerance of $\epsilon = 10^{-4}$). Tabulate the nodal values of $u(x)$ for eight linear elements and four quadratic elements.

4.12 Develop the finite element model of the nonlinear differential equation

$$-\frac{d^2u}{dx^2} - \left(\frac{du}{dx}\right)^3 = 0, \quad 0 < x < 1 \tag{1}$$

$$\left[\frac{du}{dx} + u\right]_{x=0} = \frac{3}{\sqrt{2}}, \qquad \left[\frac{du}{dx}\right]_{x=1} = 0.5 \tag{2}$$

using the finite element method, and solve the problem using direct iteration procedure. Tabulate the nodal values of $u(x)$ for four and eight linear elements and two and four quadratic elements. The exact solution is given by $u(x) = \sqrt{2(1+x)}$.

4.13 Consider simultaneous steady-state conduction and radiation in a plate. The mathematical formulation of the problem in non-dimensional form is given by

$$-\frac{d}{dx}\left(k(\theta)\frac{d\theta}{dx}\right) = 0, \quad 0 < x < x_0 \tag{1}$$

$$\theta(0) = \theta_0, \quad \theta(x_0) = 1.0 \tag{2}$$

where x is the non-dimensional thickness coordinate, θ is the non-dimensional temperature, and k is the conductivity

$$k(\theta) = k_0\left(1 + \frac{4}{3N}\theta^n\right) \tag{3}$$

and N is called the conduction-to-radiation parameter. Develop the finite element model of the problem and compute the element tangent matrix.

4.14 Analyze the nonlinear problem in Problem 4.13 using the finite element method with Newton's iteration procedure for the following two cases:

$$\text{Case 1: } k(\theta) = k_0\left(1 + \frac{4}{3N}\theta^2\right); \quad \text{Case 2: } k(\theta) = k_0\left(1 + \frac{4}{3N}\theta^3\right) \tag{1}$$

with $k_0 = 1$, $N = 0.01$, $x_0 = 1$, $\theta_0 = 0.5$, error tolerance of $\epsilon = 10^{-3}$, and maximum allowable iterations to be ten. Tabulate the nodal values of $\theta(x)$ for eight linear elements and four quadratic elements.

4.15 The explosion of a solid explosive material in the form of an infinite cylinder may be described by

$$-\frac{1}{r}\frac{d}{dr}\left(r\frac{du}{dr}\right) = 2e^u, \quad 0 < r < 1$$

subject to the boundary conditions

$$\text{At } r = 0 : \quad \frac{du}{dr} = 0; \quad \text{At } r = 1 : \quad u = 0$$

The exact solution of the nonlinear equation is

$$u(r) = \ln\frac{4}{(1+r^2)^2}$$

Analyze the nonlinear problem using the finite element method with (a) Newton's and (b) direct iteration procedure. Tabulate and plot the nodal values of $u(r)$ for eight linear elements and four quadratic elements and compare with the exact solution.

4.16 Redo Problem 4.15 when the right-hand side is replaced by e^u. This problem has two solutions

$$u_i(r) = \ln\frac{8\lambda_i}{(1+\lambda_i r^2)^2}, \quad i = 1, 2$$

where λ_i are the roots of the equation

$$\frac{8\lambda}{(1+\lambda)^2} = 1$$

4.17 Axial mixing in an isothermal tubular reactor where a second-order reaction occurs is described by

$$-\frac{1}{\text{Pe}}\frac{d^2u}{dx^2} + \frac{du}{dx} + \text{Da } u^2 = 0, \quad 0 < x < 1$$

subject to the boundary conditions

$$\text{At } x = 0 : \quad u = 1 + \frac{1}{\text{Pe}}\frac{du}{dx}; \quad \text{At } x = 1 : \quad \frac{du}{dx} = 0$$

where Pe and Da are the problem parameters. Analyze the nonlinear problem using the finite element method with the direct iteration procedure. Take Pe $= 5$ and Da $= 1$, and tabulate and plot the solution $u(x)$ for eight linear elements.

4.18 Consider the nonlinear equation

$$-\frac{d}{dx}\left(a\frac{du}{dx}\right) + cu - f = 0 \tag{1}$$

with

$$a = a_0(x) + a_1(x)u(x) + a_2(x)\frac{du}{dx}, \quad c = c_0(x) + c_1(x)u(x) + c_2(x)\frac{du}{dx} \tag{2}$$

Develop the least-squares finite element model and compute the coefficients of the tangent matrix.

4.19 Consider the following differential equation, which contains Eq. (4.1.1) as a special case:

$$-\frac{d}{dx}\left[a(x, u)\frac{du}{dx} + b_1(x, u)u\right] + b_2(x, u)\frac{du}{dx} + c(x, u)u = f(x), \quad x_0 < x < x_L \tag{1}$$

where (x_0, x_L) is the problem domain and a, b_1, b_2, c, and f are the problem data. Develop the finite element model of the equation over a typical element.

4.20 Show that the finite element model developed in Problem 3.11 is a special case of the one developed in Problem 4.19. Identify the coefficients a, b_1, b_2, c, and f in terms of μ, λ, and the radial coordinate r.

4.21 Verify that the exact solutions presented in Eqs. (11) and (15) of Example 4.5.3 are indeed the solutions that satisfy the respective differential equation and boundary conditions.

4.22 Put together a complete finite element program based on the logic presented in Fig. 4.5.1 and Boxes 4.5.1–4.5.4 that has the capability of solving the problems of this chapter.

Nonlinear Bending of Straight Beams

It is not uncommon for engineers to accept the reality of phenomena that are not yet understood, as it is very common for physicists to disbelieve the reality of phenomena that seem to contradict contemporary beliefs of physics.
——— Henry H. Bauer

5.1 Introduction

In this chapter, we consider a slightly more complicated one-dimensional nonlinear problem than that discussed in Chapter 4. We consider bending of straight beams. A *beam* is a structural member whose length to cross-sectional dimensions is very large and it undergoes not only stretching along its length but also bending about an axis transverse to the length. When the applied loads on the beam are large, the linear load–deflection relationship ceases to be valid because the beam develops internal forces that resist deformation and the magnitude of internal forces increases with loading as well as deformation. This nonlinear load–deflection response of straight beams is the topic of this section. The development presented herein is limited to static problems.

In developing a general nonlinear formulation of beams, straight or curved, one must define the measures of stress and strain consistent with the deformations accounted for in the formulation (see Chapter 2 and [1–3]). Such a formulation, called a *continuum formulation*, will be discussed in Chapter 9. The present nonlinear formulation of straight beams is based on assumptions of small strains and small or moderate rotations. Further, it is assumed that the geometry does not change significantly, so that the principle of virtual displacements for the deformed body can be written over the undeformed body. These assumptions allow us to use the stress as a measure of force per unit undeformed area and strain as a measure of change in length to the original length (and in the case of shear strain, change in the angle from $\pi/2$). The changes in the geometry are small so that no distinction between the Cauchy stress tensor and other stress measures (e.g. the Piola–Kirchhoff tensor discussed in Chapter 2) will be made. The nonlinearity in the formulation comes solely from the inclusion of the in-plane forces that are proportional to the square of the rotation of a transverse normal line in the beam (see [4–6] for details).

J.N. Reddy, *An Introduction to Nonlinear Finite Element Analysis*, Second Edition. ©J.N. Reddy 2015. Published in 2015 by Oxford University Press.

Two different theories to model the kinematic behavior of beams are considered here: (1) the Euler–Bernoulli beam theory (EBT), in which the transverse shear strain γ_{xz} is neglected and (2) the Timoshenko beam theory (TBT), in which the transverse shear strain is included in its simplest way. In each case, we begin with an assumed displacement field and compute strains that are consistent with the kinematic assumptions of the theory. Then we develop the weak forms using the principle of virtual displacements and the associated displacement finite element models for static bending. We also discuss certain computational aspects (e.g. membrane and shear locking) and review iterative methods of solution for the problems at hand. Computer implementation issues are also discussed. Other linear finite element models of the Timoshenko beam theory are also presented for completeness.

5.2 The Euler–Bernoulli Beam Theory

5.2.1 Basic Assumptions

For the sake of completeness, the governing equations of the nonlinear bending of beams are developed from basic considerations. The classical beam theory is based on the Euler–Bernoulli hypothesis that plane sections perpendicular to the axis of the beam before deformation remain (a) plane, (b) rigid (not deform), and (c) rotate such that they remain perpendicular to the (deformed) axis after deformation, as shown in Fig. 5.2.1. These assumptions amount to neglecting the Poisson effect and transverse normal and shear strains. The principle of virtual displacements will be used to formulate the variational problem and associated finite element model.

5.2.2 Displacement and Strain Fields

The bending of beams with moderately large rotations but with small strains can be derived using the displacement field

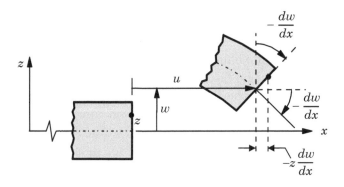

Fig. 5.2.1: Kinematics of deformation of the Euler–Bernoulli beam theory.

$$u_1 = u(x) - z\frac{dw}{dx}, \quad u_2 = 0, \quad u_3 = w(x) \tag{5.2.1}$$

where (u_1, u_2, u_3) are the total displacements of a material point occupying the location (x, y, z) in the undeformed body [usually denoted with capital letters: $(X, Y, Z) = (X_1, X_2, X_3)$], and u and w denote the axial and transverse displacements of a point on the x-axis, which is taken along the geometric centroid of the cross section of the beam. We note that once $(u, 0, w)$ are known, the total displacement field $(u_1, 0, u_3)$ can be determined.

Using the Green strain tensor components [nonlinear strain-displacement relations from Eqs. (2.4.8) and (2.4.9)], we can write

$$E_{ij} \approx \varepsilon_{ij} = \frac{1}{2}\left(\frac{\partial u_i}{\partial x_j} + \frac{\partial u_j}{\partial x_i}\right) + \frac{1}{2}\frac{\partial u_k}{\partial x_i}\frac{\partial u_k}{\partial x_j} \tag{5.2.2}$$

where sum on repeated subscripts is implied. In the present case $(u_2 = 0)$, the explicit form of ε_{11}, for example, is

$$\varepsilon_{11} = \frac{\partial u_1}{\partial x_1} + \frac{1}{2}\left[\left(\frac{\partial u_1}{\partial x_1}\right)^2 + \left(\frac{\partial u_3}{\partial x_1}\right)^2\right]$$

In the remainder of the chapter we shall use the notation $x_1 = x$, $x_2 = y$, and $x_3 = z$.

In view of the small strain assumption, the first term in the square brackets is neglected, i.e. $(\partial u_1/\partial x) = O(\epsilon)$ and hence $(\partial u_1/\partial x)^2 = O(\epsilon^2) \approx 0$. The second term, $\partial u_3/\partial x$, denotes the rotation of a line perpendicular to the beam axis. Even when the strains are small, because of the slenderness, beams can undergo moderate to large rotations. Hence, we retain the square of the rotation $\partial u_3/\partial x$, which is assumed to be of the order $O(\sqrt{\epsilon})$, in the strain component ε_{11}. Nonlinear strains in which only squares of the rotations are included while neglecting the squares of the in-plane stretching terms are known as the *von Kármán nonlinearity* in beams, plates, and shells. In view of the displacement field in Eq. (5.2.1), the only nonzero strain (assuming that $\varepsilon_{33} = \varepsilon_{13} = 0$ so that one may use uniaxial stress-strain relations) is

$$\begin{aligned}
\varepsilon_{11} = \varepsilon_{xx} &= \frac{du}{dx} - z\frac{d^2w}{dx^2} + \frac{1}{2}\left(\frac{dw}{dx}\right)^2 \\
&= \left[\frac{du}{dx} + \frac{1}{2}\left(\frac{dw}{dx}\right)^2\right] - z\frac{d^2w}{dx^2} \\
&\equiv \varepsilon_{xx}^0 + z\varepsilon_{xx}^1 \\
\varepsilon_{xx}^0 &= \frac{du}{dx} + \frac{1}{2}\left(\frac{dw}{dx}\right)^2, \quad \varepsilon_{xx}^1 = -\frac{d^2w}{dx^2}
\end{aligned} \tag{5.2.3}$$

Such strains are known as the *von Kármán strains*.

In the next section, we develop the weak form of the problem using the *principle of virtual displacements* (see Section 2.9). This is a departure from the usual method of developing weak forms (see Section 3.2.3) from governing differential equations, but it is the most common procedure used in the field of solid and structural mechanics.

5.2.3 The Principle of Virtual Displacements: Weak Form

The weak form of structural problems can be directly derived using the principle of virtual displacements, without knowing the governing differential equations. Recall from Section 2.9 that the principle can be stated as *if a body is in equilibrium, the total virtual work done by actual internal as well as external forces in moving through their respective virtual displacements is zero.* The virtual displacements are arbitrary except that they are zero where displacements are prescribed (see Section 2.9 and Reddy [3] for details). The analytical form of the principle over a typical element $\Omega^e = (x_a, x_b)$ is (see Fig. 5.2.2 for a typical beam finite element)

$$\delta W^e \equiv \delta W_I^e + \delta W_E^e = 0 \qquad (5.2.4)$$

where δW_I^e is the virtual strain energy stored in the element due to actual stresses σ_{ij} in moving through the virtual strains $\delta \varepsilon_{ij}$, and δW_E^e is the work done by external applied loads in moving through their respective virtual displacements. Here σ_{ij} and ε_{ij} denote the Cartesian components of the Cauchy or second Piola–Kirchhoff stress and the Green–Lagrange strain tensors, respectively. As stated earlier, no distinction is made between deformed and undeformed configurations and various measures of stress and strain (see Chapter 2 for a detailed discussion of various measures of stress and strain).

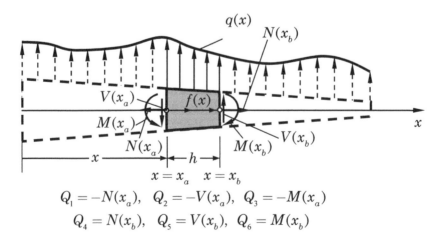

$$Q_1 = -N(x_a), \quad Q_2 = -V(x_a), \quad Q_3 = -M(x_a)$$
$$Q_4 = N(x_b), \quad Q_5 = V(x_b), \quad Q_6 = M(x_b)$$

Fig. 5.2.2: A typical beam element, $\Omega^e = (x_a, x_b)$.

The internal and external virtual work expressions for an Euler–Bernoulli beam element $\Omega^e = (x_a, x_b)$ are

$$\delta W_I^e = \int_{V^e} \delta \varepsilon_{xx} \, \sigma_{xx} \, dV = \int_{x_a}^{x_b} \int_{A^e} \left(\frac{d\delta u}{dx} + \frac{dw}{dx} \frac{d\delta w}{dx} - z \frac{d^2 \delta w}{dx^2} \right) \sigma_{xx} \, dA dx$$

$$\delta W_E^e = - \left[\int_{x_a}^{x_b} q \, \delta w \, dx + \int_{x_a}^{x_b} f \, \delta u \, dx + \sum_{i=1}^{6} Q_i^e \, \delta \Delta_i^e \right] \tag{5.2.5}$$

where V^e and A^e are the element volume and cross sectional area, respectively, $(f(x), q(x))$ are the distributed axial and transverse loads (measured per unit length), respectively, Q_i^e are the *generalized nodal forces*, and $\delta \Delta_i^e$ are the virtual *generalized nodal displacements* of the element [see Figs. 5.2.3(a) and 5.2.3(b) for the sign convention used]:

$$\Delta_1^e = u(x_a), \quad \Delta_2^e = w(x_a), \quad \Delta_3^e = \left[-\frac{dw}{dx} \right]_{x_a} \equiv \theta(x_a)$$

$$\Delta_4^e = u(x_b), \quad \Delta_5^e = w(x_b), \quad \Delta_6^e = \left[-\frac{dw}{dx} \right]_{x_b} \equiv \theta(x_b) \tag{5.2.6}$$

$$Q_1^e = -N_{xx}(x_a), \quad Q_4^e = N_{xx}(x_b), \quad Q_2^e = -\left[\frac{dw}{dx} N_{xx} + \frac{dM_{xx}}{dx} \right]_{x_a} \equiv -V_x(x_a)$$

$$Q_5^e = \left[\frac{dw}{dx} N_{xx} + \frac{dM_{xx}}{dx} \right]_{x_b} \equiv V_x(x_b), \quad Q_3^e = -M_{xx}(x_a), \quad Q_6^e = M_{xx}(x_b) \tag{5.2.7}$$

$$\Delta_2 = w(x_a) = \overline{\Delta}_1 \qquad \Delta_5 = w(x_b) = \overline{\Delta}_3$$

$$\Delta_1 = u(x_a) = u_1 \qquad \qquad \Delta_4 = u(x_b) = u_2$$

$$\Delta_3 = \theta(x_a) = \overline{\Delta}_2 \qquad \Delta_6 = \theta(x_b) = \overline{\Delta}_4$$

(a) Generalized displacement degrees of freedom

$$Q_2 = -V(x_a) \qquad Q_5 = V(x_b)$$

$$Q_1 = -N_{xx}(x_a) \qquad \qquad Q_4 = N_{xx}(x_b)$$

$$Q_3 = -M_{xx}(x_a) \qquad Q_6 = M_{xx}(x_b)$$

(b) Generalized force degrees of freedom

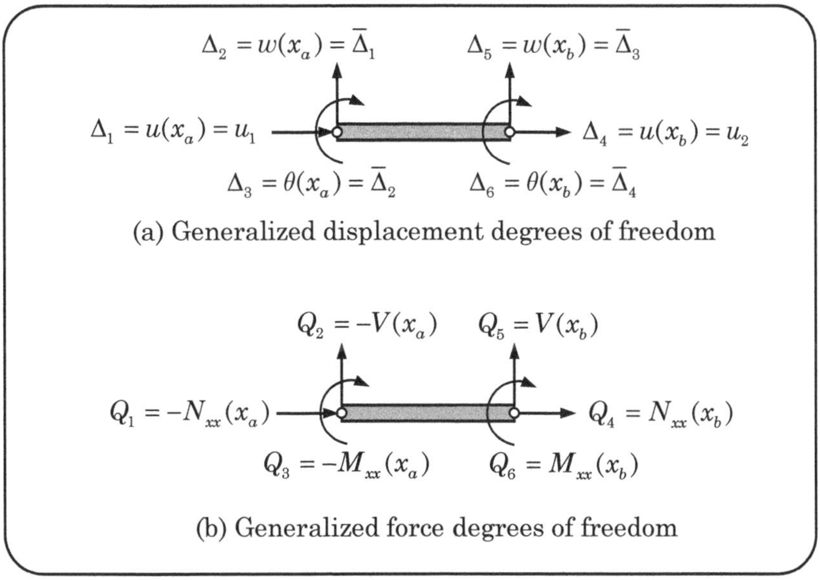

Fig. 5.2.3: The Euler–Bernoulli beam finite element with generalized displacement and force degrees of freedom.

The word *generalized* is used here to convey the meaning that rotations and moments are treated as displacements and forces, respectively; because, for example, rotation is also a displacement that is angular.

In view of the explicit nature of the assumed displacement field in Eq. (5.2.1) in the thickness coordinate z (and it is independent of coordinate y), the volume integral can be expressed as a product of integrals over the length and area of the element. Therefore, the expression for the virtual strain energy can be simplified as

$$
\begin{aligned}
\delta W_I^e &= \int_{x_a}^{x_b} \int_{A^e} \left[\left(\frac{d\delta u}{dx} + \frac{dw}{dx}\frac{d\delta w}{dx} \right) - z \left(\frac{d^2\delta w}{dx^2} \right) \right] \sigma_{xx}\, dA\, dx \\
&= \int_{x_a}^{x_b} \left[\left(\frac{d\delta u}{dx} + \frac{dw}{dx}\frac{d\delta w}{dx} \right) N_{xx} - \frac{d^2\delta w}{dx^2} M_{xx} \right] dx
\end{aligned}
\tag{5.2.8}
$$

where N_{xx} is the axial force (N) and M_{xx} is the moment (N·m)

$$
N_{xx} = \int_{A^e} \sigma_{xx}\, dA \ , \quad M_{xx} = \int_{A^e} \sigma_{xx} z\, dA
\tag{5.2.9}
$$

The virtual work statement in Eq. (5.2.4) becomes

$$
\begin{aligned}
0 = & \int_{x_a}^{x_b} \left[\left(\frac{d\delta u}{dx} + \frac{dw}{dx}\frac{d\delta w}{dx} \right) N_{xx} - \frac{d^2\delta w}{dx^2} M_{xx} \right] dx \\
& - \int_{x_a}^{x_b} \delta w\, q\, dx - \int_{x_a}^{x_b} \delta u\, f\, dx - \sum_{i=1}^{6} \delta\Delta_i^e Q_i^e
\end{aligned}
\tag{5.2.10}
$$

The virtual work statement in Eq. (5.2.10) is equivalent to the following *two* weak forms, which are obtained by collecting terms involving δu and δw separately:

$$
0 = \int_{x_a}^{x_b} \left(\frac{d\delta u}{dx} N_{xx} - \delta u f \right) dx - \delta\Delta_1^e Q_1^e - \delta\Delta_4^e Q_4^e
\tag{5.2.11}
$$

$$
0 = \int_{x_a}^{x_b} \left[\frac{d\delta w}{dx}\left(\frac{dw}{dx} N_{xx} \right) - \frac{d^2\delta w}{dx^2} M_{xx} - \delta w\, q \right] dx
$$

$$
- \delta\Delta_2^e Q_2^e - \delta\Delta_3^e Q_3^e - \delta\Delta_5^e Q_5^e - \delta\Delta_6^e Q_6^e
\tag{5.2.12}
$$

The differential equations governing nonlinear bending of straight beams can be obtained, although not needed for finite element model development, from the virtual work statement in Eq. (5.2.10), or equivalently, the weak forms in Eqs. (5.2.11) and (5.2.12), as explained next.

Integration by parts of the expressions in Eqs. (5.2.11) and (5.2.12) to relieve δu and δw of any differentiation results in the following statements:

$$
0 = \int_{x_a}^{x_b} \delta u \left(-\frac{dN_{xx}}{dx} - f \right) dx + [\delta u\, N_{xx}]_{x_a}^{x_b} - \delta\Delta_1^e Q_1^e - \delta\Delta_4^e Q_4^e
$$

$$0 = -\int_{x_a}^{x_b} \delta w \left[\frac{d}{dx}\left(\frac{dw}{dx} N_{xx} \right) + \frac{d^2 M_{xx}}{dx^2} + q \right] dx - \left[\frac{d\delta w}{dx} M_{xx} \right]_{x_a}^{x_b}$$

$$+ \left[\delta w \left(\frac{dw}{dx} N_{xx} + \frac{dM_{xx}}{dx} \right) \right]_{x_a}^{x_b} - \delta\Delta_2^e\, Q_2^e - \delta\Delta_3^e\, Q_3^e - \delta\Delta_5^e\, Q_5^e - \delta\Delta_6^e\, Q_6^e$$

Since δu and δw are arbitrary and independent of each other in $x_a < x < x_b$ as well as at $x = x_a$ and $x = x_b$, it follows that the governing equations of equilibrium, known as the *Euler equations* (see Reddy [3] for details), are

$$\delta u : \quad -\frac{dN_{xx}}{dx} = f(x) \tag{5.2.13}$$

$$\delta w : \quad -\frac{d}{dx}\left(\frac{dw}{dx} N_{xx} \right) - \frac{d^2 M_{xx}}{dx^2} = q(x) \tag{5.2.14}$$

and the natural (or force) boundary conditions are [as already defined by Eq. (5.2.7)]

$$Q_1^e = -N_{xx}(x_a), \qquad\qquad Q_4^e = N_{xx}(x_b)$$

$$Q_2^e = -\left[\frac{dw}{dx} N_{xx} + \frac{dM_{xx}}{dx} \right]_{x_a}, \qquad Q_5^e = \left[\frac{dw}{dx} N_{xx} + \frac{dM_{xx}}{dx} \right]_{x_b} \tag{5.2.15}$$

$$Q_3^e = -M_{xx}(x_a), \qquad\qquad Q_6^e = M_{xx}(x_b)$$

The vector approach that is commonly used in books on mechanics of materials involves identifying the free-body-diagram of a typical beam element of length Δx from the *deformed* configuration (in contrast to the element taken from the undeformed configuration in the linear analysis), with all its forces and moments displayed on it, setting the sums of the forces in the x and z directions and moments about the y-axis to zero separately (i.e. applying Newton's second law), and taking the limit $\Delta x \to 0$. However, the vector approach does not provide the boundary conditions. To see how this approach works, consider the free-body-diagram of the beam element shown in Fig. 5.2.4. Summing the forces in the x- and z-directions and moments about the y-axis, we obtain

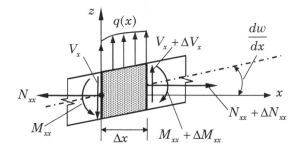

Fig. 5.2.4: A typical deformed beam element with forces and moments to derive equations of equilibrium using the vector approach.

$$-N_{xx} + (N_{xx} + \Delta N_{xx}) + f(x)\Delta x = 0$$
$$-V_x + (V_x + \Delta V_x) + q(x)\Delta x = 0$$
$$-M_{xx} + (M_{xx} + \Delta M_{xx}) - V_x\Delta x + N_{xx}\Delta x\frac{dw}{dx} - q(x)\Delta x(c\Delta x) = 0$$

where $0 < c < 1$ is a constant that depends on the nature of the load $q(x)$. Taking the limit $\Delta x \to 0$, we obtain the following three equations:

$$\frac{dN_{xx}}{dx} + f(x) = 0, \quad \frac{dV_x}{dx} + q(x) = 0, \quad \frac{dM_{xx}}{dx} - V_x + N_{xx}\frac{dw}{dx} = 0$$

which are equivalent to the two equations in Eqs. (5.2.13) and (5.2.14) (when V_x is eliminated from the second equation using the third equation). Note that V_x is the shear force on a section perpendicular to the x-axis, and it is not equal to the shear force $Q_x \equiv dM_{xx}/dx$ acting on the section perpendicular to the deformed beam. In fact, one can show that $V_x = Q_x + N_{xx}(dw/dx)$.

If one begins with Eqs. (5.2.13) and (5.2.14), their weak forms can be developed using the usual three-step procedure of Section 3.2.3:

$$0 = \int_{x_a}^{x_b} v_1\left(-\frac{dN_{xx}}{dx} - f\right)dx = \int_{x_a}^{x_b}\left(\frac{dv_1}{dx}N_{xx} - v_1 f\right)dx - [v_1 N_{xx}]_{x_a}^{x_b}$$
$$= \int_{x_a}^{x_b}\left(\frac{dv_1}{dx}N_{xx} - v_1 f\right)dx - v_1(x_a)[-N_{xx}(x_a)] - v_1(x_b)N_{xx}(x_b) \quad (5.2.16)$$

$$0 = \int_{x_a}^{x_b} v_2\left[-\frac{d}{dx}\left(\frac{dw}{dx}N_{xx}\right) - \frac{d^2 M_{xx}}{dx^2} - q\right]dx$$
$$= \int_{x_a}^{x_b}\left[\frac{dv_2}{dx}\left(\frac{dw}{dx}N_{xx}\right) - \frac{d^2 v_2}{dx^2}M_{xx} - v_2 q\right]dx$$
$$- \left[v_2\left(\frac{dw}{dx}N_{xx} + \frac{dM_{xx}}{dx}\right)\right]_{x_a}^{x_b} - \left[\left(-\frac{dv_2}{dx}\right)M_{xx}\right]_{x_a}^{x_b}$$
$$= \int_{x_a}^{x_b}\left[\frac{dv_2}{dx}\left(\frac{dw}{dx}N_{xx}\right) - \frac{d^2 v_2}{dx^2}M_{xx} - v_2 q\right]dx$$
$$- v_2(x_a)\left[-\left(\frac{dw}{dx}N_{xx} + \frac{dM_{xx}}{dx}\right)\right]_{x_a} - v_2(x_b)\left[\frac{dw}{dx}N_{xx} + \frac{dM_{xx}}{dx}\right]_{x_b}$$
$$- \left[-\frac{dv_2}{dx}\right]_{x_a}[-M_{xx}(x_a)] - \left[-\frac{dv_2}{dx}\right]_{x_b}M_{xx}(x_b) \quad (5.2.17)$$

where v_1 and v_2 are the weight functions, whose meaning is obvious if the expressions $f v_1 dx$ and $q v_2 dx$ are to represent the work done by external forces. We see that $v_1 \sim \delta u$ and $v_2 \sim \delta w$. Clearly, Eqs. (5.2.16) and (5.2.17) are the same as those in Eqs. (5.2.11) and (5.2.12) with $v_1 = \delta u$ and $v_2 = \delta w$.

The force N_{xx} and moment M_{xx} can be expressed in terms of the displacements once the constitutive behavior is assumed. Suppose that the beam is made of a *linear elastic material* with modulus of elasticity E, which can be, in general, a function of x and possibly z, and obeys Hooke's law

$$\sigma_{xx} = E\varepsilon_{xx} \tag{5.2.18}$$

Then we have

$$
\begin{aligned}
N_{xx} &= \int_{A^e} \sigma_{xx}\, dA = \int_{A^e} E^e \left[\frac{du}{dx} + \frac{1}{2}\left(\frac{dw}{dx}\right)^2 - z\frac{d^2w}{dx^2} \right] dA \\
&= A^e_{xx}\left[\frac{du}{dx} + \frac{1}{2}\left(\frac{dw}{dx}\right)^2 \right] - B^e_{xx}\frac{d^2w}{dx^2} \tag{5.2.19}
\end{aligned}
$$

$$
\begin{aligned}
M_{xx} &= \int_{A^e} \sigma_{xx}\, z\, dA = \int_{A^e} E^e \left[\frac{du}{dx} + \frac{1}{2}\left(\frac{dw}{dx}\right)^2 - z\frac{d^2w}{dx^2} \right] z\, dA \\
&= B^e_{xx}\left[\frac{du}{dx} + \frac{1}{2}\left(\frac{dw}{dx}\right)^2 \right] - D^e_{xx}\frac{d^2w}{dx^2} \tag{5.2.20}
\end{aligned}
$$

where A^e_{xx}, B^e_{xx}, and D^e_{xx} are the extensional, extensional-bending, and bending stiffnesses of the beam element

$$(A^e_{xx},\ B^e_{xx},\ D^e_{xx}) = \int_{A^e} E^e(x,z)\left(1, z, z^2\right) dA \tag{5.2.21}$$

For beams made of a material for which E^e is a function of only x and the x-axis is taken along the geometric centroid, then A^e_{xx}, B^e_{xx}, and D^e_{xx} are given by

$$A^e_{xx} = E^e A^e, \quad B^e_{xx} = 0, \quad D^e_{xx} = E^e I^e \tag{5.2.22}$$

where

$$\int_{A^e} dA = A^e, \quad \int_{A^e} z\, dA = 0, \quad \int_{A^e} z^2\, dA = I^e$$

Here A^e and I^e are the cross-sectional area and moment of inertia about the y-axis of the beam element. In general, A^e_{xx} and D^e_{xx} are functions of x whenever the modulus E^e and/or the cross-sectional area A^e is a function of x.

When $B^e_{xx} = 0$, Eqs. (5.2.11) and (5.2.12) can be expressed in terms of the generalized displacements (u, w) with the help of Eqs. (5.2.19) and (5.2.20) as

$$
\begin{aligned}
0 = \int_{x_a}^{x_b} &\left\{ A_{xx}\frac{d\delta u}{dx}\left[\frac{du}{dx} + \frac{1}{2}\left(\frac{dw}{dx}\right)^2 \right] - f\,\delta u \right\} dx \\
&- \delta u(x_a)\, Q_1 - \delta u(x_b) Q_4 \tag{5.2.23}
\end{aligned}
$$

$$
0 = \int_{x_a}^{x_b} \left\{ \frac{d\delta w}{dx} \frac{dw}{dx} A_{xx} \left[\frac{du}{dx} + \frac{1}{2} \left(\frac{dw}{dx} \right)^2 \right] + D_{xx} \frac{d\delta^2 w}{dx^2} \frac{d^2 w}{dx^2} - q \, \delta w \right\} dx
$$
$$
- \delta w(x_a) \, Q_2 - \delta\theta(x_a) \, Q_3 - \delta w(x_b) \, Q_5 - \delta\theta(x_b) \, Q_6 \qquad (5.2.24)
$$

where the element label 'e' is omitted on A_{xx}, B_{xx}, and D_{xx} for simplicity.

5.2.4   Finite Element Model

Let the axial displacement $u(x)$ and transverse deflection $w(x)$ be approximated as $[\theta = -(dw/dx)]$

$$
u(x) \approx u_h(x) = \sum_{j=1}^{2} u_j \psi_j(x) , \quad w(x) \approx w_h(x) = \sum_{J=1}^{4} \bar{\Delta}_J \varphi_J(x) \quad (5.2.25)
$$
$$
\bar{\Delta}_1 \equiv w(x_a), \quad \bar{\Delta}_2 \equiv \theta(x_a), \quad \bar{\Delta}_3 \equiv w(x_b), \quad \bar{\Delta}_4 \equiv \theta(x_b) \qquad (5.2.26)
$$

and ψ_j $(j = 1, 2)$ are the linear Lagrange interpolation functions and φ_J $(J = 1, 2, 3, 4)$ are the Hermite cubic interpolation functions:

$$
\psi_1^e(\zeta) = 1 - \zeta, \qquad \psi_2^e(\zeta) = \zeta \qquad (5.2.27)
$$

$$
\varphi_1^e(\zeta) = 1 - 3\zeta^2 + 2\zeta^3, \quad \varphi_2^e(\zeta) = -h_e \zeta \left(1 - \zeta \right)^2
$$
$$
\varphi_3^e(\zeta) = 3\zeta^2 - 2\zeta^3, \qquad \varphi_4^e(\zeta) = -h_e \zeta \left(\zeta^2 - \zeta \right) \qquad (5.2.28)
$$

where

$$
\zeta = \frac{x - x_a}{h_e}, \quad 0 \le \zeta \le 1 \qquad (5.2.29)
$$

Substituting the approximations from Eq. (5.2.25) for u and w, and $\delta u(x) = \psi_i(x)$ and $\delta w(x) = \varphi_J(x)$ (to obtain the ith algebraic equation of each weak form) into the weak forms in Eqs. (5.2.23) and (5.2.24), we obtain

$$
0 = \sum_{j=1}^{2} K_{ij}^{11} u_j + \sum_{J=1}^{4} K_{iJ}^{12} \bar{\Delta}_J - F_i^1 \quad (i = 1, 2)
$$
$$
0 = \sum_{j=1}^{2} K_{Ij}^{21} u_j + \sum_{J=1}^{4} K_{IJ}^{22} \bar{\Delta}_J - F_I^2 \quad (I = 1, 2, 3, 4) \qquad (5.2.30)
$$

where, for the case in which $B_{xx} = 0$, we have

$$
K_{ij}^{11} = \int_{x_a}^{x_b} A_{xx} \frac{d\psi_i}{dx} \frac{d\psi_j}{dx} \, dx , \qquad K_{iJ}^{12} = \frac{1}{2} \int_{x_a}^{x_b} \left(A_{xx} \frac{dw}{dx} \right) \frac{d\psi_i}{dx} \frac{d\varphi_J}{dx} \, dx
$$

$$
K_{Ij}^{21} = \int_{x_a}^{x_b} A_{xx} \frac{dw}{dx} \frac{d\varphi_I}{dx} \frac{d\psi_j}{dx} \, dx , \quad K_{Ij}^{21} = 2K_{jI}^{12}
$$

$$
K_{IJ}^{22} = \int_{x_a}^{x_b} D_{xx} \frac{d^2 \varphi_I}{dx^2} \frac{d^2 \varphi_J}{dx^2} \, dx + \frac{1}{2} \int_{x_a}^{x_b} \left[A_{xx} \left(\frac{dw}{dx} \right)^2 \right] \frac{d\varphi_I}{dx} \frac{d\varphi_J}{dx} \, dx
$$

$$F_i^1 = \int_{x_a}^{x_b} f\,\psi_i\,dx + \hat{Q}_i\,, \qquad\qquad F_I^2 = \int_{x_a}^{x_b} q\,\varphi_I\,dx + \bar{Q}_I \qquad (5.2.31)$$

for $(i, j = 1, 2)$ and $(I, J = 1, 2, 3, 4)$, where $\hat{Q}_1 = Q_1$, $\hat{Q}_2 = Q_4$, $\bar{Q}_1 = Q_2$, $\bar{Q}_2 = Q_3$, $\bar{Q}_3 = Q_5$, and $\bar{Q}_4 = Q_6$, and the definition of Q_i is given Eq. (5.2.15). The pair of equations in Eq. (5.2.30) can be expressed in matrix form as

$$\begin{bmatrix} \mathbf{K}^{11} & \mathbf{K}^{12} \\ \mathbf{K}^{21} & \mathbf{K}^{22} \end{bmatrix} \begin{Bmatrix} \boldsymbol{\Delta}^1 \\ \boldsymbol{\Delta}^2 \end{Bmatrix} = \begin{Bmatrix} \mathbf{F}^1 \\ \mathbf{F}^2 \end{Bmatrix} \quad \text{or} \quad \mathbf{K}^e \boldsymbol{\Delta}^e = \mathbf{F}^e \qquad (5.2.32)$$

where

$$\Delta_i^1 = u_i, \quad i = 1, 2; \quad \Delta_I^2 = \bar{\Delta}_I, \quad I = 1, 2, 3, 4 \qquad (5.2.33)$$

We also note that $(\mathbf{K}^{12})^T \neq \mathbf{K}^{21}$.

The coefficient matrices $\mathbf{K}^{12}$, $\mathbf{K}^{21}$, and $\mathbf{K}^{22}$ of Eq. (5.2.32) are functions of the unknown displacement $w(x)$. When the nonlinearity is not considered, we will have $\mathbf{K}^{12} = \mathbf{0}$ and $\mathbf{K}^{21} = \mathbf{0}$, and the equations involving $\mathbf{u}$ (the bar problem) are uncoupled from those involving $\bar{\boldsymbol{\Delta}}$ (the beam problem), and therefore, they can be solved independent of each other:

$$\mathbf{K}^{11}\mathbf{u} = \mathbf{F}^1, \qquad \mathbf{K}^{22}\bar{\boldsymbol{\Delta}} = \mathbf{F}^2 \qquad (5.2.34)$$

When nonlinearity is accounted for, the equations are coupled (i.e. bending and stretching deformations of the beam are coupled) and nonlinear. Therefore, an iterative method must be used to solve the assembled equations. When these matrix coefficients are evaluated using the value of $w(x) = \sum \bar{\Delta}_j \varphi_j$ from the previous iteration, we say that the equations are linearized. Hence, the element stiffness matrix and therefore, the assembled stiffness matrix, are not symmetric.

The definitions of the coefficients $K_{ij}^{\alpha\beta}$ in Eq. (5.2.31) can be different if we had accounted for the terms differently, even though the particular choice in deriving Eqs. (5.2.23) and (5.2.24) is probably the most natural. For example, all expressions involving $d\delta w/dx$ and dw/dx in Eq. (5.2.24) could have been used to define $\mathbf{K}^{22}$, leaving no term to define $\mathbf{K}^{21}$ (making $\mathbf{K}^{21} = \mathbf{0}$). Thus, it is possible to decouple the nonlinear equations (after linearization) for $\mathbf{u}$ and $\bar{\boldsymbol{\Delta}}$ and solve them iteratively, feeding the solution from one equation to the other. However, this choice of linearization for the direct iteration solution scheme is known to slow down the convergence. It does not make any difference in the case of Newton's iteration scheme because the tangent matrix will be the same independent of how the direct stiffness coefficients $\mathbf{K}^{\alpha\beta}$ are defined.

Note that the direct stiffness matrix is unsymmetric only due to the fact that $\mathbf{K}^{12}$ contains the factor $1/2$ whereas $\mathbf{K}^{21}$ does not. One way to make $(\mathbf{K}^{21})^T = \mathbf{K}^{12}$ is to split the linear strain du/dx in Eq. (5.2.24) into two equal parts and take one of the two parts as known from a previous iteration:

$$\int_{x_a}^{x_b} \left\{ A_{xx} \frac{d\delta w}{dx} \frac{dw}{dx} \left[\frac{du}{dx} + \frac{1}{2} \left(\frac{dw}{dx} \right)^2 \right] \right\} dx$$

$$= \frac{1}{2} \int_{x_a}^{x_b} A_{xx} \left\{ \frac{dw}{dx} \frac{d\delta w}{dx} \frac{du}{dx} + \left[\frac{du}{dx} + \left(\frac{dw}{dx} \right)^2 \right] \frac{d\delta w}{dx} \frac{dw}{dx} \right\} dx$$

The first term of the above equation constitutes $\mathbf{K}^{21}$ and the second one constitutes a part of $\mathbf{K}^{22}$. The resulting algebraic equations are now symmetric

$$\begin{bmatrix} \bar{\mathbf{K}}^{11} & \bar{\mathbf{K}}^{12} \\ \bar{\mathbf{K}}^{21} & \bar{\mathbf{K}}^{22} \end{bmatrix} \left\{ \begin{matrix} \mathbf{u} \\ \bar{\Delta} \end{matrix} \right\} = \left\{ \begin{matrix} \mathbf{F}^1 \\ \mathbf{F}^2 \end{matrix} \right\} \quad \text{or} \quad \bar{\mathbf{K}}^e \Delta^e = \mathbf{F}^e \tag{5.2.35}$$

where (the element label e is omitted)

$$\Delta_1 = u_1, \ \Delta_2 = \bar{\Delta}_1, \ \Delta_3 = \bar{\Delta}_2, \ \Delta_4 = u_2, \ \Delta_5 = \bar{\Delta}_3 , \ \Delta_6 = \bar{\Delta}_4$$

$$F_1 = F_1^1, \ F_2 = F_1^2, \ F_3 = F_2^2, \ F_4 = F_2^1, \ F_5 = F_3^2, \ F_6 = F_4^2$$

$$\bar{K}_{ij}^{11} = K_{ij}^{11} = \int_{x_a}^{x_b} A_{xx} \frac{d\psi_i}{dx} \frac{d\psi_j}{dx} \, dx$$

$$\bar{K}_{iJ}^{12} = K_{iJ}^{12} = \frac{1}{2} \int_{x_a}^{x_b} \left(A_{xx} \frac{dw}{dx} \right) \frac{d\psi_i}{dx} \frac{d\varphi_J}{dx} \, dx \tag{5.2.36}$$

$$\bar{K}_{Ij}^{21} = \frac{1}{2} \int_{x_a}^{x_b} \left(A_{xx} \frac{dw}{dx} \right) \frac{d\varphi_I}{dx} \frac{d\psi_j}{dx} \, dx \ , \ \ \bar{K}_{Ij}^{21} = \bar{K}_{jI}^{12}$$

$$\bar{K}_{IJ}^{22} = \int_{x_a}^{x_b} D_{xx} \frac{d^2\varphi_I}{dx^2} \frac{d^2\varphi_J}{dx^2} \, dx + \frac{1}{2} \int_{x_a}^{x_b} A_{xx} \left[\frac{du}{dx} + \left(\frac{dw}{dx} \right)^2 \right] \frac{d\varphi_I}{dx} \frac{d\varphi_J}{dx} \, dx$$

Note that in the symmetrized case, we must assume that $u(x)$ is also known, in addition to $w(x)$, from the previous iteration. As one can show, this choice of definition of the direct stiffness coefficients $\bar{\mathbf{K}}^{\alpha\beta}$ also gives the same tangent stiffness matrix as those defined in Eqs. (5.2.49)–(5.2.51).

Readers with a structural mechanics background may have seen $\mathbf{K}^e$ and $\mathbf{F}^e$ in Eq. (5.2.32) expressed in terms of matrices as

$$\mathbf{K}^e = \int_{\Omega^e} \mathbf{B}^{\mathrm{T}} \mathbf{D} \mathbf{B} \, dv, \quad \mathbf{F}^e = \int_{\Omega^e} \boldsymbol{\Psi}^{\mathrm{T}} \mathbf{f} \, dv + \oint_{\Gamma^e} \boldsymbol{\Psi}^{\mathrm{T}} \mathbf{t} \, ds \tag{5.2.37}$$

It is possible to rewrite the virtual work statement in Eq. (5.2.10), finite element approximation in Eq. (5.2.25), and subsequently the finite element model in Eq. (5.2.32) in the form of Eq. (5.2.37).

5.2.5   Iterative Solution Strategies

The direct iteration and Newton's iteration methods introduced in Chapter 4 are revisited here in connection with the nonlinear finite element equations derived in Section 5.2.4. Consider the nonlinear equations in Eq. (5.2.32)

$$\mathbf{K}^e(\mathbf{\Delta}^e)\mathbf{\Delta}^e = \mathbf{F}^e \tag{5.2.38}$$

where the sign conventions of the generalized nodal displacements and forces are shown in Fig. 5.2.3. The system of nonlinear algebraic equations in Eq. (5.2.38) can be linearized using the direct iteration method or Newton's iteration method discussed in Section 4.4. Note that the linearized equations may be symmetric or unsymmetric, depending on the formulation (i.e. weighted-residual or weak form) and the form of the original differential equations (i.e. self-adjoint or not) and, therefore, an appropriate equation solver must be used. On the other hand, an unsymmetric banded equation solver may be used for all finite element equations.

5.2.5.1 Direct iteration procedure

In the direct iteration procedure, the solution at the rth iteration is determined from the assembled set of equations

$$\mathbf{K}(\mathbf{U}^{(r-1)})\mathbf{U}^{(r)} = \mathbf{F}^{(r-1)} \tag{5.2.39}$$

where $\mathbf{U}$ is the global displacement vector and $\mathbf{K}$ is the global direct stiffness matrix, which is obtained by assembling the element stiffness matrices $\mathbf{K}^e$ evaluated at the element level using the known solution vector $\mathbf{\Delta}^e$ from the $(r-1)$st iteration. At the end of each iteration (i.e. after assembly, imposition of boundary conditions, and solution of equations), we check for convergence using the normalized global residual vector $\mathbf{R}$ or global solution vector $\mathbf{U}$:

$$\sqrt{\frac{\mathbf{R}^{(r)} \cdot \mathbf{R}^{(r)}}{\mathbf{F}^{(r)} \cdot \mathbf{F}^{(r)}}} \leq \epsilon, \quad \mathbf{R}^{(r)} \equiv \mathbf{K}(\mathbf{U}^{(r)})\mathbf{U}^{(r)} - \mathbf{F}^{(r)}$$

$$\sqrt{\frac{\mathbf{\Delta U} \cdot \mathbf{\Delta U}}{\mathbf{U}^{(r)} \cdot \mathbf{U}^{(r)}}} \leq \epsilon, \quad \mathbf{\Delta U} \equiv \mathbf{U}^{(r)} - \mathbf{U}^{(r-1)} \tag{5.2.40}$$

where ϵ denotes a preselected value of the error tolerance (e.g. $\epsilon = 10^{-3}$). Figure 4.4.1 depicts the general idea of the direct iteration procedure for a single-degree-of-freedom system, and the steps involved in the direct iteration solution scheme remain the same as those presented in Box 4.4.1.

5.2.5.2 Newton's iteration procedure

In Newton's iterative procedure, the linearized element equation at the beginning of the rth iteration is

$$\mathbf{T}^e(\mathbf{\Delta}^{(r-1)})\delta\mathbf{\Delta}^{(r)} = -\mathbf{R}^e(\mathbf{\Delta}^{(r-1)}) = (\mathbf{F}^e - \mathbf{K}^e\mathbf{\Delta}^e)^{(r-1)} \tag{5.2.41}$$

where the tangent stiffness matrix $\mathbf{T}^e$ associated with the Euler–Bernoulli beam element is calculated using the definition

$$\mathbf{T}^e \equiv \frac{\partial \mathbf{R}^e}{\partial \mathbf{\Delta}^e} \quad \text{or} \quad T_{ij}^e \equiv \frac{\partial R_i^e}{\partial \Delta_j^e} \tag{5.2.42}$$

The global incremental displacement vector $\mathbf{\Delta U}$ at the rth iteration is obtained by solving the assembled equations (after the imposition of the boundary conditions)

$$\mathbf{\Delta U} = -[\mathbf{T}(\mathbf{U}^{(r-1)})]^{-1}\mathbf{R}^{(r-1)} \tag{5.2.43}$$

and the total solution is computed from

$$\mathbf{U}^{(r)} = \mathbf{U}^{(r-1)} + \mathbf{\Delta U} \tag{5.2.44}$$

The comments made in the remarks after Box 4.4.2 about the imposition of boundary conditions in Newton's scheme also apply here.

Although the direct stiffness matrix $\mathbf{K}^e$ may be unsymmetric, it can be shown that the tangent stiffness matrix $\mathbf{T}^e$ of the Euler–Bernoulli beam is always symmetric. That is, the tangent stiffness matrix is the same whether one uses $\mathbf{K}^e$ of Eq. (5.2.32) or $\bar{\mathbf{K}}^e$ of Eq. (5.2.35).

The coefficients of the element tangent stiffness matrix $\mathbf{T}^e$ can be computed using the definition in Eq. (5.2.42). It is convenient to view $\mathbf{T}^e$ as one that has structure similar to $\mathbf{K}^e$ in Eq. (5.2.32). That is, Eq. (5.2.41) is actually has the form

$$\begin{bmatrix} \mathbf{T}^{11} & \mathbf{T}^{12} \\ \mathbf{T}^{21} & \mathbf{T}^{22} \end{bmatrix} \begin{Bmatrix} \delta\mathbf{\Delta}^1 \\ \delta\mathbf{\Delta}^2 \end{Bmatrix} = - \begin{Bmatrix} \mathbf{R}^1 \\ \mathbf{R}^2 \end{Bmatrix} \tag{5.2.45}$$

where symbol δ is used in place of Δ, for obvious reason, to denote the increment of the displacements. Also, note that $\mathbf{\Delta}^1 = \mathbf{u}^e$ and $\mathbf{\Delta}^2 = \bar{\mathbf{\Delta}}^e$. Then we can compute $\mathbf{T}^{\alpha\beta}$ from the definition

$$T_{ij}^{\alpha\beta} = \frac{\partial R_i^\alpha}{\partial \Delta_j^\beta}, \quad \alpha, \beta = 1, 2 \tag{5.2.46}$$

The components R_i^α of the residual vector $\mathbf{R}$ can be expressed as

$$R_i^\alpha = \sum_{\gamma=1}^{2}\sum_{p=1}^{n(\gamma)} K_{ip}^{\alpha\gamma}\Delta_p^\gamma - F_i^\alpha = \sum_{p=1}^{2} K_{ip}^{\alpha1}\Delta_p^1 + \sum_{P=1}^{4} K_{iP}^{\alpha2}\Delta_P^2 - F_i^\alpha$$

$$= \sum_{p=1}^{2} K_{ip}^{\alpha1} u_p + \sum_{P=1}^{4} K_{iP}^{\alpha2}\bar{\Delta}_P - F_i^\alpha \tag{5.2.47}$$

where $n(\gamma)$ denotes the number of element degrees of freedom $[n(1) = 2$ and $n(2) = 4]$. We have

$$T_{ij}^{\alpha\beta} = \frac{\partial R_i^\alpha}{\partial \Delta_j^\beta} = \frac{\partial}{\partial \Delta_j^\beta}\left(\sum_{\gamma=1}^{2}\sum_{p=1}^{n(\gamma)} K_{ip}^{\alpha\gamma}\Delta_p^\gamma - F_i^\alpha\right)$$

$$= \sum_{\gamma=1}^{2} \sum_{p=1}^{n(\gamma)} \left(K_{ip}^{\alpha\gamma} \frac{\partial \Delta_p^{\gamma}}{\partial \Delta_j^{\beta}} + \frac{\partial K_{ip}^{\alpha\gamma}}{\partial \Delta_j^{\beta}} \Delta_p^{\gamma} \right) - \frac{\partial F_i^{\alpha}}{\partial \Delta_j^{\beta}}$$

$$= K_{ij}^{\alpha\beta} + \sum_{p=1}^{2} \frac{\partial}{\partial \Delta_j^{\beta}} \left(K_{ip}^{\alpha 1} \right) u_p + \sum_{P=1}^{4} \frac{\partial}{\partial \Delta_j^{\beta}} \left(K_{iP}^{\alpha 2} \right) \bar{\Delta}_P - \frac{\partial F_i^{\alpha}}{\partial \Delta_j^{\beta}} \quad (5.2.48)$$

Since F_i^{α} is independent of $\boldsymbol{\Delta}$ (unless the boundary conditions are nonlinear) and $K_{ij}^{\alpha\beta}$ depend, at the most, on $\boldsymbol{\Delta}^2 = \bar{\boldsymbol{\Delta}}$ and not on $\boldsymbol{\Delta}^1 = \mathbf{u}$, we have

$$T_{ij}^{\alpha 1} = K_{ij}^{\alpha 1} \quad \text{for } \alpha = 1, 2 \quad (5.2.49)$$

$$T_{iJ}^{12} = K_{iJ}^{12} + \sum_{P=1}^{4} \frac{\partial}{\partial \bar{\Delta}_J} \left(K_{iP}^{12} \right) \bar{\Delta}_P$$

$$= K_{iJ}^{12} + \sum_{P=1}^{4} \left[\int_{x_a}^{x_b} \frac{1}{2} A_{xx}^{e} \frac{\partial}{\partial \bar{\Delta}_J} \left(\sum_{K=1}^{4} \bar{\Delta}_K \frac{d\varphi_K}{dx} \right) \frac{d\psi_i}{dx} \frac{d\varphi_P}{dx} \, dx \right] \bar{\Delta}_P$$

$$= K_{iJ}^{12} + \sum_{P=1}^{4} \left[\int_{x_a}^{x_b} \frac{1}{2} A_{xx}^{e} \frac{d\varphi_J}{dx} \frac{d\psi_i}{dx} \frac{d\varphi_P}{dx} \, dx \right] \bar{\Delta}_P$$

$$= K_{iJ}^{12} + \int_{x_a}^{x_b} \frac{1}{2} A_{xx}^{e} \frac{d\psi_i}{dx} \frac{d\varphi_J}{dx} \left(\sum_{P=1}^{4} \frac{d\varphi_P}{dx} \bar{\Delta}_P \right) dx$$

$$= K_{iJ}^{12} + \int_{x_a}^{x_b} \left(\frac{1}{2} A_{xx}^{e} \frac{dw}{dx} \right) \frac{d\psi_i}{dx} \frac{d\varphi_J}{dx} \, dx = 2 K_{iJ}^{12} = K_{Ji}^{21} \quad (5.2.50)$$

$$T_{IJ}^{22} = K_{IJ}^{22} + \sum_{p=1}^{2} \frac{\partial}{\partial \bar{\Delta}_J} \left(K_{Ip}^{21} \right) u_p + \sum_{P=1}^{4} \frac{\partial}{\partial \bar{\Delta}_J} \left(K_{IP}^{22} \right) \bar{\Delta}_P$$

$$= K_{IJ}^{22} + \sum_{p=1}^{2} \left[\int_{x_a}^{x_b} A_{xx}^{e} \frac{\partial}{\partial \bar{\Delta}_J} \left(\sum_{K=1}^{4} \bar{\Delta}_K \frac{d\varphi_K}{dx} \right) \frac{d\varphi_I}{dx} \frac{d\psi_p}{dx} \, dx \right] u_p$$

$$+ \sum_{P=1}^{4} \left[\int_{x_a}^{x_b} \frac{1}{2} A_{xx}^{e} \frac{\partial}{\partial \bar{\Delta}_J} \left(\frac{dw}{dx} \right)^2 \frac{d\varphi_I}{dx} \frac{d\varphi_P}{dx} \, dx \right] \bar{\Delta}_P$$

$$= K_{IJ}^{22} + \int_{x_a}^{x_b} A_{xx}^{e} \frac{d\varphi_I}{dx} \frac{d\varphi_J}{dx} \left(\sum_{p=1}^{2} \frac{d\psi_p}{dx} u_p \right) dx$$

$$+ \int_{x_a}^{x_b} A_{xx}^{e} \left(\frac{dw}{dx} \right) \frac{d\varphi_I}{dx} \frac{d\varphi_J}{dx} \left(\sum_{P=1}^{4} \bar{\Delta}_P \frac{d\varphi_P}{dx} \right) dx$$

$$= K_{IJ}^{22} + \int_{x_a}^{x_b} A_{xx}^{e} \left(\frac{du}{dx} + \frac{dw}{dx} \frac{dw}{dx} \right) \frac{d\varphi_I}{dx} \frac{d\varphi_J}{dx} \, dx \quad (5.2.51)$$

5.2.6   Load Increments

Examining the expression in Eq. (5.2.19) for the internal axial force N_{xx}, it is clear that the rotation of a transverse normal contributes to the tensile component of N_{xx}, irrespective of the sign of the applied bending loads. As a result, the beam becomes increasingly stiff as it experiences bending deformation. Hence, for large bending loads the nonlinearity may be too large for the numerical scheme to yield convergent solution. Therefore, it is necessary to divide the applied total load, say F, into several smaller load increments $\delta F_1, \delta F_2, \ldots, \delta F_N$ such that

$$F = \sum_{i=1}^{N} \delta F_i \qquad (5.2.52)$$

For the first load step, we use the initial guess of $\mathbf{U}^{(0)} = \mathbf{0}$ so that we can obtain the linear solution of the problem for the first load, F_1, at the end of the first iteration. One of the iterative procedures outlined earlier is used to determine the solution, $\mathbf{U}^{(1)}$. If the iterative scheme does not converge within a pre-specified number of iterations (ITMAX), it may be necessary to further reduce the size of the load increment $F_1 = \delta F_1$. Once the solution for the first load increment is obtained, it is used as the initial "guess" vector for the next load, $F_2 = \delta F_1 + \delta F_2$. This is continued until the total load is reached.

Another way to accelerate the convergence is to use a weighted average of the solutions from the last two iterations in evaluating the stiffness matrix at the rth iteration, as discussed in Eq. (4.4.6):

$$\bar{\mathbf{U}}^{(r-1)} = \beta \mathbf{U}^{(r-2)} + (1 - \beta)\mathbf{U}^{(r-1)}, \qquad 0 \le \beta \le 1 \qquad (5.2.53)$$

where β is the acceleration parameter. A value of $\beta = 0.5$ is suggested, but one may experiment with the value of β when the iterative scheme experiences non-convergence. Otherwise, one should use $\beta = 0$.

5.2.7   Membrane Locking

For the linear case, the axial displacement u is uncoupled from the bending deflection w, and they can be determined independently from the finite element equations [see Eq. (5.2.34)]

$$\mathbf{K}^{11}\mathbf{u} = \mathbf{F}^1, \qquad K_{ij}^{11} = \int_{x_a}^{x_b} A_{xx}^e \frac{d\psi_i}{dx}\frac{d\psi_j}{dx}\, dx \qquad (5.2.54)$$

$$\mathbf{K}^{22(L)}\boldsymbol{\Delta} = \mathbf{F}^2, \qquad K_{IJ}^{22(L)} = \int_{x_a}^{x_b} D_{xx}^e \frac{d^2\varphi_I}{dx^2}\frac{d^2\varphi_J}{dx^2}\, dx \qquad (5.2.55)$$

respectively. Here the superscript L signifies the linear stiffness coefficients. Under the assumptions of linearity, if a beam is subjected to only bending loads and no axial loads, then $u(x) = 0$ [when u is specified to be zero at least at one point]. In other words, a hinged–hinged beam and a pinned–pinned beam [see Figs. 5.2.5(a) and 5.2.5(b), respectively] will have the same deflection $w(x)$ under the same loads and $u(x) = 0$ for all x. However, this is not the case when the von Kármán nonlinearity is included. The coupling between u and w will cause the beam to undergo axial displacement even when there are no axial forces, and the solution (u, w) will be different for the two cases shown in Figs. 5.2.5(a) and 5.2.5(b).

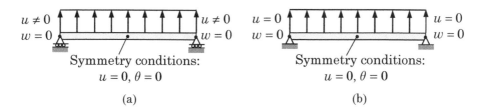

$$u \neq 0 \qquad u \neq 0 \qquad u = 0 \qquad u = 0$$
$$w = 0 \qquad w = 0 \qquad w = 0 \qquad w = 0$$

Symmetry conditions: Symmetry conditions:
$$u = 0, \theta = 0 \qquad\qquad u = 0, \theta = 0$$

(a)   (b)

Fig. 5.2.5: (a) Hinged–hinged and (b) pinned–pinned beams.

First, we note that the hinged–hinged beam does not have any end constraints on u [one may set $u(L/2) = 0$ to remove the rigid-body mode], and a transverse load does not induce axial strain, that is, $\varepsilon_{xx}^0 = 0$ because the beam is free to slide on the rollers to accommodate transverse deflection. On the other hand, the pinned–pinned beam is constrained from axial movement at $x = 0$ and $x = L$. As a result, it will develop axial strain to accommodate the transverse deflection. Thus, a hinged–hinged beam will have larger transverse deflection than a pinned–pinned beam because the latter offers axial stiffness to stretching. The axial stiffness increases in the form $N_{xx}(dw/dx)$ with any transverse load.

Let us examine the requirement that for a hinged–hinged beam and, hence, every finite element not to have an axial strain:

$$\varepsilon_{xx}^0 \equiv \frac{du}{dx} + \frac{1}{2}\left(\frac{dw}{dx}\right)^2 = 0 \text{ (membrane strain)} \tag{5.2.56}$$

In order to satisfy the constraint in (5.2.56), we must have

$$-\frac{du}{dx} \sim \left(\frac{dw}{dx}\right)^2 \tag{5.2.57}$$

The similarity is in the sense of having the same degree of polynomial variation of du/dx and $(dw/dx)^2$. In order to satisfy the constraint in Eq. (5.2.57), both (du/dx) and $(dw/dx)^2$ should be a polynomial of the same degree. For example, when w is interpolated using a Hermite cubic polynomial, then $(dw/dx)^2$ is a fourth-order polynomial. Hence, (du/dx) should be a fourth-order polynomial (or u should be a fifth-order polynomial). Approximation of u with any polynomial less than the fifth-order and w with Hermite cubic polynomials, the constraint in Eq. (5.2.57) is clearly *not* satisfied and the resulting element stiffness matrix will be excessively stiff; hence, results in almost zero displacement field, and the element is said to lock, and this phenomenon is known as the *membrane locking*.

A practical way to satisfy the constraint in Eq. (5.2.57) is to use the minimum interpolation of u and w (i.e. linear interpolation of u and Hermite cubic interpolation of w) but treat ε_{xx}^0 as a constant. This may seem impossible, but it is possible from a numerical point of view. Since du/dx is constant, it is necessary to treat $(dw/dx)^2$ as a constant in the numerical evaluation of the element stiffness coefficients. Thus, if A_{xx} is a constant, all nonlinear stiffness coefficients should be evaluated using one-point Gauss quadrature, that is, one has to use the *reduced integration* to evaluate all nonlinear terms of the stiffness coefficients. These coefficients include K_{iJ}^{12}, K_{Ij}^{21}, T_{iJ}^{12}, T_{Ij}^{21}, and the nonlinear parts of K_{IJ}^{22} and T_{IJ}^{22}. All other terms may be evaluated exactly using two-point quadrature for constant values of A_{xx} and D_{xx}.

5.2.8 Computer Implementation

The flow chart for nonlinear bending of beams is presented in Fig. 5.2.6. Note that there is an outer loop on load increments (NLS = number of load steps). Except for the definition of stiffness coefficients, much of the logic remains the same as that shown in Box 4.5.2.

5.2.8.1 Rearrangement of equations and computation of element coefficients

While the particular form in Eq. (5.2.32) of the finite element model of multi-degree-of-freedom systems is natural, that is, the element stiffness matrix is defined by submatrices $\mathbf{K}^{11}$, $\mathbf{K}^{12}$, $\mathbf{K}^{21}$, and $\mathbf{K}^{22}$ and the solution vector $\boldsymbol{\Delta}$ is partitioned into vectors of extensional and bending degrees of freedom, $\mathbf{u}$ and $\bar{\boldsymbol{\Delta}}$, respectively. However, in practice, it is desirable to rearrange (e.g. to reduce the bandwidth of the assembled system of equations) the solution vector as

$$\boldsymbol{\Delta} = \{u_1, w_1 = \bar{\Delta}_1, \theta_1 = \bar{\Delta}_2, u_2, w_2 = \bar{\Delta}_3, \theta_2 = \bar{\Delta}_4\}^{\mathrm{T}} \qquad (5.2.58)$$

This, in turn, requires rearrangement of the stiffness coefficients such that the original symmetry, if any, of the stiffness matrix is preserved. For linear interpolation of $u(x)$ and Hermite cubic interpolation of $w(x)$, the order of various submatrices of Eq. (5.2.32) is as follows: $\mathbf{K}^{11}$ is of the order 2×2, $\mathbf{K}^{12}$ is 2×4, $\mathbf{K}^{21}$ is 4×2, and $\mathbf{K}^{22}$ is 4×4. The total size of the stiffness matrix is 6×6. The vectors $\boldsymbol{\Delta}^1$ and $\mathbf{F}^1$ are 2×1 and $\boldsymbol{\Delta}^2$ and $\mathbf{F}^2$ are 4×1.

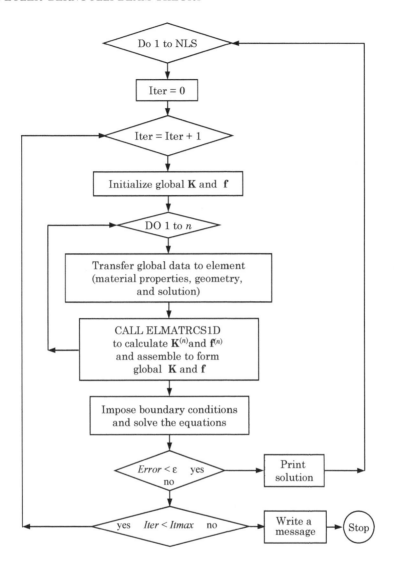

Fig. 5.2.6: A computer flow chart for the nonlinear finite element analysis of beams.

In terms of the submatrix coefficients, Eq. (5.2.32) can be expressed as

$$
\begin{bmatrix}
K_{11}^{11} & K_{12}^{11} & K_{11}^{12} & K_{12}^{12} & K_{13}^{12} & K_{14}^{12} \\
K_{21}^{11} & K_{22}^{11} & K_{21}^{12} & K_{22}^{12} & K_{23}^{12} & K_{24}^{12} \\
K_{11}^{21} & K_{12}^{21} & K_{11}^{22} & K_{12}^{22} & K_{13}^{22} & K_{14}^{22} \\
K_{21}^{21} & K_{22}^{21} & K_{21}^{22} & K_{22}^{22} & K_{23}^{22} & K_{24}^{22} \\
K_{31}^{21} & K_{32}^{21} & K_{31}^{22} & K_{32}^{22} & K_{33}^{22} & K_{34}^{22} \\
K_{41}^{21} & K_{42}^{21} & K_{41}^{22} & K_{42}^{22} & K_{43}^{22} & K_{44}^{22}
\end{bmatrix}
\begin{Bmatrix}
u_1 \\ u_2 \\ \bar{\Delta}_1 \\ \bar{\Delta}_2 \\ \bar{\Delta}_3 \\ \bar{\Delta}_4
\end{Bmatrix}
=
\begin{Bmatrix}
F_1^1 \\ F_2^1 \\ F_1^2 \\ F_2^2 \\ F_3^2 \\ F_4^2
\end{Bmatrix}
\tag{5.2.59}
$$

Rearranging the equations as per the components of the displacement vector in Eq. (5.2.58), we obtain

$$
\begin{bmatrix}
K^{11}_{11} & K^{12}_{11} & K^{12}_{12} & K^{11}_{12} & K^{12}_{13} & K^{12}_{14} \\
K^{21}_{11} & K^{22}_{11} & K^{22}_{12} & K^{21}_{12} & K^{22}_{13} & K^{22}_{14} \\
K^{21}_{21} & K^{22}_{21} & K^{22}_{22} & K^{21}_{23} & K^{22}_{24} & K^{22}_{24} \\
K^{11}_{21} & K^{12}_{21} & K^{12}_{22} & K^{11}_{22} & K^{12}_{23} & K^{12}_{24} \\
K^{21}_{31} & K^{22}_{31} & K^{22}_{32} & K^{21}_{32} & K^{22}_{33} & K^{22}_{34} \\
K^{21}_{41} & K^{22}_{41} & K^{22}_{42} & K^{21}_{42} & K^{22}_{43} & K^{22}_{44}
\end{bmatrix}
\begin{Bmatrix}
u^e_1 \\ \bar{\Delta}^e_1 \\ \bar{\Delta}^e_2 \\ u^e_2 \\ \bar{\Delta}^e_3 \\ \bar{\Delta}^e_4
\end{Bmatrix}
=
\begin{Bmatrix}
F^1_1 \\ F^2_1 \\ F^2_2 \\ F^1_2 \\ F^2_3 \\ F^2_4
\end{Bmatrix}
\tag{5.2.60}
$$

The computer implementation of such rearrangement of elements of the stiffness matrix and force vector is made easier if we rewrite (5.2.32) as

$$
\begin{bmatrix}
\mathbf{K}^{11} & \hat{\mathbf{K}}^{12} & \hat{\mathbf{K}}^{13} \\
\hat{\mathbf{K}}^{21} & \hat{\mathbf{K}}^{22} & \hat{\mathbf{K}}^{23} \\
\hat{\mathbf{K}}^{31} & \hat{\mathbf{K}}^{32} & \hat{\mathbf{K}}^{33}
\end{bmatrix}
\begin{Bmatrix}
\mathbf{u} \\ \mathbf{w} \\ \theta
\end{Bmatrix}
=
\begin{Bmatrix}
\mathbf{F}^1 \\ \mathbf{F}^2 \\ \mathbf{F}^3
\end{Bmatrix}
\tag{5.2.61}
$$

$$
\begin{aligned}
\hat{K}^{12}_{ij} &= K^{12}_{iJ}, & \hat{K}^{21}_{ij} &= K^{21}_{Ij}, & \hat{K}^{13}_{ij} &= K^{12}_{iL}, & \hat{K}^{31}_{ij} &= K^{21}_{Kj} \\
\hat{K}^{22}_{ij} &= K^{22}_{IJ}, & \hat{K}^{23}_{ij} &= \hat{K}^{22}_{IL}, & \hat{K}^{32}_{ij} &= K^{22}_{KJ}, & \hat{K}^{33}_{ij} &= K^{22}_{KL}
\end{aligned}
\tag{5.2.62}
$$

for $I = 2i-1$, $J = 2j-1$, $K = 2i$, $L = 2j$, $i, j = 1, 2$, with similar arrangement of Fs. Then the rearranged system of equations is given by

$$
\begin{bmatrix}
K^{11}_{11} & \hat{K}^{12}_{11} & \hat{K}^{13}_{11} & K^{11}_{12} & \hat{K}^{12}_{12} & \hat{K}^{13}_{12} \\
K^{21}_{11} & \hat{K}^{22}_{11} & \hat{K}^{23}_{11} & \hat{K}^{21}_{12} & \hat{K}^{22}_{12} & \hat{K}^{23}_{12} \\
\hat{K}^{31}_{11} & \hat{K}^{32}_{11} & \hat{K}^{33}_{11} & \hat{K}^{31}_{12} & \hat{K}^{32}_{12} & \hat{K}^{33}_{12} \\
K^{11}_{21} & \hat{K}^{12}_{21} & \hat{K}^{13}_{21} & K^{11}_{22} & \hat{K}^{12}_{22} & \hat{K}^{13}_{22} \\
\hat{K}^{21}_{21} & \hat{K}^{22}_{21} & \hat{K}^{23}_{21} & \hat{K}^{21}_{22} & \hat{K}^{22}_{22} & \hat{K}^{23}_{22} \\
\hat{K}^{31}_{21} & \hat{K}^{32}_{21} & \hat{K}^{33}_{21} & \hat{K}^{31}_{22} & \hat{K}^{32}_{22} & \hat{K}^{33}_{22}
\end{bmatrix}
\begin{Bmatrix}
u^e_1 \\ w^e_1 \\ \theta^e_1 \\ u^e_2 \\ w_2 \\ \theta^e_2
\end{Bmatrix}
=
\begin{Bmatrix}
F^1_1 \\ F^2_1 \\ F^2_2 \\ F^1_2 \\ F^2_3 \\ F^2_4
\end{Bmatrix}
\tag{5.2.63}
$$

Box 5.2.1 contains Fortran statements to rearrange a system of equations of the form in Eq. (5.2.61) to the system in Eq. (5.2.63). The logic is good for any number of degrees of freedom per node (NDF).

The element coefficients $\mathbf{K}^{\alpha\beta}$ and $\mathbf{F}^{\alpha}$ are computed inside the loops on Gauss quadrature. We recall that the stiffness coefficients must be computed in two separate integration loops, one on full integration for the linear terms and another on reduced integration for the nonlinear terms (see Box 5.2.2). The number of full Gauss points (NGP) is determined by the highest polynomial degree p of all integrands of the linear stiffness coefficients: NGP$=(p+1)/2$. For example, if linear interpolation of u and Hermite cubic interpolation of w is used, the integrands of the stiffness coefficients defined in Eq. (5.2.31) have the following polynomial degrees (as if the nonlinear terms are constant):

$$
\begin{aligned}
K^{11}_{ij} &= \text{degree of } A^e_{xx}, & K^{12}_{iJ} &= \text{degree of } A^e_{xx} \\
K^{22(1)}_{IJ} &= \text{degree of } D^e_{xx} + 2, & K^{22(2)}_{IJ} &= \text{degree of } A^e_{xx} \\
F^1_i &= \text{degree of } f(x) + 1, & F^2_I &= \text{degree of } q(x) + 3
\end{aligned}
$$

Box 5.2.1: Fortran statements to rearrange stiffness coefficients.

```
C    REARRANGE THE ELEMENT COEFFICIENTS
C
        II=1
        DO 220 I=1,NPE
           ELF(II)    = ELF1(I)
           ELF(II+1)  = ELF2(I)
           ELF(II+2)  = ELF3(I)
           JJ=1
           DO 210 J=1,NPE
              ELK(II,JJ)     = ELK11(I,J)
              ELK(II,JJ+1)   = ELK12(I,J)
              ELK(II,JJ+2)   = ELK13(I,J)
              ELK(II+1,JJ)   = ELK21(I,J)
              ELK(II+2,JJ)   = ELK31(I,J)
              ELK(II+1,JJ+1) = ELK22(I,J)
              ELK(II+1,JJ+2) = ELK23(I,J)
              ELK(II+2,JJ+1) = ELK32(I,J)
              ELK(II+2,JJ+2) = ELK33(I,J)
 210          JJ=NDF*J+1
 220    II=NDF*I+1
```

In particular, for constant values of $AXX = A_{xx}$, $DXX = D_{xx}$, $FX = f$, and $QX = q$ (F_0 is a scale factor used for load increment), we have $NGP = (3+1)/2 = 2$ (dictated by F_I^2) and the number of reduced integration points is $LGP = 1$.

The meaning of the variables used in Box 5.2.2 is described in the following (NPE is the number of nodes per element):

$$SFL(i) = \psi_i, \qquad SFH(I) = \varphi_I, \qquad ELU(i) = \Delta_i^e$$

$$GDSFL(i) = \frac{d\psi_i}{dx}, \quad GDSFH(I) = \frac{d\varphi_I}{dx}, \quad GDDSFH(I) = \frac{d^2\varphi_I}{dx^2}$$

$$ELK11(i,j) = K_{ij}^{11}, \quad ELK12(i,J) = \hat{K}_{iJ}^{12}, \quad ELK21(J,i) = \hat{K}_{Ji}^{21}$$

$$ELK13(i,J) = \hat{K}_{iJ}^{13}, \quad ELK31(I,j) = \hat{K}_{Ij}^{31}, \quad ELK22(I,J) = \hat{K}_{IJ}^{22}$$

$$ELK23(I,J) = \hat{K}_{IJ}^{23}, \quad ELK32(I,J) = \hat{K}_{IJ}^{32}, \quad ELK33(I,J) = \hat{K}_{IJ}^{33}$$

$$ELF1(i) = F_i^1, \qquad ELF2(I) = F_I^2, \qquad ELF3(I) = F_I^3$$

$$DW = \frac{dw}{dx}, \quad DU = \frac{du}{dx}, \quad AXX = A_{xx}, \quad DXX = D_{xx}$$

$$AXW = AXX * DW, \quad AXWH = 0.5 * AXW$$

$$AN0 = AXWH * DW, \quad AN1 = AXX * (DU + DW * DW)$$

One must extract the nodal degrees of freedom $\Delta_1 = u_1$ and $\Delta_4 = u_2$ from the solution vector $\{\Delta\}$ to compute $u(x)$ and its derivative DU; similarly, one must isolate $\Delta_2 = \bar{\Delta}_1$, $\Delta_3 = \bar{\Delta}_2$, $\Delta_5 = \bar{\Delta}_3$, and $\Delta_6 = \bar{\Delta}_4$ to compute $w(x)$ and its derivative DW (see Box 5.2.2 for the logic).

Box 5.2.2: Fortran statements to compute the element stiffness coefficients.

```
C      Element matrices for the EULER-BERNOULLI beam element
C      Linear terms to be included in the FULL integration loop
C
       DO 50 I=1,NPE
       IO=2*I-1
       ELF1(I)=ELF1(I)+FO*FX*SFL(I)*CNST
       ELF2(I)=ELF2(I)+FO*QX*SFH(IO)*CNST
       ELF3(I)=ELF3(I)+FO*QX*SFH(IO+1)*CNST
       DO 50 J=1,NPE
       JO=2*J-1
       S11=GDSFL(I)*GDSFL(J)*CNST
       H22=GDDSFH(IO)*GDDSFH(JO)*CNST
       H23=GDDSFH(IO)*GDDSFH(JO+1)*CNST
       H32=GDDSFH(IO+1)*GDDSFH(JO)*CNST
       H33=GDDSFH(IO+1)*GDDSFH(JO+1)*CNST
       ELK11(I,J)=ELK11(I,J)+AXX*S11
       ELK22(I,J)=ELK22(I,J)+DXX*H22
       ELK23(I,J)=ELK23(I,J)+DXX*H23
       ELK32(I,J)=ELK32(I,J)+DXX*H32
       ELK33(I,J)=ELK33(I,J)+DXX*H33
    50 CONTINUE
              .
              .
              .

C
C      Nonlinear terms to be included in the REDUCED integration loop
C                loop for the EULER-BERNOULLI beam element
C
           DO 130 I=1,NPE
           IO=2*I-1
           DO 130 J=1,NPE
           JO=2*J-1
           ELK12(I,J)=ELK12(I,J)+AXWH*GDSFL(I)*GDSFH(JO)*CNST
           ELK13(I,J)=ELK13(I,J)+AXWH*GDSFL(I)*GDSFH(JO+1)*CNST
           ELK21(I,J)=ELK21(I,J)+AXW*GDSFL(J)*GDSFH(IO)*CNST
           ELK31(I,J)=ELK31(I,J)+AXW*GDSFL(J)*GDSFH(IO+1)*CNST
           ELK22(I,J)=ELK22(I,J)+ANO*GDSFH(IO)*GDSFH(JO)*CNST
           ELK23(I,J)=ELK23(I,J)+ANO*GDSFH(IO)*GDSFH(JO+1)*CNST
           ELK32(I,J)=ELK32(I,J)+ANO*GDSFH(IO+1)*GDSFH(JO)*CNST
           ELK33(I,J)=ELK33(I,J)+ANO*GDSFH(IO+1)*GDSFH(JO+1)*CNST
           IF(NONLIN.GT.1)THEN
               TAN12(I,J)=TAN12(I,J)+AXWH*GDSFL(I)*GDSFH(JO)*CNST
               TAN13(I,J)=TAN13(I,J)+AXWH*GDSFL(I)*GDSFH(JO+1)*CNST
               TAN22(I,J)=TAN22(I,J)+AN1*GDSFH(IO)*GDSFH(JO)*CNST
               TAN23(I,J)=TAN23(I,J)+AN1*GDSFH(IO)*GDSFH(JO+1)*CNST
               TAN32(I,J)=TAN32(I,J)+AN1*GDSFH(IO+1)*GDSFH(JO)*CNST
               TAN33(I,J)=TAN33(I,J)+AN1*GDSFH(IO+1)*GDSFH(JO+1)*CNST
           ENDIF
   130     CONTINUE
```

5.2.8.2 Computation of strains and stresses

The postprocessing of the solution to compute strains and stresses in structures is of considerable interest because their design is often based on whether the strains and stresses produced are less than the allowable values for the material used. It is well-known that the strains and stresses are closest to the actual values when they are computed at the Gauss points of the reduced integration rule. In the case of the Euler–Bernoulli beam element, the axial displacement is approximated using the linear functions whereas the transverse displacement is approximated using the Hermite cubic polynomials. Thus, the membrane strain ε_{xx}^0, being constant within each element (recall that the nonlinear part must be treated as if it is a constant), should be evaluated using the one-point Gauss rule; and the bending strain ε_{xx}^1, which is linear within each element, may also be evaluated using one-point Gauss rule. Thus, the reduced (i.e. one-point) Gauss-point rule is consistent with the spatial variation of the strains in the element.

The extensional and bending components of strain can be computed from the known finite element solution as

$$
\varepsilon_{xx}^0 = \frac{du}{dx} + \frac{1}{2}\left(\frac{dw}{dx}\right)^2 \approx \sum_{j=1}^{2} u_j^e \frac{d\psi_j^e}{dx} + \frac{1}{2}\left(\sum_{J=1}^{4} \bar{\Delta}_J^e \frac{d\varphi_J^e}{dx}\right)^2
$$

$$
\varepsilon_{xx}^1 = -\frac{d^2 w}{dx^2} \approx -\sum_{J=1}^{4} \bar{\Delta}_J^e \frac{d^2\varphi_J^e}{dx^2}
$$

(5.2.64)

The membrane and bending components of the axial stress $\sigma_{xx} = \sigma_{xx}^0 + z\sigma_{xx}^1$ are given by

$$
\sigma_{xx}^0 = E\varepsilon_{xx}^0 \approx E\left[\sum_{j=1}^{2} u_j^e \frac{d\psi_j^e}{dx} + \frac{1}{2}\left(\sum_{J=1}^{4} \bar{\Delta}_J^e \frac{d\varphi_J^e}{dx}\right)^2\right]
$$

$$
\sigma_{xx}^1 = E\varepsilon_{xx}^1 = -E\frac{d^2 w}{dx^2} \approx -E\sum_{J=1}^{4} \bar{\Delta}_J^e \frac{d^2\varphi_J^e}{dx^2}
$$

(5.2.65)

In theory, all of the expressions can be evaluated at any point in the element, including at the nodes. However, they are the most accurate when evaluated at the center of the element (i.e. using one-point Gauss rule). Using the variable names already introduced previously, we can compute, for example, ε_{xx}^0 as

$$
EXX0 = DU + 0.5 * DW * DW, \quad EXX1 = -DDW, \quad \text{where} \quad DDW = \frac{d^2 w}{dx^2}
$$

$$
SXX0 = E * EXX0, \quad SXX1 = E * EXX1
$$

The Fortran statements to compute strains and stresses are presented in Box 5.2.3.

Box 5.2.3: Fortran statements to compute strains and stresses.

```
          DO 100 NI=1,LGP
          XI = GAUSPT(NI,LGP)
          CALL INTERPLN1D(ELX,GJ,IEL,MODEL,NPE,XI)
          X  = 0.0
          DO 10 I=1,NPE
10        X = X+SFL(I)*ELX(I)
          IF(MODEL.EQ.1)THEN
             U  = 0.0
             DU = 0.0
             DO 20 I=1,NPE
             U  = U  + ELU(I)*SFL(I)
             DU = DU + ELU(I)*GDSFL(I)
20           CONTINUE
             WRITE(IT,200)X,U,DX
          ELSE
             DW =0.0
             DDW=0.0
             PHI=0.0
             DPHI=0.0
             DO 30 I=1,NPE
             L=I*NDF-1
             J=I*(NDF-1)-1
             DU=DU+GDSFL(I)*ELU(L-1)
             IF(MODEL.EQ.2)THEN
                 DW  = DW  + GDSFH(J)*ELU(L)+ GDSFH(J+1)*ELU(L+1)
                 DDW = DDW + GDDSFH(J)*ELU(L)+GDDSFH(J+1)*ELU(L+1)
             ELSE
                 DW   = DW   + GDSFL(I)*ELU(L)
                 PHI  = PHI  + SFL(I)*ELU(L+1)
                 DPHI = DPHI + GDSFL(I)*ELU(L+1)
             ENDIF
30           CONTINUE
             EXXO=DU+0.5*DW*DW
             IF(MODEL.EQ.2)THEN
                 EXX1=-DDW
             ELSE
                 EXX1=DPHI
             ENDIF
             SXXO=AU1*EXXO
             SXX1=AU1*EXX1
             IF(MODEL.EQ.2)THEN
                 WRITE(IT,200) X,EXXO,EXX1,SXXO,SXX1
             ELSE
                 EXZO=PHI+DW
                 SXZO=AU2*EXZO
                 WRITE(IT,200) X,EXXO,EXX1,SXXO,SXX1,EXZO,SXZO
             ENDIF
          ENDIF
100   CONTINUE
200   FORMAT(5X,8E13.4)
```

5.2.9 Numerical Examples

In this section, a number of examples are included to illustrate the effect of boundary conditions, the integration rule used to evaluate the stiffness coefficients, and the iterative solution procedure on the accuracy of the solutions. The notation "$M \times N$ Gauss rule" means M Gauss points are used to evaluate the linear stiffness coefficients and N Gauss points are used to evaluate the nonlinear stiffness coefficients.

Example 5.2.1

Consider a beam of length $L = 100$ in, 1 in $\times$ 1 in cross-sectional dimensions, *hinged* at both ends, made of steel ($E = 30 \times 10^6$ psi and $\nu = 0.3$), and subjected to uniformly distributed transverse load of intensity q_0 lb/in, as shown in Fig. 5.2.5(a). Using the symmetry about $x = L/2$, investigate the effect of integration rule on the transverse deflection. Use uniform meshes of four and eight elements in half beam.

Solution: The geometric boundary conditions for the left half of the beam are

$$w(0) = u(\tfrac{L}{2}) = \tfrac{dw}{dx}(\tfrac{L}{2}) = 0 \tag{1}$$

The load is divided into load increments of equal size $\Delta q_0 = 1$ lb/in. A tolerance of $\epsilon = 10^{-3}$ and maximum allowable iterations of 25 (per load step) are used in the analysis. The initial solution vector is chosen to be the zero vector, so that the first iteration solution corresponds to the linear solution. The linear analytical solution is

$$u(x) = 0, \quad w(x) = \frac{q_0 L^4}{24 D_{xx}}\left(\frac{x}{L} - 2\frac{x^3}{L^3} + \frac{x^4}{L^4}\right) \tag{2}$$

In particular, the center deflection for $q_0 = 1$ lb/in is

$$w(0.5L) = \frac{5 q_0 L^4}{384 D_{xx}} = 0.5208 \text{ in} \tag{3}$$

and the linear finite element solution matches with the exact solution.

For the mesh of four elements in half beam, the element linear stiffness matrix and force vector (during the first iteration) for element 1 are

$$\mathbf{K}^1 = 10^5 \begin{bmatrix} 24 & 0.0000 & 0.00 & -24 & 0.0000 & 0.00 \\ 0 & 0.1536 & -0.96 & 0 & -0.1536 & -0.96 \\ 0 & -0.9600 & 8.00 & 0 & 0.9600 & 4.00 \\ -24 & 0.0000 & 0.00 & 24 & 0.0000 & 0.00 \\ 0 & -0.1536 & 0.96 & 0 & 0.1536 & 0.96 \\ 0 & -0.9600 & 4.00 & 0 & 0.9600 & 8.00 \end{bmatrix}, \quad \mathbf{F}^1 = \begin{Bmatrix} 0.000 \\ 6.250 \\ -13.021 \\ 0.000 \\ 6.250 \\ 13.021 \end{Bmatrix} \tag{4}$$

Because of the uniform mesh and data used, linear element matrices and force vectors are the same for all elements in the mesh. However, $\mathbf{K}^{(e)}$ will be different for each element Ω^e after the first iteration because of the nonlinearity that varies with x. The direct stiffness matrix for element 1 at load $q_0 = 1$ lb/in, using one-point Gauss rule for the nonlinear terms, after the first iteration (i.e. at the beginning of the second iteration) is

$$\mathbf{K}^1 = 10^5 \begin{bmatrix} 24.0000 & 0.2933 & -0.6110 & -24.0000 & -0.2933 & -0.6110 \\ 0.5865 & 0.1608 & -0.9749 & -0.5865 & -0.1608 & -0.9749 \\ -1.2219 & -0.9749 & 8.0311 & 1.2219 & 0.9749 & 4.0311 \\ -24.0000 & -0.2933 & 0.6110 & 24.0000 & 0.2933 & 0.6110 \\ -0.5865 & -0.1608 & 0.9749 & 0.5865 & 0.1608 & 0.9749 \\ -1.2219 & -0.9749 & 4.0311 & 1.2219 & 0.9749 & 8.0311 \end{bmatrix} \tag{5}$$

The global linear solution, which coincides with the exact solution, and the converged solution in the direct iteration scheme (error= 0.1419×10^{-12}) for $q_0 = 1$ lb/in are

$$
\mathbf{U} = \begin{Bmatrix} U_1 \\ U_3 \\ U_4 \\ U_5 \\ U_6 \\ U_7 \\ U_8 \\ U_9 \\ U_{10} \\ U_{11} \\ U_{12} \\ U_{14} \end{Bmatrix} = \begin{Bmatrix} 0.00000 \\ -0.01666 \\ 0.00000 \\ 0.20223 \\ -0.01523 \\ 0.00000 \\ 0.37109 \\ -0.01146 \\ 0.00000 \\ 0.48218 \\ -0.00612 \\ 0.52083 \end{Bmatrix}_{\text{Linear}}, \quad
\mathbf{U} = \begin{Bmatrix} U_1 \\ U_3 \\ U_4 \\ U_5 \\ U_6 \\ U_7 \\ U_8 \\ U_9 \\ U_{10} \\ U_{11} \\ U_{12} \\ U_{14} \end{Bmatrix} = \begin{Bmatrix} 0.00337 \\ -0.01666 \\ 0.00171 \\ 0.20223 \\ -0.01523 \\ 0.00056 \\ 0.37109 \\ -0.01146 \\ 0.00006 \\ 0.48218 \\ -0.00612 \\ 0.52083 \end{Bmatrix}_{\text{3rd iter.}} \tag{6}
$$

For the four-element mesh in half beam, the element tangent stiffness matrix for element 1, using one-point Gauss rule for the nonlinear terms, after the first iteration in Newton's iteration scheme is

$$
\mathbf{T}^1 = 10^5 \begin{bmatrix}
24.0000 & 0.5865 & -1.2219 & -24.0000 & -0.5865 & -1.2219 \\
0.5865 & 0.1751 & -1.0048 & -0.5865 & -0.1751 & -1.0048 \\
-1.2219 & -1.0048 & 8.0933 & 1.2219 & 1.0048 & 4.0933 \\
-24.0000 & -0.5865 & 1.2219 & 24.0000 & 0.5865 & 1.2219 \\
-0.5865 & -0.1751 & 1.0048 & 0.5865 & 0.1751 & 1.0048 \\
-1.2219 & -1.0048 & 4.0933 & 1.2219 & 1.0048 & 8.0933
\end{bmatrix} \tag{7}
$$

and element 1 residual force vector after the first iteration and the converged global solution increment vector after third iteration, which is the same as the total global solution vector because of the zero initial guess vector, in Newton's iteration scheme (error= 0.1480×10^{-13}) are

$$
-\mathbf{R}^1 = 10^3 \begin{Bmatrix} 3.9816 \\ 0.1473 \\ -0.2027 \\ -3.9816 \\ -0.1348 \\ -0.7496 \end{Bmatrix}, \quad
\Delta\mathbf{U} = \begin{Bmatrix} \Delta U_1 \\ \Delta U_3 \\ \Delta U_4 \\ \Delta U_5 \\ \Delta U_6 \\ \Delta U_7 \\ \Delta U_8 \\ \Delta U_9 \\ \Delta U_{10} \\ \Delta U_{11} \\ \Delta U_{12} \\ \Delta U_{14} \end{Bmatrix} = \begin{Bmatrix} 0.00337 \\ -0.01666 \\ 0.00171 \\ 0.20223 \\ -0.01523 \\ 0.00056 \\ 0.37109 \\ -0.01146 \\ 0.00006 \\ 0.48218 \\ -0.00612 \\ 0.52083 \end{Bmatrix}_{\text{3rd iter.}} \tag{8}
$$

Table 5.2.1 contains the results of the nonlinear analysis obtained with the direct iteration and the Newton iteration schemes. As discussed earlier, the problem should not exhibit any nonlinearity. The correct linear solution, given by Eq. (2), is predicted using 2×1 Gauss rule (see the last column of Table 5.2.1). Both four- and eight-element meshes and the direct and Newton's iteration schemes predicted the same result. The 2×2 Gauss rule not only yields incorrect results, but also takes more iterations to converge.

Post-computation of the strains and stresses show that the element predicts the extensional strains and stresses that are essentially zero. The bending strains and stresses are predicted correctly and they remain linear with the load.

Table 5.2.1: Finite element results for the transverse deflections, $w(L/2)$, of a beam with both ends *hinged* and subjected to uniformly distributed load (half-beam model).

Load q_0	Direct iteration (DI) (2×2 Gauss rule)		Newton iteration (NI) (2×2 Gauss rule)		DI–NI 2×1
	4 elements	8 elements	4 elements	8 elements	4 & 8 elements
1.0	0.5108 (3)*	0.5182 (3)	0.5108 (4)	0.5182 (4)	0.5208 (3)
2.0	0.9739 (5)	1.0213 (3)	0.9739 (4)	1.0213 (4)	1.0417 (3)
3.0	1.3763 (6)	1.4986 (4)	1.3764 (4)	1.4986 (4)	1.5625 (3)
4.0	1.7269 (7)	1.9451 (4)	1.7265 (4)	1.9453 (4)	2.0833 (3)
5.0	2.0356 (9)	2.3609 (5)	2.0351 (4)	2.3607 (4)	2.6042 (3)
6.0	2.3122 (11)	2.7471 (5)	2.3115 (3)	2.7466 (4)	3.1250 (3)
7.0	2.5617 (14)	3.1054 (6)	2.5624 (3)	3.1059 (4)	3.6458 (3)
8.0	2.7936 (17)	3.4418 (7)	2.7926 (3)	3.4414 (3)	4.1667 (3)
9.0	3.0049 (22)	3.7570 (7)	3.0060 (3)	3.7560 (3)	4.6875 (3)
10.0	3.2063 (29)	4.5013 (8)	3.2051 (3)	4.0522 (3)	5.2083 (3)

* Number of iterations taken to converge (with acceleration parameter, $\beta = 0$).

Example 5.2.2

Consider the straight beam of Example 5.2.1 but with two different sets of boundary conditions. Both ends either (1) *pinned*, as shown in Fig. 5.2.5(b) or (2) *clamped*. Assume uniformly distributed transverse load of intensity q_0 and use symmetry of the solution about $x = L/2$ to investigate the convergence characteristics and effect of integration rule on the transverse deflection. Use uniform meshes of four and eight elements in half beam.

Solution: The geometric boundary conditions for the computational domain of the two problems are

$$\text{pinned: } u(0) = w(0) = u(\tfrac{L}{2}) = \frac{dw}{dx}\Big|_{x=\frac{L}{2}} = 0 \tag{1}$$

$$\text{clamped: } u(0) = w(0) = \frac{dw}{dx}\Big|_{x=0} = u(\tfrac{L}{2}) = \frac{dw}{dx}\Big|_{x=\frac{L}{2}} = 0 \tag{2}$$

The load increments of $\Delta q_0 = 1.0$ lb/in, a tolerance of $\epsilon = 10^{-3}$, and maximum allowable iterations of 25 (per load step) are used in the analysis. The initial solution vector is chosen to be the zero vector.

The linear nodal displacements obtained using a mesh of four elements in a half beam for the two different boundary conditions are (independent of the iterative scheme used)

$$\mathbf{U} = \begin{Bmatrix} U_3 \\ U_4 \\ U_5 \\ U_6 \\ U_7 \\ U_8 \\ U_9 \\ U_{10} \\ U_{11} \\ U_{12} \\ U_{14} \end{Bmatrix} = \begin{Bmatrix} -0.01666 \\ 0.00000 \\ 0.20223 \\ -0.01523 \\ 0.00000 \\ 0.37109 \\ -0.11458 \\ 0.00000 \\ 0.48218 \\ -0.00612 \\ 0.52083 \end{Bmatrix}_{\text{pinned}} , \quad \mathbf{U} = \begin{Bmatrix} U_4 \\ U_5 \\ U_6 \\ U_7 \\ U_8 \\ U_9 \\ U_{10} \\ U_{11} \\ U_{12} \\ U_{14} \end{Bmatrix} = \begin{Bmatrix} 0.00000 \\ 0.01994 \\ -0.00273 \\ 0.00000 \\ 0.05859 \\ -0.00313 \\ 0.00000 \\ 0.09155 \\ -0.00195 \\ 0.10417 \end{Bmatrix}_{\text{clamped}} \tag{3}$$

The exact solutions to the linear problems are

$$pinned: \ u(x) = 0, \quad w(x) = \frac{q_0 L^4}{24 D_{xx}} \left(\frac{x}{L} - 2\frac{x^3}{L^3} + \frac{x^4}{L^4} \right) \tag{4}$$

$$clamped: \ u(x) = 0, \quad w(x) = \frac{q_0 L^4}{24 D_{xx}} \frac{x^2}{L^2} \left(1 - \frac{x}{L} \right)^2 \tag{5}$$

and the maximum transverse deflections occur at $L/2$. For $q_0 = 1$ lb/in, $L = 100$ in, and $E = 30 \times 10^6$ psi $(\nu = 0.3)$, they are given by $(D_{xx} = EH^3/12, \ H = 1)$

$$pinned: \ w(\tfrac{L}{2}) = \frac{5 q_0 L^4}{384 D_{xx}} = 0.5208 \text{ in}; \quad clamped: \ w(\tfrac{L}{2}) = \frac{q_0 L^4}{384 D_{xx}} = 0.1042 \text{ in} \tag{6}$$

The linear finite element solutions at the nodes coincide with the exact solutions.

Table 5.2.2 contains the center transverse deflections as a function of the load intensity for the beam with pinned ends for four- and eight-element meshes and $\Delta q_0 = 1.0$ lb/in. The direct iteration scheme does not converge even for 100 iterations per load step when $\beta = 0.0$. However, it converges for $\beta = 0.25$ whereas Newton's scheme yields convergent results for no more than four iterations per load step and with $\beta = 0.0$. The effect of the integration rule is not significant for this problem; the full integration rule also gives good results, slightly under-predicting the deflections.

Table 5.2.2: The center transverse deflection $w(L/2)$ versus load q_0 for a beam with both ends *pinned* and subjected to a uniformly distributed transverse load (half-beam model).

	2 × 2 Gauss rule				2 × 1 Gauss rule	
Load	4 elements		8 elements		8 elements	
q_0	DI	NI	DI	NI	DI	NI
1	0.3668 (9*)	0.3669 (5)	0.3680 (9)	0.3680 (5)	0.3685 (9)	0.3685 (5)
2	0.5425 (11)	0.5424(4)	0.5447 (11)	0.5446 (4)	0.5458 (11)	0.5457 (4)
3	0.6598 (10)	0.6601 (3)	0.6626 (10)	0.6629 (3)	0.6642 (10)	0.6645 (3)
4	0.7512 (11)	0.7510 (3)	0.7545 (11)	0.7543 (3)	0.7565 (11)	0.7564 (3)
5	0.8263 (12)	0.8263 (3)	0.8299 (12)	0.8299 (3)	0.8324 (12)	0.8324 (3)
6	0.8914 (12)	0.8912 (3)	0.8952 (12)	0.8950 (3)	0.8981 (12)	0.8979 (3)
7	0.9484 (13)	0.9485 (3)	0.9524 (13)	0.9525 (3)	0.9557 (13)	0.9558 (3)
8	1.0013 (8)	1.0002 (3)	1.0055 (8)	1.0043 (3)	1.0093 (8)	1.0080 (3)
9	1.0475 (14)	1.0473 (3)	1.0518 (14)	1.0516 (3)	1.0559 (14)	1.0557 (3)
10	1.0903 (9)	1.0908 (3)	1.0946 (9)	1.0952 (3)	1.0992 (9)	1.0997 (3)

* Number of iterations taken to converge; $\epsilon = 10^{-3}$.

Table 5.2.3 contains the center transverse deflections as a function of the load intensity for the beam with clamped ends for four- and eight-element meshes and $\Delta q_0 = 1.0$ lb/in. Both direct iteration and Newton's iteration methods converge for $\beta = 0.0$, direct iteration taking more number of iterations than Newton's method. There is no significant difference between the solutions obtained with the two integration rules; again, the fully-integrated element yields slightly lower deflections as compared to the reduced-integration element.

Figure 5.2.7 contains load-deflection curves for the two different boundary conditions using the eight-element mesh with 2×1 Gauss quadrature. The difference between the results obtained with the two iterative schemes cannot be seen on the graph. The clamped beam, being stiffer than the pinned beam, exhibits less degree of nonlinearity.

If the axial displacement degrees of freedom are suppressed (i.e. set $u = 0$ at every point of the beam) in the nonlinear analysis of beams, the beam will behave very stiff and the deflections experienced will be less than those shown in Tables 5.2.2 and 5.2.3. Figure 5.2.8 contains plots of bending stresses σ_{xx}^1 at $x = 46.875$ versus load for the two boundary conditions [see Eq. (5.2.65) for the expressions used for stresses; $\sigma_{xx}(x, z) = \sigma_{xx}^0(x) + z\sigma_{xx}^1(x)$]. For the clamped case, the maximum stress occurs at $x = 0$ and it is compressive.

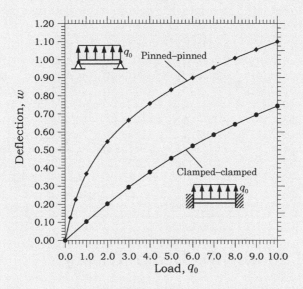

Fig. 5.2.7: Load versus center transverse deflection for pinned and clamped beams under a uniform load.

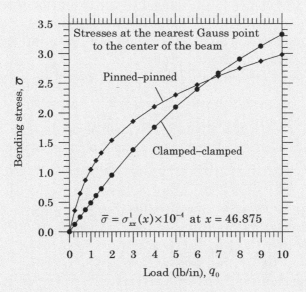

Fig. 5.2.8: Load versus center bending stress for pinned and clamped beams under a uniform load.

Table 5.2.3: The center transverse deflection $w(L/2)$ versus load q_0 for a beam with both ends *clamped* and subjected to a uniformly distributed transverse load (half-beam model).

	Direct iteration		Newton's iteration		
Load q_0	4 elements 2×1	8 elements 2×1	4 elements 2×1	8 elements 2×1	8 elements 2×2
1.0	0.1033 (3)*	0.1034 (3)	0.1034 (3)	0.1034 (3)	0.1033 (3)
2.0	0.2022 (4)	0.2023 (4)	0.2022 (3)	0.2023 (3)	0.2022 (3)
3.0	0.2938 (4)	0.2939 (4)	0.2939 (3)	0.2939 (3)	0.2936 (3)
4.0	0.3773 (5)	0.3774 (5)	0.3773 (3)	0.3774 (3)	0.3767 (3)
5.0	0.4529 (5)	0.4531 (5)	0.4528 (3)	0.4530 (3)	0.4519 (3)
6.0	0.5213 (6)	0.5215 (6)	0.5214 (3)	0.5216 (3)	0.5200 (3)
7.0	0.5840 (7)	0.5842 (7)	0.5839 (3)	0.5841 (3)	0.5821 (3)
8.0	0.6412 (8)	0.6412 (8)	0.6413 (3)	0.6414 (3)	0.6389 (3)
9.0	0.6945 (9)	0.6944 (9)	0.6943 (3)	0.6943 (3)	0.6913 (3)
10.0	0.7433 (10)	0.7431 (10)	0.7435 (3)	0.7433 (3)	0.7399 (3)

* Number of iterations taken to converge; $\epsilon = 10^{-3}$.

5.3   The Timoshenko Beam Theory

5.3.1   Displacement and Strain Fields

The Euler–Bernoulli beam theory (EBT) is based on the assumption that a straight line transverse to the axis of the beam before deformation remains: (1) straight, (2) inextensible, and (3) normal to the mid-plane after deformation. In the Timoshenko beam theory (TBT), the first two assumptions are kept but the normality condition is relaxed by assuming that the rotation is independent of the slope $[\theta = -(dw/dx)]$ of the beam.

The displacement field in the TBT can be expressed as

$$u_1 = u(x) + z\phi_x(x), \quad u_2 = 0, \quad u_3 = w(x) \tag{5.3.1}$$

where (u_1, u_2, u_3) are the displacements of a point along the (x, y, z) coordinates, (u, w) are the displacements of a point on the mid-plane of an undeformed beam, and ϕ_x is the rotation (about the y-axis) of a transverse straight line (see Fig. 5.3.1). The only non-zero strains are

$$\varepsilon_{xx} = \frac{\partial u_1}{\partial x} + \frac{1}{2}\left(\frac{\partial u_3}{\partial x}\right)^2 = \frac{du}{dx} + \frac{1}{2}\left(\frac{dw}{dx}\right)^2 + z\frac{d\phi_x}{dx} \equiv \varepsilon_{xx}^0 + z\varepsilon_{xx}^1 \tag{5.3.2}$$

$$\gamma_{xz} = \frac{\partial u_1}{\partial z} + \frac{\partial u_3}{\partial x} = \phi_x + \frac{dw}{dx} \equiv \gamma_{xz}^0 \tag{5.3.3}$$

The virtual strains are

$$\delta\varepsilon_{xx}^0 = \frac{d\delta u}{dx} + \frac{dw}{dx}\frac{d\delta w}{dx}, \quad \delta\varepsilon_{xx}^1 = \frac{d\delta\phi_x}{dx}, \quad \delta\gamma_{xz}^0 = \delta\phi_x + \frac{d\delta w}{dx} \tag{5.3.4}$$

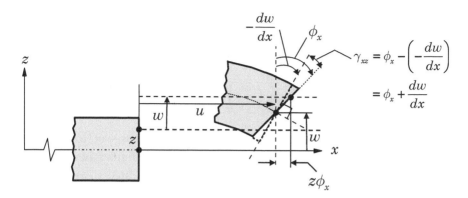

Fig. 5.3.1: Kinematics of a beam in the Timoshenko beam theory.

5.3.2 Weak Forms

Next, we use the principle of virtual displacements to develop the weak forms of the TBT. We have

$$0 = \delta W^e \equiv \delta W_I^e + \delta W_E^e \tag{5.3.5}$$

$$\delta W_I^e = \int_{x_a}^{x_b} \int_{A^e} \left(\sigma_{xx} \delta \varepsilon_{xx} + \sigma_{xz} \delta \gamma_{xz} \right) dA\, dx$$

$$= \int_{x_a}^{x_b} \int_{A^e} \left[\sigma_{xx} \left(\delta \varepsilon_{xx}^0 + z \delta \varepsilon_{xx}^1 \right) + \sigma_{xz} \delta \gamma_{xz}^0 \right] dA\, dx$$

$$= \int_{x_a}^{x_b} \left(N_{xx} \delta \varepsilon_{xx}^0 + M_{xx} \delta \varepsilon_{xx}^1 + Q_x \delta \gamma_{xz}^0 \right) dx \tag{5.3.6a}$$

$$\delta W_E^e = - \left[\int_{x_a}^{x_b} q\, \delta w\, dx + \int_{x_a}^{x_b} f\, \delta u\, dx + \sum_{i=1}^{6} Q_i^e \delta \Delta_i^e \right] \tag{5.3.6b}$$

where q is the distributed transverse load, f is the distributed axial load, Q_i^e ($i = 1, 2, \ldots, 6$) are the element generalized forces, Δ_i^e ($i = 1, 2, \ldots, 6$) are the element generalized displacements, and

$$N_{xx} = \int_A \sigma_{xx}\, dA, \quad M_{xx} = \int_A \sigma_{xx} z\, dA, \quad Q_x = \int_A \sigma_{xz}\, dA$$

$$\sigma_{xx} = E \varepsilon_{xx}, \quad \sigma_{xz} = K_s G \gamma_{xz} \tag{5.3.7}$$

and E is the modulus of elasticity, G is the shear modulus, and K_s is the *shear correction coefficient* introduced into the constitutive equation to account for the difference between the transverse shear energy calculated using equilibrium stresses and that predicted by the Timoshenko beam theory on account of constant state of shear stress through the beam height.

For example, consider a homogeneous beam with rectangular cross section, with width B and height H. The shear stress distribution through the thickness of the beam according to the theory of elasticity is given by

$$\sigma_{xz}^e = \frac{3V}{2A}\left[1 - \left(\frac{2z}{H}\right)^2\right], \quad -\frac{H}{2} \le z \le \frac{H}{2}$$

where V is the transverse force and $A = BH$ is the cross-sectional area. The transverse shear stress and shear strain in the Timoshenko beam theory are independent of z, $\sigma_{xz}^T = V/A$ and $\gamma_{xz}^T = V/(GAK_s)$. Thus, the strain energies due to the transverse shear stress in the two theories are

$$U_s^e = \frac{1}{2}\int_A \sigma_{xz}^e \gamma_{xz}^e \, dA = \frac{1}{2G}\int_A (\sigma_{xz}^e)^2 \, dA = \frac{3V^2}{5GA}$$

$$U_s^T = \frac{1}{2}\int_A \sigma_{xz}^T \gamma_{xz}^T \, dA = \frac{A}{2}\sigma_{xz}^T \gamma_{xz}^T = \frac{V^2}{2GAK_s}$$

Equating the two energies, we find that the shear correction factor is equal to $K_s = 5/6$. The shear correction factor, in general, depends on the beam cross section and material properties.

The statement of the principle of virtual displacements in Eq. (5.3.5) can be used to derive the governing equations, known as the *Euler–Lagrange equations*

$$\delta u: \quad -\frac{dN_{xx}}{dx} = f(x) \tag{5.3.8}$$

$$\delta w: \quad -\frac{dQ_x}{dx} - \frac{d}{dx}\left(N_{xx}\frac{dw}{dx}\right) = q(x) \tag{5.3.9}$$

$$\delta \phi_x: \quad -\frac{dM_{xx}}{dx} + Q_x = 0 \tag{5.3.10}$$

It is clear that (u, w, ϕ_x) are the primary variables and

$$N_{xx}, \quad Q_x, \quad \text{and} \quad M_{xx} \tag{5.3.11}$$

are the secondary variables. Thus, the pairing of the primary and secondary variables is as follows:

$$(u, N_{xx}), \quad (w, Q_x), \quad (\phi_x, M_{xx}) \tag{5.3.12}$$

Only one member of each pair may be specified at a point in the beam.

The axial force N_{xx}, shear force Q_x, and bending moment M_{xx} can be expressed in terms of the generalized displacements (u, w, ϕ_x) using the definitions in Eq. (5.3.7). We obtain

$$N_{xx} = A_{xx}\left[\frac{du}{dx} + \frac{1}{2}\left(\frac{dw}{dx}\right)^2\right] + B_{xx}\frac{d\phi_x}{dx}$$

$$M_{xx} = B_{xx} \left[\frac{du}{dx} + \frac{1}{2} \left(\frac{dw}{dx} \right)^2 \right] + D_{xx} \frac{d\phi_x}{dx} \tag{5.3.13}$$

$$Q_x = S_{xx} \left(\frac{dw}{dx} + \phi_x \right)$$

where A_{xx}, B_{xx}, and D_{xx} are defined in Eq. (5.2.21), and S_{xx} is the shear stiffness

$$S_{xx} = K_s \int_A G \, dA = K_s GA = \frac{K_s EA}{2(1+\nu)} \tag{5.3.14}$$

As discussed earlier, the beam stiffness coefficients A_{xx}, B_{xx}, and D_{xx} are functions of x whenever the modulus E and/or cross-sectional area A is a function of x. For homogeneous (through the thickness) beams, we have $B_{xx} = 0$.

The equations of equilibrium of the TBT for a *homogeneous beam* can be expressed in terms of the generalized displacements as

$$-\frac{d}{dx} \left\{ A_{xx} \left[\frac{du}{dx} + \frac{1}{2} \left(\frac{dw}{dx} \right)^2 \right] \right\} = f \tag{5.3.15}$$

$$-\frac{d}{dx} \left[S_{xx} \left(\frac{dw}{dx} + \phi_x \right) \right] - \frac{d}{dx} \left\{ A_{xx} \frac{dw}{dx} \left[\frac{du}{dx} + \frac{1}{2} \left(\frac{dw}{dx} \right)^2 \right] \right\} = q \tag{5.3.16}$$

$$-\frac{d}{dx} \left(D_{xx} \frac{d\phi_x}{dx} \right) + S_{xx} \left(\frac{dw}{dx} + \phi_x \right) = 0 \tag{5.3.17}$$

5.3.3 General Finite Element Model

The finite element model of the Timoshenko beam equations can be constructed using the virtual work statement in Eq. (5.3.5), where the axial force N_{xx}, the shear force Q_x, and the bending moment M_{xx} are known in terms of the generalized displacements (u, w, ϕ_x) by Eq. (5.3.13). The virtual work statement in Eq. (5.3.5) is equivalent to the following three statements, which can also be obtained from the governing equations in Eqs. (5.3.15)–(5.3.17):

$$0 = \int_{x_a}^{x_b} \left\{ A_{xx} \frac{d\delta u}{dx} \left[\frac{du}{dx} + \frac{1}{2} \left(\frac{dw}{dx} \right)^2 \right] - f \delta u \right\} dx$$
$$- Q_1^e \, \delta u(x_a) - Q_4^e \, \delta u(x_b) \tag{5.3.18}$$

$$0 = \int_{x_a}^{x_b} \frac{d\delta w}{dx} \left\{ S_{xx} \left(\frac{dw}{dx} + \phi_x \right) + A_{xx} \frac{dw}{dx} \left[\frac{du}{dx} + \frac{1}{2} \left(\frac{dw}{dx} \right)^2 \right] \right\} dx$$
$$- \int_{x_a}^{x_b} \delta w q \, dx - Q_2^e \, \delta w(x_a) - Q_5^e \, \delta w(x_b) \tag{5.3.19}$$

$$0 = \int_{x_a}^{x_b} \left[D_{xx} \frac{d\delta\phi_x}{dx} \frac{d\phi_x}{dx} + S_{xx} \, \delta\phi_x \left(\frac{dw}{dx} + \phi_x \right) \right] dx$$
$$- Q_3^e \, \delta\phi_x(x_a) - Q_6^e \, \delta\phi_x(x_b) \tag{5.3.20}$$

where δu, δw, and $\delta\phi_x$ are the virtual displacements, which take the roles of the weight functions in the weak-form Galerkin model (or Ritz model). The Q_i^e ($i = 1, 2, \ldots, 6$) have the same physical meaning as in the Euler–Bernoulli beam element, and their relationships to the horizontal displacement u, transverse deflection w, and rotation ϕ_x, are

$$Q_1^e = -N_{xx}(x_a), \qquad\qquad Q_4^e = N_{xx}(x_b)$$

$$Q_2^e = -\left[Q_x + N_{xx}\frac{dw}{dx}\right]_{x=x_a}, \quad Q_5^e = \left[Q_x + N_{xx}\frac{dw}{dx}\right]_{x=x_b} \qquad (5.3.21)$$

$$Q_3^e = -M_{xx}(x_a), \qquad\qquad Q_6^e = M_{xx}(x_b)$$

An examination of the virtual work statements in Eqs. (5.3.18)–(5.3.20) suggests that $u(x)$, $w(x)$, and $\phi_x(x)$ are the primary variables and, therefore, must be carried as nodal degrees of freedom. In general, u, w, and ϕ_x need not be approximated by polynomials of the same degree. However, the approximations should be such that possible deformation modes (i.e. kinematics) are represented correctly. We will return to this point shortly.

Suppose that the displacements are approximated as

$$u(x) = \sum_{j=1}^{m} u_j^e \psi_j^{(1)}, \quad w(x) = \sum_{j=1}^{n} w_j^e \psi_j^{(2)}, \quad \phi_x(x) = \sum_{j=1}^{p} s_j^e \psi_j^{(3)} \qquad (5.3.22)$$

where $\psi_j^{(\alpha)}(x)$ ($\alpha = 1, 2, 3$) are Lagrange interpolation functions of degree $(m-1)$, $(n-1)$, and $(p-1)$, respectively. At the moment, the values of m, n, and p are arbitrary, that is arbitrary degree of polynomial approximations of u, w, and ϕ_x may be used. Substitution of Eq. (5.3.22) for u, w, and ϕ_x, and $\delta u = \psi_i^{(1)}$, $\delta w = \psi_i^{(2)}$, and $\delta\phi_x = \psi_i^{(3)}$ into Eqs. (5.3.18)–(5.3.20) yields the finite element model

$$0 = \sum_{j=1}^{m} K_{ij}^{11} u_j^e + \sum_{j=1}^{n} K_{ij}^{12} w_j^e + \sum_{j=1}^{p} K_{ij}^{13} s_j^e - F_i^1 \qquad (5.3.23)$$

$$0 = \sum_{j=1}^{m} K_{ij}^{21} u_j^e + \sum_{j=1}^{n} K_{ij}^{22} w_j^e + \sum_{j=1}^{p} K_{ij}^{23} s_j^e - F_i^2 \qquad (5.3.24)$$

$$0 = \sum_{j=1}^{m} K_{ij}^{31} u_j^e + \sum_{j=1}^{n} K_{ij}^{32} w_j^e + \sum_{j=1}^{p} K_{ij}^{33} s_j^e - F_i^3 \qquad (5.3.25)$$

The stiffness and force coefficients are

$$K_{ij}^{11} = \int_{x_a}^{x_b} A_{xx} \frac{d\psi_i^{(1)}}{dx} \frac{d\psi_j^{(1)}}{dx}\, dx, \quad K_{ij}^{12} = \frac{1}{2}\int_{x_a}^{x_b} A_{xx} \frac{dw}{dx} \frac{d\psi_i^{(1)}}{dx} \frac{d\psi_j^{(2)}}{dx}\, dx$$

$$K_{ij}^{21} = \int_{x_a}^{x_b} A_{xx} \frac{dw}{dx} \frac{d\psi_i^{(2)}}{dx} \frac{d\psi_j^{(1)}}{dx}\, dx, \quad K_{ij}^{13} = 0, \quad K_{ij}^{31} = 0$$

$$K_{ij}^{22} = \int_{x_a}^{x_b} S_{xx} \frac{d\psi_i^{(2)}}{dx} \frac{d\psi_j^{(2)}}{dx}\, dx + \frac{1}{2} \int_{x_a}^{x_b} A_{xx} \left(\frac{dw}{dx}\right)^2 \frac{d\psi_i^{(2)}}{dx} \frac{d\psi_j^{(2)}}{dx}\, dx$$

$$K_{ij}^{23} = \int_{x_a}^{x_b} S_{xx} \frac{d\psi_i^{(2)}}{dx} \psi_j^{(3)}\, dx = K_{ji}^{32} \tag{5.3.26}$$

$$K_{ij}^{33} = \int_{x_a}^{x_b} \left(D_{xx} \frac{d\psi_i^{(3)}}{dx} \frac{d\psi_j^{(3)}}{dx} + S_{xx}\psi_i^{(3)}\psi_j^{(3)}\right) dx$$

$$F_i^1 = \int_{x_a}^{x_b} \psi_i^{(1)} f\, dx + Q_1^e \psi_i^{(1)}(x_a) + Q_4^e \psi_i^{(1)}(x_b)$$

$$F_i^2 = \int_{x_a}^{x_b} \psi_i^{(2)} q\, dx + Q_2^e \psi_i^{(2)}(x_a) + Q_5^e \psi_i^{(2)}(x_b)$$

$$F_i^3 = Q_3^e \psi_i^{(3)}(x_a) + Q_6^e \psi_i^{(3)}(x_b)$$

The element equations, Eqs. (5.3.23)–(5.3.25), can be expressed in matrix form as

$$\begin{bmatrix} \mathbf{K}^{11} & \mathbf{K}^{12} & \mathbf{K}^{13} \\ \mathbf{K}^{21} & \mathbf{K}^{22} & \mathbf{K}^{23} \\ \mathbf{K}^{31} & \mathbf{K}^{32} & \mathbf{K}^{33} \end{bmatrix} \begin{Bmatrix} \mathbf{u} \\ \mathbf{w} \\ \mathbf{s} \end{Bmatrix} = \begin{Bmatrix} \mathbf{F}^1 \\ \mathbf{F}^2 \\ \mathbf{F}^3 \end{Bmatrix} \tag{5.3.27}$$

The choice of the approximation functions $\psi_i^{(\alpha)}$ ($\alpha = 1, 2, 3$) dictates different finite element models. For the pure bending case (i.e. linear analysis), the use of linear polynomials to approximate both w and ϕ_x (i.e. $\psi_i^{(2)} = \psi_i^{(3)}$) is known to yield a stiffness matrix that is nearly singular and the phenomenon is known as *shear locking*. When $\psi_i^{(2)}$ and $\psi_i^{(3)}$ are higher-order (than linear), with $n \geq p$, shear locking disappears.

5.3.4   Shear and Membrane Locking

A number of Timoshenko beam finite elements for the linear case (i.e. without von Kármán nonlinearity) have appeared in the literature. They differ from each other in the choice of approximation functions used for the transverse deflection w and rotation ϕ_x, or in the variational form used to develop the finite element model. Some are based on equal interpolation and others on unequal interpolation of w and ϕ_x.

The Timoshenko beam finite element with linear interpolation of both w and ϕ_x is the simplest element. Linear interpolation of w means that the slope dw/dx is constant (see Fig. 5.3.2). In thin beam limit, that is, as the length-to-thickness ratio becomes large (say, 100), the slope should be equal to $-\phi_x$, which is also represented as linear as opposed to being a constant. On the other hand, a constant representation of ϕ_x results in zero bending energy while the transverse shear is nonzero. This inconsistency in the representation of the kinematics through linear approximation of both w and ϕ_x results in zero displacements and rotations, which trivially satisfy the Kirchhoff constraint $\phi_x = -dw/dx$, and the element is said to be very stiff in the thin-beam limit and experience shear locking.

To overcome the locking, one may use equal interpolation for both w and ϕ_x but treat ϕ_x as a constant in the evaluation of the shear strain, $\gamma_{xz} = (dw/dx) + \phi_x$. This is often realized by using selective integration, in which one-point (reduced) Gauss quadrature is used to evaluate the stiffness coefficients associated with the transverse shear strain, and all other coefficients of the stiffness matrix are evaluated using the exact (full) integration.

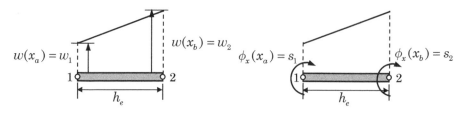

(a) Linear approximation of $w(x)$ (b) Linear approximation of $\phi_x(x)$

Fig. 5.3.2: Kinematics of deformation of the Timoshenko beam element when both (a) w and (b) ϕ_x are interpolated linearly.

Since applied distributed loads are represented as point loads in the finite element method, Eq. (5.3.16) with $q = 0$ and constant S_{xx} implies that

$$\text{(shear strain)} \quad \gamma_{xz}^0 \equiv \phi_x + \frac{dw}{dx} = \text{constant} \tag{5.3.28}$$

Similarly, for a problem that involves only bending deformation, the element should experience no stretching [see Eq. (5.3.15) with $f = 0$]:

$$\text{(membrane strain)} \quad \varepsilon_{xx}^0 \equiv \frac{du}{dx} + \frac{1}{2}\left(\frac{dw}{dx}\right)^2 = 0 \tag{5.3.29}$$

In order to satisfy the above constraints, we must have

$$\phi_x \sim \frac{dw}{dx} \tag{5.3.30}$$

$$\frac{du}{dx} \sim \left(\frac{dw}{dx}\right)^2 \tag{5.3.31}$$

The similarity is in the sense of having the same degree of polynomial variation. For example, when ϕ_x is linear and w is quadratic, the constraint in Eq. (5.3.28) is clearly met. Similarly, when both u and w are linear, the constraint in Eq. (5.3.29) is automatically met; however, when quadratic interpolation is used for both u and w, then du/dx is linear and $(dw/dx)^2$ is quadratic and the resulting element experiences the membrane locking. If u is interpolated using a cubic polynomial, w with a quadratic polynomial, and ϕ_x with a linear polynomial, then we have

$$\phi_x \text{ (linear)} \quad \sim \quad \frac{dw}{dx} \text{ (linear)} \tag{5.3.32}$$

$$\frac{du}{dx} \text{ (quadratic)} \quad \sim \quad \left(\frac{dw}{dx}\right)^2 \text{ (quadratic)} \tag{5.3.33}$$

In summary, the constraints in Eqs. (5.3.28) and (5.3.29) are satisfied for the following two choices of approximations of u, w, and ϕ_x:

1. u, ϕ_x, and w are all linear, with constant representation of γ_{xz}; the latter can be accomplished by using reduced integration to evaluate the shear stiffness coefficients.

2. u is cubic, w is quadratic, and ϕ_x is linear. However, this will result in a 9×9 stiffness matrix with different degrees of freedom at different nodes, making it difficult (but not impossible) to implement in a computer program.

Membrane locking can also be avoided, in addition to using appropriate interpolation of the variables, by using selective Gauss quadrature, as discussed in Section 5.2.7. In the present study, we shall use equal linear or quadratic approximation of the variables (u, w, ϕ_x) with reduced integration of all nonlinear terms (to avoid membrane locking) and shear terms, that is, stiffness coefficients involving S_{xx} (to avoid shear locking) of the stiffness matrix.

5.3.5 Tangent Stiffness Matrix

Returning to the nonlinear finite element model in Eqs. (5.3.23)–(5.3.25) of Section 5.3.3, we compute the tangent stiffness matrix of the Timoshenko beam element. Much of the computer implementation discussion presented in Section 5.2.8 is also valid for the Timoshenko beam element.

The submatrices $\mathbf{T}^{\alpha\beta}$ of the tangent matrix $\mathbf{T}$, in the same form as the stiffness matrix $\mathbf{K}$ in Eq. (5.3.27), are defined by

$$T_{ij}^{\alpha\beta} = K_{ij}^{\alpha\beta} + \sum_{\gamma=1}^{3}\sum_{k=1}^{n} \frac{\partial}{\partial\Delta_j^\beta}\left(K_{ik}^{\alpha\gamma}\right)\Delta_k^\gamma - \frac{\partial F_i^\alpha}{\partial\Delta_j^\beta} \tag{5.3.34}$$

In view of the fact that $\mathbf{F}^\alpha$ is independent of u, w, and ϕ_x, and $\mathbf{K}^{\alpha\beta}$ are only dependent on w, we have (when the same degree of interpolation is used for all variables)

$$\mathbf{T}^{\alpha 1} = \mathbf{K}^{\alpha 1}, \quad \mathbf{T}^{\alpha 3} = \mathbf{K}^{\alpha 3} \quad \text{for } \alpha = 1, 2, 3$$

$$T_{ij}^{12} = K_{ij}^{12} + \frac{1}{2} \int_{x_a}^{x_b} A_{xx} \frac{dw}{dx} \frac{d\psi_i}{dx} \frac{d\psi_j}{dx} dx = 2K_{ij}^{12}$$

$$T_{ij}^{22} = K_{ij}^{22} + \int_{x_a}^{x_b} A_{xx} \left[\frac{du}{dx} + \left(\frac{dw}{dx} \right)^2 \right] \frac{d\psi_i}{dx} \frac{d\psi_j}{dx} dx \qquad (5.3.35)$$

$$T_{ij}^{32} = K_{ij}^{32} + 0 = K_{ij}^{32}, \quad T_{ij}^{33} = K_{ij}^{33} + 0 = K_{ij}^{33}$$

where the direct stiffness coefficients $K_{ij}^{\alpha\beta}$ are defined in Eq. (5.3.26).

As discussed earlier, we must use reduced integration on all nonlinear stiffness coefficients as well as all shear stiffness coefficients, while the remaining stiffness coefficients are evaluated using full integration. For example, consider the following integral expression:

$$\int_{x_a}^{x_b} A_{xx} \left[\frac{du}{dx} \frac{d\psi_i}{dx} \frac{d\psi_j}{dx} + \frac{3}{2} \left(\frac{dw}{dx} \right)^2 \frac{d\psi_i}{dx} \frac{d\psi_j}{dx} \right] dx \qquad (5.3.36)$$

If quadratic approximation of both u and w is used (see Fig. 5.3.3) and A_{xx} is constant, then the first term in the integrand is a cubic polynomial while the second term is a fourth-order polynomial. Thus, exact evaluation of the first term requires two-point Gauss quadrature while the second term requires three-point Gauss quadrature. If we use two-point Gauss quadrature to evaluate T_{ij}^{22} (and K_{ij}^{22}), the first term in the coefficient of A_{xx} is integrated exactly while the second term is integrated approximately. This amounts to approximating $(dw/dx)^2$ as a linear polynomial while du/dx as a linear polynomial. Consequently, the constraint $\varepsilon_{xx}^0 = 0$ [see Eq. (5.3.29)] is satisfied.

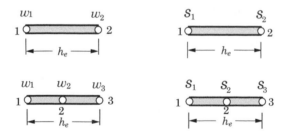

Fig. 5.3.3: Linear and quadratic Timoshenko beam finite elements.

Computer implementation of the Timoshenko beam element follows along the same lines as discussed for the Euler–Bernoulli beam element and, therefore, no additional discussion is presented here. The rearrangement of the elements of nodal displacement vector, as shown in Eq. (5.2.63), requires rearrangement of $K_{ij}^{\alpha\beta}$ in Eq. (5.3.27).

5.3.6 Numerical Examples

Example 5.3.1 ——————————————————————————

Consider the hinged–hinged beam of Example 5.2.1. Using the symmetry about $x = L/2$ and the Timoshenko beam element, investigate the convergence characteristics and effect of integration rule on the transverse deflection.

Solution: The geometric boundary conditions for the computational domain are

$$w(0) = u(0.5L) = \phi_x(0.5L) = 0 \tag{1}$$

The load is divided into load increments of equal size $\Delta q_0 = 1$ lb/in. A tolerance of $\epsilon = 10^{-3}$ is used in the analysis. The initial solution vector is chosen to be the zero vector, so that the first iteration solution corresponds to the linear problem. The exact linear solution is $[u(x) = 0]$

$$w(x) = \frac{q_0 L^4}{24 D_{xx}} \left(\frac{x}{L} - 2\frac{x^3}{L^3} + \frac{x^4}{L^4} \right) + \frac{q_0 L^2}{2 S_{xx}} \left(\frac{x}{L} - \frac{x^2}{L^2} \right)$$

$$\phi_x(x) = \frac{q_0 L^3}{24 D_{xx}} \left(1 - 6\frac{x^2}{L^2} + 4\frac{x^3}{L^3} \right) \tag{2}$$

The center deflection of the linear problem, for the data $q_0 = 1$ lb/in, $L = 100$ in, $E = 30 \times 10^6$ psi, $\nu = 0.3$, and $K_s = 5/6$, is

$$w(0.5L) = \frac{5q_0 L^4}{384 D_{xx}} + \frac{q_0 L^2}{8 S_{xx}} = 0.52083 + 0.013 \times 10^{-2} = 0.52096 \text{ in} \tag{3}$$

Thus, the effect of shear deformation on the deflection is not significant (differs from the Euler–Bernoulli beam solution by only 0.013%!). It should be noted that the reduced-integration (RI) Timoshenko beam element, in general, does not give exact nodal values even for the linear problem, unless a sufficiently large number of elements are used. For a refined mesh of linear elements or higher-order elements, one may expect the element to yield the exact linear solution. In the present case, the mesh of four quadratic elements (with 3×2 Gauss rule) gives the exact linear solution in Eq. (3).

For the mesh of four RI linear elements, the linear stiffness matrix and force vector are

$$\mathbf{K}^1 = 10^5 \begin{bmatrix} 24 & 0.000 & 0.000 & -24 & 0.000 & 0.000 \\ 0 & 7.692 & -48.077 & 0 & -7.692 & -48.077 \\ 0 & -48.077 & 302.480 & 0 & 48.077 & 298.480 \\ -24 & 0.000 & 0.000 & 24 & 0.000 & 0.000 \\ 0 & -7.692 & 48.077 & 0 & 7.692 & 48.077 \\ 0 & -48.077 & 298.480 & 0 & 48.077 & 302.480 \end{bmatrix}, \quad \mathbf{F}^1 = \begin{Bmatrix} 0.00 \\ 6.25 \\ 0.00 \\ 0.00 \\ 6.25 \\ 0.00 \end{Bmatrix} \tag{4}$$

The linear solution (i.e. the solution after the first iteration) for the specified boundary conditions $U_2 = 0$, $U_{13} = 0$, and $U_{15} = 0$ is

$$\begin{Bmatrix} U_1 \\ U_3 \\ U_4 \\ U_5 \\ U_6 \\ U_7 \\ U_8 \\ U_9 \\ U_{10} \\ U_{11} \\ U_{12} \\ U_{14} \end{Bmatrix} = \begin{Bmatrix} 0.00000 \\ -0.01641 \\ 0.00000 \\ 0.19659 \\ -0.01504 \\ 0.00000 \\ 0.36143 \\ -0.01133 \\ 0.00000 \\ 0.47009 \\ -0.00605 \\ 0.50794 \end{Bmatrix} \tag{5}$$

The element 1 tangent stiffness matrix $\mathbf{T}^1$ and residual vector $-\mathbf{R}^1$ at the beginning of the second iteration are

$$
\mathbf{T}^1 = 10^5 \begin{bmatrix}
24.000 & 0.377 & 0.000 & -24.000 & 0.377 & 0.000 \\
0.377 & 7.701 & -48.077 & 0.377 & -7.701 & -48.077 \\
0.000 & -48.077 & 302.480 & 0.000 & 48.077 & 298.480 \\
-24.000 & 0.377 & 0.000 & 24.000 & 0.377 & 0.000 \\
0.377 & -7.701 & 48.077 & 0.377 & 7.701 & 48.077 \\
0.000 & -48.077 & 298.480 & 0.000 & 48.077 & 302.480
\end{bmatrix}, \quad -\mathbf{R}^1 = 10^3 \begin{Bmatrix}
3.710 \\
0.108 \\
0.000 \\
-3.710 \\
-0.096 \\
-0.547
\end{Bmatrix}
$$

$$\tag{6}$$

The information provided in Eqs. (4)–(6) should help the reader in checking their computer program.

Table 5.3.1 contains the results obtained with various meshes in half beam and the Newton iteration scheme. Recall that NGP denotes the number of integration points used for the evaluation of all terms except the shear and nonlinear terms and LGP is the number of integration points used for the shear and nonlinear terms; FI denotes full integration (NGP = LGP = 2 for linear elements and NGP = LGP = 3 for quadratic elements) and RI denotes reduced/selective integration (NGP = 2 and LGP = 1 for linear elements and NGP = 3 and LGP = 2 for quadratic elements). Convergence is achieved in three or less number of iterations. It is clear that the quadratic elements are not as sensitive as the linear elements to locking and reduced integration is a must for linear elements. Also, the effect of locking on the solution becomes less with refined meshes. The convergence of the solution with mesh refinement is also clear.

Table 5.3.1: Transverse deflections of a *hinged–hinged* beam under uniform transverse load according to the Timoshenko beam theory (TBT).

Load q_0	Integration rule	4L	2Q	8L	4Q	16L	8Q
1.0	RI	0.5079	0.5210	0.5177	0.5210	0.5201	0.5210
	FI	0.0101	0.4943	0.0384	0.5150	0.1260	0.5198
2.0	RI	1.0159	1.0419	1.0354	1.0419	1.0403	1.0419
	FI	0.0201	0.9817	0.0768	1.0294	0.2521	1.0395
3.0	RI	1.5238	1.5629	1.5531	1.5629	1.5604	1.5629
	FI	0.0302	1.4560	0.1152	1.5429	0.3781	1.5592
4.0	RI	2.0318	2.0838	2.0708	2.0838	2.0806	2.0838
	FI	0.0403	1.9130	0.1537	2.0549	0.5042	2.0788
5.0	RI	2.5397	2.6048	2.5885	2.6048	2.6007	2.6048
	FI	0.0504	2.3502	0.1921	2.5656	0.6302	2.5983
10.0	RI	5.0794	5.2096	5.1770	5.2096	5.2015	5.2096
	FI	0.1007	4.2312	0.3841	5.0731	1.2604	5.1927

4L = 4 linear elements, 2Q = 2 quadratic elements, etc., in half beam.

Example 5.3.2

Consider the pinned–pinned beam of Example 5.2.2. Using the symmetry about $x = L/2$ and the Timoshenko beam element, investigate the convergence characteristics and effect of integration rule on the transverse deflection at the center of the beam.

Solution: The geometric boundary conditions for the computational domain of the problem are $u(0) = w(0) = u(0.5L) = \phi_x(0.5L) = 0$. Load increments of $\Delta q_0 = 1.0$ lb/in and a tolerance of $\epsilon = 10^{-3}$ are used in the analysis. The initial solution vector is chosen to be $\mathbf{u}^0 = \mathbf{0}$. Table 5.3.2 contains the results for pinned–pinned beam; the results were obtained using the Newton iteration (NI) scheme with $\beta = 0$ and the direct iteration (DI) scheme with $\beta = 0.25$ (and with reduced integration). The direct iteration scheme would not converge for $\beta = 0$. Convergence of the solutions with mesh refinement and higher-order elements is apparent from the results. Also, higher-order elements and refined meshes are less sensitive to shear locking.

Table 5.3.2: Transverse deflections as a function of load for a *pinned–pinned* beam under uniformly distributed transverse load (TBT).

Load q_0	4L-DI	4L-NI	2Q-DI	2Q-NI	8L-NI	4Q-NI
1.0	0.3654 (9)	0.3654 (5)	0.3687 (9)	0.3687 (5)	0.3677 (5)	0.3685 (5)
2.0	0.5435 (8)	0.5439 (4)	0.5458 (11)	0.5458 (4)	0.5451 (4)	0.5454 (4)
3.0	0.6633 (10)	0.6637 (3)	0.6641 (10)	0.6644 (3)	0.6639 (3)	0.6640 (3)
4.0	0.7564 (11)	0.7562 (3)	0.7562 (11)	0.7560 (3)	0.7557 (3)	0.7555 (3)
5.0	0.8356 (6)	0.8327 (3)	0.8317 (12)	0.8318 (3)	0.8316 (3)	0.8312 (3)
6.0	0.8986 (12)	0.8985 (3)	0.8972 (12)	0.8970 (3)	0.8969 (3)	0.8964 (3)
7.0	0.9566 (13)	0.9567 (3)	0.9545 (13)	0.9546 (3)	0.9546 (3)	0.9540 (3)
8.0	1.0088 (13)	1.0090 (3)	1.0077 (8)	1.0065 (3)	1.0066 (3)	1.0058 (3)
9.0	1.0564 (13)	1.0568 (3)	1.0540 (14)	1.0538 (3)	1.0540 (3)	1.0531 (3)
10.0	1.1000 (9)	1.1009 (3)	1.0969 (9)	1.0975 (3)	1.0977 (3)	1.0967 (3)

Table 5.3.3 contains deflection parameter $\bar{w} = w(L/2)(EH^3/L^4)$ of pinned–pinned beams, obtained using four quadratic elements with the 3×2 Gauss rule and the Newton iteration scheme for various length-to-thickness ratios, L/H. The effect of shear deformation is clear from the results; the thicker the beams, the larger the shear strains and deflections $\bar{w}$. Of course, the maximum deflection $w(L/2)$ will be smaller with a smaller L/H ratio (or thicker beams), because thicker beams have larger stiffness. Therefore, they also exhibit less geometric nonlinearity, as can be seen from Fig. 5.3.4, where the deflection $\bar{w}$ is plotted as a function of the intensity of the transverse load q_0 for various values of L/H ratio. The deflections predicted by the TBT are the same as those predicted by the EBT when $L/H \geq 25$, indicating that the effect of shear deformation is negligible for $L/H \geq 25$.

Table 5.3.3: The effect of L/H ratio on the deflection parameter $\bar{w} = w(L/2)EH^3/L^4$ of a *pinned–pinned* beam under uniformly distributed load (4Q element mesh; TBT).

Load	Length-to-thickness ratio, L/H					
q_0	10	20	25	50	80	100
1.0	0.160 (2)	0.157 (2)	0.157 (2)	0.156 (3)	0.141 (4)	0.111 (5)
2.0	0.320 (2)	0.314 (2)	0.314 (2)	0.309 (3)	0.237 (4)	0.164 (4)
3.0	0.480 (2)	0.472 (2)	0.471 (2)	0.457 (3)	0.306 (3)	0.199 (3)
4.0	0.641 (2)	0.629 (2)	0.627 (2)	0.598 (3)	0.360 (3)	0.227 (3)
5.0	0.801 (2)	0.786 (2)	0.784 (2)	0.731 (3)	0.404 (3)	0.249 (3)
6.0	0.961 (2)	0.943 (2)	0.941 (2)	0.856 (3)	0.443 (3)	0.269 (3)
7.0	1.121 (2)	1.100 (2)	1.097 (2)	0.973 (3)	0.476 (3)	0.286 (3)
8.0	1.281 (2)	1.258 (2)	1.254 (2)	1.083 (3)	0.506 (3)	0.302 (3)
9.0	1.441 (2)	1.415 (2)	1.410 (2)	1.187 (3)	0.534 (3)	0.316 (3)
10.0	1.601 (2)	1.572 (2)	1.567 (2)	1.284 (3)	0.559 (3)	0.329 (3)

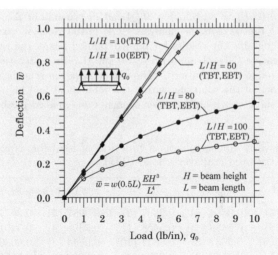

Fig. 5.3.4: Load versus deflection curves for pinned–pinned beam.

Example 5.3.3

Consider the *clamped–clamped* beam of Example 5.2.2. Using the symmetry about $x = L/2$ and the Timoshenko beam element, investigate the convergence characteristics, effect of shear deformation, and effect of integration rule on the transverse deflection.

Solution: The geometric boundary conditions for the computational domain of the problem are $u(0) = w(0) = \phi_x(0) = u(L/2) = \phi_x(L/2) = 0$. The data used are the same as that used in Example 5.3.2. Table 5.3.4 contains the nonlinear analysis results for the clamped–clamped beam; the results were obtained with the direct iteration scheme as well as the Newton iteration scheme. The results obtained with both schemes and meshes are virtually the same, but the direct iteration scheme takes more iterations than the Newton iteration scheme to converge. Table 5.3.5 contains the strains and stresses calculated after convergence at load $q_0 = 10$ lb/in.

Table 5.3.4: Finite element results for the deflections $w(L/2)$ of a *clamped–clamped* beam under a uniformly distributed load (TBT).

Load	Direct iteration		Newton's iteration		
	8L	4Q	8L	4Q	4Q
q_0	$a/h = 100$	$a/h = 100$	$a/h = 100$	$a/h = 100$	$a/h = 10^\dagger$
1.0	0.1019 (3)*	0.1035 (3)	0.1019 (3)	0.1035 (3)	0.1172 (2)
2.0	0.1997 (4)	0.2025 (4)	0.1997 (3)	0.2025 (3)	0.2343 (2)
3.0	0.2906 (4)	0.2943 (4)	0.2906 (3)	0.2943 (3)	0.3515 (2)
4.0	0.3738 (5)	0.3778 (5)	0.3738 (3)	0.3778 (3)	0.4687 (2)
5.0	0.4493 (5)	0.4535 (5)	0.4493 (3)	0.4535 (3)	0.5858 (2)
6.0	0.5178 (6)	0.5219 (6)	0.5179 (3)	0.5220 (3)	0.7030 (2)
7.0	0.5806 (7)	0.5846 (7)	0.5805 (3)	0.5845 (3)	0.8202 (2)
8.0	0.6379 (8)	0.6417 (8)	0.6380 (3)	0.6418 (3)	0.9373 (2)
9.0	0.6908 (8)	0.6947 (9)	0.6911 (3)	0.6946 (3)	1.0545 (2)
10.0	0.7406 (9)	0.7434 (10)	0.7403 (3)	0.7436 (3)	1.1717 (2)

* Number of iterations taken to converge. † The values are multiplied by 10^3.

Table 5.3.5: Strains and stresses in a clamped–clamped beam subjected to a uniformly distributed load, $q_0 = 10$ lb/in (TBT).

x	$\varepsilon_{xx}^0 \times 10^4$	$\varepsilon_{xx}^1 \times 10^2$	$\sigma_{xx}^0 \times 10^{-4}$	$\sigma_{xx}^1 \times 10^{-5}$	$\gamma_{xz}^0 \times 10^4$	$\sigma_{xz}^0 \times 10^{-3}$
2.64	1.348	−0.2184	0.4044	−0.6551	0.4658	0.5374
9.86	2.493	−0.1063	0.7479	−0.3190	0.3415	0.3940
15.14	2.208	−0.0450	0.6625	−0.1350	0.2707	0.3123
22.36	1.174	0.0192	0.3521	0.0575	0.1917	0.2212
27.64	−0.067	0.0530	−0.0201	0.1589	0.1455	0.1678
34.86	−0.859	0.0859	−0.2577	0.2576	0.0915	0.1056
40.14	−0.739	0.1008	−0.2218	0.3025	0.0578	0.0667
47.36	0.044	0.1109	0.0133	0.3327	0.0149	0.0172

Figure 5.3.5 contains plots of the deflection parameter, $\bar{w} = w(L/2)EH^3/L^4$, against q_0 for the EBT and TBT. The results were obtained using the direct iteration scheme, and four quadratic TBT elements and eight EBT elements. Note that, for $L/H = 10$, the beam does not exhibit nonlinearity but the effect of shear deformation is clear; it increases the deflection compared to the Euler–Bernoulli beam theory. For $L/H = 100$, the effect of shear deformation is negligible, and both theories predict essentially the same load-deflection response.

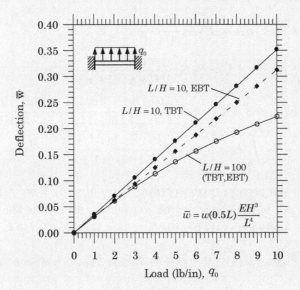

Fig. 5.3.5: Load–deflection response predicted by the EBT and TBT for clamped–clamped, thin $(L/H = 100)$ and thick $(L/H = 10)$ beams.

This completes the discussion of nonlinear bending analysis of homogeneous Timoshenko beams. In the next section, we consider Timoshenko beams with inhomogeneous material variation through the beam height. Such beams are called *functionally graded material* (FGM) beams (see Reddy [53]).

5.3.7 Functionally Graded Material Beams

5.3.7.1 Material variation and stiffness coefficients

Consider a rectangular cross-section beam made of two homogeneous isotropic materials that are combined according to the following power-law distribution:

$$E(x, z) = [E_1(x) - E_2(x)] f(z) + E_2(x), \quad f(z) = \left(\frac{1}{2} + \frac{z}{H} \right)^n \quad (5.3.37)$$

where E_1 and E_2 are moduli of the two materials that are being combined, and H is the total height (thickness) of the beam. Equation (5.3.37) describes a beam whose top surface is 100% material 1 while bottom is 100% material 2, and n is the power-law index that dictates the volume fraction $f(z)$ of material 1 [and $1 - f(z)$ is the volume fraction of material 2], as shown in Fig. 5.3.6. When $n = 0$, the beam is a homogeneous beam made of material 1 and when n is large, then the beam is a homogeneous beam made of material 2.

The beam stiffnesses (A_{xx}, B_{xx}, D_{xx}) of Eq. (5.2.21) and S_{xx} of Eq. (5.3.14) are given by (Poisson's ratio is assumed to be constant)

$$A_{xx} = \int_A E(x, z)\, dA = E_2 A_0 \frac{M + n}{1 + n}$$

$$B_{xx} = \int_A E(x, z) z\, dA = E_2 B_0 \frac{n(M - 1)}{2(1 + n)(2 + n)} \quad (5.3.38)$$

$$D_{xx} = \int_A E(x, z) z^2\, dA = E_2 I_0 \frac{(6 + 3n + 3n^2)M + (8n + 3n^2 + n^3)}{6 + 11n + 6n^2 + n^3}$$

$$S_{xx} = \frac{K_s}{2(1 + \nu)} \int_A E(x, z)\, dA = \frac{K_s E_2 A_0}{2(1 + \nu)} \frac{M + n}{1 + n}$$

where $A_0 = bh$ is the area of cross section, $B_0 = bh^2$, $I_0 = bh^3/12$ is the second moment of area, K_s is the shear correction factor, and M is the modulus ratio $M = E_1/E_2$. For homogeneous beams (i.e. $n = 0$ or $E_1 = E_2 = E$), we have

$$A_{xx} = EA_0, \quad B_{xx} = 0, \quad D_{xx} = EI_0, \quad S_{xx} = K_s GA_0 \quad (5.3.39)$$

and the stress resultants (N_{xx}, M_{xx}, Q_x) are defined by Eq. (5.3.13).

Figure 5.3.7 contains the variation of the non-dimensional axial stiffness $\bar{A}_{xx} = A_{xx}/E_2 A_0$ and bending stiffness $\bar{D}_{xx} = D_{xx}/E_2 I_0$ as functions of the power-law index n for various values of the modulus ratio $M = E_1/E_2 \geq 1$ and Fig. 5.3.8 contains similar plots for the non-dimensional axial-bending coupling stiffness $\bar{B}_{xx} = B_{xx}/E_2 B_0$. It is clear that $\bar{A}_{xx}$ is the maximum at $n = 0$ and decreases with increasing value of n; $\bar{B}_{xx}$ is zero at $n = 0$, increases to a maximum at $n = \sqrt{2}$, and then decreases with increasing value of n; $\bar{D}_{xx}$ is the maximum at $n = 0$ and decreases with increasing value of n.

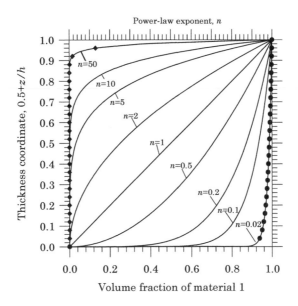

Fig. 5.3.6: Volume fraction of material 1 as a function of the power-law index n.

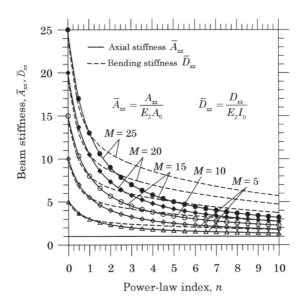

Fig. 5.3.7: Variation of $\bar{A}_{xx}$ and $\bar{D}_{xx}$ as a function of the power-law index n.

5.3.7.2 Equations of equilibrium

The equations of equilibrium expressed in terms of the stress resultants (N_{xx}, M_{xx}, Q_x) remain the same as those in Eqs. (5.3.8)–(5.3.10). The equations of equilibrium of the TBT for an *FGM beam*, when expressed in terms of the

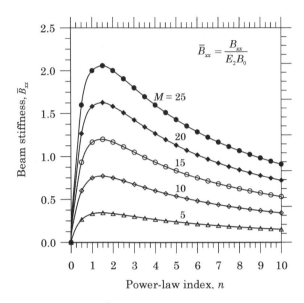

Fig. 5.3.8: Variation of $\bar{B}_{xx}$ as a function of the power-law index n.

generalized displacements, take the form [make use of Eqs. (5.3.8)–(5.3.10) and (5.3.13)]

$$-\frac{d}{dx}\left\{A_{xx}\left[\frac{du}{dx}+\frac{1}{2}\left(\frac{dw}{dx}\right)^2\right]+B_{xx}\frac{d\phi_x}{dx}\right\}=f \quad (5.3.40)$$

$$-\frac{d}{dx}\left\{A_{xx}\frac{dw}{dx}\left[\frac{du}{dx}+\frac{1}{2}\left(\frac{dw}{dx}\right)^2\right]+B_{xx}\frac{dw}{dx}\frac{d\phi_x}{dx}\right\}$$
$$-\frac{d}{dx}\left[S_{xx}\left(\frac{dw}{dx}+\phi_x\right)\right]=q \quad (5.3.41)$$

$$-\frac{d}{dx}\left\{D_{xx}\frac{d\phi_x}{dx}+B_{xx}\left[\frac{du}{dx}+\frac{1}{2}\left(\frac{dw}{dx}\right)^2\right]\right\}+S_{xx}\left(\frac{dw}{dx}+\phi_x\right)=0 \quad (5.3.42)$$

In contrast to Eqs. (5.3.15)–(5.3.17), the dependent unknowns (u, w, ϕ_x) for an FGM beam are coupled even when the geometric nonlinearity is not accounted for. This is due to the fact that the stiffness coefficient B_{xx} couples extensional deformation to bending deformation.

5.3.7.3   Finite element model

The finite element model of the through-thickness functionally graded Timoshenko beam equations can be constructed in the same way as in Section 5.3.3.

The weak forms associated with Eqs. (5.3.40)–(5.3.42) are

$$
0 = \int_{x_a}^{x_b} \left\{ A_{xx} \frac{d\delta u}{dx} \left[\frac{du}{dx} + \frac{1}{2} \left(\frac{dw}{dx} \right)^2 \right] + B_{xx} \frac{d\delta u}{dx} \frac{d\phi_x}{dx} - f \delta u \right\} dx
$$
$$
- Q_1^e \, \delta u(x_a) - Q_4^e \, \delta u(x_b) \tag{5.3.43}
$$

$$
0 = \int_{x_a}^{x_b} \frac{d\delta w}{dx} \left\{ S_{xx} \left(\frac{dw}{dx} + \phi_x \right) + A_{xx} \frac{dw}{dx} \left[\frac{du}{dx} + \frac{1}{2} \left(\frac{dw}{dx} \right)^2 \right] \right.
$$
$$
\left. + B_{xx} \frac{dw}{dx} \frac{d\phi_x}{dx} \right\} dx - \int_{x_a}^{x_b} \delta w \, q \, dx
$$
$$
- Q_2^e \, \delta w(x_a) - Q_5^e \, \delta w(x_b) \tag{5.3.44}
$$

$$
0 = \int_{x_a}^{x_b} \left\{ D_{xx} \frac{d\delta\phi_x}{dx} \frac{d\phi_x}{dx} + B_{xx} \frac{d\delta\phi_x}{dx} \left[\frac{du}{dx} + \frac{1}{2} \left(\frac{dw}{dx} \right)^2 \right] \right.
$$
$$
\left. + S_{xx} \delta\phi_x \left(\frac{dw}{dx} + \phi_x \right) \right\} dx - Q_3^e \, \delta\phi_x(x_a) - Q_6^e \, \delta\phi_x(x_b) \tag{5.3.45}
$$

where Q_i^e ($i = 1, 2, \ldots, 6$) have the same physical meaning as before and they are defined by Eq. (5.3.21).

Using the finite element approximations in Eq. (5.3.22), we obtain the finite element model in Eq. (5.3.27), and the only coefficients that are different from those listed in Eq. (5.3.26) are

$$
K_{ij}^{13} = \int_{x_a}^{x_b} B_{xx} \frac{d\psi_i^{(1)}}{dx} \frac{d\psi_j^{(3)}}{dx} \, dx = K_{ji}^{31}
$$

$$
K_{ij}^{23} = \int_{x_a}^{x_b} \left(S_{xx} \frac{d\psi_i^{(2)}}{dx} \psi_j^{(3)} + B_{xx} \frac{dw}{dx} \frac{d\psi_i^{(2)}}{dx} \frac{d\psi_j^{(3)}}{dx} \right) dx \tag{5.3.46}
$$

$$
K_{ij}^{32} = \int_{x_a}^{x_b} \left(S_{xx} \psi_i^{(3)} \frac{d\psi_j^{(2)}}{dx} + \frac{1}{2} B_{xx} \frac{dw}{dx} \frac{d\psi_i^{(3)}}{dx} \frac{d\psi_j^{(2)}}{dx} \right) dx
$$

The tangent stiffness coefficients also remain the same as in Eq. (5.3.35), except for the following coefficients:

$$
T_{ij}^{22} = K_{ij}^{22} + \int_{x_a}^{x_b} \left\{ A_{xx} \left[\frac{du}{dx} + \left(\frac{dw}{dx} \right)^2 \right] + B_{xx} \frac{d\phi_x}{dx} \right\} \frac{d\psi_i^{(2)}}{dx} \frac{d\psi_j^{(2)}}{dx} \, dx
$$

$$
T_{ij}^{32} = K_{ij}^{32} + \frac{1}{2} \int_{x_a}^{x_b} B_{xx} \frac{dw}{dx} \frac{d\psi_i^{(3)}}{dx} \frac{d\psi_j^{(2)}}{dx} \, dx = K_{ji}^{23} \tag{5.3.47}
$$

In the implementation of the element, we use the same degree of interpolation functions for all three variables. The computational aspects with regard to the shear locking and membrane locking remain the same (i.e. use reduced integration for the nonlinear and shear terms) as for the homogeneous Timoshenko beam finite element.

Example 5.3.4

Consider a functionally graded beam of length $L = 100$ in, height $H = 1$ in, and width $B = 1$ in. Suppose that the beam is fixed at the left end and pinned at the right end, and subjected to uniformly distributed load of intensity q_0. The FGM beam is made of two materials with the following values:

$$E^{\text{top}} = E_1 = 30 \times 10^6 \text{ psi}, \quad E^{\text{bot}} = E_2 = 10 \times 10^6 \text{ psi}, \quad \nu = 0.3, \quad K_s = \frac{5}{6}$$

Using a uniform mesh of eight linear Timoshenko beam elements, investigate the parametric effect of the power-law index, n, on the transverse deflection. In particular, plot the deflection $\bar{w} = w(0.5L)E_1BH^3/L^4 = w(0.5L)E_1H^3/L^4$ versus q_0, with $\Delta q_0 = 0.25$, for three different values of $n = 0, 1$, and 10.

Solution: The geometric boundary conditions for the problem are

$$u(0) = w(0) = \phi_x(0) = 0, \quad u(L) = w(L) = 0$$

The program developed earlier for homogeneous beams can be modified for two-constituent functionally graded beams. Figure 5.3.9 shows the load-deflection curves for three values of the power-law index, $n = 0, 1$, and 10. The results were obtained with the Newton iteration scheme with a tolerance of $\epsilon = 10^{-3}$. The value $n = 0$ corresponds to the homogeneous beam made of material 1. For increasing values of n the beam material approaches that of material 2 (with smaller modulus) and, therefore, the beam experiences larger deflections.

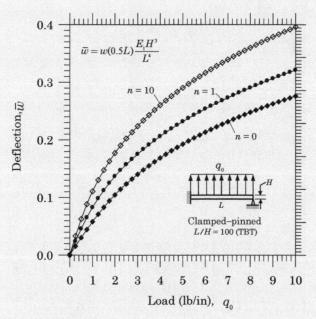

Fig. 5.3.9: Load versus dimensionless deflection of clamped–pinned FGM beams for three different values of the power-law index n.

5.4 Summary

Two different beam theories, namely, the Euler–Bernoulli beam theory and the Timoshenko beam theory, are discussed and their finite element models are developed. Geometric nonlinearity in the form of small strains and moderate rotations (i.e. the von Kármán nonlinearity) is included in simplifying the Green strain component $E_{xx} \approx \varepsilon_{xx}$. The principle of virtual displacements is used to derive the weak forms of the two beam theories. Displacement finite element models are developed using arbitrary and independent degree of approximations of the generalized displacements.

Linear approximation of both transverse deflection w and rotation ϕ_x in the Timoshenko beam finite element make it experience shear locking. Membrane locking occurs in both theories when inconsistent interpolation of the axial and transverse displacements is used. Both types of locking are eliminated by use of reduced integration rule in the numerical evaluation stiffness coefficients involving all nonlinear terms and all shear terms.

Computer implementation of the finite element models developed herein is discussed and numerical examples are presented to illustrate the effects of the integration rule, the degree of approximation, and the number of elements on the accuracy of the results. Postcomputation of stresses is also discussed.

The ideas presented in this chapter for straight beams can be extended to study axisymmetric bending of circular plates (see Problems 5.6 and 5.7). In fact, the computer program developed for straight beams can be readily modified for axisymmetric bending of circular plates.

Problems

WEAK FORMS AND FINITE ELEMENT MODELS

5.1 Give the expressions for the stiffness coefficients $\mathbf{K}^{\alpha\beta}$ $(\alpha,\beta = 1,2)$ for the Euler–Bernoulli beam theory for the case in which $B_{xx} \neq 0$.

5.2 Give the expressions for the tangent matrix coefficients $\mathbf{T}^{\alpha\beta}$ $(\alpha,\beta = 1,2)$ for the Euler–Bernoulli beam theory for the case in which $B_{xx} \neq 0$.

5.3 Consider the following pair of nonlinear equations:

$$-\frac{dN_{xx}}{dx} = f(x) \qquad (1)$$

$$\frac{d^2}{dx^2}\left(D_{xx}\frac{d^2w}{dx^2}\right) - \frac{d}{dx}\left(N_{xx}\frac{dw}{dx}\right) = q(x) \qquad (2)$$

where $q(x)$ is the distributed transverse force (positive upward), and

$$N_{xx} = A_{xx}\left[\frac{du}{dx} + \frac{1}{2}\left(\frac{dw}{dx}\right)^2\right] \qquad (3)$$

Rewrite the equations, by introducing the bending moment $M(x)$ as a dependent variable, as a set of second-order equations

$$-\frac{d}{dx}\left\{A_{xx}\left[\frac{du}{dx} + \frac{1}{2}\left(\frac{dw}{dx}\right)^2\right]\right\} - f(x) = 0 \qquad (4)$$

$$-\frac{d^2w}{dx^2} - \frac{M_{xx}}{D_{xx}} = 0 \tag{5}$$

$$-\frac{d^2M_{xx}}{dx^2} - \frac{d}{dx}\left\{\frac{dw}{dx}A_{xx}\left[\frac{du}{dx} + \frac{1}{2}\left(\frac{dw}{dx}\right)^2\right]\right\} = q(x) \tag{6}$$

Develop: (a) the weak form and (b) the finite element model using interpolation of the form

$$u = \sum_{j=1}^{m} u_j^e\psi_j^{(1)}(x), \quad w = \sum_{j=1}^{n} w_j^e\psi_j^{(2)}(x), \quad M_{xx} = \sum_{j=1}^{r} M_j^e\psi_j^{(3)}(x) \tag{7}$$

5.4 Consider the problem of bending of beams according to the Euler–Bernoulli beam theory. The principle of minimum total potential energy states that if the beam is in equilibrium, then the total potential energy associated with the equilibrium configuration is the minimum; that is, the equilibrium displacements are those which make the total potential energy a minimum. Thus, solving the equations governing the equilibrium of the Euler–Bernoulli beam is equivalent to minimizing the total potential energy

$$\Pi(u,w) = \int_{x_a}^{x_b}\left\{\frac{A_{xx}}{2}\left[\frac{du}{dx} + \frac{1}{2}\left(\frac{dw}{dx}\right)^2\right]^2 + \frac{D_{xx}}{2}\left(\frac{d^2w}{dx^2}\right)^2\right\}dx$$

$$- \int_{x_a}^{x_b}(fu + qw)\,dx \tag{1}$$

where $A_{xx} = EA$ and $D_{xx} = EI$ are the extensional and bending stiffness coefficients, respectively. The necessary condition for the minimum of a functional is that its first variation be zero: $\delta\Pi = 0$, which yields the governing equations of equilibrium. As you know, the statement $\delta\Pi = 0$ is the same as the weak forms of the governing equations of the Euler–Bernoulli beam theory. The weak form requires Hermite cubic interpolation of the transverse deflection w. Now suppose that we wish to relax the continuity required of the interpolation used for $w(x)$ by introducing the relation

$$-\frac{dw}{dx} = \phi_x(x) \tag{2}$$

Then the total potential energy functional takes the form

$$\Pi(u,w,\phi) = \int_{x_a}^{x_b}\left\{\frac{A_{xx}}{2}\left[\frac{du}{dx} + \frac{1}{2}\left(\frac{dw}{dx}\right)^2\right]^2 + \frac{D_{xx}}{2}\left(\frac{d\phi_x}{dx}\right)^2\right\}dx$$

$$- \int_{x_a}^{x_b}(fu + qw)\,dx - \sum_{i=1}^{6}\Delta_i^e Q_i^e \tag{3}$$

Since the functional now contains only the first derivative of u, w, and ϕ_x, linear Lagrange (minimum) interpolation can be used. Thus the original problem is replaced with the following equivalent problem: minimize $\Pi(u,w,\phi_x)$ in Eq. (3) subjected to the constraint

$$\frac{dw}{dx} + \phi_x(x) = 0 \tag{4}$$

In the penalty method, the constraint is included in the functional to be minimized in a least-squares sense, that is construct Π_p such that

$$\Pi_p = \Pi + \frac{\gamma}{2}\int_{x_a}^{x_b}\left[\frac{dw}{dx} + \phi_x(x)\right]^2 dx \tag{5}$$

where γ is known as the penalty parameter. Develop the penalty function formulation of the constrained problem by deriving (a) the weak form and (b) the finite element model.

5.5 Develop the weak forms of the governing equations, Eqs. (5.3.15)–(5.3.17); make use of the definitions of Q_i in Eq. (5.3.21).

5.6 Beginning with the displacement field in axisymmetric bending of a circular plate

$$u_r = u(r) - z\frac{dw}{dr}, \quad u_\theta = 0, \quad u_z = w(r) \tag{1}$$

(a) develop the virtual work statement and (b) the finite element model for nonlinear bending of circular plates. The strain–displacement relations for small strains and moderate rotation case are given by (see Reddy [1])

$$\varepsilon_{rr} = \frac{\partial u_r}{\partial r} + \frac{1}{2}\left(\frac{\partial u_z}{\partial r}\right)^2, \quad \varepsilon_{\theta\theta} = \frac{u_r}{r} \tag{2}$$

and the (plane–stress) stress–strain relations are

$$\left\{\begin{matrix} \sigma_{rr} \\ \sigma_{\theta\theta} \end{matrix}\right\} = \frac{E}{1-\nu^2}\begin{bmatrix} 1 & \nu \\ \nu & 1 \end{bmatrix}\left\{\begin{matrix} \varepsilon_{rr} \\ \varepsilon_{\theta\theta} \end{matrix}\right\} \tag{3}$$

where E and ν are Young's modulus and Poisson's ratio, respectively.

5.7 The displacement field of circular plates according to the first-order shear deformation plate theory is

$$u_r = u(r) + z\phi_r(r), \quad u_\theta = 0, \quad u_z = w(r) \tag{1}$$

Develop (a) the virtual work statement and (b) finite element model accounting for the von Kármán nonlinear strains. See the statement of Problem 5.6 for additional information.

5.8 Develop the *mixed* finite element model of the following set of nonlinear equations among the variables (u, w, M) of the Euler–Bernoulli beam theory with the von Kármán strains:

$$-\frac{dN}{dx} = f \tag{1}$$

$$-\frac{d^2 M}{dx^2} - \frac{d}{dx}\left(N\frac{dw}{dx}\right) = q \tag{2}$$

$$\frac{M}{EI} + \frac{d^2 w}{dx^2} = 0 \tag{3}$$

where $f(x)$ and $q(x)$ are the distributed axial and transverse forces, respectively, EA and EI are the axial and bending stiffnesses, and N is known in terms of u and w as

$$N = EA\left[\frac{du}{dx} + \frac{1}{2}\left(\frac{dw}{dx}\right)^2\right] \tag{4}$$

Also, compute the tangent stiffness coefficients. Present the finite element model in the standard format [see, for example, Eq. (5.2.32)] and define all coefficients.

5.9 Develop a mixed least-squares finite element model of the equations in Problem 5.8.

5.10 Develop a least-squares finite element model of the following set of nonlinear equations that arise in connection with the Timoshenko beam theory (assume that EA, EI, and GAK_s are constant):

$$-\frac{dN}{dx} = f \tag{1}$$

$$-\frac{d}{dx}\left[GAK_s\left(\phi + \frac{dw}{dx}\right)\right] - \frac{d}{dx}\left(N\frac{dw}{dx}\right) = q \tag{2}$$

$$-\frac{d}{dx}\left(EI\frac{d\phi}{dx}\right) + GAK_s\left(\phi + \frac{dw}{dx}\right) = 0 \tag{3}$$

COMPUTATIONAL PROBLEMS (see Problem 5.15 for the development of a computer program to solve the problems of this chapter)

5.11 Analyze a propped cantilever beam (clamped at one end and pinned at the other end) under uniformly distributed load (of intensity, $q = q_0$) using the (a) direct iteration as well as the (b) Newton's iteration methods. Use eight EBT or eight linear TBT elements, with $E = 30 \times 10^6$ psi, ($\nu = 0.3$), $L = 100$ in, and $H = W = 1$ in. You may begin with a load of $q_0 = 0.25$ and increment the load until the total load reaches $q_0 = 10$. Use a convergence tolerance of $\epsilon = 10^{-3}$ and maximum number of iterations to be $ITMAX = 50$. Tabulate and plot the deflection $w(L/2)$ versus load q_0.

5.12 Analyze a beam (EA, EI, L) with one end fixed and the other end vertically supported by a linear elastic spring with spring constant k (lb/in), as shown in Fig. P5.12. The beam is subjected to uniformly distributed transverse load of intensity q_0 (lb/in). Use the Euler–Bernoulli beam element with Newton's iteration method. Use eight elements in the beam (1×1 in^2 cross section, $E = 30 \times 10^6$ psi, and $L = 100$ in) and $k = 0, 25$, and 250 (i.e. three different cases). Tabulate and plot the deflection $w(L)$ versus the load q_0, with maximum load of $q_0 = 10$ (lb/in), for two separate cases: (a) horizontal movement of the free end is not restricted, and (b) the horizontal movement of the free end is restricted to zero. A load increment of $\Delta q_0 = 0.5$ (lb/in) is suggested. Use a convergence tolerance of $\epsilon = 10^{-3}$ and the maximum number of iterations to be $ITMAX = 25$.

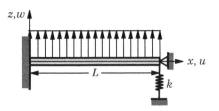

Fig. P5.12

5.13 Repeat Problem 5.12 with (a) eight linear elements and (b) four quadratic elements of the Timoshenko beam theory.

5.14 Determine the nonlinear response of a clamped (all around its circumference) circular plate of radius $a = 100$ in, thickness $h = 10$ in, modulus $E = 10^6$ psi, Poisson's ratio $\nu = 0.3$, and subjected to uniformly distributed load of intensity q_0. Use five quadratic elements, Newton's iteration scheme, error tolerance of $\epsilon = 10^{-3}$, maximum allowable number of iterations to be 10, and 20 load steps of $\Delta q_0 = 250$ lb to obtain the load-deflection response. Plot w/h versus load parameter $P = q_0 a^4 / E h^4 = q_0 \times 10^{-2}$.

5.15 Put together a complete finite element program based on the logic presented in Fig. 5.2.6 and Boxes 5.2.1–5.2.3 that has the capability of solving the Euler–Bernoulli beam and the Timoshenko beam problems of this chapter.

Two-Dimensional Problems Involving a Single Variable

I know that most men, including those at ease with problems of the greatest complexity, can seldom accept even the simplest and most obvious truth if it be such as would oblige them to admit the falsity of conclusions which they have delighted in explaining to colleagues, which they have proudly taught to others, and which they have woven, thread by thread, into the fabric of their lives. —— Leo Nikolayevich Tolstoy

6.1 Model Equation

The finite element analysis of nonlinear two-dimensional problems involves the same basic steps as those described for one-dimensional problems in Chapter 3. Finite element formulation of a model second-order equation was presented in Section 3.3. Here, we extend that development to problems in which the coefficients of the differential equation are possibly functions of the dependent variable and its derivatives.

Consider the problem of finding the solution $u(x, y)$ of the following second-order partial differential equation (a more general equation than the one considered in Section 3.3)

$$-\frac{\partial}{\partial x}\left(a_{xx}\frac{\partial u}{\partial x} + a_{xy}\frac{\partial u}{\partial y}\right) - \frac{\partial}{\partial y}\left(a_{yx}\frac{\partial u}{\partial x} + a_{yy}\frac{\partial u}{\partial y}\right) + a_{00}u = f(x, y) \ \text{ in } \ \Omega$$

$$(6.1.1)$$

where a_{xx}, a_{yy}, a_{xy}, a_{yx}, and a_{00} are known functions of position (x, y) as well as the dependent unknown u and its derivatives, and f is a known function of position in a two-dimensional domain Ω with boundary Γ [see Fig. 6.1.1(a)]. In general, $a_{xy} \neq a_{yx}$ and a_{00}, in most cases, is zero. For example, a_{xx} may be assumed to be of the form

$$a_{xx} = a_{xx}\left(x, y, u, \frac{\partial u}{\partial x}, \frac{\partial u}{\partial y}\right)$$

$$(6.1.2)$$

The same type of dependence of other coefficients on u, $\partial u/\partial x$, and $\partial u/\partial y$ is assumed. The equation is subject to certain boundary conditions, whose form will be apparent from the weak formulation.

J.N. Reddy, *An Introduction to Nonlinear Finite Element Analysis*, Second Edition. ©J.N. Reddy 2015. Published in 2015 by Oxford University Press.

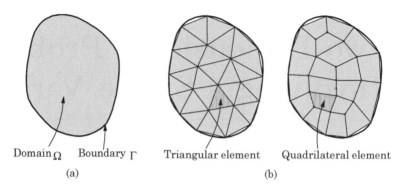

(a)　　　　　　　　　　　　　　　　　(b)

Fig. 6.1.1: (a) Actual domain and (b) finite element discretization of the domain into a mesh of triangles or quadrilaterals.

In the finite element method, the domain $\bar{\Omega}$ is discretized into a mesh of elements Ω^e, as shown in Fig. 6.1.1(b):

$$\bar{\Omega} \approx \bar{\Omega}_h = \bigcup_{e=1}^{N} \bar{\Omega}^e, \quad \bar{\Omega} = \Omega \cup \Gamma, \quad \bar{\Omega}^e = \Omega^e \cup \Gamma^e \tag{6.1.3}$$

where N is the total number of elements in the mesh. The residual due to the approximation of u by u_h over Ω^e is

$$R(u_h) = -\frac{\partial}{\partial x}\left(a_{xx}\frac{\partial u_h}{\partial x} + a_{xy}\frac{\partial u_h}{\partial y}\right) - \frac{\partial}{\partial y}\left(a_{yx}\frac{\partial u_h}{\partial x} + a_{yy}\frac{\partial u_h}{\partial y}\right) + a_{00}u_h - f(x,y)$$

6.2　Weak Form

We use a representative element Ω^e to derive the weak form of the model equation, Eq. (6.1.1). Following the steps of Section 3.3, the first step is to multiply the residual R with the ith weight function $w_i(x,y)$, which is assumed to be differentiable once with respect to x and y, and then set the integral of the product $w_i R$ over the element domain Ω^e to zero:

$$0 = \int_{\Omega^e} w_i\left[-\frac{\partial}{\partial x}\left(a_{xx}\frac{\partial u_h}{\partial x} + a_{xy}\frac{\partial u_h}{\partial y}\right) - \frac{\partial}{\partial y}\left(a_{yx}\frac{\partial u_h}{\partial x} + a_{yy}\frac{\partial u_h}{\partial y}\right) + a_{00}u_h - f\right] dx\, dy \tag{6.2.1}$$

In the second step, we distribute the differentiation among u_h and w_i equally in the first two terms of Eq. (6.2.1) using the Green–Gauss theorem:

$$0 = \int_{\Omega^e}\left[\frac{\partial w_i}{\partial x}\left(a_{xx}\frac{\partial u_h}{\partial x} + a_{xy}\frac{\partial u_h}{\partial y}\right) + \frac{\partial w_i}{\partial y}\left(a_{yx}\frac{\partial u_h}{\partial x} + a_{yy}\frac{\partial u_h}{\partial y}\right)\right.$$
$$\left. + a_{00}w_i u_h - w_i f\right] dx dy$$

$$-\oint_{\Gamma^e} w_i \left[\left(a_{xx}\frac{\partial u_h}{\partial x} + a_{xy}\frac{\partial u_h}{\partial y} \right) n_x + \left(a_{yx}\frac{\partial u_h}{\partial x} + a_{yy}\frac{\partial u_h}{\partial y} \right) n_y \right] ds \quad (6.2.2)$$

where n_x and n_y are the components (i.e. the direction cosines) of the unit normal vector

$$\hat{\mathbf{n}} = n_x\,\hat{\mathbf{i}} + n_y\,\hat{\mathbf{j}} = \cos\alpha\,\hat{\mathbf{i}} + \sin\alpha\,\hat{\mathbf{j}} \quad (6.2.3)$$

on the boundary Γ^e, and ds is the arc length of an infinitesimal line element along the boundary. Here α denotes the angle between the x-axis and the unit normal (see Fig. 6.2.1). The circle on the boundary integral denotes integration over the closed boundary Γ^e. From an inspection of the boundary term in Eq. (6.2.2), we note that u_h is the primary variable. The coefficient of the weight function w_i in the boundary expression is

$$\left(a_{xx}\frac{\partial u_h}{\partial x} + a_{xy}\frac{\partial u_h}{\partial y} \right) n_x + \left(a_{yx}\frac{\partial u_h}{\partial x} + a_{yy}\frac{\partial u_h}{\partial y} \right) n_y \equiv q_n \quad (6.2.4)$$

and it constitutes the secondary variable. Thus, the weak form of Eq. (6.1.1) is

$$0 = \int_{\Omega^e} \left[\frac{\partial w_i}{\partial x}\left(a_{xx}\frac{\partial u_h}{\partial x} + a_{xy}\frac{\partial u_h}{\partial y} \right) + \frac{\partial w_i}{\partial y}\left(a_{yx}\frac{\partial u_h}{\partial x} + a_{yy}\frac{\partial u_h}{\partial y} \right) \right.$$
$$\left. + a_{00}w_i u_h - w_i f \right] dx dy - \oint_{\Gamma^e} w_i\, q_n\, ds \quad (6.2.5)$$

The function $q_n = q_n(s)$ denotes the outward flux normal to the boundary as we move counter-clockwise along the boundary Γ^e. The secondary variable q_n is of physical interest in most problems. For example, in the case of the heat transfer through an anisotropic medium, a_{ij} denotes the conductivities of the medium, and q_n denotes the heat flux normal to the boundary of the element. The weak form in Eq. (6.2.5) will be the basis of the weak-form Galerkin finite element model, in which w_i will be replaced by the approximation functions used to represent u_h.

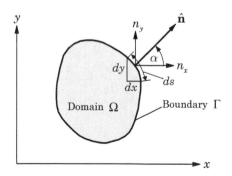

Fig. 6.2.1: A typical two-dimensional domain with a curved boundary.

6.3   Finite Element Model

The weak form in Eq. (6.2.5) requires that the approximation chosen for u should be at least linear in both x and y so that every term in Eq. (6.2.5) has a non-zero contribution to the integral. Since the primary variable is just u, which must be made continuous between elements, the Lagrange family of interpolation functions is admissible. Hence, u is approximated over a typical finite element Ω^e by the expression

$$u(x, y) \approx u_h^e(x, y) = \sum_{j=1}^{n} u_j^e \psi_j^e(x, y) \tag{6.3.1}$$

where u_j^e is the value of u_h^e at the jth node of the element, and ψ_j^e are the Lagrange interpolation functions derived in Section 3.3.5, and they have the following properties:

$$\psi_i^e(x_j^e, y_j^e) = \delta_{ij}, \quad \sum_{j=1}^{n} \psi_j^e(x, y) = 1 \tag{6.3.2}$$

where (x_j^e, y_j^e) denote the global coordinates of the jth node of element Ω^e.

Substituting the finite element approximation in Eq. (6.3.1) for u_h into the weak form, Eq. (6.2.5), we obtain

$$0 = \sum_{j=1}^{n} u_j^e \int_{\Omega^e} \left[\frac{\partial w_i}{\partial x} \left(a_{xx} \frac{\partial \psi_j^e}{\partial x} + a_{xy} \frac{\partial \psi_j^e}{\partial y} \right) + \frac{\partial w_i}{\partial y} \left(a_{yx} \frac{\partial \psi_j^e}{\partial x} + a_{yy} \frac{\partial \psi_j^e}{\partial y} \right) \right.$$
$$\left. + a_{00} w_i \psi_j^e \right] dx dy - \int_{\Omega^e} w_i f \, dx dy - \oint_{\Gamma^e} w_i q_n \, ds \tag{6.3.3}$$

For the weak-form Galerkin model, we replace the weight function with ψ_i^e and obtain

$$\sum_{j=1}^{n} K_{ij}^e u_j^e - f_i^e - Q_i^e = 0 \quad \text{or} \quad \mathbf{K}^e \mathbf{u}^e = \mathbf{f}^e + \mathbf{Q}^e \tag{6.3.4}$$

where

$$K_{ij}^e = \int_{\Omega^e} \left[\frac{\partial \psi_i^e}{\partial x} \left(a_{xx} \frac{\partial \psi_j^e}{\partial x} + a_{xy} \frac{\partial \psi_j^e}{\partial y} \right) + \frac{\partial \psi_i^e}{\partial y} \left(a_{yx} \frac{\partial \psi_j^e}{\partial x} + a_{yy} \frac{\partial \psi_j^e}{\partial y} \right) \right.$$
$$\left. + a_{00} \psi_i^e \psi_j^e \right] dx dy \tag{6.3.5}$$

$$f_i^e = \int_{\Omega^e} \psi_i^e f \, dx \, dy, \quad Q_i^e = \oint_{\Gamma^e} \psi_i^e q_n \, ds$$

Note that $K_{ij}^e \neq K_{ji}^e$ (i.e. $\mathbf{K}^e$ is not symmetric) unless $a_{xy} = a_{yx}$. Equation (6.3.4) represents a set of n nonlinear algebraic equations.

Equation (6.3.4) represents the finite element model of Eq. (6.1.1). This completes the finite element model development. The usual tasks of assembly of element equations, imposition of boundary conditions, and solution of linear algebraic equations (after an iterative method is applied) are standard, which were discussed in Chapter 3. Therefore, they are not discussed here again.

6.4 Solution of Nonlinear Equations

6.4.1 Direct Iteration Scheme

The assembled form of the nonlinear equation, Eq. (6.3.4) is

$$\mathbf{K}(\mathbf{U})\mathbf{U} = \mathbf{F} \tag{6.4.1}$$

where $\mathbf{K}$ is the assembled coefficient matrix, $\mathbf{U}$ the vector of global nodal values, and $\mathbf{F}$ the assembled source vector. In the direct iteration scheme, we assume that the solution at the $(r-1)$st iteration is known and seek the solution at the rth iteration from the equation

$$\mathbf{K}(\mathbf{U}^{(r-1)})\mathbf{U}^{(r)} = \mathbf{F}^{(r-1)} \tag{6.4.2}$$

where the direct coefficient matrix $\mathbf{K}$ and the right-hand side vector $\mathbf{F}$, if depends on $\mathbf{U}$, are evaluated using the known solution from the $(r-1)$st iteration. As applied to the element equation in Eq. (6.3.4), the direct iteration scheme has the form

$$\mathbf{K}^e(\mathbf{u}^{(r-1)})\mathbf{u}(r) = \mathbf{F}^e \tag{6.4.3}$$

Thus, we evaluate element matrices using the solution known from $(r-1)$st iteration and assemble them to obtain Eq. (6.4.2). The assembled system of equations is then solved after applying the boundary conditions.

6.4.2 Newton's Iteration Scheme

In the Newton procedure, we solve the assembled system of equations for the solution increment $\Delta\mathbf{U}$

$$\mathbf{T}(\mathbf{U}^{(r-1)})\Delta\mathbf{U} = -\mathbf{R}(\mathbf{U}^{(r-1)}) \tag{6.4.4}$$

where $\mathbf{R}$ is the assembled residual vector

$$\mathbf{R}(\mathbf{U}) = \mathbf{K}(\mathbf{U})\mathbf{U} - \mathbf{F} \tag{6.4.5}$$

and $\mathbf{T}$ is the assembled tangent matrix

$$\mathbf{T}(\mathbf{U}) \equiv \frac{\partial\mathbf{R}}{\partial\mathbf{U}} \tag{6.4.6}$$

The total solution at the end of the rth iteration is given by

$$\mathbf{U}^{(r)} = \mathbf{U}^{(r-1)} + \Delta\mathbf{U} \tag{6.4.7}$$

The element-wise computation of the tangent matrix coefficients were discussed in Chapters 4 and 5. The coefficients of the element tangent matrix are computed using the definition

$$T_{ij}^e \equiv \frac{\partial R_i^e}{\partial u_j^e} = K_{ij}^e + \sum_{m=1}^{n} \frac{\partial K_{im}^e}{\partial u_j^e} u_m^e - \frac{\partial F_i^e}{\partial u_j^e} \tag{6.4.8}$$

At the element level, Eq. (6.4.4) takes the form

$$[\mathbf{T}^e(\mathbf{u}^e)]^{(r-1)}\Delta\mathbf{u}^{(r)} = \mathbf{F}^e(\mathbf{u}^{(r-1)}) - [\mathbf{K}^e(\mathbf{u}^e)]^{(r-1)}(\mathbf{u}^e)^{(r-1)} \tag{6.4.9}$$

Example 6.4.1

Suppose that a_{xy}, a_{yx}, and a_{00} are functions of x and y, and a_{xx}^e and a_{yy}^e have the form

$$\begin{aligned}
a_{xx}^e &= a_{11}^e + a_{1u}^e \cdot u + a_{1ux}^e \cdot \frac{\partial u}{\partial x} + a_{1uy}^e \cdot \frac{\partial u}{\partial y} \\
a_{yy}^e &= a_{22}^e + a_{2u}^e \cdot u + a_{2ux}^e \cdot \frac{\partial u}{\partial x} + a_{2uy}^e \cdot \frac{\partial u}{\partial y}
\end{aligned} \tag{1}$$

where a_{11}, a_{1u}, and so on are functions of only x and y. Determine the tangent coefficient matrix.

Solution: We have

$$T_{ij}^e = K_{ij}^e + \sum_{m=1}^{n} \frac{\partial K_{im}^e}{\partial u_j} u_m^e \tag{2}$$

where

$$\begin{aligned}
\sum_{m=1}^{n} \frac{\partial K_{im}^e}{\partial u_j^e} u_m^e &= \sum_{m=1}^{n} \left[\int_{\Omega^e} \left(\frac{\partial a_{xx}^e}{\partial u_j^e} \frac{\partial \psi_i^e}{\partial x} \frac{\partial \psi_m^e}{\partial x} + \frac{\partial a_{yy}^e}{\partial u_j^e} \frac{\partial \psi_i^e}{\partial y} \frac{\partial \psi_m^e}{\partial y} \right) dx\,dy \right] u_m^e \\
&= \sum_{m=1}^{n} \left\{ \int_{\Omega^e} \left[\left(a_{1u}^e \psi_j^e + a_{1ux}^e \frac{\partial \psi_j^e}{\partial x} + a_{1uy}^e \frac{\partial \psi_j^e}{\partial y} \right) \frac{\partial \psi_i^e}{\partial x} \frac{\partial \psi_m^e}{\partial x} \right. \right. \\
&\qquad \left. \left. + \left(a_{2u}^e \psi_j^e + a_{2ux}^e \frac{\partial \psi_j^e}{\partial x} + a_{2uy}^e \frac{\partial \psi_j^e}{\partial y} \right) \frac{\partial \psi_i^e}{\partial y} \frac{\partial \psi_m^e}{\partial y} \right] dx\,dy \right\} u_m^e \\
&= \int_{\Omega^e} \left[\frac{\partial u}{\partial x} \frac{\partial \psi_i^e}{\partial x} \left(a_{1u}^e \psi_j^e + a_{1ux}^e \frac{\partial \psi_j^e}{\partial x} + a_{1uy}^e \frac{\partial \psi_j^e}{\partial y} \right) \right. \\
&\qquad \left. + \frac{\partial u}{\partial y} \frac{\partial \psi_i^e}{\partial y} \left(a_{2u}^e \psi_j^e + a_{2ux}^e \frac{\partial \psi_j^e}{\partial x} + a_{2uy}^e \frac{\partial \psi_j^e}{\partial y} \right) \right] dx\,dy
\end{aligned} \tag{3}$$

The symmetry of $\mathbf{T}^e$ depends on the nature of the nonlinearity and $\mathbf{K}^e$.

6.5 Axisymmetric Problems

6.5.1 Introduction

Consider the differential equation in the cylindrical coordinate system (r, θ, z)

$$-\frac{1}{r}\frac{\partial}{\partial r}\left(r a_{rr}\frac{\partial u}{\partial r}\right) - \frac{1}{r^2}\frac{\partial}{\partial \theta}\left(a_{\theta\theta}\frac{\partial u}{\partial \theta}\right) - \frac{\partial}{\partial z}\left(a_{zz}\frac{\partial u}{\partial z}\right) = f \qquad (6.5.1)$$

where, in general, u, f, a_{rr}, $a_{\theta\theta}$, and a_{zz} are functions of r, θ, and z. When the domain of the phenomena is a circular cylinder, the problem may be reduced to a two-dimensional or even one-dimensional problem, depending on the material properties (a_{rr}, $a_{\theta\theta}$, and a_{zz}), boundary conditions, and applied *load*, f. The problems described by Eq. (6.5.1) and posed on a circular cylinder, hollow or solid, can be simplified as follows.

When the cylinder is very long and material properties, boundary conditions, and applied loads do not vary along the length of the cylinder (i.e. independent of the coordinate z), it can be modeled as a circular disk of arbitrary thickness, that is, two-dimensional domain in the $r\theta$-plane, as shown in Fig. 6.5.1. In addition, if the material properties, boundary conditions, and applied loads do not vary with the circumferential coordinate θ, the disc can be replaced with a radial line and the problem becomes a one-dimensional one.

The governing equation for the two-dimensional case is given by

$$-\frac{1}{r}\frac{\partial}{\partial r}\left(r a_{rr}\frac{\partial u}{\partial r}\right) - \frac{1}{r^2}\frac{\partial}{\partial \theta}\left(a_{\theta\theta}\frac{\partial u}{\partial \theta}\right) = f(r, \theta) \qquad (6.5.2)$$

and for the one-dimensional case, the governing equation is

$$-\frac{1}{r}\frac{d}{dr}\left(r a_{rr}\frac{du}{dr}\right) = f(r) \qquad (6.5.3)$$

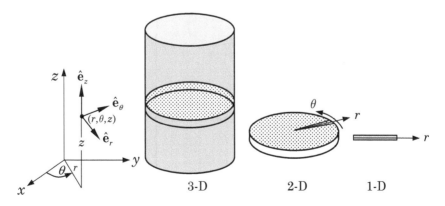

Fig. 6.5.1: Simplification of three-dimensional problems posed on a cylinder to two- and one-dimensional problems.

6.5.2   Governing Equation and the Finite Element Model

When the cylinder is of finite length and material properties, boundary conditions, and applied loads vary along the length of the cylinder but independent of the circumferential coordinate θ, we can use any typical plane (a wedge) of the domain, as shown in Fig. 6.5.2. The governing equation in this case becomes

$$-\frac{1}{r}\frac{\partial}{\partial r}\left(ra_{rr}\frac{\partial u}{\partial r}\right) - \frac{\partial}{\partial z}\left(a_{zz}\frac{\partial u}{\partial z}\right) = f(r,z) \tag{6.5.4}$$

The weak form of Eq. (6.5.4) is given by

$$
\begin{aligned}
0 &= \int_{\Omega^e} w_i\left[-\frac{1}{r}\frac{\partial}{\partial r}\left(ra_{rr}\frac{\partial u_h}{\partial r}\right) - \frac{\partial}{\partial z}\left(a_{zz}\frac{\partial u_h}{\partial z}\right) - f(r,z)\right]rdrdz \\
&= \int_{\Omega^e}\left[a_{rr}(r,z,u_h)\frac{\partial w_i}{\partial r}\frac{\partial u_h}{\partial r} + a_{zz}(r,z,u_h)\frac{\partial w_i}{\partial z}\frac{\partial u_h}{\partial z}\right]rdrdz \\
&\quad - \int_{\Omega^e} w_i f(r,z)\,rdrdz - \oint_{\Gamma^e} w_i q_n\,ds
\end{aligned}
\tag{6.5.5}
$$

where

$$q_n(s) = r\left[a_{rr}(r,z,u_h)\frac{\partial u_h}{\partial r}n_r + a_{zz}(r,z,u_h)\frac{\partial u_h}{\partial z}n_z\right] \tag{6.5.6}$$

and (n_r, n_z) are the direction cosines of the unit normal $\hat{n}$ to the boundary Γ^e. The finite element model is given by

$$\mathbf{K}^e\mathbf{u}^e = \mathbf{f}^e + \mathbf{Q}^e \equiv \mathbf{F}^e \tag{6.5.7}$$

where

$$
\begin{aligned}
K_{ij}^e &= \int_{\Omega^e}\left[a_{rr}(r,z,u_h)\frac{\partial \psi_i^e}{\partial r}\frac{\partial \psi_j^e}{\partial r} + a_{zz}(r,z,u_h)\frac{\partial \psi_i^e}{\partial z}\frac{\partial \psi_j^e}{\partial z}\right]rdrdz \\
f_i^e &= \int_{\Omega^e} f(r,z)\psi_i^e\,rdrdz, \quad Q_i^e = \oint_{\Gamma^e} q_n\psi_i^e\,ds
\end{aligned}
\tag{6.5.8}
$$

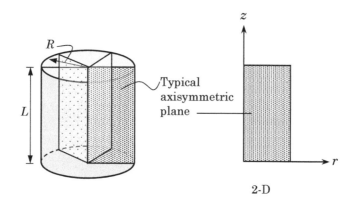

Fig. 6.5.2: An axisymmetric two-dimensional problem.

and $\psi_i^e(r, z)$ are the Lagrange interpolation functions of Section 3.3.5 [with (x, y) replaced by (r, z)]. This completes the finite element formulation of axisymmetric problems. All other aspects are the same as those discussed for plane problems.

6.6 Computer Implementation

6.6.1 Introduction

An accurate representation of non-rectangular domains and domains with curved boundaries can be accomplished by the use of refined meshes and/or higher-order elements (i.e. elements with higher-order interpolation functions). The geometry of the elements used to represent an irregular domain tend to be non-rectangular and, therefore, evaluation of the integrals involved in the computation of K_{ij}^e, f_i^e, and T_{ij}^e over the irregularly-shaped elements is difficult. A coordinate transformation between the coordinates (x, y) used in the formulation of the problem, called *global coordinates*, and another coordinate system (ξ, η), called a *local coordinate system*, which is convenient in deriving the interpolation functions as well as evaluating the integrals is needed. The choice of the local coordinate system is dictated by the choice of the numerical integration method. As discussed in Section 3.6, we shall use, as was done in one-dimensional problems, the Gauss quadrature rule to evaluate integrals defined over two-dimensional elements. The main steps in the numerical evaluation of finite element matrices are reviewed here from a computational view point.

6.6.2 Numerical Integration

The transformation between Ω^e and $\hat{\Omega}$ is accomplished by a coordinate transformation of the form [see Eqs. (3.6.1) and (3.6.2)]

$$x = \sum_{j=1}^m x_j^e \hat{\psi}_j^e(\xi, \eta) \,, \qquad y = \sum_{j=1}^m y_j^e \hat{\psi}_j^e(\xi, \eta) \tag{6.6.1}$$

while a typical dependent variable $u(x, y)$ is approximated by

$$u(x, y) = \sum_{j=1}^n u_j^e \psi_j^e(x, y) = \sum_{j=1}^n u_j^e \psi_j^e(x(\xi, \eta), y(\xi, \eta)) \tag{6.6.2}$$

where $\hat{\psi}_j^e$ denote the interpolation functions of the master element $\hat{\Omega}$ and ψ_j^e are the interpolation functions of a typical element Ω^e over which u is approximated. The transformation in Eq. (6.6.1) maps a point (x, y) in a typical element Ω^e of the mesh to a point (ξ, η) in the master element $\hat{\Omega}$ and vice versa, if the Jacobian J_e of the transformation is positive-definite [see Eqs. (3.6.5)–(3.6.8)].

The integral expressions involved in the calculation of $\mathbf{K}^e$ and $\mathbf{f}^e$ [see Eqs. (6.3.5) and (6.5.8)] are, in general, complicated algebraically due to the geometry of the element, spatial variation of the parameters and their dependence on the solution and possibly its derivatives, and the coordinate transformations. Therefore, the integrals are evaluated numerically, which requires evaluation of the integrand at a selective number of points in the domain, multiplying their values with suitable weights, and summing. Here we use the Gauss quadrature discussed in Section 3.6.3 to evaluate integrals over quadrilateral elements.

We recall from Section 3.6.3 the essential elements of the Gauss quadrature by considering the following integral expression:

$$K_{ij}^e = \int_{\Omega^e} \left(a_{xx} \frac{\partial \psi_i^e}{\partial x} \frac{\partial \psi_j^e}{\partial x} + a_{yy} \frac{\partial \psi_i^e}{\partial y} \frac{\partial \psi_j^e}{\partial y} + a_{00} \psi_i^e \psi_j^e \right) dx\, dy \qquad (6.6.3)$$

We wish to transform the integral from Ω^e to the master element $\hat{\Omega} = \{(\xi, \eta) : -1 \leq \xi \leq 1, -1 \leq \eta \leq 1\}$ so that the Gauss quadrature can be used. The derivatives of ψ_i^e with respect to the global coordinates (x, y) in K_{ij}^e must be expressed in terms of the derivative with respect to (ξ, η). This relationship is given by Eq. (3.6.6)

$$\left\{ \begin{array}{c} \frac{\partial \psi_i^e}{\partial x} \\ \frac{\partial \psi_i^e}{\partial y} \end{array} \right\} = (\mathbf{J}^e)^{-1} \left\{ \begin{array}{c} \frac{\partial \psi_i^e}{\partial \xi} \\ \frac{\partial \psi_i^e}{\partial \eta} \end{array} \right\} \qquad (6.6.4)$$

where $\mathbf{J}^e$ is the Jacobian matrix defined in Eq. (3.6.5), and it is calculated from

$$\mathbf{J}^e = \begin{bmatrix} \frac{\partial x}{\partial \xi} & \frac{\partial y}{\partial \xi} \\ \frac{\partial x}{\partial \eta} & \frac{\partial y}{\partial \eta} \end{bmatrix} = \begin{bmatrix} \sum_{i=1}^m x_i \frac{\partial \hat{\psi}_i}{\partial \xi} & \sum_{i=1}^m y_i \frac{\partial \hat{\psi}_i}{\partial \xi} \\ \sum_{i=1}^m x_i \frac{\partial \hat{\psi}_i}{\partial \eta} & \sum_{i=1}^m y_i \frac{\partial \hat{\psi}_i}{\partial \eta} \end{bmatrix} = \begin{bmatrix} \frac{\partial \hat{\psi}_1}{\partial \xi} & \frac{\partial \hat{\psi}_2}{\partial \xi} & \cdots & \frac{\partial \hat{\psi}_m}{\partial \xi} \\ \frac{\partial \hat{\psi}_1}{\partial \eta} & \frac{\partial \hat{\psi}_2}{\partial \eta} & \cdots & \frac{\partial \hat{\psi}_m}{\partial \eta} \end{bmatrix} \begin{bmatrix} x_1 & y_1 \\ x_2 & y_2 \\ \vdots & \vdots \\ x_m & y_m \end{bmatrix}$$
$$(6.6.5)$$

Thus, given the global coordinates (x_i^e, y_i^e) of element nodes and the interpolation functions $\hat{\psi}_i^e$ used for geometry, the Jacobian matrix can be evaluated using Eq. (6.6.5). Note that $\hat{\psi}_i^e$ are different, in general, from ψ_i^e used in the approximation of the dependent variables. Whenever $\hat{\psi}_i^e = \psi_i^e$, we say that an *isoparametric formulation* is adopted. The Jacobian is given by

$$J_e = |\mathbf{J}^e| = J_{11}^e J_{22}^e - J_{12}^e J_{21}^e \qquad (6.6.6)$$

We have from Eq. (6.6.4)

$$\left\{ \begin{array}{c} \frac{\partial \psi_i^e}{\partial x} \\ \frac{\partial \psi_i^e}{\partial y} \end{array} \right\} = (\mathbf{J}^e)^{-1} \left\{ \begin{array}{c} \frac{\partial \psi_i^e}{\partial \xi} \\ \frac{\partial \psi_i^e}{\partial \eta} \end{array} \right\} \equiv \mathbf{J}^* \left\{ \begin{array}{c} \frac{\partial \psi_i^e}{\partial \xi} \\ \frac{\partial \psi_i^e}{\partial \eta} \end{array} \right\} \qquad (6.6.7)$$

where

$$J_{11}^* = J_{22}^e / J_e, \quad J_{12}^* = -J_{12}^e / J_e, \quad J_{22}^* = J_{11}^e / J_e, \quad J_{21}^* = -J_{21}^e / J_e \qquad (6.6.8)$$

Returning to the coefficients K_{ij}^e in Eq. (6.6.3), we can write it now in terms of the natural coordinates (ξ, η) as

$$
\begin{aligned}
K_{ij}^e = \int_{\hat{\Omega}} \Bigg\{ & \hat{a}_{xx}(\xi,\eta) \left(J_{11}^* \frac{\partial \psi_i^e}{\partial \xi} + J_{12}^* \frac{\partial \psi_i^e}{\partial \eta} \right) \left(J_{11}^* \frac{\partial \psi_j^e}{\partial \xi} + J_{12}^* \frac{\partial \psi_j^e}{\partial \eta} \right) \\
& + \hat{a}_{yy}(\xi,\eta) \left(J_{21}^* \frac{\partial \psi_i^e}{\partial \xi} + J_{22}^* \frac{\partial \psi_i^e}{\partial \eta} \right) \left(J_{21}^* \frac{\partial \psi_j^e}{\partial \xi} + J_{22}^* \frac{\partial \psi_j^e}{\partial \eta} \right) \\
& + \hat{a}_{00}(\xi,\eta) \psi_i^e \psi_j^e \Bigg\} J_e \, d\xi \, d\eta \\
\equiv & \int_{\hat{\Omega}} F_{ij}^e(\xi,\eta) \, d\xi \, d\eta
\end{aligned}
\tag{6.6.9}
$$

where the element area $dA = dxdy$ in element Ω^e is transformed to $J_e \, d\xi \, d\eta$ in the master element $\hat{\Omega}$, and $\hat{a}_{xx} = a_{xx}(x(\xi,\eta), y(\xi,\eta), u(\xi,\eta))$, and so on.

Using the $M \times N$ Gauss quadrature to evaluate integrals defined over a rectangular master element $\hat{\Omega}$, we obtain

$$
\begin{aligned}
\int_{\hat{\Omega}} F_{ij}^e(\xi,\eta) \, d\xi \, d\eta &= \int_{-1}^1 \left[\int_{-1}^1 F_{ij}^e(\xi,\eta) \, d\eta \right] d\xi \approx \int_{-1}^1 \left[\sum_{J=1}^N F_{ij}^e(\xi, \eta_J) W_J \right] d\xi \\
&\approx \sum_{I=1}^M \sum_{J=1}^N F_{ij}^e(\xi_I, \eta_J) W_I W_J
\end{aligned}
\tag{6.6.10}
$$

where M and N denote the number of Gauss quadrature points in the ξ and η directions, respectively, (ξ_I, η_J) denote the Gauss points, and W_I and W_J denote the corresponding Gauss weights, as listed in Table 3.6.1.

As already discussed, if the integrand is a polynomial of degree p in a coordinate direction, it is integrated exactly by employing $NGP \equiv N = \text{int}[\frac{1}{2}(p+1)]$ (the nearest equal or larger integer number) Gauss points in that direction. In most cases, the interpolation functions are of the same degree in both ξ and η, and we take $N = M$. For example, consider the expression involving a_{00} in K_{ij}^e of Eq. (6.6.3). When a_{00} is a linear function of ξ and η, it requires a 2×2 Gauss rule when ψ_i^e are linear and a 3×3 Gauss rule when ψ_i^e are quadratic to be evaluated exactly. When a_{00} is quadratic or cubic, it requires 3×3 and 4×4 Gauss rules for linear and quadratic *rectangular* elements, respectively.

6.6.3 Element Calculations

Calculation of element coefficients require evaluation of interpolation functions and their derivatives. The statements for the calculation of the approximation functions $\psi_i^e(\xi,\eta)$, their derivatives with respect to (ξ,η), the Jacobian matrix and its determinant, and derivatives of ψ_i^e with respect to (x,y) using Eq. (6.6.7) are included in the subroutine INTERPLN2D, which is listed in Box 3.7.3. The notation used in INTERPLN2D is recalled briefly here (also see Section 3.7.3 for more details):

$$\text{SFL}(i) = \psi_i^e$$

$$\text{DSFL}(1, i) = \frac{\partial \psi_i^e}{\partial \xi}, \quad \text{DSFL}(2, i) = \frac{\partial \psi_i^e}{\partial \eta}$$

$$[\text{ELXY}] = \text{array of the global coordinates of element nodes}$$

$$\text{ELXY}(i, 1) = x_i, \quad \text{ELXY}(i, 2) = y_i$$

$$[\text{GJ}] = \text{the Jacobian matrix}, \mathbf{J}^e; \quad [\text{GJ}] = [\text{DSFL}][\text{ELXY}]$$

$$[\text{GJINV}] = \text{inverse of the Jacobian matrix}, [\text{GJ}]^{-1}$$

$$\text{DET} = \text{the determinant of the Jacobian matrix}, J_e = |\mathbf{J}^e|$$

$$\text{GDSFL}(1, i) = \frac{\partial \psi_i^e}{\partial x}, \quad \text{GDSFL}(2, i) = \frac{\partial \psi_i^e}{\partial y}$$

$$[\text{GDSFL}] = [\text{GJINV}][\text{DSFL}]$$

A Fortran listing of subroutine ELEMATRCS2D, in which matrices $\mathbf{K}^e$ and $\mathbf{f}^e$ for the model problem (with $a_{xy} = a_{yx} = 0$) of this chapter are calculated, is listed in Box 6.6.1. The following notation is used ($n = \text{NPE}$):

$$\text{ELF}(i) = f_i^e, \text{ the } i\text{th component of the source vector}$$

$$\text{ELK}(i, j) = K_{ij}^e, \text{ the } (i, j) \text{ component of the element coefficient matrix}$$

$$\text{TANG}(i, j) = \sum_{m=1}^{n} \frac{\partial K_{im}^e}{\partial u_j} u_m^e, \text{ addition to } K_{ij}^e \text{ to obtain } T_{ij}^e$$

$$\text{GAUSPT}(I, J) = I\text{th Gauss point in the } J\text{th Gauss-point rule } (I \leq J)$$

$$\text{GAUSWT}(I, J) = I\text{th Gauss weight in the } J\text{th Gauss-point rule } (I \leq J)$$

$$\text{ELU}(I) = u_i^e \text{ from the previous iteration}$$

$$\text{NGPF} = \text{number of Gauss points - full integration}$$

$$\text{NONLIN} = \text{Flag for nonlinear analysis; it is also used for type of}$$

$$\text{the iteration method: } = 1, \text{ Picard; } > 1, \text{ Newton}$$

The element matrix $\mathbf{K}^e$ can be expressed in terms of the submatrices $\mathbf{S}^{\alpha\beta}$ $[S^{\alpha\beta}(i, j) = S_{ij}^{\alpha\beta}]$ defined in Eq. (3.7.10), and therefore one can use them to verify the values of $\mathbf{K}^e$ calculated in a computer program.

Box 6.6.1: Subroutine ELMATRCS2D for the calculation of element matrices $\mathbf{K}^e$ and $\mathbf{f}^e$ of Eq. (6.3.5) with $a_{xy} = a_{yx} = 0$.

```
      SUBROUTINE ELMATRCS2D(MODEL,NPE,NN,NONLIN)
C     ----------------------------------------------------------------
C     Element calculations based on linear and quadratic rectangular
C               elements with isoparametric formulation.
C     ----------------------------------------------------------------
C
      IMPLICIT REAL*8(A-H,O-Z)
      COMMON/STF/ELF(9),ELK(9,9),ELXY(9,2),ELU(9)
      COMMON/PST/A10,A1X,A1Y,A20,A2X,A2Y,A00,F0,FX,FY,
     *           A1U,A1UX,A1UY,A2U,A2UX,A2UY
      COMMON/SHP/SFL(9),GDSFL(2,9)
      DIMENSION GAUSPT(5,5),GAUSWT(5,5),TANG(9,9)
      COMMON/IO/IN,IT
C
      DATA GAUSPT/5*0.0D0, -0.57735027D0, 0.57735027D0, 3*0.0D0,
     2  -0.77459667D0, 0.0D0, 0.77459667D0, 2*0.0D0, -0.86113631D0,
     3  -0.33998104D0, 0.33998104D0, 0.86113631D0, 0.0D0, -0.90617984D0,
     4  -0.53846931D0,0.0D0,0.53846931D0,0.90617984D0/
C
      DATA GAUSWT/2.0D0, 4*0.0D0, 2*1.0D0, 3*0.0D0, 0.55555555D0,
     2   0.88888888D0, 0.55555555D0, 2*0.0D0, 0.34785485D0,
     3 2*0.65214515D0, 0.34785485D0, 0.0D0, 0.23692688D0,
     4   0.47862867D0, 0.56888888D0, 0.47862867D0, 0.23692688D0/
C
C     Initialize the arrays
C
      DO 100 I = 1,NPE
         ELF(I)  = 0.0
      DO 100 J = 1,NPE
         IF(NONLIN.GT.1)THEN
            TANG(I,J)=0.0
         ENDIF
  100    ELK(I,J)= 0.0
C
C     Do-loops on numerical (Gauss) integration begin here.
C     Subroutine INTERPLN2D is called here
C
      DO 200 NI = 1,NGPF
      DO 200 NJ = 1,NGPF
         XI  = GAUSPT(NI,NGPF)
         ETA = GAUSPT(NJ,NGPF)
         CALL INTERPLN2D(NPE,XI,ETA,DET,ELXY)
         CNST = DET*GAUSWT(NI,NGPF)*GAUSWT(NJ,NGPF)
         X=0.0
         Y=0.0
         U=0.0
         UX=0.0
         UY=0.0
```

Box 6.6.1: Subroutine ELMATRCS2D (continued).

```
      DO 140 I=1,NPE
      IF(NONLIN.GT.0)THEN
         U=U+ELU(I)*SFL(I)
         UX=UX+ELU(I)*GDSFL(1,I)
         UY=UY+ELU(I)*GDSFL(2,I)
      ENDIF
         X=X+ELXY(I,1)*SFL(I)
  140    Y=Y+ELXY(I,2)*SFL(I)
C *** Define the coefficients of the differential equation ***
         FXY=F0+FX*X+FY*Y
         A11=A10+A1X*X+A1Y*Y
         A22=A20+A2X*X+A2Y*Y
         IF(NONLIN.GT.0)THEN
            AXX=A11+A1U*U+A1UX*UX+A1UY*UY
            AYY=A22+A2U*U+A2UX*UX+A2UY*UY
         ENDIF
C *** Define the element source vector and the coefficient matrix ***
         DO 180 I=1,NPE
            ELF(I)=ELF(I)+FXY*SFL(I)*CNST
            DO 160 J=1,NPE
               S00=SFL(I)*SFL(J)*CNST
               S11=GDSFL(1,I)*GDSFL(1,J)*CNST
               S22=GDSFL(2,I)*GDSFL(2,J)*CNST
               ELK(I,J)=ELK(I,J)+A00*S00+AXX*S11+AYY*S22
C *** Define the part needed to be added to [K] in order to define [T] ***
               IF(NONLIN.GT.1)THEN
                  S10=GDSFL(1,I)*SFL(J)*CNST
                  S20=GDSFL(2,I)*SFL(J)*CNST
                  S12=GDSFL(1,I)*GDSFL(2,J)*CNST
                  S21=GDSFL(2,I)*GDSFL(1,J)*CNST
                  TANG(I,J)=TANG(I,J)
     *                     +UX*(A1U*S10+A1UX*S11+A1UY*S12)
     *                     +UY*(A2U*S20+A2UX*S21+A2UY*S22)
               ENDIF
  160       CONTINUE
  180    CONTINUE
  200 CONTINUE
C
C ** Write statements to compute the residual vector and tangent matrix **
C
      IF(NONLIN.GT.1)THEN
         DO 220 I=1,NPE
            DO 220 J=1,NPE
  220          ELF(I)=ELF(I)-ELK(I,J)*ELU(J)
C
         DO 240 I=1,NPE
            DO 240 J=1,NPE
  240          ELK(I,J)=ELK(I,J)+TANG(I,J)
      ENDIF
      RETURN
      END
```

Next, we consider several examples of applications of the ideas presented.

Example 6.6.1

Consider heat conduction in a rectangular, isotropic medium with conductivity $a_{xx} = a_{yy} = k$ [and $a_{xy} = a_{yx} = a_{00} = 0$ in Eq. (6.1.1)]. The domain is of dimensions $a \times b$. The conductivity k is assumed to vary according to the relation

$$k = k_0 \left[1 + k_1 (T - T_0) \right] \tag{1}$$

where k_0 is the constant thermal conductivity, k_1 is the temperature coefficient of thermal conductivity, T_0 is a reference temperature, and T is the temperature. Determine the finite element solution when there is no internal heat generation (i.e. $f = 0$) and the problem data and boundary conditions are

$$a = 0.18 \text{ m}, \ \ b = 0.1 \text{ m}, \ \ T_0 = 0^\circ \text{ K}, \ \ k_0 = 0.2 \text{ W/(m }^\circ\text{K)}, \ \ k_1 = 2 \times 10^{-3} \, ^\circ\text{K}^{-1}$$

$$T(0, y) = 500^\circ \text{ K} \ , \ \ T(a, y) = 300^\circ \text{ K} \ , \ \ \frac{\partial T}{\partial y} = 0 \text{ at } y = 0, b \text{ for any } x \tag{2}$$

where the origin of the coordinate system (x, y) is taken at the lower left corner of the domain. Consider the following uniform meshes: 4×2 and 8×2 meshes of linear rectangular elements and 2×1 and 4×1 meshes of nine-node quadratic elements, and use both direct and Newton's iteration schemes with error tolerance of 10^{-3} and maximum iterations of 10.

Solution: Comparing $a_{xx} = a_{yy} = k$ from Eq. (1) to that in Eq. (1) of Example 6.4.1, we have

$$a_{11} = a_{22} = k_0 - k_0 \, k_1 \, T_0 = k_0, \ \ a_{1u} = a_{2u} = k_0 \, k_1, \ \ a_{1ux} = a_{1uy} = a_{2ux} = a_{2uy} = 0 \tag{3}$$

and the tangent coefficient matrix $\mathbf{T}^e$ is unsymmetric while the direct coefficient matrix $\mathbf{K}^e$ is symmetric. Hence, one must use an equation solver suitable for solving the banded unsymmetric equations when Newton's iteration method is used. Further, in Newton's iteration scheme, one must make sure that the values of the nonzero specified values on the temperature are made zero immediately after the first iteration. In the present analysis, the 2×2 and 3×3 Gauss quadratures are used for linear and quadratic elements, respectively, while the gradient of the solution is computed at 1×1 and 2×2 Gauss points, respectively. For this set of boundary conditions, the temperature T is only a function of x (because there is repeated symmetry about a horizontal line) and the problem can be solved as such (see Table 4.5.1).

The linear solution may be naturally obtained at the end of the first iteration if the initial guess vector is taken as zero and the actual boundary conditions are imposed through the boundary subroutine. The element matrix at the beginning of the first iteration for the 4×2L4 mesh (with the equal subdivision in each coordinate direction) of linear elements, with zero guess vector, is

$$\mathbf{K}^{(1)} = \begin{bmatrix} 0.13407 & -0.04407 & -0.06704 & -0.02296 \\ -0.04407 & 0.13407 & -0.02296 & -0.06704 \\ -0.06704 & -0.02296 & 0.13407 & -0.04407 \\ -0.02296 & -0.06704 & -0.04407 & 0.13407 \end{bmatrix} \tag{4}$$

The element matrix at the beginning of the second iteration in the direct iteration scheme is

$$\mathbf{K}^{(1)} = \begin{bmatrix} 0.26294 & -0.08594 & -0.13072 & -0.04628 \\ -0.08594 & 0.25994 & -0.04328 & -0.01307 \\ -0.13072 & -0.04328 & 0.25944 & -0.08594 \\ -0.04628 & -0.01307 & -0.08594 & 0.26294 \end{bmatrix} \tag{5}$$

Linear and nonlinear solutions $T(x, y)$ obtained with the direct iteration and Newton's iteration procedures gave essentially the same results; results obtained with various meshes are presented in Table 6.6.1 as a function of x for $y = 0$ or $y = b$. Both iterative schemes took three iterations to converge for error tolerance of $\epsilon = 10^{-3}$. The solution is independent of the mesh in the y-direction. The present results were found to be identical, as expected, to those obtained with the one-dimensional model of Example 4.5.1.

Table 6.6.1: Finite element solutions of a nonlinear heat conduction equation [$k_0 = 0.2$ W/(m °K) and $k_1 = 2 \times 10^{-3}$ (°K^{-1})].

x	Linear	$4 \times$ 2L4*	$2 \times$ 1Q9	$8 \times$ 2L4	$4 \times$ 1Q9
0.0000	500.00	500.00	500.00	500.00	500.00
0.0225	475.00	$--$	$--$	477.24	477.24
0.0450	450.00	453.94	453.94	453.94	453.94
0.0675	425.00	$--$	$--$	430.06	430.06
0.0900	400.00	405.54	405.54	405.54	405.54
0.1125	375.00	$--$	$--$	380.35	380.35
0.1350	350.00	354.40	354.40	354.40	354.40
0.1575	325.00	$--$	$--$	327.65	327.65
0.1800	300.00	300.00	300.00	300.00	300.00

* $m \times n$-L4, for example, denotes a mesh of m by n (m along x and n along y) four-node linear (L4) elements, with equal subdivisions in each coordinate direction.

Example 6.6.2

Analyze the problem in Example 6.6.1 with the following data:

$$a = 0.2 \text{ m}, \quad b = 0.1 \text{ m}, \quad T_0 = 0° \text{ K}, \quad k_0 = 0.2 \text{ W/(m °K)}, \quad k_1 = 100 \text{ °K}^{-1}$$
$$T(0, y) = 500° \text{ K}, \quad T(a, y) = 300° \text{ K}$$
$$\frac{\partial T}{\partial y} = 0 \text{ at } y = 0, \quad T(x, b) = 500(1 - 10x^2) \text{ °K} \tag{1}$$

Use a uniform mesh of 4×4 nine-node quadratic elements with direct iteration.

Solution: In this case, the problem no longer exhibits any symmetry, and $T(x, y)$ has a two-dimensional distribution. Figure 6.6.1 shows the temperature distributions for different values of y. The dashed lines indicate the corresponding linear solutions. The direct iteration procedure with a tolerance of $\epsilon = 10^{-3}$ is used. Only three iterations were taken for convergence.

Example 6.6.3

Consider a unit square membrane, fixed on all its sides, and subjected to uniformly distributed load of intensity $f_0 = 1$. Suppose that the tensions a_{xx} and a_{yy} in the membrane are nonlinearly dependent on u according to the relations

$$a_{xx} = 1 + a_{1ux}\frac{\partial u}{\partial x} + a_{1uy}\frac{\partial u}{\partial y}, \quad a_{yy} = a_{xx} \tag{1}$$

Exploit the biaxial symmetry and use 4×4Q9 uniform mesh to analyze the problem.

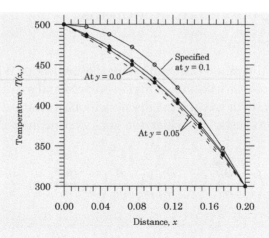

Fig. 6.6.1: Temperature distribution in a rectangular region with temperature dependent conductivity.

Solution: The Newton iteration scheme with $\epsilon = 10^{-3}$ is used. The deflections $u(x, 0)$ along the symmetry line $y = 0$ are presented in Table 6.6.2 (also see Fig. 6.6.2) for $a_{1ux} = -0.4, -0.2, 0,$ and 0.2. The deflections increase with a_{1ux} increasing from -0.4 to 0.2.

Table 6.6.2: Deflections of a square membrane subjected to uniform load.

x	$a_{1ux} = -0.4$	$a_{1ux} = -0.2$	$a_{1ux} = 0.0$	$a_{1ux} = 0.2$
0.0000	0.068440	0.070848	0.073670	0.077077
0.0625	0.067473	0.069875	0.072691	0.076092
0.1250	0.064573	0.066940	0.069719	0.073081
0.1875	0.059696	0.061973	0.064656	0.067913
0.2500	0.052723	0.054834	0.057334	0.060387
0.3125	0.043488	0.045328	0.047523	0.050229
0.3750	0.031776	0.033208	0.034931	0.037083
0.4375	0.017358	0.018196	0.019218	0.020515

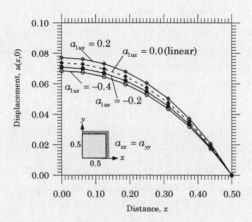

Fig. 6.6.2: Displacement $u(x, 0)$ versus x for various values of $a_{1ux} = a_{1uy}$ of Eq. (1).

6.7   Time-Dependent Problems

6.7.1   Introduction

In this section, we develop the finite element models of 2-D time-dependent problems with nonlinearities and describe some standard time approximation schemes. We begin with the general discussion of the finite element modeling of time-dependent problems. Consider the time-dependent version of the model equation, Eq. (6.1.1)

$$-\frac{\partial}{\partial x}\left(a_{xx}\frac{\partial u}{\partial x} + a_{xy}\frac{\partial u}{\partial y}\right) - \frac{\partial}{\partial y}\left(a_{yx}\frac{\partial u}{\partial x} + a_{yy}\frac{\partial u}{\partial y}\right)$$

$$+a_{00}u + c_0\frac{\partial u}{\partial t} + c_1\frac{\partial^2 u}{\partial t^2} = f(x, y, t) \qquad (6.7.1)$$

where a_{ij} and c_i are, in general, functions of position $\mathbf{x}$, time t, u, $\partial u/\partial x$, and $\partial u/\partial y$. The equation is subjected to appropriate boundary and initial conditions.

The finite element formulation of time-dependent problems involves following two steps:

1. *Spatial approximation*, where the solution $u(\mathbf{x}, t)$ of the equation under consideration is approximated by expressions of the form

$$u(\mathbf{x}, t) \approx u_h^e(\mathbf{x}, t) = \sum_{j=1}^{n} u_j^e(t)\psi_j^e(\mathbf{x}) \qquad (6.7.2)$$

and the spatial finite element model of the equation is developed using the procedures of static or steady-state problems, while carrying all time-dependent terms in the formulation. This step results in a set of ordinary differential equations in time for the nodal variables $u_j^e(t)$ of the element $\mathbf{K}^e\mathbf{u}^e + \mathbf{C}^e\dot{\mathbf{u}}^e + \mathbf{M}^e\ddot{\mathbf{u}}^e = \mathbf{F}^e$. When the solution is separable into functions of time only and space only, that is, $u(\mathbf{x}, t) = T(t)X(\mathbf{x})$, approximation of the type in Eq. (6.7.2) is justified for the overall transient response of a system, in contrast to wave propagation type solutions. Since this step only addresses spatial approximation, it is termed *semidiscretization*.

2. *Temporal approximation*, where the system of ordinary differential equations in time are further approximated to obtain a system of algebraic equations, called *fully discretized equations*. Depending on the goal of the study, one may reduce the ordinary differential equations to either an algebraic eigenvalue problem $(\mathbf{K} - \lambda\mathbf{M})\bar{\mathbf{u}} = \mathbf{Q}$, or an algebraic system of equations $\hat{\mathbf{K}}\mathbf{u}^{s+1} = \hat{\mathbf{F}}^{s,s+1}$ that can be solved for various fixed times by marching in time to obtain the *transient response*; here λ denotes an eigenvalue and $\mathbf{u}^{s+1} \equiv \mathbf{u}(t_{s+1})$.

6.7.2 Semidiscretization

The spatial approximation (i.e. semidiscretization) follows the same two steps as in the steady-state case discussed in Sections 6.2 and 6.3, namely, the development of the weak form and approximation of the field variables to obtain the algebraic relations. The weak form of Eq. (6.7.1) is given by

$$
0 = \int_{\Omega^e} \left[\frac{\partial w_i}{\partial x} \left(a_{xx} \frac{\partial u_h}{\partial x} + a_{xy} \frac{\partial u_h}{\partial y} \right) + \frac{\partial w_i}{\partial y} \left(a_{yx} \frac{\partial u_h}{\partial x} + a_{yy} \frac{\partial u_h}{\partial y} \right) + a_{00} w_i u_h \right.
$$

$$
\left. + c_0 w_i \frac{\partial u_h}{\partial t} + c_1 w_i \frac{\partial^2 u_h}{\partial t^2} - w_i f \right] dx\, dy - \oint_{\Gamma^e} w_i q_n\, ds \tag{6.7.3}
$$

The finite element approximation is assumed to be of the form

$$
u(\mathbf{x}, t) \approx u_h^e(\mathbf{x}, t) = \sum_{j=1}^{n} u_j^e(t) \psi_j^e(\mathbf{x}) \tag{6.7.4}
$$

where the nodal values u_j^e are now assumed to be functions of time. Substitution of Eq. (6.7.4) into Eq. (6.7.3) gives rise to the finite element equations

$$
\mathbf{M}^e \ddot{\mathbf{u}}^e + \mathbf{C}^e \dot{\mathbf{u}}^e + \mathbf{K}^e \mathbf{u}^e = \mathbf{f}^e + \mathbf{Q}^e \equiv \mathbf{F}^e \tag{6.7.5}
$$

where

$$
C_{ij}^e = \int_{\Omega^e} c_0 \psi_i^e \psi_j^e\, dx\, dy, \quad M_{ij}^e = \int_{\Omega^e} c_1 \psi_i^e \psi_j^e\, dx\, dy
$$

$$
K_{ij}^e = \int_{\Omega^e} \left[\frac{\partial \psi_i^e}{\partial x} \left(a_{xx} \frac{\partial \psi_j^e}{\partial x} + a_{xy} \frac{\partial \psi_j^e}{\partial y} \right) + \frac{\partial \psi_i^e}{\partial y} \left(a_{yx} \frac{\partial \psi_j^e}{\partial x} + a_{yy} \frac{\partial \psi_j^e}{\partial y} \right) \right.
$$

$$
\left. + a_{00} \psi_i^e \psi_j^e \right] dx\, dy
$$

$$
f_i^e = \int_{\Omega^e} \psi_i^e f\, dx\, dy, \quad Q^e = \oint_{\Gamma^e} \psi_i^e q_n\, ds \tag{6.7.6}
$$

In Sections 6.7.3 and 6.7.4, we consider time-approximation schemes to reduce Eq. (6.7.5) to a set of algebraic equations among the nodal values, completing the discretization in space and time. As a matter of information, the reduction of an equation of motion to the associated eigenvalue problem is also discussed. The case of $\mathbf{M} = \mathbf{0}$ arises in heat transfer and fluid dynamics problems, and the equation is known as the *parabolic equation*. Equation (6.7.5), in its general form is known as a hyperbolic equation, where $\mathbf{C}$ denotes the damping matrix and $\mathbf{M}$ the mass matrix, and it arises in structural dynamics with damping $(\mathbf{C} \neq \mathbf{0})$ and without damping $(\mathbf{C} = \mathbf{0})$. The time approximation of Eq. (6.7.5) for parabolic and hyperbolic equations will be derived separately. An equation of the type in Eq. (6.7.5) is also obtained by other spatial approximations methods like the finite difference method and the boundary element method, among others. Therefore, the discussion of converting matrix equations of the type in Eq. (6.7.5) is equally valid for other numerical methods used to approximate the spatial variation of the solution.

6.7.3   Full Discretization of Parabolic Equations

6.7.3.1   Eigenvalue problem

Consider Eq. (6.7.5), with $\mathbf{M}^e$ and $\mathbf{f}^e$ set to zero:

$$\mathbf{C}^e \dot{\mathbf{u}}^e + \mathbf{K}^e \mathbf{u}^e = \mathbf{Q}^e \tag{6.7.7}$$

which arises in heat transfer and fluid dynamics problems. The global solution vector is subject to the initial condition

$$\mathbf{u}(0) = \mathbf{u}_0 \tag{6.7.8}$$

where $\mathbf{u}_0$ denotes the value of the enclosed quantity $\mathbf{u}$ at time $t = 0$.

The *linear* eigenvalue problem associated with Eq. (6.7.7) is obtained by assuming that the problem is linear (i.e. $\mathbf{K}^e$ and $\mathbf{C}^e$ are independent of $\mathbf{u}^e$) and that the solution $\mathbf{u}(t)$ decays with time

$$\mathbf{u} = \bar{\mathbf{u}}\, e^{-\lambda t}, \quad \mathbf{Q} = \bar{\mathbf{Q}}\, e^{-\lambda t} \tag{6.7.9}$$

where $\bar{\mathbf{u}}$ is the vector of amplitudes (independent of time) and λ is the eigenvalue, which represents wave speed in heat transfer and fluid mechanics type problems. Substitution of Eq. (6.7.9) into Eq. (6.7.7) gives

$$(-\lambda \mathbf{C}^e + \mathbf{K}^e)\, \bar{\mathbf{u}}^e = \bar{\mathbf{Q}}^e \tag{6.7.10}$$

Assembly of element equations and imposition of boundary conditions follows along the same lines as for static problems.

6.7.3.2   Time (α-family of) approximations

The most commonly used method for reducing Eq. (6.7.7) to a set of algebraic equations is the α-family of approximation, in which a weighted average of the time derivative of a dependent variable is approximated at two consecutive time steps by linear approximation of the values of the variable at the two steps:

$$(1 - \alpha)\, \dot{\mathbf{u}}_s + \alpha\, \dot{\mathbf{u}}_{s+1} \approx \frac{\mathbf{u}_{s+1} - \mathbf{u}_s}{t_{s+1} - t_s} \quad \text{for } 0 \le \alpha \le 1 \tag{6.7.11}$$

where $(\cdot)_s$, for example, refers to the value of the enclosed quantity at time $t = t_s$. In the interest of brevity, the element label e on various quantities is omitted (i.e. the time approximation scheme is used at the element level). Equation (6.7.11) can be expressed alternatively as

$$\begin{aligned}
\mathbf{u}_{s+1} &= \mathbf{u}_s + \Delta t\left[(1 - \alpha)\dot{\mathbf{u}}_s + \alpha\dot{\mathbf{u}}_{s+1}\right] \\
&= \mathbf{u}_s + a_2\dot{\mathbf{u}}_s + a_1\dot{\mathbf{u}}_{s+1}, \quad \text{for } 0 \le \alpha \le 1
\end{aligned} \tag{6.7.12}$$

where $\Delta t = t_{s+1} - t_s$, $a_1 = \alpha\Delta t$, and $a_2 = (1 - \alpha)\Delta t$.

Note that when $\alpha = 0$, Eq. (6.7.11) reduces, for any vector of nodal values $\mathbf{u}(t)$, to

$$\dot{\mathbf{u}}_s = \left(\frac{d\mathbf{u}}{dt}\right)_{t=t_s} \approx \frac{\mathbf{u}(t_{s+1}) - \mathbf{u}(t_s)}{t_{s+1} - t_s} \tag{6.7.13}$$

Clearly, it amounts to replacing the time derivative of $\mathbf{u}(t)$ at $t = t_s$ with the *finite difference* of its values at $t = t_{s+1}$ (i.e. the value from a time step ahead) and $t = t_s$. Equation (6.7.13) is nothing but an approximation of the derivative of a function, since $\Delta t = t_{s+1} - t_s$ is finite (i.e. Δt cannot approach zero). The approximation in Eq. (6.7.13) is known as the *forward difference* method because it uses the function value ahead of the current position in computing the slope [see Fig. 6.7.1 for the approximation of the slope of a typical nodal value $u_j(t)$]. One may also use the value of the function from a time step behind

$$\left(\frac{d\mathbf{u}}{dt}\right)_{t=t_s} \approx \frac{\mathbf{u}(t_s) - \mathbf{u}(t_{s-1})}{t_s - t_{s-1}} \tag{6.7.14a}$$

or

$$\left(\frac{d\mathbf{u}}{dt}\right)_{t=t_{s+1}} \approx \frac{\mathbf{u}(t_{s+1}) - \mathbf{u}(t_s)}{t_{s+1} - t_s} \tag{6.7.14b}$$

which is the same as that in Eq. (6.7.11) when $\alpha = 1$. Equation (6.7.14b) is known as the *backward difference* method. If we use the values ahead and behind in computing the slope

$$\left(\frac{d\mathbf{u}}{dt}\right)_{t=t_s} \approx \frac{\mathbf{u}(t_{s+1}) - \mathbf{u}(t_{s-1})}{t_{s+1} - t_{s-1}} \tag{6.7.15}$$

it is called the *centered difference* method, which is not a special case of the α-family of approximation in Eq. (6.7.11).

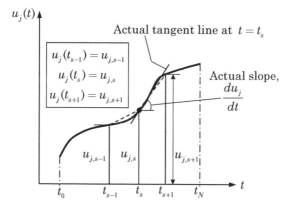

Fig. 6.7.1: Approximation of the first derivative of a function.

For different values of α, we obtain the following well-known numerical integration schemes from Eq. (6.7.12):

$$\alpha = \begin{cases} 0, \text{ the forward difference scheme (conditionally stable);} \\ \quad\quad \text{Order of accuracy } = O(\Delta t) \\ \frac{1}{2}, \text{ the Crank–Nicolson scheme (stable);} \quad O((\Delta t)^2) \\ \frac{2}{3}, \text{ the Galerkin method (stable);} \quad O((\Delta t)^2) \\ 1, \text{ the backward difference scheme (stable);} \quad O(\Delta t) \end{cases} \quad (6.7.16)$$

The meaning of the phrases "stability" and "conditional stability" will be discussed in Section 6.7.5.

6.7.3.3 Fully discretized equations

Equation (6.7.12) can be used to reduce ordinary differential equations in Eq. (6.7.7) to algebraic equations among the nodal values u_j at time t_{s+1}. Assuming that $\mathbf{C}$ is independent of time t (hence, independent of $\mathbf{u}$), we can use the approximation in Eq. (6.7.12) to eliminate $\dot{\mathbf{u}}$ in Eq. (6.7.7). Premultiplying Eq. (6.7.12) with $\mathbf{C}$ and replacing $\mathbf{C}\dot{\mathbf{u}}_s$ and $\mathbf{C}\dot{\mathbf{u}}_{s+1}$ with the aid of Eq. (6.7.7) evaluated at times t_s and t_{s+1}, respectively, we obtain

$$\hat{\mathbf{K}}(\mathbf{u}_{s+1})\mathbf{u}_{s+1} = \bar{\mathbf{K}}(\mathbf{u}_s)\mathbf{u}_s + \bar{\mathbf{F}}_{s,s+1} \equiv \hat{\mathbf{F}} \quad (6.7.17)$$

where

$$\begin{aligned} \hat{\mathbf{K}}(\mathbf{u}_{s+1}) &= \mathbf{C} + a_1\mathbf{K}(\mathbf{u}_{s+1}), \quad \bar{\mathbf{K}}(\mathbf{u}_s) = \mathbf{C} - a_2\mathbf{K}(\mathbf{u}_s) \\ \bar{\mathbf{F}}_{s,s+1} &= a_1\mathbf{F}_{s+1} + a_2\mathbf{F}_s, \quad a_1 = \alpha\Delta t, \quad a_2 = (1-\alpha)\Delta t \end{aligned} \quad (6.7.18)$$

6.7.3.4 Direct iteration scheme

When the direct iteration is used to solve the nonlinear equations in Eq. (6.7.17), at the $(r+1)$st iteration we solve the equation

$$\hat{\mathbf{K}}(\mathbf{u}_{s+1}^r)\mathbf{u}_{s+1}^{r+1} = \hat{\mathbf{F}}(\mathbf{u}_s, \mathbf{F}_s, \mathbf{F}_{s+1}) \quad (6.7.19)$$

with

$$\hat{\mathbf{F}} = \left[\mathbf{C} - a_2\mathbf{K}(\mathbf{u}_s)\right]\mathbf{u}_s + a_1\mathbf{F}_{s+1} + a_2\mathbf{F}_s \quad (6.7.20)$$

Note that $\hat{\mathbf{F}}$ remains unchanged during the nonlinear iteration for a given time step, whereas $\hat{\mathbf{K}}$ changes during the iteration due to its dependence on the latest known solution $\mathbf{u}_{s+1}^r$. Equation (6.7.19) provides a means to compute for $\mathbf{u}_{s+1}^{r+1}$ whenever $\mathbf{u}_s$ and $\mathbf{u}_{s+1}^r$ are known. Of course, $\mathbf{C}$, $\mathbf{K}$, and $\mathbf{F}$ are known for all times (parts of $\mathbf{F}$ may not be known at the element level, but after assembly they are known whenever the corresponding $\mathbf{u}$ is unknown). Equations (6.7.19) and (6.7.20) are valid for a typical element.

6.7.3.5 Newton's iteration scheme

When the Newton iteration procedure is used, we solve for the incremental solution vector $\Delta \mathbf{u}$ from

$$\Delta \mathbf{u} = -[\hat{\mathbf{T}}(\mathbf{u}_{s+1}^r)]^{-1} \mathbf{R}_{s+1}^r \qquad (6.7.21)$$

where $\hat{\mathbf{T}}$ is the tangent matrix associated with Eq. (6.7.17)

$$\hat{\mathbf{T}}(\mathbf{u}_{s+1}^r) \equiv \left[\frac{\partial \mathbf{R}_{s+1}}{\partial \mathbf{u}_{s+1}} \right]^r, \quad \mathbf{R}_{s+1}^r = \hat{\mathbf{K}}(\mathbf{u}_{s+1}^r) \mathbf{u}_{s+1}^r - \hat{\mathbf{F}} \qquad (6.7.22)$$

The total solution is obtained from

$$\mathbf{u}_{s+1}^{r+1} = \mathbf{u}_{s+1}^r + \Delta \mathbf{u} \qquad (6.7.23)$$

6.7.3.6 Explicit and implicit formulations and mass lumping

Note that for $\alpha = 0$ (the forward difference scheme), Eq. (6.7.18) gives $\hat{\mathbf{K}}^e = \mathbf{C}^e$. If the matrix $\mathbf{C}^e$ is diagonal, then Eq. (6.7.17) becomes *explicit* in the sense that the solution of Eq. (6.7.17) for $\mathbf{u}_{s+1}$ does not involve inverting a matrix. However, in the spatial approximation by the finite element method, matrix $\mathbf{C}^e$ is derived using a weak form and it is never a diagonal matrix. The matrix $\mathbf{C}^e$ derived using a weak form is called the *consistent matrix*. Thus, the finite element equations with consistent matrix $\mathbf{C}^e$ can never be explicit. In a finite difference method, the matrix $\mathbf{C}^e$ is diagonal, and therefore a time integration scheme like the forward difference method results in explicit set of equations, which are quite inexpensive to solve at each time step. To have the advantage of less computational time in dynamic/transient analysis by the finite element method, it is desirable to have $\mathbf{C}^e$ diagonalized. Thus, explicit finite element equations can be obtained only when: (a) the time approximation scheme is such that $\hat{\mathbf{K}}$ is equal to or a multiple of $\mathbf{C}$ and (b) the matrix $\mathbf{C}$ is diagonal. Otherwise, the formulation is called *implicit*.

There are two ways one may diagonalize the consistent matrix $\mathbf{C}$. The diagonalized matrices are also called *lumped matrices*. The *row-sum lumping* and *proportional lumping* techniques are the two ways to compute diagonal matrices. In row-sum lumping, the sum of the coefficients of each row of the consistent matrix $\mathbf{C}$ is used as the diagonal element and the off-diagonal elements are set to zero (no sum on i):

$$C_{ii} = \sum_{j=1}^{n} \int_{\Omega^e} c_0 \psi_i^e \psi_j^e \, d\mathbf{x} = \int_{\Omega^e} c_0 \psi_i^e \, d\mathbf{x}, \quad C_{ij} = 0 \text{ for } i \neq j \qquad (6.7.24)$$

where the property $\sum_{j=1}^{n} \psi_j^e = 1$ of the interpolation functions is used. In proportional lumping the diagonal elements are computed to be proportional to the diagonal elements of the consistent matrix while conserving the total integral of c_0 over the element (no sum on i):

$$C_{ii} = \alpha \int_{\Omega^e} c_0 \psi_i^e \psi_i^e \, dx, \quad C_{ij} = 0 \text{ for } i \neq j, \quad \alpha = \frac{\int_{\Omega^e} c_0 \, dx}{\sum_{i=1}^{n} \int_{\Omega^e} c_0 \psi_i^e \psi_i^e \, dx} \quad (6.7.25)$$

As an example consider the one-dimensional linear element. The consistent matrix is

$$\mathbf{C} = \frac{c_0 h}{6} \begin{bmatrix} 2 & 1 \\ 1 & 2 \end{bmatrix}$$

The row-sum lumping (RSL) yields the diagonal matrix

$$\mathbf{C}_{RSL} = \frac{c_0 h}{6} \begin{bmatrix} 2+1 & 0 \\ 0 & 1+2 \end{bmatrix} = \frac{c_0 h}{2} \begin{bmatrix} 1 & 0 \\ 0 & 1 \end{bmatrix}$$

The proportional lumping (PRL) gives

$$\alpha = \frac{c_0 h}{(c_0 h/3) + (c_0 h/3)} = \frac{3}{2}$$

and

$$\mathbf{C}_{PRL} = \frac{3}{2} \frac{c_0 h}{6} \begin{bmatrix} 2 & 0 \\ 0 & 2 \end{bmatrix} = \frac{c_0 h}{2} \begin{bmatrix} 1 & 0 \\ 0 & 1 \end{bmatrix}$$

Similarly, for the one-dimensional quadratic element, the consistent matrix is

$$\mathbf{C} = \frac{c_0 h}{30} \begin{bmatrix} 4 & 2 & -1 \\ 2 & 16 & 2 \\ -1 & 2 & 4 \end{bmatrix}$$

The row-sum lumping yields the diagonal matrix

$$\mathbf{C}_{RSL} = \frac{c_0 h}{30} \begin{bmatrix} 4+2-1 & 0 & 0 \\ 0 & 2+16+2 & 0 \\ 0 & 0 & -1+2+4 \end{bmatrix} = \frac{c_0 h}{6} \begin{bmatrix} 1 & 0 & 0 \\ 0 & 4 & 0 \\ 0 & 0 & 1 \end{bmatrix}$$

The proportional lumping gives

$$\alpha = \frac{c_0 h}{(4 c_0 h/30) + (16 c_0 h/30) + (4 c_0 h/30)} = \frac{3}{2}$$

and

$$\mathbf{C}_{PRL} = \frac{5}{4} \frac{c_0 h}{30} \begin{bmatrix} 4 & 0 & 0 \\ 0 & 16 & 0 \\ 0 & 0 & 4 \end{bmatrix} = \frac{c_0 h}{6} \begin{bmatrix} 1 & 0 & 0 \\ 0 & 4 & 0 \\ 0 & 0 & 1 \end{bmatrix}$$

6.7.4 Full Discretization of Hyperbolic Equations

6.7.4.1 Newmark's scheme

Consider the original second-order equation in Eq. (6.7.5)

$$\mathbf{M}^e \ddot{\mathbf{u}}^e + \mathbf{C}^e \dot{\mathbf{u}}^e + \mathbf{K}^e \mathbf{u}^e = \mathbf{F}^e \tag{6.7.26}$$

which arises in structural dynamics, where $\mathbf{C}^e$ denotes the damping matrix, $\mathbf{M}^e$ the mass matrix, and $\mathbf{K}^e$ the stiffness matrix. The global displacement vector $\mathbf{u}$ is subject to the initial conditions that the displacement and velocity are known at time $t = 0$

$$\mathbf{u}(0) = \mathbf{u}_0, \quad \dot{\mathbf{u}}(0) = \mathbf{v}_0 \tag{6.7.27}$$

There are several numerical integration methods available to integrate second-order (i.e. hyperbolic) equations [6–10]. Among these, the Newmark family of time integration schemes [9] is widely used in structural dynamics. Other methods, such as the Wilson method and the Houbolt method [10], can be used to develop the algebraic equations from the second-order differential equations.

In the Newmark method, the function and its time derivatives are approximated according to

$$\begin{aligned}
\mathbf{u}_{s+1} &= \mathbf{u}_s + \Delta t\, \dot{\mathbf{u}}_s + \tfrac{1}{2}(\Delta t)^2 \left[(1-\gamma)\ddot{\mathbf{u}}_s + \gamma\, \ddot{\mathbf{u}}_{s+1}\right] \\
\dot{\mathbf{u}}_{s+1} &= \dot{\mathbf{u}}_s + a_2 \ddot{\mathbf{u}}_s + a_1 \ddot{\mathbf{u}}_{s+1}, \quad a_1 = \alpha\, \Delta t, \quad a_2 = \Delta t\, (1-\alpha)
\end{aligned} \tag{6.7.28}$$

and α and γ are parameters that determine the stability and accuracy of the scheme. For $\alpha = 0.5$, the following values of γ define various well-known schemes:

$$\gamma = \begin{cases}
\tfrac{1}{2}, \text{ the constant-average acceleration method (stable)} \\
\tfrac{1}{3}, \text{ the linear acceleration method (conditionally stable)} \\
0, \text{ the central difference method (conditionally stable)} \\
\tfrac{8}{5}, \text{ the Galerkin method (stable)} \\
2, \text{ the backward difference method (stable)}
\end{cases} \tag{6.7.29}$$

6.7.4.2 Fully discretized equations

The set of ordinary differential equations in Eq. (6.7.26) can be reduced, with the help of Eq. (6.7.28), to a set of algebraic equations relating $\mathbf{u}_{s+1}$ to $\mathbf{u}_s$, as explained next.

Solving the first equation in Eq. (6.7.28) for $\ddot{\mathbf{u}}_{s+1}$, we obtain

$$\begin{aligned}
\ddot{\mathbf{u}}_{s+1} &= \frac{2}{\gamma(\Delta t)^2}\left[\mathbf{u}_{s+1} - \mathbf{u}_s - \Delta t \dot{\mathbf{u}}_s - \frac{(\Delta t)^2}{2}(1-\gamma)\ddot{\mathbf{u}}_s\right] \\
&= a_3\,(\mathbf{u}_{s+1} - \mathbf{u}_s) - a_4 \dot{\mathbf{u}}_s - a_5 \ddot{\mathbf{u}}_s
\end{aligned} \tag{6.7.30}$$

where

$$a_3 = \frac{2}{\gamma(\Delta t)^2}, \quad a_4 = \Delta t\, a_3, \quad a_5 = \frac{(\Delta t)^2}{2}(1-\gamma)a_3 = \frac{1-\gamma}{\gamma}$$

Substituting the result into the second equation in Eq. (6.7.28), we obtain

$$\dot{\mathbf{u}}_{s+1} = \dot{\mathbf{u}}_s + a_2\ddot{\mathbf{u}}_s + a_1\left[a_3\left(\mathbf{u}_{s+1} - \mathbf{u}_s\right) - a_4\dot{\mathbf{u}}_s - a_5\ddot{\mathbf{u}}_s\right]$$
$$= a_6\left(\mathbf{u}_{s+1} - \mathbf{u}_s\right) - a_7\dot{\mathbf{u}}_s - a_8\ddot{\mathbf{u}}_s \qquad (6.7.31)$$

where

$$a_6 = \frac{2\alpha}{\gamma\Delta t}, \quad a_7 = \frac{2\alpha}{\gamma} - 1, \quad a_8 = \left(\frac{\alpha}{\gamma} - 1\right)\Delta t$$

Premultiplying Eq. (6.7.30) with $\mathbf{M}_{s+1}$ and substituting for $\mathbf{M}_{s+1}\ddot{\mathbf{u}}_{s+1}$ from Eq. (6.7.26), we obtain the result,

$$\mathbf{F}_{s+1} - \mathbf{K}_{s+1}\mathbf{u}_{s+1} - \mathbf{C}_{s+1}\dot{\mathbf{u}}_{s+1} = \mathbf{M}_{s+1}\left[a_3\left(\mathbf{u}_{s+1} - \mathbf{u}_s\right) - a_4\dot{\mathbf{u}}_s - a_5\ddot{\mathbf{u}}_s\right]$$

and replacing $\mathbf{C}_{s+1}\dot{\mathbf{u}}_{s+1}$ using Eq. (6.7.31), we arrive at

$$\mathbf{F}_{s+1} - \mathbf{K}_{s+1}\mathbf{u}_{s+1} - \mathbf{C}_{s+1}\left[a_6\left(\mathbf{u}_{s+1} - \mathbf{u}_s\right) - a_7\dot{\mathbf{u}}_s - a_8\ddot{\mathbf{u}}_s\right]$$
$$= \mathbf{M}_{s+1}\left[a_3\left(\mathbf{u}_{s+1} - \mathbf{u}_s\right) - a_4\dot{\mathbf{u}}_s - a_5\ddot{\mathbf{u}}_s\right]$$

Collecting the terms involving $\mathbf{u}_{s+1}$ on one side and the remaining terms on the other side, we obtain the result:

$$\left(\mathbf{K}_{s+1} + a_3\mathbf{M}_{s+1} + a_6\mathbf{C}_{s+1}\right)\mathbf{u}_{s+1}$$
$$= \mathbf{M}_{s+1}\left(a_3\mathbf{u}_s + a_4\dot{\mathbf{u}}_s + a_5\ddot{\mathbf{u}}_s\right) + \mathbf{C}_{s+1}\left(a_6\mathbf{u}_s + a_7\dot{\mathbf{u}}_s + a_8\ddot{\mathbf{u}}_s\right) + \mathbf{F}_{s+1}$$

or

$$\hat{\mathbf{K}}_{s+1}(\mathbf{u}_{s+1})\mathbf{u}_{s+1} = \hat{\mathbf{F}}_{s,s+1} \qquad (6.7.32)$$

where

$$\hat{\mathbf{K}}_{s+1} = \mathbf{K}_{s+1} + a_3\mathbf{M}_{s+1} + a_6\mathbf{C}_{s+1}$$
$$\hat{\mathbf{F}}_{s,s+1} = \mathbf{F}_{s+1} + \mathbf{M}_{s+1}\mathbf{A}_s + \mathbf{C}_{s+1}\mathbf{B}_s \qquad (6.7.33)$$
$$\mathbf{A}_s = a_3\mathbf{u}_s + a_4\dot{\mathbf{u}}_s + a_5\ddot{\mathbf{u}}_s, \quad \mathbf{B}_s = a_6\mathbf{u}_s + a_7\dot{\mathbf{u}}_s + a_8\ddot{\mathbf{u}}_s$$

and a_i $(i = 1, 2, \ldots, 8)$ are defined as

$$a_1 = \alpha\Delta t, \quad a_2 = (1-\alpha)\Delta t$$
$$a_3 = \frac{2}{\gamma(\Delta t)^2}, \quad a_4 = a_3\Delta t, \quad a_5 = \frac{1}{\gamma} - 1 \qquad (6.7.34)$$
$$a_6 = \frac{2\alpha}{\gamma\Delta t}, \quad a_7 = \frac{2\alpha}{\gamma} - 1, \quad a_8 = \Delta t\left(\frac{\alpha}{\gamma} - 1\right)$$

The direct iteration and Newton iteration procedures for the hyperbolic equations are the same as those given in Eqs. (6.7.19)–(6.7.23), except for the definitions of the matrices involved.

The following remarks concerning the Newmark scheme are in order:

1. The calculation of $\hat{\mathbf{K}}$ and $\hat{\mathbf{F}}$ in Newmark's scheme requires knowledge of the initial conditions $\mathbf{u}_0$, $\dot{\mathbf{u}}_0$, and $\ddot{\mathbf{u}}_0$. In practice, one does not know $\ddot{\mathbf{u}}_0$. As an approximation, it can be calculated from the assembled system of equations associated with Eq. (6.7.26) using initial conditions on $\mathbf{u}$, $\dot{\mathbf{u}}$, and $\mathbf{F}$ (often $\mathbf{F}$ is assumed to be zero at $t = 0$):

$$\ddot{\mathbf{u}}_0 = \mathbf{M}^{-1}\left(\mathbf{F}_0 - \mathbf{K}\mathbf{u}_0 - \mathbf{C}\dot{\mathbf{u}}_0\right) \qquad (6.7.35)$$

2. At the end of each time step, the new velocity vector $\dot{\mathbf{u}}_{s+1}$ and acceleration vector $\ddot{\mathbf{u}}_{s+1}$ are computed using Eqs. (6.7.30) and (6.7.31):

$$
\begin{aligned}
\ddot{\mathbf{u}}_{s+1} &= a_3\left(\mathbf{u}_{s+1} - \mathbf{u}_s\right) - a_4\dot{\mathbf{u}}_s - a_5\ddot{\mathbf{u}}_s \\
\dot{\mathbf{u}}_{s+1} &= \dot{\mathbf{u}}_s + a_2\ddot{\mathbf{u}}_s + a_1\ddot{\mathbf{u}}_{s+1}
\end{aligned} \qquad (6.7.36)
$$

where a_1 through a_5 are defined in Eq. (6.7.34).

3. Equation (6.7.32) is not valid for the *centered difference* scheme ($\gamma = 0$), as some of the parameters a_i are not defined for this scheme. An alternative algebraic manipulation of Eqs. (6.7.26) and (6.7.28) is required. It can be shown that (see Problem 6.9)

$$\mathbf{H}_{s+1}\ddot{\mathbf{u}}_{s+1} = \mathbf{F}_{s+1} - \mathbf{K}_{s+1}\mathbf{A}_s - \mathbf{C}_{s+1}\mathbf{B}_s \qquad (6.7.37)$$

where

$$
\begin{aligned}
\mathbf{H}_{s+1} &= 0.5\gamma(\Delta t)^2\mathbf{K}_{s+1} + a_1\,\mathbf{C}_{s+1} + \mathbf{M}_{s+1} \\
\mathbf{A}_s &= \mathbf{u}_s + \Delta t\,\dot{\mathbf{u}}_s + \frac{1-\gamma}{2}(\Delta t)^2\,\ddot{\mathbf{u}}_s \\
\mathbf{B}_s &= \dot{\mathbf{u}}_s + a_2\,\ddot{\mathbf{u}}_s
\end{aligned} \qquad (6.7.38)
$$

and the displacements and velocities are updated using Eq. (6.7.28):

$$
\begin{aligned}
\mathbf{u}_{s+1} &= \mathbf{u}_s + \Delta t\,\dot{\mathbf{u}}_s + \frac{(\Delta t)^2}{2}\left[(1-\gamma)\ddot{\mathbf{u}}_s + \gamma\ddot{\mathbf{u}}_{s+1}\right] \\
\dot{\mathbf{u}}_{s+1} &= \dot{\mathbf{u}}_s + \Delta t\left[(1-\alpha)\ddot{\mathbf{u}}_s + \alpha\ddot{\mathbf{u}}_{s+1}\right]
\end{aligned}
$$

4. The centered difference scheme ($\gamma = 0$) with $\alpha = 0$ yields [see Eqs. (6.7.37) and (6.7.38)]

$$\mathbf{M}_{s+1}\ddot{\mathbf{u}}_{s+1} = \mathbf{F}_{s+1} - \mathbf{K}_{s+1}\left[\mathbf{u}_s + \Delta t\dot{\mathbf{u}}_s + \tfrac{1}{2}(\Delta t)^2\ddot{\mathbf{u}}_s\right] - \mathbf{C}_{s+1}\left(\dot{\mathbf{u}}_s + \Delta t\ddot{\mathbf{u}}_s\right) \qquad (6.7.39)$$

Thus, if the mass matrix is made diagonal, the system in Eq. (6.7.37) becomes explicit (no inversion of the coefficient matrix is required).

5. For natural vibration, the forces and the solution are assumed to be periodic ($\mathbf{F} = \mathbf{Q}$ when the source term is set to zero)

$$\mathbf{u} = \mathbf{u}^0 \, e^{i\omega t}, \quad \mathbf{Q} = \mathbf{Q}^0 \, e^{i\omega t}, \quad i = \sqrt{-1}$$

where $\mathbf{u}^0$ is the vector of amplitudes (independent of time) and ω is the frequency of natural vibration of the system. Substitution of the periodic motion into Eq. (6.7.26), assuming that it is linear, yields

$$\left(-\omega^2 \mathbf{M} + i\omega \mathbf{C} + \mathbf{K}\right)\mathbf{u}^0 = \mathbf{Q}^0 \tag{6.7.40}$$

where $\mathbf{Q}^0$ is the vector of secondary variables. Equation (6.7.40) is called an eigenvalue problem, which may have complex eigenvalues when damping $\mathbf{C}$ is included.

6.7.5   Stability and Accuracy

6.7.5.1   Preliminary comments

In general, the application of a time approximation scheme to an initial-value problem results in an equation of the type

$$\hat{\mathbf{K}}\mathbf{u}_{s+1} = \bar{\mathbf{K}}\mathbf{u}_s \quad \text{or} \quad \mathbf{u}_{s+1} = \mathbf{A}\mathbf{u}_s \tag{6.7.41}$$

where $\mathbf{A} = \hat{\mathbf{K}}^{-1}\bar{\mathbf{K}}$ is called the *amplification* matrix, and $\hat{\mathbf{K}}$ and $\bar{\mathbf{K}}$ are matrix operators that depend on the problem parameters [see Eqs. (6.7.33) and (6.7.34)], for example, geometric and material properties, time step, and mesh parameter, and $\mathbf{u}_{s+1}$ is the solution vector at time t_{s+1}.

Since Eq. (6.7.11), for example, represents an approximation that is used to derive an equation of the type in Eq. (6.7.41), error is introduced into the solution $\mathbf{u}_{s+1}$ at each time step. Since the solution $\mathbf{u}_{s+1}$ at time t_{s+1} depends on the solution $\mathbf{u}_s$ at time t_s, the error can grow with time. The time approximation scheme is said to be *stable* if the error introduced in $\mathbf{u}_s$ does not grow unbounded as Eq. (6.7.41) is solved repeatedly for $s = 0, 1, \ldots,$. In order for the error to remain bounded, it is necessary and sufficient that the largest eigenvalue of the amplification matrix $\mathbf{A}$ be less than or equal to unity:

$$|\lambda_{\max}| \leq 1 \tag{6.7.42}$$

where $\lambda_{\max}$ is the largest value that satisfies the equation

$$\left(\mathbf{A} - \lambda_A \mathbf{I}\right)\mathbf{u} = \mathbf{0} \tag{6.7.43}$$

We note that λ_A depends on the characteristic length of the mesh, time step, and material parameters (such as the density and modulus). Since modulus and density are fixed for a given problem, the stability criterion essentially provides a relation between the characteristic element size and the time step used. Equation (6.7.43) represents a nonlinear eigenvalue problem, if one can be formulated. If the condition in Eq. (6.7.42) is satisfied for any value of Δt, independent of the mesh size, the scheme is said to be *unconditionally stable*, or simply *stable*. If Eq. (6.7.42) places a restriction, for a given mesh, on the size of the time step Δt, the scheme is said to be *conditionally stable*. As the mesh is refined, the value of the maximum eigenvalue increases and, hence, the value of the critical time step decreases.

Accuracy of a numerical scheme is a measure of the closeness between the approximate solution and the exact solution, whereas *stability* of a solution is a measure of the boundedness of the approximate solution with time. As one might expect, the size of the time step can influence both accuracy and stability. When we construct an approximate solution, we like it to converge to the true solution when the number of elements or the degree of approximation is increased and the time step Δt is decreased. A time approximation scheme is said to be *convergent* if, for fixed t_s and Δt, the numerical value $\mathbf{u}_s$ converges to its true value $\mathbf{u}(t_s)$ as $\Delta t \to 0$. Accuracy is measured in terms of the rate at which the approximate solution converges. If a numerical scheme is stable and consistent, it is also convergent.

6.7.5.2 Stability criteria

The discussion here is limited only to linear problems, as there are no proofs available for stability criteria for nonlinear problems. The α-family of approximations applied to linear problems can be shown to be stable for all numerical schemes for which $\alpha \geq \frac{1}{2}$; for $\alpha < \frac{1}{2}$, the scheme is conditionally stable if the time step Δt is such that the following (stability) condition is satisfied:

$$\Delta t < \Delta t_{\mathrm{cr}} \equiv \frac{2}{(1 - 2\alpha)\lambda_{\max}} \qquad (6.7.44)$$

where $\lambda_{\max}$ is the largest eigenvalue of the finite element equations after assembly and imposition of boundary conditions

$$(\mathbf{K} - \lambda\mathbf{C})\,\mathbf{u}^0 = \mathbf{0} \qquad (6.7.45)$$

Note that the same mesh as that used for the transient analysis must be used to calculate the eigenvalues of the assembled system; after assembly and imposition of boundary conditions, the eigenvalue problem becomes homogeneous, as given in Eq. (6.7.45).

The stability criterion in Eq. (6.7.44) is arrived using Eq. (6.7.42). The amplification matrix for the α-family of approximations as applied to linear problems is given by

$$\mathbf{A} = (\hat{\mathbf{K}})^{-1}\,\bar{\mathbf{K}} = (\mathbf{C} + a_1\mathbf{K})^{-1}(\mathbf{C} - a_2\mathbf{K}) \qquad (6.7.46)$$

Let $\lambda_{\max}$ be the maximum eigenvalue of Eq. (6.7.45). Then it can be shown that (using spectral decomposition of $\mathbf{A}$) the maximum eigenvalue of $\mathbf{A}$ is equal to

$$(\lambda_A)_{\max} = \left| \frac{1 - (1-\alpha)\Delta t\, \lambda_{\max}}{1 + \alpha\, \Delta t\, \lambda_{\max}} \right| \le 1 \tag{6.7.47}$$

from which it follows that the α-family of approximations is unconditionally stable if $\alpha \ge \frac{1}{2}$. In the case $\alpha < \frac{1}{2}$, the method is stable only if the condition in Eq. (6.7.44) is satisfied.

Similarly, for all Newmark schemes in which $\gamma < \alpha$ and $\alpha \ge \frac{1}{2}$, the stability requirement is (for linear problems)

$$\Delta t \le \Delta t_{cr} = \left[\frac{1}{2}\omega_{\max}^2(\alpha - \gamma) \right]^{-1/2} \tag{6.7.48}$$

where $\omega_{\max}$ is the maximum natural frequency of the undamped system (6.7.40)

$$\left(\mathbf{K} - \omega^2 \mathbf{M} \right) \mathbf{u} = \mathbf{0} \tag{6.7.49}$$

For all nonlinear problems, one often uses the time step restrictions imposed by linear stability criteria.

6.7.6   Computer Implementation

Computer implementation of linear or nonlinear time-dependent problems is complicated by the fact that one must keep track of the solution vectors at different times and iterations. In general, there are three loops. The outermost one is on the *load*, which can be a force or source term whose influence on the system is investigated. While it is the load placed on the system in the case of a solid mechanics problem, it may be the Reynolds number in fluid flow problems and heat flux in heat transfer problems. The outer loop on the number of load steps is followed by a loop on the number of time steps, and the inner most loop being on nonlinear equilibrium iterations. The flow chart shown in Fig. 6.7.2 illustrates the general idea.

One must keep track of the solution vector from the previous time step $\mathbf{u}_s$ and the solution vector from the previous iteration $\mathbf{u}_{s+1}^r$. This is because the formulation requires solution $\mathbf{u}_s$ to compute $\hat{\mathbf{F}}$ [see Eqs. (6.7.18) and (6.7.20) in the case of parabolic equations and Eqs. (6.7.33) and (6.7.34) for hyperbolic equations], and the latest known solution $\mathbf{u}_{s+1}^r$ is required to update the coefficient matrix $\hat{\mathbf{K}}_{s+1}$ during the equilibrium iteration.

Another point one must note is that the source vector $\mathbf{F}$ due to the body force $f(\mathbf{x}, t)$ [see Eqs. (6.7.1) and (6.7.6)] appears in the fully discretized equation of the parabolic equations [see Eq. (6.7.18)] as a vector multiplied by Δt. Therefore, when nonzero secondary variables are read as input through the boundary conditions, they must be multiplied by Δt (assuming that **f** is independent of time) inside the program. This is not required in the case of hyperbolic equations, as can be seen from Eq. (6.7.33). Although it can be done, we will not consider time-dependent boundary conditions here.

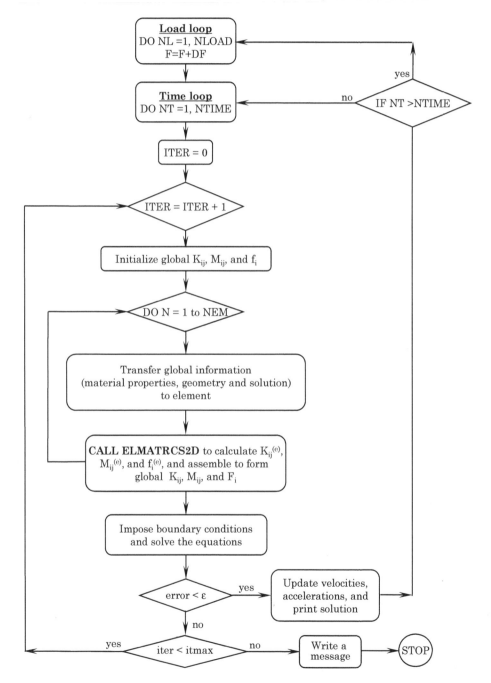

Fig. 6.7.2: Flow chart of the nonlinear transient analysis of a typical problem.

Fortran statements showing the input data to be read for transient problems and the transfer of global solution vectors from the previous time step as well as from the latest iteration of the current time step to subroutine ELMATRCS2D (used to calculate element matrices and residual vector) are presented in Box 6.7.1. Here, variable ITEM denotes a flag for static (ITEM = 0) or transient (ITEM > 0) analysis. Also, when ITEM > 0, ITEM = 1 is used for parabolic equations and ITEM $\geq$ 2 is used for hyperbolic equations. Damping is not included in the program and the variable ELM is used for **C** or **M**, depending whether a parabolic or hyperbolic equation is solved. Coefficient c_0 or c_1 is represented as $CXY = C0 + CX \cdot x + CY \cdot y$.

Box 6.7.1: Fortran statements showing the transfer of global solution vectors to element solution vectors.

```
C
C      Read the necessary data for time-dependent problems
C
        IF(ITEM.NE.0) THEN
             READ(IN,*) C0,CX,CY
             WRITE(ITT,820)
             WRITE(ITT,540) C0,CX,CY
             READ(IN,*) NTIME
             READ(IN,*) DT,ALFA,GAMA,EPSLN
             A1=ALFA*DT
             A2=(1.0-ALFA)*DT
             DO 40 I=1,NEQ
  40            GLU(I)=0.0
             IF(ITEM.EQ.1) THEN
                  IF(NSSV.NE.0) THEN
                       DO 50 I=1,NSSV
  50                    VSSV(I)=VSSV(I)*DT
                  ENDIF
             ELSE
                  DT2=DT*DT
                  A3=2.0/GAMA/DT2
                  A4=A3*DT
                  A5=1.0/GAMA-1.0
C
C ***The initial conditions are assumed to be homogeneous***
C
                  DO 70 I=1,NEQ
                  GLV(I)=0.0
  70              GLA(I)=0.0
             ENDIF
        ENDIF
C
C ***Initialize the arrays***
        . . . . . . . . . . .
C
```

Box 6.7.1: Fortran statements showing the transfer of global solution vectors to element solution vectors (continued).

```
        DO 250 N=1,NEM
        DO 200 I=1,NPE
        NI=NOD(N,I)
        IF(NONLIN.GT.0 .OR. ITEM.GT.0)THEN
            ELU(I)=GLU(NI)          !Transfer of the current solution
            IF(ITEM.GT.0)THEN
                ELUO(I)=GLP(NI) !Transfer of previous time step solution
            ENDIF
        ENDIF
        IF(ITEM.EQ.2) THEN
            ELU1(I)=GLV(NI)    !Transfer of previous first time derivative
            ELU2(I)=GLA(NI)    !Transfer of previous second time derivative
        ENDIF
        ELXY(I,1)=GLXY(NI,1)
        ELXY(I,2)=GLXY(NI,2)
200 CONTINUE
C
C     Call Subroutine ELMATRCS2D to compute ELK, ELK-HAT, etc.
C     and assemble them into global matrices GLK
        . . . . . . . . .
C
250 CONTINUE
```

The following variables are used (the meaning of other variables used is obvious by their names):

NTIME = number of time steps

ELU0(i) = element solution vector at time t_s

ELU1(i) = first time derivative of the element solution vector at time t_s

ELU2(i) = second time derivative of the element solution vector at time t_s

ELU(i) = element solution vector at t_{s+1} in the latest iteration, r

The corresponding global vectors are denoted by GPU, GLV, GLA, and GLU, respectively. In the case of hyperbolic equations, these vectors must be updated at the end of each time step (i.e. once convergence is reached) using Eqs. (6.7.30) and (6.7.31).

Fortran statements showing the calculation of $\hat{\mathbf{K}}$ and $\hat{\mathbf{F}}$ inside the subroutine are presented in Box 6.7.2. The matrix $\mathbf{K}_s = \mathbf{K}(\mathbf{u}_s)$ is denoted with ELK0(i, j). When $\mathbf{K}$ is independent of time t and $\mathbf{u}$ (i.e. for the linear problems), we have $\mathbf{K}_s = \mathbf{K}_{s+1} = \mathbf{K}$ and hence ELK0(i, j) = ELK(i, j). Of course, even when the problem is linear, $\mathbf{K}$ is a function of time if the coefficients appearing in the definition of K_{ij}^e are time dependent; then $\mathbf{K}_s \neq \mathbf{K}_{s+1}$.

Box 6.7.2: Fortran statements for the calculation of $\hat{\mathbf{K}}$ and $\hat{\mathbf{F}}$.

```
         DO 200 NI = 1,NGPF      ! Full Gauss integration loop
         DO 200 NJ = 1,NGPF
                . . . . . . . .
C
C    Define linear and nonlinear coefficients of the equation
                . . . . . . .
         IF(ITEM.GT.0)THEN
             CXY=C0+CX*X+CY*Y
         ENDIF
                . . . . . . . .
C
         IF(ITEM.GT.0)THEN
             UP =0.0
             UPX=0.0
             UPY=0.0
             DO 140 I=1,NPE
             UP =UP +ELU0(I)*SF(I)
             UPX=UPX+ELU0(I)*GDSF(1,I)
140          UPY=UPY+ELU0(I)*GDSF(2,I)
             APXX=A11+A1U*UP+A1UX*UPX+A1UY*UPY
             APYY=A22+A2U*UP+A2UX*UPX+A2UY*UPY
         ENDIF
C
C    Define the element coefficient matrices ELK, ELF, and ELM
              . . . . . . . .
C
         IF(ITEM.GT.0)THEN
             ELM(I,J)=ELM(I,J)+CXY*S00
             IF(NONLIN.GT.0)THEN            ! Define
                 ELK0(I,J)=ELK0(I,J)+APXX*SXX+APYY*SYY+A00*S00
             ENDIF
           ENDIF
C        . . . . . . . .
C
200 CONTINUE
C
C    Compute  effective [K] and {F}
C
         IF(ITEM.EQ.1) THEN   ! For parabolic equations
             DO 220 I=1,NN
             SUM=0.0
             DO 210 J=1,NN
             IF(NONLIN.GT.0)THEN
                 SUM = SUM+(ELM(I,J)-A2*ELK0(I,J))*ELU0(J)
             ELSE
                 SUM = SUM+(ELM(I,J)-A2*ELK(I,J))*ELU0(J)
             ENDIF
210          ELK(I,J) = ELM(I,J)+A1*ELK(I,J)
220          ELF(I) = (A1+A2)*ELF(I)+SUM
         ENDIF
```

Box 6.7.2: Fortran statements for the calculation of $\hat{\mathbf{K}}$ and $\hat{\mathbf{F}}$ (continued).

```
         IF(ITEM.GT.1) THEN   ! For hyperbolic equations
             DO 270 I = 1,NN
             SUM  = 0.0
             DO 260 J = 1,NN
             SUM  = SUM+ELM(I,J)*(A3*ELU0(J)+A4*ELU1(J)+A5*ELU2(J))
260      ELK(I,J) = ELK(I,J)+A3*ELM(I,J)
270      ELF(I)   = ELF(I)+SUM
         ENDIF
```

6.7.7   Numerical Examples

Here we consider several representative examples of time-dependent problems. We begin with a linear heat conduction problem.

Example 6.7.1

Consider the transient heat conduction equation

$$\frac{\partial \theta}{\partial t} - \left(\frac{\partial^2 \theta}{\partial x^2} + \frac{\partial^2 \theta}{\partial y^2} \right) = 1 \text{ in } \Omega; \quad \theta = 0 \text{ on } \Gamma \tag{1}$$

where θ is the non-dimensional temperature, Ω is the square domain of side 2, and Γ is its boundary (see Fig. 6.7.3). Determine the temperature field inside the domain for $t > 0$ for the initial condition $\theta(x, y, 0) = 0$. Exploit the biaxial symmetry and use a uniform mesh of 8×8 linear rectangular elements to model the domain. Also, investigate the stability and accuracy of various schemes.

Solution: The boundary conditions along the lines of symmetry require that the heat flux be zero there. Thus, the boundary conditions of the computational domain are

$$\frac{\partial \theta}{\partial x}(0, y, t) = 0, \quad \frac{\partial \theta}{\partial y}(x, 0, t) = 0, \quad \theta(1, y, t) = 0, \quad \theta(x, 1, t) = 0 \tag{2}$$

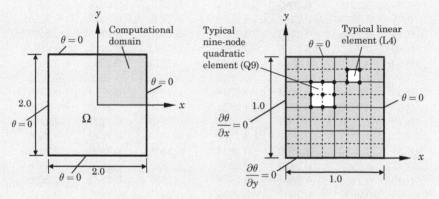

Fig. 6.7.3: Actual and computational domains of the transient heat transfer problem.

Since the Crank–Nicolson ($\alpha = 0.5$) and backward difference ($\alpha = 1.0$) methods are unconditionally stable schemes, one can choose any value of Δt. However, if Δt is too large, the solution may not be accurate even when it is stable. In order to estimate the time step, one must calculate the maximum eigenvalue for the mesh used in the transient analysis. The solution of the eigenvalue problem associated with the 8×8 mesh of linear elements yields 64 eigenvalues, of which the maximum eigenvalue is $\lambda_{\max} = 1492.56$. Therefore, the critical time step for the forward difference scheme ($\alpha = 0.0$) is given by $\Delta t_{cr} = (2/1492.56) = 0.00134$.

Figure 6.7.4 shows plots of the temperature $\theta(0, 0, t)$ versus time t for $\Delta t = 0.002$, which is greater than the critical time step. For very small times, $\theta(0, 0, t) \approx t$, and both backward difference and Crank–Nicolson schemes show stable behavior while the forward difference scheme is unstable. For $\Delta t = 0.001$, the forward difference scheme also gives the same result as the stable schemes using $\Delta t = 0.05$. Table 6.7.1 shows the numerical values of $\theta(0, 0, t)$ predicted by various schemes and two different meshes. One must note that the critical time step for the $4 \times 4Q9$ mesh is different as the maximum eigenvalue is different (Δt_{cr} is likely to be smaller for the mesh of quadratic elements). Figure 6.7.5 contains plots of the evolution of $\theta(0, 0, t)$ with time, reaching a steady-state at around $t = 1.25$. Both meshes give results that cannot be distinguished in the graph. The Crank–Nicolson scheme ($\Delta t = 0.05$) and forward difference scheme ($\Delta t = 0.001$) essentially give the same results, whereas the backward difference scheme ($\Delta t = 0.05$) differs slightly at initial times but develops to the same steady-state solution. Finally, $\theta(x, 0, t)$ versus x, obtained with the Crank–Nicolson and backward difference schemes and 8×8 mesh of linear elements, for $t = 0.25, 0.5, 1.0,$ and 1.25 are presented in Fig. 6.7.6.

Table 6.7.1: Variation of $\theta(0, 0, t)$ with time t, obtained with various time approximation schemes.

	8 × 8L			4 × 4Q9		
Time, t	Crank–Nicolson $\Delta t = 0.05$	Backward difference $\Delta t = 0.05$	Forward difference $\Delta t = 0.001$	Crank–Nicolson $\Delta t = 0.05$	Backward difference $\Delta t = 0.05$	Forward difference $\Delta t = 0.001$
0.05	0.0497	0.0480	0.0500	0.0496	0.0479	0.0500
0.10	0.0975	0.0916	0.0983	0.0971	0.0913	0.0979
0.15	0.1398	0.1294	0.1400	0.1390	0.1288	0.1393
0.20	0.1740	0.1611	0.1737	0.1730	0.1604	0.1728
0.25	0.2006	0.1873	0.2004	0.1996	0.1864	0.1994
0.30	0.2215	0.2085	0.2213	0.2205	0.2075	0.2202
0.35	0.2379	0.2257	0.2376	0.2368	0.2247	0.2365
0.40	0.2506	0.2395	0.2503	0.2495	0.2385	0.2493
0.45	0.2605	0.2506	0.2603	0.2594	0.2496	0.2592
0.50	0.2682	0.2595	0.2680	0.2672	0.2585	0.2670
0.55	0.2743	0.2667	0.2741	0.2732	0.2656	0.2731
0.60	0.2790	0.2724	0.2788	0.2779	0.2714	0.2778
0.65	0.2826	0.2770	0.2825	0.2816	0.2760	0.2815
0.70	0.2855	0.2807	0.2854	0.2845	0.2797	0.2844
0.75	0.2877	0.2837	0.2876	0.2867	0.2827	0.2866
0.80	0.2894	0.2860	0.2894	0.2885	0.2850	0.2884
0.85	0.2908	0.2879	0.2908	0.2898	0.2870	0.2898
0.90	0.2919	0.2894	0.2918	0.2909	0.2885	0.2909
0.95	0.2927	0.2907	0.2926	0.2917	0.2897	0.2917
1.00	0.2933	0.2916	0.2933	0.2924	0.2907	0.2924
1.25	0.2949	0.2943	0.2949	0.2940	0.2934	0.2940

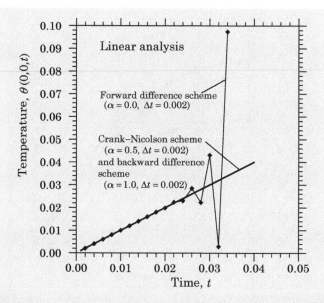

Fig. 6.7.4: Transient response predicted by various schemes.

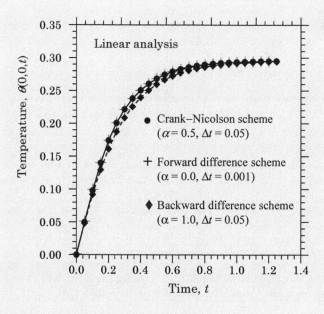

Fig. 6.7.5: Evolution of the temperature $\theta(0, 0, t)$ with time t.

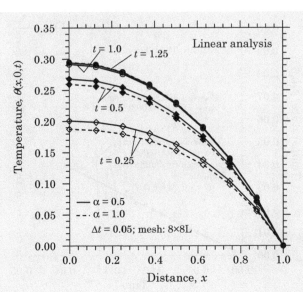

Fig. 6.7.6: Variation of temperature field $\theta(x, 0, t)$ with x for various values of time t of the heat conduction problem of Example 6.7.1 (the Crank–Nicolson and backward difference schemes are used).

Example 6.7.2

Consider the equation

$$\frac{\partial T}{\partial t} - \frac{\partial}{\partial x}\left(k\frac{\partial T}{\partial x}\right) - \frac{\partial}{\partial y}\left(k\frac{\partial T}{\partial y}\right) = 0 \text{ in } \Omega \tag{1}$$

where T is the temperature and k is the conductivity. Domain Ω is a rectangle of dimensions $a = 0.18$ m and $b = 0.1$ m along the x and y coordinates, respectively; conductivity k is of the form

$$k = k_0 \left(1 + \beta T\right) \tag{2}$$

where k_0 is the constant thermal conductivity, and β is the temperature coefficient of thermal conductivity. Take $k_0 = 0.2$ W/(m °C) and $\beta = 2 \times 10^{-3}$ (°C^{-1}); the boundary conditions are

$$T(0, y, t) = 500°\text{C} , \quad T(a, y, t) = 300°\text{C} , \quad \frac{\partial T}{\partial y} = 0 \text{ at } y = 0, b \tag{3}$$

and the initial condition is

$$T(x, y, 0) = 0°\text{C} \tag{4}$$

Determine the temperature distribution $T(x, y, t)$ using the 4×2 uniform mesh of nine-node quadratic elements and the Crank–Nicolson scheme ($\alpha = 0.5$).

Solution: As noted in Example 6.6.1, this is essentially a one-dimensional problem; that is, the solution $T(x, y, t)$ is independent of the y-coordinate. Table 6.7.2 contains numerical results obtained for various times as a function of position x. Figure 6.7.7 shows the evolution of the temperature $T(x, y, t)$ with x for various times, while Fig. 6.7.8 shows the temperature $T(x, y, t)$ as a function of time for different values of x.

Table 6.7.2: The temperature field $T(x, y, t)$ (for any fixed y) of the heat transfer problem in Example 6.7.2 (4×2Q9 uniform mesh, $\Delta t = 0.005$, $\alpha = 0.5$).

x	$t = 0.005$	$t = 0.01$	$t = 0.02$	$t = 0.03$	$t = 0.05$	$t = 0.08$	$t = 0.1$	S-State
0.0225	245.11	426.98	456.58	471.48	477.94	477.91	477.58	477.31
0.0450	107.47	289.91	386.74	427.69	448.56	448.57	453.09	454.03
0.0675	45.03	178.65	343.35	401.46	427.61	427.63	430.44	430.12
0.0900	24.16	119.18	311.87	375.39	402.23	402.23	405.01	405.57
0.1125	26.78	118.01	289.07	350.37	377.43	377.39	380.55	380.32
0.1350	56.82	167.10	278.43	326.88	350.15	350.13	354.00	354.34
0.1575	134.67	250.20	293.20	316.45	327.11	327.12	327.79	327.58

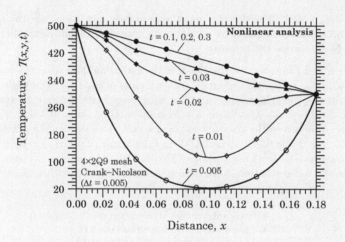

Fig. 6.7.7: Temperature $T(x, y, t)$ versus x for different values of t.

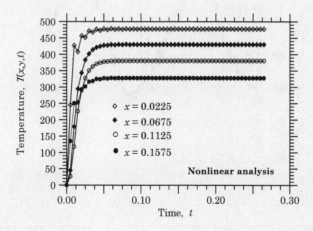

Fig. 6.7.8: Temperature $T(x, y, t)$ versus time t for different values of x.

Example 6.7.3

Consider the linear and nonlinear transient response $u(x, y, t)$ of a square membrane of dimensions 2×2, fixed on its boundary (i.e. $u = 0$), and subjected to uniform pressure of intensity $f_0 = 1$. The governing equation is

$$\frac{\partial^2 u}{\partial t^2} - \left[\frac{\partial}{\partial x} \left(a_{xx} \frac{\partial u}{\partial x} \right) + \frac{\partial}{\partial y} \left(a_{yy} \frac{\partial u}{\partial y} \right) \right] = f(x, y) \quad \text{in} \quad \Omega = (-1, 1) \times (-1, 1) \tag{1}$$

where the coefficients a_{xx} and a_{yy} are defined by Eq. (1) of Example 6.6.3:

$$a_{xx} = 1 + a_{1ux} \frac{\partial u}{\partial x} + a_{1uy} \frac{\partial u}{\partial y}, \quad a_{yy} = a_{xx} \tag{2}$$

where $a_{1ux} = a_{1uy} = 0.2$. The domain and the boundary conditions are the same as those shown in Fig. 6.7.3 with θ replaced by u. Use a 8×8 mesh of linear elements in the quarter of the domain to determine the response.

Solution: The critical time step for the linear acceleration method ($\alpha = 0.5$ and $\gamma = 1/3$) is ($\lambda_{\max} = 1492.56$) $\Delta t_{\text{cr}} = 0.0897$. Figure 6.7.9 shows plots of the transverse deflection $u(0, 0, t)$ versus time t for the linear case (i.e. $a_{1ux} = a_{1uy} = 0$) with $\Delta t = 0.1$, which is greater than the critical time step. The linear acceleration method ($\alpha = 0.5$ and $\gamma = 1/3$) gives almost the same response as the constant-average acceleration method ($\alpha = 0.5$ and $\gamma = 0.5$) when $\Delta t = 0.05$ (see Table 6.7.3). The linear and nonlinear (with $a_{1ux} = a_{1uy} = 0.2$) center deflections as functions of time are shown in Fig. 6.7.10. The nonlinear deflection starts to drift away from the linear solution, both in amplitude and period with time.

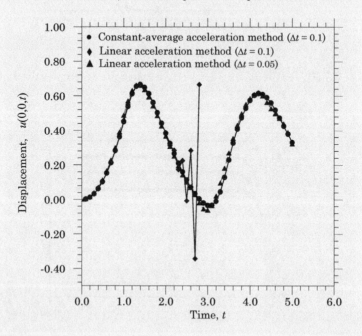

Fig. 6.7.9: Evolution of the center deflection $u(0, 0, t)$ of the membrane.

Table 6.7.3: Deflection $u(0,0,t)$ versus time t for a square membrane fixed on its edges and subjected to uniform load $(8 \times 8L4)$.

	$u(0,0,t)$				$u(0,0,t)$		
t	CAM*	LAM**	LAM†	t	CAM*	LAM**	LAM†
0.1	0.0025	0.0017	0.0029	1.7	0.5623	0.5578	0.5424
0.2	0.0125	0.0117	0.0154	1.8	0.5025	0.5023	0.4858
0.3	0.0325	0.0317	0.0379	1.9	0.4419	0.4397	0.4260
0.4	0.0625	0.0617	0.0704	2.0	0.3833	0.3872	0.3661
0.5	0.1025	0.1017	0.1129	2.1	0.3243	0.3122	0.3069
0.6	0.1525	0.1517	0.1657	2.2	0.2655	0.2841	0.2511
0.7	0.2125	0.2117	0.2269	2.3	0.2105	0.1740	0.1959
0.8	0.2825	0.2812	0.2989	2.4	0.1601	0.2250	0.1427
0.9	0.3624	0.3626	0.3896	2.5	0.1131	−0.0078	0.1014
1.0	0.4500	0.4565	0.4876	2.6	0.0706	0.2833	0.0693
1.1	0.5378	0.5482	0.5701	2.7	0.0343	−0.3422	0.0289
1.2	0.6110	0.6161	0.6301	2.8	0.0038	0.6653	−0.0196
1.3	0.6550	0.6546	0.6619	2.9	−0.0204	−1.1816	−0.0553
1.4	0.6656	0.6658	0.6608	3.0	−0.0348	1.9454	−0.0625
1.5	0.6492	0.6482	0.6359	3.1	−0.0339	−3.5058	−0.0358
1.6	0.6133	0.6079	0.5939	3.2	−0.0102	5.9399	0.0207

* CAM = constant-average acceleration method ($\alpha = 0.5$, $\gamma = 0.5$, $\Delta t = 0.1$);
** LAM = linear acceleration method ($\alpha = 0.5$, $\gamma = 1/3$, $\Delta t = 0.1$).
† LAM with $\Delta t = 0.05$.

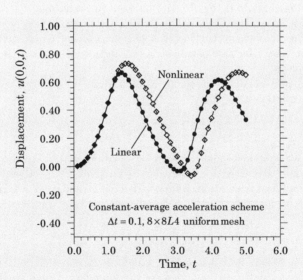

Fig. 6.7.10: Comparison of the linear and nonlinear center deflections $u(0,0,t)$ of the membrane.

6.8 Summary

In this chapter, the weak-form Galerkin finite element model of a generalized, nonlinear Poisson equation in two dimensions is presented for both steady-state and transient response. Computer implementation of the formulation is discussed. Several numerical examples of heat conduction with temperature-dependent conductivity and deflections of membranes with tensions that are functions of the displacement are presented. The finite element formulations developed in this chapter for steady-state and transient analysis of nonlinear two-dimensional problems involving a single unknown provide the necessary tools for the two-dimensional problems of solid and fluid mechanics in Chapters 7 through 11.

Problems

WEAK FORMS AND FINITE ELEMENT MODELS

6.1 The energy equation for simultaneous conduction and radiation in a participating medium can be expressed by

$$-\nabla \cdot [k_e(T)\nabla T] = g$$

where

$$k_e(T) = k + \frac{16\sigma n^2 T^3}{3\beta}$$

Here T is the temperature, g is the internal heat generation, n denotes the refractive index of the medium, σ is the Stefan–Boltzman constant, and β is the Roseland mean extinction coefficient (see Özisik [54]). Develop the finite element model of the equation and determine the tangent coefficient matrix for a planar (two-dimensional) domain.

6.2 Repeat Problem 6.1 for an axisymmetric (i.e. independent of the circumferential coordinate) problem.

6.3 Consider the problem of finding temperature field $u(x, y)$ such that the following partial differential equation is satisfied:

$$-\left[\frac{\partial}{\partial x}\left(a_{xx}\frac{\partial u}{\partial x} \right) + \frac{\partial}{\partial y}\left(a_{yy}\frac{\partial u}{\partial y} \right) \right] = f(x, y) \quad \text{in } \Omega \tag{1}$$

where Ω is a two-dimensional (orthotropic) medium with boundary Γ. Here a_{xx} and a_{yy} are thermal conductivities in the x and y directions, respectively, and $f(x, y)$ is the known internal heat generation. Suppose that the boundary of a typical finite element is subject to both convective and enclosed radiation heat transfer:

$$\left(a_{xx}\frac{\partial u}{\partial x}n_x + a_{yy}\frac{\partial u}{\partial y}n_y \right) + h_c(u - u_c) + \sigma\epsilon(u^4 - u_c^4) = q_n \tag{2}$$

where σ is the Stefan–Boltzmann constant and ϵ is the emissivity of the surface. Assuming that σ, ϵ, h_c and u_c are functions of the boundary coordinate s and the conductivities are independent of the temperature u, formulate the finite element equations to account for the radiation term. *Hint:* $\sigma\epsilon(u^4 - u_c^4) = \sigma\epsilon(u^2 + u_c^2)(u + u_c)(u - u_c) \equiv h_r(u)(u - u_c)$.

6.4 Compute the tangent coefficient matrix associated with the nonlinear radiation boundary condition of Problem 6.3.

COMPUTATIONAL PROBLEMS

6.5 Solve the problem of Example 6.6.1 for the following data:

$$a = 0.2 \text{ m}, \quad b = 0.1 \text{ m}, \quad \beta = 0.2 \ (^\circ\text{K}^{-1}) \tag{1}$$

$$T(0, y) = 500^\circ \text{ K} \ , \quad T(a, y) = 300^\circ \text{ K} \ , \quad \frac{\partial T}{\partial y} = 0 \text{ at } y = 0, \quad T(x, b) = 500 - 1000x \ ^\circ \text{ K} \tag{2}$$

All other data remain the same. Use 4×4 and 8×8 linear element meshes, and 2×2 and 4×4 nine-node quadratic element meshes to analyze the two problems using the direct iteration and Newton's iteration schemes. Use a convergence tolerance of $\varepsilon = 10^{-3}$ and maximum allowable iterations to be 10.

6.6 Consider the nonlinear problem of Example 6.6.1 (heat transfer in two dimensions). Use the uniform 4×4 nine-node quadratic element mesh to analyze the problem using the following data and boundary conditions (see Fig. P6.6) and the direct iteration procedure:

$$a = 0.18 \text{ m}, \quad b = 0.1 \text{ m}, \quad f_0 = 0 \text{ W/m}^3 \tag{1}$$

$$k = k_0 \left(1 + \beta T\right), \quad k_0 = 25 \text{ W/(m } ^\circ\text{C)} \tag{2}$$

$$T(0, y) = 100 \ ^\circ\text{C} \ , \quad T(a, y) = 50 \ ^\circ\text{C}$$

$$k \frac{\partial T}{\partial n} + h_c(T - T_\infty) = 0 \text{ at } y = 0, b \tag{3}$$

Use $\beta = 0.2$, $T_\infty = 10^\circ\text{C}$, and $h_c = 50 \text{ W/(m}^2 \ ^\circ\text{C)}$, and $\epsilon = 10^{-3}$ and ITMAX $= 10$. Plot linear and nonlinear solutions $T(x, y)$ as functions of x for fixed $y = 0$ and $y = 0.05$. *Hint*: See Section 3.3.4 for the treatment of convection heat transfer boundary conditions, and Chapter 8 of Reddy [2] for computational details of such problems.

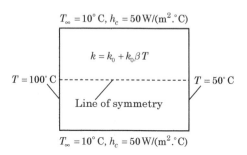

$$T_\infty = 10^\circ \text{ C}, \ h_c = 50 \text{ W/(m}^2.^\circ\text{C)}$$

$$k = k_0 + k_0 \beta T$$

$$T = 100^\circ\text{ C} \qquad T = 50^\circ\text{ C}$$

Line of symmetry

$$T_\infty = 10^\circ \text{ C}, \ h_c = 50 \text{ W/(m}^2.^\circ\text{C)}$$

Fig. P6.6

6.7 Analyze Problem 6.6 using the Newton iteration procedure.

TIME-DEPENDENT PROBLEMS

6.8 Derive Eq. (6.7.17) for parabolic equations when $\mathbf{C}$ is a function of time.

6.9 Show that Eq. (6.7.27) can be expressed in the alternative form

$$\mathbf{H}_{s+1} \ddot{\mathbf{u}}_{s+1} = \tilde{\mathbf{F}}_{s+1}$$

and define $\mathbf{H}_{s+1}$ and $\tilde{\mathbf{F}}_{s+1}$.

6.10 The α-family of approximation and a general class of time-approximation schemes based on truncated Taylor's series can also be derived using the finite element method in time domain. Consider a first-order differential equation of the form

$$C\frac{du}{dt} + Ku(t) = F(t) \tag{1}$$

where, in general, C, K, and F are functions of time, t. Suppose that $u(t)$ is approximated as

$$u(t) \approx \sum_{j=1}^{n} u_j \psi_j(t) \tag{2}$$

where u_j denotes the value of $u(t)$ at time $t = t_j$ and $\psi_j(t)$ is the associated approximation function. The weighted-integral method seeks the solution of Eq. (1) over a typical time interval (t_s, t_{s+1}) by requiring

$$0 = \int_{t_s}^{t_{s+1}} w_j(t)\,(C\dot{u} + Ku - F)\,dt \tag{3}$$

where $w_j(t)$ is the jth weight function $(j = 1, 2, \cdots, n)$. Obtain the (Petrov–Galerkin) finite element model using the weighted-integral statement, Eq. (3), and then specialize the result for the choice $\phi_j = \psi_j$ (i.e. using the Galerkin method) as the linear interpolation functions. Assume that $F(t)$ can also be approximated as

$$F(t) = F_1\psi_1(t) + F_2\psi_2(t) \tag{4}$$

and derive the equation for $u_{s+1} = u_2$ in terms of $u_s = u_1$, $F_s = F_1$, $F_{s+1} = F_2$, and $\Delta t = t_{s+1} - t_s$ using constant values of C and K.

6.11 Use quadratic approximation of $u(t)$ with $\phi_i = \psi_i$ (i.e. Galerkin's method) in the weighted-integral form in Eq. (3) of Problem 6.10 and arrive at the finite element equations

$$\left(\frac{C}{6}\begin{bmatrix} -3 & 4 & -1 \\ -4 & 0 & 4 \\ 1 & -4 & 3 \end{bmatrix} + \frac{K\Delta t}{30}\begin{bmatrix} 4 & 2 & -1 \\ 2 & 16 & 2 \\ -1 & 2 & 4 \end{bmatrix} \right) \left\{ \begin{array}{c} U_1 \\ U_2 \\ U_3 \end{array} \right\}$$

$$= \frac{\Delta t}{30}\begin{bmatrix} 4 & 2 & -1 \\ 2 & 16 & 2 \\ -1 & 2 & 4 \end{bmatrix} \left\{ \begin{array}{c} F_1 \\ F_2 \\ F_3 \end{array} \right\} \tag{1}$$

where

$$U_1 = u_s, \quad U_2 = u_{s+\frac{1}{2}}, \quad U_3 = u_{s+1} \tag{2}$$

and similar definition holds for F_i. Determine the values of U_2 and U_3 in terms of U_1.

6.12 Consider the equation

$$c\frac{\partial u}{\partial t} - \frac{\partial}{\partial x}\left(a_{xx}\frac{\partial u}{\partial x} \right) - \frac{\partial}{\partial y}\left(a_{yy}\frac{\partial u}{\partial y} \right) = f(x, y, t) \text{ in } \Omega \tag{1}$$

where u is the dependent unknown and a_{xx} and a_{yy} are known functions of position (x, y) and u and its derivatives

$$a_{xx} = a_{yy} = c_1(x, y) + \tfrac{1}{2}c_2(x, y)\left[\left(\frac{\partial u}{\partial x}\right)^2 + \left(\frac{\partial u}{\partial y}\right)^2 \right] \tag{2}$$

Develop a fully discretized finite element model and determine the tangent coefficient matrix of a typical element.

6.13 Consider the transient heat conduction equation

$$\frac{\partial\theta}{\partial t} - \hat{k}\left(\frac{\partial^2\theta}{\partial x^2} + \frac{\partial^2\theta}{\partial y^2}\right) = 1 \text{ in } \Omega; \quad \theta = 0 \text{ on } \Gamma \tag{1}$$

where θ is the non-dimensional temperature, Ω is the square domain of side 2, and Γ is its boundary with the boundary conditions shown in Fig. 6.7.3. Determine the temperature field inside the domain as a function of position and time when $\hat{k} = k_0(1+\beta\theta)$ with $k_0 = 1.0$ and $\beta = 0.4\times10^{-3}$. Assume zero initial condition $\theta(x, y, 0) = 0$, exploit the biaxial symmetry, and use a uniform mesh of 4×4 nine-node rectangular elements in a quadrant, the Crank–Nicolson scheme (i.e. $\alpha = 0.5$) with $\Delta t = 0.05$, and error tolerance of $\epsilon = 10^{-3}$ for the direct iteration method. Plot the temperature $\theta(x_0, 0, t)$ as a function of time for $x_0 = 0.0, 0.5$, and 0.75. The total number of time steps are such that the temperature reaches a steady state.

6.14 Repeat Problem 6.13 with $\beta = 0.4$ and all other data being the same.

6.15 Solve the problem in Example 6.7.3 with

$$a_{xx} = a_{yy} = 1 + a_u \cdot u$$

with $a_u = 0.1$ and plot the solution $u(0, 0, t)$ versus time.

6.16 Solve the problem in Example 6.7.3 with

$$a_{xx} = a_{yy} = 1 + a_u \cdot u$$

with $a_u = 1.0$ and plot the solution $u(0, 0, t)$ versus time.

6.17 Put together a complete finite element program based on the logic presented in Box 6.6.1 that has the capability of solving two-dimensional nonlinear problems involving a single unknown.

Nonlinear Bending of Elastic Plates

Science is the search for the truth – it is not a game in which one tries to beat his opponent, to do harm to others.
———— Linus Pauling

7.1 Introduction

A *plate* is a flat structural element with planform dimensions that are large compared to its thickness and is subjected to loads that cause bending deformation in addition to stretching in the plane [3, 55, 56]. In most cases, the thickness is no greater than one-tenth of the smallest in-plane dimension. Because of the smallness of the thickness dimension, it is often not necessary to model them using three-dimensional elasticity equations. Simple two-dimensional plate theories, much like the beam theories considered in Chapter 5, can be developed to study the deformation and stresses in plate structures undergoing small strains, small to moderate rotations, and large displacements (i.e. $w/h > 1$).

As in the case of beams, we consider both classical and shear deformation plate theories, which are extensions of the Euler–Bernoulli and Timoshenko beam theories, respectively, to two dimensions. The extension of the Euler–Bernoulli beam theory to two dimensions is known as the *classical* or *Kirchhoff* plate theory, while the extension of the Timoshenko beam theory to two dimensions is termed the *first-order shear deformation plate theory*. Geometric nonlinearity is considered via the von Kármán strains, as in the case of beams.

In nonlinear continuum mechanics (see Reddy [1] and Chapter 2), one uses the notation $\mathbf{X} = (X_1, X_2, X_3)$ for the material coordinates and $\mathbf{x}$ for the spatial coordinates. As discussed previously in connection with beams, the present study of plates is limited to small strains and moderate rotations, and therefore no distinction is made here between the material coordinates $\mathbf{X}$ and spatial coordinates $\mathbf{x} = (x_1, x_2, x_3)$. In fact, $\mathbf{x}$ is used to denote the material coordinates, i.e. $x = x_1 = X_1$, $y = x_2 = X_2$, and $z = x_3 = X_3$.

The weak forms of each plate theory are developed using the principle of virtual displacements. Then the displacement finite element models of both theories are developed, their computer implementation is discussed, and numerical examples of some representative plate problems are presented.

J.N. Reddy, *An Introduction to Nonlinear Finite Element Analysis*, Second Edition. ©J.N. Reddy 2015. Published in 2015 by Oxford University Press.

7.2   The Classical Plate Theory

7.2.1   Assumptions of the Kinematics

The *classical plate theory* (CPT) is one in which the displacement field is selected so as to satisfy the *Kirchhoff hypothesis,* which is an extension of the Euler–Bernoulli hypothesis of beams to two dimensions. The Kirchhoff hypothesis consists of the following three assumptions (see Fig. 7.2.1):

(1) Straight lines perpendicular to the mid-surface (i.e. transverse normals) before deformation, remain straight after deformation.
(2) The transverse normals do not experience elongation (i.e. inextensible).
(3) The transverse normals rotate such that they remain perpendicular to the mid-surface after deformation.

7.2.2   Displacement and Strain Fields

Let us denote the undeformed mid-plane of the plate with symbol Ω_0. The total domain of the plate is the tensor product $\Omega_0 \times (-h/2, h/2)$. The boundary of the total domain consists of surfaces $S_t(z = h/2)$ and $S_b(z = -h/2)$, and the edge $\bar{\Gamma} \equiv \Gamma \times (-h/2, h/2)$. In general, Γ is a curved surface, with outward normal $\hat{\mathbf{n}} = n_x \hat{\mathbf{e}}_x + n_y \hat{\mathbf{e}}_y$, where n_x and n_y are the direction cosines of the unit normal.

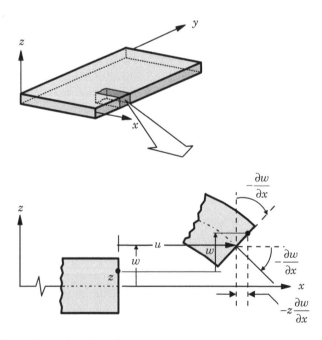

Fig. 7.2.1: Undeformed and deformed geometries of an edge of a plate under the Kirchhoff assumptions.

The Kirchhoff hypothesis implies the following form of the displacement field (see Fig. 7.2.1 for the kinematics of bending)

$$u_1(x, y, z, t) = u(x, y, t) - z\frac{\partial w}{\partial x}$$

$$u_2(x, y, z, t) = v(x, y, t) - z\frac{\partial w}{\partial y} \qquad (7.2.1)$$

$$u_3(x, y, z, t) = w(x, y, t)$$

where (u, v, w) denote the displacements of a material point occupying the position $(x, y, 0)$ in the undeformed configuration for any time t, whereas (u_1, u_2, u_3) are the total displacement components of a material point located at an arbitrary point $\mathbf{x} = (x, y, z)$ in the undeformed body.

The definition of the components of the Green–Lagrange strain tensor $\mathbf{E}$ in terms of the components of the total displacement vector $\mathbf{u} = \mathbf{x}(\mathbf{X}, t) - \mathbf{X}$ is [see Eqs. (2.4.8) and (2.4.9); shortly, X_i will be replaced with x_i]

$$E_{11} = \frac{\partial u_1}{\partial X_1} + \frac{1}{2}\left[\left(\frac{\partial u_1}{\partial X_1}\right)^2 + \left(\frac{\partial u_2}{\partial X_1}\right)^2 + \left(\frac{\partial u_3}{\partial X_1}\right)^2\right]$$

$$E_{22} = \frac{\partial u_2}{\partial X_2} + \frac{1}{2}\left[\left(\frac{\partial u_1}{\partial X_2}\right)^2 + \left(\frac{\partial u_2}{\partial X_2}\right)^2 + \left(\frac{\partial u_3}{\partial X_2}\right)^2\right]$$

$$E_{33} = \frac{\partial u_3}{\partial X_3} + \frac{1}{2}\left[\left(\frac{\partial u_1}{\partial X_3}\right)^2 + \left(\frac{\partial u_2}{\partial X_3}\right)^2 + \left(\frac{\partial u_3}{\partial X_3}\right)^2\right] \qquad (7.2.2)$$

$$E_{12} = \frac{1}{2}\left(\frac{\partial u_1}{\partial X_2} + \frac{\partial u_2}{\partial X_1} + \frac{\partial u_1}{\partial X_1}\frac{\partial u_1}{\partial X_2} + \frac{\partial u_2}{\partial X_1}\frac{\partial u_2}{\partial X_2} + \frac{\partial u_3}{\partial X_1}\frac{\partial u_3}{\partial X_2}\right)$$

$$E_{13} = \frac{1}{2}\left(\frac{\partial u_1}{\partial X_3} + \frac{\partial u_3}{\partial X_1} + \frac{\partial u_1}{\partial X_1}\frac{\partial u_1}{\partial X_3} + \frac{\partial u_2}{\partial X_1}\frac{\partial u_2}{\partial X_3} + \frac{\partial u_3}{\partial X_1}\frac{\partial u_3}{\partial X_3}\right)$$

$$E_{23} = \frac{1}{2}\left(\frac{\partial u_2}{\partial X_3} + \frac{\partial u_3}{\partial X_2} + \frac{\partial u_1}{\partial X_2}\frac{\partial u_1}{\partial X_3} + \frac{\partial u_2}{\partial X_2}\frac{\partial u_2}{\partial X_3} + \frac{\partial u_3}{\partial X_2}\frac{\partial u_3}{\partial X_3}\right)$$

If the components of the displacement gradients are of the order ϵ (a very small parameter, $\epsilon << 1$)

$$\frac{\partial u_1}{\partial X_1}, \frac{\partial u_1}{\partial X_2}, \frac{\partial u_1}{\partial X_3}, \frac{\partial u_2}{\partial X_1}, \frac{\partial u_2}{\partial X_2}, \frac{\partial u_2}{\partial X_3}, \cdots, \frac{\partial u_3}{\partial X_3} = O(\epsilon) \qquad (7.2.3)$$

then the small strain assumption implies that terms of order ϵ^2 are omitted in the strains. If the rotations of transverse normals are moderate (say $10°$–$15°$), then the following terms are small but *not* negligible compared to ϵ

$$\left(\frac{\partial u_3}{\partial X_1}\right)^2, \left(\frac{\partial u_3}{\partial X_2}\right)^2, \frac{\partial u_3}{\partial X_1}\frac{\partial u_3}{\partial X_2} \qquad (7.2.4)$$

Thus for small strains and moderate rotations, the strain-displacement relations
(7.2.2) take the form [here we switch from X_i to x_i and subsequently to (x, y, z)]

$$E_{xx} \approx \varepsilon_{xx} = \frac{\partial u_1}{\partial x} + \frac{1}{2}\left(\frac{\partial u_3}{\partial x}\right)^2, \quad E_{xy} \approx \varepsilon_{xy} = \frac{1}{2}\left(\frac{\partial u_1}{\partial y} + \frac{\partial u_2}{\partial x} + \frac{\partial u_3}{\partial x}\frac{\partial u_3}{\partial y}\right)$$

$$E_{xz} \approx \varepsilon_{xz} = \frac{1}{2}\left(\frac{\partial u_1}{\partial z} + \frac{\partial u_3}{\partial x}\right), \quad E_{yy} \approx \varepsilon_{yy} = \frac{\partial u_2}{\partial y} + \frac{1}{2}\left(\frac{\partial u_3}{\partial y}\right)^2$$

$$E_{yz} \approx \varepsilon_{yz} = \frac{1}{2}\left(\frac{\partial u_2}{\partial z} + \frac{\partial u_3}{\partial y}\right), \quad E_{zz} \approx \varepsilon_{zz} = \frac{\partial u_3}{\partial z}$$

$$(7.2.5)$$

where, for this special case of geometric nonlinearity (i.e. small strains but
moderate rotations), notation ε_{ij} is used in place of E_{ij}. The corresponding
Cartesian components of the second Piola–Kirchhoff stress tensor will be denoted σ_{ij} (notation σ_{ij} is usually reserved for the Cauchy stress tensor; see
Reddy [1] and Chapter 2).

For the displacement field in Eq. (7.2.1), we have $\varepsilon_{zz} = \frac{\partial u_3}{\partial z} = \frac{\partial w}{\partial z} = 0$ and
the strains in Eq. (7.2.5) reduce to

$$\varepsilon_{xx} = \frac{\partial u}{\partial x} + \frac{1}{2}\left(\frac{\partial w}{\partial x}\right)^2 - z\frac{\partial^2 w}{\partial x^2}$$

$$\varepsilon_{yy} = \frac{\partial v}{\partial y} + \frac{1}{2}\left(\frac{\partial w}{\partial y}\right)^2 - z\frac{\partial^2 w}{\partial y^2}$$

$$2\varepsilon_{xy} = \frac{\partial u}{\partial y} + \frac{\partial v}{\partial x} + \frac{\partial w}{\partial x}\frac{\partial w}{\partial y} - 2z\frac{\partial^2 w}{\partial x \partial y} \equiv \gamma_{xy} \qquad (7.2.6)$$

$$2\varepsilon_{xz} = -\frac{\partial w}{\partial x} + \frac{\partial w}{\partial x} \equiv \gamma_{xz} = 0$$

$$2\varepsilon_{yz} = -\frac{\partial w}{\partial y} + \frac{\partial w}{\partial y} \equiv \gamma_{yz} = 0$$

The strains in Eq. (7.2.6) are called the *von Kármán strains*, and the associated
plate theory is termed the *classical plate theory with the von Kármán strains*.
Note that the transverse strains $(\varepsilon_{xz}, \varepsilon_{yz}, \varepsilon_{zz})$ are identically zero in the classical
plate theory. In matrix notation, the total strains in Eq. (7.2.6) can be written
as the sum of membrane strains $\{\varepsilon^{(0)}\}$ and bending strains $\{\varepsilon^{(1)}\}$

$$\begin{Bmatrix} \varepsilon_{xx} \\ \varepsilon_{yy} \\ \gamma_{xy} \end{Bmatrix} = \begin{Bmatrix} \varepsilon_{xx}^0 \\ \varepsilon_{yy}^0 \\ \gamma_{xy}^0 \end{Bmatrix} + z \begin{Bmatrix} \varepsilon_{xx}^1 \\ \varepsilon_{yy}^1 \\ \gamma_{xy}^1 \end{Bmatrix} \qquad (7.2.7)$$

$$\begin{Bmatrix} \varepsilon_{xx}^0 \\ \varepsilon_{yy}^0 \\ \gamma_{xy}^0 \end{Bmatrix} = \begin{Bmatrix} \frac{\partial u}{\partial x} + \frac{1}{2}\left(\frac{\partial w}{\partial x}\right)^2 \\ \frac{\partial v}{\partial y} + \frac{1}{2}\left(\frac{\partial w}{\partial y}\right)^2 \\ \frac{\partial u}{\partial y} + \frac{\partial v}{\partial x} + \frac{\partial w}{\partial x}\frac{\partial w}{\partial y} \end{Bmatrix}, \quad \begin{Bmatrix} \varepsilon_{xx}^1 \\ \varepsilon_{yy}^1 \\ \gamma_{xy}^1 \end{Bmatrix} = -\begin{Bmatrix} \frac{\partial^2 w}{\partial x^2} \\ \frac{\partial^2 w}{\partial y^2} \\ 2\frac{\partial^2 w}{\partial x \partial y} \end{Bmatrix} \qquad (7.2.8)$$

7.3 Weak Formulation of the CPT

7.3.1 Virtual Work Statement

The principle of virtual displacements (see Section 2.9 and Reddy [3] for a discussion of the principle of virtual displacements) is used to derive the weak form of the classical plate theory for a typical plate finite element Ω^e. In the derivation, we account for thermal (and hence, moisture) effects only with the understanding that the material properties do not change with temperature and that temperature T is a known function of position (hence, $\delta T = 0$). Thus, temperature enters the formulation only through constitutive equations.

Suppose that the total domain (i.e. midplane) Ω of the plate is represented by a set of finite elements, and let Ω^e be a typical plate element with boundary Γ^e. Suppose that $q(x, y)$ is the distributed transverse force at the top surface, $z = h/2$, and $(\sigma_{nn}, \sigma_{ns}, \sigma_{nz})$ are the stress components on the boundary $\bar{\Gamma}^e = \Gamma^e \times (-\frac{h}{2}, \frac{h}{2})$ of the plate element (see Fig. 7.3.1 for a plate element with curved boundary).

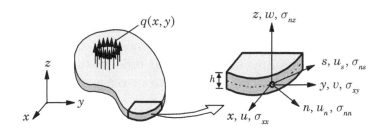

Fig. 7.3.1: Geometry of a plate element with curved boundary.

The principle of virtual displacements requires

$$0 = \delta W^e \equiv \delta W_I^e + \delta W_E^e \tag{7.3.1}$$

where δW_I^e denotes the internal virtual work due to the straining of the element (i.e. virtual strain energy stored) and δW_E^e is the virtual work done by externally applied forces on the element Ω^e.

As noted earlier, the transverse strains (γ_{xz}, γ_{yz}, and ε_{zz}) are identically zero in the classical plate theory. Consequently, the transverse stresses ($\sigma_{xz}, \sigma_{yz}, \sigma_{zz}$) do not enter the formulation because the virtual strain energy of these stresses is zero (due to the fact that kinematically consistent virtual strains must be zero):

$$\delta\gamma_{xz} = 0, \quad \delta\gamma_{yz} = 0, \quad \delta\varepsilon_{zz} = 0 \tag{7.3.2}$$

Whether the transverse shear stresses are or are not accounted for in a theory, in reality they exist in the plate to keep it in equilibrium. In addition, these stress components may be specified on the boundary. Thus, the transverse stresses do not enter the virtual strain energy expression but must be accounted for in equilibrium equations.

The virtual strain energy in element Ω^e is given by

$$
\delta W_I^e = \int_{\Omega^e} \int_{-\frac{h}{2}}^{\frac{h}{2}} \left(\sigma_{xx} \delta \varepsilon_{xx} + \sigma_{yy} \delta \varepsilon_{yy} + 2\sigma_{xy} \delta \varepsilon_{xy} \right) \, dz \, dx \, dy
$$

$$
= \int_{\Omega^e} \left(N_{xx} \delta \varepsilon_{xx}^0 + M_{xx} \delta \varepsilon_{xx}^1 + N_{yy} \delta \varepsilon_{yy}^0 + M_{yy} \delta \varepsilon_{yy}^1 \right.
$$

$$
\left. + N_{xy} \delta \gamma_{xy}^0 + M_{xy} \delta \gamma_{xy}^1 \right) dx \, dy \tag{7.3.3}
$$

where (N_{xx}, N_{yy}, N_{xy}) are the *forces per unit length* and (M_{xx}, M_{yy}, M_{xy}) are the *moments per unit length* as defined by

$$
\begin{Bmatrix} N_{xx} \\ N_{yy} \\ N_{xy} \end{Bmatrix} = \int_{-\frac{h}{2}}^{\frac{h}{2}} \begin{Bmatrix} \sigma_{xx} \\ \sigma_{yy} \\ \sigma_{xy} \end{Bmatrix} dz \,, \quad \begin{Bmatrix} M_{xx} \\ M_{yy} \\ M_{xy} \end{Bmatrix} = \int_{-\frac{h}{2}}^{\frac{h}{2}} \begin{Bmatrix} \sigma_{xx} \\ \sigma_{yy} \\ \sigma_{xy} \end{Bmatrix} z \, dz \tag{7.3.4}
$$

Figure 7.3.2 shows various stress resultants on a plate edge with unit normal $\hat{\mathbf{n}} = n_x \hat{\mathbf{i}} + n_y \hat{\mathbf{j}}$; an alternative sign convention for moments, based on the right-hand screw rule, is displayed in Fig. 7.3.3. We note that the subscripts on N's and M's refer to the subscripts of stresses that produce them.

The virtual work done by the distributed transverse load $q(x, y)$, the transverse reaction force of an elastic foundation at $z = -h/2$, in-plane normal stress σ_{nn}, in-plane tangential stress σ_{ns}, and transverse shear stress σ_{nz} is

$$
\delta W_E^e = - \left\{ \int_{\Omega^e} q(x, y) \delta w(x, y, \frac{h}{2}) \, dx \, dy + \int_{\Omega^e} F_s(x, y) \, \delta w(x, y, -\frac{h}{2}) \, dx \, dy \right.
$$

$$
\left. + \oint_{\Gamma^e} \int_{-\frac{h}{2}}^{\frac{h}{2}} \left[\sigma_{nn} \left(\delta u_n - z \frac{\partial \delta w}{\partial n} \right) + \sigma_{ns} \left(\delta u_s - z \frac{\partial \delta w}{\partial s} \right) + \sigma_{nz} \delta w \right] dz \, ds \right\} \tag{7.3.5}
$$

$$
= - \left[\oint_{\Gamma^e} \left(N_{nn} \delta u_n - M_{nn} \frac{\partial \delta w}{\partial n} + N_{ns} \delta u_s - M_{ns} \frac{\partial \delta w}{\partial s} + Q_n \delta w \right) ds \right.
$$

$$
\left. + \int_{\Omega^e} (q - kw) \, \delta w \, dx \, dy \right] \tag{7.3.6}
$$

where $F_s = -kw$ is the foundation force, k is the foundation modulus, and u_n, u_s, and w are the displacements along the normal, tangential, and transverse directions, respectively (see Fig. 7.3.1), and

$$
\begin{Bmatrix} N_{nn} \\ N_{ns} \end{Bmatrix} = \int_{-\frac{h}{2}}^{\frac{h}{2}} \begin{Bmatrix} \sigma_{nn} \\ \sigma_{ns} \end{Bmatrix} dz, \quad \begin{Bmatrix} M_{nn} \\ M_{ns} \end{Bmatrix} = \int_{-\frac{h}{2}}^{\frac{h}{2}} \begin{Bmatrix} \sigma_{nn} \\ \sigma_{ns} \end{Bmatrix} z \, dz, \quad Q_n = \int_{-\frac{h}{2}}^{\frac{h}{2}} \sigma_{nz} \, dz \tag{7.3.7}
$$

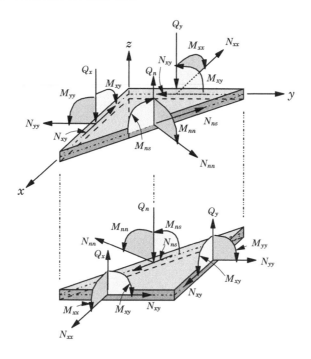

Fig. 7.3.2: Forces and moments per unit length on a plate element.

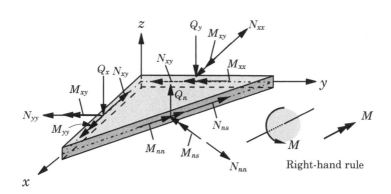

Fig. 7.3.3: Alternative sign convention for moments on a plate element.

The stresses $(\sigma_{nn}, \sigma_{ns})$ on the boundary Γ^e are related to $(\sigma_{xx}, \sigma_{yy}, \sigma_{xy})$ in the interior of Ω^e by the stress transformation equations

$$
\left\{ \begin{array}{c} \sigma_{nn} \\ \sigma_{ns} \end{array} \right\} = \left[\begin{array}{ccc} n_x^2 & n_y^2 & 2n_x n_y \\ -n_x n_y & n_x n_y & n_x^2 - n_y^2 \end{array} \right] \left\{ \begin{array}{c} \sigma_{xx} \\ \sigma_{yy} \\ \sigma_{xy} \end{array} \right\}
\tag{7.3.8}
$$

7.3.2   Weak Forms

Substituting for δW_I^e and δW_E^e from Eqs. (7.3.3) and (7.3.6) into the virtual work statement in Eq. (7.3.1), we obtain the weak form

$$
\begin{aligned}
0 = & \int_{\Omega^e} \Big(N_{xx}\delta\varepsilon_{xx}^0 + M_{xx}\delta\varepsilon_{xx}^1 + N_{yy}\delta\varepsilon_{yy}^0 + M_{yy}\delta\varepsilon_{yy}^1 + N_{xy}\delta\gamma_{xy}^0 \\
& + M_{xy}\delta\gamma_{xy}^1 + kw\delta w - q\delta w \Big)dx\,dy \\
& - \oint_{\Gamma^e}\Big(N_{nn}\delta u_n + N_{ns}\delta u_s - M_{nn}\frac{\partial\delta w}{\partial n} - M_{ns}\frac{\partial\delta w}{\partial s} + Q_n\delta w \Big)ds \\
= & \int_{\Omega^e}\Bigg[\Big(\frac{\partial\delta u}{\partial x} + \frac{\partial w}{\partial x}\frac{\partial\delta w}{\partial x}\Big)N_{xx} + \Big(\frac{\partial\delta v}{\partial y} + \frac{\partial w}{\partial y}\frac{\partial\delta w}{\partial y}\Big)N_{yy} \\
& + \Big(\frac{\partial\delta u}{\partial y} + \frac{\partial\delta v}{\partial x} + \frac{\partial w}{\partial x}\frac{\partial\delta w}{\partial y} + \frac{\partial w}{\partial y}\frac{\partial\delta w}{\partial x}\Big)N_{xy} \\
& - \frac{\partial^2\delta w}{\partial x^2}M_{xx} - \frac{\partial^2\delta w}{\partial y^2}M_{yy} - 2\frac{\partial^2\delta w}{\partial x\partial y}M_{xy} + k\delta ww - \delta wq\Bigg]dx\,dy \\
& - \oint_{\Gamma^e}\Big(N_{nn}\delta u_n + N_{ns}\delta u_s - M_{nn}\frac{\partial\delta w}{\partial n} - M_{ns}\frac{\partial\delta w}{\partial s} + Q_n\delta w \Big)ds \quad (7.3.9)
\end{aligned}
$$

The statement in Eq. (7.3.9) is equivalent to the following three statements:

$$
0 = \int_{\Omega^e}\Big(\frac{\partial\delta u}{\partial x}N_{xx} + \frac{\partial\delta u}{\partial y}N_{xy} \Big)dx\,dy - \oint_{\Gamma^e}N_{nn}\delta u_n\,ds \qquad (7.3.10)
$$

$$
0 = \int_{\Omega^e}\Big(\frac{\partial\delta v}{\partial x}N_{xy} + \frac{\partial\delta v}{\partial y}N_{yy} \Big)dx\,dy - \oint_{\Gamma^e}N_{ns}\delta u_s\,ds \qquad (7.3.11)
$$

$$
\begin{aligned}
0 = & \int_{\Omega^e}\Bigg[\frac{\partial\delta w}{\partial x}\Big(\frac{\partial w}{\partial x}N_{xx} + \frac{\partial w}{\partial y}N_{xy}\Big) + \frac{\partial\delta w}{\partial y}\Big(\frac{\partial w}{\partial x}N_{xy} + \frac{\partial w}{\partial y}N_{yy}\Big) \\
& - \frac{\partial^2\delta w}{\partial x^2}M_{xx} - \frac{\partial^2\delta w}{\partial y^2}M_{yy} - 2\frac{\partial^2\delta w}{\partial x\partial y}M_{xy} + k\delta ww - \delta wq\Bigg]dx\,dy \\
& - \oint_{\Gamma^e}\Big(-M_{nn}\frac{\partial\delta w}{\partial n} - M_{ns}\frac{\partial\delta w}{\partial s} + Q_n\delta w \Big)ds \qquad (7.3.12)
\end{aligned}
$$

7.3.3   Equilibrium Equations

To obtain the governing equations of the CPT, we integrate expressions in Eqs. (7.3.10)–(7.3.12) by parts to relieve the virtual displacements $(\delta u, \delta v, \delta w)$ in Ω^e of any differentiation (so that we can use the fundamental lemma of variational calculus); we obtain

$$
\begin{aligned}
0 = & \int_{\Omega^e}\Big[-(N_{xx,x} + N_{xy,y})\,\delta u - (N_{xy,x} + N_{yy,y})\,\delta v \\
& - (M_{xx,xx} + 2M_{xy,xy} + M_{yy,yy} + \mathcal{N} - kw + q)\delta w\Big]dx\,dy
\end{aligned}
$$

$$+ \oint_{\Gamma^e} \Big[(N_{xx}n_x + N_{xy}n_y)\,\delta u + (N_{xy}n_x + N_{yy}n_y)\,\delta v$$

$$+ \Big(M_{xx,x}n_x + M_{xy,y}n_x + M_{yy,y}n_y + M_{xy,x}n_y + \mathcal{P} \Big)\delta w$$

$$- (M_{xx}n_x + M_{xy}n_y)\frac{\partial \delta w}{\partial x} - (M_{xy}n_x + M_{yy}n_y)\frac{\partial \delta w}{\partial y}\Big]\,ds$$

$$- \oint_{\Gamma^e} \Big(N_{nn}\delta u_n + N_{ns}\delta u_s - M_{nn}\frac{\partial \delta w}{\partial n} - M_{ns}\frac{\partial \delta w}{\partial s} + Q_n\delta w \Big)ds \qquad (7.3.13)$$

where a comma followed by subscripts denotes differentiation with respect to the subscripts: $N_{xx,x} = \partial N_{xx}/\partial x$, and so on, and

$$\mathcal{N}(u,v,w) = \frac{\partial}{\partial x}\Big(N_{xx}\frac{\partial w}{\partial x} + N_{xy}\frac{\partial w}{\partial y} \Big) + \frac{\partial}{\partial y}\Big(N_{xy}\frac{\partial w}{\partial x} + N_{yy}\frac{\partial w}{\partial y} \Big)$$

$$\mathcal{P}(u,v,w) = \Big(N_{xx}\frac{\partial w}{\partial x} + N_{xy}\frac{\partial w}{\partial y} \Big)n_x + \Big(N_{xy}\frac{\partial w}{\partial x} + N_{yy}\frac{\partial w}{\partial y} \Big)n_y$$

$$(7.3.14)$$

The equations of equilibrium are obtained by setting the coefficients of δu, δv, and δw in Ω^e to zero:

$$\delta u : \qquad \frac{\partial N_{xx}}{\partial x} + \frac{\partial N_{xy}}{\partial y} = 0 \qquad\qquad\qquad (7.3.15)$$

$$\delta v : \qquad \frac{\partial N_{xy}}{\partial x} + \frac{\partial N_{yy}}{\partial y} = 0 \qquad\qquad\qquad (7.3.16)$$

$$\delta w : \qquad \frac{\partial^2 M_{xx}}{\partial x^2} + 2\frac{\partial^2 M_{xy}}{\partial y \partial x} + \frac{\partial^2 M_{yy}}{\partial y^2} + \mathcal{N}(u,v,w) - kw + q = 0 \quad (7.3.17)$$

7.3.4 Boundary Conditions

To cast the boundary conditions on an arbitrary edge whose normal is $\hat{n}$, we express all generalized displacements $(u, v, w, \frac{\partial w}{\partial x}, \frac{\partial w}{\partial x})$ from the (x, y, z) system in terms of the corresponding generalized displacements in the normal, tangential, and transverse directions. We have

$$u = u_n n_x - u_s n_y, \qquad v = u_n n_y + u_s n_x \qquad (7.3.18a)$$

$$\frac{\partial w}{\partial x} = \frac{\partial w}{\partial n}n_x - \frac{\partial w}{\partial s}n_y, \qquad \frac{\partial w}{\partial y} = \frac{\partial w}{\partial n}n_y + \frac{\partial w}{\partial s}n_x \qquad (7.3.18b)$$

The boundary expression of Eq. (7.3.13) takes the form

$$\oint_{\Gamma^e} \Big[(N_{xx}n_x + N_{xy}n_y)\,(\delta u_n n_x - \delta u_s n_y) + (N_{xy}n_x + N_{yy}n_y)\,(\delta u_n n_y + \delta u_s n_x)$$

$$+ \Big(M_{xx,x}n_x + M_{xy,y}n_x + M_{yy,y}n_y + M_{xy,x}n_y + \mathcal{P} \Big)\delta w$$

$$- (M_{xx}n_x + M_{xy}n_y) \left(\frac{\partial \delta w}{\partial n} n_x - \frac{\partial \delta w}{\partial s} n_y \right)$$

$$- (M_{xy}n_x + M_{yy}n_y) \left(\frac{\partial \delta w}{\partial n} n_y + \frac{\partial \delta w}{\partial s} n_x \right) \Bigg] ds$$

$$- \oint_{\Gamma^e} \left(N_{nn} \delta u_n + N_{ns} \delta u_s - M_{nn} \frac{\partial \delta w}{\partial n} - M_{ns} \frac{\partial \delta w}{\partial s} + Q_n \delta w \right) ds$$

$$= \oint_{\Gamma^e} \Bigg\{ (N_{xx}n_x^2 + 2N_{xy}n_x n_y + N_{yy}n_y^2 - N_{nn}) \, \delta u_n$$

$$+ \left[(N_{yy} - N_{xx})n_x n_y + N_{xy}(n_x^2 - n_y^2) - N_{ns} \right] \delta u_s$$

$$+ (M_{xx,x}n_x + M_{xy,y}n_x + M_{yy,y}n_y + M_{xy,x}n_y + \mathcal{P} - Q_n) \, \delta w$$

$$- \left(M_{xx}n_x^2 + 2M_{xy}n_x n_y + M_{yy}n_y^2 - M_{nn} \right) \frac{\partial \delta w}{\partial n}$$

$$- \left[(M_{yy} - M_{xx})n_x n_y + M_{xy}(n_x^2 - n_y^2) - M_{ns} \right] \frac{\partial \delta w}{\partial s} \Bigg\} ds \qquad (7.3.19)$$

The natural boundary conditions are obtained by setting the coefficients of δu_n, δu_s, δw, $\frac{\partial \delta w}{\partial n}$ and $\frac{\partial \delta w}{\partial s}$ on Γ^e to zero:

$$\delta u_n : \; N_{nn} = N_{xx}n_x^2 + 2N_{xy}n_x n_y + N_{yy}n_y^2$$

$$\delta u_s : \; N_{ns} = (N_{yy} - N_{xx})n_x n_y + N_{xy}(n_x^2 - n_y^2) \qquad (7.3.20a)$$

$$\delta w : \; Q_n = M_{xx,x}n_x + M_{xy,y}n_x + M_{yy,y}n_y + M_{xy,x}n_y + \mathcal{P}$$

$$\frac{\partial \delta w}{\partial n} : \; M_{nn} = M_{xx}n_x^2 + 2M_{xy}n_x n_y + M_{yy}n_y^2$$

$$\frac{\partial \delta w}{\partial s} : \; M_{ns} = (M_{yy} - M_{xx})n_x n_y + M_{xy}(n_x^2 - n_y^2) \qquad (7.3.20b)$$

From Eq. (7.3.20), it is clear that the primary variables (i.e. generalized displacements) and secondary variables (i.e. generalized forces) of the theory are:

$$\text{Primary variables:} \quad u_n, \quad u_s, \quad w, \quad \frac{\partial w}{\partial n}, \quad \frac{\partial w}{\partial s}$$

$$\text{Secondary variables:} \quad N_{nn}, \; N_{ns}, \; Q_n, \; M_{nn}, \; M_{ns} \qquad (7.3.21)$$

7.3.4.1   The Kirchhoff free-edge condition

We note that the equations of equilibrium in Eqs. (7.3.15)–(7.3.17) have the combined spatial differential order of eight. In other words, if the governing equations are expressed in terms of displacements (u, v, w), they would contain second-order spatial derivatives of u and v and fourth-order spatial derivatives of w. This implies that there should be only eight (four essential and four natural) boundary conditions, whereas Eq. (7.3.21) shows five essential and five natural boundary conditions, giving a total of ten boundary conditions. To eliminate this discrepancy, one may integrate the tangential derivative term by parts to obtain the boundary term

$$-\oint_\Gamma M_{ns}\frac{\partial \delta w}{\partial s}\,ds = \oint_\Gamma \frac{\partial M_{ns}}{\partial s}\delta w\,ds - [M_{ns}\delta w]_\Gamma \qquad (7.3.22)$$

The term $[M_{ns}\delta w]_\Gamma$ is zero when the end points of a closed curve coincide or when $M_{ns} = 0$. If $M_{ns} = 0$ is not specified at corners of the boundary Γ of a polygonal plate, concentrated forces of magnitude $F_c = -2M_{ns}$ will be produced at the corners. The factor of 2 appears because M_{ns} from two sides of the corner are added there.

The remaining boundary term in Eq. (7.3.22) is added to the shear force Q_n (because it is a coefficient of δw on Γ^e) to obtain the effective shear force

$$V_n \equiv Q_n + \frac{\partial M_{ns}}{\partial s} \qquad (7.3.23)$$

The specification of this effective shear force V_n is known as the *Kirchhoff free-edge condition*. Finally, the correct boundary conditions of the CPT involve specifying the following quantities:

$$\text{Generalized displacements: } u_n, \ u_s, \ w, \ \frac{\partial w}{\partial n} \qquad (7.3.24)$$
$$\text{Generalized forces: } N_{nn}, \ N_{ns}, \ V_n, \ M_{nn}$$

Thus, at every boundary point one must know u_n or N_{nn}, u_s or N_{ns}, w or V_n, and $\partial w/\partial n$ or M_{nn}. On an edge parallel to the x-axis (i.e. $s = x$ and $n = y$), for example, the above boundary conditions become

$$u_n = v, \quad u_s = u, \quad w, \quad \frac{\partial w}{\partial n} = \frac{\partial w}{\partial y} \qquad (7.3.25a)$$
$$N_{nn} = N_{yy}, \quad N_{ns} = N_{yx}, \quad V_n = V_y, \quad M_{nn} = M_{yy} \qquad (7.3.25b)$$

7.3.4.2 Typical edge conditions

Next we discuss some common types of boundary conditions for the linear bending of a rectangular plate with edges parallel to the x and y coordinates. Here we use the edge with normal $\hat{n}$ (with components n_x and n_y) to discuss the boundary conditions (see Fig. 7.3.2). The force boundary conditions may be expressed in terms of the generalized displacements using the plate constitutive equations discussed in the next section.

Free edge with normal $\hat{n}$: A free edge is one that is geometrically not restrained in any way. Hence, we have

$$u_n \neq 0, \quad u_s \neq 0, \quad w \neq 0, \quad \frac{\partial w}{\partial n} \neq 0 \qquad (7.3.26a)$$

However, the edge may have applied forces and/or moments

$$N_{nn} = \hat{N}_{nn}, \quad N_{ns} = \hat{N}_{ns}, \quad V_n \equiv Q_n + \frac{\partial M_{ns}}{\partial s} = \hat{V}_n, \quad M_{nn} = \hat{M}_{nn} \quad (7.3.26b)$$

where quantities with a hat are specified forces/moments.

Fixed (or clamped) with normal n̂: A fixed edge is one that is geometrically fully restrained

$$u_n = 0, \quad u_s = 0, \quad w = 0, \quad \frac{\partial w}{\partial n} = 0 \quad\quad\quad (7.3.27)$$

Therefore, the forces and moments on a fixed edge are not known *a priori* (i.e. they are reactions to be determined as part of the analysis).

Simply-supported edge with normal n̂: The phrase "simply-supported" does not uniquely define the boundary conditions and one must indicate what it means, especially when both in-plane and bending deflections are coupled. Here we define two types of simply-supported boundary conditions:

$$\text{SS--1:} \quad u_s = 0, \quad w = 0; \quad N_{nn} = \hat{N}_{nn}, \quad M_{nn} = \hat{M}_{nn} \quad (7.3.28)$$

$$\text{SS--2:} \quad u_n = 0, \quad w = 0; \quad N_{ns} = \hat{N}_{ns}, \quad M_{nn} = \hat{M}_{nn} \quad (7.3.29)$$

7.3.5   Stress Resultant–Deflection Relations

To express the forces and moments $(\mathbf{N}, \mathbf{M})$ per unit length in terms of the generalized displacements (u, v, w), we must invoke appropriate stress–strain relations. In the CPT, all three transverse strain components $(\varepsilon_{zz}, \varepsilon_{xz}, \varepsilon_{yz})$ are zero by definition. Since $\varepsilon_{zz} = 0$, the transverse normal stress σ_{zz}, though not zero identically, does not appear in the virtual work statement and hence in the equations of motion. Consequently, it amounts to neglecting the transverse normal stress. Thus, in theory, we have a case of both plane strain and plane stress. However, from practical considerations, a thin to moderately thick plate is in a state of plane stress because the thickness is small as compared to the in-plane dimensions. Hence, the plane stress-reduced constitutive relations are used.

For an orthotropic material with principal material axes (x_1, x_2, x_3) coinciding with the plate coordinates (x, y, z), the plane stress-reduced thermoelastic constitutive equations can be expressed as (see Reddy [55] for the discussion of constitutive relations for orthotropic and laminated plates)

$$\begin{Bmatrix} \sigma_{xx} \\ \sigma_{yy} \\ \sigma_{xy} \end{Bmatrix} = \begin{bmatrix} Q_{11} & Q_{12} & 0 \\ Q_{12} & Q_{22} & 0 \\ 0 & 0 & Q_{66} \end{bmatrix} \begin{Bmatrix} \varepsilon_{xx} - \alpha_1\, \Delta T \\ \varepsilon_{yy} - \alpha_2\, \Delta T \\ \gamma_{xy} \end{Bmatrix} \quad (7.3.30)$$

where Q_{ij} are the plane stress-reduced stiffnesses

$$Q_{11} = \frac{E_1}{1 - \nu_{12}\nu_{21}}, \quad Q_{12} = \frac{\nu_{12}E_2}{1 - \nu_{12}\nu_{21}} = \frac{\nu_{21}E_1}{1 - \nu_{12}\nu_{21}}$$

$$Q_{22} = \frac{E_2}{1 - \nu_{12}\nu_{21}}, \quad Q_{66} = G_{12} \tag{7.3.31}$$

and $(\sigma_i, \varepsilon_i)$ are the stress and strain components, respectively, α_1 and α_2 are the coefficients of thermal expansion, and ΔT is the temperature increment from a reference state, $\Delta T = T - T_0$. The moisture strains are similar to thermal strains (i.e. for moisture strains, replace ΔT and α_i with the moisture concentration increment and coefficients of hygroscopic expansion, respectively).

The plate constitutive equations relate the forces and moments per unit length in Eq. (7.3.4) to the strains in Eq. (7.2.8) of the plate theory. For a plate made of a single or multiple orthotropic layers, the plate constitutive relations are obtained using the definitions in Eq. (7.3.4). For plates laminated of multiple orthotropic layers whose material axes are arbitrarily oriented with respect to the plate axes, the plate constitutive relations couple the in-plane displacements to the out-of-plane displacements even for linear problems (see Reddy [55, 56] for details). For a single orthotropic layer, the plate constitutive relations are greatly simplified. They are

$$\left\{ \begin{matrix} N_{xx} \\ N_{yy} \\ N_{xy} \end{matrix} \right\} = \int_{-\frac{h}{2}}^{\frac{h}{2}} \begin{bmatrix} Q_{11} & Q_{12} & 0 \\ Q_{12} & Q_{22} & 0 \\ 0 & 0 & Q_{66} \end{bmatrix} \left\{ \begin{matrix} \varepsilon_{xx}^0 + z\varepsilon_{xx}^1 - \alpha_1 \Delta T \\ \varepsilon_{yy}^0 + z\varepsilon_{yy}^1 - \alpha_2 \Delta T \\ \gamma_{xy}^0 + z\gamma_{xy}^1 \end{matrix} \right\} dz$$

$$= \begin{bmatrix} A_{11} & A_{12} & 0 \\ A_{12} & A_{22} & 0 \\ 0 & 0 & A_{66} \end{bmatrix} \left\{ \begin{matrix} \varepsilon_{xx}^0 \\ \varepsilon_{yy}^0 \\ \gamma_{xy}^0 \end{matrix} \right\} - \left\{ \begin{matrix} N_{xx}^T \\ N_{yy}^T \\ 0 \end{matrix} \right\} \tag{7.3.32}$$

$$\left\{ \begin{matrix} M_{xx} \\ M_{yy} \\ M_{xy} \end{matrix} \right\} = \int_{-\frac{h}{2}}^{\frac{h}{2}} \begin{bmatrix} Q_{11} & Q_{12} & 0 \\ Q_{12} & Q_{22} & 0 \\ 0 & 0 & Q_{66} \end{bmatrix} \left\{ \begin{matrix} \varepsilon_{xx}^0 + z\varepsilon_{xx}^1 - \alpha_1 \Delta T \\ \varepsilon_{yy}^0 + z\varepsilon_{yy}^1 - \alpha_2 \Delta T \\ \gamma_{xy}^0 + z\gamma_{xy}^1 \end{matrix} \right\} z\, dz$$

$$= \begin{bmatrix} D_{11} & D_{12} & 0 \\ D_{12} & D_{22} & 0 \\ 0 & 0 & D_{66} \end{bmatrix} \left\{ \begin{matrix} \varepsilon_{xx}^1 \\ \varepsilon_{yy}^1 \\ \gamma_{xy}^1 \end{matrix} \right\} - \left\{ \begin{matrix} M_{xx}^T \\ M_{yy}^T \\ 0 \end{matrix} \right\} \tag{7.3.33}$$

where A_{ij} are *extensional stiffnesses* and D_{ij} are *bending stiffnesses*, which are defined in terms of the elastic stiffnesses Q_{ij} as

$$(A_{ij}, D_{ij}) = \int_{-\frac{h}{2}}^{\frac{h}{2}} Q_{ij} \left(1, z^2 \right) dz \quad \text{or} \quad A_{ij} = Q_{ij}\, h, \quad D_{ij} = Q_{ij} \frac{h^3}{12} \tag{7.3.34}$$

and $\mathbf{N}^T$ and $\mathbf{M}^T$ are thermal stress resultants

$$\left\{ \begin{matrix} N_{xx}^T \\ N_{yy}^T \end{matrix} \right\} = \left\{ \begin{matrix} Q_{11}\alpha_1 + Q_{12}\alpha_2 \\ Q_{12}\alpha_1 + Q_{22}\alpha_2 \end{matrix} \right\} \int_{-\frac{h}{2}}^{\frac{h}{2}} \Delta T(x, y, z)\, dz \qquad (7.3.35a)$$

$$\left\{ \begin{matrix} M_{xx}^T \\ M_{yy}^T \end{matrix} \right\} = \left\{ \begin{matrix} Q_{11}\alpha_1 + Q_{12}\alpha_2 \\ Q_{12}\alpha_1 + Q_{22}\alpha_2 \end{matrix} \right\} \int_{-\frac{h}{2}}^{\frac{h}{2}} \Delta T(x, y, z)\, z\, dz \qquad (7.3.35b)$$

where α_1 and α_2 are the thermal coefficients of expansion, and ΔT is the temperature change (above a stress-free temperature), which is a known function of position. For isotropic plates, Eqs. (7.3.35a,b) simplify to $N_{xx}^T = N_{yy}^T = N^T$ and $M_{xx}^T = M_{yy}^T = M^T$, where

$$(N^T, M^T) = \frac{E\alpha}{1 - \nu} \int_{-\frac{h}{2}}^{\frac{h}{2}} \Delta T(1, z)\, dz \qquad (7.3.36)$$

7.4   Finite Element Models of the CPT

7.4.1   General Formulation

In this section, the displacement finite element model of Eqs. (7.3.15)–(7.3.17) governing plates according to the CPT is developed. The virtual work statements of the CPT over a typical orthotropic plate finite element Ω^e are given by [from Eqs. (7.3.10)–(7.3.12)]

$$0 = \int_{\Omega^e} \left(\frac{\partial \delta u}{\partial x} \left\{ A_{11} \left[\frac{\partial u}{\partial x} + \frac{1}{2}\left(\frac{\partial w}{\partial x}\right)^2 \right] + A_{12} \left[\frac{\partial v}{\partial y} + \frac{1}{2}\left(\frac{\partial w}{\partial y}\right)^2 \right] \right\} \right.$$
$$\left. + A_{66} \frac{\partial \delta u}{\partial y} \left[\frac{\partial u}{\partial y} + \frac{\partial v}{\partial x} + \frac{\partial w}{\partial x}\frac{\partial w}{\partial y} \right] \right) dx\, dy - \int_{\Omega^e} \frac{\partial \delta u}{\partial x} N_{xx}^T \, dxdy$$
$$- \oint_{\Gamma^e} N_{nn} \delta u_n \, ds \qquad (7.4.1)$$

$$0 = \int_{\Omega^e} \left(\frac{\partial \delta v}{\partial y} \left\{ A_{12} \left[\frac{\partial u}{\partial x} + \frac{1}{2}\left(\frac{\partial w}{\partial x}\right)^2 \right] + A_{22} \left[\frac{\partial v}{\partial y} + \frac{1}{2}\left(\frac{\partial w}{\partial y}\right)^2 \right] \right\} \right.$$
$$\left. + A_{66} \frac{\partial \delta v}{\partial x} \left[\frac{\partial u}{\partial y} + \frac{\partial v}{\partial x} + \frac{\partial w}{\partial x}\frac{\partial w}{\partial y} \right] \right) dx\, dy - \int_{\Omega^e} \frac{\partial \delta v}{\partial y} N_{yy}^T \, dxdy$$
$$- \oint_{\Gamma^e} N_{ns} \delta u_s \, ds \qquad (7.4.2)$$

$$0 = \int_{\Omega^e} \left\{ \frac{\partial \delta w}{\partial x} \frac{\partial w}{\partial x} \left\{ A_{11} \left[\frac{\partial u}{\partial x} + \frac{1}{2}\left(\frac{\partial w}{\partial x}\right)^2 \right] + A_{12} \left[\frac{\partial v}{\partial y} + \frac{1}{2}\left(\frac{\partial w}{\partial y}\right)^2 \right] \right\} \right.$$
$$+ A_{66} \frac{\partial w}{\partial y} \left(\frac{\partial u}{\partial y} + \frac{\partial v}{\partial x} + \frac{\partial w}{\partial x}\frac{\partial w}{\partial y} \right) + \frac{\partial \delta w}{\partial y} \frac{\partial w}{\partial y} \left\{ A_{12} \left[\frac{\partial u}{\partial x} + \frac{1}{2}\left(\frac{\partial w}{\partial x}\right)^2 \right] \right.$$
$$\left. + A_{22} \left[\frac{\partial v}{\partial y} + \frac{1}{2}\left(\frac{\partial w}{\partial y}\right)^2 \right] \right\} + A_{66} \frac{\partial w}{\partial x} \left(\frac{\partial u}{\partial y} + \frac{\partial v}{\partial x} + \frac{\partial w}{\partial x}\frac{\partial w}{\partial y} \right) \right]$$

$$+ \frac{\partial^2 \delta w}{\partial x^2}\left(D_{11}\frac{\partial^2 w}{\partial x^2} + D_{12}\frac{\partial^2 w}{\partial y^2}\right) + \frac{\partial^2 \delta w}{\partial y^2}\left(D_{12}\frac{\partial^2 w}{\partial x^2} + D_{22}\frac{\partial^2 w}{\partial y^2}\right)$$

$$+ 4D_{66}\frac{\partial^2 \delta w}{\partial x \partial y}\frac{\partial^2 w}{\partial x \partial y} + k\,\delta w\,w - \delta w\,q\Big\}dx\,dy$$

$$+ \int_{\Omega^e}\left(\frac{\partial^2 \delta w}{\partial x^2}M_{xx}^T + \frac{\partial^2 \delta w}{\partial y^2}M_{yy}^T - \frac{\partial \delta w}{\partial x}\frac{\partial w}{\partial x}N_{xx}^T - \frac{\partial \delta w}{\partial y}\frac{\partial w}{\partial y}N_{yy}^T\right)dx\,dy$$

$$- \oint_{\Gamma^e}\left(V_n\delta w - M_{nn}\frac{\partial \delta w}{\partial n}\right)ds \tag{7.4.3}$$

where (N_{nn}, N_{ns}), (M_{nn}, M_{ns}), and V_n are defined in Eqs. (7.3.20a), (7.3.20b), and (7.3.23), respectively, and (n_x, n_y) denote the direction cosines of the unit normal on the element boundary Γ^e. We note from the boundary terms in Eqs. (7.4.1)–(7.4.3) that u_n, u_s, w, and $\partial w/\partial n$ are used as the primary variables (or generalized displacements), and $\hat{N}_{nn}$, $\hat{N}_{ns}$, $\hat{V}_n$, and $\hat{M}_{nn}$ as the secondary degrees of freedom (or generalized forces). Thus, finite elements based on the CPT require continuity of the transverse deflection w and its derivatives across element boundaries (i.e. C^1-continuity of w). Also, to satisfy the constant displacement (rigid body mode) and constant strain requirements, the polynomial expansion for w should be a complete quadratic. The in-plane displacements u_n and u_s need only be C^0 continuous. In the present study, attention is focused only on rectangular finite elements and, therefore, we shall use $(u, v, w, \partial w/\partial x, \partial w/\partial y)$ as the generalized displacements.

We assume finite element approximation of the form

$$u(x, y) = \sum_{j=1}^{m=4} u_j^e \psi_j^e(x, y), \quad v(x, y) = \sum_{j=1}^{m=4} v_j^e \psi_j^e(x, y), \quad w(x, y) = \sum_{j=1}^{n} \bar{\Delta}_j^e \varphi_j^e(x, y)$$

$$\tag{7.4.4}$$

where ψ_j^e are the linear Lagrange interpolation functions, $\bar{\Delta}_j^e$ are the values of w and its derivatives at the nodes, and φ_j^e are the interpolation functions, the specific form of which depends on the geometry of the element and the nodal degrees of freedom interpolated. In the present study, we shall use rectangular elements with two different sets of degrees of freedom: $(u, v, w, w_{,x}, w_{,y})$ and $(u, v, w, w_{,x}, w_{,y}, w_{,xy})$, as shown in Figs. 7.4.1(a) and 7.4.1(b), respectively (i.e. $m = 4$ and $n = 12$ or $m = 4$ and $n = 16$). The first one is known as a *non-conforming element* and the latter is called a *conforming element*. Additional details on these two elements are presented in Section 7.4.3.

Substituting the approximations in Eq. (7.4.4) for (u, v, w) and $(\psi_i^e, \psi_i^e, \varphi_i^e)$ for the virtual displacements $(\delta u, \delta v, \delta w)$ into Eqs. (7.4.1)–(7.4.3), we obtain

$$\begin{bmatrix} \mathbf{K}^{11} & \mathbf{K}^{12} & \mathbf{K}^{13} \\ \mathbf{K}^{21} & \mathbf{K}^{22} & \mathbf{K}^{23} \\ \mathbf{K}^{31} & \mathbf{K}^{32} & \mathbf{K}^{33} \end{bmatrix}\begin{Bmatrix} \mathbf{u} \\ \mathbf{v} \\ \bar{\boldsymbol{\Delta}} \end{Bmatrix} = \begin{Bmatrix} \mathbf{F}^1 \\ \mathbf{F}^2 \\ \mathbf{F}^3 \end{Bmatrix} + \begin{Bmatrix} \mathbf{F}^{1T} \\ \mathbf{F}^{2T} \\ \mathbf{F}^{3T} \end{Bmatrix} \quad \text{or} \quad \mathbf{K\Delta} = \mathbf{F} \tag{7.4.5}$$

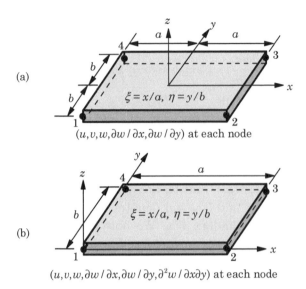

Fig. 7.4.1: (a) The non-conforming and (b) the conforming rectangular elements.

The stiffness matrix coefficients $K_{ij}^{\alpha\beta}$ and force vector components F_i^α and $F_i^{\alpha T}$ $(\alpha, \beta = 1, 2, 3)$ are defined as follows:

$$K_{ij}^{11} = \int_{\Omega^e} \left(A_{11} \frac{\partial \psi_i^e}{\partial x} \frac{\partial \psi_j^e}{\partial x} + A_{66} \frac{\partial \psi_i^e}{\partial y} \frac{\partial \psi_j^e}{\partial y} \right) dx\, dy$$

$$K_{ij}^{12} = \int_{\Omega^e} \left(A_{12} \frac{\partial \psi_i^e}{\partial x} \frac{\partial \psi_j^e}{\partial y} + A_{66} \frac{\partial \psi_i^e}{\partial y} \frac{\partial \psi_j^e}{\partial x} \right) dx\, dy = K_{ji}^{21}$$

$$K_{ij}^{13} = \frac{1}{2} \int_{\Omega^e} \left[\frac{\partial \psi_i^e}{\partial x} \left(A_{11} \frac{\partial w}{\partial x} \frac{\partial \varphi_j^e}{\partial x} + A_{12} \frac{\partial w}{\partial y} \frac{\partial \varphi_j^e}{\partial y} \right) \right.$$
$$\left. + A_{66} \frac{\partial \psi_i^e}{\partial y} \left(\frac{\partial w}{\partial x} \frac{\partial \varphi_j^e}{\partial y} + \frac{\partial w}{\partial y} \frac{\partial \varphi_j^e}{\partial x} \right) \right] dx\, dy$$

$$K_{ij}^{22} = \int_{\Omega^e} \left(A_{66} \frac{\partial \psi_i^e}{\partial x} \frac{\partial \psi_j^e}{\partial x} + A_{22} \frac{\partial \psi_i^e}{\partial y} \frac{\partial \psi_j^e}{\partial y} \right) dx\, dy$$

$$K_{ij}^{23} = \frac{1}{2} \int_{\Omega^e} \left[\frac{\partial \psi_i^e}{\partial y} \left(A_{12} \frac{\partial w}{\partial x} \frac{\partial \varphi_j^e}{\partial x} + A_{22} \frac{\partial w}{\partial y} \frac{\partial \varphi_j^e}{\partial y} \right) \right.$$
$$\left. + A_{66} \frac{\partial \psi_i^e}{\partial x} \left(\frac{\partial w}{\partial x} \frac{\partial \varphi_j^e}{\partial y} + \frac{\partial w}{\partial y} \frac{\partial \varphi_j^e}{\partial x} \right) \right] dx\, dy$$

$$K_{ij}^{31} = \int_{\Omega^e} \left[\frac{\partial \varphi_i^e}{\partial x} \left(A_{11} \frac{\partial w}{\partial x} \frac{\partial \psi_j^e}{\partial x} + A_{66} \frac{\partial w}{\partial y} \frac{\partial \psi_j^e}{\partial y} \right) \right.$$
$$\left. + \frac{\partial \varphi_i^e}{\partial y} \left(A_{66} \frac{\partial w}{\partial x} \frac{\partial \psi_j^e}{\partial y} + A_{12} \frac{\partial w}{\partial y} \frac{\partial \psi_j^e}{\partial x} \right) \right] dx\, dy$$

$$K_{ij}^{32} = \int_{\Omega^e} \left[\frac{\partial \varphi_i^e}{\partial x} \left(A_{12} \frac{\partial w}{\partial x} \frac{\partial \psi_j^e}{\partial y} + A_{66} \frac{\partial w}{\partial y} \frac{\partial \psi_j^e}{\partial x} \right) \right. $$
$$ \left. + \frac{\partial \varphi_i^e}{\partial y} \left(A_{66} \frac{\partial w}{\partial x} \frac{\partial \psi_j^e}{\partial x} + A_{22} \frac{\partial w}{\partial y} \frac{\partial \psi_j^e}{\partial y} \right) \right] dx\,dy$$

$$K_{ij}^{33} = \int_{\Omega^e} \left[D_{11} \frac{\partial^2 \varphi_i^e}{\partial x^2} \frac{\partial^2 \varphi_j^e}{\partial x^2} + D_{22} \frac{\partial^2 \varphi_i^e}{\partial y^2} \frac{\partial^2 \varphi_j^e}{\partial y^2} \right. $$
$$ + D_{12} \left(\frac{\partial^2 \varphi_i^e}{\partial x^2} \frac{\partial^2 \varphi_j^e}{\partial y^2} + \frac{\partial^2 \varphi_i^e}{\partial y^2} \frac{\partial^2 \varphi_j^e}{\partial x^2} \right) $$
$$ \left. + 4 D_{66} \frac{\partial^2 \varphi_i^e}{\partial x \partial y} \frac{\partial^2 \varphi_j^e}{\partial x \partial y} + k \varphi_i^e \varphi_j^e \right] dx\,dy$$

$$ + \frac{1}{2} \int_{\Omega^e} \left\{ \left[A_{11} \left(\frac{\partial w}{\partial x} \right)^2 + A_{66} \left(\frac{\partial w}{\partial y} \right)^2 \right] \frac{\partial \varphi_i^e}{\partial x} \frac{\partial \varphi_j^e}{\partial x} \right. $$
$$ + \left[A_{66} \left(\frac{\partial w}{\partial x} \right)^2 + A_{22} \left(\frac{\partial w}{\partial y} \right)^2 \right] \frac{\partial \varphi_i^e}{\partial y} \frac{\partial \varphi_j^e}{\partial y} $$
$$ \left. + (A_{12} + A_{66}) \frac{\partial w}{\partial x} \frac{\partial w}{\partial y} \left(\frac{\partial \varphi_i^e}{\partial x} \frac{\partial \varphi_j^e}{\partial y} + \frac{\partial \varphi_i^e}{\partial y} \frac{\partial \varphi_j^e}{\partial x} \right) \right\} dx\,dy$$

$$ - \int_{\Omega^e} \left(N_{xx}^T \frac{\partial \varphi_i^e}{\partial x} \frac{\partial \varphi_j^e}{\partial x} + N_{yy}^T \frac{\partial \varphi_i^e}{\partial y} \frac{\partial \varphi_j^e}{\partial y} \right) dx\,dy$$

$$F_i^1 = \oint_{\Gamma^e} N_{nn} \psi_i^e \, ds, \quad F_i^2 = \oint_{\Gamma^e} N_{ns} \psi_i^e \, ds$$

$$F_i^3 = \int_{\Omega^e} q \, \varphi_i^e \, dx\,dy + \oint_{\Gamma^e} \left(V_n \varphi_i^e - M_{nn} \frac{\partial \varphi_i^e}{\partial n} \right) ds$$

$$F_i^{1T} = \int_{\Omega^e} N_{xx}^T \frac{\partial \psi_i^e}{\partial x} \, dx\,dy, \quad F_i^{2T} = \int_{\Omega^e} N_{yy}^T \frac{\partial \psi_i^e}{\partial y} \, dx\,dy$$

$$F_i^{3T} = - \int_{\Omega^e} \left(\frac{\partial^2 \varphi_i^e}{\partial x^2} M_{xx}^T + \frac{\partial^2 \varphi_i^e}{\partial y^2} M_{yy}^T \right) dx\,dy \qquad (7.4.6)$$

where N_{xx}^T, M_{xx}^T, etc., are the thermal forces and moments in Eqs. (7.3.35a,b). Hats ^ on the stress resultants N_{nn}, N_{ns}, M_{nn}, and V_n are removed because they are now defined on the element boundary. Note that the thermal resultant term included in K_{ij}^{33} is due to the von Kármán nonlinearity; alternatively, it could have been included in F_i^{3T} as a nonlinear term. The finite element model in Eq. (7.4.5) is called a *displacement finite element model*.

7.4.2 Tangent Stiffness Coefficients

In the solution of nonlinear algebraic equations arising in the analysis of structural problems, one often uses the Newton method or its improvements (see Appendix 2). To this end it is useful to derive the tangent stiffness coefficients associated with the CPT. The Newton iterative method involves solving equations of the form

$$
\begin{bmatrix} \mathbf{T}^{11} & \mathbf{T}^{12} & \mathbf{T}^{13} \\ \mathbf{T}^{21} & \mathbf{T}^{22} & \mathbf{T}^{23} \\ \mathbf{T}^{31} & \mathbf{T}^{32} & \mathbf{T}^{33} \end{bmatrix} \begin{Bmatrix} \delta\boldsymbol{\Delta}^1 \\ \delta\boldsymbol{\Delta}^2 \\ \delta\boldsymbol{\Delta}^3 \end{Bmatrix} = - \begin{Bmatrix} \mathbf{R}^1 \\ \mathbf{R}^2 \\ \mathbf{R}^3 \end{Bmatrix} \quad \text{or} \quad \mathbf{T}\,\delta\boldsymbol{\Delta} = -\mathbf{R} \tag{7.4.7}
$$

where

$$
\Delta_i^1 = u_i, \quad \Delta_i^2 = v_i, \quad \Delta_i^3 = \bar{\Delta}_i^3 \tag{7.4.8}
$$

The coefficients of the submatrices $\mathbf{T}^{\alpha\beta}$ of the tangent stiffness matrix $\mathbf{T}$ and the components of the residual vector $\mathbf{R}^\alpha$ are defined by

$$
T_{ij}^{\alpha\beta} = \frac{\partial R_i^\alpha}{\partial \Delta_j^\beta}, \quad R_i^\alpha = \sum_{\gamma=1}^{3}\sum_{k=1}^{n(\gamma)} K_{ik}^{\alpha\gamma}\Delta_k^\gamma - F_i^\alpha \tag{7.4.9}
$$

where $n(\gamma)$ denotes m or n, depending on γ value $[n(1) = n(2) = m = 4$ and $n(3) = n]$. We obtain

$$
T_{ij}^{\alpha\beta} = \frac{\partial}{\partial \Delta_j^\beta}\left(\sum_{\gamma=1}^{3}\sum_{k=1}^{n(\gamma)} K_{ik}^{\alpha\gamma}\Delta_k^\gamma - F_i^\alpha \right) = \sum_{\gamma=1}^{3}\sum_{k=1}^{n(\gamma)} \frac{\partial K_{ik}^{\alpha\gamma}}{\partial \Delta_j^\beta}\Delta_k^\gamma + K_{ij}^{\alpha\beta} \tag{7.4.10}
$$

Since the nonlinearity involved in the problem is only a function of w (or, equivalently, only a function of the nodal values $\bar{\boldsymbol{\Delta}} = \boldsymbol{\Delta}_3$) and the only coefficients that depend on w are K_{ij}^{13}, K_{ij}^{23}, K_{ij}^{31}, K_{ij}^{32}, and K_{ij}^{33}, the derivatives of all submatrices $\mathbf{K}^{\alpha\beta}$ with respect to $\Delta_j^1 = u_j$ and $\Delta_j^2 = v_j$ are zero. Hence, we have (assuming that $\mathbf{F}^\alpha$ are independent of $\boldsymbol{\Delta}$)

$$
\mathbf{T}^{\alpha\beta} = \mathbf{K}^{\alpha\beta} \quad \text{for} \quad \alpha = 1, 2, 3; \ \beta = 1, 2 \tag{7.4.11}
$$

The remaining coefficients, $\mathbf{T}^{13}$, $\mathbf{T}^{23}$, and $\mathbf{T}^{33}$, are computed as follows:

$$
T_{ij}^{13} = \sum_{\gamma=1}^{3}\sum_{k=1}^{n(\gamma)} \frac{\partial K_{ik}^{1\gamma}}{\partial \bar{\Delta}_j^3}\Delta_k^\gamma + K_{ij}^{13} = \sum_{k=1}^{n} \frac{\partial K_{ik}^{13}}{\partial \bar{\Delta}_j^3}\bar{\Delta}_k^3 + K_{ij}^{13}
$$

$$
= \frac{1}{2}\sum_{k=1}^{n}\bar{\Delta}_k^3 \frac{\partial}{\partial \bar{\Delta}_j^3}\left\{ \int_{\Omega^e}\left[\frac{\partial \psi_i^e}{\partial x}\left(A_{11}\frac{\partial w}{\partial x}\frac{\partial \varphi_k^e}{\partial x} + A_{12}\frac{\partial w}{\partial y}\frac{\partial \varphi_k^e}{\partial y} \right) \right.\right.
$$

$$
\left.\left. + A_{66}\frac{\partial \psi_i^e}{\partial y}\left(\frac{\partial w}{\partial x}\frac{\partial \varphi_k^e}{\partial y} + \frac{\partial w}{\partial y}\frac{\partial \varphi_k^e}{\partial x} \right) \right] dx\, dy \right\} + K_{ij}^{13}
$$

$$
= \frac{1}{2}\sum_{k=1}^{n}\bar{\Delta}_k^3\left\{ \int_{\Omega^e}\left[\frac{\partial \psi_i^e}{\partial x}\left(A_{11}\frac{\partial \varphi_j^e}{\partial x}\frac{\partial \varphi_k^e}{\partial x} + A_{12}\frac{\partial \varphi_j^e}{\partial y}\frac{\partial \varphi_k^e}{\partial y} \right) \right.\right.
$$

$$
\left.\left. + A_{66}\frac{\partial \psi_i^e}{\partial y}\left(\frac{\partial \varphi_j^e}{\partial x}\frac{\partial \varphi_k^e}{\partial y} + \frac{\partial \varphi_j^e}{\partial y}\frac{\partial \varphi_k^e}{\partial x} \right) \right] dx\, dy \right\} + K_{ij}^{13}
$$

$$= \frac{1}{2} \int_{\Omega^e} \left[\frac{\partial \psi_i^e}{\partial x} \left(A_{11} \frac{\partial w}{\partial x} \frac{\partial \varphi_j^e}{\partial x} + A_{12} \frac{\partial w}{\partial y} \frac{\partial \varphi_j^e}{\partial y} \right) \right.$$

$$\left. + A_{66} \frac{\partial \psi_i^e}{\partial y} \left(\frac{\partial w}{\partial x} \frac{\partial \varphi_j^e}{\partial y} + \frac{\partial w}{\partial y} \frac{\partial \varphi_j^e}{\partial x} \right) \right] dx \, dy + K_{ij}^{13}$$

$$= K_{ij}^{13} + K_{ij}^{13} = 2K_{ij}^{13} \ (= T_{ji}^{31}); \quad (\mathbf{T}^{13})^{\mathrm{T}} = \mathbf{T}^{31} = \mathbf{K}^{31} \qquad (7.4.12)$$

$$T_{ij}^{23} = \sum_{\gamma=1}^{3} \sum_{k=1}^{n(\gamma)} \frac{\partial K_{ik}^{2\gamma}}{\partial \bar{\Delta}_j^3} \Delta_k^\gamma + K_{ij}^{23} = \sum_{k=1}^{n} \frac{\partial K_{ik}^{23}}{\partial \bar{\Delta}_j^3} \bar{\Delta}_k^3 + K_{ij}^{23}$$

$$= \frac{1}{2} \sum_{k=1}^{n} \bar{\Delta}_k^3 \frac{\partial}{\partial \bar{\Delta}_j^3} \left\{ \int_{\Omega^e} \left[\frac{\partial \psi_i^e}{\partial y} \left(A_{12} \frac{\partial w}{\partial x} \frac{\partial \varphi_k^e}{\partial x} + A_{22} \frac{\partial w}{\partial y} \frac{\partial \varphi_k^e}{\partial y} \right) \right. \right.$$

$$\left. \left. + A_{66} \frac{\partial \psi_i^e}{\partial x} \left(\frac{\partial w}{\partial x} \frac{\partial \varphi_k^e}{\partial y} + \frac{\partial w}{\partial y} \frac{\partial \varphi_k^e}{\partial x} \right) \right] dx \, dy \right\} + K_{ij}^{23}$$

$$= \frac{1}{2} \sum_{k=1}^{n} \bar{\Delta}_k^3 \left\{ \int_{\Omega^e} \left[\frac{\partial \psi_i^e}{\partial y} \left(A_{12} \frac{\partial \varphi_j^e}{\partial x} \frac{\partial \varphi_k^e}{\partial x} + A_{22} \frac{\partial \varphi_j^e}{\partial y} \frac{\partial \varphi_k^e}{\partial y} \right) \right. \right.$$

$$\left. \left. + A_{66} \frac{\partial \psi_i^e}{\partial x} \left(\frac{\partial \varphi_j^e}{\partial x} \frac{\partial \varphi_k^e}{\partial y} + \frac{\partial \varphi_j^e}{\partial y} \frac{\partial \varphi_k^e}{\partial x} \right) \right] dx \, dy \right\} + K_{ij}^{23}$$

$$= \frac{1}{2} \int_{\Omega^e} \left[\frac{\partial \psi_i^e}{\partial y} \left(A_{12} \frac{\partial w}{\partial x} \frac{\partial \varphi_j^e}{\partial x} + A_{22} \frac{\partial w}{\partial y} \frac{\partial \varphi_j^e}{\partial y} \right) \right.$$

$$\left. + A_{66} \frac{\partial \psi_i^e}{\partial x} \left(\frac{\partial w}{\partial y} \frac{\partial \varphi_j^e}{\partial x} + \frac{\partial w}{\partial x} \frac{\partial \varphi_j^e}{\partial y} \right) \right] dx \, dy + K_{ij}^{23}$$

$$= K_{ij}^{23} + K_{ij}^{23} = 2K_{ij}^{23} \ (= T_{ji}^{32}); \quad (\mathbf{T}^{23})^{\mathrm{T}} = \mathbf{T}^{32} = \mathbf{K}^{32} \qquad (7.4.13)$$

The computation of T_{ij}^{33} requires the calculation of three parts:

$$T_{ij}^{33} = K_{ij}^{33} + \sum_{k=1}^{m=4} \frac{\partial K_{ik}^{31}}{\partial \bar{\Delta}_j^3} u_k + \sum_{k=1}^{m=4} \frac{\partial K_{ik}^{32}}{\partial \bar{\Delta}_j^3} v_k + \sum_{k=1}^{n} \frac{\partial K_{ik}^{33}}{\partial \bar{\Delta}_j^3} \bar{\Delta}_k^3 \qquad (7.4.14)$$

We shall compute these three terms first. We have

$$\sum_{k=1}^{m=4} u_k \frac{\partial K_{ik}^{31}}{\partial \bar{\Delta}_j^3} = \sum_{k=1}^{m=4} u_k \frac{\partial}{\partial \bar{\Delta}_j^3} \left\{ \int_{\Omega^e} \left[\frac{\partial \varphi_i^e}{\partial x} \left(A_{11} \frac{\partial w}{\partial x} \frac{\partial \psi_k^e}{\partial x} + A_{66} \frac{\partial w}{\partial y} \frac{\partial \psi_k^e}{\partial y} \right) \right. \right.$$

$$\left. \left. + \frac{\partial \varphi_i^e}{\partial y} \left(A_{66} \frac{\partial w}{\partial x} \frac{\partial \psi_k^e}{\partial y} + A_{12} \frac{\partial w}{\partial y} \frac{\partial \psi_k^e}{\partial x} \right) \right] dx \, dy \right\}$$

$$= \sum_{k=1}^{m=4} u_k \left\{ \int_{\Omega^e} \left[\frac{\partial \varphi_i^e}{\partial x} \left(A_{11} \frac{\partial \varphi_j^e}{\partial x} \frac{\partial \psi_k^e}{\partial x} + A_{66} \frac{\partial \varphi_j^e}{\partial y} \frac{\partial \psi_k^e}{\partial y} \right) \right. \right.$$

$$\left. \left. + \frac{\partial \varphi_i^e}{\partial y} \left(A_{66} \frac{\partial \varphi_j^e}{\partial x} \frac{\partial \psi_k^e}{\partial y} + A_{12} \frac{\partial \varphi_j^e}{\partial y} \frac{\partial \psi_k^e}{\partial x} \right) \right] dx \, dy \right\}$$

$$
= \int_{\Omega^e} \left[\frac{\partial \varphi_i^e}{\partial x} \left(A_{11} \frac{\partial u}{\partial x} \frac{\partial \varphi_j^e}{\partial x} + A_{66} \frac{\partial u}{\partial y} \frac{\partial \varphi_j^e}{\partial y} \right) \right.
$$
$$
\left. + \frac{\partial \varphi_i^e}{\partial y} \left(A_{66} \frac{\partial u}{\partial y} \frac{\partial \varphi_j^e}{\partial x} + A_{12} \frac{\partial u}{\partial x} \frac{\partial \varphi_j^e}{\partial y} \right) \right] dx\, dy \tag{7.4.15}
$$

$$
\sum_{k=1}^{m=4} v_k \frac{\partial K_{ik}^{32}}{\partial \bar{\Delta}_j^3} = \sum_{k=1}^{m=4} v_k \frac{\partial}{\partial \bar{\Delta}_j^3} \left\{ \int_{\Omega^e} \left[\frac{\partial \varphi_i^e}{\partial x} \left(A_{12} \frac{\partial w}{\partial x} \frac{\partial \psi_k^e}{\partial y} + A_{66} \frac{\partial w}{\partial y} \frac{\partial \psi_k^e}{\partial x} \right) \right. \right.
$$
$$
\left. \left. + \frac{\partial \varphi_i^e}{\partial y} \left(A_{66} \frac{\partial w}{\partial x} \frac{\partial \psi_k^e}{\partial x} + A_{22} \frac{\partial w}{\partial y} \frac{\partial \psi_k^e}{\partial y} \right) \right] dx\, dy \right\}
$$
$$
= \sum_{k=1}^{m} v_k \left\{ \int_{\Omega^e} \left[\frac{\partial \varphi_i^e}{\partial x} \left(A_{12} \frac{\partial \varphi_j^e}{\partial x} \frac{\partial \psi_k^e}{\partial y} + A_{66} \frac{\partial \varphi_j^e}{\partial y} \frac{\partial \psi_k^e}{\partial x} \right) \right. \right.
$$
$$
\left. \left. + \frac{\partial \varphi_i^e}{\partial y} \left(A_{66} \frac{\partial \varphi_j^e}{\partial x} \frac{\partial \psi_k^e}{\partial x} + A_{22} \frac{\partial \varphi_j^e}{\partial y} \frac{\partial \psi_k^e}{\partial y} \right) \right] dx\, dy \right\}
$$
$$
= \int_{\Omega^e} \left[\frac{\partial \varphi_i^e}{\partial x} \left(A_{12} \frac{\partial v}{\partial y} \frac{\partial \varphi_j^e}{\partial x} + A_{66} \frac{\partial v}{\partial x} \frac{\partial \varphi_j^e}{\partial y} \right) \right.
$$
$$
\left. + \frac{\partial \varphi_i^e}{\partial y} \left(A_{66} \frac{\partial v}{\partial x} \frac{\partial \varphi_j^e}{\partial x} + A_{22} \frac{\partial v}{\partial y} \frac{\partial \varphi_j^e}{\partial y} \right) \right] dx\, dy \tag{7.4.16}
$$

$$
\sum_{k=1}^{n} \bar{\Delta}_k^3 \frac{\partial K_{ik}^{33}}{\partial \bar{\Delta}_j^3} = \frac{1}{2} \sum_{k=1}^{n} \bar{\Delta}_k^3 \frac{\partial}{\partial \bar{\Delta}_j^3} \left[\int_{\Omega^e} \left\{ \left[A_{11} \left(\frac{\partial w}{\partial x} \right)^2 + A_{66} \left(\frac{\partial w}{\partial y} \right)^2 \right] \frac{\partial \varphi_i^e}{\partial x} \frac{\partial \varphi_k^e}{\partial x} \right. \right.
$$
$$
+ \left[A_{66} \left(\frac{\partial w}{\partial x} \right)^2 + A_{22} \left(\frac{\partial w}{\partial y} \right)^2 \right] \frac{\partial \varphi_i^e}{\partial y} \frac{\partial \varphi_k^e}{\partial y}
$$
$$
\left. \left. + \bar{A}_{12} \frac{\partial w}{\partial x} \frac{\partial w}{\partial y} \left(\frac{\partial \varphi_i^e}{\partial x} \frac{\partial \varphi_k^e}{\partial y} + \frac{\partial \varphi_i^e}{\partial y} \frac{\partial \varphi_k^e}{\partial x} \right) \right\} dx\, dy \right]
$$
$$
= \frac{1}{2} \sum_{k=1}^{n} \bar{\Delta}_k^3 \left\{ \int_{\Omega^e} \left[\left(2 A_{11} \frac{\partial w}{\partial x} \frac{\partial \varphi_j^e}{\partial x} + 2 A_{66} \frac{\partial w}{\partial y} \frac{\partial \varphi_j^e}{\partial y} \right) \frac{\partial \varphi_i^e}{\partial x} \frac{\partial \varphi_k^e}{\partial x} \right. \right.
$$
$$
+ \left(2 A_{66} \frac{\partial w}{\partial x} \frac{\partial \varphi_j^e}{\partial x} + 2 A_{22} \frac{\partial w}{\partial y} \frac{\partial \varphi_j^e}{\partial y} \right) \frac{\partial \varphi_i^e}{\partial y} \frac{\partial \varphi_k^e}{\partial y}
$$
$$
\left. \left. + \bar{A}_{12} \left(\frac{\partial w}{\partial x} \frac{\partial \varphi_j^e}{\partial y} + \frac{\partial w}{\partial y} \frac{\partial \varphi_j^e}{\partial x} \right) \left(\frac{\partial \varphi_i^e}{\partial x} \frac{\partial \varphi_k^e}{\partial y} + \frac{\partial \varphi_i^e}{\partial y} \frac{\partial \varphi_k^e}{\partial x} \right) \right] dx\, dy \right\}
$$
$$
= \int_{\Omega^e} \left[A_{11} \left(\frac{\partial w}{\partial x} \right)^2 \frac{\partial \varphi_i^e}{\partial x} \frac{\partial \varphi_j^e}{\partial x} + A_{22} \left(\frac{\partial w}{\partial y} \right)^2 \frac{\partial \varphi_i^e}{\partial y} \frac{\partial \varphi_j^e}{\partial y} \right.
$$
$$
+ A_{66} \frac{\partial w}{\partial x} \frac{\partial w}{\partial y} \left(\frac{\partial \varphi_i^e}{\partial x} \frac{\partial \varphi_j^e}{\partial y} + \frac{\partial \varphi_i^e}{\partial y} \frac{\partial \varphi_j^e}{\partial x} \right)
$$
$$
\left. + \frac{\bar{A}_{12}}{2} \left(\frac{\partial \varphi_i^e}{\partial x} \frac{\partial w}{\partial y} + \frac{\partial \varphi_i^e}{\partial y} \frac{\partial w}{\partial x} \right) \left(\frac{\partial w}{\partial x} \frac{\partial \varphi_j^e}{\partial y} + \frac{\partial w}{\partial y} \frac{\partial \varphi_j^e}{\partial x} \right) \right] dx\, dy \tag{7.4.17}
$$

where $\bar{A}_{12} = A_{12} + A_{66}$. Combining the expressions in Eqs. (7.4.15)–(7.4.17), we obtain

$$
\bar{T}_{ij}^{33} = \int_{\Omega^e} \left[\bar{N}_{xx} \frac{\partial \varphi_i^e}{\partial x} \frac{\partial \varphi_j^e}{\partial x} + \bar{N}_{yy} \frac{\partial \varphi_i^e}{\partial y} \frac{\partial \varphi_j^e}{\partial y} + \bar{N}_{xy} \left(\frac{\partial \varphi_i^e}{\partial x} \frac{\partial \varphi_j^e}{\partial y} + \frac{\partial \varphi_i^e}{\partial y} \frac{\partial \varphi_j^e}{\partial x} \right) \right] dx\, dy
$$
(7.4.18)

where

$$
\bar{N}_{xx} = N_{xx} + \frac{1}{2} \left[A_{11} \left(\frac{\partial w}{\partial x} \right)^2 + A_{66} \left(\frac{\partial w}{\partial y} \right)^2 \right]
$$

$$
\bar{N}_{yy} = N_{yy} + \frac{1}{2} \left[A_{66} \left(\frac{\partial w}{\partial x} \right)^2 + A_{22} \left(\frac{\partial w}{\partial y} \right)^2 \right]
$$
(7.4.19)

$$
\bar{N}_{xy} = N_{xy} + \left(\frac{A_{12} + A_{66}}{2} \right) \frac{\partial w}{\partial x} \frac{\partial w}{\partial y}
$$

Therefore, $\mathbf{T}^{33} = \mathbf{K}^{33} + \bar{\mathbf{T}}^{33}$ is given by

$$
T_{ij}^{33} = \int_{\Omega^e} \left[D_{11} \frac{\partial^2 \varphi_i^e}{\partial x^2} \frac{\partial^2 \varphi_j^e}{\partial x^2} + D_{22} \frac{\partial^2 \varphi_i^e}{\partial y^2} \frac{\partial^2 \varphi_j^e}{\partial y^2} + D_{12} \left(\frac{\partial^2 \varphi_i^e}{\partial x^2} \frac{\partial^2 \varphi_j^e}{\partial y^2} + \frac{\partial^2 \varphi_i^e}{\partial y^2} \frac{\partial^2 \varphi_j^e}{\partial x^2} \right) \right.
$$

$$
+ 4 D_{66} \frac{\partial^2 \varphi_i^e}{\partial x \partial y} \frac{\partial^2 \varphi_j^e}{\partial x \partial y} + k \varphi_i^e \varphi_j^e + \hat{N}_{xx} \frac{\partial \varphi_i^e}{\partial x} \frac{\partial \varphi_j^e}{\partial x} + \hat{N}_{yy} \frac{\partial \varphi_i^e}{\partial y} \frac{\partial \varphi_j^e}{\partial y}
$$

$$
\left. + \hat{N}_{xy} \left(\frac{\partial \varphi_i^e}{\partial x} \frac{\partial \varphi_j^e}{\partial y} + \frac{\partial \varphi_i^e}{\partial y} \frac{\partial \varphi_j^e}{\partial x} \right) \right] dx\, dy
$$
(7.4.20)

where

$$
\hat{N}_{xx} = N_{xx} + A_{11} \left(\frac{\partial w}{\partial x} \right)^2 + A_{66} \left(\frac{\partial w}{\partial y} \right)^2
$$

$$
\hat{N}_{yy} = N_{yy} + A_{66} \left(\frac{\partial w}{\partial x} \right)^2 + A_{22} \left(\frac{\partial w}{\partial y} \right)^2
$$
(7.4.21)

$$
\hat{N}_{xy} = N_{xy} + \left(A_{12} + A_{66} \right) \frac{\partial w}{\partial x} \frac{\partial w}{\partial y}
$$

and (N_{xx}, N_{yy}, N_{xy}) are defined by Eq. (7.3.32). Clearly, the tangent stiffness matrix of the CPT element is symmetric, while the original element stiffness is *not* symmetric.

7.4.3 Non-Conforming and Conforming Plate Elements

A non-conforming rectangular element has w, $\partial w/\partial x$, and $\partial w/\partial y$ as the nodal variables, in addition to the in-plane displacement degrees of freedom (u, v), as shown in Fig. 7.4.1(a). The in-plane displacements are approximated using linear interpolation functions [see Eqs. (3.3.38)]

$$
u \approx u_h = \sum_{j=1}^{4} u_j \psi_j(\xi, \eta), \quad v \approx v_h = \sum_{j=1}^{4} v_j \psi_j(\xi, \eta)
$$

The transverse deflection w is approximated by (see Melosh [57] and Zienkiewicz and Cheung [58])

$$w \approx w_h(\xi, \eta) = c_1 + c_2\xi + c_3\eta + c_4\xi\eta + c_5\xi^2 + c_6\eta^2 + c_7\xi^3 + c_8\xi^2\eta + c_9\xi\eta^2$$
$$+ c_{10}\eta^3 + c_{11}\xi^3\eta + c_{12}\xi\eta^3 = \sum_{j=1}^{12} \bar{\Delta}_j \varphi_j(\xi, \eta) \quad (7.4.22)$$

where φ_i are the cubic interpolation functions associated with the degrees of freedom $(w, \partial w/\partial x, \partial w/\partial y)$

$$\varphi_i = g_{i1} \ (i = 1, 4, 7, 10); \quad \varphi_i = g_{i2} \ (i = 2, 5, 8, 11); \quad \varphi_i = g_{i3} \ (i = 3, 6, 9, 12)$$

$$g_{i1} = \frac{1}{8}(1 + \xi_i\xi)(1 + \eta_i\eta)(2 + \xi_i\xi + \eta_i\eta - \xi^2 - \eta^2)$$

$$g_{i2} = \frac{1}{8}a\xi_i(\xi_0 - 1)(1 + \eta_0)(1 + \xi_0)^2, \quad \xi = \frac{x - x_c}{a}, \quad \eta = \frac{y - y_c}{b} \quad (7.4.23)$$

$$g_{i3} = \frac{1}{8}b\eta_i(\eta_0 - 1)(1 + \xi_0)(1 + \eta_0)^2, \quad \xi_0 = \xi_i\xi, \quad \eta_0 = \eta_i\eta$$

where $2a$ and $2b$ are the sides of the rectangle, (x_c, y_c) are the global coordinates of the center of the rectangle, and (ξ_i, η_i) are the coordinates of the nodes in the (ξ, η) coordinate system [e.g. $(\xi_1, \eta_1) = (-1, -1)$, $(\xi_2, \eta_2) = (1, -1)$, and so on]. The variation of the normal slope $\partial w/\partial n$ is cubic along an edge, whereas there are only two values of it available on the edge. Therefore, the cubic polynomials for the normal derivatives of w are not the same on an edge common to two elements; hence, it is said to be *non-conforming*. The size of the non-conforming element stiffness matrix is 20×20.

A conforming rectangular element with w, $\partial w/\partial x$, $\partial w/\partial y$, and $\partial^2 w/\partial x \partial y$ as the bending degrees of freedom was developed by Bogner *et al.* [59]. The deflection w is approximated by the 16-term complete polynomial

$$w \approx w_h(\xi, \eta) = c_1 + c_2\xi + c_3\eta + c_4\xi\eta + c_5\xi^2 + c_6\eta^2 + c_7\xi^3 + c_8\xi^2\eta + c_9\xi\eta^2$$
$$+ c_{10}\eta^3 + c_{11}\xi^3\eta + c_{12}\xi\eta^3 + c_{13}\xi^2\eta^2$$
$$+ c_{14}\xi^3\eta^2 + c_{15}\xi^2\eta^3 + c_{16}\xi^3\eta^3 = \sum_{j=1}^{16} \bar{\Delta}_j \varphi_j(\xi, \eta) \quad (7.4.24)$$

where

$$\varphi_i^e = g_{i1} \ (i = 1, 5, 9, 13); \quad \varphi_i^e = g_{i2} \ (i = 2, 6, 10, 14)$$
$$\varphi_i^e = g_{i3} \ (i = 3, 7, 11, 15); \quad \varphi_i^e = g_{i4} \ (i = 4, 8, 12, 16)$$
$$g_{i1} = \frac{1}{16}(\xi + \xi_i)^2(\xi_i\xi - 2)(\eta + \eta_i)^2(\eta_i\eta - 2)$$
$$g_{i2} = \frac{1}{16}a\xi_i(\xi + \xi_i)^2(1 - \xi_i\xi)(\eta + \eta_i)^2(\eta_i\eta - 2) \quad (7.4.25)$$
$$g_{i3} = \frac{1}{16}b\eta_i(\xi + \xi_i)^2(\xi_i\xi - 2)(\eta + \eta_i)^2(1 - \eta_i\eta)$$

$$g_{i4} = \frac{1}{16} ab\xi_i\eta_i(\xi + \xi_i)^2(1 - \xi_i\xi)(\eta + \eta_i)^2(1 - \eta_i\eta)$$

Here a and b are the sides of the rectangular element, and $\xi = x/a$ and $\eta = y/b$. In this case the normal slope continuity between elements is satisfied, hence termed a *conforming element* (also see [60,61]). The conforming element has 6 degrees of freedom per node $(u, v, w, w_{,x}, w_{,y}, w_{,xy})$, as shown in Fig. 7.4.1(b). For the conforming rectangular element, the total number of nodal degrees of freedom per element is $6 \times 4 = 24$, and hence the size of the stiffness matrix is 24×24.

7.5 Computer Implementation of the CPT Elements

7.5.1 General Remarks

The conforming and non-conforming rectangular finite elements developed in this chapter can be implemented into a computer program in the same way as the Euler–Bernoulli beam element. Here we use a bilinear interpolation of (u, v) and a Hermite cubic interpolation of w, as discussed in Section 7.4.3. The element geometry is represented using bilinear interpolation functions. In view of the different interpolation of the in-plane displacements and the transverse deflection, one must compute both types of interpolation functions and their derivatives. Since the stiffness coefficients of the CPT element involve both first- and second-order derivatives of $\varphi_i(x, y)$ with respect to the global coordinates (x, y), the associated formulas, similar to that for the first-order derivatives [see Eq. (3.6.6) or Eq. (6.6.4)], must be developed.

Let the transformation between the global coordinates (x, y) and local normalized coordinates (ξ, η) in a Lagrange element Ω_e be

$$x = \sum_{i=1}^{m} x_i \hat{\psi}_i(\xi, \eta), \qquad y = \sum_{i=1}^{m} y_i \hat{\psi}_i(\xi, \eta) \tag{7.5.1}$$

where (x_i^e, y_i^e) denote the global coordinates of the element nodes. As shown in Sections 3.6.2 and 6.6.2, the derivatives of the interpolation function φ_i^e with respect to the global coordinates (x, y) are related to their derivatives with respect to the local coordinates (ξ, η) by

$$\left\{ \begin{array}{c} \frac{\partial \varphi_i^e}{\partial x} \\ \frac{\partial \varphi_i^e}{\partial y} \end{array} \right\} = \mathbf{J}^{-1} \left\{ \begin{array}{c} \frac{\partial \varphi_i^e}{\partial \xi} \\ \frac{\partial \varphi_i^e}{\partial \eta} \end{array} \right\} \tag{7.5.2}$$

where the Jacobian matrix $\mathbf{J}$ is defined in Eq. (6.6.5). in terms of $\hat{\psi}_i^e$ and (x_i^e, y_i^e). The same procedure as that used to derive the result in Eq. (7.5.2) (i.e. chain rule of differentiation) can be used to develop the relationship between the second-order derivatives of φ_i with respect to (x, y) and the derivatives of φ_i with respect to (ξ, η).

We begin with

$$\frac{\partial \varphi_i}{\partial \xi} = \frac{\partial \varphi_i}{\partial x} \frac{\partial x}{\partial \xi} + \frac{\partial \varphi_i}{\partial y} \frac{\partial y}{\partial \xi}, \qquad \frac{\partial \varphi_i}{\partial \eta} = \frac{\partial \varphi_i}{\partial x} \frac{\partial x}{\partial \eta} + \frac{\partial \varphi_i}{\partial y} \frac{\partial y}{\partial \eta}$$

and find the second derivative

$$
\begin{aligned}
\frac{\partial^2 \varphi_i}{\partial \xi^2} &= \frac{\partial}{\partial \xi} \left(\frac{\partial \varphi_i}{\partial x} \frac{\partial x}{\partial \xi} + \frac{\partial \varphi_i}{\partial y} \frac{\partial y}{\partial \xi} \right) \\
&= \frac{\partial}{\partial \xi} \left(\frac{\partial \varphi_i}{\partial x} \right) \frac{\partial x}{\partial \xi} + \frac{\partial \varphi_i}{\partial x} \frac{\partial^2 x}{\partial \xi^2} + \frac{\partial}{\partial \xi} \left(\frac{\partial \varphi_i}{\partial y} \right) \frac{\partial y}{\partial \xi} + \frac{\partial \varphi_i}{\partial y} \frac{\partial^2 y}{\partial \xi^2} \\
&= \frac{\partial^2 \varphi_i}{\partial x^2} \left(\frac{\partial x}{\partial \xi} \right)^2 + \frac{\partial^2 \varphi_i}{\partial x \partial y} \frac{\partial x}{\partial \xi} \frac{\partial y}{\partial \xi} + \frac{\partial \varphi_i}{\partial x} \frac{\partial^2 x}{\partial \xi^2} \\
&\quad + \frac{\partial^2 \varphi_i}{\partial y^2} \left(\frac{\partial y}{\partial \xi} \right)^2 + \frac{\partial^2 \varphi_i}{\partial x \partial y} \frac{\partial x}{\partial \xi} \frac{\partial y}{\partial \xi} + \frac{\partial \varphi_i}{\partial y} \frac{\partial^2 y}{\partial \xi^2}
\end{aligned}
\tag{7.5.3a}
$$

Similarly, the second derivative with respect to η and the mixed derivative can be computed as

$$
\begin{aligned}
\frac{\partial^2 \varphi_i}{\partial \eta^2} &= \frac{\partial^2 \varphi_i}{\partial x^2} \left(\frac{\partial x}{\partial \eta} \right)^2 + 2 \frac{\partial^2 \varphi_i}{\partial x \partial y} \frac{\partial x}{\partial \eta} \frac{\partial y}{\partial \eta} + \frac{\partial \varphi_i}{\partial x} \frac{\partial^2 x}{\partial \eta^2} + \frac{\partial \varphi_i}{\partial y} \frac{\partial^2 y}{\partial \eta^2} + \frac{\partial^2 \varphi_i}{\partial y^2} \left(\frac{\partial y}{\partial \eta} \right)^2 \\
\frac{\partial^2 \varphi_i}{\partial \eta \partial \xi} &= \frac{\partial^2 \varphi_i}{\partial x^2} \frac{\partial x}{\partial \xi} \frac{\partial x}{\partial \eta} + \frac{\partial^2 \varphi_i}{\partial x \partial y} \left(\frac{\partial x}{\partial \eta} \frac{\partial y}{\partial \xi} + \frac{\partial x}{\partial \xi} \frac{\partial y}{\partial \eta} \right) + \frac{\partial^2 \varphi_i}{\partial y^2} \frac{\partial y}{\partial \xi} \frac{\partial y}{\partial \eta} \\
&\quad + \frac{\partial \varphi_i}{\partial x} \frac{\partial^2 x}{\partial \eta \partial \xi} + \frac{\partial \varphi_i}{\partial y} \frac{\partial^2 y}{\partial \eta \partial \xi}
\end{aligned}
\tag{7.5.3b}
$$

Since we need to write the global derivatives in terms of the local derivatives, we invert Eqs. (7.5.3a,b) to write the global derivatives in terms of the local derivatives and obtain

$$
\left\{
\begin{array}{c}
\frac{\partial^2 \varphi_i^e}{\partial x^2} \\[4pt]
\frac{\partial^2 \varphi_i^e}{\partial y^2} \\[4pt]
\frac{\partial^2 \varphi_i^e}{\partial x \partial y}
\end{array}
\right\}
=
\left[
\begin{array}{ccc}
\left(\frac{\partial x_e}{\partial \xi} \right)^2 & \left(\frac{\partial y_e}{\partial \xi} \right)^2 & 2 \frac{\partial x_e}{\partial \xi} \frac{\partial y_e}{\partial \xi} \\[6pt]
\left(\frac{\partial x_e}{\partial \eta} \right)^2 & \left(\frac{\partial y_e}{\partial \eta} \right)^2 & 2 \frac{\partial x_e}{\partial \eta} \frac{\partial y_e}{\partial \eta} \\[6pt]
\frac{\partial x_e}{\partial \xi} \frac{\partial x_e}{\partial \eta} & \frac{\partial y_e}{\partial \xi} \frac{\partial y_e}{\partial \eta} & \frac{\partial x_e}{\partial \eta} \frac{\partial y_e}{\partial \xi} + \frac{\partial x_e}{\partial \xi} \frac{\partial y_e}{\partial \eta}
\end{array}
\right]^{-1}
$$

$$
\times \left(
\left\{
\begin{array}{c}
\frac{\partial^2 \varphi_i^e}{\partial \xi^2} \\[4pt]
\frac{\partial^2 \varphi_i^e}{\partial \eta^2} \\[4pt]
\frac{\partial^2 \psi_i^e}{\partial \xi \partial \eta}
\end{array}
\right\}
-
\left[
\begin{array}{cc}
\frac{\partial^2 x_e}{\partial \xi^2} & \frac{\partial^2 y_e}{\partial \xi^2} \\[4pt]
\frac{\partial^2 x_e}{\partial \eta^2} & \frac{\partial^2 y_e}{\partial \eta^2} \\[4pt]
\frac{\partial^2 x_e}{\partial \xi \partial \eta} & \frac{\partial^2 y_e}{\partial \xi \partial \eta}
\end{array}
\right]
\left\{
\begin{array}{c}
\frac{\partial \varphi_i^e}{\partial x} \\[4pt]
\frac{\partial \varphi_i^e}{\partial y}
\end{array}
\right\}
\right)
\tag{7.5.4}
$$

$$
= \begin{bmatrix} \left(\dfrac{\partial x_e}{\partial \xi}\right)^2 & \left(\dfrac{\partial y_e}{\partial \xi}\right)^2 & 2\dfrac{\partial x_e}{\partial \xi}\dfrac{\partial y_e}{\partial \xi} \\[2mm] \left(\dfrac{\partial x_e}{\partial \eta}\right)^2 & \left(\dfrac{\partial y_e}{\partial \eta}\right)^2 & 2\dfrac{\partial x_e}{\partial \eta}\dfrac{\partial y_e}{\partial \eta} \\[2mm] \dfrac{\partial x_e}{\partial \xi}\dfrac{\partial x_e}{\partial \eta} & \dfrac{\partial y_e}{\partial \xi}\dfrac{\partial y_e}{\partial \eta} & \dfrac{\partial x_e}{\partial \eta}\dfrac{\partial y_e}{\partial \xi} + \dfrac{\partial x_e}{\partial \xi}\dfrac{\partial y_e}{\partial \eta} \end{bmatrix}^{-1}
$$

$$
\times \left(\left\{ \begin{matrix} \dfrac{\partial^2 \varphi_i^e}{\partial \xi^2} \\[2mm] \dfrac{\partial^2 \varphi_i^e}{\partial \eta^2} \\[2mm] \dfrac{\partial^2 \varphi_i^e}{\partial \xi \partial \eta} \end{matrix} \right\} - \begin{bmatrix} \dfrac{\partial^2 x_e}{\partial \xi^2} & \dfrac{\partial^2 y_e}{\partial \xi^2} \\[2mm] \dfrac{\partial^2 x_e}{\partial \eta^2} & \dfrac{\partial^2 y_e}{\partial \eta^2} \\[2mm] \dfrac{\partial^2 x_e}{\partial \xi \partial \eta} & \dfrac{\partial^2 y_e}{\partial \xi \partial \eta} \end{bmatrix} \mathbf{J}^{-1} \left\{ \begin{matrix} \dfrac{\partial \varphi_i^e}{\partial \xi} \\[2mm] \dfrac{\partial \varphi_i^e}{\partial \eta} \end{matrix} \right\} \right) \tag{7.5.5}
$$

where the first- and second-derivatives of x and y with respect to ξ and η are determined from the transformation equations in Eq. (7.5.1). These operations are carried out in the subroutine INTERPLN2D, which must now contain the Hermite cubic interpolation functions and their derivatives with respect to ξ and η. Box 7.5.1 contains a listing of the subroutine INTERPLN2D with Hermite interpolation functions. In addition, one must rearrange the stiffness coefficients such that the finite element nodal displacement vector of the conforming element, for example, is of the form (to minimize the bandwidth of the stiffness matrix)

$$
\mathbf{\Delta} = \{u_1, v_1, w_1, w_{,x1}, w_{,y1}, w_{,xy1}, u_2, v_2, w_2, w_{,x2}, w_{,y2}, w_{,xy2}, \ldots\}^{\mathrm{T}}
$$

where $(u_i, v_i, w_i, w_{,xi}, w_{,yi}, w_{,xyi})$ are the generalized displacements at node i. Box 5.2.1 contains the logic used to rearrange the element stiffness coefficients.

7.5.2 Programming Aspects

As in the case of beams, reduced integration of the nonlinear terms in the stiffness matrices is required to eliminate the membrane locking. Suppose that the elastic foundation term is absent (i.e. $k = 0$) and all the plate stiffnesses A_{ij} and D_{ij} are constant. For a rectangular element, the Jacobian and hence its inverse are independent of (ξ, η). Then the exact evaluation of $\mathbf{K}^{11}$, $\mathbf{K}^{12}$, $\mathbf{K}^{21}$, and $\mathbf{K}^{22}$ requires only the 2×2 Gauss rule (2 Gauss points along the ξ-axis and 2 Gauss points along the η-axis) for linear element and 3×3 for a quadratic element. When the element is a quadrilateral, the Jacobian is a linear function of ξ and η. The exact evaluation of the linear terms in $\mathbf{K}^{33}$ may require the 3×3 or no more than the 4×4 Gauss rule because its integrand is a fifth-degree polynomial in ξ as well as η when quadrilateral elements are involved, not counting the contribution of J_{ij}^* (coefficients of the inverse $\mathbf{J}^{-1}$) to the polynomial degree. When a mesh of rectangular elements is used, the nonlinear terms, being linear in ξ and η, should be evaluated using the 1×1 Gauss rule. If quadrilateral elements are used, it requires the 2×2 Gauss rule. Thus, full integration requires the 4×4 Gauss rule and the nonlinear stiffnesses may be evaluated

Box 7.5.1: Listing of subroutine **INTERPLN2D** with Hermite cubic interpolation functions.

```
C
C       CONFORMING Hermite functions (four-node element; NPE=4)
C
        IF(ITYPE.LE.1) THEN
           IF(ITYPE.EQ.1) THEN
              NET=4*NPE
              II = 1
              DO 60 I = 1, NPE
              XP   = XNODE(I,1)
              YP   = XNODE(I,2)
              XI1  = XI*XP-1.0
              XI2  = XI1-1.0
              ETA1 = ETA*YP-1.0
              ETA2 = ETA1-1.0
              XIO  = (XI+XP)**2
              ETAO = (ETA+YP)**2
              XIPO = XI+XP
              XIP1 = 3.0*XI*XP+XP*XP
              XIP2 = 3.0*XI*XP+2.0*XP*XP
              YIPO = ETA+YP
              YIP1 = 3.0*ETA*YP+YP*YP
              YIP2 = 3.0*ETA*YP+2.0*YP*YP
C
              SFH(II)       = 0.0625*FNC(ETAO,ETA2)*FNC(XIO,XI2)
              DSFH(1,II)    = 0.0625*FNC(ETAO,ETA2)*XIPO*(XIP1-4.0)
              DSFH(2,II)    = 0.0625*FNC(XIO,XI2)*YIPO*(YIP1-4.0)
              DDSFH(1,II)   = 0.125*FNC(ETAO,ETA2)*(XIP2-2.0)
              DDSFH(2,II)   = 0.125*FNC(XIO,XI2)*(YIP2-2.0)
              DDSFH(3,II)   = 0.0625*(XIP1-4.0)*(YIP1-4.0)*XIPO*YIPO
              SFH(II+1)     = -0.0625*XP*FNC(XIO,XI1)*FNC(ETAO,ETA2)
              DSFH(1,II+1)  = -0.0625*FNC(ETAO,ETA2)*XP*XIPO*(XIP1-2.0)
              DSFH(2,II+1)  = -0.0625*FNC(XIO,XI1)*XP*YIPO*(YIP1-4.0)
              DDSFH(1,II+1) = -0.125*FNC(ETAO,ETA2)*XP*(XIP2-1.0)
              DDSFH(2,II+1) = -0.125*FNC(XIO,XI1)*(YIP2-2.0)*XP
              DDSFH(3,II+1) = -0.0625*XP*XIPO*(XIP1-2.0)*(YIP1-4.0)*YIPO
              SFH(II+2)     = -0.0625*YP*FNC(XIO,XI2)*FNC(ETAO,ETA1)
              DSFH(1,II+2)  = -0.0625*FNC(ETAO,ETA1)*YP*XIPO*(XIP1-4.0)
              DSFH(2,II+2)  = -0.0625*FNC(XIO,XI2)*YP*YIPO*(YIP1-2.0)
              DDSFH(1,II+2) = -0.125*FNC(ETAO,ETA1)*YP*(XIP2-2.0)
              DDSFH(2,II+2) = -0.125*FNC(XIO,XI2)*YP*(YIP2-1.0)
              DDSFH(3,II+2) = -0.0625*YP*YIPO*(YIP1-2.0)*(XIP1-4.0)*XIPO
              SFH(II+3)     = 0.0625*XP*YP*FNC(XIO,XI1)*FNC(ETAO,ETA1)
              DSFH(1,II+3)  = 0.0625*FNC(ETAO,ETA1)*XP*YP*(XIP1-2.0)*XIPO
              DSFH(2,II+3)  = 0.0625*FNC(XIO,XI1)*XP*YP*(YIP1-2.0)*YIPO
              DDSFH(1,II+3) = 0.125*FNC(ETAO,ETA1)*XP*YP*(XIP2-1.0)
              DDSFH(2,II+3) = 0.125*FNC(XIO,XI1)*XP*YP*(YIP2-1.0)
              DDSFH(3,II+3) = 0.0625*XP*YP*YIPO*XIPO*(YIP1-2.0)*(XIP1-2.)
60            II = I*4 + 1
           ELSE
```

Box 7.5.1: Listing of **INTERPLN2D** (continued).

```
C
C       NON-CONFORMING Hermite functions (Four-node element; NPE=4)
              NET=3*NPE
              II = 1
              DO 80 I = 1, NPE
              XP    = XNODE(I,1)
              YP    = XNODE(I,2)
              XIO   = XI*XP
              ETAO  = ETA*YP
              XIP1  = 1.0+XIO
              ETAP1 = 1.0+ETAO
              XIM1  = XIO-1.0
              ETAM1 = ETAO-1.0
              XID   = 3.0+2.0*XIO+ETAO-3.0*XI*XI-ETA*ETA-2.0*XI/XP
              ETAD  = 3.0+XIO+2.0*ETAO-XI*XI-3.0*ETA*ETA-2.0*ETA/YP
              ETAXI = 4.0+2.0*(XIO+ETAO)-3.0*(XI*XI+ETA*ETA)
     *                                  -2.0*(ETA/YP+XI/XP)
              SFH(II) = 0.125*XIP1*ETAP1*(2.0+XIO+ETAO-XI*XI-ETA*ETA)
              DSFH(1,II)    = 0.125*XP*ETAP1*XID
              DSFH(2,II)    = 0.125*YP*XIP1*ETAD
              DDSFH(1,II)   = 0.250*XP*ETAP1*(XP-3.0*XI-1.0/XP)
              DDSFH(2,II)   = 0.250*YP*XIP1*(YP-3.0*ETA-1.0/YP)
              DDSFH(3,II)   = 0.125*XP*YP*ETAXI
              SFH(II+1)     = 0.125*XP*XIP1*XIP1*XIM1*ETAP1
              DSFH(1,II+1)  = 0.125*XP*XP*ETAP1*(3.0*XIO-1.0)*XIP1
              DSFH(2,II+1)  = 0.125*XP*YP*XIP1*XIP1*XIM1
              DDSFH(1,II+1) = 0.250*XP*XP*XP*ETAP1*(3.0*XIO+1.0)
              DDSFH(2,II+1) = 0.0
              DDSFH(3,II+1) = 0.125*XP*XP*YP*(3.0*XIO-1.0)*XIP1
              SFH(II+2)     = 0.125*YP*XIP1*ETAP1*ETAP1*ETAM1
              DSFH(1,II+2)  = 0.125*XP*YP*ETAP1*ETAP1*ETAM1
              DSFH(2,II+2)  = 0.125*YP*YP*XIP1*(3.0*ETAO-1.0)*ETAP1
              DDSFH(1,II+2) = 0.0
              DDSFH(2,II+2) = 0.250*YP*YP*YP*XIP1*(3.0*ETAO+1.0)
              DDSFH(3,II+2) = 0.125*XP*YP*YP*(3.0*ETAO-1.0)*ETAP1
  80          II = I*3 + 1
          ENDIF
C
C       The geometry is approximated using the linear functions
              DDSF(1,1) =    0.0D0
              DDSF(2,1) =    0.0D0
              DDSF(3,1) =    0.250D0
              DDSF(1,2) =    0.0D0
              DDSF(2,2) =    0.0D0
              DDSF(3,2) =   -0.250D0
              DDSF(1,3) =    0.0D0
              DDSF(2,3) =    0.0D0
              DDSF(3,3) =    0.250D0
              DDSF(1,4) =    0.0D0
              DDSF(2,4) =    0.0D0
              DDSF(3,4) =   -0.250D0
```

Box 7.5.1: Listing of **INTERPLN2D** (continued).

```
C
C      Compute global first derivatives of Hermite functions
           DO 110 I = 1, 2
           DO 100 J = 1, NET
           SUM = 0.0D0
           DO 90 K = 1, 2
   90  SUM = SUM + GJINV(I,K)*DSFH(K,J)
             GDSFH(I,J) = SUM
  100  CONTINUE
  110  CONTINUE
C      Compute global second derivatives of Hermite functions
           DO 140 I = 1, 3
           DO 130 J = 1, 2
           SUM = 0.0D0
           DO 120 K = 1, NPE
           SUM = SUM + DDSF(I,K)*ELXY(K,J)
  120  CONTINUE
             DJCB(I,J) = SUM
  130  CONTINUE
  140  CONTINUE
           DO 170 K = 1, 3
           DO 160 J = 1, NET
           SUM = 0.0D0
           DO 150 L = 1, 2
           SUM = SUM + DJCB(K,L)*GDSFH(L,J)
  150  CONTINUE
             DDSJ(K,J) = SUM
  160  CONTINUE
  170  CONTINUE
C      Compute the Jacobian of the transformation
           GGJ(1,1)=GJ(1,1)*GJ(1,1)
           GGJ(1,2)=GJ(1,2)*GJ(1,2)
           GGJ(1,3)=2.0*GJ(1,1)*GJ(1,2)
           GGJ(2,1)=GJ(2,1)*GJ(2,1)
           GGJ(2,2)=GJ(2,2)*GJ(2,2)
           GGJ(2,3)=2.0*GJ(2,1)*GJ(2,2)
           GGJ(3,1)=GJ(2,1)*GJ(1,1)
           GGJ(3,2)=GJ(2,2)*GJ(1,2)
           GGJ(3,3)=GJ(2,1)*GJ(1,2)+GJ(1,1)*GJ(2,2)
           CALL INVRSE(GGJ,GGINV)
C
           DO 200 I = 1, 3
           DO 190 J = 1, NET
           SUM = 0.0D0
           DO 180 K = 1, 3
           SUM = SUM + GGINV(I,K)*(DDSFH(K,J)-DDSJ(K,J))
  180  CONTINUE
             GDDSFH(I,J) = SUM
  190  CONTINUE
  200  CONTINUE
       ENDIF
```

using (reduced integration) the 1×1 Gauss rule when the element is rectangular and 2×2 when the element is quadrilateral.

One may use separate arrays, say, EXT13(I,J), EXT23(I,J), and EXT33(I,J), to store the extra terms to be added to K_{ij}^{13}, K_{ij}^{23}, K_{ij}^{33}, respectively, to calculate the respective tangent stiffness matrix coefficients

$$-\mathbf{R} = \mathbf{F} - \mathbf{K}\boldsymbol{\Delta} \rightarrow \text{ELF(I)} = \text{ELF(I)} - \text{ELK(I,J)}*\text{ELU(J)}$$
$$\mathbf{K}^{\text{tan}} = \mathbf{K} + \mathbf{EXT} \rightarrow \text{ELK(I,J)} = \text{ELK(I,J)} + \text{EXT(I,J)}$$

where **EXT** in the present discussion is used only for the *extra* terms that are added to **K** to obtain the tangent stiffness matrix **T**. The above two operations must be carried out sequentially in separate do-loops.

7.5.3 Post-Computation of Stresses

Here we discuss the evaluation of stresses from the known displacement expansions. Once the nodal values of generalized displacements (u, v, w) have been obtained by solving the assembled equations of a problem, the strains are evaluated in each element by differentiating the displacement expansions in Eq. (7.4.4). Then the stresses σ_{xx}, σ_{yy}, and σ_{xy} are computed using the stress–strain relations. Since the displacements in the finite element models are referred to the global coordinates (x, y, z), the stresses are computed in the global coordinates using the constitutive relations

$$\begin{Bmatrix} \sigma_{xx} \\ \sigma_{yy} \\ \sigma_{xy} \end{Bmatrix} = \begin{bmatrix} Q_{11} & Q_{12} & 0 \\ Q_{12} & Q_{22} & 0 \\ 0 & 0 & Q_{66} \end{bmatrix} \begin{Bmatrix} \varepsilon_{xx} \\ \varepsilon_{yy} \\ \gamma_{xy} \end{Bmatrix} \tag{7.5.6a}$$

where the material stiffnesses Q_{ij} are defined in Eqs. (7.3.31a,b)

$$Q_{11} = \frac{E_1}{1 - \nu_{12}\nu_{21}}, \quad Q_{12} = \frac{\nu_{12}E_2}{1 - \nu_{12}\nu_{21}}, \quad Q_{22} = \frac{E_2}{1 - \nu_{12}\nu_{21}}, \quad Q_{66} = G_{12}$$
$$\tag{7.5.6b}$$

and

$$\begin{Bmatrix} \varepsilon_{xx} \\ \varepsilon_{yy} \\ \gamma_{xy} \end{Bmatrix} = \begin{Bmatrix} \frac{\partial u}{\partial x} + \frac{1}{2}\left(\frac{\partial w}{\partial x}\right)^2 \\ \frac{\partial v}{\partial y} + \frac{1}{2}\left(\frac{\partial w}{\partial y}\right)^2 \\ \frac{\partial u}{\partial y} + \frac{\partial v}{\partial x} + \frac{\partial w}{\partial x}\frac{\partial w}{\partial y} \end{Bmatrix} - z \begin{Bmatrix} \frac{\partial^2 w}{\partial x^2} \\ \frac{\partial^2 w}{\partial y^2} \\ \frac{\partial^2 w}{\partial x \partial y} \end{Bmatrix} \tag{7.5.7}$$

Typically the strains and stresses are computed at points (x, y) corresponding to the reduced Gauss points of the element, called *Barlow points*, as they are found to be more accurate there (see Barlow [62, 63] for a discussion). The stresses can be evaluated for any desired value of z, say at the top $(z = +h/2)$ and bottom $(z = -h/2)$ of the element. For example, the values of σ_{xx} at the top and bottom of the element at a point (x_c, y_c) corresponding to a Gauss point are computed using

$$
\sigma_{xx}^{\text{top}} = \sigma_{xx}(x_c, y_c, h/2) = Q_{11} \left\{ \left[\frac{\partial u}{\partial x} + \frac{1}{2} \left(\frac{\partial w}{\partial x} \right)^2 \right] - \frac{h}{2} \frac{\partial^2 w}{\partial x^2} \right\}_{(x_c, y_c)}
$$

$$
+ Q_{12} \left\{ \left[\frac{\partial v}{\partial y} + \frac{1}{2} \left(\frac{\partial w}{\partial y} \right)^2 \right] - \frac{h}{2} \frac{\partial^2 w}{\partial y^2} \right\}_{(x_c, y_c)} \tag{7.5.8}
$$

$$
\sigma_{xx}^{\text{bottom}} = \sigma_{xx}(x_c, y_c, -h/2) = Q_{11} \left\{ \left[\frac{\partial u}{\partial x} + \frac{1}{2} \left(\frac{\partial w}{\partial x} \right)^2 \right] + \frac{h}{2} \frac{\partial^2 w}{\partial x^2} \right\}_{(x_c, y_c)}
$$

$$
+ Q_{12} \left\{ \left[\frac{\partial v}{\partial y} + \frac{1}{2} \left(\frac{\partial w}{\partial y} \right)^2 \right] + \frac{h}{2} \frac{\partial^2 w}{\partial y^2} \right\}_{(x_c, y_c)} \tag{7.5.9}
$$

Similar expressions hold for σ_{yy} and σ_{xy}.

7.6 Numerical Examples using the CPT Elements

7.6.1 Preliminary Comments

Solution symmetries available in a problem should be taken advantage of to identify the computational domain because they reduce computational effort. A solution is symmetric about a line only if (a) the geometry, including boundary conditions, (b) the material properties, and (c) the loading are symmetric about the line. The boundary conditions along the edges and the symmetry lines of a simply-supported rectangular plate are shown in Fig. 7.6.1. In the case of the conforming element, we may also set $\partial^2 w / \partial x \partial y = 0$ at the center of the plate. When one is not sure of the solution symmetry, it is advised that the whole plate be modeled.

   All meshes used in the examples are uniform (i.e. the same size elements are used). At every boundary node, one is required to specify one element in each of the four pairs (i.e. either the generalized displacement or generalized force) given in Eq. (7.3.24). When no displacement or force is specified at a node, then it is understood that the specified generalized force is specified to be zero at that node. The range of loads used in the examples has the sole purpose of bringing out the geometric nonlinearity exhibited in the problem, and very seldom plate structures are loaded into the nonlinear range.

7.6.2 Results of Linear Analysis

We consider the bending of rectangular plates with various edge conditions to evaluate the non-conforming and conforming rectangular plate elements based

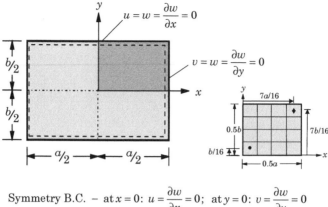

$$\text{Symmetry B.C.} - \text{at } x = 0:\ u = \frac{\partial w}{\partial x} = 0;\quad \text{at } y = 0:\ v = \frac{\partial w}{\partial y} = 0$$

Fig. 7.6.1: Simply-supported (SS–1) boundary and symmetry conditions for rectangular plates, and the solid circle and diamond denote locations of maximum σ_{xx} (σ_{yy}) and σ_{xy}, respectively.

on CPT. The plate in-plane dimensions along the x and y directions are denoted by a and b, respectively, and the total thickness is denoted by h.

The foundation modulus k is set to zero, and the linear stiffness coefficients are evaluated using the 4×4 Gauss rule, although the 3×3 Gauss rule also produced the same results for the problems discussed here; the stresses are computed at the center of the elements (i.e. the one-point Gauss rule is used).

Example 7.6.1

Consider a simply-supported (SS–1) square plate ($a = b$) under uniformly distributed transverse load of intensity q_0. The geometric boundary conditions of the computational domain (see the shaded quadrant in Fig. 7.6.1) are

$$u = \frac{\partial w}{\partial x} = 0 \text{ at } x = 0;\quad v = \frac{\partial w}{\partial y} = 0 \text{ at } y = 0$$

$$v = w = \frac{\partial w}{\partial y} = 0 \text{ at } x = \frac{a}{2};\quad u = w = \frac{\partial w}{\partial x} = 0 \text{ at } y = \frac{b}{2}$$

(1)

$$\frac{\partial^2 w}{\partial x \partial y} = 0 \text{ at } x = y = 0 \text{ (for conforming element only)}$$

(2)

Analyze the plate for deflections and stresses and investigate convergence of the results using 2×2, 4×4, and 8×8 uniform meshes of conforming and non-conforming elements.

Solution: The numerical results are presented in dimensionless form; therefore, the actual values of the geometric and material parameters used in the computation are not needed to compare the solutions. However, to aid the reader in verifying his/her computer program, the following data are used:

$$a = b = 1.0,\quad h = 0.01,\quad E = 10^6,\quad \nu = 0.25,\quad q_0 = 1$$

(3)

The axial and bending stiffnesses are $A_{11} = A_{22} = 0.10667 \times 10^5$, $A_{12} = 0.26667 \times 10^4$, $A_{66} = 0.4 \times 10^4$, $D_{11} = D_{22} = 0.088889$, $D_{12} = 0.022222$, and $D_{66} = 0.033333$. The first five rows and the last two rows, denoted by the subscripts, of the element stiffness matrix, and the element force vector for the non-conforming element in the 2×2 mesh are

$$\mathbf{K}_1^{(1)} = \{0.48889 \times 10^4 \quad 0.16667 \times 10^4 \ 0.0 \ 0.0 \quad 0.0$$
$$-0.28889 \times 10^4 \ -0.33333 \times 10^3 \ 0.0 \ 0.0 \quad 0.0$$
$$-0.24444 \times 10^4 \ -0.16667 \times 10^4 \ 0.0 \ 0.0 \quad 0.0$$
$$0.44444 \times 10^3 \quad 0.33333 \times 10^3 \ 0.0 \ 0.0 \quad 0.0\}$$

$$\mathbf{K}_2^{(1)} = \{0.16667 \times 10^4 \quad 0.48889 \times 10^4 \ 0.0 \ 0.0 \quad 0.0$$
$$0.33333 \times 10^3 \quad 0.44444 \times 10^3 \ 0.0 \ 0.0 \quad 0.0$$
$$-0.16667 \times 10^4 \ -0.24444 \times 10^4 \ 0.0 \ 0.0 \quad 0.0$$
$$-0.33333 \times 10^3 \ -0.28889 \times 10^4 \ 0.0 \ 0.0 \quad 0.0\}$$

$$\mathbf{K}_3^{(1)} = \{0.0 \ 0.0 \quad 0.60302 \times 10^2 \ 0.27307 \times 10^2 \quad 0.27307 \times 10^2$$
$$0.0 \ 0.0 \ -0.26169 \times 10^2 \ 0.24462 \times 10^2 \quad 0.68267 \times 10^1$$
$$0.0 \ 0.0 \ -0.79644 \times 10^1 \ 0.96711 \times 10^1 \quad 0.96711 \times 10^1$$
$$0.0 \ 0.0 \ -0.26169 \times 10^2 \ 0.68267 \times 10^1 \quad 0.24462 \times 10^2\}$$

$$\mathbf{K}_4^{(1)} = \{0.0 \ 0.0 \quad 0.27307 \times 10^2 \ 0.34892 \times 10^2 \quad 0.56889 \times 10^1$$
$$0.0 \ 0.0 \ -0.24462 \times 10^2 \ 0.14033 \times 10^2 \ -0.55708 \times 10^{-8}$$
$$0.0 \ 0.0 \ -0.96711 \times 10^1 \ 0.87230 \times 10^1 \ -0.22204 \times 10^{-15}$$
$$0.0 \ 0.0 \quad 0.68267 \times 10^1 \ 0.10619 \times 10^2 \quad 0.55708 \times 10^{-8}\}$$

$$\mathbf{K}_5^{(1)} = \{0.0 \ 0.0 \quad 0.27307 \times 10^2 \quad 0.56889 \times 10^1 \quad 0.34892 \times 10^2$$
$$0.0 \ 0.0 \quad 0.68267 \times 10^1 \quad 0.55708 \times 10^{-8} \quad 0.10619 \times 10^2$$
$$0.0 \ 0.0 \ -0.96711 \times 10^1 \ -0.16653 \times 10^{-15} \ 0.87230 \times 10^1$$
$$0.0 \ 0.0 \ -0.24462 \times 10^2 \ -0.55708 \times 10^{-8} \quad 0.14033 \times 10^2\}$$

$$\mathbf{K}_{19}^{(1)} = \{0.0 \ 0.0 \quad 0.68267 \times 10^1 \quad 0.10619 \times 10^2 \ -0.42595 \times 10^{-8}$$
$$0.0 \ 0.0 \ -0.96711 \times 10^1 \ 0.87230 \times 10^1 \ -0.27756 \times 10^{-15}$$
$$0.0 \ 0.0 \ -0.24462 \times 10^2 \ 0.14033 \times 10^2 \quad 0.42595 \times 10^{-8}$$
$$0.0 \ 0.0 \quad 0.27307 \times 10^2 \ 0.34892 \times 10^2 \ -0.56889 \times 10^1\}$$

$$\mathbf{K}_{20}^{(1)} = \{0.0 \ 0.0 \quad 0.24462 \times 10^2 \quad 0.42595 \times 10^{-8} \quad 0.14033 \times 10^2$$
$$0.0 \ 0.0 \quad 0.96711 \times 10^1 \ -0.22204 \times 10^{-15} \ 0.87230 \times 10^1$$
$$0.0 \ 0.0 \ -0.68267 \times 10^1 \ -0.42595 \times 10^{-8} \quad 0.10619 \times 10^2$$
$$0.0 \ 0.0 \ -0.27307 \times 10^2 \ -0.56889 \times 10^1 \quad 0.34892 \times 10^2\}$$

$$\mathbf{f}^{(1)} = \{0.0 \quad 0.0 \ 0.39063 \times 10^{-2} \quad 0.13021 \times 10^{-2} \quad 0.13021 \times 10^{-2}$$
$$0.0 \quad 0.0 \ 0.39063 \times 10^{-2} \ -0.13021 \times 10^{-2} \quad 0.13021 \times 10^{-2}$$
$$0.0 \quad 0.0 \ 0.39063 \times 10^{-2} \ -0.13021 \times 10^{-2} \ -0.13021 \times 10^{-2} \tag{4}$$
$$0.0 \quad 0.0 \ 0.39063 \times 10^{-2} \quad 0.13021 \times 10^{-2} \ -0.13021 \times 10^{-2}\}^{\mathrm{T}}$$

The non-dimensional finite element solutions for center transverse displacement and maximum normal and shear stresses along with the analytical solutions (see Reddy [3, 55, 56]) of isotropic ($\nu = 0.25$) and orthotropic ($E_1/E_2 = 25$, $G_{12} = G_{13} = 0.5E_2$, $\nu_{12} = 0.25$) square plates are presented in Table 7.6.1. The (x, y, z) locations in the plate where the normal stresses are computed are $(a/8, a/8, h/2)$, $(a/16, a/16, h/2)$, and $(a/32, a/32, h/2)$ for 2×2, 4×4, and 8×8 uniform meshes, respectively, while those of σ_{xy} are $(3a/8, 3a/8, -h/2)$, $(7a/16, 7a/16, -h/2)$, and $(15a/32, 15a/32, -h/2)$ for the same three meshes. The analytical solutions were evaluated using $m, n = 1, 3, \ldots, 19$. The exact maximum deflection occurs at $x = y = 0$, maximum stresses σ_{xx} and σ_{yy} occur at $(0, 0, h/2)$, and the maximum shear stress σ_{xy} occurs at $(a/2, a/2, -h/2)$. The conforming element is slightly more accurate than the non-conforming element in predicting the deflections and both converge to the analytical solutions as the mesh is refined.

Table 7.6.1: Maximum transverse deflections and stresses* of simply-supported square plates under a uniformly distributed load q_0.

Variable	Non-conforming			Conforming			Analytical
	2×2	4×4	8×8	2×2	4×4	8×8	solution
Isotropic plates							
$\bar{w} \times 10^2$	4.8571	4.6425	4.5883	4.7619	4.5952	4.5739	4.5698
$\bar{\sigma}_{xx}$	0.2405	0.2673	0.2740	0.2239	0.2637	0.2731	0.2762
$\bar{\sigma}_{xy}$	0.1713	0.1964	0.2050	0.1688	0.1935	0.2040	0.2085
Orthotropic plates							
$\bar{w} \times 10^2$	0.7082	0.6635	0.6531	0.7710	0.6651	0.6522	0.6497
$\bar{\sigma}_{xx}$	0.7148	0.7709	0.7828	0.5560	0.7388	0.7743	0.7866
$\bar{\sigma}_{yy}$	0.0296	0.0253	0.0246	0.0278	0.0249	0.0245	0.0244
$\bar{\sigma}_{xy}$	0.0337	0.0421	0.0444	0.0375	0.0416	0.0448	0.0463

*$\bar{w} = w E_2 h^3 / (q_0 a^4)$, $\bar{\sigma} = \sigma h^2 / (q_0 a^2)$.

Example 7.6.2

Consider a clamped square plate under uniformly distributed transverse load of intensity q_0. Consider both isotropic ($\nu = 0.25$) and orthotropic ($E_1/E_2 = 25$, $G_{12} = G_{13} = 0.5E_2$, $\nu_{12} = 0.25$) plates. The boundary conditions are [the origin of the coordinate system (x, y, z) is taken at the center of the plate]:

$$u = \frac{\partial w}{\partial x} = 0 \text{ at } x = 0; \quad v = \frac{\partial w}{\partial y} = 0 \text{ at } y = 0 \tag{1}$$

$$u = v = w = \frac{\partial w}{\partial x} = \frac{\partial w}{\partial y} = 0 \text{ at } x = \frac{a}{2}; \quad u = v = w = \frac{\partial w}{\partial x} = \frac{\partial w}{\partial y} = 0 \text{ at } y = \frac{b}{2} \tag{2}$$

$$\frac{\partial^2 w}{\partial x \partial y} = 0 \text{ on clamped edges (for conforming element only)} \tag{3}$$

Analyze the plate for deflections and stresses and investigate convergence of the results using 2×2, 4×4, and 8×8 uniform meshes of conforming and non-conforming elements.

Solution: Table 7.6.2 contains the non-dimensional deflections and stresses. The (x, y, z) locations of the normal stresses reported for the three meshes are:

$$2 \times 2 : \left(\frac{a}{8}, \frac{b}{8}, -\frac{h}{2}\right); \quad 4 \times 4 : \left(\frac{a}{16}, \frac{b}{16}, -\frac{h}{2}\right); \quad 8 \times 8 : \left(\frac{a}{32}, \frac{b}{32}, -\frac{h}{2}\right) \tag{4}$$

and shear stresses reported for the three meshes are

$$2 \times 2 : \left(\frac{3a}{8}, \frac{3b}{8}, -\frac{h}{2}\right); \quad 4 \times 4 : \left(\frac{7a}{16}, \frac{7b}{16}, -\frac{h}{2}\right); \quad 8 \times 8 : \left(\frac{15a}{32}, \frac{15b}{32}, -\frac{h}{2}\right) \tag{5}$$

These stresses are not necessarily the maximum ones in the plate. For example, for an 8×8 mesh, the maximum normal stress in the isotropic plate is found to be 0.2300 at $(0.46875a, 0.03125b, -h/2)$ and the maximum shear stress is 0.0226 at $(0.28125a, 0.09375b, -0.5h)$ for the non-conforming element. The conforming element yields slightly different solutions than the non-conforming element for deflections but not for the stresses, and both elements show good convergence.

Table 7.6.2: Maximum transverse deflections and stresses* of clamped (CCCC), isotropic and orthotropic, square plates ($a = b$) under a uniformly distributed load q_0 (linear analysis).

Variable	Non-conforming			Conforming		
	2×2	4×4	8×8	2×2	4×4	8×8
Isotropic plate						
$\bar{w} \times 10^2$	1.5731	1.4653	1.4342	1.4778	1.4370	1.4249
$\bar{\sigma}_{xx}$	0.0987	0.1238	0.1301	0.0861	0.1197	0.1288
$\bar{\sigma}_{xy}$	0.0497	0.0222	0.0067	0.0489	0.0224	0.0068
Orthotropic plate						
$\bar{w} \times 10^2$	0.1434	0.1332	0.1314	0.1402	0.1330	0.1311
$\bar{\sigma}_{xx}$	0.1962	0.2491	0.2598	0.1559	0.2358	0.2576
$\bar{\sigma}_{yy}$	0.0085	0.0046	0.0042	0.0066	0.0047	0.0043
$\bar{\sigma}_{xy}$	0.0076	0.0046	0.0019	0.0083	0.0048	0.0020

*$\bar{w} = wE_2h^3/(q_0a^4)$, $\bar{\sigma} = \sigma h^2/(q_0a^2)$.

7.6.3   Results of Nonlinear Analysis

Here we investigate geometrically nonlinear response of plates using the conforming and non-conforming plate finite elements. The nonlinear terms are evaluated using reduced integration. Full integration (F) means 4×4 Gauss rule and reduced integration (R) means 1×1 Gauss rule.

Example 7.6.3 ——————————————————————————————————

Consider nonlinear bending of an isotropic ($\nu = 0.3$) square plate under uniformly distributed transverse load, q_0. This problem was analyzed by Lévy [64], Wang [65], and Kawai and Yoshimura [66]. The following simply-supported (SS–3) geometric boundary conditions are used:

$$u = v = w = 0 \quad \text{on all four edges} \tag{1}$$

Since $(\partial w/\partial x)$ and $(\partial w/\partial y)$ are not specified in SS–3, it follows that the following force boundary conditions are satisfied in the integral sense [see Eq. (7.4.6)]:

$$\text{on } y = 0, b: \quad M_{xy} = M_{yy} = 0; \quad \text{on } x = 0, a: \quad M_{xx} = M_{xy} = 0 \tag{2}$$

Use the biaxial symmetry to model only a quadrant of the plate with uniform 4×4 and 8×8 meshes of rectangular elements to obtain deflections and stresses as functions of the load.

Solution: The boundary conditions along the lines of symmetry are shown in Fig. 7.6.1. The following geometric and material parameters are used, although the non-dimensional deflections and stresses presented here are independent of them (but may depend on ν):

$$a = b = 10\,\text{in}, \quad h = 0.1\,\text{in}, \quad E = 30 \times 10^6\,\text{psi}, \quad \nu = 0.3 \tag{3}$$

A load increment of $\Delta q = 7.5$, which is equal to the increment of load parameter, $\Delta P \equiv a^4\Delta q/Eh^4 = 25$, is used along with the convergence tolerance of $\epsilon = 10^{-2}$. Except for the first load step, which took five iterations, the convergence was achieved for two or three iterations.

Plots of the load parameter P versus the deflection $\bar{w}$ and P versus various stresses are presented in Fig. 7.6.2. Although the linear and nonlinear values of $\bar{\sigma}_{xx}$ for the load parameter $P = 25$ are maximum at the center of the plate, the location of the maximum normal stress $\bar{\sigma}_{xx}$ in the nonlinear analysis changes with the load. For example, the maximum value of $\bar{\sigma}_{xx}$ at $P = 250$ in the finite element analysis occurs at $(x,y)=(2.8125,0.3125)$, and its value is found to be 21.177. Note that membrane stresses are a significant part of total stresses.

The center deflection, $\bar{w} = w/h$, and total stresses (i.e. sum of membrane and flexural contributions), $\bar{\sigma}_{xx} = \sigma_{xx}(A,A,h/2)(a^2/Eh^2)$ and $\bar{\sigma}_{xy} = \sigma_{xy}(B,B,-h/2)(a^2/Eh^2)$, as functions of the load parameter, $P = q_0 a^4/Eh^4$ are presented in Table 7.6.3. The location $(A,A,h/2)$ refers to the Gauss point nearest to the center ($x = y = 0$) but at the top of the plate, while $(B,B,-h/2)$ refers to the Gauss point nearest to the corner $x = y = a/2$, at the bottom of the

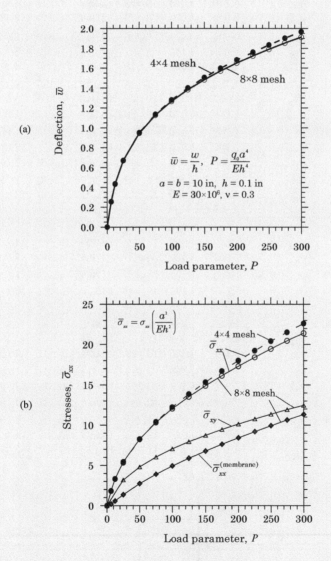

Fig. 7.6.2: (a) Load versus deflection and (b) load versus stress plots for simply-supported (SS–3) isotropic ($\nu = 0.3$) square plates under uniformly distributed transverse load q_0.

Table 7.6.3: Maximum transverse deflections and stresses* of simply-supported (SS–3), isotropic ($\nu = 0.3$) square plates under a uniformly distributed transverse load q_0 ($\bar{w} = w/h$ and $P \equiv a^4 q_0 / E h^4$; *nonlinear analysis*).

P		Non-conforming			Conforming		
		4F–F	4F–R	8F–R	4F–F	4F–R	8F–R
25 Linear $\bar{w} \rightarrow$		1.127	1.127	1.113	1.116	1.116	1.110
25	$\bar{w}$	0.670	0.673	0.670	0.669	0.670	0.669
	$\bar{\sigma}_{xx}$	5.279	5.321	5.423	5.252	5.324	5.426
	$\bar{\sigma}_{xy}$	3.014	3.028	3.179	2.994	3.019	3.172
50	$\bar{w}$	0.944	0.951	0.946	0.943	0.949	0.945
	$\bar{\sigma}_{xx}$	8.035	8.142	8.227	8.018	8.219	8.247
	$\bar{\sigma}_{xy}$	4.510	4.552	4.818	4.484	4.566	4.810
75	$\bar{w}$	1.124	1.136	1.128	1.123	1.135	1.127
	$\bar{\sigma}_{xx}$	10.057	10.231	10.271	10.048	10.390	10.309
	$\bar{\sigma}_{xy}$	5.600	5.674	6.036	5.569	5.719	6.029
100	$\bar{w}$	1.262	1.280	1.268	1.262	1.279	1.267
	$\bar{\sigma}_{xx}$	11.731	11.974	11.961	11.729	12.217	12.017
	$\bar{\sigma}_{xy}$	6.493	6.599	7.049	6.457	6.679	7.042
125	$\bar{w}$	1.377	1.400	1.383	1.377	1.400	1.383
	$\bar{\sigma}_{xx}$	13.197	13.509	13.440	13.201	13.838	13.513
	$\bar{\sigma}_{xy}$	7.265	7.403	7.933	7.224	7.521	7.927
150	$\bar{w}$	1.475	1.504	1.483	1.476	1.505	1.483
	$\bar{\sigma}_{xx}$	14.519	14.902	14.777	14.529	15.317	14.867
	$\bar{\sigma}_{xy}$	7.951	8.122	8.726	7.905	8.280	8.722
175	$\bar{w}$	1.563	1.597	1.571	1.563	1.598	1.571
	$\bar{\sigma}_{xx}$	15.737	16.191	16.009	15.752	16.693	16.117
	$\bar{\sigma}_{xy}$	8.573	8.778	9.451	8.524	8.979	9.449
200	$\bar{w}$	1.641	1.681	1.651	1.642	1.683	1.651
	$\bar{\sigma}_{xx}$	16.874	17.401	17.162	16.894	17.989	17.287
	$\bar{\sigma}_{xy}$	9.147	9.385	10.123	9.094	9.631	10.123
225	$\bar{w}$	1.713	1.758	1.724	1.713	1.761	1.724
	$\bar{\sigma}_{xx}$	17.946	18.546	18.251	17.970	19.222	18.393
	$\bar{\sigma}_{xy}$	9.681	9.952	10.751	9.625	10.244	10.754
250	$\bar{w}$	1.779	1.830	1.791	1.779	1.834	1.791
	$\bar{\sigma}_{xx}$	18.965	19.638	19.287	18.993	20.401	19.446
	$\bar{\sigma}_{xy}$	10.182	10.486	11.344	10.122	10.827	11.349

*$\bar{w} = w/h$, $\bar{\sigma} = \sigma(a^2/Eh^2)$.

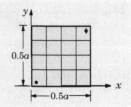

plate. The notation F–F means full integration (4×4 Gauss rule) is used for the numerical evaluation of all coefficients, while F–R means full integration is used for all but nonlinear terms and reduced integration (1×1 Gauss rule) is used for the nonlinear terms. The number in front of F–F or F–R stands for the mesh type (for example, 4 refers to 4×4 mesh). Stresses are evaluated at the center of the element. There is very little difference between the results obtained with reduced and full integrations of the nonlinear stiffness coefficients.

Example 7.6.4

Consider the bending of (a) simply-supported (SS–1) and (b) clamped orthotropic square plates under a uniformly distributed transverse load q_0. Use the following geometric and material properties:

$$a = b = 12 \text{ in}, \quad h = 0.138 \text{ in}, \quad E_1 = 3 \times 10^6 \text{ psi}, \quad E_2 = 1.28 \times 10^6 \text{ psi}$$
$$G_{12} = G_{23} = G_{13} = 0.37 \times 10^6 \text{ psi}, \quad \nu_{12} = \nu_{23} = \nu_{13} = 0.32 \tag{1}$$

Use a load increment of $\Delta q = 0.2$ psi and a uniform mesh of 4×4 non-conforming elements in a quarter plate for the simply-supported plate. Assume the boundary conditions of a clamped edge to be $u = v = w = \frac{\partial w}{\partial x} = \frac{\partial w}{\partial y} = \frac{\partial^2 w}{\partial x \partial y} = 0$, use a uniform mesh of 8×8 non-conforming elements in a quarter plate, and load increments of $\{\Delta q\} = \{0.05, 0.05, 0.1, 0.2, 0.2, \ldots, 0.2\}$ psi. Plot the center deflection versus load for the two problems.

Solution: Numerical solutions to large deflection bending of orthotropic rectangular plates by the method of differential quadrature can be found in a paper by Bert, Jang, and Striz [67]. Figure 7.6.3 shows a plot of the center deflection versus the intensity of the distributed load for the simply-supported (SS–1) plate.

The linear solution for the clamped plate at $q_0 = 0.05$ is found to be $w(0,0) = 0.00302$ in. Figure 7.6.3 also contains a plot of the center deflection versus the intensity of the distributed load for the clamped orthotropic plate (see [68–80]).

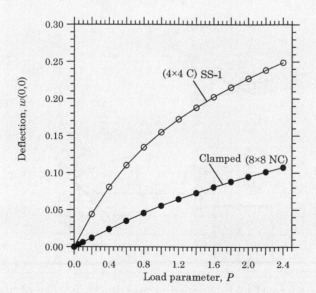

Fig. 7.6.3: Load-deflection curves for simply-supported and clamped orthotropic square plates under uniform transverse load.

7.7 The First-Order Shear Deformation Plate Theory

7.7.1 Introduction

The preceding sections of the book were devoted to the study of bending of plates using the classical plate theory (CPT), in which transverse normal and shear stresses were neglected. The first-order shear deformation theory (FSDT) extends the kinematics of the CPT by relaxing the normality restriction (see Section 7.2) and allowing for arbitrary but constant rotation of transverse normals, as shown in Fig. 7.7.1. In this section, we develop displacement finite element models of the FSDT. As we shall see in the following sections, the formulation requires only C^0 interpolation of all generalized displacements.

7.7.2 Displacement Field

Under the same assumptions and restrictions as in the classical laminate theory but relaxing the normality condition, the displacement field of the FSDT can be expressed in the form

$$
\begin{aligned}
u_1(x, y, z) &= u(x, y) + z\phi_x(x, y) \\
u_2(x, y, z) &= v(x, y) + z\phi_y(x, y) \\
u_3(x, y, z) &= w(x, y)
\end{aligned}
\tag{7.7.1}
$$

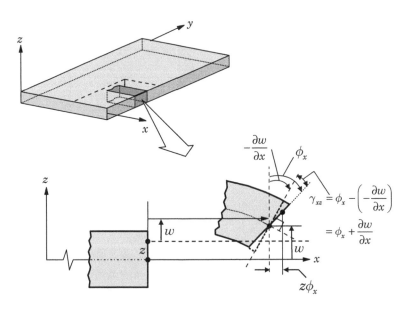

Fig. 7.7.1: Undeformed and deformed geometries of an edge of a plate under the assumptions of the FSDT.

where $(u, v, w, \phi_x, \phi_y)$ are unknown functions to be determined. As before, (u, v, w) denote the displacements of a point on the plane $z = 0$ and ϕ_x and ϕ_y are the rotations of a transverse normal line about the y- and x-axes, respectively (see Fig. 7.7.1). The quantities $(u, v, w, \phi_x, \phi_y)$ are called the *generalized displacements*.

The notation that ϕ_x denotes the rotation of a transverse normal about the y-axis and ϕ_y denotes the rotation about the x-axis may be confusing to some because they do not follow the right-hand rule. However, the notation has been used extensively in the literature and we will not depart from it. If (β_x, β_y) denote the rotations about the x- and y-axes, respectively, that follow the right-hand rule, then

$$\beta_x = -\phi_y , \quad \beta_y = \phi_x \tag{7.7.2}$$

For thin plates, that is, when the plate in-plane characteristic dimension to thickness ratio is on the order of 50 or greater, the rotation functions ϕ_x and ϕ_y should approach the respective slopes of the transverse deflection [21]:

$$\phi_x = -\frac{\partial w}{\partial x} , \quad \phi_y = -\frac{\partial w}{\partial y} \tag{7.7.3}$$

However, this equality is not achieved in the discrete model, resulting in shear locking, as in the case of the Timoshenko beam theory, when the same lower-order approximations are used for the transverse deflection w as well as the rotations (ϕ_x and ϕ_y).

The von Kármán nonlinear strains associated with the displacement field in Eq. (7.7.1) are ($\varepsilon_{zz} = 0$)

$$\begin{Bmatrix} \varepsilon_{xx} \\ \varepsilon_{yy} \\ \gamma_{yz} \\ \gamma_{xz} \\ \gamma_{xy} \end{Bmatrix} = \begin{Bmatrix} \varepsilon_{xx}^0 \\ \varepsilon_{yy}^0 \\ \gamma_{yz}^0 \\ \gamma_{xz}^0 \\ \gamma_{xy}^0 \end{Bmatrix} + z \begin{Bmatrix} \varepsilon_{xx}^1 \\ \varepsilon_{yy}^1 \\ 0 \\ 0 \\ \gamma_{xy}^1 \end{Bmatrix} = \begin{Bmatrix} \frac{\partial u}{\partial x} + \frac{1}{2}\left(\frac{\partial w}{\partial x}\right)^2 \\ \frac{\partial v}{\partial y} + \frac{1}{2}\left(\frac{\partial w}{\partial y}\right)^2 \\ \frac{\partial w}{\partial y} + \phi_y \\ \frac{\partial w}{\partial x} + \phi_x \\ \frac{\partial u}{\partial y} + \frac{\partial v}{\partial x} + \frac{\partial w}{\partial x}\frac{\partial w}{\partial y} \end{Bmatrix} + z \begin{Bmatrix} \frac{\partial \phi_x}{\partial x} \\ \frac{\partial \phi_y}{\partial y} \\ 0 \\ 0 \\ \frac{\partial \phi_x}{\partial y} + \frac{\partial \phi_y}{\partial x} \end{Bmatrix}$$

$$\tag{7.7.4}$$

Note that the strains ($\varepsilon_{xx}, \varepsilon_{yy}, \gamma_{xy}$) are linear through the plate thickness, while the transverse shear strains (γ_{xz}, γ_{yz}) are constant.

7.7.3 Weak Forms using the Principle of Virtual Displacements

The weak forms of the FSDT can be derived using the principle of virtual displacements

$$0 = \delta W^e \equiv \delta W_I^e + \delta W_E^e \tag{7.7.5}$$

where the virtual strain energy δW_I^e and the virtual work done by applied forces δW_E^e in an element Ω^e are given by

$$\delta W_I^e = \int_{\Omega^e} \left\{ \int_{-\frac{h}{2}}^{\frac{h}{2}} \left[\sigma_{xx} \left(\delta\varepsilon_{xx}^0 + z\delta\varepsilon_{xx}^1 \right) + \sigma_{yy} \left(\delta\varepsilon_{yy}^0 + z\delta\varepsilon_{yy}^1 \right) \right. \right.$$

$$\left. \left. + \sigma_{xy} \left(\delta\gamma_{xy}^0 + z\delta\gamma_{xy}^1 \right) + K_s\sigma_{xz}\delta\gamma_{xz}^0 + K_s\sigma_{yz}\delta\gamma_{yz}^0 \right] dz \right\} dx\, dy \quad (7.7.6)$$

$$\delta W_E^e = -\left\{ \oint_{\Gamma^e} \int_{-\frac{h}{2}}^{\frac{h}{2}} \left[\sigma_{nn} \left(\delta u_n + z\delta\phi_n \right) + \sigma_{ns} \left(\delta u_s + z\delta\phi_s \right) + \sigma_{nz}\delta w \right] dz\, ds \right.$$

$$\left. + \int_{\Omega^e} (q - kw)\, \delta w\ dx\, dy \right\} \quad (7.7.7)$$

where Ω^e denotes the undeformed mid-plane of a typical plate element, h is the total thickness, k is the modulus of the elastic foundation (if any), $(\sigma_{nn}, \sigma_{ns}, \sigma_{nz})$ are the edge stresses along the (n, s, z) coordinates, and K_s is the shear correction coefficient ($K_s = 5/6$).

Substituting for δW_I^e and δW_E^e from Eqs. (7.7.6) and (7.7.7) into the virtual work statement in Eq. (7.7.5) and integrating through the thickness, we obtain

$$0 = \int_{\Omega^e} \left[N_{xx}\delta\varepsilon_{xx}^0 + M_{xx}\delta\varepsilon_{xx}^1 + N_{yy}\delta\varepsilon_{yy}^0 + M_{yy}\delta\varepsilon_{yy}^1 + N_{xy}\delta\gamma_{xy}^0 \right.$$

$$\left. + M_{xy}\delta\gamma_{xy}^1 + Q_x\delta\gamma_{xz}^0 + Q_y\delta\gamma_{yz}^0 + kw\delta w - q\delta w \right] dx\, dy$$

$$- \oint_{\Gamma^e} (N_{nn}\delta u_n + N_{ns}\delta u_s + M_{nn}\delta\phi_n + M_{ns}\delta\phi_s + Q_n\delta w)\, ds \quad (7.7.8)$$

where ϕ_n and ϕ_s are the rotations of a transverse normal about s and $-n$ coordinates, respectively, $(N_{xx}, N_{yy}, N_{xy}, M_{xx}, M_{yy}, M_{xy})$ are defined in Eq. (7.3.4), and $(N_{nn}, N_{ns}, M_{nn}, M_{ns}, Q_n)$ are defined in Eq. (7.3.7). The *transverse shear forces per unit length* (Q_x, Q_y) are defined by

$$\left\{ \begin{array}{c} Q_x \\ Q_y \end{array} \right\} = K_s \int_{-\frac{h}{2}}^{\frac{h}{2}} \left\{ \begin{array}{c} \sigma_{xz} \\ \sigma_{yz} \end{array} \right\} dz \quad (7.7.9)$$

7.7.4   Governing Equations

Equation (7.7.8) contains five different statements associated with the five virtual displacements $(\delta u, \delta v, \delta w, \delta\phi_x, \delta\phi_y)$, and they form the basis of the finite element model. The governing equations of equilibrium of the FSDT, although they are not needed to develop the finite element model, can be derived by first expressing the virtual strains in Eq. (7.7.8) in terms of the virtual displacements and then integrating-by-parts all expressions to relieve the virtual displacements $(\delta u, \delta v, \delta w, \delta \phi_x, \delta \phi_y)$ in Ω^e of any differentiation. Then the fundamental lemma of calculus of variations is used to declare that the coefficients of $(\delta u, \delta v, \delta w, \delta \phi_x, \delta \phi_y)$ in Ω^e must be zero, obtaining

$$
\begin{aligned}
0 = \int_{\Omega^e} \Big[& -(N_{xx,x} + N_{xy,y})\,\delta u - (N_{xy,x} + N_{yy,y})\,\delta v \\
& -(M_{xx,x} + M_{xy,y} - Q_x)\,\delta\phi_x - (M_{xy,x} + M_{yy,y} - Q_y)\,\delta\phi_y \\
& -(Q_{x,x} + Q_{y,y} - kw + \mathcal{N} + q)\,\delta w \Big] dx\, dy \\
+ \oint_{\Gamma^e} \Big[& (N_{xx}n_x + N_{xy}n_y)\,\delta u + (N_{xy}n_x + N_{yy}n_y)\,\delta v \\
& + (M_{xx}n_x + M_{xy}n_y)\,\delta\phi_x + (M_{xy}n_x + M_{yy}n_y)\,\delta\phi_y \\
& + \Big(Q_x n_x + Q_y n_y + \mathcal{P}\Big)\delta w \Big] ds \\
- \oint_{\Gamma^e} & (N_{nn}\delta u_n + N_{ns}\delta u_s + M_{nn}\delta\phi_n + M_{ns}\delta\phi_s + Q_n \delta w)\, ds \qquad (7.7.10)
\end{aligned}
$$

where $\mathcal{N}$ and $\mathcal{P}$ are defined by Eq. (7.3.14). The boundary terms can be expressed in terms of the normal and tangential components u_n, u_s, ϕ_n, and ϕ_s using Eqs. (7.3.18a) and

$$
\phi_x = n_x \phi_n - n_y \phi_s\,, \quad \phi_y = n_y \delta\phi_n + n_x \delta\phi_s \qquad (7.7.11)
$$

This will yield the natural boundary conditions given in Eqs. (7.3.20a, b), which relate the forces and moments on an arbitrary edge to those on edges parallel to the coordinates (x, y, z).

The Euler–Lagrange equations are

$$
\delta u: \quad \frac{\partial N_{xx}}{\partial x} + \frac{\partial N_{xy}}{\partial y} = 0 \qquad (7.7.12)
$$

$$
\delta v: \quad \frac{\partial N_{xy}}{\partial x} + \frac{\partial N_{yy}}{\partial y} = 0 \qquad (7.7.13)
$$

$$
\delta w: \quad \frac{\partial Q_x}{\partial x} + \frac{\partial Q_y}{\partial y} - kw + \mathcal{N}(u, v, w, \phi_x, \phi_y) + q = 0 \qquad (7.7.14)
$$

$$
\delta\phi_x: \quad \frac{\partial M_{xx}}{\partial x} + \frac{\partial M_{xy}}{\partial y} - Q_x = 0 \qquad (7.7.15)
$$

$$
\delta\phi_y: \quad \frac{\partial M_{xy}}{\partial x} + \frac{\partial M_{yy}}{\partial y} - Q_y = 0 \qquad (7.7.16)
$$

The primary and secondary variables of the theory are

$$
\begin{aligned}
\text{primary variables:} \quad & u_n,\ u_s,\ w,\ \phi_n,\ \phi_s \\
\text{secondary variables:} \quad & N_{nn},\ N_{ns},\ Q_n,\ M_{nn},\ M_{ns}
\end{aligned} \qquad (7.7.17)
$$

The plate constitutive equations in Eqs. (7.3.32) and (7.3.33) are valid also for the first-order plate theory. Since the transverse shear strains are represented as constant through the plate thickness, it follows that the transverse shear stresses will also be constant. It is well known from elementary theory of homogeneous beams that the transverse shear stress variation is parabolic through the beam thickness, as noted in connection with the Timoshenko beam theory. This discrepancy between the actual stress state and the constant stress state predicted by the FSDT is often corrected in computing the transverse shear forces (Q_x, Q_y) by multiplying the transverse shear stiffnesses (Q_{44}, Q_{55}) with the shear correction factor K_s. Thus, the constitutive equations for transverse shear forces of an orthotropic plate are

$$\left\{ \begin{array}{c} Q_y \\ Q_x \end{array} \right\} = K_s \int_{-\frac{h}{2}}^{\frac{h}{2}} \left\{ \begin{array}{c} \sigma_{yz} \\ \sigma_{xz} \end{array} \right\} dz = \left[\begin{array}{cc} A_{44} & 0 \\ 0 & A_{55} \end{array} \right] \left\{ \begin{array}{c} \gamma_{yz} \\ \gamma_{xz} \end{array} \right\} \tag{7.7.18}$$

where the transverse shear stiffnesses A_{44} and A_{55} are defined by

$$(A_{44}, A_{55}) = K_s \int_{-\frac{h}{2}}^{\frac{h}{2}} (Q_{44}, Q_{55}) \, dz, \quad Q_{44} = G_{23}, \quad Q_{55} = G_{13} \tag{7.7.19}$$

In summary, the stress resultants in an orthotropic plate are related to the generalized displacements $(u, v, w, \phi_x, \phi_y)$ by [see Eqs. (7.3.32) and (7.3.33)]

$$\left\{ \begin{array}{c} N_{xx} \\ N_{yy} \\ N_{xy} \end{array} \right\} = \left[\begin{array}{ccc} A_{11} & A_{12} & 0 \\ A_{12} & A_{22} & 0 \\ 0 & 0 & A_{66} \end{array} \right] \left\{ \begin{array}{c} \varepsilon_{xx}^0 \\ \varepsilon_{yy}^0 \\ \gamma_{xy}^0 \end{array} \right\} - \left\{ \begin{array}{c} N_{xx}^T \\ N_{yy}^T \\ 0 \end{array} \right\} \tag{7.7.20}$$

$$\left\{ \begin{array}{c} M_{xx} \\ M_{yy} \\ M_{xy} \end{array} \right\} = \left[\begin{array}{ccc} D_{11} & D_{12} & 0 \\ D_{12} & D_{22} & 0 \\ 0 & 0 & D_{66} \end{array} \right] \left\{ \begin{array}{c} \varepsilon_{xx}^1 \\ \varepsilon_{yy}^1 \\ \gamma_{xy}^1 \end{array} \right\} - \left\{ \begin{array}{c} M_{xx}^T \\ M_{yy}^T \\ 0 \end{array} \right\} \tag{7.7.21}$$

$$\left\{ \begin{array}{c} Q_y \\ Q_x \end{array} \right\} = \left[\begin{array}{cc} A_{44} & 0 \\ 0 & A_{55} \end{array} \right] \left\{ \begin{array}{c} \gamma_{yz}^0 \\ \gamma_{xz}^0 \end{array} \right\} \tag{7.7.22}$$

where $(i, j = 1, 2, 6)$

$$A_{ij} = Q_{ij} h, \quad D_{ij} = Q_{ij} \frac{h^3}{12}; \quad A_{44} = K_s G_{23} h, \quad A_{55} = K_s G_{13} h \tag{7.7.23}$$

$$Q_{11} = \frac{E_1}{1 - \nu_{12}\nu_{21}}, \quad Q_{22} = Q_{11} \frac{E_2}{E_1}, \quad Q_{12} = \nu_{12} Q_{22}, \quad Q_{66} = G_{12} \tag{7.7.24}$$

7.8 Finite Element Models of the FSDT

7.8.1 Weak Forms

Using the weak forms, Eq. (7.7.8), developed in Section 7.7, we can construct the finite element models of the equations governing the FSDT. The stress resultants in Eq. (7.7.8) are understood to be known in terms of the generalized displacements $(u, v, w, \phi_x, \phi_y)$ via Eqs. (7.7.20)–(7.7.22). As stated earlier, the virtual work statement of Eq. (7.7.8) is equivalent to (collecting the terms involving δu, δv, δw, $\delta\phi_x$, and $\delta\phi_y$ separately) the following five weak forms:

$$0 = \int_{\Omega^e} \left(\frac{\partial \delta u}{\partial x} N_{xx} + \frac{\partial \delta u}{\partial y} N_{xy} \right) dx\,dy - \oint_{\Gamma^e} (N_{xx} n_x + N_{xy} n_y)\, \delta u\; ds \qquad (7.8.1)$$

$$0 = \int_{\Omega^e} \left(\frac{\partial \delta v}{\partial x} N_{xy} + \frac{\partial \delta v}{\partial y} N_{yy} \right) dx\,dy - \oint_{\Gamma^e} (N_{xy} n_x + N_{yy} n_y)\, \delta v\; ds \qquad (7.8.2)$$

$$0 = \int_{\Omega^e} \left[\frac{\partial \delta w}{\partial x} Q_x + \frac{\partial \delta w}{\partial y} Q_y + \frac{\partial \delta w}{\partial x} \left(N_{xx} \frac{\partial w}{\partial x} + N_{xy} \frac{\partial w}{\partial y} \right) \right.$$
$$+ \frac{\partial \delta w}{\partial y} \left(N_{xy} \frac{\partial w}{\partial x} + N_{yy} \frac{\partial w}{\partial y} \right) - \delta w q + k w \delta w \bigg] dx\,dy$$
$$- \oint_{\Gamma^e} \left[\left(Q_x + N_{xx} \frac{\partial w}{\partial x} + N_{xy} \frac{\partial w}{\partial y} \right) n_x \right.$$
$$+ \left(Q_y + N_{xy} \frac{\partial w}{\partial x} + N_{yy} \frac{\partial w}{\partial y} \right) n_y \bigg]\, \delta w\; ds \qquad (7.8.3)$$

$$0 = \int_{\Omega^e} \left(\frac{\partial \delta\phi_x}{\partial x} M_{xx} + \frac{\partial \delta\phi_x}{\partial y} M_{xy} + \delta\phi_x Q_x \right) dx\,dy$$
$$- \oint_{\Gamma^e} (M_{xx} n_x + M_{xy} n_y)\, \delta\phi_x\; ds \qquad (7.8.4)$$

$$0 = \int_{\Omega^e} \left(\frac{\partial \delta\phi_y}{\partial x} M_{xy} + \frac{\partial \delta\phi_y}{\partial y} M_{yy} + \delta\phi_y Q_y \right) dx\,dy$$
$$- \oint_{\Gamma^e} (M_{xy} n_x + M_{yy} n_y)\, \delta\phi_y\; ds \qquad (7.8.5)$$

We note from the boundary terms in Eqs. (7.8.1)–(7.8.5) that $(u, v, w, \phi_x, \phi_y)$ are used as the primary variables (or generalized displacements) as opposed to $(u_n, u_s, w, \phi_n, \phi_s)$. We identify the secondary variables [dual to $(u, v, w, \phi_x, \phi_y)$] of the formulation as

$$\bar{N}_{nn} \equiv N_{xx} n_x + N_{xy} n_y, \qquad \bar{N}_{ns} \equiv N_{xy} n_x + N_{yy} n_y \qquad (7.8.6a)$$
$$\bar{M}_{nn} \equiv M_{xx} n_x + M_{xy} n_y, \qquad \bar{M}_{ns} \equiv M_{xy} n_x + M_{yy} n_y \qquad (7.8.6b)$$
$$Q_n \equiv \left(Q_x + N_{xx} \frac{\partial w}{\partial x} + N_{xy} \frac{\partial w}{\partial y} \right) n_x + \left(Q_y + N_{xy} \frac{\partial w}{\partial x} + N_{yy} \frac{\partial w}{\partial y} \right) n_y \qquad (7.8.7)$$

Recall that the rotations (ϕ_x, ϕ_y) are independent of w. Since no derivatives of $(u, v, w, \phi_x, \phi_y)$ appear in the list of the primary variables, all generalized displacements may be interpolated using Lagrange interpolation functions. Hence, the element is called C^0 element with respect to all dependent unknowns.

7.8.2   The Finite Element Model

The virtual work statements in Eqs. (7.8.1)–(7.8.5) contain at the most only the first derivatives of the dependent variables $(u, v, w, \phi_x, \phi_y)$. Therefore, they can all be approximated using the Lagrange interpolation functions. In principle, (u, v), w, and (ϕ_x, ϕ_y) can be approximated with differing degrees of Lagrange interpolation functions. Let

$$u(x, y) = \sum_{j=1}^{m} u_j \psi_j^{(1)}(x, y), \quad v(x, y) = \sum_{j=1}^{m} v_j \psi_j^{(1)}(x, y) \tag{7.8.8}$$

$$w(x, y) = \sum_{j=1}^{n} w_j \psi_j^{(2)}(x, y) \tag{7.8.9}$$

$$\phi_x(x, y) = \sum_{j=1}^{p} S_j^1 \psi_j^{(3)}(x, y), \quad \phi_y(x, y) = \sum_{j=1}^{p} S_j^2 \psi_j^{(3)}(x, y) \tag{7.8.10}$$

where $\psi_j^{(\alpha)}$ $(\alpha = 1, 2, 3)$ are Lagrange interpolation functions. One can use linear, quadratic, or higher-order interpolations of these variables. Although the development is general, in the implementation of this element, we shall use equal interpolation of all variables (see Reddy [55, 56]).

Substituting Eqs. (7.8.8)–(7.8.10) for $(u, v, w, \phi_x, \phi_y)$ into Eqs. (7.8.1)–(7.8.5), we obtain the following finite element model:

$$\begin{bmatrix} \mathbf{K}^{11} & \mathbf{K}^{12} & \mathbf{K}^{13} & \mathbf{K}^{14} & \mathbf{K}^{15} \\ \mathbf{K}^{21} & \mathbf{K}^{22} & \mathbf{K}^{23} & \mathbf{K}^{24} & \mathbf{K}^{25} \\ \mathbf{K}^{31} & \mathbf{K}^{32} & \mathbf{K}^{33} & \mathbf{K}^{34} & \mathbf{K}^{35} \\ \mathbf{K}^{41} & \mathbf{K}^{42} & \mathbf{K}^{43} & \mathbf{K}^{44} & \mathbf{K}^{45} \\ \mathbf{K}^{51} & \mathbf{K}^{52} & \mathbf{K}^{53} & \mathbf{K}^{54} & \mathbf{K}^{55} \end{bmatrix} \begin{Bmatrix} \mathbf{u}^e \\ \mathbf{v}^e \\ \mathbf{w}^e \\ \mathbf{S}^1 \\ \mathbf{S}^2 \end{Bmatrix} = \begin{Bmatrix} \mathbf{F}^1 \\ \mathbf{F}^2 \\ \mathbf{F}^3 \\ \mathbf{F}^4 \\ \mathbf{F}^5 \end{Bmatrix} + \begin{Bmatrix} \mathbf{F}^{1T} \\ \mathbf{F}^{2T} \\ \mathbf{0} \\ \mathbf{F}^{4T} \\ \mathbf{F}^{5T} \end{Bmatrix} \tag{7.8.11}$$

or, in generic matrix form

$$\mathbf{K}^e \mathbf{\Delta}^e = \mathbf{F}^e \tag{7.8.12}$$

where the coefficients of the submatrices $\mathbf{K}^{\alpha\beta}$ and vectors $\mathbf{F}^\alpha$ and $\mathbf{F}^{\alpha T}$ are defined, for $\alpha, \beta = 1, 2, 3, 4, 5$, by the expressions

$$K_{ij}^{11} = \int_{\Omega^e} \left(A_{11} \frac{\partial \psi_i^{(1)}}{\partial x} \frac{\partial \psi_j^{(1)}}{\partial x} + A_{66} \frac{\partial \psi_i^{(1)}}{\partial y} \frac{\partial \psi_j^{(1)}}{\partial y} \right) dx\, dy$$

$$K_{ij}^{12} = \int_{\Omega^e} \left(A_{12} \frac{\partial \psi_i^{(1)}}{\partial x} \frac{\partial \psi_j^{(1)}}{\partial y} + A_{66} \frac{\partial \psi_i^{(1)}}{\partial y} \frac{\partial \psi_j^{(1)}}{\partial x} \right) dx\, dy$$

$$K_{ij}^{13} = \frac{1}{2} \int_{\Omega^e} \left[\frac{\partial \psi_i^{(1)}}{\partial x} \left(A_{11} \frac{\partial w}{\partial x} \frac{\partial \psi_j^{(2)}}{\partial x} + A_{12} \frac{\partial w}{\partial y} \frac{\partial \psi_j^{(2)}}{\partial y} \right) \right.$$

$$\left. + A_{66} \frac{\partial \psi_i^{(1)}}{\partial y} \left(\frac{\partial w}{\partial x} \frac{\partial \psi_j^{(2)}}{\partial y} + \frac{\partial w}{\partial y} \frac{\partial \psi_j^{(2)}}{\partial x} \right) \right] dx\, dy$$

$$K_{ij}^{22} = \int_{\Omega^e} \left(A_{66} \frac{\partial \psi_i^{(1)}}{\partial x} \frac{\partial \psi_j^{(1)}}{\partial x} + A_{22} \frac{\partial \psi_i^{(1)}}{\partial y} \frac{\partial \psi_j^{(1)}}{\partial y} \right) dx\, dy$$

$$K_{ij}^{23} = \frac{1}{2} \int_{\Omega^e} \left[\frac{\partial \psi_i^{(1)}}{\partial y} \left(A_{12} \frac{\partial w}{\partial x} \frac{\partial \psi_j^{(2)}}{\partial x} + A_{22} \frac{\partial w}{\partial y} \frac{\partial \psi_j^{(2)}}{\partial y} \right) \right.$$
$$\left. + A_{66} \frac{\partial \psi_i^{(1)}}{\partial x} \left(\frac{\partial w}{\partial x} \frac{\partial \psi_j^{(2)}}{\partial y} + \frac{\partial w}{\partial y} \frac{\partial \psi_j^{(2)}}{\partial x} \right) \right] dx\, dy$$

$$K_{ij}^{31} = \int_{\Omega^e} \left[\frac{\partial \psi_i^{(2)}}{\partial x} \left(A_{11} \frac{\partial w}{\partial x} \frac{\partial \psi_j^{(1)}}{\partial x} + A_{66} \frac{\partial w}{\partial y} \frac{\partial \psi_j^{(1)}}{\partial y} \right) \right.$$
$$\left. + \frac{\partial \psi_i^{(2)}}{\partial y} \left(A_{66} \frac{\partial w}{\partial x} \frac{\partial \psi_j^{(1)}}{\partial y} + A_{12} \frac{\partial w}{\partial y} \frac{\partial \psi_j^{(1)}}{\partial x} \right) \right] dx\, dy$$

$$K_{ij}^{32} = \int_{\Omega^e} \left[\frac{\partial \psi_i^{(2)}}{\partial x} \left(A_{12} \frac{\partial w}{\partial x} \frac{\partial \psi_j^{(1)}}{\partial y} + A_{66} \frac{\partial w}{\partial y} \frac{\partial \psi_j^{(1)}}{\partial x} \right) \right.$$
$$\left. + \frac{\partial \psi_i^{(2)}}{\partial y} \left(A_{66} \frac{\partial w}{\partial x} \frac{\partial \psi_j^{(1)}}{\partial x} + A_{22} \frac{\partial w}{\partial y} \frac{\partial \psi_j^{(1)}}{\partial y} \right) \right] dx\, dy$$

$$K_{ij}^{33} = \int_{\Omega^e} \left(A_{55} \frac{\partial \psi_i^{(2)}}{\partial x} \frac{\partial \psi_j^{(2)}}{\partial x} + A_{44} \frac{\partial \psi_i^{(2)}}{\partial y} \frac{\partial \psi_j^{(2)}}{\partial y} + k \psi_i^{(2)} \psi_j^{(2)} \right) dx\, dy$$
$$+ \frac{1}{2} \int_{\Omega^e} \left\{ \left[A_{11} \left(\frac{\partial w}{\partial x} \right)^2 + A_{66} \left(\frac{\partial w}{\partial y} \right)^2 \right] \frac{\partial \psi_i^{(2)}}{\partial x} \frac{\partial \psi_j^{(2)}}{\partial x} \right.$$
$$+ \left[A_{66} \left(\frac{\partial w}{\partial x} \right)^2 + A_{22} \left(\frac{\partial w}{\partial y} \right)^2 \right] \frac{\partial \psi_i^{(2)}}{\partial y} \frac{\partial \psi_j^{(2)}}{\partial y}$$
$$\left. + (A_{12} + A_{66}) \frac{\partial w}{\partial x} \frac{\partial w}{\partial y} \left(\frac{\partial \psi_i^{(2)}}{\partial x} \frac{\partial \psi_j^{(2)}}{\partial y} + \frac{\partial \psi_i^{(2)}}{\partial y} \frac{\partial \psi_j^{(2)}}{\partial x} \right) \right\} dx\, dy$$
$$- \int_{\Omega^e} \left(N_{xx}^T \frac{\partial \psi_i^{(2)}}{\partial x} \frac{\partial \psi_j^{(2)}}{\partial x} + N_{yy}^T \frac{\partial \psi_i^{(2)}}{\partial y} \frac{\partial \psi_j^{(2)}}{\partial y} \right) dx\, dy$$

$$K_{ij}^{34} = \int_{\Omega^e} A_{55} \frac{\partial \psi_i^{(2)}}{\partial x} \psi_j^{(3)} \, dx\, dy, \quad K_{ij}^{35} = \int_{\Omega^e} A_{44} \frac{\partial \psi_i^{(2)}}{\partial y} \psi_j^{(3)} \, dx\, dy$$

$$K_{ij}^{44} = \int_{\Omega^e} \left(D_{11} \frac{\partial \psi_i^{(3)}}{\partial x} \frac{\partial \psi_j^{(3)}}{\partial x} + D_{66} \frac{\partial \psi_i^{(3)}}{\partial y} \frac{\partial \psi_j^{(3)}}{\partial y} + A_{55} \psi_i^{(3)} \psi_j^{(3)} \right) dx\, dy$$

$$K_{ij}^{45} = \int_{\Omega^e} \left(D_{12} \frac{\partial \psi_i^{(3)}}{\partial x} \frac{\partial \psi_j^{(3)}}{\partial y} + D_{66} \frac{\partial \psi_i^{(3)}}{\partial y} \frac{\partial \psi_j^{(3)}}{\partial x} \right) dx\, dy$$

$$K_{ij}^{55} = \int_{\Omega^e} \left(D_{66} \frac{\partial \psi_i^{(3)}}{\partial x} \frac{\partial \psi_j^{(3)}}{\partial x} + D_{22} \frac{\partial \psi_i^{(3)}}{\partial y} \frac{\partial \psi_j^{(3)}}{\partial y} + A_{44} \psi_i^{(3)} \psi_j^{(3)} \right) dx\, dy$$

$$F_i^1 = \oint_{\Gamma^e} \bar{N}_{nn}\,\psi_i^{(1)}\,ds, \quad F_i^2 = \oint_{\Gamma^e} \bar{N}_{ns}\,\psi_i^{(1)}\,ds$$

$$F_i^3 = \int_{\Omega^e} q\psi_i^{(2)}\,dx\,dy + \oint_{\Gamma^e} Q_n\,\psi_i^{(2)}\,ds, \quad F_i^4 = \oint_{\Gamma^e} \bar{M}_{nn}\,\psi_i^{(3)}\,ds$$

$$F_i^5 = \oint_{\Gamma^e} \bar{M}_{ns}\,\psi_i^{(3)}\,ds, \quad F_i^{1T} = \int_{\Omega^e} \frac{\partial \psi_i^{(1)}}{\partial x} N_{xx}^T\,dxdy, \quad F_i^{2T} = \int_{\Omega^e} \frac{\partial \psi_i^{(1)}}{\partial y} N_{yy}^T\,dxdy$$

$$F_i^{4T} = \int_{\Omega^e} \frac{\partial \psi_i^{(3)}}{\partial x} M_{xx}^T\,dxdy, \quad F_i^{5T} = \int_{\Omega^e} \frac{\partial \psi_i^{(3)}}{\partial y} M_{yy}^T\,dxdy$$

$$\mathbf{K}^{21} = (\mathbf{K}^{12})^{\mathrm{T}}; \quad \mathbf{K}^{43} = (\mathbf{K}^{34})^{\mathrm{T}}; \quad \mathbf{K}^{53} = (\mathbf{K}^{35})^{\mathrm{T}}; \quad \mathbf{K}^{54} = (\mathbf{K}^{45})^{\mathrm{T}}$$
$$\mathbf{K}^{14} = \mathbf{K}^{15} = \mathbf{K}^{24} = \mathbf{K}^{25} = \mathbf{K}^{41} = \mathbf{K}^{42} = \mathbf{K}^{51} = \mathbf{K}^{52} = \mathbf{0} \qquad (7.8.13)$$

Here N_{xx}^T and N_{yy}^T are the thermal forces and M_{xx}^T and M_{yy}^T are the thermal moments. We note that the element stiffness matrix is *not* symmetric. When the bilinear rectangular element is used for all generalized displacements (u, v, w, ϕ_x, ϕ_y), the element stiffness matrices are of the order 20×20; and for the nine-node quadratic element they are 45×45 (see Fig. 7.8.1). The plate bending element of Eq. (7.8.11) is often referred to in the finite element literature as the *Mindlin plate element*, which is labeled in this book as the FSDT element.

7.8.3   Tangent Stiffness Coefficients

The tangent stiffness matrix coefficients for the FSDT can be computed using the definition in Eq. (7.4.9), with $\gamma = 1, 2, \ldots, 5$. Since the source of nonlinearity in the CPT and FSDT is the same, the nonlinear parts of the tangent stiffness coefficients derived for the CPT are also valid for the FSDT.

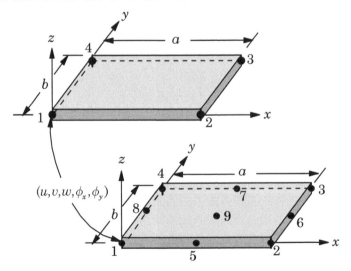

Fig. 7.8.1: Linear and nine-node quadratic rectangular elements for the FSDT.

Suppose that the tangent stiffness matrix is of the same form as the direct stiffness matrix in Eq. (7.8.11). Then the coefficients of the submatrices $\mathbf{T}^{\alpha\beta}$ are defined by

$$T_{ij}^{\alpha\beta} = \frac{\partial R_i^\alpha}{\partial \Delta_j^\beta} \tag{7.8.14}$$

where the components of the residual vector $\mathbf{R}^\alpha$ are given by

$$R_i^\alpha = \sum_{\gamma=1}^{5} \sum_{k=1}^{n(\gamma)} K_{ik}^{\alpha\gamma} \Delta_k^\gamma - F_i^\alpha \tag{7.8.15}$$

$$\Delta_i^1 = u_i, \quad \Delta_i^2 = v_i, \quad \Delta_i^3 = w_i, \quad \Delta_i^4 = S_i^1, \quad \Delta_i^5 = S_i^2 \tag{7.8.16}$$

and $n(\gamma)$ denotes n, m, or p, depending on the nodal degree of freedom. Thus, we have

$$T_{ij}^{\alpha\beta} = \frac{\partial}{\partial \Delta_j^\beta} \left(\sum_{\gamma=1}^{5} \sum_{k=1}^{n(\gamma)} K_{ik}^{\alpha\gamma} \Delta_k^\gamma - F_i^\alpha \right) = \sum_{\gamma=1}^{5} \sum_{k=1}^{n(\gamma)} \frac{\partial K_{ik}^{\alpha\gamma}}{\partial \Delta_j^\beta} \Delta_k^\gamma + K_{ij}^{\alpha\beta} \tag{7.8.17}$$

We note that the only coefficients that depend on the solution are K_{ij}^{13}, K_{ij}^{23}, K_{ij}^{31}, K_{ij}^{32}, and K_{ij}^{33}. Further, they are functions of only w (or functions of w_j). Hence, derivatives of all submatrices with respect to u_j, v_j, S_j^1, and S_j^2 are zero. Thus, we have (for $\alpha = 1, 2, 3, 4, 5$)

$$\begin{aligned}
\mathbf{T}^{\alpha 1} = \mathbf{K}^{\alpha 1} = (\mathbf{K}^{1\alpha})^{\mathrm{T}}, \quad \mathbf{T}^{\alpha 2} = \mathbf{K}^{\alpha 2} = (\mathbf{K}^{2\alpha})^{\mathrm{T}} \\
\mathbf{T}^{\alpha 4} = \mathbf{K}^{\alpha 4} = (\mathbf{K}^{4\alpha})^{\mathrm{T}}, \quad \mathbf{T}^{\alpha 5} = \mathbf{K}^{\alpha 5} = (\mathbf{K}^{5\alpha})^{\mathrm{T}}
\end{aligned} \tag{7.8.18}$$

and $\mathbf{T}^{\alpha 3}$, for $\alpha = 1, 2, 3$, are calculated as follows:

$$\begin{aligned}
T_{ij}^{13} &= \sum_{\gamma=1}^{5} \sum_{k=1}^{n(\gamma)} \frac{\partial K_{ik}^{1\gamma}}{\partial w_j} \Delta_k^\gamma + K_{ij}^{13} = \sum_{k=1}^{n} \frac{\partial K_{ik}^{13}}{\partial w_j} w_k + K_{ij}^{13} \\
&= \frac{1}{2} \int_{\Omega^e} \left[\frac{\partial \psi_i^{(1)}}{\partial x} \left(A_{11} \frac{\partial w}{\partial x} \frac{\partial \psi_j^{(2)}}{\partial x} + A_{12} \frac{\partial w}{\partial y} \frac{\partial \psi_j^{(2)}}{\partial y} \right) \right. \\
&\quad \left. + A_{66} \frac{\partial \psi_i^{(1)}}{\partial y} \left(\frac{\partial w}{\partial x} \frac{\partial \psi_j^{(2)}}{\partial y} + \frac{\partial w}{\partial y} \frac{\partial \psi_j^{(2)}}{\partial x} \right) \right] dx\, dy + K_{ij}^{13} \\
&= K_{ij}^{13} + K_{ij}^{13} = 2K_{ij}^{13} = T_{ji}^{31}, \quad (\mathbf{T}^{13})^{\mathrm{T}} = \mathbf{T}^{31} = \mathbf{K}^{31}
\end{aligned} \tag{7.8.19}$$

$$\begin{aligned}
T_{ij}^{23} &= \sum_{\gamma=1}^{5} \sum_{k=1}^{n(\gamma)} \frac{\partial K_{ik}^{2\gamma}}{\partial w_j} \Delta_k^\gamma + K_{ij}^{23} = \sum_{k=1}^{n} \frac{\partial K_{ik}^{23}}{\partial w_j} w_k + K_{ij}^{23} \\
&= \frac{1}{2} \int_{\Omega^e} \left[\frac{\partial \psi_i^{(1)}}{\partial y} \left(A_{12} \frac{\partial w}{\partial x} \frac{\partial \psi_j^{(2)}}{\partial x} + A_{22} \frac{\partial w}{\partial y} \frac{\partial \psi_j^{(2)}}{\partial y} \right) \right.
\end{aligned}$$

$$+ A_{66} \frac{\partial \psi_i^{(1)}}{\partial x} \left(\frac{\partial w}{\partial y} \frac{\partial \psi_j^{(2)}}{\partial x} + \frac{\partial w}{\partial x} \frac{\partial \psi_j^{(2)}}{\partial y} \right) \Bigg] dx\, dy + K_{ij}^{23}$$

$$= K_{ij}^{23} + K_{ij}^{23} = 2K_{ij}^{23} = T_{ji}^{32}, \quad (\mathbf{T}^{23})^{\mathrm{T}} = \mathbf{T}^{32} = \mathbf{K}^{32} \qquad (7.8.20)$$

$$T_{ij}^{33} = \sum_{\gamma=1}^{5} \sum_{k=1}^{n(\gamma)} \frac{\partial K_{ik}^{3\gamma}}{\partial w_j} \Delta_k^{\gamma} + K_{ij}^{33} = \sum_{k=1}^{n(\gamma)} \left(\frac{\partial K_{ik}^{31}}{\partial w_j} u_k + \frac{\partial K_{ik}^{32}}{\partial w_j} v_k + \frac{\partial K_{ik}^{33}}{\partial w_j} w_k \right) + K_{ij}^{33}$$

$$= \int_{\Omega^e} \left(A_{55} \frac{\partial \psi_i^{(2)}}{\partial x} \frac{\partial \psi_j^{(2)}}{\partial x} + A_{44} \frac{\partial \psi_i^{(2)}}{\partial y} \frac{\partial \psi_j^{(2)}}{\partial y} + k \psi_i^{(2)} \psi_j^{(2)} \right) dx\, dy$$

$$+ \int_{\Omega^e} \left\{ N_{xx} \frac{\partial \psi_i^{(2)}}{\partial x} \frac{\partial \psi_j^{(2)}}{\partial x} + N_{yy} \frac{\partial \psi_i^{(2)}}{\partial y} \frac{\partial \psi_j^{(2)}}{\partial y} + N_{xy} \frac{\partial \psi_i^{(2)}}{\partial x} \frac{\partial \psi_j^{(2)}}{\partial y} \right.$$

$$+ N_{xy} \frac{\partial \psi_i^{(2)}}{\partial y} \frac{\partial \psi_j^{(2)}}{\partial x} + \left(A_{12} + A_{66} \right) \frac{\partial w}{\partial x} \frac{\partial w}{\partial y} \left(\frac{\partial \psi_i^{(2)}}{\partial x} \frac{\partial \psi_j^{(2)}}{\partial y} + \frac{\partial \psi_i^{(2)}}{\partial y} \frac{\partial \psi_j^{(2)}}{\partial x} \right)$$

$$+ \left[A_{11} \left(\frac{\partial w}{\partial x} \right)^2 + A_{66} \left(\frac{\partial w}{\partial y} \right)^2 \right] \frac{\partial \psi_i^{(2)}}{\partial x} \frac{\partial \psi_j^{(2)}}{\partial x}$$

$$+ \left[A_{66} \left(\frac{\partial w}{\partial x} \right)^2 + A_{22} \left(\frac{\partial w}{\partial y} \right)^2 \right] \frac{\partial \psi_i^{(2)}}{\partial y} \frac{\partial \psi_j^{(2)}}{\partial y} \right\} dx\, dy \qquad (7.8.21)$$

where (N_{xx}, N_{yy}, N_{xy}) are given by Eq. (7.7.20). We note that the tangent stiffness matrix of the FSDT element is symmetric.

7.8.4   Shear and Membrane Locking

The C^0 plate bending elements based on the FSDT are among the simplest available in the literature. Unfortunately, when lower order (quadratic or less) equal interpolation of the generalized displacements is used, the elements become excessively stiff in the thin plate limit, yielding displacements that are too small compared to the true solution. As discussed earlier for beams, this type of behavior is known as *shear locking*. There are a number of papers on the subject of shear locking and elements developed to alleviate the problem (see [23–30]). A commonly used technique to avoid shear locking and membrane locking is to use selective integration: use full integration to evaluate all linear stiffness coefficients and use reduced integration to evaluate the transverse shear stiffnesses, i.e. all coefficients in $K_{ij}^{\alpha\beta}$ that contain A_{44} and A_{55} and nonlinear stiffnesses. Higher-order elements or refined meshes of lower-order elements experience relatively less locking, but sometimes at the expense of the rate of convergence. With the suggested Gauss rule, highly distorted elements tend to have slower rates of convergence but they give sufficiently accurate results.

7.9 Computer Implementation and Numerical Results of the FSDT Elements

7.9.1 Computer Implementation

The FSDT element is quite simple to implement, and the implementation follows the same ideas as discussed in Chapter 6 for single-variable problems in two dimensions and for CPT elements discussed in Section 7.5. The element equations are rearranged using the Fortran statements included in Box 5.2.1. The element displacement vector is of the form

$$\mathbf{\Delta}^{\mathrm{T}} = \{u_1, v_1, w_1, \phi_{x1}, \phi_{y1}, u_2, v_2, w_2, \phi_{x2}, \phi_{y2}, \ldots\}$$

Two separate do-loops on Gauss quadrature are required to compute all the force and stiffness coefficients. The full integration loop is used to evaluate all force components and all linear stiffnesses except for the transverse shear stiffnesses. The reduced integration loop has two parts: one for the transverse shear terms and the other for nonlinear terms. One may use separate arrays (say, $\mathrm{EXT13}(I, J)$, $\mathrm{EXT23}(I, J)$, $\mathrm{EXT33}(I, J)$,) to store the extra terms to those of $K_{ij}^{\alpha\beta}$ to obtain the tangent stiffness coefficients.

The stresses in the FSDT are computed in the same way as discussed in Section 7.5.3, except that the strains are defined by Eq. (7.7.4). Also note that in the FSDT the transverse shear stresses computed through the constitutive equations are not zero.

7.9.2 Results of Linear Analysis

In this section, the effect of the integration rule and the convergence characteristics of the FSDT plate element based on equal interpolation (linear or quadratic elements) of all variables is illustrated through several examples. All meshes used consist of either linear or quadratic elements, and they are uniform (i.e. same size elements are used). A shear correction factor of $K_s = 5/6$ is used.

We note that the submatrices $\mathbf{K}^{\alpha\beta}$ appearing in Eq. (7.8.11) for the linear case are readily available, for verification of the element stiffness matrices, from $\mathbf{S}^{\alpha\beta}$ [see Eq. (3.7.10)]. For example, we have $\mathbf{K}^{11} = A_{11}\,\mathbf{S}^{11} + A_{66}\,\mathbf{S}^{22}$, $\mathbf{K}^{12} = A_{12}\,\mathbf{S}^{12} + A_{66}\,\mathbf{S}^{21}$, $\mathbf{K}^{22} = A_{66}\,\mathbf{S}^{11} + A_{22}\,\mathbf{S}^{22}$, $\mathbf{K}^{34} = A_{55}\,\mathbf{S}^{10}$, $\mathbf{K}^{35} = A_{44}\,\mathbf{S}^{20}$, and so on. Also at every boundary node, one element of each of the following pairs should be known [see Eq. (7.7.17)]: (u_n, N_{nn}), (u_s, N_{ns}), (w, Q_n), (ϕ_n, M_{nn}), and (ϕ_s, M_{ns}).

Example 7.9.1 ───

Consider a simply-supported (SS–1) isotropic ($\nu = 0.25$ and $K_s = 5/6$) square plate under uniformly distributed transverse load q_0. The geometric boundary conditions of the computational domain (see the shaded quadrant in Fig. 7.9.1) are

$$u = \phi_x = 0 \text{ at } x = 0; \quad v = \phi_y = 0 \text{ at } y = 0 \quad \text{(along the lines of symmetry)} \tag{1}$$

$$v = w = \phi_y = 0 \text{ at } x = a/2; \quad u = w = \phi_x = 0 \text{ at } y = b/2 \tag{2}$$

Investigate the effect of integration rules and mesh on shear locking.

Solution: The following non-dimensional quantities are used:

$$\bar{w} = w(0,0)\frac{E_2 h^3}{a^4 q_0}, \qquad \bar{\sigma}_{xx} = \sigma_{xx}(0,0,\frac{h}{2})\frac{h^2}{b^2 q_0}, \qquad \bar{\sigma}_{yy} = \sigma_{yy}(0,0,\frac{h}{2})\frac{h^2}{b^2 q_0} \tag{3}$$

$$\bar{\sigma}_{xy} = \sigma_{xy}(\frac{a}{2},\frac{b}{2},-\frac{h}{2})\frac{h^2}{b^2 q_0}, \qquad \bar{\sigma}_{xz} = \sigma_{xz}(\frac{a}{2},0,-\frac{h}{2})\frac{h}{b q_0}, \qquad \bar{\sigma}_{yz} = \sigma_{yz}(0,\frac{b}{2},\frac{h}{2})\frac{h}{b q_0}$$

where the origin of the coordinate system is taken at the center of the plate, $0 \le x \le a/2, 0 \le y \le b/2$, and $-h/2 \le z \le h/2$. The stresses in the finite element analysis are computed at the reduced Gauss points, irrespective of the Gauss rule used for the evaluation of the element stiffness coefficients. The Gauss point coordinates A and B are shown in Table 7.9.1.

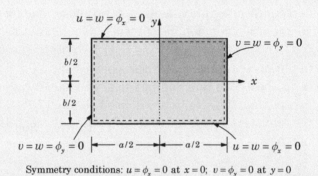

Symmetry conditions: $u = \phi_x = 0$ at $x = 0$; $v = \phi_y = 0$ at $y = 0$

Fig. 7.9.1: Geometric boundary conditions used in the SS–1 type simply-supported rectangular plates.

Table 7.9.1: The Gauss point locations at which the stresses are computed in the finite element analysis of simply-supported plates.

Point	2L	4L	8L	1Q9	2Q9	4Q9
A	0.125a	0.0625a	0.03125a	0.10566a	0.05283a	0.02642a
B	0.375a	0.4375a	0.46875a	0.39434a	0.44717a	0.47358a

The finite element solutions are compared in Table 7.9.2 with the analytical solutions from [3, 56] for two side-to-thickness ratios $a/h = 10$ and 100. The notation nL stands for $n \times n$ uniform mesh of linear rectangular elements and $nQ9$ for $n \times n$ uniform mesh of nine-node quadratic elements in a quarter plate. The notation used in Table 7.9.2 is as follows: F = full integration; R = reduced integration; S = selective integration: full integration of all except the transverse shear coefficients, which are evaluated using reduced integration rule. The CPT solution is independent of side-to-thickness ratio, a/h. The stresses are evaluated at the Gauss points

$$\sigma_{xx}(A,A,\frac{h}{2}), \quad \sigma_{xy}(B,B,-\frac{h}{2}), \quad \sigma_{xz}(B,A,-\frac{h}{2}) \tag{4}$$

Thus, as mesh is refined or higher-order elements are used, the Gauss point locations get closer to the points at which the analytical solutions for stresses are evaluated.

Although the results are presented in non-dimensional form (i.e. the actual values of a, h, q_0, and E_2 do not affect the non-dimensional results), the following data are used to compute

Table 7.9.2: Effect of integration rule on the linear deflections $\bar{w}$ and stresses $\bar{\sigma}$ of simply-supported isotropic square plates under a uniform transverse load q_0.

a/h	Mesh	$\bar{w} \times 10^2$	$\bar{\sigma}_{xx}$	$\bar{\sigma}_{xy}$	$\bar{\sigma}_{xz}$
	2L–F	2.4742	0.1185	0.0727	0.2627
	2L–S	4.7120	0.2350	0.1446	0.2750
	2L–R	4.8887	0.2441	0.1504	0.2750
	1Q–F	4.5304	0.2294	0.1610	0.2813
	1Q–S	4.9426	0.2630	0.1639	0.2847
	1Q–R	4.9711	0.2645	0.1652	0.2886
Finite	4L–F	3.8835	0.2160	0.1483	0.3366
element	4L–S	4.7728	0.2661	0.1850	0.3356
solution	4L–R	4.8137	0.2684	0.1869	0.3356
(FSDT)	2Q–F	4.7707	0.2699	0.1930	0.3437
10	2Q–S	4.7989	0.2715	0.1939	0.3424
	2Q–R	4.8005	0.2716	0.1943	0.3425
	8L–F	4.5268	0.2590	0.1891	0.3700
	8L–S	4.7966	0.2743	0.2743	0.2014
	8L–R	4.7866	0.2737	0.2737	0.2008
	4Q–F	4.7897	0.2749	0.2044	0.3737
	4Q–S	4.7916	0.2750	0.2043	0.3735
	4Q–R	4.7917	0.2750	0.2044	0.3735
FSDT anal. soln. [56]		4.7914	0.2762	0.2085	0.3927
	2L–F	0.0469	0.0024	0.0014	0.2635
	2L–S	4.4645	0.2350	0.1446	0.2750
	2L–R	4.6412	0.2441	0.1504	0.2750
	1Q–F	4.0028	0.2040	0.1591	0.2733
	1Q–S	4.7196	0.2629	0.1643	0.2837
	1Q–R	4.7483	0.2645	0.1652	0.2886
Finite	4L–F	0.1819	0.0108	0.0071	0.3462
element	4L–S	4.5481	0.2661	0.1850	0.3356
solution	4L–R	4.5890	0.2684	0.1869	0.3356
(FSDT)	2Q–F	4.4822	0.2644	0.1893	0.3485
100	2Q–S	4.5799	0.2715	0.1941	0.3414
	2Q–R	4.5815	0.2716	0.1943	0.3425
	8L–F	0.6497	0.0401	0.0275	0.3847
	8L–S	4.5664	0.2737	0.2008	0.3691
	8L–R	4.5764	0.2743	0.2014	0.3691
	4Q–F	4.5530	0.2741	0.2020	0.3749
	4Q–S	4.5728	0.2750	0.2044	0.3734
	4Q–R	4.5729	0.2750	0.2044	0.3735
(CPT)	4 × 4C	4.5952	0.2637	0.1935	–
	8 × 8C	4.5734	0.2732	0.2040	–
FSDT anal. soln. [56]		4.5698	0.2762	0.2085	0.3927

the plate stiffnesses (A_{ij}, D_{ij}), element matrix coefficients $(\mathbf{K}^{(1)}, \mathbf{f}^{(1)})$ for 4×4 mesh of linear elements in a quadrant (using 2×2 for full integration and one-point for reduced integration rules), which may be helpful to the readers in verifying their computer programs:

$$a = 1.0, \quad h = 0.01, \quad E = 10^6, \quad \nu = 0.25, \quad K_s = 5/6, \quad q_0 = 1 \tag{5}$$

The values of the extensional and bending stiffnesses are $A_{11} = A_{22} = 1.0667 \times 10^4$, $A_{12} = 0.2667 \times 10^4$, $A_{44} = A_{55} = 0.4 \times 10^4 K_s$, $A_{66} = 0.4 \times 10^4$, $D_{11} = D_{22} = 8.8889 \times 10^{-2}$, $D_{12} = 2.2222 \times 10^{-2}$, and $D_{66} = 3.3333 \times 10^{-2}$; selective nonzero coefficients of the stiffness matrix and force vector of element 1 are

$$K_{11}^{(1)} = 0.48889 \times 10^4, \quad K_{12}^{(1)} = 0.16667 \times 10^4, \quad K_{16}^{(1)} = -0.28889 \times 10^4$$

$$K_{17}^{(1)} = -0.33333 \times 10^3, \quad K_{1(11)}^{(1)} = -0.24444 \times 10^4, \quad K_{1(12)}^{(1)} = -0.16667 \times 10^4$$

$$K_{1(16)}^{(1)} = 0.44444 \times 10^3, \quad K_{1(17)}^{(1)} = 0.33333 \times 10^3, \quad K_{22}^{(1)} = K_{11}^{(1)}, \quad K_{26}^{(1)} = -K_{17}^{(1)}$$

$$K_{27}^{(1)} = K_{1(16)}^{(1)}, \quad K_{2(11)}^{(1)} = K_{1(12)}^{(1)}, \quad K_{2(12)}^{(1)} = -0.24444 \times 10^4, \quad K_{2(16)}^{(1)} = K_{17}^{(1)}$$

$$K_{2(17)}^{(1)} = K_{16}^{(1)}, \quad K_{33}^{(1)} = K_{12}^{(1)}, \quad K_{34}^{(1)} = -0.52083 \times 10^2, \quad K_{44}^{(1)} = 3.2959 \tag{6}$$

$$K_{(15)(10)}^{(1)} = 3.2311, \quad K_{(15)5}^{(1)} = 3.2348 = K_{(14)4}^{(1)}, \quad K_{(20)(14)}^{(1)} = -0.27778 \times 10^{-2}$$

$$K_{(20)(15)}^{(1)} = 3.2589, \quad K_{(20)(18)}^{(1)} = 0.52083 \times 10^2, \quad K_{(20)(19)}^{(1)} = -0.013889$$

$$K_{(20)(20)}^{(1)} = 3.2959, \quad f_3^{(1)} = f_8^{(1)} = f_{13}^{(1)} = f_{18}^{(1)} = 0.39063 \times 10^{-2}$$

The results of Table 7.9.2 indicate that the FSDT finite element with equal interpolation of all generalized displacements does not experience shear locking for thick plates even when the full integration rule is used for the evaluation of all stiffness coefficients. Shear locking is evident when the element is used to model thin plates ($a/h \geq 100$) with the full integration rule (F). Also, higher-order elements are less sensitive to locking but exhibit slower convergence. The element behaves well for thin and thick plates when the reduced (R) or selectively reduced integration (S) rule is used, with the selective integration rule being the best.

Example 7.9.2

Analyze a clamped isotropic ($\nu = 0.25$ and $K_s = 5/6$) square plate under uniformly distributed transverse load, using various types of meshes.

Solution: The non-dimensional quantities used are the same as in Eq. (3) of Example 7.9.1, except for the location of the stresses. The stresses were non-dimensionalized as follows:

$$\bar{\sigma}_{xx} = \sigma_{xx}(a/2, 0, -h/2)(h^2/b^2 q_0), \quad \bar{\sigma}_{xz} = \sigma_{xz}(a/2, 0, -h/2)(h/b q_0) \tag{1}$$

$$\bar{\sigma}_{xx} = \sigma_{xx}(A, B, -h/2)(h^2/b^2 q_0), \quad \bar{\sigma}_{xz} = \sigma_{xz}(A, B, -h/2)(h/b q_0) \tag{2}$$

where the values of A and B for different meshes are given in Table 7.9.3. Table 7.9.4 contains the non-dimensionalized displacements and stresses. The FSDT element with selective integration or reduced integration is accurate in predicting the bending response.

Table 7.9.3: The Gauss point locations at which the stresses are computed in the finite element analysis of clamped plates.

Point	4L	8L	2Q9	4Q9
A	$0.4375a$	$0.46875a$	$0.44717a$	$0.47358a$
B	$0.0625a$	$0.03125a$	$0.05283a$	$0.02642a$

Table 7.9.4: Effect of integration rule on the linear deflections $\bar{w}$ and stresses $\bar{\sigma}$ of clamped, isotropic, square plates under uniform transverse load q_0; the CPT(C) solution for $a/h = 100$ and 4×4 mesh is: $\bar{w} \times 10^2 = 1.4370$ and $\bar{\sigma}_{xx} = 0.1649$.

a/h	Integ.	Variable	4×4L	2×2Q9	8×8L	4×4Q9
10	F	$\bar{w} \times 10^2$	1.2593	1.5983	1.5447	1.6685
		$\bar{\sigma}_{xx}$	0.1190	0.1568	0.2054	0.2301
		$\bar{\sigma}_{xz}$	0.3890	0.4193	0.4463	0.4578
10	S	$\bar{w} \times 10^2$	1.6632	1.6880	1.6721	1.6758
		$\bar{\sigma}_{xx}$	0.1689	0.1813	0.2275	0.2357
		$\bar{\sigma}_{xz}$	0.4056	0.4118	0.4511	0.4566
10	R	$\bar{w} \times 10^2$	1.6854	1.6903	1.6776	1.6760
		$\bar{\sigma}_{xx}$	0.1718	0.1817	0.2204	0.2358
		$\bar{\sigma}_{xz}$	0.4045	0.4120	0.4509	0.4566
100	F	$\bar{w} \times 10^2$	0.0386	1.1222	1.3982	1.3546
		$\bar{\sigma}_{xx}$	0.0041	0.0947	0.0204	0.1825
		$\bar{\sigma}_{xz}$	0.3734	0.4732	0.4229	0.4862
100	S	$\bar{w} \times 10^2$	1.4093	1.4382	1.4219	1.4268
		$\bar{\sigma}_{xx}$	0.1731	0.1846	0.2337	0.2417
		$\bar{\sigma}_{xz}$	0.4236	0.4255	0.4729	0.4776
100	R	$\bar{w} \times 10^2$	1.4334	1.4417	1.4279	1.4271
		$\bar{\sigma}_{xx}$	0.1762	0.1853	0.2346	0.2418
		$\bar{\sigma}_{xz}$	0.4234	0.4284	0.4727	0.4779

7.9.3   Results of Nonlinear Analysis

Here we consider several examples of nonlinear bending of rectangular plates using the nonlinear FSDT element. The effect of the integration rule to evaluate the nonlinear and transverse shear stiffness coefficients is investigated in Example 7.9.3. Unless stated otherwise, a uniform mesh of 4×4 nine-node quadratic elements is used in a quarter plate for the FSDT. For this choice of mesh, full integration (F) means the 3×3 Gauss rule and reduced integration (R) means the 2×2 Gauss rule. Stresses are calculated at the center of the element.

Example 7.9.3 ——————————————————————————

Consider a square isotropic plate with

$$a = b = 10 \text{ in}, \quad h = 1 \text{ in}, \quad E = 7.8 \times 10^6 \text{ psi}, \quad \nu = 0.3, \quad K_s = 5/6 \qquad (1)$$

and subjected to a uniformly distributed transverse load of intensity q_0. Two types of simply-supported boundary conditions, namely, SS–1 and SS–3, are to be studied (shown in Figs. 7.9.1 and 7.9.2, respectively). The geometric boundary conditions are

$$\text{SS–1: At } x = a/2: \quad v = w = \phi_y = 0; \quad \text{At } y = b/2: \quad u = w = \phi_x = 0 \qquad (2)$$
$$\text{SS–3: } u = v = w = 0 \text{ on simply-supported edges} \qquad (3)$$

The boundary conditions along the symmetry lines for both cases are given in Eq. (1) of Example 7.9.1.

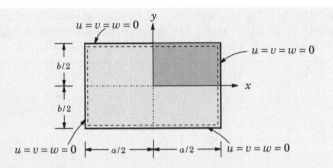

Symmetry conditions: $u = \phi_x = 0$ at $x = 0$; $v = \phi_y = 0$ at $y = 0$

Fig. 7.9.2: Geometric boundary conditions used in the SS–3 type simply-supported rectangular plates.

Solution: Using the load parameter introduced earlier, $P \equiv q_0 a^4 / E h^4$, the incremental load vector is chosen to be

$$\{\Delta P\} = \{6.25,\ 6.25,\ 12.5,\ 25.0,\ 25.0,\ \ldots,\ 25.0\}$$

A tolerance of $\epsilon = 10^{-2}$ is used for convergence in the Newton iteration scheme to check for convergence of the nodal displacements.

Table 7.9.5 contains the deflections $w(0,0)$ obtained with a uniform mesh of 4×4Q9 FSDT elements for various integration rules (also see Fig. 7.9.3). The number of iterations taken for convergence are listed in parentheses. The linear FSDT plate solution for load $q_0 = 4875$ psi (or $P = 6.25$) is $w = 0.2917$ in for SS–1 and $w = 0.3151$ in for SS–3 (for the R-R integration rule). The linear solution, being independent of the boundary conditions on u and v, of the SS–1 plate is smaller than that of the SS–3 plate. In the nonlinear case, the SS–3 type boundary conditions provide more edge restraint than the SS–1 boundary conditions and, therefore, would produce smaller transverse deflections. As discussed earlier, the 4×4Q9 meshes are not sensitive to shear or membrane locking, and therefore the results obtained with various integration rules are essentially the same.

Table 7.9.5: Center deflection w of simply-supported (SS–1 and SS–3) plates under uniformly distributed load q_0.

	SS–3			SS–1		
P	R–R*	F–R	F–F	R–R	F–R	F–F
6.25	0.2790 (4)	0.2790 (4)	0.2780 (3)	0.2813 (3)	0.2813 (3)	0.2812 (3)
12.5	0.4630 (3)	0.4630 (3)	0.4619 (3)	0.5186 (3)	0.5186 (3)	0.5185 (3)
25.0	0.6911 (3)	0.6911 (3)	0.6902 (3)	0.8673 (4)	0.8673 (4)	0.8672 (4)
50.0	0.9575 (3)	0.9575 (3)	0.9570 (3)	1.3149 (4)	1.3149 (4)	1.3147 (4)
75.0	1.1333 (3)	1.1333 (3)	1.1330 (3)	1.6241 (3)	1.6239 (3)	1.6237 (3)
100.0	1.2688 (3)	1.2688 (3)	1.2686 (3)	1.8687 (3)	1.8683 (3)	1.8679 (3)
125.0	1.3809 (2)	1.3809 (2)	1.3808 (2)	2.0758 (2)	2.0751 (2)	2.0746 (2)
150.0	1.4774 (2)	1.4774 (2)	1.4774 (2)	2.2567 (2)	2.2556 (2)	2.2549 (2)
175.0	1.5628 (2)	1.5629 (2)	1.5629 (2)	2.4194 (2)	2.4177 (2)	2.4168 (2)
200.0	1.6398 (2)	1.6399 (2)	1.6399 (2)	2.5681 (2)	2.5657 (2)	2.5645 (2)

* The first letter refers to the integration rule used for the nonlinear terms while the second letter refers to the integration rule used for the shear terms; R = reduced; F = full integration.

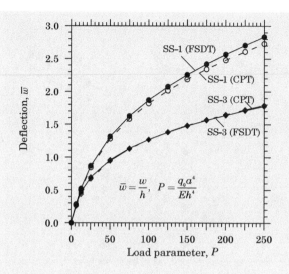

Fig. 7.9.3: Plots of center deflection w versus load parameter P for isotropic ($\nu = 0.3$), simply-supported square plates under uniformly distributed transverse load (4×4Q9 for FSDT and 8×8C for CPT).

Table 7.9.6 contains the normal stresses $\bar{\sigma}_{xx} = \sigma_{xx}(a^2/Eh^2)$ obtained with a uniform mesh of 4×4Q9 FSDT elements for various integration rules (also see Fig. 7.9.4).

Table 7.9.6: Center stresses $\bar{\sigma}_{xx}$ of simply-supported (SS–1 and SS–3) plates under uniformly distributed load q_0.

	SS–3			SS–1		
P	R–R	F–R	F–F	R–R	F–R	F–F
6.25	1.861	1.861	1.856	1.779	1.779	1.780
12.5	3.305	3.305	3.300	3.396	3.396	3.398
25.0	5.319	5.320	5.317	5.882	5.882	5.885
50.0	8.001	8.002	8.001	9.159	9.162	9.165
75.0	9.983	9.984	9.983	11.458	11.462	11.465
100.0	11.633	11.634	11.634	13.299	13.307	13.308
125.0	13.084	13.085	13.085	14.878	14.890	14.889
150.0	14.396	14.398	14.398	16.278	16.293	16.290
175.0	15.608	15.610	15.610	17.553	17.572	17.567
200.0	16.741	16.743	16.743	18.733	18.755	18.748

Example 7.9.4

Consider an orthotropic square plate subjected to a uniformly distributed transverse load of intensity q_0. The geometric and material parameters to be used are

$$a = b = 12 \text{ in}, \quad h = 0.138 \text{ in}, \quad E_1 = 3 \times 10^6 \text{ psi}, \quad E_2 = 1.28 \times 10^6 \text{ psi}$$
$$G_{12} = G_{13} = G_{23} = 0.37 \times 10^6 \text{psi}, \quad \nu_{12} = 0.32, \quad K_s = 5/6 \tag{1}$$

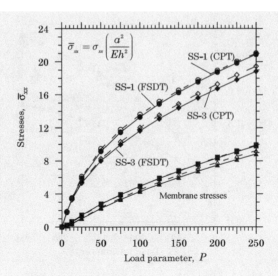

Fig. 7.9.4: Plots of center normal stress $\bar{\sigma}_{xx}$ versus load parameter P for isotropic ($\nu = 0.3$), simply-supported square plates under uniformly distributed transverse load ($4 \times 4Q9$ for FSDT and $8 \times 8C$ for CPT).

Use a uniform mesh of $4 \times 4Q9$ elements in a quadrant, and analyze the problem for both SS–1 and SS–3 boundary conditions. Choose the incremental load vector to be $\{\Delta q_0\} = \{0.05, 0.05, 0.1, 0.2, 0.2, \ldots, 0.2\}$ with twelve load steps and error tolerance of $\epsilon = 0.01$.

Solution: Table 7.9.7 contains the center deflection and total normal stress as a function of the load for the two boundary conditions. The linear FSDT solution for load $q_0 = 0.05$ is $w = 0.01132$ for SS–1 and $w = 0.01140$ for SS–3. Plots of load q_0 (psi) versus center deflection w (in) and q_0 versus normal stress (total as well as membrane) $\bar{\sigma}_{xx} = \sigma_{xx}(a^2/E_2h^2)$ are shown in Figs. 7.9.5 and 7.9.6 for SS–1 and SS–3 plates, respectively. The figures also show the results obtained using 8×8 mesh of conforming CPT elements.

Table 7.9.7: Center deflection w and normal stress $\bar{\sigma}_{xx}$ for simply-supported orthotropic square plates under uniformly distributed load ($4 \times 4Q9$).

	SS–1			SS–3		
q_0	CPT	FSDT	FSDT	CPT	FSDT	FSDT
	w	w	$\bar{\sigma}_{xx}$	w	w	$\bar{\sigma}_{xx}$
0.05	0.0113 (2)	0.0113	1.034	0.0112	0.0113	1.056
0.10	0.0224 (2)	0.0224	2.070	0.0217	0.0218	2.116
0.20	0.0438 (3)	0.0439	4.092	0.0395	0.0397	4.058
0.40	0.0812 (3)	0.0815	7.716	0.0648	0.0650	7.103
0.60	0.1116 (3)	0.1122	10.702	0.0823	0.0824	9.406
0.80	0.1367 (3)	0.1377	13.169	0.0957	0.0959	11.284
1.00	0.1581 (2)	0.1594	15.255	0.1068	0.1069	12.894
1.20	0.1767 (2)	0.1783	17.050	0.1162	0.1162	14.316
1.40	0.1932 (2)	0.1951	18.631	0.1245	0.1244	15.602
1.60	0.2081 (2)	0.2103	20.044	0.1318	0.1318	16.783
1.80	0.2217 (2)	0.2241	21.324	0.1385	0.1384	17.880
2.00	0.2343 (2)	0.2370	22.495	0.1447	0.1445	18.909

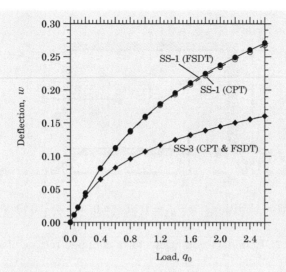

Fig. 7.9.5: Center deflection $w(0,0)$ as a function of the load q_0 for simply-supported, orthotropic, square plates under uniformly distributed transverse load.

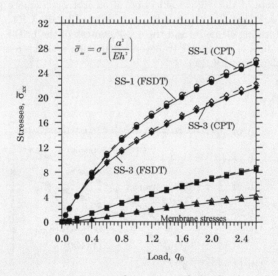

Fig. 7.9.6: Maximum stresses $\bar{\sigma}_{xx}$ as functions of the load q_0 for simply-supported, orthotropic, square plates under uniformly distributed transverse load.

Example 7.9.5

Analyze an orthotropic plate with clamped edges (see Fig. 7.9.7). Assume that the boundary conditions of a clamped edge are

$$u = v = w = \phi_x = \phi_y = 0 \tag{1}$$

Use the geometric and material parameters given in Eq. (1) of Example 7.9.4, and assume a uniformly distributed transverse load of intensity q_0 on the plate. Use shear correction factor of $K_s = 5/6$.

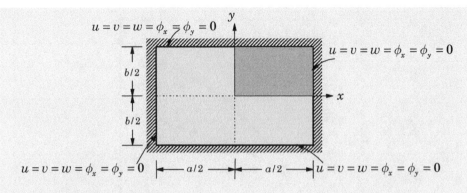

Symmetry conditions: $u = \phi_x = 0$ at $x = 0$; $v = \phi_y = 0$ at $y = 0$

Fig. 7.9.7: Boundary conditions for a clamped rectangular plate.

Solution: The linear transverse deflection at the center of the plate is $w(0,0) = 0.0301$ for load $q_0 = 0.5$. Figure 7.9.8 contains a plot of load q_0 versus center deflection $w(0,0)$. Figure 7.9.8 also contains plots of the center deflection of an isotropic square plate ($h = 0.138$ in, $E = 1.28 \times 10^6$ psi, and $\nu = 0.3$). The results are predicted by the CPT element using a 8×8 mesh of the non-conforming elements and the 4×4Q9 mesh of the FSDT element. Table 7.9.8 contains numerical values of the center transverse deflection w and stress $\bar{\sigma}_{xx}$ as a function of q_0 for the same problem (see [18–20]).

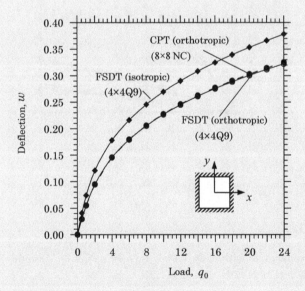

Fig. 7.9.8: Nonlinear center deflection $w(0,0)$ versus load parameter q_0 for clamped, isotropic and orthotropic, square plates under uniform load.

Table 7.9.8: Center deflection w and normal stress $\bar{\sigma}_{xx}$ for clamped orthotropic square plates under uniformly distributed load (4×4Q9).

q_0	w	$\bar{\sigma}_{xx}$	q_0	w	$\bar{\sigma}_{xx}$
0.5	0.0294 (3)	4.317	12.0	0.2450 (2)	46.001
1.0	0.0552 (3)	8.467	14.0	0.2610 (2)	49.851
2.0	0.0948 (3)	15.309	16.0	0.2754 (2)	53.431
4.0	0.1456 (3)	24.811	18.0	0.2886 (2)	56.800
6.0	0.1795 (3)	31.599	20.0	0.3006 (2)	59.998
8.0	0.2054 (3)	37.078	22.0	0.3119 (2)	63.053
10.0	0.2268 (2)	41.793	24.0	0.3224 (2)	65.986

Example 7.9.6

Consider the nonlinear bending of a clamped, isotropic ($E = 10^6$ and $\nu = 0.3$), circular plate (radius $a = 100$ in and thickness $h = 10$ in) under uniformly distributed transverse load of intensity q_0. Due to the symmetry, the plate experiences axisymmetric deformation; that is, every radial line experiences the same deformation. However, to use the plate elements, one may use any sector of the plate. The geometric boundary condition along a radial line oriented at an angle other than $0°$ or $90°$ to the x or y axis is that the inplane displacement normal to the edge (i.e. u_n) be zero (due to the symmetry), which presents difficulty because u_n is not one of the degrees of freedom of the element. Analyze the problem using the computational domain (a quadrant of the plate) and the finite element mesh and boundary conditions shown in Fig. 7.9.9, with a load step of $\Delta q_0 = 2.5$ lb and a total load of $q_0 = 120$ lb. Plot the center deflection w and stress σ_{xx} as a function of the load.

Solution: The exact center deflection for the linear case, for the two theories, is given by (see Reddy [3,56])

$$w^{\text{FSDT}}(0) = \frac{q_0 a^4}{64D}\left(1 + \frac{8}{3(1-\nu)K_s}\frac{h^2}{a^2}\right), \quad w^{\text{CPT}}(0) = \frac{q_0 a^4}{64D} \tag{1}$$

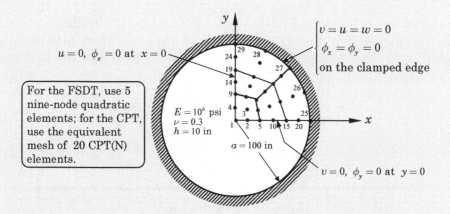

Fig. 7.9.9: Mesh and boundary conditions used for a clamped circular plate.

where a denotes the radius of the plate. For $\nu = 0.3$, $K_s = 5/6$, and $a/h = 10$, the linear FSDT center deflection is given by $w(0) = 1.0457(q_0 a^4/64D)$. The linear FSDT solution obtained with the mesh of five nine-node quadratic elements is $w(0) = 1.0659(q_0 a^4/64D)$ ($< 2\%$ error). Figure 7.9.10 shows the nonlinear load-deflection curve and load versus $\bar{\sigma}_{xx} = \sigma_{xx}(A, A)(a/h)^2/10E$ curve for the problem. The same results can be obtained with an axisymmetric one-dimensional element (see Problem 5.14).

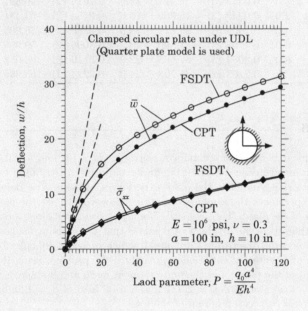

Fig. 7.9.10: Load-deflection curve for clamped isotropic circular plate under uniform load.

7.10   Transient Analysis of the FSDT

7.10.1   Equations of Motion

The equations of motion of FSDT with the von Kármán nonlinearity can be derived using the dynamic version of the principle of virtual displacements (see Reddy [3, 55, 56]). They are the same as Eqs. (7.7.12)–(7.7.16) with inertia terms added, as follows:

$$I_0\frac{\partial^2 u}{\partial t^2} - \frac{\partial N_{xx}}{\partial x} - \frac{\partial N_{xy}}{\partial y} = 0 \qquad (7.10.1)$$

$$I_0\frac{\partial^2 v}{\partial t^2} - \frac{\partial N_{xy}}{\partial x} - \frac{\partial N_{yy}}{\partial y} = 0 \qquad (7.10.2)$$

$$I_0\frac{\partial^2 w}{\partial t^2} - \frac{\partial}{\partial x}\left(N_{xx}\frac{\partial w}{\partial x} + N_{xy}\frac{\partial w}{\partial y}\right) - \frac{\partial}{\partial y}\left(N_{xy}\frac{\partial w}{\partial x} + N_{yy}\frac{\partial w}{\partial y}\right)$$
$$-\frac{\partial Q_x}{\partial x} - \frac{\partial Q_y}{\partial y} - q = 0 \qquad (7.10.3)$$

$$I_2 \frac{\partial^2 \phi_x}{\partial t^2} - \frac{\partial M_{xx}}{\partial x} - \frac{\partial M_{xy}}{\partial y} + Q_x = 0 \qquad (7.10.4)$$

$$I_2 \frac{\partial^2 \phi_y}{\partial t^2} - \frac{\partial M_{xy}}{\partial x} - \frac{\partial M_{yy}}{\partial y} + Q_y = 0 \qquad (7.10.5)$$

where I_0 and I_2 are the principal and rotatory inertias, respectively,

$$I_0 = \rho h, \quad I_2 = \frac{\rho h^3}{12} \qquad (7.10.6)$$

and the stress resultants (N, M, Q), are known in terms of the displacements through Eqs. (7.7.20)–(7.7.22).

7.10.2   The Finite Element Model

The finite element model is developed using equal interpolation of all generalized displacements $(u, v, w, \phi_x, \phi_y)$ [see Eqs. (7.8.8)–(7.8.10)]

$$u(x, y, t) = \sum_{j=1}^{n} \Delta_j^1(t)\psi_j(x, y), \quad v(x, y, t) = \sum_{j=1}^{n} \Delta_j^2(t)\psi_j(x, y)$$

$$w(x, y, t) = \sum_{j=1}^{n} \Delta_j^3(t)\psi_j(x, y) \qquad (7.10.7)$$

$$\phi_x(x, y, t) = \sum_{j=1}^{n} \Delta_j^4(t)\psi_j(x, y), \quad \phi_y(x, y, t) = \sum_{j=1}^{n} \Delta_j^5(t)\psi_j(x, y)$$

where Δ_i^α denotes the value of the αth variable at the ith node with $\Delta_i^1 = u_i$, $\Delta_i^2 = v_i$, $\Delta_i^3 = w_i$, $\Delta_i^4 = (\phi_x)_i = S_i^1$, and $\Delta_i^5 = (\phi_y)_i = S_i^2$.

The weak forms of Eqs. (7.10.1)–(7.10.5), when the weight functions are replaced with the approximation functions ψ_i, are given by

$$0 = \int_{\Omega^e} \left(\frac{\partial \psi_i}{\partial x} N_{xx} + \frac{\partial \psi_i}{\partial y} N_{xy} + I_0 \psi_i \frac{\partial^2 u}{\partial t^2} \right) dx\, dy - \oint_{\Gamma^e} \bar{N}_{nn} \psi_i\, ds$$

$$0 = \int_{\Omega^e} \left(\frac{\partial \psi_i}{\partial x} N_{xy} + \frac{\partial \psi_i}{\partial y} N_{yy} + I_0 \psi_i \frac{\partial^2 v}{\partial t^2} \right) dx\, dy - \oint_{\Gamma^e} \bar{N}_{ns} \psi_i\, ds$$

$$0 = \int_{\Omega^e} \left[\frac{\partial \psi_i}{\partial x} \left(Q_x + N_{xx} \frac{\partial w}{\partial x} + N_{xy} \frac{\partial w}{\partial y} \right) + \frac{\partial \psi_i}{\partial y} \left(Q_y + N_{xy} \frac{\partial w}{\partial x} + N_{yy} \frac{\partial w}{\partial y} \right) \right.$$

$$\left. + I_0 \psi_i \frac{\partial^2 w}{\partial t^2} - \psi_i q \right] dx\, dy - \oint_{\Gamma^e} Q_n \psi_i\, ds \qquad (7.10.8)$$

$$0 = \int_{\Omega^e} \left(\frac{\partial \psi_i}{\partial x} M_{xx} + \frac{\partial \psi_i}{\partial y} M_{xy} + \psi_i Q_x + I_2 \psi_i \frac{\partial^2 \phi_x}{\partial t^2} \right) dx\, dy - \oint_{\Gamma^e} \bar{M}_{nn} \psi_i\, ds$$

$$0 = \int_{\Omega^e} \left(\frac{\partial \psi_i}{\partial x} M_{xy} + \frac{\partial \psi_i}{\partial y} M_{yy} + \psi_i Q_y + I_2 \psi_i \frac{\partial^2 \phi_y}{\partial t^2} \right) dx\, dy - \oint_{\Gamma^e} \bar{M}_{ns} \psi_i\, ds$$

where $(\bar{N}_{nn}, \bar{N}_{ns}, \bar{M}_{nn}, \bar{M}_{ns}, Q_n)$ are the stress resultants on an edge with unit normal $\hat{n}$, and they are defined in Eqs. (7.8.6a,b) and (7.8.7).

Note that the nonlinearity in Eq. (7.10.8) is solely due to w, and the non-linear terms are present only in (N_{xx}, N_{xy}, N_{yy}). For convenience of writing the finite element equations, the linear and nonlinear parts of the inplane forces (N_{xx}, N_{xy}, N_{yy}) are denoted as

$$\mathbf{N} = \mathbf{N}^0 + \mathbf{N}^1 \tag{7.10.9}$$

where $\mathbf{N}^0$ is the linear part and $\mathbf{N}^1$ is the nonlinear part. For an orthotropic plate (with the principal material axes coinciding with the plate axes), the forces and moments are related to the generalized displacements $(u, v, w, \phi_x, \phi_y)$ as follows:

$$N^0_{xx} = A_{11}\frac{\partial u}{\partial x} + A_{12}\frac{\partial v}{\partial y}, \qquad N^0_{yy} = A_{12}\frac{\partial u}{\partial x} + A_{22}\frac{\partial v}{\partial y}$$

$$N^0_{xy} = A_{66}\left(\frac{\partial u}{\partial y} + \frac{\partial v}{\partial x}\right), \qquad M_{xx} = D_{11}\frac{\partial \phi_x}{\partial x} + D_{12}\frac{\partial \phi_y}{\partial y}$$

$$M_{yy} = D_{12}\frac{\partial \phi_x}{\partial x} + D_{22}\frac{\partial \phi_y}{\partial y}, \qquad M_{xy} = D_{66}\left(\frac{\partial \phi_x}{\partial y} + \frac{\partial \phi_y}{\partial x}\right)$$

$$Q_x = A_{55}\left(\phi_x + \frac{\partial w}{\partial x}\right), \qquad Q_y = A_{44}\left(\phi_y + \frac{\partial w}{\partial y}\right) \tag{7.10.10}$$

$$N^1_{xx} = \frac{1}{2}\left[A_{11}\left(\frac{\partial w}{\partial x}\right)^2 + A_{12}\left(\frac{\partial w}{\partial y}\right)^2\right]$$

$$N^1_{yy} = \frac{1}{2}\left[A_{12}\left(\frac{\partial w}{\partial x}\right)^2 + A_{22}\left(\frac{\partial w}{\partial y}\right)^2\right]$$

$$N^1_{xy} = A_{66}\frac{\partial w}{\partial x}\frac{\partial w}{\partial y}$$

and A_{ij} and D_{ij} are the plate stiffnesses [see Eq. (7.7.23)]

$$A_{ij} = hQ_{ij}, \qquad D_{ij} = \frac{h^3}{12}Q_{ij} \tag{7.10.11}$$

and Q_{ij} are defined in Eq. (7.7.24).

By substituting the expansions in Eq. (7.10.10) into the weak forms in Eq. (7.10.8), we obtain the finite element model

$$0 = \sum_{\beta=1}^{5}\sum_{j=1}^{n} M_{ij}^{\alpha\beta}\ddot{\Delta}_j^{\beta} + \sum_{\beta=1}^{5}\sum_{j=1}^{n} K_{ij}^{\alpha\beta}\Delta_j^{\beta} - F_i^{\alpha} \equiv R_i^{\alpha} \tag{7.10.12}$$

for $\alpha = 1, 2, \ldots, 5$, where Δ_i^{α} denotes the value of the αth variable, in the order $(u, v, w, \phi_x, \phi_y)$, at the ith node $(i = 1, 2, \ldots, n)$ of the element. The *non-zero* coefficients of Eq. (7.10.12) are defined by

$$M_{ij}^{11} = \int_{\Omega^e} I_0 \psi_i \psi_j \, dx \, dy; \quad M_{ij}^{44} = \int_{\Omega^e} I_2 \psi_i \psi_j \, dx \, dy$$

$$M_{ij}^{22} = M_{ij}^{33} = M_{ij}^{11}, \quad M_{ij}^{55} = M_{ij}^{44}$$

$$K_{ij}^{1\alpha} = \int_{\Omega^e} \left(\frac{\partial \psi_i}{\partial x} N_{1j}^{\alpha} + \frac{\partial \psi_i}{\partial y} N_{6j}^{\alpha} \right) dx \, dy$$

$$K_{ij}^{2\alpha} = \int_{\Omega^e} \left(\frac{\partial \psi_i}{\partial x} N_{6j}^{\alpha} + \frac{\partial \psi_i}{\partial y} N_{2j}^{\alpha} \right) dx \, dy$$

$$K_{ij}^{3\alpha} = \int_{\Omega^e} \left[\frac{\partial \psi_i}{\partial x} \left(Q_{1j}^{\alpha} + \frac{\partial w}{\partial x} N_{1j}^{\alpha} + \frac{\partial w}{\partial y} N_{6j}^{\alpha} \right) \right.$$

$$\left. + \frac{\partial \psi_i}{\partial y} \left(Q_{2j}^{\alpha} + \frac{\partial w}{\partial x} N_{6j}^{\alpha} + \frac{\partial w}{\partial y} N_{2j}^{\alpha} \right) \right] dx \, dy \qquad (7.10.13)$$

$$K_{ij}^{4\alpha} = \int_{\Omega^e} \left(Q_{1j}^{\alpha} \psi_i + \frac{\partial \psi_i}{\partial x} M_{1j}^{\alpha} + \frac{\partial \psi_i}{\partial y} M_{6j}^{\alpha} \right) dx \, dy$$

$$K_{ij}^{5\alpha} = \int_{\Omega^e} \left(Q_{2j}^{\alpha} \psi_i + \frac{\partial \psi_i}{\partial x} M_{6j}^{\alpha} + \frac{\partial \psi_i}{\partial y} M_{2j}^{\alpha} \right) dx \, dy$$

$$F_i^1 = \oint_{\Gamma^e} \bar{N}_{nn} \psi_i \, ds, \quad F_i^2 = \oint_{\Gamma^e} \bar{N}_{ns} \psi_i \, ds, \quad F_i^4 = \oint_{\Gamma^e} \bar{M}_{nn} \psi_i \, ds$$

$$F_i^3 = \int_{\Omega^e} q \psi_i \, dx \, dy + \oint_{\Gamma^e} Q_n \psi_i \, ds, \quad F_i^5 = \oint_{\Gamma^e} \bar{M}_{ns} \psi_i \, ds$$

and $K_{ij}^{\alpha\beta}$ denotes the coefficient matrix of the αth variable in the βth equation $(\alpha, \beta = 1, 2, 3, 4, 5)$. Thus, N_{ij}^I denotes the contribution from $N_i (= N_i^0 + N_i^1)$ $(i = 1, 2, \ldots, 6)$ to the Ith variable, $(I = 1, 2, \ldots, 5)$ where j denotes the node number. The *non-zero* terms appearing in the definition of $K_{ij}^{\alpha\beta}$ are:

$$M_{1j}^4 = D_{11} \frac{\partial \psi_j}{\partial x}, \quad M_{6j}^4 = D_{66} \frac{\partial \psi_j}{\partial y}, \quad M_{2j}^4 = D_{12} \frac{\partial \psi_j}{\partial x}$$

$$M_{1j}^5 = D_{12} \frac{\partial \psi_j}{\partial y}, \quad M_{6j}^5 = D_{66} \frac{\partial \psi_j}{\partial x}, \quad M_{2j}^5 = D_{22} \frac{\partial \psi_j}{\partial y} \qquad (7.10.14)$$

$$N_{1j}^1 = A_{11} \frac{\partial \psi_j}{\partial x}, \quad N_{6j}^1 = A_{66} \frac{\partial \psi_j}{\partial y}, \quad N_{1j}^2 = A_{12} \frac{\partial \psi_j}{\partial y}$$

$$N_{6j}^2 = A_{66} \frac{\partial \psi_j}{\partial x}, \quad N_{1j}^3 = \frac{1}{2} \left(A_{11} \frac{\partial w}{\partial x} \frac{\partial \psi_j}{\partial x} + A_{12} \frac{\partial w}{\partial y} \frac{\partial \psi_j}{\partial y} \right)$$

$$N_{6j}^3 = \frac{A_{66}}{2} \left(\frac{\partial w}{\partial x} \frac{\partial \psi_j}{\partial y} + \frac{\partial w}{\partial y} \frac{\partial \psi_j}{\partial x} \right), \quad N_{2j}^1 = A_{12} \frac{\partial \psi_j}{\partial x} \qquad (7.10.15)$$

$$N_{2j}^2 = A_{22} \frac{\partial \psi_j}{\partial y}, \quad N_{2j}^3 = \frac{1}{2} \left(A_{12} \frac{\partial w}{\partial x} \frac{\partial \psi_j}{\partial x} + A_{22} \frac{\partial w}{\partial y} \frac{\partial \psi_j}{\partial y} \right)$$

$$Q_{1j}^3 = A_{55} \frac{\partial \psi_j}{\partial x}, \quad Q_{2j}^3 = A_{44} \frac{\partial \psi_j}{\partial y}, \quad Q_{1j}^4 = A_{55} \psi_j, \quad Q_{2j}^5 = A_{44} \psi_j \qquad (7.10.16)$$

7.10.3 Time Approximation

The finite element equations in Eq. (7.10.12) can be cast in the general matrix form

$$\mathbf{M}\ddot{\mathbf{\Delta}} + \mathbf{K}(\mathbf{\Delta})\mathbf{\Delta} = \mathbf{F} \tag{7.10.17}$$

which is of the same general form as one in Eq. (6.7.27) with $\mathbf{C} = \mathbf{0}$. Therefore, the discussion presented in Section 6.7.4 is valid. The fully discretized equations are [see Eq. (6.7.33)]

$$\hat{\mathbf{K}}(\mathbf{\Delta}_{s+1})\mathbf{\Delta}_{s+1} = \hat{\mathbf{F}}_{s,s+1} \tag{7.10.18}$$

where

$$\begin{aligned}
\hat{\mathbf{K}}(\mathbf{\Delta}_{s+1}) &= \mathbf{K}(\mathbf{\Delta}_{s+1}) + a_3\,\mathbf{M}_{s+1} \\
\hat{\mathbf{F}}_{s,s+1} &= \mathbf{F}_{s+1} + \mathbf{M}_{s+1}\big(a_3\mathbf{\Delta}_s + a_4\dot{\mathbf{\Delta}}_s + a_5\ddot{\mathbf{\Delta}}_s\big)
\end{aligned} \tag{7.10.19}$$

and a_i are defined as

$$a_1 = \alpha\Delta t, \quad a_2 = (1-\alpha)\Delta t, \quad a_3 = \frac{2}{\gamma(\Delta t)^2}, \quad a_4 = a_3\Delta t, \quad a_5 = \frac{1}{\gamma}-1 \tag{7.10.20}$$

At the end of each time step, the new velocity vector $\dot{\mathbf{\Delta}}_{s+1}$ and acceleration vector $\ddot{\mathbf{\Delta}}_{s+1}$ are computed using the equations

$$\begin{aligned}
\ddot{\mathbf{\Delta}}_{s+1} &= a_3\big(\mathbf{\Delta}_{s+1} - \mathbf{\Delta}_s\big) - a_4\,\dot{\mathbf{\Delta}} - a_5\,\ddot{\mathbf{\Delta}}_s \\
\dot{\mathbf{\Delta}}_{s+1} &= \dot{\mathbf{\Delta}}_s + a_2\,\ddot{\mathbf{\Delta}}_s + a_1\,\ddot{\mathbf{\Delta}}_{s+1}
\end{aligned} \tag{7.10.21}$$

Solution of Eq. (7.10.18) by the Newton iteration method results in the following linearized equations for the incremental solution at the $(r + 1)$st iteration:

$$\delta\mathbf{\Delta} = -(\hat{\mathbf{T}}(\mathbf{\Delta}_{s+1}^r))^{-1}\mathbf{R}_{s+1}^r \tag{7.10.22}$$

$$\hat{\mathbf{T}}(\mathbf{\Delta}_{s+1}^r) \equiv \left[\frac{\partial \mathbf{R}}{\partial \mathbf{\Delta}}\right]_{s+1}^r, \quad \mathbf{R}_{s+1}^r = \hat{\mathbf{K}}(\mathbf{\Delta}_{s+1}^r)\mathbf{\Delta}_{s+1}^r - \hat{\mathbf{F}} \tag{7.10.23}$$

The total solution is obtained from

$$\mathbf{\Delta}_{s+1}^{r+1} = \mathbf{\Delta}_{s+1}^r + \delta\mathbf{\Delta} \tag{7.10.24}$$

Note that the tangent stiffness matrix $\hat{\mathbf{T}}_{s+1}^r$, the coefficient matrix $\hat{\mathbf{K}}_{s+1}^r$, and the residual vector $\mathbf{R}_{s+1}^r$ are evaluated using the latest known solution $\mathbf{\Delta}_{s+1}^r$. The right-hand side vector $\hat{\mathbf{F}}$ only depends on the known load vector (after assembly and imposition of boundary conditions) $\mathbf{F}_{s,s+1}$ and known solutions $(\mathbf{\Delta}_s, \dot{\mathbf{\Delta}}_s, \ddot{\mathbf{\Delta}}_s)$ from the previous time step. At the end of each time step (i.e. after nonlinear convergence is reached), the velocity and acceleration vectors are updated using the formulas in Eq. (7.10.21). Computer implementation aspects were discussed in Section 6.7.6, and the flow chart in Fig. 6.7.2 and Fortran statements presented in Boxes 6.7.1 and 6.7.2 are also applicable here.

7.10.4 Numerical Examples

Here we present numerical examples to illustrate the implementation of the ideas discussed in the preceding sections. First two examples, one linear and one nonlinear, deal with the transient response of rectangular plates. The third example deals with a clamped circular plate. In all cases, a shear correction coefficient of $K_s = 5/6$ is used. We begin with linear transient response of a simply-supported (SS–1) rectangular plate under uniformly distributed load.

Example 7.10.1 ──

Obtain the transient response $w(x, y, t)$ of a simply-supported (SS–1), isotropic, square plate under uniform load of intensity $q(x, y, t) = q_0 H(t)$, where $H(t)$ denotes the Heaviside step function. Exploit the problem symmetry and use the uniform 4×4Q9 mesh in a quadrant of the domain and the constant-average acceleration method $[\alpha = 0.5, \gamma = 0.5$; see Eq. (6.7.30)] with $\Delta t = 20 \, \mu s = 20 \times 10^{-6}$ s. Take the geometric and material parameters to be

$$a = b = 25 \text{ cm}, \ h = 1 \text{ or } 2.5 \text{ cm}, \ E_1 = E_2 = 2.1 \times 10^6 \text{ N/cm}^2$$
$$\nu_{12} = 0.25, \ \rho = 8 \times 10^{-6} \text{ N.s}^2/\text{cm}^4 \tag{1}$$

Solution: Figure 7.10.1 shows plots of the non-dimensionalized transverse deflection $\bar{w} = 10^3 w(0,0,t) E_2 h^3/q_0 a^4$ versus time t (in μs) for thin ($h = 1$ cm) and thick ($h = 2.5$ cm), simply-supported, isotropic ($\nu = 0.25$) plates (4×4Q9 mesh; $\Delta t = 20 \, \mu s$). Comparing the results of thin and thick plates, we conclude that the effect of shear deformation is to increase the amplitude and reduce the period of the transverse deflection (also see Table 7.10.1 for the numerical results).

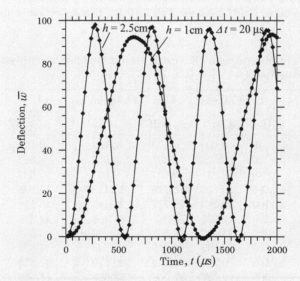

Fig. 7.10.1: Evolution of the transverse deflection $\bar{w}$ of a simply-supported, isotropic plate under uniform load ($\alpha = \gamma = 0.5$).

Table 7.10.1: Non-dimensionalized center deflection $\bar{w}$ versus time t for simply-supported, isotropic, square plates ($\alpha = \gamma = 0.5$).

t	$\bar{w}$		t	$\bar{w}$	
(μs)	$h = 1$	$h = 2.5$	(μs)	$h = 1$	$h = 2.5$
20	0.067	0.426	360	50.661	76.739
40	0.334	2.193	380	54.768	68.263
60	0.860	6.074	400	58.922	58.352
80	1.626	12.813	420	63.078	46.337
100	2.674	22.770	440	67.238	33.574
120	4.189	34.696	460	71.368	22.403
140	6.405	46.408	480	75.488	13.751
160	9.348	56.971	500	79.533	7.423
180	12.871	66.677	520	83.218	3.213
200	16.819	75.708	540	86.276	0.605
220	21.044	84.185	560	88.635	−0.511
240	25.459	91.721	580	90.348	0.728
260	30.002	96.806	600	91.472	5.332
280	34.532	98.062	620	92.084	13.441
300	38.861	95.669	640	92.323	23.727
320	42.878	90.826	660	92.251	34.641

Example 7.10.2

Obtain the nonlinear transient response of the plate considered in Example 7.10.1. Use the same data as given in Eq. (1) of Example 7.10.1, except $h = 2.5$ cm and $\Delta t = 10$ μs.

Solution: Table 7.10.2 contains the linear and nonlinear center deflections versus time for $q_0 = 10^3$, $q_0 = 5 \times 10^3$, and $q_0 = 10^4$ (also see Fig. 7.10.2). The deflection is non-dimensionalized as $\bar{w} = 10^3 w(0,0,t) E_2 h^3 / q_0 a^4$.

Table 7.10.2: Center deflections $\bar{w}$ versus time t (μs) for a simply-supported square plate under uniform load ($\alpha = \gamma = 0.5$).

t	Linear	Nonlinear		
(μs)	$q_0 = 1$	$q_0 = 10^3$	$q_0 = 5 \times 10^3$	$q_0 = 10^4$
10	0.105	0.105	0.105	0.105
20	0.525	0.525	0.525	0.525
40	2.637	2.637	2.637	2.637
80	15.119	15.119	15.116	15.108
100	26.282	26.277	26.147	25.746
200	78.605	77.565	61.334	42.258
300	93.457	86.558	30.263	4.352
400	53.509	41.715	−2.343	7.872
500	6.760	2.214	25.179	42.854
600	9.727	17.908	62.625	18.889
700	59.089	69.535	33.608	0.115
800	95.215	90.271	0.151	36.143
900	73.660	51.497	23.921	37.009
1,000	20.968	5.232	60.393	1.060

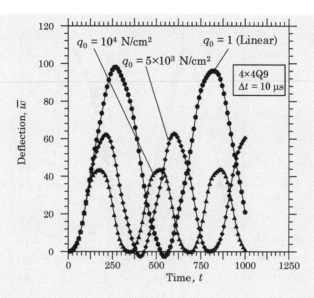

Fig. 7.10.2: Plots of the linear and nonlinear center deflections $\bar{w}$ versus time t (μs) for a simply-supported square plate under uniform load (4×4Q9 mesh, $\Delta t = 10\mu$s, and $\alpha = \gamma = 0.5$).

Example 7.10.3

Obtain the transient response of a clamped circular plate under uniformly distributed load. Use the following geometric, material, and load parameters (may not represent any real situation):

$$a = 100 \text{ in}, \quad h = 2 \text{ in}, \quad E = 10^7 \text{ psi}, \quad \nu = 0.3, \quad \rho = 0.1 \text{ slugs/in}^3, \quad q_0 = 1 \text{ psi}, \quad \Delta t = 10^{-2}\text{s} \quad (1)$$

Once again, use a quadrant of the plate with the mesh shown in Fig. 7.9.9 to determine linear and nonlinear response with the constant-average acceleration method ($\alpha = \gamma = 0.5$).

Solution: Table 7.10.3 contains the linear solutions for $q_0 = 1$ psi, whereas Fig. 7.10.3 shows a center deflection $w(0,0)$ normalized by q_0 versus time t. The linear and nonlinear solutions for $q_0 = 1$ do not differ that much, but as the load is increased to $q_0 = 5$ psi, the nonlinearity decreases the amplitude as well as the period of oscillation.

Table 7.10.3: Center deflections $\bar{w} = w/q_0$ versus time t for a clamped circular plate under uniform load ($\alpha = \gamma = 0.5$).

$t(\times 10^{-2})$	Linear	Nonlinear		
	$q_0 = 1$	$q_0 = 1$	$q_0 = 2$	$q_0 = 5$
10	0.0186	0.0186	0.0186	0.0186
20	0.1369	0.1369	0.1369	0.1367
30	0.2828	0.2825	0.2814	0.2742
40	0.3821	0.3803	0.3751	0.3431
50	0.4574	0.4520	0.4368	0.3560
60	0.4097	0.3989	0.3694	0.2372
70	0.3066	0.2927	0.2567	0.1193
80	0.1808	0.1659	0.1289	0.0190
100	−0.0004	−0.0012	0.0011	0.0650

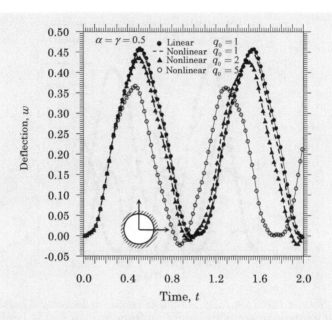

Fig. 7.10.3: Plots of the linear and nonlinear center deflections $\bar{w} = w/q_0$ versus time t for a clamped circular plate under uniform load (4Q9 mesh, $\Delta t = 10^{-2}$).

7.11   Summary

In this chapter the classical plate theory (CPT) and first-order shear deformation plate theory (FSDT) are developed based on assumed displacement fields and accounting for the von Kármán nonlinearity. The governing equations and the weak forms for the two theories are derived using the principle of virtual displacements and the displacement finite element models are developed. Computational as well as computer programming aspects are discussed. For the CPT, C^1–approximation of the transverse displacement w while C^0 Lagrange approximations of (u, v) are admissible, whereas the FSDT requires C^0–approximation of all variables $(u, v, w, \phi_x, \phi_y)$.

Like the beam elements, both plate elements suffer from membrane locking when geometric nonlinearity is included, while the FSDT suffers from shear locking. The membrane locking is a result of using the same low-order approximations of the in-plane displacements (u, v) and the transverse displacement w, while the membrane locking is due to the use of the same low-order approximations of the rotation variables (ϕ_x, ϕ_y) and the transverse displacement w. Both types of locking are avoided using reduced integration of shear and nonlinear terms of the stiffness coefficients. In addition, finite element model for the transient response of FSDT is presented. Numerical examples of static and transient response of rectangular and circular plates with different boundary conditions are presented.

Problems

WEAK FORMS AND FINITE ELEMENT MODELS

7.1 Develop the weak forms of the following linear equations governing the classical plate theory:

$$-\left(\frac{\partial^2 M_{xx}}{\partial x^2} + 2\frac{\partial^2 M_{xy}}{\partial x \partial y} + \frac{\partial^2 M_{yy}}{\partial y^2}\right) = q \tag{1}$$

$$-\frac{\partial^2 w}{\partial x^2} - \left(\bar{D}_{22}M_{xx} - \bar{D}_{12}M_{yy}\right) = 0 \tag{2}$$

$$-\frac{\partial^2 w}{\partial y^2} - \left(\bar{D}_{12}M_{xx} - \bar{D}_{11}M_{yy}\right) = 0 \tag{3}$$

$$-2\frac{\partial^2 w}{\partial x \partial y} - (D_{66})^{-1}M_{xy} = 0 \tag{4}$$

where

$$\bar{D}_{ij} = \frac{D_{ij}}{D_0}, \quad D_0 = D_{11}D_{22} - D_{12}^2, \quad D_{ij} = Q_{ij}\frac{h^3}{12}, \quad (i = 1, 2, 6) \tag{5}$$

and Q_{ij} are the elastic stiffnesses defined in terms of the principal moduli (E_1, E_2), shear modulus G_{12}, and Poisson's ratio ν_{12} as

$$Q_{11} = \frac{E_1}{1 - \nu_{12}\nu_{21}}, \quad Q_{12} = \frac{\nu_{12}E_2}{1 - \nu_{12}\nu_{21}}, \quad Q_{22} = \frac{E_2}{1 - \nu_{12}\nu_{21}}$$

$$Q_{66} = \frac{1}{S_{66}} = G_{12}, \quad \nu_{21} = \nu_{12}\frac{E_2}{E_1} \tag{6}$$

7.2 Develop the (mixed) finite element model associated with the equations in Problem 7.1. Assume approximation of the form

$$w = \sum_{i=1}^{r} w_i \psi_i^{(1)}, \quad M_{xx} = \sum_{i=1}^{s} M_{xi}\psi_i^{(2)}$$

$$M_{yy} = \sum_{i=1}^{p} M_{yi}\psi_i^{(3)}, \quad M_{xy} = \sum_{i=1}^{q} M_{xyi}\psi_i^{(4)}$$

where $\psi_i^{(\alpha)}$, $(\alpha = 1, 2, 3, 4)$ are appropriate interpolation functions. Discuss the minimum requirements of the interpolation functions.

7.3 Develop the weak forms of the following nonlinear equations governing the classical plate theory:

$$-\left(\frac{\partial N_{xx}}{\partial x} + \frac{\partial N_{xy}}{\partial y}\right) = 0 \tag{1}$$

$$-\left(\frac{\partial N_{xy}}{\partial x} + \frac{\partial N_{yy}}{\partial y}\right) = 0 \tag{2}$$

$$-\left(\frac{\partial^2 M_{xx}}{\partial x^2} + 2\frac{\partial^2 M_{xy}}{\partial x \partial y} + \frac{\partial^2 M_{yy}}{\partial y^2}\right) - \mathcal{N}(u, v, w) = q \tag{3}$$

$$-\frac{\partial^2 w}{\partial x^2} - \left(\bar{D}_{22}M_{xx} - \bar{D}_{12}M_{yy}\right) = 0 \tag{4}$$

$$-\frac{\partial^2 w}{\partial y^2} - \left(\bar{D}_{12}M_{xx} - \bar{D}_{11}M_{yy}\right) = 0 \tag{5}$$

$$-2\frac{\partial^2 w}{\partial x \partial y} - (D_{66})^{-1}M_{xy} = 0 \tag{6}$$

where $(N_{xx}, N_{yy}, N_{xy}, \mathcal{N})$ are known in terms of (u, v, w) and their derivatives by Eqs. (7.3.32) and (7.3.14). Note that the dependent unknowns are u, v, w, M_{xx}, M_{yy}, and M_{xy}.

7.4 Develop the (mixed) nonlinear finite element model of the equations in Problem 7.3. Assume interpolation of the form

$$u = \sum_{i=1}^{r} u_i \psi_i^{(1)}, \quad v = \sum_{i=1}^{r} v_i \psi_i^{(1)}, \quad w = \sum_{i=1}^{r} w_i \psi_i^{(2)} \tag{1}$$

$$M_{xx} = \sum_{i=1}^{s} M_{xi} \psi_i^{(3)}, \quad M_{yy} = \sum_{i=1}^{p} M_{yi} \psi_i^{(4)}, \quad M_{xy} = \sum_{i=1}^{q} M_{xyi} \psi_i^{(5)} \tag{2}$$

7.5 Develop the weak forms of the following nonlinear equations governing the classical plate theory:

$$-\left(\frac{\partial N_{xx}}{\partial x} + \frac{\partial N_{xy}}{\partial y}\right) = 0 \tag{1}$$

$$-\left(\frac{\partial N_{xy}}{\partial x} + \frac{\partial N_{yy}}{\partial y}\right) = 0 \tag{2}$$

$$-\frac{\partial^2 M_{xx}}{\partial x^2} + 4D_{66}\frac{\partial^4 w}{\partial x^2 \partial y^2} - \frac{\partial^2 M_{yy}}{\partial y^2} - \frac{\partial}{\partial x}\left(N_{xx}\frac{\partial w}{\partial x} + N_{xy}\frac{\partial w}{\partial y}\right)$$

$$-\frac{\partial}{\partial y}\left(N_{xy}\frac{\partial w}{\partial x} + N_{yy}\frac{\partial w}{\partial y}\right) = q \tag{3}$$

$$-\frac{\partial^2 w}{\partial x^2} - \left(\bar{D}_{22}M_{xx} + \bar{D}_{12}M_{yy}\right) = 0 \tag{4}$$

$$-\frac{\partial^2 w}{\partial y^2} - \left(\bar{D}_{12}M_{xx} + \bar{D}_{11}M_{yy}\right) = 0 \tag{5}$$

where (N_{xx}, N_{yy}, N_{xy}) are known in terms of (u, v, w) and their derivatives by Eq. (7.3.32). Note that the dependent unknowns are u, v, w, M_{xx}, and M_{yy}.

7.6 Develop the (mixed) nonlinear finite element model of the equations in Problem 7.5 by assuming approximations of the form

$$u = \sum_{i=1}^{r} u_i \psi_i^{(1)}, \quad v = \sum_{i=1}^{r} v_i \psi_i^{(1)}, \quad w = \sum_{i=1}^{r} w_i \psi_i^{(2)} \tag{1}$$

$$M_{xx} = \sum_{i=1}^{s} M_{xi} \psi_i^{(3)}, \quad M_{yy} = \sum_{i=1}^{p} M_{yi} \psi_i^{(4)} \tag{2}$$

7.7 Consider the following alternative form of nonlinear equations governing the first-order shear deformation plate theory:

$$-\left(\frac{\partial N_{xx}}{\partial x} + \frac{\partial N_{xy}}{\partial y}\right) = 0 \tag{1}$$

$$-\left(\frac{\partial N_{xy}}{\partial x} + \frac{\partial N_{yy}}{\partial y}\right) = 0, \tag{2}$$

$$-A_{55}\frac{\partial}{\partial x}\left(\frac{\partial w}{\partial x} + \phi_x\right) - A_{44}\frac{\partial}{\partial y}\left(\frac{\partial w}{\partial y} + \phi_y\right)$$

$$-\frac{\partial}{\partial x}\left(N_{xx}\frac{\partial w}{\partial x} + N_{xy}\frac{\partial w}{\partial y}\right) - \frac{\partial}{\partial y}\left(N_{xy}\frac{\partial w}{\partial x} + N_{yy}\frac{\partial w}{\partial y}\right) = q, \tag{3}$$

$$-\frac{\partial M_{xx}}{\partial x} - D_{66}\frac{\partial}{\partial y}\left(\frac{\partial \phi_x}{\partial y} + \frac{\partial \phi_y}{\partial x}\right) + K_s A_{55}\left(\frac{\partial w}{\partial x} + \phi_x\right) = 0, \tag{4}$$

$$-\frac{\partial M_{yy}}{\partial y} - D_{66}\frac{\partial}{\partial x}\left(\frac{\partial \phi_x}{\partial y} + \frac{\partial \phi_y}{\partial x}\right) + K_s A_{44}\left(\frac{\partial w}{\partial y} + \phi_y\right) = 0, \tag{5}$$

$$\frac{\partial \phi_x}{\partial x} - \left(\bar{D}_{11} M_{xx} + \bar{D}_{12} M_{yy}\right) = 0, \tag{6}$$

$$\frac{\partial \phi_y}{\partial y} - \left(\bar{D}_{12} M_{xx} + \bar{D}_{22} M_{yy}\right) = 0. \tag{7}$$

where (N_{xx}, N_{yy}, N_{xy}) are known in terms of (u, v, w) and their derivatives by [see Eqs. (7.7.20)–(7.7.22)] and

$$\bar{D}_{ij} = \frac{D_{ij}}{D_0}, \quad D_0 = D_{11} D_{22} - D_{12}^2 \tag{8}$$

The dependent unknowns of the formulation (note the order) are $(u, v, w, \phi_x, \phi_y, M_{xx}, M_{yy})$. Develop the weak forms of Eqs. (4)–(7) such that they are symmetric in $(\phi_x, \phi_y, M_{xx}, M_{yy})$, and identify the primary and secondary variables.

7.8 Develop the (mixed) nonlinear finite element model of the equations in Problem 7.7 by assuming approximations of the form

$$u = \sum_{i=1}^{r} u_i \psi_i^{(1)}, \quad v = \sum_{i=1}^{r} v_i \psi_i^{(1)}, \quad w = \sum_{i=1}^{r} w_i \psi_i^{(2)}$$

$$\phi_x = \sum_{i=1}^{r} S_i^1 \psi_i^{(3)}, \quad \phi_y = \sum_{i=1}^{r} S_i^2 \psi_i^{(3)}, \quad M_{xx} = \sum_{i=1}^{s} M_{xi} \psi_i^{(4)}, \quad M_{yy} = \sum_{i=1}^{p} M_{yi} \psi_i^{(5)}$$

7.9 Consider the governing equations of equilibrium of elastic plates according to the first-order shear deformation plate theory with the von Kármán nonlinearity. Assume that the material of the plate is isotropic and obeys the generalized Hooke's law. However, the elastic modulus E varies nonlinearly with the displacement gradients according to the equation

$$E = E_0 + E_1\left(\frac{\partial u}{\partial x} + \frac{\partial v}{\partial y}\right) \tag{1}$$

Assuming that Poisson's ratio is constant, develop (a) the finite element model of the equations and (b) compute the tangent stiffness coefficients, T_{ij}^{11} and T_{ij}^{33}.

7.10 Consider the governing equations of equilibrium of elastic plates according to the first-order shear deformation plate theory with the von Kármán nonlinearity. Assume that the plate is made of two homogeneous isotropic materials that are combined according to the following power-law distribution [see Eq. (5.3.37)]:

$$E(z) = (E_1 - E_2) f(z) + E_2, \quad f(z) = \left(\frac{1}{2} + \frac{z}{h}\right)^n \tag{1}$$

where E_1 and E_2 are moduli of the two materials that are being mixed, h is the total thickness of the plate, and n is the power-law index that dictates the volume fraction $f(z)$ of material 1. Develop the displacement finite element model of the functionally graded plate and determine the tangent stiffness coefficients.

PROGRAMMING PROBLEMS (Development of finite element computer programs based on Problems 7.1–7.8 and Problem 7.10 can also be assigned)

7.11 Write a computer subroutine to calculate stresses for the CPT, based on the discussion of Section 7.5.3.

7.12 Write a computer subroutine to calculate stresses for FSDT, following the discussion of Section 7.5.3.

7.13 Write a computer subroutine to implement the time-approximation schemes of Section 7.10.

COMPUTATIONAL PROBLEMS

7.14 Analyze an orthotropic square plate using $a = 10$ in, $h = 1$ in, $E_1 = 7.8 \times 10^6$ psi, $E_2 = 2.6 \times 10^6$ psi, $\nu_{12}=0.25$, $G_{12} = G_{13} = G_{23} = 1.3 \times 10^6$ psi, and 8×8 uniform mesh of conforming CPT(C) elements in a quadrant of the plate for SS–1 and SS–3 boundary conditions. The material coordinates (x_1, x_2) coincide with the plate coordinates (x, y). Use load parameter $P = q_0 a^4/(E_2 h^4)$ equal to 25 for the first four steps, 50 for the next two steps, and 100 for the last ten steps. Use a convergence tolerance of $\varepsilon = 10^{-3}$ and a maximum of ten allowable iterations.

7.15 Analyze Problem 7.14 using the FSDT elements with 4×4 mesh of Q9 elements.

7.16 Consider a clamped square plate, subjected to uniformly distributed transverse load $q = q_0 = $ constant. The geometric and material parameters are $a = 10$ in, $h = 1$ in, $E_1 = 7.8 \times 10^6$ psi, $E_2 = 2.6 \times 10^6$ psi, $\nu_{12}=0.25$, $G_{12} = G_{13} = G_{23} = 1.3 \times 10^6$ psi. Use 8×8 and 4×4 uniform meshes of linear and equivalent meshes of nine-node quadratic rectangular elements in a quadrant of the plate to analyze the problem. Use load steps such that the *load parameter* $\Delta P \equiv q_0 a^4/E_2 h^4$ is equal to 5, and use 32 load steps to obtain the nonlinear response. Plot (a) load versus center deflection (w/h versus P), and (b) load versus maximum stress (σ_{xx} and σ_{xy}). Use a convergence tolerance of $\varepsilon = 10^{-3}$ and maximum number of iterations for convergence to be ten.

7.17 Determine the load–deflection behavior of a square plate with two opposite edges simply-supported (SS–1) and the other two edges free, and subjected to uniformly distributed transverse load. The material coordinates (x_1, x_2) coincide with the plate coordinates (x, y), and assume that the edges parallel to the x-axis are simply-supported. Use $a = 10$ in, $h = 1$ in, $E_1 = 7.8 \times 10^6$ psi, $E_2 = 2.6 \times 10^6$ psi, $\nu_{12}=0.25$, $G_{12} = G_{13} = G_{23} = 1.3 \times 10^6$ psi, and 8×8 uniform mesh of conforming CPT elements in a quadrant of the plate. Use load steps such that the *load parameter* $P \equiv q_0 a^4/E_2 h^4$ is equal to 20, and use 12 load steps. Tabulate results for the maximum transverse deflection w and stress σ_{xx}. Use a tolerance of $\varepsilon = 10^{-2}$ and ten maximum iterations.

7.18 Repeat Problem 7.17 for a square plate with two opposite edges simply-supported (SS–1) and the other two edges clamped (CC). The material coordinates (x_1, x_2) coincide with the plate coordinates (x, y), and assume that the edges parallel to the x-axis are clamped and the edges parallel to the y-axis are simply-supported. Use 8×8 mesh of (a) conforming CPT element, and (b) non-conforming CPT element. Present the load-deflection curves in graphical form.

7.19 Determine the load–deflection behavior of a square plate with two opposite edges clamped and the other two edges free, and subjected to uniformly distributed transverse load. The material coordinates (x_1, x_2) coincide with the plate coordinates (x, y), and assume that the edges parallel to the x-axis are clamped. Use $a = 10$ in, $h = 1$ in, $E_1 = 7.8 \times 10^6$ psi, $E_2 = 2.6 \times 10^6$ psi, $\nu_{12}=0.25$, $G_{12} = G_{13} = G_{23} = 1.3 \times 10^6$ psi, and 8×8 uniform mesh of (a) non-conforming CPT elements, and (b) conforming CPT elements. Use load steps such that the *load parameter* $P \equiv q_0 a^4/E_2 h^4$ is equal to 20, and use 12 load steps with $\epsilon = 10^{-2}$. Plot load versus center deflection (P versus w). Use a convergence tolerance of $\varepsilon = 10^{-2}$ and ten maximum iterations.

7.20 Repeat Problem 7.17 using 4×4Q9 mesh of the FSDT elements.

7.21 Repeat Problem 7.18 using 4×4Q9 mesh of the FSDT elements.

7.22 Repeat Problem 7.19 using 4×4Q9 mesh of the FSDT elements.

7.23 Analyze the circular plate problem of Example 7.9.6 using an equivalent mesh of the nonconforming plate elements. Use all other parameters as in Example 7.9.6.

7.24 Analyze the circular plate problem of Example 7.10.3 when the edge is simply-supported (SS–3). Use the 5Q9 mesh of FSDT elements and parameters of Example 7.10.3.

7.25 Find the nonlinear transient response of the circular plate problem in Problem 7.24 for $q_0 = 10, 20$, and 100. Use all parameters as in Example 7.10.3.

7.26 Find the nonlinear transient response of a simply-supported (SS–1) orthotropic square plate under uniformly distributed load of intensity $q_0 = 10^3$ psi. Use the following geometric and material parameters:

$$a = b = 25 \text{ in}, \quad h = 2.5 \text{ in}, \quad E_1 = 7.5 \times 10^6 \text{ psi}, \quad E_2 = 2 \times 10^6 \text{ psi}$$

$$G_{12} = 1.25 \times 10^6 \text{ psi}, \quad G_{13} = G_{23} = 0.625 \times 10^6 \text{ psi}$$

$$\nu_{12} = 0.25, \quad \rho = 1.0 \times 10^{-6} \text{ lb-s}^2/\text{in}^4, \quad \alpha = \gamma = 0.5$$

and a time step of $\Delta t = 10^{-5}$ to capture at least the first period of the response.

7.27 Repeat Problem 7.26 for the SS–3 boundary conditions.

Nonlinear Bending of Elastic Shells

The man who cannot occasionally imagine events and conditions of existence that are contrary to the causal principle as he knows it will never enrich his science by the addition of a new idea. —— Max Planck

8.1 Introduction

In Chapter 7, we considered plates whose planform dimensions are large compared to their thicknesses and subjected to loads that cause bending deformation in addition to stretching. Shells are much like plates in terms of their thicknesses being considerably smaller than the other two dimensions, except that they are curved. Shells are common in nature as well as in many engineering structures, including pressure vessels, submarine and ship hulls, airplane wings and fuselages, containment structures, pipes, exteriors of rockets, missiles, automobile tires, concrete roofs, chimneys, cooling towers, liquid storage tanks, and many others. In nature, they can be found in the form of sea shells, eggs, inner ear, skull, and geological formations. Shell structures have been built by humans for centuries, mostly in the form of masonry and stone domes, roofs, and arches; some of which still exist around the world. They are the most efficient structures because loads applied in any direction are resisted by a combination of transverse as well as surface forces even when the deformation is small. This is primarily due to its curved geometry.

Finite element models of shells are developed using either (1) a shell theory or (2) the 2-D equations obtained from a *degenerated* 3-D elasticity model. The latter is more often known as the *continuum shell model*, which will be discussed in Chapter 9. The present chapter is devoted to the development of shell theory-based finite element models.

A number of shell theories exist in the literature, and many of these theories were developed originally for thin shells and are based on the Kirchhoff–Love kinematic hypothesis, similar to the Kirchhoff hypothesis for plates, that straight lines normal to the undeformed midsurface remain straight, inextensible, and normal to the middle surface after deformation. A detailed study of thin isotropic shells can be found in the monographs by Ambartsumyan [81–83],

J.N. Reddy, *An Introduction to Nonlinear Finite Element Analysis*, Second Edition. ©J.N. Reddy 2015. Published in 2015 by Oxford University Press.

Flügge [84], Kraus [85], Timoshenko and Woinowsky-Krieger [86], Leissa [87], Dym [88], Ugural [89], and Farshad [90]. The first-order shear deformation theory of shells, also known as the *Sanders' shell theory* [91, 92], can be found in Kraus [85], Palazotto and Dennis [93], and Reddy [55, 56]. Other shell theories can be found in the works of Vlasov [94] and Koiter [95], among others.

Naghdi [96, 97] presented a complete treatise of shell theories in the well-known Encyclopedia of Physics. There, he derived two approaches for nonlinear shell theories: the direct approach based on the Cosserat continuum theory and the derivation from the 3D continuum theory, or the so-called single-layer theories. A general shear deformation theory for geometrically nonlinear laminated shells was given Librescu [98], which is well-known as the moderate rotation shell theory. The first-order version of theory was developed by Schmidt and Reddy [99], and it was utilized by Palmerio, Reddy, and Schmidt [100, 101] and Kreja, Schmidt, and Reddy [102] for finite element analysis of laminated composite shells.

Simo and co-workers [103–107] proposed the stress-resultant geometrically exact shell model which is formulated entirely in terms of stress resultants and is essentially equivalent to the single director inextensible Cosserat surface. The phrase "geometrically exact" was used by Simo because there was no approximation of the geometry.

Another group of shell theories is based on the order-of-magnitude assumptions on strains and rotations in full nonlinear equations, and they are called the *finite rotation theories.* Pietraszkiewicz and his colleagues [108–110] developed these theories using both total Lagrangian and update Lagrangian formulations for geometrically nonlinear shells based on the Kirchhoff–Love assumptions. Strains and rotations about the normal to the surface are assumed to be of the order $\epsilon \ll 1$. Rotations about tangents to the surface were organized in a consistent classification where for each range of magnitude of rotations specific shell equations were obtained. Refined high-order shear deformation models for composite shells using finite rotations can be found in Başar [111, 112], Başar, Ding, and Schultz [113] and Balah and Al-Ghamedy [114]. Both approaches utilize the third-order shell theory for the analysis of composite shells via the finite element method.

The objective of this chapter is to review the governing equations of shells based on the first-order shear deformation shell theory and develop its finite element model. Following this introduction, a description of the geometry, kinematic relations, and equilibrium equations for shells are presented. The presentation is similar to that presented for plates in Chapter 7, except for the description of shell geometry, which is more involved than that of a plate. Much of the theoretical development comes from this author's book on plates and shells [56]. Finite element models of doubly-curved shells are presented in Section 8.3.

8.2 Governing Equations

8.2.1 Geometric Description

Consider a curved shell element of uniform thickness, as shown in Fig. 8.2.1(a). Here (ξ_1, ξ_2, ζ) denote the curvilinear coordinates such that ξ_1 and ξ_2 curves are the lines of curvature on the middle surface $(\zeta = 0)$. The position vector of a typical point $(\xi_1, \xi_2, 0)$ on the middle surface is denoted by $\mathbf{r}$, and the position of an arbitrary point (ξ_1, ξ_2, ζ) is denoted by $\mathbf{R}$ [see Figs. 8.2.1(b) and (c)]. A differential line element on the middle surface can be written as

$$dr = \frac{\partial \mathbf{r}}{\partial \xi_\alpha} d\xi_\alpha = \mathbf{g}_\alpha d\xi_\alpha, \quad \mathbf{g}_\alpha = \frac{\partial \mathbf{r}}{\partial \xi_\alpha} \quad (\alpha = 1, 2) \tag{8.2.1}$$

where the vectors $\mathbf{g}_1$ and $\mathbf{g}_2$ are tangent to the ξ_1 and ξ_2 coordinate lines, as shown in Fig. 8.2.1(b). The components of the *surface metric tensor* $g_{\alpha\beta}$ $(\alpha, \beta = 1, 2)$ are

$$\mathbf{g}_\alpha \cdot \mathbf{g}_\beta \equiv g_{\alpha\beta}, \quad a_1 = \sqrt{g_{11}}, \quad a_2 = \sqrt{g_{22}}, \quad \mathbf{g}_1 \cdot \mathbf{g}_2 = a_1 a_2 \cos \chi \tag{8.2.2}$$

where χ denotes the angle between the coordinate curves. The unit vector normal to the middle surface (hence, normal to both $\mathbf{g}_1$ and $\mathbf{g}_2$) can be determined from

$$\hat{\mathbf{n}} = \frac{\mathbf{g}_1 \times \mathbf{g}_2}{a_1 a_2 \sin \chi} \tag{8.2.3}$$

In general, $\hat{\mathbf{n}}$, a_1 and a_2 are functions of ξ_1 and ξ_2. The square of the distance ds between points $(\xi_1, \xi_2, 0)$ and $(\xi_1 + d\xi_1, \xi_2 + d\xi_2, 0)$ on the middle surface is determined by

$$(ds)^2 = d\mathbf{r} \cdot d\mathbf{r} = g_{\alpha\beta} \, d\xi_\alpha \, d\xi_\beta = a_1^2 \, (d\xi_1)^2 + a_2^2 \, (d\xi_2)^2 + 2a_1 a_2 \cos \chi \, d\xi_1 \, d\xi_2 \tag{8.2.4}$$

The right-hand side of the above equation is known as the *first quadratic form of the surface*, which allows us to determine the infinitesimal lengths, the angle between the curves, and area of the surface. The terms a_1^2, a_2^2, and $a_1 a_2 \cos \chi$ are called the *first fundamental quantities*.

Let $\mathbf{r} = \mathbf{r}(s)$ be the equation of a curve s on the surface. The unit vector tangent to the curve is

$$\hat{\mathbf{t}} = \frac{d\mathbf{r}}{ds} = \frac{\partial \mathbf{r}}{\partial \xi_\alpha} \frac{d\xi_\alpha}{ds} = \mathbf{g}_\alpha \frac{d\xi_\alpha}{ds} \tag{8.2.5}$$

According to Frenet's formula, the derivative of this unit vector is

$$\frac{d\hat{\mathbf{t}}}{ds} = \frac{\hat{\mathbf{N}}}{\rho} \tag{8.2.6}$$

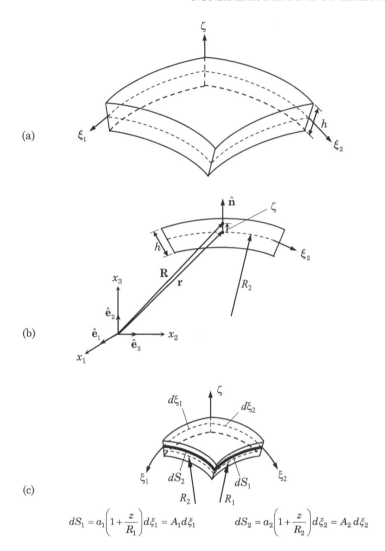

$$dS_1 = a_1\left(1 + \frac{z}{R_1}\right)d\xi_1 = A_1 d\xi_1 \qquad dS_2 = a_2\left(1 + \frac{z}{R_2}\right)d\xi_2 = A_2 d\xi_2$$

Fig. 8.2.1: (a) Shell geometry. (b) Position vector of a point and coordinates on the middle surface. (c) Position vectors of points on the middle surface and above the middle surface.

where $1/\rho$ is the curvature of the curve, and $\hat{\mathbf{N}}$ is the unit vector of the principal normal to the curve. From Eqs. (8.2.5) and (8.2.6), we obtain

$$\frac{\hat{\mathbf{N}}}{\rho} = \frac{d}{ds}\left(\frac{d\mathbf{r}}{ds}\right) = \frac{\partial \mathbf{g}_\alpha}{\partial \xi_\beta}\frac{d\xi_\alpha}{ds}\frac{d\xi_\beta}{ds} + \mathbf{g}_\alpha\frac{d^2\xi_\alpha}{ds^2} \tag{8.2.7}$$

Let φ be the angle between the normal to the surface, $\hat{\mathbf{n}}$, and the principal normal to the curve, $\hat{\mathbf{N}}$; i.e. $\cos\varphi = \hat{\mathbf{n}}\cdot\hat{\mathbf{N}}$. Then, taking the dot product of both sides of Eq. (8.2.7), we obtain

$$\frac{\cos\varphi}{\rho} = \frac{L\,(d\xi_1)^2 + 2M\,d\xi_1\,d\xi_2 + N\,(d\xi_2)^2}{ds^2} \tag{8.2.8}$$

where

$$L = \hat{\mathbf{n}} \cdot \frac{\partial^2 \mathbf{r}}{\partial \xi_1^2}, \quad M = \hat{\mathbf{n}} \cdot \frac{\partial^2 \mathbf{r}}{\partial \xi_1 \partial \xi_2}, \quad N = \hat{\mathbf{n}} \cdot \frac{\partial^2 \mathbf{r}}{\partial \xi_2^2} \tag{8.2.9}$$

The numerator on the right-hand side of Eq. (8.2.8) is called the *second quadratic form* of the surface, and the quantities L, M, and N are the coefficients of the quadratic form.

Equation (8.2.8) can be used to compute the curvatures of the curves obtained by intersecting the surface with normal planes. For the curve generated by a normal plane, $\hat{\mathbf{n}}$ and $\hat{\mathbf{N}}$ have opposite directions ($\varphi = \pi$) because the outward normal (to the right of the curve) is taken positive as we traverse along $d\mathbf{r}$. Thus, Eq. (8.2.8) yields

$$\frac{1}{R} = -\frac{L\,(d\xi_1)^2 + 2M\,d\xi_1\,d\xi_2 + N\,(d\xi_2)^2}{a_1^2\,(d\xi_1)^2 + 2a_1 a_2\,\cos\chi\,d\xi_1\,d\xi_2 + a_2^2\,(d\xi_2)^2} \tag{8.2.10}$$

In particular, the curvatures of the ξ_1 curve ($\xi_2 = $ constant) and ξ_2 curve ($\xi_1 = $ constant) are

$$\frac{1}{R_1} = -\frac{L}{a_1^2}, \quad \frac{1}{R_2} = -\frac{N}{a_2^2} \tag{8.2.11}$$

From this point onwards, we assume that ξ_1 and ξ_2 are orthogonal, that is, curves $\xi_1 = $ constant and $\xi_2 = $ constant are lines of principal curvatures of the undeformed middle surface; then $\cos\chi = 0$ and $M = 0$. The derivatives of the vectors $\hat{\mathbf{n}}$ and $\hat{\mathbf{e}}_\alpha \equiv \mathbf{g}_\alpha / a_\alpha$ (no sum on α) are given by (see Kraus [85])

$$\frac{\partial \hat{\mathbf{n}}}{\partial \xi_\alpha} = \frac{a_\alpha}{R_\alpha}\,\hat{\mathbf{e}}_\alpha = \frac{\mathbf{g}_\alpha}{R_\alpha} \quad (\alpha = 1, 2)$$

$$\frac{\partial \hat{\mathbf{e}}_\alpha}{\partial \xi_\beta} = \frac{1}{a_\alpha}\frac{\partial a_\beta}{\partial \xi_\alpha}\,\hat{\mathbf{e}}_\beta \quad (\alpha \neq \beta) \tag{8.2.12}$$

$$\frac{\partial \hat{\mathbf{e}}_\alpha}{\partial \xi_\alpha} = -\frac{1}{a_\beta}\frac{\partial a_\alpha}{\partial \xi_\beta}\,\hat{\mathbf{e}}_\beta - \frac{a_\alpha}{R_\alpha}\,\hat{\mathbf{n}} \quad (\alpha \neq \beta)$$

In Eq. (8.2.12), no sum on repeated indices is implied. The first equation in Eq. (8.2.12) is known as the *Theorem of Rodrigues*, and the second and third relations are known as the *Weingarten formulas*.

In addition, we have the identities

$$\frac{\partial}{\partial \xi_1}\left(\frac{1}{a_1}\frac{\partial a_2}{\partial \xi_1}\right) + \frac{\partial}{\partial \xi_2}\left(\frac{1}{a_2}\frac{\partial a_1}{\partial \xi_2}\right) = -\frac{a_1 a_2}{R_1 R_2} = -\frac{a_1 a_2}{K} \tag{8.2.13}$$

$$\frac{\partial}{\partial \xi_2}\left(\frac{a_1}{R_1}\right) = \frac{1}{R_2}\frac{\partial a_1}{\partial \xi_2}, \quad \frac{\partial}{\partial \xi_1}\left(\frac{a_2}{R_2}\right) = \frac{1}{R_1}\frac{\partial a_2}{\partial \xi_1} \tag{8.2.14}$$

where $1/K = 1/R_1 R_2$ is called the Gaussian curvature. Equation (8.2.13) is known as the *Gauss characteristic equation*, and the formulas given in Eq. (8.2.14) are called *Mainardi–Codazzi relations*.

The position vector $\mathbf{R}$ of an arbitrary point (ξ_1, ξ_2, ζ) in the shell can be expressed as [see Fig. 8.2.1(c)]

$$\mathbf{R} = \mathbf{r} + \zeta \hat{\mathbf{n}} \tag{8.2.15}$$

where ζ is the distance of the point measured from the middle surface along $\hat{\mathbf{n}}$. The differential line element on a surface away from the middle surface is

$$d\mathbf{R} = \frac{\partial \mathbf{R}}{\partial \xi_i} d\xi_i = \frac{\partial \mathbf{R}}{\partial \xi_\alpha} d\xi_\alpha + \frac{\partial \mathbf{R}}{\partial \zeta} d\zeta = \mathbf{G}_\alpha \, d\xi_\alpha + \hat{\mathbf{n}} \, d\zeta \tag{8.2.16}$$

where

$$\mathbf{G}_\alpha = \frac{\partial \mathbf{R}}{\partial \xi_\alpha} = \frac{\partial \mathbf{r}}{\partial \xi_\alpha} + \zeta \frac{\partial \hat{\mathbf{n}}}{\partial \xi_\alpha} = \left(1 + \frac{\zeta}{R_\alpha}\right) \mathbf{g}_\alpha \quad \text{(no sum on } \alpha) \tag{8.2.17}$$

where the Theorem of Rodrigues [i.e. the first equation in Eq. (8.2.12)] is used in arriving at the last step. The components of the metric tensor $G_{\alpha\beta}$ are defined by

$$G_{\alpha\beta} = \mathbf{G}_\alpha \cdot \mathbf{G}_\beta, \quad A_1 = \sqrt{G_{11}}, \quad A_2 = \sqrt{G_{22}}, \quad A_3 = 1 \tag{8.2.18}$$

where A_1, A_2, and A_3 are known as the *Lamé coefficients*, and they can be expressed in terms of a_α as

$$A_\alpha = a_\alpha \left(1 + \frac{\zeta}{R_\alpha}\right) \quad \text{(no sum on } \alpha) \tag{8.2.19}$$

The square of the distance dS between points (ξ_1, ξ_2, ζ) and $(\xi_1 + d\xi_1, \xi_2 + d\xi_2, \zeta + d\zeta)$ is given by

$$(dS)^2 = d\mathbf{R} \cdot d\mathbf{R} = G_{\alpha\beta} \, d\xi_\alpha d\xi_\beta + (d\zeta)^2$$
$$= A_1^2 (d\xi_1)^2 + A_2^2 (d\xi_2)^2 + A_3^2 (d\zeta)^2 \tag{8.2.20}$$

The Gauss characteristic equation, Eq. (8.2.13), and the Mainardi–Codazzi equations in Eq. (8.2.14) can be generalized for a surface at distance ζ from the middle surface. We have

$$\frac{\partial}{\partial \xi_1} \left(\frac{1}{A_1} \frac{\partial A_2}{\partial \xi_1}\right) + \frac{\partial}{\partial \xi_2} \left(\frac{1}{A_2} \frac{\partial A_1}{\partial \xi_2}\right) = -\frac{a_1 a_2}{R_1 R_2} \tag{8.2.21}$$

$$\frac{\partial^2 A_1}{\partial \xi_2 \partial \zeta} = \frac{1}{A_2} \frac{\partial A_1}{\partial \xi_2} \frac{\partial A_2}{\partial \zeta}, \quad \frac{\partial^2 A_2}{\partial \xi_1 \partial \zeta} = \frac{1}{A_1} \frac{\partial A_2}{\partial \xi_1} \frac{\partial A_1}{\partial \zeta} \tag{8.2.22}$$

In view of the Mainardi–Codazzi conditions in Eq. (8.2.14), Eq. (8.2.22) reduces to

$$\frac{1}{A_2} \frac{\partial A_1}{\partial \xi_2} = \frac{1}{a_2} \frac{\partial a_1}{\partial \xi_2}, \quad \frac{1}{A_1} \frac{\partial A_2}{\partial \xi_1} = \frac{1}{a_1} \frac{\partial a_2}{\partial \xi_1} \tag{8.2.23}$$

Under the assumption that (ξ_1, ξ_2, ζ) is a curvilinear orthogonal system with two principal curvatures, the cross-sectional areas of the edge faces of the shell element shown in Fig. 8.2.2 are

$$dS_1 d\zeta = A_1\, d\xi_1 d\zeta = a_1 \left(1 + \frac{\zeta}{R_1}\right) d\xi_1\, d\zeta$$

$$dS_2 d\zeta = A_2\, d\xi_2 d\zeta = a_2 \left(1 + \frac{\zeta}{R_2}\right) d\xi_2\, d\zeta$$

(8.2.24)

An elemental area of the middle surface $(\zeta = 0)$ is determined by [see Fig. 8.2.3(a)]

$$dA_0 = d\mathbf{r}_1 \times d\mathbf{r}_2 \cdot \hat{\mathbf{n}} = \left(\frac{\partial \mathbf{r}}{\partial \xi_1} \times \frac{\partial \mathbf{r}}{\partial \xi_2} \cdot \hat{\mathbf{n}}\right) d\xi_1\, d\xi_2 = a_1 a_2\, d\xi_1\, d\xi_2 \qquad (8.2.25)$$

and an elemental area of the surface at ζ is given by [see Fig. 8.2.3(b)]

$$dA_\zeta = d\mathbf{R}_1 \times d\mathbf{R}_2 \cdot \hat{\mathbf{n}} = \left(\frac{\partial \mathbf{R}}{\partial \xi_1} \times \frac{\partial \mathbf{R}}{\partial \xi_2} \cdot \hat{\mathbf{n}}\right) d\xi_1\, d\xi_2 = A_1 A_2\, d\xi_1\, d\xi_2 \qquad (8.2.26)$$

The volume of a differential shell element is given by

$$dV = d\mathbf{R}_1 \times d\mathbf{R}_2 \cdot \hat{\mathbf{n}}\, d\zeta = dA_\zeta\, d\zeta = A_1 A_2\, d\xi_1\, d\xi_2\, d\zeta$$

$$= \left[a_1 \left(1 + \frac{\zeta}{R_1}\right)\right]\left[a_2 \left(1 + \frac{\zeta}{R_2}\right)\right] d\xi_1\, d\xi_2\, d\zeta \qquad (8.2.27)$$

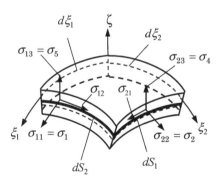

$$dS_1 = a_1\left(1 + \frac{\zeta}{R_1}\right)d\xi_1 = A_1 d\xi_1, \quad dS_2 = a_2\left(1 + \frac{\zeta}{R_2}\right)d\xi_2 = A_2\, d\xi_2$$

Fig. 8.2.2: A differential element of the shell (dS_1 and dS_2 denote the arc lengths).

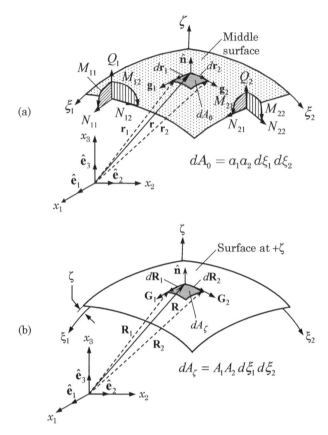

Fig. 8.2.3: (a) Area element on the middle surface ($\zeta = 0$). (b) Area element on a surface at $+\zeta$.

8.2.2   General Strain-Displacement Relations

Let **u** be the displacement vector, with components $u_i(\xi_1, \xi_2, \zeta)$ ($i = 1, 2, 3$) in an orthogonal curvilinear coordinate system ($\xi_1, \xi_2, \xi_3 = \zeta$). Then, the normal ($\varepsilon_{ij}$ for $i = j$ is denoted with ε_i) and shear ($2\varepsilon_{ij} = \gamma_{ij}$ for $i \neq j$) components of the Green–Lagrange strain tensor $E_{ij} = \varepsilon_{ij}$ are given by (no sum on repeated indices; see Palazotto and Dennis [93])

$$
\varepsilon_i = \frac{\partial}{\partial \xi_i}\left(\frac{u_i}{A_i}\right) + \frac{1}{A_i}\sum_{k=1}^{3} \frac{u_k}{A_k}\frac{\partial A_i}{\partial \xi_k} + \frac{1}{2}\left[\frac{\partial}{\partial \xi_i}\left(\frac{u_i}{A_i}\right) + \frac{1}{A_i}\sum_{k=1}^{3}\frac{u_k}{A_k}\frac{\partial A_i}{\partial \xi_k}\right]^2
$$

$$
+ \frac{1}{2A_i^2}\sum_{k=1, k\neq i}^{3}\left(\frac{\partial u_k}{\partial \xi_i} - \frac{u_i}{A_k}\frac{\partial A_i}{\partial \xi_k}\right)^2 \quad (i = 1, 2, 3) \tag{8.2.28}
$$

$$
\gamma_{ij} = \frac{A_i}{A_j}\frac{\partial}{\partial \xi_j}\left(\frac{u_i}{A_i}\right) + \frac{A_j}{A_i}\frac{\partial}{\partial \xi_i}\left(\frac{u_j}{A_j}\right)
$$

$$+ \sum_{k=1, k\neq i, k\neq j}^{3} \frac{1}{A_i A_j} \left(\frac{\partial u_k}{\partial \xi_i} - \frac{u_i}{A_k} \frac{\partial A_i}{\partial \xi_k} \right) \left(\frac{\partial u_k}{\partial \xi_j} - \frac{u_j}{A_k} \frac{\partial A_j}{\partial \xi_k} \right)$$

$$+ \frac{1}{A_j} \left(\frac{\partial u_i}{\partial \xi_j} - \frac{u_j}{A_i} \frac{\partial A_j}{\partial \xi_i} \right) \left[\frac{\partial}{\partial \xi_i} \left(\frac{u_i}{A_i} \right) + \frac{1}{A_i} \sum_{k=1}^{3} \frac{u_k}{A_k} \frac{\partial A_i}{\partial \xi_k} \right]$$

$$+ \frac{1}{A_i} \left(\frac{\partial u_j}{\partial \xi_i} - \frac{u_i}{A_j} \frac{\partial A_i}{\partial \xi_j} \right) \left[\frac{\partial}{\partial \xi_j} \left(\frac{u_j}{A_j} \right) + \frac{1}{A_j} \sum_{k=1}^{3} \frac{u_k}{A_k} \frac{\partial A_j}{\partial \xi_k} \right] \tag{8.2.29}$$

where $i \neq j$ $(i, j = 1, 2, 3)$ in Eq. (8.2.29), and

$$\xi_3 = \zeta, \quad A_1 = a_1 \left(1 + \frac{\zeta}{R_1} \right), \quad A_2 = a_2 \left(1 + \frac{\zeta}{R_2} \right), \quad A_3 = a_3 = 1 \tag{8.2.30}$$

Substituting Eq. (8.2.30) into Eqs. (8.2.28) and (8.2.29) and making use of conditions in Eqs. (8.2.12)–(8.2.14) and Eq. (8.2.23), one obtains

$$\varepsilon_1 = \varepsilon_{11} = \frac{1}{A_1} \left(\frac{\partial u_1}{\partial \xi_1} + \frac{u_2}{a_2} \frac{\partial a_1}{\partial \xi_2} + \frac{a_1}{R_1} u_3 \right) + \frac{1}{2A_1^2} \left[\left(\frac{\partial u_1}{\partial \xi_1} + \frac{u_2}{a_2} \frac{\partial a_1}{\partial \xi_2} + \frac{a_1}{R_1} u_3 \right)^2 \right.$$

$$\left. + \left(\frac{\partial u_2}{\partial \xi_1} - \frac{u_1}{a_2} \frac{\partial a_1}{\partial \xi_2} \right)^2 + \left(\frac{\partial u_3}{\partial \xi_1} - \frac{a_1}{R_1} u_1 \right)^2 \right]$$

$$\varepsilon_2 = \varepsilon_{22} = \frac{1}{A_2} \left(\frac{\partial u_2}{\partial \xi_2} + \frac{u_1}{a_1} \frac{\partial a_2}{\partial \xi_1} + \frac{a_2}{R_2} u_3 \right) + \frac{1}{2A_2^2} \left[\left(\frac{\partial u_2}{\partial \xi_2} + \frac{u_1}{a_1} \frac{\partial a_2}{\partial \xi_1} + \frac{a_2}{R_2} u_3 \right)^2 \right.$$

$$\left. + \left(\frac{\partial u_1}{\partial \xi_2} - \frac{u_2}{a_1} \frac{\partial a_2}{\partial \xi_1} \right)^2 + \left(\frac{\partial u_3}{\partial \xi_2} - \frac{a_2}{R_2} u_2 \right)^2 \right]$$

$$\varepsilon_3 = \varepsilon_{33} = \frac{\partial u_3}{\partial \zeta} + \frac{1}{2} \left[\left(\frac{\partial u_1}{\partial \zeta} \right)^2 + \left(\frac{\partial u_2}{\partial \zeta} \right)^2 + \left(\frac{\partial u_3}{\partial \zeta} \right)^2 \right]$$

$$\varepsilon_4 = 2\varepsilon_{23} = \frac{1}{A_2} \frac{\partial u_3}{\partial \xi_2} + A_2 \frac{\partial}{\partial \zeta} \left(\frac{u_2}{A_2} \right) + \frac{1}{A_2} \left[\frac{\partial u_2}{\partial \zeta} \left(\frac{\partial u_2}{\partial \xi_2} + \frac{u_1}{a_1} \frac{\partial a_2}{\partial \xi_1} + \frac{a_2}{R_2} u_3 \right) \right.$$

$$\left. + \frac{\partial u_1}{\partial \zeta} \left(\frac{\partial u_1}{\partial \xi_2} - \frac{u_2}{a_1} \frac{\partial a_2}{\partial \xi_1} \right) + \frac{\partial u_3}{\partial \zeta} \left(\frac{\partial u_3}{\partial \xi_2} - \frac{a_2}{R_2} u_2 \right) \right]$$

$$\varepsilon_5 = 2\varepsilon_{13} = \frac{1}{A_1} \frac{\partial u_3}{\partial \xi_1} + A_1 \frac{\partial}{\partial \zeta} \left(\frac{u_1}{A_1} \right) + \frac{1}{A_1} \left[\frac{\partial u_1}{\partial \zeta} \left(\frac{\partial u_1}{\partial \xi_1} + \frac{u_2}{a_2} \frac{\partial a_1}{\partial \xi_2} + \frac{a_1}{R_1} u_3 \right) \right.$$

$$\left. + \frac{\partial u_2}{\partial \zeta} \left(\frac{\partial u_2}{\partial \xi_1} - \frac{u_1}{a_2} \frac{\partial a_1}{\partial \xi_2} \right) + \frac{\partial u_3}{\partial \zeta} \left(\frac{\partial u_3}{\partial \xi_1} - \frac{a_1}{R_1} u_1 \right) \right]$$

$$\varepsilon_6 = 2\varepsilon_{12} = \frac{A_2}{A_1} \frac{\partial}{\partial \xi_1} \left(\frac{u_2}{A_2} \right) + \frac{A_1}{A_2} \frac{\partial}{\partial \xi_2} \left(\frac{u_1}{A_1} \right)$$

$$+ \frac{1}{A_1 A_2} \left[\left(\frac{\partial u_1}{\partial \xi_2} - \frac{u_2}{a_1} \frac{\partial a_2}{\partial \xi_1} \right) \left(\frac{\partial u_1}{\partial \xi_1} + \frac{1}{a_2} \frac{\partial a_1}{\partial \xi_2} u_2 + \frac{a_1}{R_1} u_3 \right) \right.$$

$$+ \left(\frac{\partial u_2}{\partial \xi_1} - \frac{u_1}{a_2} \frac{\partial a_1}{\partial \xi_2} \right) \left(\frac{\partial u_2}{\partial \xi_2} + \frac{1}{a_1} \frac{\partial a_2}{\partial \xi_1} u_1 + \frac{a_2}{R_2} u_3 \right)$$

$$\left. + \left(\frac{\partial u_3}{\partial \xi_1} - \frac{a_1}{R_1} u_1 \right) \left(\frac{\partial u_3}{\partial \xi_2} - \frac{a_2}{R_2} u_2 \right) \right] \tag{8.2.31}$$

The linear strain–displacement relations are obtained by setting the terms in the square brackets of Eq. (8.2.31) to zero.

8.2.3 Stress Resultants

Next, we introduce the stress resultants acting on a shell element. The tensile force measured per unit length along a ξ_2-coordinate line on a cross section perpendicular to a ξ_1 coordinate line [see Fig. 8.2.3(a)] is $\sigma_{11} \, dS_2$. The total tensile force on the differential element in the ξ_1-direction can be computed by integrating over the entire thickness of the shell:

$$\int_{-h/2}^{h/2} \sigma_{11} \, dS_2 \, d\zeta = a_2 \left[\int_{-h/2}^{h/2} \sigma_{11} \left(1 + \frac{\zeta}{R_2} \right) d\zeta \right] d\xi_2 \equiv N_{11} \, a_2 \, d\xi_2 \tag{8.2.32}$$

where h is the total thickness of the shell, $\zeta = -h/2$ and $\zeta = h/2$ denote the bottom and top surfaces of the shell, and N_{11} is the membrane force per unit length in the ξ_1-direction, acting on a surface perpendicular to the ξ_1-coordinate (see Fig. 8.2.3):

$$N_{11} = \int_{-h/2}^{h/2} \sigma_{11} \left(1 + \frac{\zeta}{R_2} \right) d\zeta$$

Similarly, the moment of the force $\sigma_{11} \, dS_2$ about the ξ_2-axis is

$$\int_{-h/2}^{h/2} \sigma_{11} \, dS_2 \, \zeta \, d\zeta = a_2 \left[\int_{-h/2}^{h/2} \sigma_{11} \left(1 + \frac{\zeta}{R_2} \right) \zeta \, d\zeta \right] d\xi_2 \equiv M_{11} \, a_2 \, d\xi_2 \tag{8.2.33}$$

Thus, the stress resultants per unit length (see Fig. 8.2.3) can be defined as

$$\begin{Bmatrix} N_{11} \\ N_{12} \\ Q_1 \end{Bmatrix} = \int_{-h/2}^{h/2} \begin{Bmatrix} \sigma_{11} \\ \sigma_{12} \\ K_s \sigma_{13} \end{Bmatrix} \left(1 + \frac{\zeta}{R_2} \right) d\zeta$$

$$\begin{Bmatrix} N_{22} \\ N_{21} \\ Q_2 \end{Bmatrix} = \int_{-h/2}^{h/2} \begin{Bmatrix} \sigma_{22} \\ \sigma_{21} \\ K_s \sigma_{23} \end{Bmatrix} \left(1 + \frac{\zeta}{R_1} \right) d\zeta \tag{8.2.34}$$

$$\begin{Bmatrix} M_{11} \\ M_{12} \end{Bmatrix} = \int_{-h/2}^{h/2} \begin{Bmatrix} \sigma_{11} \\ \sigma_{12} \end{Bmatrix} \left(1 + \frac{\zeta}{R_2}\right) \zeta \, d\zeta$$

$$\begin{Bmatrix} M_{22} \\ M_{21} \end{Bmatrix} = \int_{-h/2}^{h/2} \begin{Bmatrix} \sigma_{22} \\ \sigma_{21} \end{Bmatrix} \left(1 + \frac{\zeta}{R_1}\right) \zeta \, d\zeta$$

(8.2.35)

where K_s is the shear correction factor. Note that, in general, $N_{\alpha\beta} \neq N_{\beta\alpha}$ and $M_{\alpha\beta} \neq M_{\beta\alpha}$ for $\alpha \neq \beta$ $(\alpha, \beta = 1, 2)$. However, if we assume that the thickness of the shell is small as compared to the smallest radius of curvature of the middle surface, one can neglect ζ/R_1 and ζ/R_2. Such shells are called *thin shells*, and for thin shells we have $N_{\alpha\beta} = N_{\beta\alpha}$ and $M_{\alpha\beta} = M_{\beta\alpha}$, as in a plate theory.

8.2.4 Displacement and Strain Fields

In developing a moderately thick (i.e. shear deformation) shell theory with the von Kármán nonlinear strains (i.e. small strains but moderate rotations), the following assumptions (as in the case of plates) are made:

1. The transverse normal is inextensible (i.e. $\varepsilon_3 \approx 0$) and the transverse normal stress is small as compared with the other normal stress components and may be neglected.

2. Normals to the undeformed middle surface of the shell before deformation remain straight, but not necessarily normal after deformation.

3. The deflections and strains are sufficiently small so that the quantities of second- and higher-order magnitudes, except for second-order rotations about the transverse normals, may be neglected in comparison with the first-order terms.

4. The rotations about the ξ_1 and ξ_2 axes are moderate so that we retain second-order terms [i.e. terms that are products and squares of the terms $(\partial u_3/\partial \xi_\alpha) - (a_\alpha u_\alpha/R_\alpha)$] in the strain–displacement relations (the von Kármán nonlinearity).

Love's *first approximation theory* further assumes that: (1) the thickness of the shell is small as compared with the other dimensions, (2) the transverse normals to the undeformed middle surface remain not only straight but also normal to the deformed middle surface after deformation, and (3) the strains are infinitesimal so that all nonlinear terms are neglected. The first assumption allows us to neglect the higher powers of ζ/R and h/R in the derivation of shell theories. The second assumption (also known as the Kirchhoff hypothesis) results in the neglect of transverse shear strains.

Consistent with the assumptions (1)–(4) of a moderately thick shell theory, we assume the following form of the displacement components u_i:

$$u_1(\xi_1, \xi_2, \zeta, t) = u(\xi_1, \xi_2, t) + \zeta\varphi_x(\xi_1, \xi_2, t)$$
$$u_2(\xi_1, \xi_2, \zeta, t) = v(\xi_1, \xi_2, t) + \zeta\varphi_y(\xi_1, \xi_2, t) \qquad (8.2.36)$$
$$u_3(\xi_1, \xi_2, \zeta, t) = w(\xi_1, \xi_2, t)$$

where (u, v, w) are the displacements of a point $(\xi_1, \xi_2, 0)$ on the middle surface of the shell in the ξ_i-direction, and (ϕ_x, ϕ_y) are the rotations of a normal to the reference surface about the ξ_2 and ξ_1 axes, respectively

$$\phi_x = \frac{\partial u_1}{\partial \zeta}, \quad \phi_y = \frac{\partial u_2}{\partial \zeta} \qquad (8.2.37)$$

The five variables $(u, v, w, \phi_x, \phi_y)$ completely determine the displacements of an arbitrary point (ξ_1, ξ_2, ζ) in the shell.

In writing the von Kármán strains, we make approximations of the type $\zeta^2\varphi_\alpha^2 \approx 0$, $\zeta u_\alpha^0 \varphi_\alpha \approx 0$, and $\zeta u_{3,\alpha}\varphi_\alpha \approx 0$ (no sum on α) and

$$\left[\frac{\partial u_3}{\partial \xi_\alpha} - \frac{a_\alpha}{R_\alpha}(u_\alpha^0 + \zeta\varphi_\alpha)\right]^2 \approx \left(\frac{\partial u_3^0}{\partial \xi_\alpha} - \frac{a_\alpha u_\alpha^0}{R_\alpha}\right)^2 \qquad (8.2.38)$$

and use the identity

$$A_\alpha \frac{\partial}{\partial \zeta}\left(\frac{u_\alpha}{A_\alpha}\right) = \frac{\partial u_\alpha}{\partial \zeta} + A_\alpha u_\alpha \frac{\partial}{\partial \zeta}\left(\frac{1}{A_\alpha}\right) = \frac{\partial u_\alpha}{\partial \zeta} - \frac{u_\alpha}{A_\alpha}\frac{a_\alpha}{R_\alpha} = \frac{a_\alpha}{A_\alpha}\left(\varphi_\alpha - \frac{u_\alpha^0}{R_\alpha}\right)$$
$$(8.2.39)$$

The von Kármán nonlinear strains can be expressed as the sum of membrane strains ε_i^0 and flexural strains ε_i^1 (see [56] for details)

$$\{\varepsilon\} = \{\varepsilon^0\} + \zeta\{\varepsilon^1\} \qquad (8.2.40a)$$

where

$$\left\{\begin{array}{c}\varepsilon_1^0 \\ \varepsilon_2^0 \\ \varepsilon_4^0 \\ \varepsilon_5^0 \\ \varepsilon_6^0\end{array}\right\} = \left\{\begin{array}{c}\frac{1}{A_1}\left[\frac{\partial u}{\partial \xi_1} + \frac{v}{a_2}\frac{\partial a_1}{\partial \xi_2} + \frac{a_1 w}{R_1} + \frac{1}{2A_1}\left(\frac{\partial w}{\partial \xi_1} - \frac{a_1 u}{R_1}\right)^2\right] \\[2mm] \frac{1}{A_2}\left[\frac{\partial v}{\partial \xi_2} + \frac{u}{a_1}\frac{\partial a_2}{\partial \xi_1} + \frac{a_2 w}{R_2} + \frac{1}{2A_2}\left(\frac{\partial w}{\partial \xi_2} - \frac{a_2 v}{R_2}\right)^2\right] \\[2mm] \frac{a_2}{A_2}\left(\frac{1}{a_2}\frac{\partial w}{\partial \xi_2} - \frac{v}{R_2} + \phi_y\right) \\[2mm] \frac{a_1}{A_1}\left(\frac{1}{a_1}\frac{\partial w}{\partial \xi_1} - \frac{u}{R_1} + \phi_x\right) \\[2mm] \frac{1}{A_1}\left[\frac{\partial v}{\partial \xi_1} - \frac{u}{a_2}\frac{\partial a_1}{\partial \xi_2} + \frac{1}{2A_2}\left(\frac{\partial w}{\partial \xi_1} - \frac{a_1 u}{R_1}\right)\left(\frac{\partial w}{\partial \xi_2} - \frac{a_2 v}{R_2}\right)\right] \\[2mm] +\frac{1}{A_2}\left[\frac{\partial u}{\partial \xi_2} - \frac{v}{a_1}\frac{\partial a_2}{\partial \xi_1} + \frac{1}{2A_1}\left(\frac{\partial w}{\partial \xi_1} - \frac{a_1 u}{R_1}\right)\left(\frac{\partial w}{\partial \xi_2} - \frac{a_2 v}{R_2}\right)\right]\end{array}\right\} \qquad (8.2.40b)$$

$$
\left\{
\begin{array}{c}
\varepsilon_1^1 \\
\varepsilon_2^1 \\
\varepsilon_4^1 \\
\varepsilon_5^1 \\
\varepsilon_6^1
\end{array}
\right\}
=
\left\{
\begin{array}{c}
\dfrac{1}{A_1}\left(\dfrac{\partial \phi_x}{\partial \xi_1} + \dfrac{\phi_y}{a_2}\dfrac{\partial a_1}{\partial \xi_2}\right) \\[2mm]
\dfrac{1}{A_2}\left(\dfrac{\partial \phi_y}{\partial \xi_2} + \dfrac{\phi_x}{a_1}\dfrac{\partial a_2}{\partial \xi_1}\right) \\[2mm]
0 \\[1mm]
0 \\[2mm]
\dfrac{1}{A_1}\left(\dfrac{\partial \phi_y}{\partial \xi_1} - \dfrac{\phi_x}{a_2}\dfrac{\partial a_1}{\partial \xi_2}\right) + \dfrac{1}{A_2}\left(\dfrac{\partial \phi_x}{\partial \xi_2} - \dfrac{\phi_y}{a_1}\dfrac{\partial a_2}{\partial \xi_1}\right)
\end{array}
\right\}
\tag{8.2.40c}
$$

8.2.5 Equations of Equilibrium

The equations governing generalized Sanders shell theory can be derived using the principle of virtual displacements. The generalization is to include the von Kármán strains. The displacement field in Eq. (8.2.36) and strains in Eq. (8.2.40) are used to derive the governing equations. In addition, the vanishing of the moments about the normal to the differential element (Fig. 8.2.3) yields the following additional relation among the twisting moments and surface shear forces:

$$
\frac{M_{21}}{R_2} - \frac{M_{12}}{R_1} + N_{21} - N_{12} = 0 \tag{8.2.41}
$$

which must be accounted for in the formulation; otherwise, it will lead to inconsistency associated with rigid body rotations (i.e. a rigid body rotation gives a non-vanishing torsion except for flat plates and spherical shells). The equations of equilibrium thus obtained are (see [56] for details)

$$
\frac{1}{a_2 a_1}\left\{\frac{\partial}{\partial \xi_1}(a_2 N_{11}) + \frac{\partial}{\partial \xi_2}(a_1 N_{21}) + N_{12}\frac{\partial a_1}{\partial \xi_2} - N_{22}\frac{\partial a_2}{\partial \xi_1} + C_0 a_1 \frac{\partial \tilde{M}_{12}}{\partial \xi_2}\right\}
$$
$$
+ \frac{\hat{N}_{11}}{a_1 R_1}\left(\frac{\partial w}{\partial \xi_1} - \frac{a_1 u}{R_1}\right) + \frac{\tilde{N}_{12}}{a_2 R_1}\left(\frac{\partial w}{\partial \xi_2} - \frac{a_2 v}{R_2}\right) + \frac{Q_1}{R_1} + f_1 = 0 \tag{8.2.42}
$$

$$
\frac{1}{a_1 a_2}\left\{\frac{\partial}{\partial \xi_2}(a_1 N_{22}) + \frac{\partial}{\partial \xi_1}(a_2 N_{12}) + N_{21}\frac{\partial a_2}{\partial \xi_1} - N_{11}\frac{\partial a_1}{\partial \xi_2} - C_0 a_2 \frac{\partial \tilde{M}_{12}}{\partial \xi_1}\right\}
$$
$$
+ \frac{\hat{N}_{22}}{a_2 R_2}\left(\frac{\partial w}{\partial \xi_2} - \frac{a_2 v}{R_2}\right) + \frac{\tilde{N}_{12}}{a_1 R_2}\left(\frac{\partial w}{\partial \xi_1} - \frac{a_1 u}{R_1}\right) + \frac{Q_2}{R_2} + f_2 = 0 \tag{8.2.43}
$$

$$
\frac{1}{a_1 a_2}\left\{\frac{\partial}{\partial \xi_1}\left[\frac{a_2}{a_1}\hat{N}_{11}\left(\frac{\partial w}{\partial \xi_1} - \frac{a_1 u}{R_1}\right) + \tilde{N}_{12}\left(\frac{\partial w}{\partial \xi_2} - \frac{a_2 v}{R_2}\right)\right]\right.
$$
$$
+ \frac{\partial}{\partial \xi_2}\left[\frac{a_1}{a_2}\hat{N}_{22}\left(\frac{\partial w}{\partial \xi_2} - \frac{a_2 v}{R_2}\right) + \tilde{N}_{12}\left(\frac{\partial w}{\partial \xi_1} - \frac{a_1 u}{R_1}\right)\right]
$$
$$
\left. + \frac{\partial}{\partial \xi_1}(a_2 Q_1) + \frac{\partial}{\partial \xi_2}(a_1 Q_2)\right\} - \left(\frac{N_{11}}{R_1} + \frac{N_{22}}{R_2}\right) + f_3 = 0 \tag{8.2.44}
$$

$$
\frac{1}{a_1 a_2}\left[\frac{\partial}{\partial \xi_1}(a_2 M_{11}) + \frac{\partial}{\partial \xi_2}(a_1 M_{21}) - M_{22}\frac{\partial a_2}{\partial \xi_1} + M_{12}\frac{\partial a_1}{\partial \xi_2}\right] - Q_1 = 0 \tag{8.2.45}
$$

$$\frac{1}{a_2 a_1}\left[\frac{\partial}{\partial \xi_1}(a_2 M_{12}) + \frac{\partial}{\partial \xi_2}(a_1 M_{22}) + M_{21}\frac{\partial a_2}{\partial \xi_1} - M_{11}\frac{\partial a_1}{\partial \xi_2}\right] - Q_2 = 0 \quad (8.2.46)$$

where

$$\hat{N}_{11} = \int_{-h/2}^{h/2} \sigma_1 \left(1 + \frac{\zeta}{R_2}\right)\left(1 + \frac{\zeta}{R_1}\right)^{-1} d\zeta$$

$$\hat{N}_{22} = \int_{-h/2}^{h/2} \sigma_2 \left(1 + \frac{\zeta}{R_1}\right)\left(1 + \frac{\zeta}{R_2}\right)^{-1} d\zeta \qquad (8.2.47)$$

$$\tilde{N}_{12} = \tilde{N}_{21} = \int_{-h/2}^{h/2} \sigma_6 \, d\zeta, \quad \tilde{M}_{12} = \tilde{M}_{21} = \int_{-h/2}^{h/2} \zeta \sigma_6 \, d\zeta$$

and

$$C_0 = \frac{1}{2}\left(\frac{1}{R_1} - \frac{1}{R_2}\right) \qquad (8.2.48)$$

We note, even for isotropic shells, the surface (or membrane) equations of motion, Eqs. (8.2.42) and (8.2.43), are coupled to the bending (flexural) equations of motion, Eqs. (8.2.44)–(8.2.46), due to the presence of the bending stress resultants Q_1, Q_2, and $C_0\tilde{M}_{21}$ in Eqs. (8.2.42) and (8.2.43), and membrane stress resultants N_{11}, N_{22}, $\hat{N}_{11}$, $\hat{N}_{22}$ and $\tilde{N}_{12}$ in Eq. (8.2.44). This coupling is responsible for the extraordinary capacity of an egg, despite its thinness, to withstand normal forces. Thus, a shell is a more efficient structure than a flat plate.

In the remainder of this development, we omit the term ζ/R in the definition of the stress resultants (amounts assuming that the shell is thin)

$$\left(1 + \frac{\zeta}{R_\alpha}\right) \approx 1, \quad A_\alpha = a_\alpha \qquad (8.2.49)$$

and we have $N_{12} = N_{21}$ and $M_{12} = M_{21}$, and the equations of equilibrium of the generalized Sanders' shell theory are given by

$$-\left(\frac{\partial N_{xx}}{\partial x} + \frac{\partial N_{xy}}{\partial y}\right) - \frac{Q_x + \mathcal{N}_1}{R_1} - f_x = 0 \qquad (8.2.50)$$

$$-\left(\frac{\partial N_{xy}}{\partial x} + \frac{\partial N_{yy}}{\partial y}\right) - \frac{Q_y + \mathcal{N}_2}{R_2} - f_y = 0 \qquad (8.2.51)$$

$$-\left(\frac{\partial Q_x}{\partial x} + \frac{\partial Q_y}{\partial y}\right) + \frac{N_{xx}}{R_1} + \frac{N_{yy}}{R_2} - \mathcal{N}_3 - q = 0 \qquad (8.2.52)$$

$$-\left(\frac{\partial M_{xx}}{\partial x} + \frac{\partial M_{xy}}{\partial y}\right) + Q_x = 0 \qquad (8.2.53)$$

$$-\left(\frac{\partial M_{xy}}{\partial x} + \frac{\partial M_{yy}}{\partial y}\right) + Q_y = 0 \qquad (8.2.54)$$

where we have used the notation

$$
\begin{aligned}
&dx = dx_1 = a_1 d\xi_1, \quad dy = dx_2 = a_2 d\xi_2 \\
&N_{xx} = N_{11}, \quad N_{xy} = N_{12} = N_{21}, \quad N_{yy} = N_{22} \\
&M_{xx} = M_{11}, \quad M_{xy} = M_{12} = M_{21}, \quad M_{yy} = N_{22} \\
&Q_x = Q_1, \quad Q_y = Q_2, \quad f_x = f_1, \quad f_y = f_2, \quad q = f_3
\end{aligned}
\tag{8.2.55}
$$

$$
\mathcal{N}_1(u, v, w, \phi_x, \phi_y) = N_{xx}\left(\frac{\partial w}{\partial x} - \frac{u}{R_1}\right) + N_{xy}\left(\frac{\partial w}{\partial y} - \frac{v}{R_2}\right)
$$

$$
\mathcal{N}_2(u, v, w, \phi_x, \phi_y) = N_{xy}\left(\frac{\partial w}{\partial x} - \frac{u}{R_1}\right) + N_{yy}\left(\frac{\partial w}{\partial y} - \frac{v}{R_2}\right)
\tag{8.2.56}
$$

$$
\mathcal{N}_3(u, v, w, \phi_x, \phi_y) = \frac{\partial \mathcal{N}_1}{\partial x} + \frac{\partial \mathcal{N}_2}{\partial y}
$$

8.2.6 Shell Constitutive Relations

The stress resultants are related to the strains, for an orthotropic shell in the absence of thermal and other influences, as follows:

$$
\begin{Bmatrix} N_{xx} \\ N_{yy} \\ N_{xy} \end{Bmatrix} = \begin{bmatrix} A_{11} & A_{12} & 0 \\ A_{12} & A_{22} & 0 \\ 0 & 0 & A_{66} \end{bmatrix} \begin{Bmatrix} \frac{\partial u}{\partial x} + \frac{w}{R_1} + \frac{1}{2}\left(\frac{\partial w}{\partial x} - \frac{u}{R_1}\right)^2 \\ \frac{\partial v}{\partial y} + \frac{w}{R_2} + \frac{1}{2}\left(\frac{\partial w}{\partial y} - \frac{v}{R_2}\right)^2 \\ \frac{\partial u}{\partial y} + \frac{\partial v}{\partial x} + \left(\frac{\partial w}{\partial x} - \frac{u}{R_1}\right)\left(\frac{\partial w}{\partial y} - \frac{v}{R_2}\right) \end{Bmatrix}
\tag{8.2.57}
$$

$$
\begin{Bmatrix} M_{xx} \\ M_{yy} \\ M_{xy} \end{Bmatrix} = \begin{bmatrix} D_{11} & D_{12} & 0 \\ D_{12} & D_{22} & 0 \\ 0 & 0 & D_{66} \end{bmatrix} \begin{Bmatrix} \frac{\partial \phi_x}{\partial x} \\ \frac{\partial \phi_y}{\partial y} \\ \frac{\partial \phi_x}{\partial y} + \frac{\partial \phi_y}{\partial x} \end{Bmatrix}
\tag{8.2.58}
$$

$$
\begin{Bmatrix} Q_y \\ Q_x \end{Bmatrix} = K_s \begin{bmatrix} A_{44} & 0 \\ 0 & A_{55} \end{bmatrix} \begin{Bmatrix} \frac{\partial w}{\partial y} - \frac{v}{R_2} + \phi_y \\ \frac{\partial w}{\partial x} - \frac{u}{R_1} + \phi_x \end{Bmatrix}
\tag{8.2.59}
$$

where $K_s = 5/6$ is the shear correction coefficient and

$$
A_{ij} = Q_{ij} h \quad (i, j = 1, 2, 4, 5, 6), \quad D_{ij} = Q_{ij}\frac{h^3}{12} \quad (i, j = 1, 2, 6)
$$

$$
Q_{11} = \frac{E_1}{1 - \nu_{12}\nu_{21}}, \quad Q_{22} = Q_{11}\frac{E_2}{E_1}, \quad Q_{12} = \nu_{12}Q_{22}, \quad Q_{66} = G_{12}
\tag{8.2.60}
$$

8.3 Finite Element Formulation

8.3.1 Weak Forms

The displacement finite element model of the equations governing doubly-curved shells, Eqs. (8.2.50)–(8.2.54), can be derived in a manner similar to that of the first-order shear deformation plate theory (FSDT) discussed in Chapter

7. In fact, the finite element model of doubly-curved shells is identical to that of FSDT with additional terms in the stiffness coefficients. It will reduce to FSDT model when $1/R_1$ and $1/R_2$ are set to zero. For the sake of completeness, the main equations are presented here.

We begin with the weak forms of Eqs. (8.2.50)–(8.2.54):

$$0 = \int_{\Omega^e} \left[\frac{\partial \delta u}{\partial x} N_{xx} + \frac{\partial \delta u}{\partial y} N_{xy} - \delta u \frac{Q_x + \mathcal{N}_1}{R_1} - \delta u f_x \right] dx\, dy - \oint_{\Gamma^e} P_x \delta u\, ds \quad (8.3.1)$$

$$0 = \int_{\Omega^e} \left[\frac{\partial \delta v}{\partial x} N_{xy} + \frac{\partial \delta v}{\partial y} N_{yy} - \delta v \frac{Q_y + \mathcal{N}_2}{R_2} - \delta v f_y \right] dx\, dy - \oint_{\Gamma^e} P_y \delta v\, ds \quad (8.3.2)$$

$$0 = \int_{\Omega^e} \left[\frac{\partial \delta w}{\partial x} Q_x + \frac{\partial \delta w}{\partial y} Q_y + \delta w \left(\frac{N_{xx}}{R_1} + \frac{N_{yy}}{R_2} \right) - \delta w q \right.$$

$$\left. + \frac{\partial \delta w}{\partial x} \mathcal{N}_1 + \frac{\partial \delta w}{\partial y} \mathcal{N}_2 \right] dx\, dy - \oint_{\Gamma^e} V_n \delta w\, ds \quad (8.3.3)$$

$$0 = \int_{\Omega^e} \left(\frac{\partial \delta \phi_x}{\partial x} M_{xx} + \frac{\partial \delta \phi_x}{\partial y} M_{xy} + \delta \phi_x Q_x \right) dx\, dy - \oint_{\Gamma^e} T_x \delta \phi_x\, ds \quad (8.3.4)$$

$$0 = \int_{\Omega^e} \left(\frac{\partial \delta \phi_y}{\partial x} M_{xy} + \frac{\partial \delta \phi_y}{\partial y} M_{yy} + \delta \phi_y Q_y \right) dx\, dy - \oint_{\Gamma^e} T_y \delta \phi_y\, ds \quad (8.3.5)$$

where the stress resultants $(\mathbf{N}, \mathbf{M}, \mathbf{Q})$ are defined by Eqs. (8.2.57)–(8.2.59). We note from the boundary terms in Eqs. (8.3.1)–(8.3.5) that $(u, v, w, \phi_x, \phi_y)$ are the primary variables. Therefore, we can use the C^0-interpolation of the displacements. The secondary variables are

$$\begin{aligned}
P_x &\equiv N_{xx} n_x + N_{xy} n_y, & P_y &\equiv N_{xy} n_x + N_{yy} n_y \\
T_x &\equiv M_{xx} n_x + M_{xy} n_y, & T_y &\equiv M_{xy} n_x + M_{yy} n_y \\
V_n &\equiv (Q_x + \mathcal{N}_1) n_x + (Q_y + \mathcal{N}_2) n_y
\end{aligned} \quad (8.3.6)$$

where (n_x, n_y) are the direction cosines of the unit normal to the surface. Note that in the case of shells, surface displacements are coupled to the transverse displacement even for linear analysis of isotropic shells.

8.3.2 Finite Element Model

Here we use equal interpolation of all variables:

$$u \approx \sum_{j=1}^{p} u_j \psi_j^e(x, y), \quad v \approx \sum_{j=1}^{p} v_j \psi_j^e(x, y), \quad w \approx \sum_{j=1}^{p} w_j \psi_j^e(x, y)$$

$$\phi_x \approx \sum_{j=1}^{p} S_j^1 \psi_j^e(x, y), \quad \phi_y \approx \sum_{j=1}^{p} S_j^2 \psi_j^e(x, y) \quad (8.3.7)$$

where ψ_j^e are Lagrange interpolation functions. In the present study, equal interpolation of five displacements, with $p = 1, 2, \ldots$, is used. Note that the finite element model developed here for doubly-curved shells contains the FSDT plate element as a special case (set $1/R_1 = 0$ and $1/R_2 = 0$).

Substituting Eq. (8.3.7) for $(u, v, w, \phi_x, \phi_y)$ into the weak forms in Eqs. (8.3.1)–(8.3.5), we obtain the finite element model of the first-order shear deformation shell theory:

$$
\begin{bmatrix}
\mathbf{K}^{11} & \mathbf{K}^{12} & \mathbf{K}^{13} & \mathbf{K}^{14} & \mathbf{K}^{15} \\
\mathbf{K}^{21} & \mathbf{K}^{22} & \mathbf{K}^{23} & \mathbf{K}^{24} & \mathbf{K}^{25} \\
\mathbf{K}^{31} & \mathbf{K}^{32} & \mathbf{K}^{33} & \mathbf{K}^{34} & \mathbf{K}^{35} \\
\mathbf{K}^{41} & \mathbf{K}^{42} & \mathbf{K}^{43} & \mathbf{K}^{44} & \mathbf{K}^{45} \\
\mathbf{K}^{51} & \mathbf{K}^{52} & \mathbf{K}^{53} & \mathbf{K}^{54} & \mathbf{K}^{55}
\end{bmatrix}
\begin{Bmatrix}
\mathbf{u} \\
\mathbf{v} \\
\mathbf{w} \\
\mathbf{S}^1 \\
\mathbf{S}^2
\end{Bmatrix}
=
\begin{Bmatrix}
\mathbf{F}^1 \\
\mathbf{F}^2 \\
\mathbf{F}^3 \\
\mathbf{F}^4 \\
\mathbf{F}^5
\end{Bmatrix}
\tag{8.3.8}
$$

where the coefficients of the submatrices $\mathbf{K}^{\alpha\beta}$ are

$$
K_{ij}^{1\alpha} = \int_{\Omega^e} \left(\frac{\partial \psi_i^e}{\partial x} N_{1j}^\alpha + \frac{\partial \psi_i^e}{\partial y} N_{6j}^\alpha - \psi_i^e \frac{Q_{1j}^\alpha + N_{1j}^\alpha}{R_1} \right) dx\, dy
$$

$$
K_{ij}^{2\alpha} = \int_{\Omega^e} \left(\frac{\partial \psi_i^e}{\partial x} N_{6j}^\alpha + \frac{\partial \psi_i^e}{\partial y} N_{2j}^\alpha - \psi_i^e \frac{Q_{2j}^\alpha + N_{2j}^\alpha}{R_2} \right) dx\, dy
$$

$$
K_{ij}^{3\alpha} = \int_{\Omega^e} \left[\frac{\partial \psi_i^e}{\partial x} Q_{1j}^\alpha + \frac{\partial \psi_i^e}{\partial y} Q_{2j}^\alpha + \psi_i^e \left(\frac{N_{1j}^\alpha}{R_1} + \frac{N_{2j}^\alpha}{R_2} \right) \right.
$$

$$
\left. + \frac{\partial \psi_i^e}{\partial x} \mathcal{N}_{1j}^\alpha + \frac{\partial \psi_i^e}{\partial y} \mathcal{N}_{2j}^\alpha \right] dx\, dy
$$

$$
K_{ij}^{4\alpha} = \int_{\Omega^e} \left(\frac{\partial \psi_i^e}{\partial x} M_{1j}^\alpha + \frac{\partial \psi_i^e}{\partial y} M_{6j}^\alpha + \psi_i^e Q_{1j}^\alpha \right) dx\, dy
$$

$$
K_{ij}^{5\alpha} = \int_{\Omega^e} \left(\frac{\partial \psi_i^e}{\partial x} M_{6j}^\alpha + \frac{\partial \psi_i^e}{\partial y} M_{2j}^\alpha + \psi_i^e Q_{2j}^\alpha \right) dx\, dy
$$

$$
F_i^1 = \int_{\Omega^e} f_x \psi_i^e \, dx dy + \oint_{\Gamma^e} P_x \psi_i^e \, dx dy
$$

$$
F_i^2 = \int_{\Omega^e} f_y \psi_i^e \, dx dy + \oint_{\Gamma^e} P_y \psi_i^e \, dx\, dy
$$

$$
F_i^3 = \int_{\Omega^e} q \psi_i^e \, dx\, dy + \oint_{\Gamma^e} V_n \psi_i^e \, ds
$$

$$
F_i^4 = \oint_{\Gamma^e} T_x \psi_i^e \, dx\, dy, \quad F_i^5 = \oint_{\Gamma^e} T_y \psi_i^e \, dx dy
\tag{8.3.9}
$$

The coefficients N_{Ij}^α, M_{Ij}^α, Q_{Ij}^α, and $\mathcal{N}_{Ij}^\alpha$ for $\alpha = 1, 2, \ldots, 5$ and $I = 1, 2, 6$ are

$$
N_{1j}^1 = A_{11} \left[\frac{\partial \psi_j^e}{\partial x} + \left(\frac{u}{2R_1^2} - \frac{1}{R_1} \frac{\partial w}{\partial x} \right) \psi_j^e \right]
$$

$$
N_{1j}^2 = A_{12} \left[\frac{\partial \psi_j^e}{\partial y} + \left(\frac{v}{2R_2^2} - \frac{1}{R_2} \frac{\partial w}{\partial y} \right) \psi_j^e \right]
$$

$$N_{1j}^3 = \left(\frac{A_{11}}{R_1} + \frac{A_{12}}{R_2} \right) \psi_j^e + \frac{1}{2} \left(A_{11} \frac{\partial w}{\partial x} \frac{\partial \psi_j^e}{\partial x} + A_{12} \frac{\partial w}{\partial y} \frac{\partial \psi_j^e}{\partial y} \right)$$

$$N_{6j}^1 = A_{66} \left[\frac{\partial \psi_j^e}{\partial y} - \frac{1}{R_1} \left(\frac{\partial w}{\partial y} - \frac{1}{2} \frac{v}{R_2} \right) \psi_j^e \right]$$

$$N_{6j}^2 = A_{66} \left[\frac{\partial \psi_j^e}{\partial x} - \frac{1}{R_2} \left(\frac{\partial w}{\partial x} - \frac{1}{2} \frac{u}{R_1} \right) \psi_j^e \right]$$

$$N_{6j}^3 = \frac{A_{66}}{2} \left(\frac{\partial w}{\partial x} \frac{\partial \psi_j^e}{\partial y} + \frac{\partial w}{\partial y} \frac{\partial \psi_j^e}{\partial x} \right)$$

$$N_{2j}^1 = A_{12} \left[\frac{\partial \psi_j^e}{\partial x} + \left(\frac{u}{2R_1^2} - \frac{1}{R_1} \frac{\partial w}{\partial x} \right) \psi_j^e \right]$$

$$N_{2j}^2 = A_{22} \left[\frac{\partial \psi_j^e}{\partial y} + \left(\frac{v}{2R_2^2} - \frac{1}{R_2} \frac{\partial w}{\partial y} \right) \psi_j^e \right]$$

$$N_{2j}^3 = \left(\frac{A_{12}}{R_1} + \frac{A_{22}}{R_2} \right) \psi_j^e + \frac{1}{2} \left(A_{12} \frac{\partial w}{\partial x} \frac{\partial \psi_j^e}{\partial x} + A_{22} \frac{\partial w}{\partial y} \frac{\partial \psi_j^e}{\partial y} \right)$$

$$N_{6j}^4 = 0, \quad N_{6j}^5 = 0, \quad N_{1j}^4 = 0, \quad N_{1j}^5 = 0, \quad N_{2j}^4 = 0, \quad N_{2j}^5 = 0$$

$$Q_{1j}^1 = -\frac{K_s A_{55}}{R_1} \psi_j^e, \quad Q_{1j}^3 = K_s A_{55} \frac{\partial \psi_j^e}{\partial x}, \quad Q_{1j}^4 = K_s A_{55} \psi_j^e$$

$$Q_{2j}^2 = -\frac{K_s A_{44}}{R_2} \psi_j^e, \quad Q_{2j}^3 = K_s A_{44} \frac{\partial \psi_j^e}{\partial y}, \quad Q_{2j}^5 = K_s A_{44} \psi_j^e$$

$$Q_{1j}^2 = 0, \quad Q_{1j}^5 = 0, \quad Q_{2j}^1 = 0, \quad Q_{2j}^4 = 0$$

$$\mathcal{N}_{1j}^1 = -\frac{N_{xx}}{R_1} \psi_j^e, \quad \mathcal{N}_{1j}^2 = -\frac{N_{xy}}{R_2} \psi_j^e, \quad \mathcal{N}_{1j}^3 = N_{xx} \frac{\partial \psi_j^e}{\partial x} + N_{xy} \frac{\partial \psi_j^e}{\partial y}$$

$$\mathcal{N}_{2j}^1 = -\frac{N_{xy}}{R_1} \psi_j^e, \quad \mathcal{N}_{2j}^2 = -\frac{N_{yy}}{R_2} \psi_j^e, \quad \mathcal{N}_{2j}^3 = N_{xy} \frac{\partial \psi_j^e}{\partial x} + N_{yy} \frac{\partial \psi_j^e}{\partial y}$$

$$\mathcal{N}_{1j}^4 = 0, \quad \mathcal{N}_{1j}^5 = 0, \quad \mathcal{N}_{2j}^4 = 0, \quad \mathcal{N}_{2j}^5 = 0$$

$$M_{1j}^1 = M_{1j}^2 = M_{1j}^3 = 0, \quad M_{1j}^4 = D_{11} \frac{\partial \psi_j^e}{\partial x}, \quad M_{1j}^5 = D_{12} \frac{\partial \psi_j^e}{\partial y}$$

$$M_{2j}^1 = M_{2j}^2 = M_{2j}^3 = 0, \quad M_{2j}^4 = D_{12} \frac{\partial \psi_j^e}{\partial x}, \quad M_{2j}^5 = D_{22} \frac{\partial \psi_j^e}{\partial y}$$

$$M_{6j}^1 = M_{6j}^2 = M_{6j}^3 = 0, \quad M_{6j}^4 = D_{66} \frac{\partial \psi_j^e}{\partial y}, \quad M_{6j}^5 = D_{66} \frac{\partial \psi_j^e}{\partial x} \qquad (8.3.10)$$

8.3.3   Linear Analysis

Numerical results are presented here for linear analysis of some benchmark problems from the literature. Quadrilateral elements are used with selective integration rule to evaluate the stiffness coefficients: full integration for bending terms and reduced integration for bending-membrane coupling terms as well as transverse shear terms.

Example 8.3.1

Consider the deformation of a cylindrical shell with internal pressure. The shell is clamped at its ends (see Fig. 8.3.1). The geometric and material parameters used are:

$$R_1 = 10^{30} \ (\text{or } 1/R_1 \approx 0), \quad R_2 = R = 20\,\text{in}, \quad a = 20\,\text{in}, \quad h = 1\,\text{in} \tag{1}$$

$$E_1 = 7.5 \times 10^6 \,\text{psi}, \ E_2 = 2 \times 10^6 \,\text{psi}, \ G_{12} = 1.25 \times 10^6 \,\text{psi}$$
$$G_{13} = G_{23} = 0.625 \times 10^6 \,\text{psi}, \ \nu_{12} = 0.25 \tag{2}$$

Assume a transverse pressure of $p_0 = (6.41/\pi)$ ksi. Using 4×4 mesh of four-node rectangular elements (4×4L4) and 2×2 mesh of nine-node (quadratic) rectangular elements (2×2Q9) in an octant ($u = \phi_x = 0$ at $x = 0$; $v = \phi_y = 0$ at $y = 0, \pi R/2$; and $u = v = w = \phi_x = \phi_y = 0$ at $x = a/2$) of the shell, obtain the maximum transverse deflection.

Solution: The numerical results obtained are presented in Table 8.3.1. The reference solutions by Rao [115] and Timoshenko and Woinowsky-Krieger [86] did not account for the transverse shear strains.

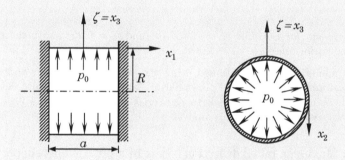

Fig. 8.3.1: Clamped cylindrical shell with internal pressure.

Table 8.3.1: Maximum radial deflection (w in) of a clamped cylindrical shell with internal pressure.

Laminate	Present solution		Ref. [115]	Ref. [86]
	4×4L4	2×2Q9		
$0°$	0.3754	0.3727	0.3666	0.367

Example 8.3.2

Consider a doubly-curved spherical shell panel ($R_1 = R_2 = R$) under a central point load $F_0 = 100$ lbs. The shell panel is simply supported at edges (see Fig. 8.3.2). The geometric and material parameters to be used are:

$$R_1 = R_2 = R = 96\,\text{in}, \quad a = b = 32\,\text{in}, \quad h = 0.1\,\text{in}$$
$$E_1 = 25E_2, \quad E_2 = 10^6 \,\text{psi}, \quad G_{12} = G_{13} = 0.5E_2, \quad G_{23} = 0.2E_2, \quad \nu_{12} = 0.25 \tag{1}$$

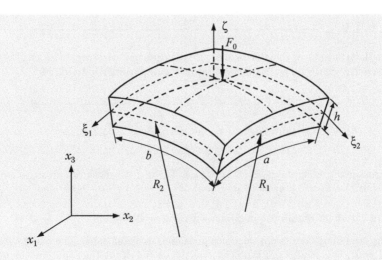

Fig. 8.3.2: Simply supported spherical shell panel under central point load.

Analyze the problem for isotropic ($E = 25 \times 10^6$ and $\nu = 0.25$) and orthotropic material properties using various meshes.

Solution: The numerical results obtained using various meshes of linear and quadratic elements in a quadrant of the shell are presented in Table 8.3.2. The finite element solution converges with refinement of the mesh to the series solution of Vlasov [94], who did not consider transverse shear strains in his analysis.

Table 8.3.2: Maximum radial deflection ($-w \times 10$ in) of a simply supported spherical shell panel under central point load.

Material	Present results				Ref. [115]	Ref. [94]
	4×4L4 Uniform	2×2Q9 Uniform	4×4Q9 Uniform	4×4Q9 Nonuniform		
Isotropic	0.3506	0.3726	0.3904	0.3935	0.3866	0.3956
Orthotropic	0.9373	1.0349	–	1.2644	–	–

The remaining example problems of this chapter are analyzed using various p levels [see Eq. (8.3.7)]. With five degrees of freedom at each node, the number of degrees of freedom per element for different p values are presented in Table 8.3.3.

The numerical integration rule (Gauss quadrature) used is $I \times J \times K$, where K denotes the Gauss rule (i.e. $K \times K$) used to evaluate the transverse shear terms (i.e. those containing A_{44} and A_{55}), J denotes the Gauss rule used to evaluate the bending-membrane coupling terms (which are zero for the linear analysis of plates), and I denotes the Gauss rule used to evaluate all remaining terms in the stiffness matrix. One may use full integration for all terms, reduced

Table 8.3.3: Number of degrees of freedom per element for different p.

Element type	p level	DoF per element
L4	1	20
Q9	2	45
Q25	4	125
Q49	6	245
Q81	8	405

integration for all terms, or selective integration where reduced integration for transverse shear and coupling terms and full integration for all other terms in the stiffness matrix. The values of I, J, and K used in the present study for different p levels and integration rules are listed in Table 8.3.4.

Table 8.3.4: The Gauss quadrature rule used for various terms.

p	Full integration	Selective integration	Reduced integration
1	$2 \times 2 \times 2$	$2 \times 1 \times 1$	$1 \times 1 \times 1$
2	$3 \times 3 \times 3$	$3 \times 2 \times 2$	$2 \times 2 \times 2$
4	$5 \times 5 \times 5$	$5 \times 4 \times 4$	$4 \times 4 \times 4$
6	$7 \times 7 \times 7$	$7 \times 6 \times 6$	$6 \times 6 \times 6$
8	$9 \times 9 \times 9$	$9 \times 8 \times 8$	$8 \times 8 \times 8$

Example 8.3.3

Consider a clamped isotropic cylindrical shell panel with the following geometric and material parameters and subjected to uniformly distributed transverse (normal to the surface) load q (see Fig. 8.3.3):

$$\alpha = 0.1 \, \text{rad}, \quad R = 100 \, \text{in}, \quad a = 20 \, \text{in}, \quad h = 0.125 \, \text{in}$$
$$E = 0.45 \times 10^6 \, \text{psi}, \quad \nu = 0.3, \quad q = 0.04 \, \text{psi} \tag{1}$$

Use two sets of uniform meshes, one with 81 nodes (405 DoF) and the other with 289 nodes (1,445 DoF), in a quadrant of the shell with different p levels to analyze the problem.

Solution: For $p = 1$ the mesh is 8×8Q4, for $p = 2$ the mesh is 4×4Q9, and for $p = 8$ the mesh is 1×1Q81 – all meshes have a total of 81 nodes. Doubling the above meshes will have 289 nodes. The vertical displacement at the center of the shell obtained with various meshes and integration rules are presented in Table 8.3.5. The results obtained with selective and reduced integrations are in close agreement with those of Palazotto and Dennis [93] and Brebbia and Connor [116].

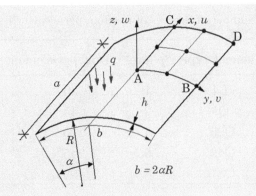

Fig. 8.3.3: Clamped cylindrical shell panel under uniform transverse load.

Table 8.3.5: Vertical deflection $(-w_A \times 10^2 \text{ in})^\dagger$ at the center of the clamped cylindrical panel under uniform transverse load.

p	Mesh of 81 nodes			Mesh of 289 nodes		
	Full integ.	Selective integ.	Reduced integ.	Full integ.	Selective integ.	Reduced integ.
1	0.3378	1.1562	1.1577	0.7456	1.1401	1.1404
2	1.1721	1.1351	1.1352	1.1427	1.1349	1.1349
4	1.1347	1.1349	1.1349	1.1349	1.1349	1.1349
8	1.1349	1.1348	1.1348	1.1348	1.1349	1.1349

† Palazotto and Dennis [93] reported 1.144×10^{-2} in while Brebbia and Connor [116] reported a value of 1.1×10^{-2} in.

The next example deals with the well-known benchmark problem of Scordelis–Lo roof (see Scordelis and Lo [117]). A solution to this problem was first discussed by Cantin and Clough [118] (who used $\nu = 0.3$).

Example 8.3.4

The *barrel vault* problem consists of a cylindrical roof with "rigid" supports at edges $x = \pm a/2$ while edges at $y = \pm b/2$ are free. The shell is assumed to deform under its own weight (i.e. q acts vertically down, not perpendicular to the surface of the shell). The geometric and material data of the problem are (see Fig. 8.3.4)

$$\alpha = 40° \ (0.698 \text{ rad}), \quad R = 300 \text{ in}, \quad a = 600 \text{ in}, \quad h = 3 \text{ in}$$

$$E = 3 \times 10^6 \text{ psi}, \quad \nu = 0.0, \quad q_y = q \sin \frac{y}{R}, \quad q_z = -q \cos \frac{y}{R}, \quad q = 0.625 \text{ psi} \tag{1}$$

The boundary conditions on the computational domain are

$$\text{At } x = 0: \quad u = \phi_x = 0, \quad \text{At } x = a/2: \quad v = w = \phi_y = 0$$
$$\text{At } y = 0: \quad v = \phi_y = 0, \quad \text{At } y = b/2: \quad \text{Free} \tag{2}$$

Use two sets of uniform meshes, one with 289 nodes (1,445 DoF) and the other with 1,089 nodes (5,445 DoF), in a quadrant of the shell with different p levels to determine the displacement at $x = 0$ and $y = \pm b/2$ (middle of the free edge) of the shell.

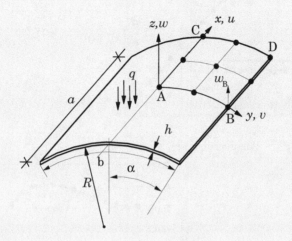

Fig. 8.3.4: A cylindrical shell roof under its own weight.

Solution: The numerical results obtained with various meshes and integration rules are presented in Table 8.3.6. To avoid shear and membrane locking one must use at least a mesh of 4×4Q25 ($p = 4$). The results obtained with selective and reduced integrations are in close agreement with those reported by Simo, Fox, and Rifai [104].

Table 8.3.6: Vertical deflection $(-w_B$ in$)^{\dagger}$ at the center of the free edge of a cylindrical roof panel under its own weight.

p	Mesh of 289 nodes			Mesh of 1089 nodes		
	Full integ.	Selective integ.	Reduced integ.	Full integ.	Selective integ.	Reduced integ.
1	0.9002	3.2681	3.6434	1.8387	3.5415	3.6431
2	3.6170	3.6393	3.6430	3.6367	3.6425	3.6428
4	3.6374	3.6430	3.6430	3.6399	3.6428	3.6428
8	3.6392	3.6429	3.6429	3.6419	3.6429	3.6429

† Simo, Fox, and Rifai [104] reported $w_{\text{ref}} = 3.6288$ in for deep shells.

Figures 8.3.5(a) and 8.3.5(b) show the variation of the vertical deflections $w(0, y)$ and $u(a/2, y)$, respectively, as a function of y, while Fig. 8.3.6 shows the convergence of the vertical displacement w_B for $p = 1, 2, 4, 8$. Figure 8.3.5 also contains the results of Zienkiewicz [119].

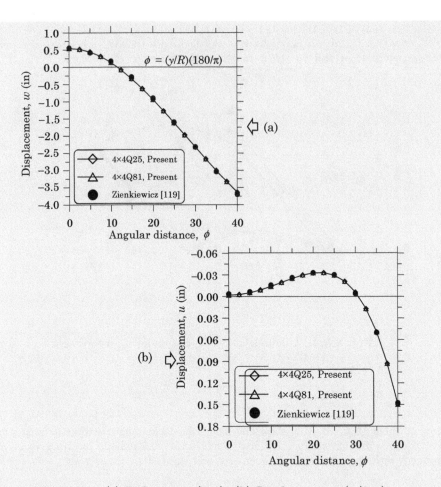

Fig. 8.3.5: (a) Deflection $w(0, y)$. (b) Displacement $u(a/2, y)$.

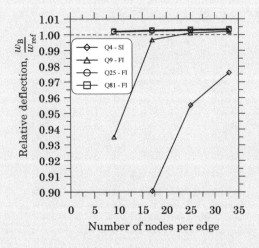

Fig. 8.3.6: Convergence of the relative vertical deflection, w_B/w_{ref}.

Example 8.3.5

Consider the problem of a *pinched cylinder*, a well-known benchmark problem considered by Flügge [84], Kreja, Schmidt, and Reddy [102], and Cho and Roh [120]. The circular cylinder with "rigid" end diaphragms is subjected to a point load at the center on opposite sides of the cylinder, as shown in Fig. 8.3.7. The geometric and material data are

$$\alpha = \frac{\pi}{2} \text{ rad}, \quad R = 300 \text{ in}, \quad a = 600 \text{ in}, \quad h = 3 \text{ in}$$
$$E = 3 \times 10^6 \text{ psi}, \quad \nu_{12} = 0.3 \tag{1}$$

The boundary conditions for the computational domain are

$$\text{At } x = 0: \quad u = \phi_x = 0, \quad \text{At } x = a/2: \quad v = w = \phi_y = 0$$
$$\text{At } y = 0, b/2: \quad v = \phi_y = 0 \tag{2}$$

Use three different meshes with 81 nodes, 289 nodes, and 1,089 nodes (with different p values) in the octant of the cylinder to analyze the problem.

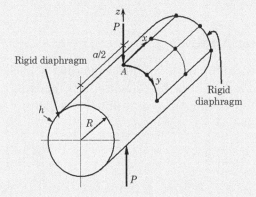

Fig. 8.3.7: Geometry of the pinched circular cylinder problem.

Solution: Table 8.3.7 contains radial displacement at the point of load application. The solution of Flügge [84] is based on classical shell theory. It is clear that the problem requires a high p level to overcome shear and membrane locking. Figure 8.3.8 shows the convergence characteristics of the problem.

Table 8.3.7: Radial displacement $(-w_A \times 10^5)^{\dagger}$ at node 1 of the pinched circular cylinder problem.

p	Mesh of 81 nodes			Mesh of 289 nodes			Mesh of 1,089 nodes		
	Full	Selec.	Reduc.	Full	Selec.	Reduc.	Full	Selec.	Reduc.
1	0.1282	1.5784	1.8453	0.2785	1.7724	1.8600	0.6017	1.8432	1.8690
2	0.4184	1.7247	1.8451	1.2238	1.8395	1.8596	1.6844	1.8636	1.8677
4	1.1814	1.8108	1.8438	1.7574	1.8510	1.8586	1.8335	1.8648	1.8667
8	1.7562	1.8309	1.8415	1.8325	1.8548	1.8579	1.8471	1.8653	1.8661

† The analytical solution of Flügge [84] is -1.8248×10^{-5} in; the value given by Cho and Roh [120] is $w_{\text{ref}} = -1.8541 \times 10^{-5}$ in.

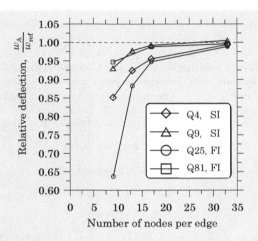

Fig. 8.3.8: Convergence of the relative radial deflection, w_A/w_{ref}.

8.3.4   Nonlinear Analysis

Here we present few examples of nonlinear bending of shells. The thin shell approximation is used and the results presented are based on the nonlinear strains in Eqs. (8.2.40)(a)–(c). The tangent stiffness coefficients can be computed as in the case of plates.

Example 8.3.6

Consider the shallow shell panel, with all its four sides clamped, shown in Fig. 8.3.9. The geometric and material parameters are:

$$E = 0.45 \times 10^6 \text{ psi}, \quad \nu = 0.3, \quad a = 20 \text{ in}, \quad R = 100 \text{ in}, \quad h = 0.125 \text{ in}, \quad \alpha = 0.1 \text{ rad} \quad (1)$$

Use a uniformly distributed load, with a load step of $\Delta q_0 = -0.02$ psi (for a total of 20 load steps) and exploit the biaxial symmetry of the panel to determine the nonlinear response. Use various uniform meshes and full integration for all stiffness coefficients.

Solution: The boundary conditions of the computational domain are

$$\text{At } \xi_1 = 0: \quad u = \phi_x = 0; \quad \text{At } \xi_2 = 0: \quad v = \phi_y = 0 \text{ (symmetry)}$$
$$\text{At } \xi_1 = \frac{a}{2} \text{ and } \xi_2 = \alpha: \quad u = v = w = \phi_x = \phi_y = 0 \quad (2)$$

Figure 8.3.10 shows the center deflection versus applied load for various meshes. The results are in close agreement with those of Palazzotto and Dennis [93].

Example 8.3.7

Consider a shallow shell panel, hinged on straight edges and free on curved edges, as shown in Fig. 8.3.11. Use the geometric and material parameters from Eq. (1) of Example 8.3.6,

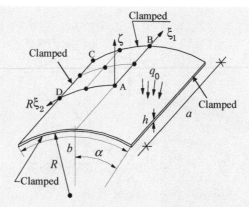

Fig. 8.3.9: Geometry and computational domain of the cylindrical shell panel.

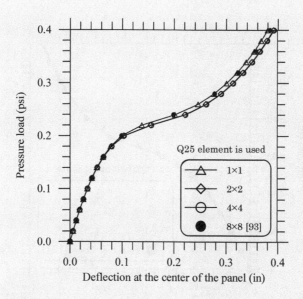

Fig. 8.3.10: Center deflection versus load for the clamped cylindrical shell panel.

except for the shell thickness, which is $h = 1.0$ in. Assume a point load P at the center of the panel with a load step of $\Delta P = -100$ lbs (for a total of 12 load steps) and use various meshes in the quadrant of the panel with full integration for all stiffness coefficients.

Solution: The boundary conditions of the computational domain are

$$\text{At } \xi_1 = 0: \quad u = \phi_x = 0; \quad \text{At } \xi_2 = 0: \quad v = \phi_y = 0 \text{ (symmetry)}$$
$$\text{At } \xi_2 = \alpha: \quad u = v = w = \phi_x = 0; \quad \text{At } \xi_1 = a/2: \quad \text{Free} \tag{1}$$

Figure 8.3.12 shows the center deflection versus applied load for various meshes. The results are in close agreement with those of Sabir and Lock [121].

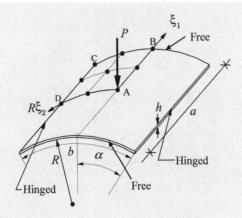

Fig. 8.3.11: Geometry and computational domain of the shell panel.

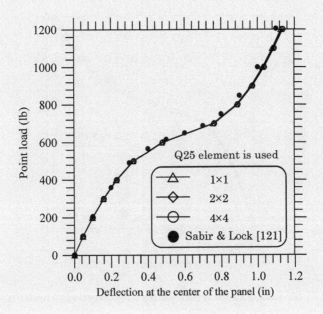

Fig. 8.3.12: Deflection versus load for the hinged shell panel.

Example 8.3.8

Analyze the *barrel vault problem* of Example 8.3.4 for nonlinear response (see Fig. 8.3.4). Use the geometric and material parameters and the boundary conditions from Eqs. (1) and (2) of Example 8.3.4 and use 16 load steps of $\Delta q_0 = -0.625$ psi.

Solution: Figure 8.3.13 shows the center deflection versus applied load for various meshes. For the total load (10 psi) considered, the shell experiences no snap-through behavior.

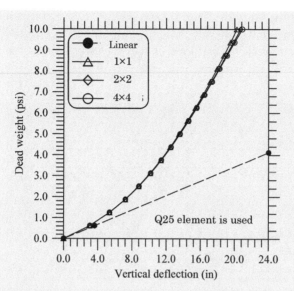

Fig. 8.3.13: Center deflection versus load for a cylindrical shell panel under its own weight.

Example 8.3.9

Consider the nonlinear bending of an orthotropic cylinder clamped at its ends as shown in Fig. 8.3.14. Use the following geometric and material parameters (those of glass–epoxy fiber-reinforced composite material):

$$E_1 = 7.5 \times 10^6 \text{ psi}, \quad E_2 = 2.0 \times 10^6 \text{ psi}, \quad G_{12} = 1.25 \times 10^6 \text{ psi}, \quad G_{13} = G_{23} = 0.625 \times 10^6 \text{ psi}$$

$$\nu_{12} = 0.25, \quad a = 10 \text{ in}, \quad R = 20 \text{ in}, \quad h = 1.0 \text{ in}, \quad \alpha = \frac{\pi}{8} \text{ rad}$$

$$(1)$$

Exploit the symmetry and use twenty load steps of $\Delta q_0 = 500$ psi.

Solution: The boundary conditions of the computational domain are

$$\text{At } \xi_1 = 0: \quad u = \phi_x = 0; \quad \text{At } \xi_2 = 0, \alpha: \quad v = \phi_y = 0 \text{ (symmetry)}$$

$$\text{At } \xi_1 = \frac{a}{2}: \quad u = v = w = \phi_x = \phi_y = 0$$

$$(2)$$

Figure 8.3.15 shows the center deflection versus applied load for various meshes. The results are in close agreement with those of Kreja, Schmidt, and Reddy [102].

8.4 Summary

In this chapter the first-order shear deformation shell theory and its displacement finite element model are presented. Beginning with the geometry description of a doubly-curved shell, kinematics of deformation are presented. Using the principle of virtual displacements, the weak forms of the shell governing equations are derived. Finite element model, which resembles that of the first-order shear deformation plate theory, is developed and some numerical examples are presented. The shell finite element developed here contains the FSDT plate element as a special case. Therefore, it can be used to analyze plate problems as well.

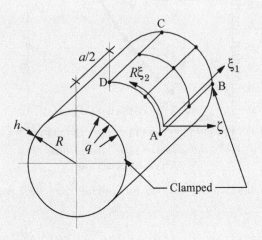

Fig. 8.3.14: Geometry and computational domain of a clamped circular cylinder.

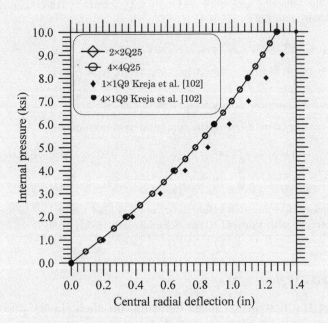

Fig. 8.3.15: Center deflection versus load for the clamped orthotropic cylinder.

Problems

THEORETICAL ASPECTS

8.1 Verify the strain-displacement relations in Eq. (8.2.31).

8.2 Verify that the strain-displacement relations associated with the displacement field in Eq. (8.2.36) are given by Eqs. (8.2.40a–c).

8.3 Derive the equations of motion in Eqs. (8.2.42)–(8.2.46).

8.4 Simplify the equations of motion in Eqs. (8.2.42)–(8.2.46) for linear analysis of cylindrical shells.

COMPUTATIONAL PROBLEMS

8.5 Derive the tangent stiffness coefficients associated with the finite element model in Eqs. (8.3.7)–(8.3.10).

8.6 Analyze the circular cylinder problem of Example 8.3.9 by assuming the fiber direction is ξ_2 (as opposed to ξ_1) and compare the results with those in Fig. 8.3.15.

Finite Element Formulations
of Solid Continua

The fact that an opinion has been widely held is no evidence whatever that it is not utterly absurd; indeed in view of the silliness of the majority of mankind a widespread belief is more likely to be foolish than sensible. —— Bertrand Russell

9.1 Introduction

9.1.1 Background

In chapters leading to the present one, we have assumed that the system under consideration was continuous in the sense that all field variables that enter the description of the physical phenomenon, such as temperature and displacements, are differentiable. Central to this assumption is that the discrete nature of matter can be overlooked, provided the length scales of interest are large compared to the length scales of discrete molecular structure. Thus, matter at sufficiently large length scales can be treated as a *continuum* in which all physical quantities of interest, including density, are continuously differentiable.

Another assumption that was implicit in writing the equations in the previous chapters was that changes in the geometry of the domain are negligible so that the physical principles can be applied to "undeformed" configuration. In geometrically nonlinear analysis of beams and plates that was considered in earlier chapters, the assumption of small strains allowed us to ignore the changes in the geometry of the body, and *no* distinction between various measures of stress and strain was made as we proceeded to determine the deformation for the next load.

When large changes in the original configuration occur, the physical principle should be stated in the deformed configuration, which is not known *a priori*. In this chapter, we shall study geometrically nonlinear behavior in which changes in geometry are updated. Consequently, it becomes necessary to distinguish between various measures of stress and strain and descriptions of motion, as discussed in Chapter 2. Intuition tells us that if the strain energy is based on the product of stress and strain, it is not expected to change simply because of our choice of stress or strain measure. Thus, "energetically conjugate" pairs of stress and strain that produce the same strain energy must be used [1].

J.N. Reddy, *An Introduction to Nonlinear Finite Element Analysis*, Second Edition. ©J.N. Reddy 2015. Published in 2015 by Oxford University Press.

9.1.2 Summary of Definitions and Concepts from Continuum Mechanics

The main equations of Chapter 2 that are useful in the present chapter are summarized here for ready reference (refer to Fig. 9.1.1 for the notation).

- Kinematics:

Material coordinates:	$\mathbf{X} = (X_1, X_2, X_3)$
Spatial coordinates:	$\mathbf{x} = (x_1, x_2, x_3)$
Motion mapping:	$\mathbf{x} = \chi(\mathbf{X}, t), \quad \mathbf{X} = \chi^{-1}(\mathbf{x}, t)$
Material and spatial gradients:	$\boldsymbol{\nabla}_0 = \partial/\partial\mathbf{X}, \quad \boldsymbol{\nabla} = \partial/\partial\mathbf{x}$

  (9.1.1)

- Displacement vector:

$$\mathbf{u}(\mathbf{x}, t) = \mathbf{x} - \mathbf{X}(\mathbf{x}, t) = \mathbf{x} - \chi^{-1}(\mathbf{x}, t) \tag{9.1.2}$$

- Deformation gradient:

$$\mathbf{F} = (\boldsymbol{\nabla}_0 \mathbf{x})^{\mathrm{T}} \tag{9.1.3}$$

- Green–Lagrange strain tensor:

$$\mathbf{E} = \frac{1}{2}\left(\mathbf{F}^{\mathrm{T}} \cdot \mathbf{F} - \mathbf{I}\right) = \frac{1}{2}\left[\boldsymbol{\nabla}_0\mathbf{u} + (\boldsymbol{\nabla}_0\mathbf{u})^{\mathrm{T}} + (\boldsymbol{\nabla}_0\mathbf{u}) \cdot (\boldsymbol{\nabla}_0\mathbf{u})^{\mathrm{T}}\right] \tag{9.1.4}$$

- Euler–Almansi strain tensor:

$$\mathbf{e} = \frac{1}{2}\left(\mathbf{I} - \mathbf{F}^{-\mathrm{T}} \cdot \mathbf{F}^{-1}\right) = \frac{1}{2}\left[\boldsymbol{\nabla}\mathbf{u} + (\boldsymbol{\nabla}\mathbf{u})^{\mathrm{T}} - (\boldsymbol{\nabla}\mathbf{u}) \cdot (\boldsymbol{\nabla}\mathbf{u})^{\mathrm{T}}\right] \tag{9.1.5}$$

- Velocity gradient tensor:

$$\mathbf{l} = \left(\frac{\partial\mathbf{v}(\mathbf{x}, t)}{\partial\mathbf{x}}\right)^{\mathrm{T}} = (\boldsymbol{\nabla}\mathbf{v}(\mathbf{x}, t))^{\mathrm{T}} = \mathbf{d} + \mathbf{w} \tag{9.1.6}$$

- Symmetric and skew-symmetric parts of $\mathbf{l}$:

$$\mathbf{d} = \frac{1}{2}\left[\boldsymbol{\nabla}\mathbf{v} + (\boldsymbol{\nabla}\mathbf{v})^{\mathrm{T}}\right] = \mathbf{d}^{\mathrm{T}}, \quad \mathbf{w} = \frac{1}{2}\left[(\boldsymbol{\nabla}\mathbf{v})^{\mathrm{T}} - \boldsymbol{\nabla}\mathbf{v}\right] = -\mathbf{w}^{\mathrm{T}} \tag{9.1.7}$$

- Nanson's formula:

$$\hat{\mathbf{n}}\, da = J\mathbf{F}^{-\mathrm{T}} \cdot \hat{\mathbf{N}}\, dA \tag{9.1.8}$$

- Relationships:

$$\mathbf{e} = \mathbf{F}^{-\mathrm{T}} \cdot \mathbf{E} \cdot \mathbf{F}^{-1}, \quad \mathbf{E} = \mathbf{F}^{\mathrm{T}} \cdot \mathbf{e} \cdot \mathbf{F}$$

$$\dot{\mathbf{F}} \equiv \frac{d\mathbf{F}}{dt} = (\boldsymbol{\nabla}_0\mathbf{v})^{\mathrm{T}}, \quad \mathbf{l} = \dot{\mathbf{F}} \cdot \mathbf{F}^{-1}, \quad \mathbf{d} = \mathbf{F}^{-\mathrm{T}} \cdot \dot{\mathbf{E}} \cdot \mathbf{F}^{-1} \tag{9.1.9}$$

- Stress measures ($\boldsymbol{\sigma}$ = Cauchy stress tensor; $\mathbf{P}$ = first Piola–Kirchhoff stress tensor; $\mathbf{S}$ = second Piola–Kirchhoff stress tensor):

$$\mathbf{P} = J\boldsymbol{\sigma} \cdot \mathbf{F}^{-\mathrm{T}}, \quad \mathbf{S} = \mathbf{F}^{-1} \cdot \mathbf{P} = J\mathbf{F}^{-1} \cdot \boldsymbol{\sigma} \cdot \mathbf{F}^{-\mathrm{T}} \tag{9.1.10}$$

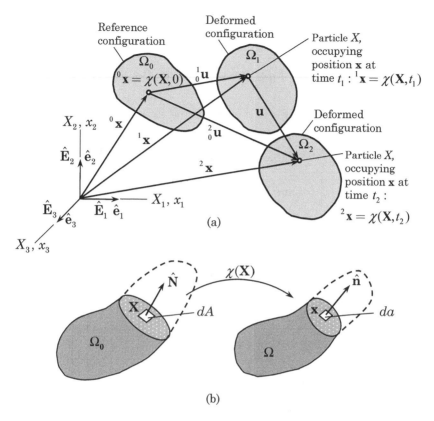

Fig. 9.1.1: (a) Kinetics and (b) kinematics of deformation.

9.1.3 Energetically-Conjugate Stresses and Strains

The rate of internal work done (power) in a continuous medium in the current configuration can be expressed as (see [1, 6])

$$W = \frac{1}{2} \int_v \boldsymbol{\sigma} : \mathbf{d}\, dv \qquad (9.1.11)$$

where $\boldsymbol{\sigma}$ is the Cauchy stress tensor and $\mathbf{d}$ is the symmetric part of the velocity gradient tensor, $\mathbf{l} = (\boldsymbol{\nabla}\mathbf{v})^{\mathrm{T}}$, and defined in Eq. (9.1.7). The pair $(\boldsymbol{\sigma}, \mathbf{d})$ is said to be *energetically-conjugate* since it produces the (strain) energy stored in the deformable medium. We can show that the first Piola–Kirchhoff stress tensor $\mathbf{P}$ is work-conjugate to the rate of the deformation gradient $\dot{\mathbf{F}}$ and the second Piola–Kirchhoff stress tensor $\mathbf{S}$ is energetically-conjugate to the rate of Green–Lagrange strain tensor $\dot{\mathbf{E}}$.

$$W = \frac{1}{2} \int_V \mathbf{P} : \dot{\mathbf{F}}\, dV = \frac{1}{2} \int_V \mathbf{S} : \dot{\mathbf{E}}\, dV \qquad (9.1.12)$$

To establish the result in Eq. (9.1.12), we begin with the expression in Eq. (9.1.11):

$$\frac{1}{2}\int_v \boldsymbol{\sigma} : \mathbf{d}\, dv = \frac{1}{2}\int_v \boldsymbol{\sigma} : \mathbf{l}\, dv \quad \text{(by symmetry of } \boldsymbol{\sigma} \text{ and } \mathbf{d)}$$

$$= \frac{1}{2}\int_V J\boldsymbol{\sigma} : \left(\dot{\mathbf{F}} \cdot \mathbf{F}^{-1}\right) dV \quad \text{[using Eqs. (2.3.7) and (9.1.9)]}$$

$$= \frac{1}{2}\int_V J\,(\sigma_{ij}\hat{\mathbf{e}}_i\hat{\mathbf{e}}_j) : \left[\left(\dot{F}_{pI}\hat{\mathbf{e}}_p\hat{\mathbf{E}}_I\right) \cdot \left((\mathbf{F}^{-1})_{Jq}\,\hat{\mathbf{E}}_J\hat{\mathbf{e}}_q\right)\right] dV$$

$$= \frac{1}{2}\int_V J\,(\sigma_{ij}\hat{\mathbf{e}}_i\hat{\mathbf{e}}_j) : \left(\dot{F}_{pI}\hat{\mathbf{e}}_p(\mathbf{F}^{-1})_{Iq}\hat{\mathbf{e}}_q\right) dV$$

$$= \frac{1}{2}\int_V J\,\sigma_{ij}\dot{F}_{iI}(\mathbf{F}^{-1})_{Ij}\, dV$$

$$= \frac{1}{2}\int_V J\,\sigma_{ij}\dot{F}_{jI}(\mathbf{F}^{-1})_{Ii}\, dV \quad \text{(by symmetry of } \sigma_{ij} = \sigma_{ji})$$

$$= \frac{1}{2}\int_V J\,\left[\left((\mathbf{F}^{-1})_{Iq}\hat{\mathbf{E}}_I\hat{\mathbf{e}}_q\right) \cdot (\sigma_{ij}\hat{\mathbf{e}}_i\hat{\mathbf{e}}_j)\right]^{\mathrm{T}} : \left(\dot{F}_{pJ}\hat{\mathbf{e}}_p\hat{\mathbf{E}}_J\right) dV$$

$$= \frac{1}{2}\int_V \left(J\boldsymbol{\sigma} \cdot \mathbf{F}^{-\mathrm{T}}\right) : \dot{\mathbf{F}}\, dV$$

$$= \frac{1}{2}\int_V \mathbf{P} : \dot{\mathbf{F}}\, dV \tag{9.1.13}$$

Similarly, we obtain

$$\frac{1}{2}\int_v \boldsymbol{\sigma} : \mathbf{d}\, dv = \frac{1}{2}\int_V J\boldsymbol{\sigma} : \left(\mathbf{F}^{-\mathrm{T}} \cdot \dot{\mathbf{E}} \cdot \mathbf{F}^{-1}\right) dV \quad \text{[using Eqs. (2.3.7) and (9.1.9)]}$$

$$= \frac{1}{2}\int_V J\,(\sigma_{ij}\hat{\mathbf{e}}_i\hat{\mathbf{e}}_j) : \left((\mathbf{F}^{-1})_{Ip}\hat{\mathbf{e}}_p\,\dot{E}_{IJ}\,(\mathbf{F}^{-1})_{Jq}\,\hat{\mathbf{e}}_q\right) dV$$

$$= \frac{1}{2}\int_V J\,\left((\mathbf{F}^{-1})_{Ii}\,\sigma_{ij}\,(\mathbf{F}^{-1})_{Jj}\right)\dot{E}_{IJ}\, dV$$

$$= \frac{1}{2}\int_V J\,\left[\left((\mathbf{F}^{-1})_{Ip}\hat{\mathbf{E}}_I\hat{\mathbf{e}}_p\right) \cdot (\sigma_{ij}\hat{\mathbf{e}}_i\hat{\mathbf{e}}_j) \cdot \left((\mathbf{F}^{-1})_{Jq}\hat{\mathbf{E}}_J\hat{\mathbf{e}}_q\right)^{\mathrm{T}}\right] : \left(\dot{E}_{PQ}\hat{\mathbf{E}}_P\hat{\mathbf{E}}_Q\right) dV$$

$$= \frac{1}{2}\int_V J\,\left(\mathbf{F}^{-1} \cdot \boldsymbol{\sigma} \cdot \mathbf{F}^{-\mathrm{T}}\right) : \dot{\mathbf{E}}\, dV = \frac{1}{2}\int_V \mathbf{S} : \dot{\mathbf{E}}\, dV \tag{9.1.14}$$

We also have the following equivalence [see Problem 2.23; follows from Problem 2.22 and Eq. (2.5.22)]

$$\int_\Omega \boldsymbol{\sigma} : \delta\mathbf{e}\, dv = \int_{\Omega_0} \mathbf{S} : \delta\mathbf{E}\, dV \tag{9.1.15}$$

which is useful in rewriting the principle of virtual displacements stated over the deformed domain Ω as one on the reference (known) domain Ω_0.

9.2 Various Strain and Stress Measures

9.2.1 Introduction

The determination of the configuration of a deformed body undergoing large deformation is not an easy task. A practical way of determining the configuration Ω from a known initial configuration Ω_0 is to assume that the total load P is applied in n suitable size increments ΔP_i,

$$P = P_n = \sum_{i=1}^{n} \Delta P_i$$

so that the body occupies n configurations, $^i\Omega$ $(i = 1, 2, \ldots, n)$, prior to occupying the final configuration $\Omega_2 \equiv {}^n\Omega$, as shown in Fig. 9.2.1. The magnitude of load increments should be such that the computational method used is capable of predicting the deformed configuration $^i\Omega$ at the end of each load $P_i = \sum_{k=1}^{i} \Delta P_k$, ΔP_i being the last load increment.

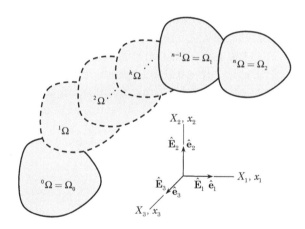

Fig. 9.2.1: Reference, intermediate, and final configurations of the body.

In the determination of an intermediate configuration $^i\Omega$, the Lagrangian description of motion can use, in principle, any of the previously known configurations Ω_0, $^1\Omega$, $\ldots$, $^{(i-1)}\Omega$, as the reference configuration. If the initial configuration is used as the reference configuration with respect to which all quantities are measured, it is called the *total Lagrangian formulation*. If the latest known configuration $^{(i-1)}\Omega$ is used as the reference configuration to determine the next configuration $^i\Omega$, it is called the *updated Lagrangian formulation*.

Knowing displacements, strains, and stresses in configurations up to and including configuration Ω_1, we wish to develop a formulation for determining the displacement field of the deformed body in the current configuration, Ω_2. It is assumed that the deformation of the body from Ω_1 to Ω_2 due to a load increment is small, and the accumulated deformation of the body from Ω_0 to Ω_1 can be arbitrarily large but continuous (i.e. neighborhoods move into neighborhoods).

9.2.2   Notation

We consider three equilibrium configurations of the body, namely, Ω_0, $^{n-1}\Omega = \Omega_1$, and $^n\Omega = \Omega_2$, which correspond to the undeformed configuration and the two configurations corresponding to the last two load increments, respectively, as shown in Fig. 9.2.1. A notation similar to one used by Bathe [4] is adopted in the present study. A left *superscript* on a quantity denotes the configuration in which the quantity *occurs*, and a left *subscript* denotes the configuration with respect to which the quantity is *measured*. Thus $^i_j Q$ indicates that the quantity Q occurs in configuration Ω_i but measured in configuration Ω_j. When the quantity under consideration is measured in the same configuration in which it occurs, the left subscript may not be used. The left superscript will be omitted on incremental quantities that occur between configurations Ω_1 and Ω_2. Of course, right subscript(s) refer to components in rectangular Cartesian coordinate systems, as will be clear in the context of the discussion. Although somewhat cumbersome, the notation conveys the meaning more directly.

Since we are dealing with three different configurations, namely, Ω_0, Ω_1, and Ω_2, it is necessary to introduce the following notation in the three configurations [see Fig. 9.1.1(a) for the notation]:

$$
\begin{array}{lccc}
\text{Configuration:} & \Omega_0 & \Omega_1 & \Omega_2 \\[4pt]
\text{Position vector:} & {}^0\mathbf{x} & {}^1\mathbf{x} & {}^2\mathbf{x} \\[4pt]
\text{Volume:} & {}^0V & {}^1V & {}^2V \\[4pt]
\text{Area:} & {}^0A & {}^1A, & {}^2A \\[4pt]
\text{Density:} & {}^0\rho & {}^1\rho & {}^2\rho \\[4pt]
\text{Displacement vector:} & & {}^1_0\mathbf{u} & {}^2_0\mathbf{u}
\end{array}
\tag{9.2.1}
$$

We use a rectangular Cartesian coordinate system to formulate the equations. We rewrite the field equations presented in Chapter 2 in a form suitable for the current objective of formulating finite element models of continua. When the body deforms under the action of external forces, a particle[1] X occupying the position $\mathbf{X} \equiv {}^0\mathbf{x} = ({}^0x_1, {}^0x_2, {}^0x_3) = (X_1, X_2, X_3)$ in configuration Ω_0 moves to a new position $^1\mathbf{x} = ({}^1x_1, {}^1x_2, {}^1x_3)$ in configuration Ω_1 and position $^2\mathbf{x} = ({}^2x_1, {}^2x_2, {}^2x_3)$ in configuration Ω_2. The total displacements of the particle X in the two configurations Ω_1 and Ω_2 can be written as

$$
{}^1_0\mathbf{u} = {}^1\mathbf{x} - {}^0\mathbf{x} \quad \text{or} \quad {}^1_0u_i = {}^1x_i - {}^0x_i \quad (i = 1, 2, 3)
\tag{9.2.2}
$$

---

[1]Recall that X denotes the name of a particle whose location in Ω_0 is $^0\mathbf{x} = \mathbf{X}$ and its current location in $\Omega = \Omega_2$ is $^2\mathbf{x}$.

$$\underset{0}{2}\mathbf{u} = {}^2\mathbf{x} - {}^0\mathbf{x} \quad \text{or} \quad \underset{0}{2}u_i = {}^2x_i - {}^0x_i \quad (i = 1, 2, 3) \tag{9.2.3}$$

and the displacement increment of the point from Ω_1 to Ω_2 is

$$\mathbf{u} = \underset{0}{2}\mathbf{u} - \underset{0}{1}\mathbf{u} \quad \text{or} \quad u_i = \underset{0}{2}u_i - \underset{0}{1}u_i \quad (i = 1, 2, 3) \tag{9.2.4}$$

9.2.3 Conservation of Mass

The relations among the mass densities ${}^0\rho$, ${}^1\rho$, and ${}^2\rho$ of the materials of configurations Ω_0, Ω_1, and Ω_2, respectively, can be established using the principle of conservation of mass. The principle requires that the mass of a material body be conserved in moving through different configurations

$$\int_{{}^2V} {}^2\rho \, d^2V = \int_{{}^1V} {}^1\rho \, d^1V = \int_{{}^0V} {}^0\rho \, d^0V \tag{9.2.5}$$

where ${}^iV$ is the volume of the body when it occupies the configuration Ω_i $(i = 0, 1, 2)$. By a change of coordinates from ${}^2x_i$ to ${}^0x_i$, we have

$$\int_{{}^2V} {}^2\rho \, d^2V = \int_{{}^0V} {}^2\rho \, \underset{0}{2}J \, d^0V \tag{9.2.6}$$

where $\underset{0}{2}J$ is the determinant of the deformation gradient $\underset{0}{2}\mathbf{F}$ (or the *Jacobian* of the transformation)

$$\underset{0}{2}J = \det\left(\underset{0}{2}\mathbf{F}\right) = \left|\frac{\partial^2 x_i}{\partial^0 x_j}\right| = \begin{vmatrix} \frac{\partial^2 x_1}{\partial^0 x_1} & \frac{\partial^2 x_1}{\partial^0 x_2} & \frac{\partial^2 x_1}{\partial^0 x_3} \\ \frac{\partial^2 x_2}{\partial^0 x_1} & \frac{\partial^2 x_2}{\partial^0 x_2} & \frac{\partial^2 x_2}{\partial^0 x_3} \\ \frac{\partial^2 x_3}{\partial^0 x_1} & \frac{\partial^2 x_3}{\partial^0 x_2} & \frac{\partial^2 x_3}{\partial^0 x_3} \end{vmatrix} \tag{9.2.7}$$

Equation (9.2.5) implies, in view of Eq. (9.2.6), that [see Eq. (2.7.7)]

$$ {}^0\rho = {}^2\rho \, \underset{0}{2}J \tag{9.2.8}$$

Similarly, we have

$$ {}^1\rho = {}^2\rho \, \underset{1}{2}J \tag{9.2.9}$$

9.2.4 Green–Lagrange Strain Tensors

The Cartesian components of the *Green–Lagrange strain tensors*, $\underset{0}{1}E_{ij}$ and $\underset{0}{2}E_{ij}$, in the two configurations Ω_1 and Ω_2 are defined by

$$\underset{0}{1}E_{ij} = \frac{1}{2}\left(\frac{\partial \underset{0}{1}u_i}{\partial^0 x_j} + \frac{\partial \underset{0}{1}u_j}{\partial^0 x_i} + \frac{\partial \underset{0}{1}u_k}{\partial^0 x_i}\frac{\partial \underset{0}{1}u_k}{\partial^0 x_j}\right) \tag{9.2.10}$$

$$\underset{0}{2}E_{ij} = \frac{1}{2}\left(\frac{\partial \underset{0}{2}u_i}{\partial^0 x_j} + \frac{\partial \underset{0}{2}u_j}{\partial^0 x_i} + \frac{\partial \underset{0}{2}u_k}{\partial^0 x_i}\frac{\partial \underset{0}{2}u_k}{\partial^0 x_j}\right) \tag{9.2.11}$$

9.2.4.1   Green–Lagrange strain increment tensor

It is useful to define the incremental strain components $_0\varepsilon_{ij}$, that is, strains induced in moving from configuration Ω_1 to configuration Ω_2. The *Green–Lagrange strain increment tensor* is defined as

$$
\begin{aligned}
2\left({_0\varepsilon_{ij}}\right) d^0x_i \, d^0x_j &= \left({^2ds}\right)^2 - \left({^1ds}\right)^2 \\
&= [({^2ds})^2 - ({^0ds})^2] - [({^1ds})^2 - ({^0ds})^2] \\
&= 2\left({_0^2E_{ij}} - {_0^1E_{ij}}\right) d^0x_i \, d^0x_j && (9.2.12) \\
&\equiv 2\left({_0e_{ij}} + {_0\eta_{ij}}\right) d^0x_i \, d^0x_j && (9.2.13)
\end{aligned}
$$

where $_0e_{ij}$ are linear components of strain increment tensor

$$
_0e_{ij} = \frac{1}{2}\left(\frac{\partial u_i}{\partial^0 x_j} + \frac{\partial u_j}{\partial^0 x_i} + \frac{\partial {_0^1u_k}}{\partial^0 x_i}\frac{\partial u_k}{\partial^0 x_j} + \frac{\partial {_0^1u_k}}{\partial^0 x_j}\frac{\partial u_k}{\partial^0 x_i}\right)
\qquad (9.2.14)
$$

and $_0\eta_{ij}$ are the nonlinear components

$$
_0\eta_{ij} = \frac{1}{2}\frac{\partial u_k}{\partial^0 x_i}\frac{\partial u_k}{\partial^0 x_j}
\qquad (9.2.15)
$$

The linearity of $_0e_{ij}$ and nonlinearity of $_0\eta_{ij}$ are understood to be in terms of the incremental displacement components u_i. Note that the displacement components $_0^1u_j$ are known, making $_0e_{ij}$ linear in the increments u_i.

For geometrically linear analysis, that is when strains are infinitesimal, only two configurations $\Omega_1 = \Omega_0$ and Ω_2 are involved, $^1u_i = 0$, $^2u_i = u_i$, and $u_{i,k}$ are small enough to neglect their products. Consequently, the linear components of strain increment tensor $_0e_{ij}$ become the same as the components of the Green–Lagrange strain tensor $_0^2E_{ij}$, and both reduce to the *infinitesimal strain components*

$$
_0e_{ij} = \frac{1}{2}\left(\frac{\partial u_i}{\partial^0 x_j} + \frac{\partial u_j}{\partial^0 x_i}\right)
\qquad (9.2.16)
$$

9.2.4.2   Updated Green–Lagrange strain tensor

The Green–Lagrange strain tensor $_0^2E_{ij}$ introduced earlier is useful in the total Lagrangian formulation. With the updated Lagrangian formulation in mind, we define the components of Green–Lagrange strain tensor with respect to configuration Ω_1. Such a strain tensor, denoted as $_1^2\varepsilon_{ij}$, is called the *updated Green–Lagrange strain tensor*. It is defined by

$$
2\left({_1^2\varepsilon_{ij}}\right) d^1x_i \, d^1x_j = \left({^2ds}\right)^2 - \left({^1ds}\right)^2
\qquad (9.2.17)
$$

Using (note that all are first-order derivatives)

$$
d^2x_i = \frac{\partial^2 x_i}{\partial^1 x_j}\, d^1x_j, \quad \left({^2ds}\right)^2 = \frac{\partial^2 x_k}{\partial^1 x_i}\frac{\partial^2 x_k}{\partial^1 x_j}\, d^1x_i \, d^1x_j
\qquad (9.2.18)
$$

$$u_i = {}^2x_i - {}^1x_i, \quad \frac{\partial\,{}^2x_k}{\partial\,{}^1x_i} = \frac{\partial\,u_k}{\partial\,{}^1x_i} + \delta_{ki} \tag{9.2.19}$$

we can write

$$\,{}^2_1\varepsilon_{ij} = \frac{1}{2}\left(\frac{\partial u_i}{\partial\,{}^1x_j} + \frac{\partial u_j}{\partial\,{}^1x_i} + \frac{\partial u_k}{\partial\,{}^1x_i}\frac{\partial u_k}{\partial\,{}^1x_j}\right) \tag{9.2.20}$$

$$\equiv \,{}_1e_{ij} + \,{}_1\eta_{ij} \tag{9.2.21}$$

where

$$\,{}_1e_{ij} = \frac{1}{2}\left(\frac{\partial u_i}{\partial\,{}^1x_j} + \frac{\partial u_j}{\partial\,{}^1x_i}\right), \quad \,{}_1\eta_{ij} = \frac{1}{2}\left(\frac{\partial u_k}{\partial\,{}^1x_i}\frac{\partial u_k}{\partial\,{}^1x_j}\right) \tag{9.2.22}$$

Note that the definition of $\,{}^2_1\varepsilon_{ij}$ involves the components of the displacement increment vector **u**. Therefore, $\,{}^2_1\varepsilon_{ij}$ are also called the *updated Green–Lagrange strain increment* components.

9.2.5 Euler–Almansi Strain Tensor

Suppose that a body occupying the undeformed configuration Ω_0 takes a number of intermediate configurations before occupying the deformed configuration Ω_1, as shown in Fig. 9.2.1. Then we wish to determine the next configuration Ω_2 in a single incremental step. Although the accumulated displacements $\,{}^1u_i$ of the body from configuration Ω_0 to configuration Ω_1 can be large, the incremental displacements u_i within the incremental step from Ω_1 to Ω_2 are assumed to be small. Hence, we may refer the strains to configuration Ω_2. The strains occurring in the body at configuration Ω_2 and measured in the same configuration, i.e. the Euler–Almansi strain tensor components defined in Eq. (2.4.14), can be expressed in terms of the components of the displacement increment u_i as

$$2\,{}^2_2\varepsilon_{ij}\,d\,{}^2x_i\,d\,{}^2x_j = \left({}^2ds\right)^2 - \left({}^1ds\right)^2$$

$$= \left(\delta_{ij} - \frac{\partial\,{}^1x_k}{\partial\,{}^2x_i}\frac{\partial\,{}^1x_k}{\partial\,{}^2x_j}\right)d\,{}^2x_i\,d\,{}^2x_j \tag{9.2.23}$$

or $\left({}^2_2\varepsilon_{ij} \equiv \,{}_2\varepsilon_{ij}\right)$

$$\,{}_2\varepsilon_{ij} = \frac{1}{2}\left(\frac{\partial u_i}{\partial\,{}^2x_j} + \frac{\partial u_j}{\partial\,{}^2x_i} - \frac{\partial u_k}{\partial\,{}^2x_i}\frac{\partial u_k}{\partial\,{}^2x_j}\right) \tag{9.2.24}$$

The linear part of $\,{}_2\varepsilon_{ij}$ is called the infinitesimal strain tensor and denoted $\,{}_2e_{ij}$

$$\,{}_2e_{ij} = \frac{1}{2}\left(\frac{\partial u_i}{\partial\,{}^2x_j} + \frac{\partial u_j}{\partial\,{}^2x_i}\right) \tag{9.2.25}$$

which is identical in form to the infinitesimal strain tensor given in Eq. (2.4.19), except that the reference configuration is changed to Ω_2.

9.2.6   Relationships Between Various Stress Tensors

The Cauchy stress components σ_{ij} in configurations Ω_1 and Ω_2 are denoted by

$$^1\sigma_{ij} = {}^1_1\sigma_{ij}, \quad {}^2\sigma_{ij} = {}^2_2\sigma_{ij} \tag{9.2.26}$$

We note that the second Piola–Kirchhoff stress tensor characterizes the current force in Ω_2 but transformed to and measured per unit area in Ω_0.

It is useful in the updated Lagrangian formulation to define another kind of stress tensor called the *updated Kirchhoff stress tensor*. Consider the infinitesimal rectangular parallelepiped containing the point P in Ω_1 enclosed by the six surfaces

$$^1x_i = \text{constant}, \quad {}^1x_i + d\,{}^1x_i = \text{constant} \quad (i = 1, 2, 3) \tag{9.2.27}$$

The Cauchy stress components on this rectangular parallelepiped are $^1\sigma_{ij}$. As the body deforms from configuration Ω_1 to configuration Ω_2, this rectangular parallelepiped will deform into, in general, a non-rectangular parallelepiped. Using Ω_1 as the reference configuration, the updated Kirchhoff stress components $^2_1S_{ij}$ are defined as the internal forces per unit area acting along the normal and two tangential directions of each of the side surfaces of the parallelepiped in Ω_2. The updated Kirchhoff stress components $^2_1S_{ij}$ can be decomposed as

$$^2_1S_{ij} = {}^1\sigma_{ij} + {}_1S_{ij} \tag{9.2.28}$$

where $^1\sigma_{ij} = {}^1_1S_{ij}$ are the Cauchy stress components in Ω_1 and $_1S_{ij}$ are the *updated Kirchhoff stress increment tensor* components.

The second Piola–Kirchhoff stress tensor components in Ω_1 and Ω_2 are denoted by $^1_0S_{ij}$ and $^2_0S_{ij}$, respectively, which are related by

$$^2_0S_{ij} = {}^1_0S_{ij} + {}_0S_{ij} \tag{9.2.29}$$

where $_0S_{ij}$ are the *Kirchhoff stress increment tensor* components.

9.2.7   Constitutive Equations

As discussed in Section 2.8, materials for which the constitutive behavior is only a function of the current state of deformation are known as *elastic*. In the special case in which the work done by the stresses during a deformation is dependent only on the initial state and the current configuration, the material is called *hyperelastic*. For hyperelastic materials we assume that there exists a stored strain energy potential, Ψ, measured per unit undeformed volume, such that the components of the *material elasticity tensor* **C** [see Section 2.8.2 and Eq. (2.8.14)] are given by

$$\frac{\partial \Psi}{\partial E_{ij}} = S_{ij} = C_{ijk\ell}E_{k\ell} \tag{9.2.30}$$

where $C_{ijk\ell}$ are the components of the fourth-order elasticity tensor.

In the derivation of finite element models of incremental nonlinear analysis of solid continua, it is necessary to specify the stress–strain relations in incremental form. In the total Lagrangian formulation, for example, the constitutive relations are expressed in terms of the Kirchhoff stress increment tensor components $_0S_{ij}$ and Green–Lagrange strain increment tensor components $_0\varepsilon_{ij}$ as

$$_0S_{ij} = {}_0C_{ijk\ell}\, {}_0\varepsilon_{k\ell} \tag{9.2.31}$$

In the updated Lagrangian formulation, the constitutive relations are expressed in terms of the updated Kirchhoff stress increment tensor components $_1S_{ij}$ and updated Green–Lagrange strain increment tensor components $_1\varepsilon_{ij}$:

$$_1S_{ij} = {}_1C_{ijk\ell}\, {}_1\varepsilon_{k\ell} \tag{9.2.32}$$

where $_0C_{ijkl}$ and $_1C_{ijkl}$ denote the components of the elasticity tensors with respect to the configurations Ω_0 and Ω_1, respectively.

Often, the incremental material laws in Eqs. (9.2.31) and (9.2.32) with identical coefficients, that is, $_0C_{ijk\ell} = {}_1C_{ijk\ell}$, are employed in the total and updated Lagrangian finite element formulations. Of course, the error introduced by the assumption can be negligible if the strains are relatively small, but the difference can be significant in large strain problems.

In single subscript notation for stresses and strains, analogous to those listed in Eq. (2.8.16), Eq. (9.2.31) can be expressed as

$$_0S_i = {}_0C_{ij}\, {}_0\varepsilon_j \tag{9.2.33}$$

where it is understood that the principal material coordinate system coincides with the coordinate system chosen to describe the motion. Otherwise, Eq. (9.2.33) has the form

$$_0S_i' = {}_0C_{ij}'\, {}_0\varepsilon_j' \tag{9.2.34}$$

where all quantities are referred to the material coordinate system (X_1', X_2', X_3'). For orthotropic materials (with respect to the principal material coordinates), $_0C_{ij}'$ can be expressed in terms of the material properties as

$$
\begin{aligned}
&_0C_{11}' = \frac{1 - \nu_{23}\nu_{23}}{E_2 E_3 \Delta}, \quad
{}_0C_{12}' = \frac{\nu_{21} + \nu_{31}\nu_{23}}{E_2 E_3 \Delta} = \frac{\nu_{12} + \nu_{32}\nu_{13}}{E_1 E_3 \Delta} \\[4pt]
&_0C_{13}' = \frac{\nu_{31} + \nu_{21}\nu_{32}}{E_2 E_3 \Delta} = \frac{\nu_{13} + \nu_{12}\nu_{23}}{E_1 E_2 \Delta} \\[4pt]
&_0C_{22}' = \frac{1 - \nu_{13}\nu_{31}}{E_1 E_3 \Delta}, \quad
{}_0C_{23}' = \frac{\nu_{32} + \nu_{12}\nu_{31}}{E_1 E_3 \Delta} = \frac{\nu_{23} + \nu_{21}\nu_{13}}{E_1 E_2 \Delta} \\[4pt]
&_0C_{33}' = \frac{1 - \nu_{12}\nu_{21}}{E_1 E_2 \Delta}, \quad
{}_0C_{44}' = G_{23}, \quad {}_0C_{55}' = G_{31}, \quad {}_0C_{66}' = G_{12} \\[4pt]
&\Delta = (1 - \nu_{12}\nu_{21} - \nu_{23}\nu_{32} - \nu_{31}\nu_{13} - 2\nu_{21}\nu_{32}\nu_{13})/E_1 E_2 E_3
\end{aligned}
\tag{9.2.35}
$$

where E_1, E_2, E_3 are Young's moduli in 1, 2, and 3 material directions, respectively, ν_{ij} is Poisson's ratio, defined as the ratio of transverse strain in the X_jth direction to the strain in the X_ith direction when stressed in the X_ith direction, and G_{23}, G_{13}, and G_{12} are shear moduli in the X_2-X_3, X_1-X_3, and X_1-X_2 planes, respectively. The following reciprocal relations hold on account of the symmetries of the stress, strain, and elasticity tensors:

$$\frac{\nu_{21}}{E_2} = \frac{\nu_{12}}{E_1}; \quad \frac{\nu_{31}}{E_3} = \frac{\nu_{13}}{E_1}; \quad \frac{\nu_{32}}{E_3} = \frac{\nu_{23}}{E_2}$$

or, in short

$$\frac{\nu_{ij}}{E_i} = \frac{\nu_{ji}}{E_j} \quad (\text{no sum on } i, j) \tag{9.2.36}$$

for $i, j = 1, 2, 3$. The nine independent material coefficients for an orthotropic material are

$$E_1, \ E_2, \ E_3, \ G_{23}, \ G_{13}, \ G_{12}, \ \nu_{12}, \ \nu_{13}, \ \nu_{23} \tag{9.2.37}$$

When the material coordinate system (X_1', X_2', X_3') does not coincide with (X_1, X_2, X_3) used to describe the motion, one must transform Eq. (9.2.34) to Eq. (9.2.33) using the direction cosines between the two coordinate systems (see Chapter 1 of Reddy [1] for detailed discussion of coordinate transformations and relations between the components of stress, strain, and elasticity tensor in the two coordinate systems).

For a two-dimensional elasticity problem, when the material coordinate system is the same as that used for the description of motion, typical stress–strain relations are given by

$$\begin{Bmatrix} \sigma_1 \\ \sigma_2 \\ \sigma_6 \end{Bmatrix} = \begin{bmatrix} C_{11} & C_{12} & 0 \\ C_{12} & C_{22} & 0 \\ 0 & 0 & C_{66} \end{bmatrix} \begin{Bmatrix} \varepsilon_1 \\ \varepsilon_2 \\ \varepsilon_6 \end{Bmatrix} \tag{9.2.38}$$

For transversely isotropic material (i.e. $E_3 = E_2$, $\nu_{23} = \nu_{32}$, and $\nu_{12} = \nu_{13}$) in a state of *plane strain* and orthotropic material in a state of *plane stress* the material stiffness C_{ij} can be expressed in terms of the engineering constants as follows:

Plane strain

$$C_{11} = \frac{E_1(1 - \nu_{23})}{(1 - \nu_{23} - 2\nu_{12}\nu_{21})}, \quad C_{12} = \frac{\nu_{21}E_1}{(1 - \nu_{23} - 2\nu_{12}\nu_{21})}$$

$$C_{22} = \frac{E_2(1 - \nu_{12}\nu_{21})}{(1 + \nu_{23})(1 - \nu_{23} - 2\nu_{12}\nu_{21})} \tag{9.2.39}$$

$$C_{44} = G_{13}, \quad C_{55} = G_{23}, \quad C_{66} = G_{12}$$

Plane stress

$$C_{11} = \frac{E_1}{(1 - \nu_{12}\nu_{21})}, \quad C_{22} = \frac{E_2}{(1 - \nu_{12}\nu_{21})}$$

$$C_{12} = \nu_{12}C_{22}, \quad C_{44} = G_{13}, \quad C_{55} = G_{23}, \quad C_{66} = G_{12} \tag{9.2.40}$$

This completes the discussion of various strain and stress measures between configurations Ω_0, Ω_1, and Ω_2. In the next section we study variational formulations based on principles of virtual displacements and total and updated Lagrangian descriptions of motion.

9.3 Total Lagrangian and Updated Lagrangian Formulations

9.3.1 Principle of Virtual Displacements

The equations of the Lagrangian incremental description of motion can be derived from the principles of virtual work (i.e. principles of virtual displacements, virtual forces, or mixed virtual displacements and forces). Since our ultimate objective is to develop the finite element model of the equations governing a deformable continuum, we will not actually derive the differential equations of motion but utilize the virtual work statements, which are the weak forms needed to develop the finite element models.

The *displacement finite element models* to be developed in this chapter are based on the principle of virtual displacements (see Section 2.9). The principle requires that the sum of the external virtual work done on and the internal virtual work stored in the body should be equal to zero:

$$\delta W \equiv \delta W_I + \delta W_E = 0 \tag{9.3.1}$$

where δW_I is the internal virtual work and δW_E is the virtual work done by external forces

$$\delta W_I = \int_{^2V} {}^2\boldsymbol{\sigma} : \delta({}_2\boldsymbol{\varepsilon}) \, d\,{}^2V = \int_{^2V} {}^2\sigma_{ij} \, \delta({}_2\varepsilon_{ij}) \, d\,{}^2V \tag{9.3.2}$$

$$\delta W_E = -\left[\int_{^2V} {}^2\mathbf{f} \cdot \delta\mathbf{u} \, d\,{}^2V + \int_{^2S} {}^2\mathbf{t} \cdot \delta\mathbf{u} \, d\,{}^2S\right]$$

$$= -\left[\int_{^2V} {}^2f_i \delta u_i \, d\,{}^2V + \int_{^2S} {}^2t_i \, \delta u_i \, d\,{}^2S\right] \equiv -\delta\,{}^2R \tag{9.3.3}$$

and $d\,{}^2S$ denotes surface element and ${}^2\mathbf{f}$ is the body force vector (measured per unit volume) and ${}^2\mathbf{t}$ is the boundary stress (or traction) vector (measured per unit surface area) in configuration Ω_2; $\boldsymbol{\sigma}$ and $\mathbf{e}$ denote the Cauchy stress tensor and Euler–Almansi strain tensors, respectively.

Equations (9.3.2) and (9.3.3) are not useful in large deformation analysis because the configuration Ω_2 is unknown and hence integration cannot be carried out on an unknown geometry. This is an important difference compared with the linear analysis in which we assume that the displacements are infinitesimally small and that the configuration of the body does not change appreciably, $\Omega_2 \approx \Omega_0$. In a large deformation analysis, special attention must be given to the fact that the configuration of the body changes continuously as load is incremented, and the configuration of the deformed body slowly drifts away from the configuration of the undeformed body, making Ω_2 to be substantially different from Ω_0 (although Ω_1 and Ω_2 may be close enough if the load increment is small). This change in configuration can be dealt with by defining appropriate stress and strain measures, as discussed in Chapter 2. The objective of their introduction into the analysis is to express the virtual work expressions in Eqs. (9.3.2) and (9.3.3) in terms of an integral over a configuration that is known. The stress and strain measures that we shall use are the second Piola–Kirchhoff stress tensor and the Green–Lagrange strain tensor, which are "energetically conjugate" to each other [see Eq. (9.1.15)].

9.3.2 Total Lagrangian Formulation

9.3.2.1 Weak form

In the total Lagrangian formulation, all quantities are measured with respect to the initial configuration Ω_0. Hence, the virtual work statement in Eq. (9.3.1) must be expressed in terms of quantities referred to the reference configuration. We use the following identities:

$$\int_{^2V} {}^2\sigma_{ij}\, \delta({}_2\varepsilon_{ij})\, d\,{}^2V = \int_{^0V} {}^2_0S_{ij}\, \delta({}^2_0E_{ij})\, d^0V$$

$$\int_{^2V} {}^2f_i\, \delta u_i\, d\,{}^2V = \int_{^0V} {}^2_0f_i\, \delta u_i\, d^0V \qquad (9.3.4)$$

$$\int_{^2S} {}^2t_i\, \delta u_i\, d\,{}^2S = \int_{^0S} {}^2_0t_i\, \delta u_i\, d^0S$$

where ${}^2_0f_i$ and ${}^2_0t_i$ are the body force and boundary traction components referred to the configuration Ω_0. Substituting the equivalence from Eq. (9.3.4) into Eqs. (9.3.2) and (9.3.3) for δW_I and δW_E, respectively, and then using them in Eq. (9.3.1), we arrive at

$$0 = \int_{^0V} {}^2_0S_{ij}\, \delta({}^2_0E_{ij})\, d^0V - \int_{^0V} {}^2_0f_i\, \delta u_i\, d^0V - \int_{^0S} {}^2_0t_i\, \delta u_i\, d^0S \qquad (9.3.5)$$

9.3.2.2 Incremental decompositions

Next we simplify the virtual work statement in Eq. (9.3.5) further. First, we note that [see Eqs. (9.2.12), (9.2.13), and (9.2.29)]

$$
\delta({}_{0}^{2}E_{ij}) = \delta({}_{0}^{1}E_{ij}) + \delta({}_{0}\varepsilon_{ij}) = \delta({}_{0}\varepsilon_{ij})
$$
$$
{}_{0}^{2}S_{ij} = {}_{0}^{1}S_{ij} + {}_{0}S_{ij}
$$

$$(9.3.6)$$

where $\delta({}_{0}^{1}E_{ij}) = 0$ because ${}_{0}^{1}E_{ij}$ is not a function of the unknown u_i [see Eq. (9.2.10)]. The virtual strains $\delta({}_{0}\varepsilon_{ij})$ are given by

$$
\delta({}_{0}\varepsilon_{ij}) = \delta({}_{0}e_{ij}) + \delta({}_{0}\eta_{ij})
$$

$$(9.3.7)$$

where $\delta({}_{0}e_{ij})$ and $\delta({}_{0}\eta_{ij})$ are [see Eqs. (9.2.14) and (9.2.15)]

$$
\delta({}_{0}e_{ij}) = \frac{1}{2}\left(\frac{\partial \delta u_i}{\partial\,{}^{0}x_j} + \frac{\partial \delta u_j}{\partial\,{}^{0}x_i} + \frac{\partial \delta u_k}{\partial\,{}^{0}x_i}\frac{\partial\,{}^{1}_{0}u_k}{\partial\,{}^{0}x_j} + \frac{\partial\,{}^{1}_{0}u_k}{\partial\,{}^{0}x_i}\frac{\partial \delta u_k}{\partial\,{}^{0}x_j}\right)
$$

$$(9.3.8)$$

$$
\delta({}_{0}\eta_{ij}) = \frac{1}{2}\left(\frac{\partial \delta u_k}{\partial\,{}^{0}x_i}\frac{\partial u_k}{\partial\,{}^{0}x_j} + \frac{\partial u_k}{\partial\,{}^{0}x_i}\frac{\partial \delta u_k}{\partial\,{}^{0}x_j}\right)
$$

$$(9.3.9)$$

Substituting Eq. (9.3.6) for $\delta({}_{0}^{2}E_{ij})$ and ${}_{0}^{2}S_{ij}$ into Eq. (9.3.5), we arrive at the expression

$$
\begin{aligned}
0 &= \int_{{}^{0}V}\left({}_{0}^{1}S_{ij} + {}_{0}S_{ij}\right)\delta({}_{0}\varepsilon_{ij})\,d\,{}^{0}V - \delta({}_{0}^{2}R)\\
&= \int_{{}^{0}V}\left\{{}_{0}^{1}S_{ij}\left[\delta({}_{0}e_{ij}) + \delta({}_{0}\eta_{ij})\right] + {}_{0}S_{ij}\,\delta({}_{0}\varepsilon_{ij})\right\}d\,{}^{0}V - \delta({}_{0}^{2}R)\\
&= \int_{{}^{0}V}{}_{0}^{1}S_{ij}\,\delta({}_{0}\eta_{ij})\,d\,{}^{0}V + \int_{{}^{0}V}{}_{0}S_{ij}\,\delta({}_{0}\varepsilon_{ij})\,d\,{}^{0}V\\
&\quad + \int_{{}^{0}V}{}_{0}^{1}S_{ij}\,\delta({}_{0}e_{ij})\,d\,{}^{0}V - \delta({}_{0}^{2}R)
\end{aligned}
$$

$$(9.3.10)$$

where

$$
\delta({}_{0}^{2}R) = \int_{{}^{0}V}{}_{0}^{2}f_i\,\delta u_i\,d{}^{0}V + \int_{{}^{0}S}{}_{0}^{2}t_i\,\delta u_i\,d{}^{0}S
$$

$$(9.3.11)$$

Since the body is in equilibrium at configuration Ω_1, by the principle of virtual work applied to configuration Ω_1 we have

$$
0 = \int_{{}^{0}V}{}_{0}^{1}S_{ij}\,\delta({}_{0}e_{ij})\,d\,{}^{0}V - \int_{{}^{0}V}{}_{0}^{1}f_i\,\delta u_i\,d\,{}^{0}V - \int_{{}^{0}S}{}_{0}^{1}t_i\delta u_i\,d\,{}^{0}S
$$

$$(9.3.12)$$

Consequently, Eq. (9.3.10) takes the form

$$
\int_{{}^{0}V}{}_{0}S_{ij}\,\delta({}_{0}\varepsilon_{ij})\,d\,{}^{0}V + \int_{{}^{0}V}{}_{0}^{1}S_{ij}\,\delta({}_{0}\eta_{ij})\,d\,{}^{0}V = \delta({}_{0}^{2}R) - \delta({}_{0}^{1}R)
$$

$$(9.3.13)$$

where

$$\delta({}_0^1 R) = \int_{{}^0 V} {}_0^1 S_{ij}\, \delta({}_0 e_{ij})\, d\,{}^0 V = \int_{{}^0 V} {}_0^1 f_i\, \delta u_i\, d\,{}^0 V + \int_{{}^0 S} {}_0^1 t_i\, \delta u_i\, d\,{}^0 S \quad (9.3.14)$$

Equation (9.3.13) forms the basis for the finite element model of the total Lagrangian formulation. The first term of Eq. (9.3.13) represents the change in the virtual strain energy due to the virtual incremental displacements u_i between configurations Ω_1 and Ω_2. The second term represents the virtual work done by forces due to initial stresses (i.e. stresses accumulated through configuration Ω_1) ${}_0^1 S_{ij}$, as shown symbolically in Fig. 9.3.1. The two terms together on the right-hand side denote the change in the virtual work done by applied body forces and surface tractions in moving from Ω_1 to Ω_2. Equation (9.3.13) represents the statement of virtual work for the incremental deformation between the configurations Ω_1 and Ω_2, and no approximations are made in arriving at it. Replacing ${}_0 S_{ij}$ in terms of the strains ${}_0 \varepsilon_{k\ell}$ using the constitutive relation, Eq. (9.2.31), we obtain

$$\int_{{}^0 V} {}_0 C_{ijk\ell}\, {}_0 \varepsilon_{k\ell}\, \delta({}_0 \varepsilon_{ij})\, d\,{}^0 V + \int_{{}^0 V} {}_0^1 S_{ij}\, \delta({}_0 \eta_{ij})\, d\,{}^0 V = \delta({}_0^2 R) - \delta({}_0^1 R) \quad (9.3.15)$$

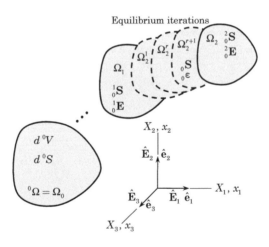

Fig. 9.3.1: Reference and final configurations of the body in the total Lagrangian formulation.

9.3.2.3 Linearization

Equation (9.3.15) is nonlinear, through ${}_0 \eta_{ij}$, in the displacement increments u_i. From a computational point of view, it is necessary to linearize the equation. Toward this end, we assume that the displacements u_i are small, which is indeed the case provided the load step is small in moving from Ω_1 to Ω_2, so that the following approximations hold:

$$_0S_{ij} = \; _0C_{ijk\ell} \; _0\varepsilon_{k\ell} \approx \; _0C_{ijk\ell} \; _0e_{k\ell} \quad \text{and} \quad \delta(_0\varepsilon_{ij}) \approx \delta(_0e_{ij}) \tag{9.3.16}$$

Note that partial linearization [i.e. linearizing only $_0\varepsilon_{k\ell}$ or $\delta(_0\varepsilon_{ij})$] or no linearization will result in unsymmetric stiffness matrices. The constitutive relation in the above equation can be expressed in the form of Eq. (9.2.33). Equation (9.3.15) now can be simplified to

$$\int_{^0V} \;_0C_{ijk\ell} \; _0e_{k\ell} \; \delta(_0e_{ij}) \; d \;^0V + \int_{^0V} \;_0^1S_{ij} \; \delta(_0\eta_{ij}) \; d \;^0V \approx \delta(_0^2R) - \delta(_0^1R) \tag{9.3.17}$$

The two expressions on the left side of the above equation are linear in the incremental displacements **u** and its first variation $\delta\mathbf{u}$.

Equation (9.3.17) is the weak form (the approximation sign will be replaced by the equal sign) used for the development of the (linear) finite element model based on the total Lagrangian formulation. The total stress components $_0^1S_{ij}$ are evaluated using the constitutive relation

$$_0^1S_{ij} = \; _0C_{ijk\ell} \; _0^1E_{k\ell} \tag{9.3.18}$$

where $_0^1E_{k\ell}$ are the Green–Lagrange strain tensor components defined in Eq. (9.2.10). In single subscript notation, we can write Eq. (9.3.18) as

$$_0^1S_i = \; _0C_{ij} \; _0^1E_j \tag{9.3.19}$$

A summary of all the pertinent equations of the total Lagrangian formulation is presented in Table 9.3.1.

9.3.3 Updated Lagrangian Formulation

9.3.3.1 Weak form

In the updated Lagrangian formulation, all quantities are referred to the latest known configuration, namely Ω_1. Hence, the virtual work statement in Eqs. (9.3.1)–(9.3.3) must be recast in terms of quantities referred to Ω_1. We use the identities

$$\int_{^2V} \;^2\sigma_{ij} \; \delta(_2\varepsilon_{ij}) \; d \;^2V = \int_{^1V} \;_1^2S_{ij} \; \delta(_1^2\varepsilon_{ij}) \; d \;^1V \tag{9.3.20}$$

$$\int_{^2V} \;^2f_i \; \delta u_i \; d \;^2V = \int_{^1V} \;_1^2f_i \; \delta u_i \; d \;^1V \tag{9.3.21}$$

$$\int_{^2S} \;^2t_i \; \delta u_i \; d \;^2S = \int_{^1S} \;_1^2t_i \; \delta u_i \; d \;^1S \tag{9.3.22}$$

Table 9.3.1: Summary of equations of the total Lagrangian formulation.

1. Weak form

$$0 = \int_{^0V} {}^2_0 S_{ij}\, \delta({}^2_0 E_{ij})\, d\,{}^0V - \int_{^0V} {}^2_0 f_i\, \delta u_i\, d\,{}^0V - \int_{^0S} {}^2_0 t_i\, \delta u_i\, d\,{}^0S \tag{1}$$

$$\delta({}^2_0 E_{ij}) = \delta({}_0\varepsilon_{ij}), \quad \delta({}_0\varepsilon_{ij}) = \delta({}_0 e_{ij}) + \delta({}_0\eta_{ij}) \tag{2}$$

2. Incremental decompositions (after configuration ${}^1\Omega \equiv {}^{n-1}\Omega$ is determined)

$$\begin{aligned}
{}^2_0 S_{ij} &= {}^1_0 S_{ij} + {}_0 S_{ij} \tag{3}\\[4pt]
{}^2_0 E_{ij} &= {}^1_0 E_{ij} + {}_0\varepsilon_{ij}, \quad {}_0\varepsilon_{ij} = {}_0 e_{ij} + {}_0\eta_{ij} \tag{4}\\[4pt]
{}^1_0 S_{ij} &= {}_0 C_{ijk\ell}\, {}^1_0 E_{k\ell}, \quad {}_0 S_{ij} = {}_0 C_{ijk\ell}\, {}_0\varepsilon_{k\ell} \tag{5}
\end{aligned}$$

$$ {}^1_0 E_{ij} = \frac{1}{2}\left(\frac{\partial\, {}^1_0 u_i}{\partial\, {}^0 x_j} + \frac{\partial\, {}^1_0 u_j}{\partial\, {}^0 x_i} + \frac{\partial\, {}^1_0 u_k}{\partial\, {}^0 x_i}\frac{\partial\, {}^1_0 u_k}{\partial\, {}^0 x_j} \right) \tag{6}$$

$$ {}_0 e_{ij} = \frac{1}{2}\left(\frac{\partial u_i}{\partial\, {}^0 x_j} + \frac{\partial u_j}{\partial\, {}^0 x_i} + \frac{\partial\, {}^1_0 u_k}{\partial\, {}^0 x_i}\frac{\partial u_k}{\partial\, {}^0 x_j} + \frac{\partial u_k}{\partial\, {}^0 x_i}\frac{\partial\, {}^1_0 u_k}{\partial\, {}^0 x_j} \right) \tag{7}$$

$$ {}_0\eta_{ij} = \frac{1}{2}\frac{\partial u_k}{\partial\, {}^0 x_i}\frac{\partial u_k}{\partial\, {}^0 x_j} \tag{8}$$

We note that quantities with left superscript 1 and left subscript 0 are known (because ${}^1_0\mathbf{u} \equiv {}^{n-1}_0\mathbf{u}$ is known). At the very first load step, we wish to determine ${}^n\Omega$ for $n = 1$; hence, ${}^{n-1}\Omega = \Omega_0$. Therefore we have ${}^1_0\mathbf{u} \equiv {}^{n-1}_0\mathbf{u} = \mathbf{0}$, and Eqs. (3)–(8) reduce to

$$ {}^1_0 S_{ij} = {}_0 S_{ij}, \quad {}^1_0 E_{ij} = {}_0\varepsilon_{ij}, \quad {}_0\varepsilon_{ij} = \frac{1}{2}\left(\frac{\partial u_i}{\partial\, {}^0 x_j} + \frac{\partial u_j}{\partial\, {}^0 x_i} + \frac{\partial u_k}{\partial\, {}^0 x_i}\frac{\partial u_k}{\partial\, {}^0 x_j} \right) \tag{9}$$

3. Weak form with incremental decompositions

$$\int_{^0V} {}^1_0 S_{ij}\, \delta({}_0\eta_{ij})\, d\,{}^0V + \int_{^0V} {}_0 S_{ij}\, \delta({}_0\varepsilon_{ij})\, d\,{}^0V = \delta({}^2_0 R) - \int_{^0V} {}^1_0 S_{ij}\, \delta({}_0 e_{ij})\, d\,{}^0V \tag{10}$$

4. Linearized weak form with incremental decompositions

    Use the approximations

$$ {}_0 S_{ij} = {}_0 C_{ijk\ell}\, {}_0\varepsilon_{k\ell} \approx {}_0 C_{ijk\ell}\, {}_0 e_{k\ell}, \quad \delta({}_0\varepsilon_{ij}) \approx \delta({}_0 e_{ij}) \tag{11}$$

to rewrite the weak form as

$$\int_{^0V} {}_0 C_{ijk\ell}\, {}_0 e_{k\ell}\, \delta({}_0 e_{ij})\, d\,{}^0V + \int_{^0V} {}^1_0 S_{ij}\, \delta({}_0\eta_{ij})\, d\,{}^0V$$

$$= \delta({}^2_0 R) - \int_{^0V} {}^1_0 S_{ij}\, \delta({}_0 e_{ij})\, d\,{}^0V \tag{12}$$

where $_1^2 f_i$ and $_1^2 t_i$ are the body force and boundary traction components referred to the configuration Ω_1, and $_1^2 \varepsilon_{ij}$ are the components of the updated Green–Lagrange strain tensor components defined in Eq. (9.2.20). Using Eqs. (9.3.20)–(9.3.22), Eqs. (9.3.1)–(9.3.3) can be expressed as

$$\int_{^1V} {}_1^2 S_{ij}\, \delta({}_1^2\varepsilon_{ij})\, d\,{}^1V - \int_{^1V} {}_1^2 f_i\, \delta u_i\, d\,{}^1V - \int_{^1S} {}_1^2 t_i\, \delta u_i\, d\,{}^1S = 0 \qquad (9.3.23)$$

The virtual strains are given by [see Eqs. (9.2.20)–(9.2.22)]

$$\delta({}_1^2\varepsilon_{ij}) = \delta(_1e_{ij}) + \delta(_1\eta_{ij})$$

$$\delta(_1e_{ij}) = \frac{1}{2}\left(\frac{\partial \delta u_i}{\partial\,{}^1x_j} + \frac{\partial \delta u_j}{\partial\,{}^1x_i}\right) \qquad (9.3.24)$$

$$\delta(_1\eta_{ij}) = \frac{1}{2}\left(\frac{\partial \delta u_k}{\partial\,{}^1x_i}\frac{\partial u_k}{\partial\,{}^1x_j} + \frac{\partial u_k}{\partial\,{}^1x_i}\frac{\partial \delta u_k}{\partial\,{}^1x_j}\right)$$

9.3.3.2 Incremental decompositions

Now using Eqs. (9.2.28) and (9.3.24), we can express Eq. (9.3.23) as

$$0 = \int_{^1V} {}_1^2 S_{ij}\, \delta({}_1^2\varepsilon_{ij})\, d\,{}^1V - \delta({}_1^2 R)$$

$$= \int_{^1V} \left({}^1\sigma_{ij} + {}_1 S_{ij}\right) \delta({}_1^2\varepsilon_{ij})\, d\,{}^1V - \delta({}_1^2 R)$$

$$= \int_{^1V} \left\{ {}_1 S_{ij}\, \delta({}_1^2\varepsilon_{ij}) + {}^1\sigma_{ij}\left[\delta(_1e_{ij}) + \delta(_1\eta_{ij})\right]\right\}\, d\,{}^1V - \delta({}_1^2 R)$$

$$= \int_{^1V} {}_1 S_{ij}\, \delta({}_1^2\varepsilon_{ij})\, d\,{}^1V + \int_{^1V} {}^1\sigma_{ij}\, \delta(_1\eta_{ij})\, d\,{}^1V + \delta({}_1^1 R) - \delta({}_1^2 R) \quad (9.3.25)$$

where **u** is the displacement from Ω_1 to Ω_2, $\delta \mathbf{u}$ is its first variation, and

$$\delta({}_1^2 R) = \int_{^1V} {}_1^2 f_i\, \delta u_i\, d\,{}^1V + \int_{^1S} {}_1^2 t_i\, \delta u_i\, d\,{}^1S$$

$$\delta({}_1^1 R) = \int_{^1V} {}^1\sigma_{ij}\, \delta(_1e_{ij})\, d\,{}^1V \qquad (9.3.26)$$

Note that $\delta({}_1^1 R)$ is the virtual internal energy stored in the body at configuration Ω_1 (see Fig. 9.3.2). Since the body is in equilibrium at configuration Ω_1, the principle of virtual work applied to configuration Ω_1 yields

$$\delta({}_1^1 R) = \int_{^1V} {}_1^1 f_i\, \delta u_i\, d\,{}^1V + \int_{^1S} {}_1^1 t_i\, \delta u_i\, d\,{}^1S \qquad (9.3.27)$$

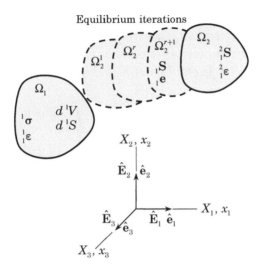

Fig. 9.3.2: Reference and final configurations of the body in the updated Lagrangian formulation.

Next, we invoke the constitutive relation, Eq. (9.2.32), to express $_1S_{ij}$ in terms of the incremental strain components $_1^2\varepsilon_{ij}$. Equation (9.3.23) takes the form

$$\int_{^1V} {_1}C_{ijk\ell} \, {_1^2}\varepsilon_{k\ell} \, \delta({_1^2}\varepsilon_{ij}) \, d\,{^1}V + \int_{^1V} {^1}\sigma_{ij} \, \delta({_1}\eta_{ij}) \, d\,{^1}V$$

$$= \delta({_1^2}R) - \int_{^1V} {^1}\sigma_{ij} \, \delta({_1}e_{ij}) \, d\,{^1}V \tag{9.3.28}$$

9.3.3.3   Linearization

Once again, we assume that the displacements u_i are small so that the following approximations hold:

$$_1S_{ij} = {_1}C_{ijk\ell} \, {_1^2}\varepsilon_{k\ell} \approx {_1}C_{ijk\ell} \, {_1}e_{k\ell} \quad \text{and} \quad \delta({_1^2}\varepsilon_{ij}) \approx \delta({_1}e_{ij}) \tag{9.3.29}$$

Then Eq. (9.3.28) takes the form

$$\int_{^1V} {_1}C_{ijk\ell} \, {_1}e_{k\ell} \, \delta({_1}e_{ij}) \, d\,{^1}V + \int_{^1V} {^1}\sigma_{ij} \, \delta({_1}\eta_{ij}) \, d\,{^1}V$$

$$= \delta({_1^2}R) - \int_{^1V} {^1}\sigma_{ij} \, \delta({_1}e_{ij}) \, d\,{^1}V \tag{9.3.30}$$

Equation (9.3.30) is the weak form for the development of the finite element model based on the updated Lagrangian formulation.

The total Cauchy stress components $^1\sigma_{ij}$ are evaluated using the constitutive relation

$$^1\sigma_{ij} = {_1}C_{ijk\ell} \, {_1^1}\varepsilon_{k\ell} \tag{9.3.31}$$

where ${_1^1}\varepsilon_{k\ell}$ are the Euler–Almansi strain tensor components defined by

$$2 \, {_1^1}\varepsilon_{ij} \, d \, {^1}x_i \, d \, {^1}x_j = \left(^1ds\right)^2 - \left(^0ds\right)^2$$

which gives

$$
{_1^1}\varepsilon_{ij} = \frac{1}{2}\left(\frac{\partial \, {_0^1}u_i}{\partial \, {^1}x_j} + \frac{\partial \, {_0^1}u_j}{\partial \, {^1}x_i} - \frac{\partial \, {_0^1}u_k}{\partial \, {^1}x_i}\frac{\partial \, {_0^1}u_k}{\partial \, {^1}x_j}\right) \tag{9.3.32}
$$

A summary of all the pertinent equations of the updated Lagrangian formulation is presented in Table 9.3.2.

9.3.4 Some Remarks on the Formulations

Let us examine the consequence of the linearizations used in arriving at the virtual work statements in Eqs. (9.3.17) and (9.3.30). First, we note that the linearization only approximated the left-hand side of the virtual work statements. Therefore, the errors in the two formulations are

$$\text{TL Formulation:} \quad \mathcal{E}_{TL} = \delta(^2_0R) - \int_{^0V} {_0^1}S_{ij}\,\delta(_0e_{ij})\,d\,{^0}V \tag{9.3.33}$$

$$\text{UL Formulation:} \quad \mathcal{E}_{UL} = \delta(^2_1R) - \int_{^1V} {^1}\sigma_{ij}\,\delta(_1e_{ij})\,d\,{^1}V \tag{9.3.34}$$

The expressions in Eqs. (9.3.33) and (9.3.34) denote the virtual work done by the "out-of-balance" forces in the two formulations. Therefore, iteration is necessary to make the virtual work done by the imbalance forces approach zero within an assumed tolerance. Even though, in theory, the displacements ${_0^1}u_i$ are assumed to be known, they have to be determined by iteration. Thus, Eqs. (9.3.17) and (9.3.30) at the beginning of the rth (Newton's) iteration are

$$\int_{^0V} {_0}C_{ijk\ell}^{(r-1)} \, {_0}e_{k\ell}^{(r)} \, \delta(_0e_{ij}) \, d\,{^0}V + \int_{^0V} {_0^1}S_{ij}^{(r-1)} \, \delta(_0\eta_{ij}^{(r)}) \, d\,{^0}V$$

$$= \delta(^2_0R) - \int_{^0V} {_0^1}S_{ij}^{(r-1)} \, \delta(_0e_{ij}^{(r-1)}) \, d\,{^0}V \tag{9.3.35}$$

$$\int_{^1V} {_1}C_{ijk\ell}^{(r-1)} \, {_1}e_{k\ell}^{(r)} \, \delta(_1e_{ij}) \, d\,{^1}V + \int_{^1V} {^1}\sigma_{ij}^{(r-1)} \, \delta(_1\eta_{ij}^{(r)}) \, d\,{^1}V$$

$$= \delta(^2_1R) - \int_{^1V} {^1}\sigma_{ij}^{(r-1)} \, \delta(_1e_{ij}^{(r-1)}) \, d\,{^1}V \tag{9.3.36}$$

Table 9.3.2: Summary of equations of the updated Lagrangian formulation.

---

1. Weak form

$$\int_{^1V} {}_1^2S_{ij}\,\delta({}_1^2\varepsilon_{ij})\,d\,{}^1V - \int_{^1V} {}_1^2f_i\,\delta u_i\,d\,{}^1V - \int_{^1S} {}_1^2t_i\,\delta u_i\,d\,{}^1S = 0 \qquad (1)$$

$$\delta({}_1^2\varepsilon_{ij}) = \delta({}_1e_{ij}) + \delta({}_1\eta_{ij})$$

$$\delta({}_1e_{ij}) = \frac{1}{2}\left(\frac{\partial\delta u_i}{\partial\,{}^1x_j} + \frac{\partial\delta u_j}{\partial\,{}^1x_i}\right)$$

$$\delta({}_1\eta_{ij}) = \frac{1}{2}\left(\frac{\partial\delta u_k}{\partial\,{}^1x_i}\frac{\partial u_k}{\partial\,{}^1x_j} + \frac{\partial u_k}{\partial\,{}^1x_i}\frac{\partial\delta u_k}{\partial\,{}^1x_j}\right) \qquad (2)$$

2. Incremental decompositions

$$\begin{aligned}
{}_1^2S_{ij} &= {}^1\sigma_{ij} + {}_1S_{ij}, \quad {}_1^1S_{ij} = {}^1\sigma_{ij} && (3)\\
{}_1^2\varepsilon_{ij} &= {}_1e_{ij} + {}_1\eta_{ij} && (4)
\end{aligned}$$

$$_1e_{ij} = \frac{1}{2}\left(\frac{\partial u_i}{\partial\,{}^1x_j} + \frac{\partial u_j}{\partial\,{}^1x_i}\right), \quad {}_1\eta_{ij} = \frac{1}{2}\left(\frac{\partial u_k}{\partial\,{}^1x_i}\frac{\partial u_k}{\partial\,{}^1x_j}\right) \qquad (5)$$

$${}^1\sigma_{ij} = {}_1C_{ijk\ell}\,{}_1^1\varepsilon_{k\ell}, \quad {}_1S_{ij} = {}_1C_{ijk\ell}\,{}_1^2\varepsilon_{k\ell} \qquad (6)\ \&\ (7)$$

$${}_1^1\varepsilon_{ij} = \frac{1}{2}\left(\frac{\partial\,{}_0^1u_i}{\partial\,{}^1x_j} + \frac{\partial\,{}_0^1u_j}{\partial\,{}^1x_i} - \frac{\partial\,{}_0^1u_k}{\partial\,{}^1x_i}\frac{\partial\,{}_0^1u_k}{\partial\,{}^1x_j}\right) \qquad (8)$$

3. Weak form with incremental decompositions

$$\int_{^1V} {}_1S_{ij}\,\delta({}_1^2\varepsilon_{ij})\,d\,{}^1V + \int_{^1V} {}^1\sigma_{ij}\,\delta({}_1\eta_{ij})\,d\,{}^1V = \delta({}_1^2R) - \int_{^1V} {}^1\sigma_{ij}\,\delta({}_1e_{ij})\,d\,{}^1V \qquad (9)$$

4. Linearized weak form with incremental decompositions

  Use the approximations

$${}_1S_{ij} = {}_1C_{ijk\ell}\,{}_1^2\varepsilon_{k\ell} \approx {}_1C_{ijk\ell}\,{}_1e_{k\ell}, \quad \delta({}_1\varepsilon_{ij}) \approx \delta({}_1e_{ij}) \qquad (10)$$

to rewrite the weak form as

$$\int_{^1V} {}_1C_{ijk\ell}\,{}_1e_{k\ell}\,\delta({}_1e_{ij})\,d\,{}^1V + \int_{^1V} {}^1\sigma_{ij}\,\delta({}_1\eta_{ij})\,d\,{}^1V$$

$$= \delta({}_1^2R) - \int_{^1V} {}^1\sigma_{ij}\,\delta({}_1e_{ij})\,d\,{}^1V \qquad (11)$$

Update the geometry at the end of each iteration

$$\mathbf{x} = \mathbf{x}^{(r)} + \mathbf{u} \qquad (12)$$

9.4 Finite Element Models of 2-D Continua

9.4.1 Introduction

Here we discuss finite element models based on the total and updated Lagrangian formulations presented in the previous section. Attention is focused here on two-dimensional problems under the assumption of linear, orthotropic, elastic material behavior. The volume integral will be replaced, at the end, with $dV = h_e\, A$, where h_e denotes the thickness of the elastic body; one may take $h_e = 1$ for plane strain problems.

9.4.2 Total Lagrangian Formulation

We introduce the notation

$$^0x_1 = x, \quad ^0x_2 = y, \quad ^1u_1 = u, \quad ^1u_2 = v, \quad u_1 = \bar{u}, \quad u_2 = \bar{v} \qquad (9.4.1)$$

We can write the first expression of Eq. (9.3.17) in matrix form

$$\int_{^0V} {}_0C_{ijk\ell}\, {}_0e_{k\ell}\, \delta({}_0e_{ij})\, d\,{}^0V = \int_{^0V} (\delta_0 \mathbf{e})^\mathrm{T}\, {}_0\mathbf{C}\, {}_0\mathbf{e}\, d\,{}^0V$$

$$= \int_{^0V} \left\{ [(\mathbf{D} + \mathbf{D_u})\delta\bar{\mathbf{u}}]^\mathrm{T}\, {}_0\mathbf{C}\, (\mathbf{D} + \mathbf{D_u})\, \bar{\mathbf{u}} \right\} d\,{}^0V \qquad (9.4.2)$$

where

$$
{}_0\mathbf{e} =
\left\{
\begin{array}{c}
{}_0e_{xx} \\
{}_0e_{yy} \\
2\,{}_0e_{xy}
\end{array}
\right\}
=
\left\{
\begin{array}{c}
\dfrac{\partial \bar{u}}{\partial x} \\[2mm]
\dfrac{\partial \bar{v}}{\partial y} \\[2mm]
\dfrac{\partial \bar{u}}{\partial y} + \dfrac{\partial \bar{v}}{\partial x}
\end{array}
\right\}
+
\left\{
\begin{array}{c}
\dfrac{\partial u}{\partial x}\dfrac{\partial \bar{u}}{\partial x} + \dfrac{\partial v}{\partial x}\dfrac{\partial \bar{v}}{\partial x} \\[2mm]
\dfrac{\partial u}{\partial y}\dfrac{\partial \bar{u}}{\partial y} + \dfrac{\partial v}{\partial y}\dfrac{\partial \bar{v}}{\partial y} \\[2mm]
\dfrac{\partial u}{\partial x}\dfrac{\partial \bar{u}}{\partial y} + \dfrac{\partial v}{\partial x}\dfrac{\partial \bar{v}}{\partial y} + \dfrac{\partial \bar{u}}{\partial x}\dfrac{\partial u}{\partial y} + \dfrac{\partial \bar{v}}{\partial x}\dfrac{\partial v}{\partial y}
\end{array}
\right\}
$$

$$
=
\left(
\begin{bmatrix}
\dfrac{\partial}{\partial x} & 0 \\[2mm]
0 & \dfrac{\partial}{\partial y} \\[2mm]
\dfrac{\partial}{\partial y} & \dfrac{\partial}{\partial x}
\end{bmatrix}
+
\begin{bmatrix}
\dfrac{\partial u}{\partial x}\dfrac{\partial}{\partial x} & \dfrac{\partial v}{\partial x}\dfrac{\partial}{\partial x} \\[2mm]
\dfrac{\partial u}{\partial y}\dfrac{\partial}{\partial y} & \dfrac{\partial v}{\partial y}\dfrac{\partial}{\partial y} \\[2mm]
\dfrac{\partial u}{\partial y}\dfrac{\partial}{\partial x} + \dfrac{\partial u}{\partial x}\dfrac{\partial}{\partial y} & \dfrac{\partial v}{\partial y}\dfrac{\partial}{\partial x} + \dfrac{\partial v}{\partial x}\dfrac{\partial}{\partial y}
\end{bmatrix}
\right)
\left\{
\begin{array}{c}
\bar{u} \\
\bar{v}
\end{array}
\right\}
$$

$$\equiv (\mathbf{D} + \mathbf{D_u})\, \bar{\mathbf{u}} \qquad (9.4.3)$$

$$
\mathbf{D} =
\begin{bmatrix}
\dfrac{\partial}{\partial x} & 0 \\[2mm]
0 & \dfrac{\partial}{\partial y} \\[2mm]
\dfrac{\partial}{\partial y} & \dfrac{\partial}{\partial x}
\end{bmatrix},
\quad
\mathbf{D_u} =
\begin{bmatrix}
\dfrac{\partial u}{\partial x}\dfrac{\partial}{\partial x} & \dfrac{\partial v}{\partial x}\dfrac{\partial}{\partial x} \\[2mm]
\dfrac{\partial u}{\partial y}\dfrac{\partial}{\partial y} & \dfrac{\partial v}{\partial y}\dfrac{\partial}{\partial y} \\[2mm]
\dfrac{\partial u}{\partial y}\dfrac{\partial}{\partial x} + \dfrac{\partial u}{\partial x}\dfrac{\partial}{\partial y} & \dfrac{\partial v}{\partial y}\dfrac{\partial}{\partial x} + \dfrac{\partial v}{\partial x}\dfrac{\partial}{\partial y}
\end{bmatrix}
\qquad (9.4.4)
$$

$$
{}_0\mathbf{C} =
\begin{bmatrix}
{}_0C_{11} & {}_0C_{12} & 0 \\
{}_0C_{12} & {}_0C_{22} & 0 \\
0 & 0 & {}_0C_{66}
\end{bmatrix},
\quad
\bar{\mathbf{u}} =
\left\{
\begin{array}{c}
\bar{u} \\
\bar{v}
\end{array}
\right\}
\qquad (9.4.5)
$$

$$\delta\,{}_0\mathbf{e} = (\mathbf{D} + \mathbf{D_u})\, \delta\bar{\mathbf{u}} \qquad (9.4.6)$$

where the coefficients $_0C_{ij}$ are defined in Eqs. (9.2.39) and (9.2.40) for plane strain and plane stress cases, respectively. The second expression of Eq. (9.3.17) can be written as

$$
\int_{^0V} {}_0^1 S_{ij} \, \delta({}_0\eta_{ij}) \, d\,{}^0V
$$

$$
= \int_{^0V} \left[{}_0^1 S_{xx} \left(\frac{\partial \delta \bar{u}}{\partial x} \frac{\partial \bar{u}}{\partial x} + \frac{\partial \delta \bar{v}}{\partial x} \frac{\partial \bar{v}}{\partial x} \right) + {}_0^1 S_{yy} \left(\frac{\partial \delta \bar{u}}{\partial y} \frac{\partial \bar{u}}{\partial y} + \frac{\partial \delta \bar{v}}{\partial y} \frac{\partial \bar{v}}{\partial y} \right) \right.
$$

$$
\left. + {}_0^1 S_{xy} \left(\frac{\partial \delta \bar{u}}{\partial x} \frac{\partial \bar{u}}{\partial y} + \frac{\partial \delta \bar{v}}{\partial x} \frac{\partial \bar{v}}{\partial y} + \frac{\partial \bar{u}}{\partial x} \frac{\partial \delta \bar{u}}{\partial y} + \frac{\partial \bar{v}}{\partial x} \frac{\partial \delta \bar{v}}{\partial y} \right) \right] d\,{}^0V
$$

$$
= \int_{^0V} \begin{Bmatrix} \frac{\partial \delta \bar{u}}{\partial x} \\ \frac{\partial \delta \bar{u}}{\partial y} \\ \frac{\partial \delta \bar{v}}{\partial x} \\ \frac{\partial \delta \bar{v}}{\partial y} \end{Bmatrix}^{\mathrm{T}} \begin{bmatrix} {}_0^1 S_{xx} & {}_0^1 S_{xy} & 0 & 0 \\ {}_0^1 S_{xy} & {}_0^1 S_{yy} & 0 & 0 \\ 0 & 0 & {}_0^1 S_{xx} & {}_0^1 S_{xy} \\ 0 & 0 & {}_0^1 S_{xy} & {}_0^1 S_{yy} \end{bmatrix} \begin{Bmatrix} \frac{\partial \bar{u}}{\partial x} \\ \frac{\partial \bar{u}}{\partial y} \\ \frac{\partial \bar{v}}{\partial x} \\ \frac{\partial \bar{v}}{\partial y} \end{Bmatrix} d\,{}^0V
$$

$$
= \int_{^0V} (\bar{\mathbf{D}} \delta \bar{\mathbf{u}})^{\mathrm{T}} {}_0^1\mathbf{S} \, (\bar{\mathbf{D}} \bar{\mathbf{u}}) \, d\,{}^0V \tag{9.4.7}
$$

where

$$
{}_0^1\mathbf{S} = \begin{bmatrix} {}_0^1 S_{xx} & {}_0^1 S_{xy} & 0 & 0 \\ {}_0^1 S_{xy} & {}_0^1 S_{yy} & 0 & 0 \\ 0 & 0 & {}_0^1 S_{xx} & {}_0^1 S_{xy} \\ 0 & 0 & {}_0^1 S_{xy} & {}_0^1 S_{yy} \end{bmatrix}, \quad \bar{\mathbf{D}} = \begin{bmatrix} \frac{\partial}{\partial x} & 0 \\ \frac{\partial}{\partial y} & 0 \\ 0 & \frac{\partial}{\partial x} \\ 0 & \frac{\partial}{\partial y} \end{bmatrix} \tag{9.4.8}
$$

Next, the total and incremental displacement fields, $\mathbf{u}$ and $\bar{\mathbf{u}}$, are interpolated as

$$
\mathbf{u} = \begin{Bmatrix} u \\ v \end{Bmatrix} = \begin{Bmatrix} \sum_{j=1}^{n} u_j \psi_j(\mathbf{x}) \\ \sum_{j=1}^{n} v_j \psi_j(\mathbf{x}) \end{Bmatrix} = \boldsymbol{\Psi} \, \boldsymbol{\Delta} \tag{9.4.9}
$$

$$
\bar{\mathbf{u}} = \begin{Bmatrix} \bar{u} \\ \bar{v} \end{Bmatrix} = \begin{Bmatrix} \sum_{j=1}^{n} \bar{u}_j \psi_j(\mathbf{x}) \\ \sum_{j=1}^{n} \bar{v}_j \psi_j(\mathbf{x}) \end{Bmatrix} = \boldsymbol{\Psi} \, \bar{\boldsymbol{\Delta}}, \qquad \delta \bar{\mathbf{u}} = \boldsymbol{\Psi} \, \delta \bar{\boldsymbol{\Delta}} \tag{9.4.10}
$$

where

$$
\boldsymbol{\Psi} = \begin{bmatrix} \psi_1 & 0 & \psi_2 & 0 & \cdots & \psi_n & 0 \\ 0 & \psi_1 & 0 & \psi_2 & \cdots & 0 & \psi_n \end{bmatrix} \tag{9.4.11}
$$

$$
\boldsymbol{\Delta}^{\mathrm{T}} = \{u_1, \; v_1, \; u_2, \; v_2, \; \cdots, \; u_n, \; v_n\} \tag{9.4.12}
$$

$$
\bar{\boldsymbol{\Delta}}^{\mathrm{T}} = \{\bar{u}_1, \; \bar{v}_1, \; \bar{u}_2, \; \bar{v}_2, \; \cdots, \; \bar{u}_n, \; \bar{v}_n\} \tag{9.4.13}
$$

Here $\{\psi_j(\mathbf{x})\}$ denote the set of Lagrange family of approximation functions and n is the number of nodes per element. The size of the element stiffness matrix is $2n \times 2n$.

The expressions in Eqs. (9.4.2) and (9.4.7) can now be expressed in matrix form as

$$\int_{0_V} \left\{ [(\mathbf{D} + \mathbf{D_u})\delta\bar{\mathbf{u}}]^{\mathrm{T}} \ _0\mathbf{C} \ (\mathbf{D} + \mathbf{D_u}) \bar{\mathbf{u}} \right\} d \ ^0V$$

$$= \int_{0_V} (\delta\bar{\mathbf{\Delta}})^{\mathrm{T}} [(\mathbf{D} + \mathbf{D_u}) \mathbf{\Psi}]^{\mathrm{T}} \ _0\mathbf{C} [(\mathbf{D} + \mathbf{D_u}) \mathbf{\Psi}] \bar{\mathbf{\Delta}} \ d \ ^0V$$

$$\equiv \int_{0_V} (\delta\bar{\mathbf{\Delta}})^{\mathrm{T}} (\mathbf{B}_1)^{\mathrm{T}} \ _0\mathbf{C} \, \mathbf{B}_1 \bar{\mathbf{\Delta}} \ d \ ^0V \qquad (9.4.14)$$

$$\int_{0_V} (\bar{\mathbf{D}}\delta\bar{\mathbf{u}})^{\mathrm{T}} \ _0^1\mathbf{S} \, (\bar{\mathbf{D}}\bar{\mathbf{u}}) \ d \ ^0V = \int_{0_V} (\delta\bar{\mathbf{\Delta}})^{\mathrm{T}} (\mathbf{B}_2)^{\mathrm{T}} \ _0^1\mathbf{S} \, \mathbf{B}_2 \bar{\mathbf{\Delta}} \, d \ ^0V \quad (9.4.15)$$

where $\mathbf{B}_1$ is the $3 \times 2n$ matrix and $\mathbf{B}_2$ is the $4 \times 2n$ matrix defined by

$$\mathbf{B}_1 = (\mathbf{D} + \mathbf{D_u}) \, \mathbf{\Psi}, \quad \mathbf{B}_2 = \bar{\mathbf{D}}\mathbf{\Psi} \qquad (9.4.16)$$

and $_0^1\mathbf{S}$ is the 4×4 matrix defined in Eq. (9.4.8). The right-hand sides of Eq. (9.3.17) can be written in terms of the finite element approximations as

$$\delta(_0^1R) = \int_{0_V} \ _0^1S_{ij} \ \delta(_0e_{ij}) \ d \ ^0V = \int_{0_V} (\delta \ _0\mathbf{e})^{\mathrm{T}} \ _0^1\mathbf{S}_v \ d \ ^0V$$

$$= \int_{0_V} (\delta\bar{\mathbf{\Delta}})^{\mathrm{T}} (\mathbf{B}_1)^{\mathrm{T}} \ _0^1\mathbf{S}_v \ d \ ^0V \qquad (9.4.17)$$

$$\delta(_0^2R) = \int_{0_V} \ _0^2f_i\delta u_i \ d \ ^0V + \int_{0_S} \ _0^2t_i\delta u_i \ d \ ^0S$$

$$= \int_{0_V} (\delta\bar{\mathbf{\Delta}})^{\mathrm{T}} \ \mathbf{\Psi}^{\mathrm{T}} \ _0^2\mathbf{f} \ d \ ^0V + \int_{0_S} (\delta\bar{\mathbf{\Delta}})^{\mathrm{T}} \ \mathbf{\Psi}^{\mathrm{T}} \ _0^2\mathbf{t} \ d \ ^0S \qquad (9.4.18)$$

where

$$_0^1\mathbf{S}_v = \left\{ \begin{matrix} _0^1S_{xx} \\ _0^1S_{yy} \\ _0^1S_{xy} \end{matrix} \right\} = \begin{bmatrix} _0C_{11} & _0C_{12} & 0 \\ _0C_{12} & _0C_{22} & 0 \\ 0 & 0 & _0C_{66} \end{bmatrix} \left\{ \begin{matrix} _0^1E_{xx} \\ _0^1E_{yy} \\ 2_0^1E_{xy} \end{matrix} \right\} \equiv \ _0\mathbf{C} \ _0^1\mathbf{E} \qquad (9.4.19)$$

$$_0^1\mathbf{E} = \left\{ \begin{matrix} _0^1E_{xx} \\ _0^1E_{yy} \\ 2_0^1E_{xy} \end{matrix} \right\} = \left\{ \begin{matrix} \frac{\partial u}{\partial x} + \frac{1}{2}\left[\left(\frac{\partial u}{\partial x}\right)^2 + \left(\frac{\partial v}{\partial x}\right)^2\right] \\ \frac{\partial v}{\partial y} + \frac{1}{2}\left[\left(\frac{\partial u}{\partial y}\right)^2 + \left(\frac{\partial v}{\partial y}\right)^2\right] \\ \frac{\partial u}{\partial y} + \frac{\partial v}{\partial x} + \left(\frac{\partial u}{\partial x}\frac{\partial u}{\partial y} + \frac{\partial v}{\partial x}\frac{\partial v}{\partial y}\right) \end{matrix} \right\} \qquad (9.4.20)$$

Substitution of Eqs. (9.4.14)–(9.4.20) into Eq. (9.3.17), and use of the fundamental lemma of calculus of variations (i.e. $\delta\bar{\mathbf{\Delta}}$ is a vector of arbitrary variations), yields the following finite element model associated with the total Lagrangian formulation of two-dimensional nonlinear continua:

$$(\mathbf{K}_1 + \mathbf{K}_2) \, \bar{\mathbf{\Delta}} = \ _0^2\mathbf{F} - \ _0^1\mathbf{F} \qquad (9.4.21)$$

where

$$\mathbf{K}_1 = \int_{0V} (\mathbf{B}_1)^{\mathrm{T}} {}_0\mathbf{C}\,\mathbf{B}_1 \, d^0V, \quad \mathbf{K}_2 = \int_{0V} (\mathbf{B}_2)^{\mathrm{T}} {}_0^1\mathbf{S}\,\mathbf{B}_2 \, d^0V$$

$$\substack{1\\0}\mathbf{F} = \int_{0V} (\mathbf{B}_1)^{\mathrm{T}} {}_0^1\mathbf{S}_v \, d^0V, \quad \substack{2\\0}\mathbf{f} = \left\{ \begin{matrix} {}_0^2 f_x \\ {}_0^2 f_y \end{matrix} \right\}, \quad \substack{2\\0}\mathbf{t} = \left\{ \begin{matrix} {}_0^2 t_x \\ {}_0^2 t_y \end{matrix} \right\} \qquad (9.4.22)$$

$$\substack{2\\0}\mathbf{F} = \int_{0V} \boldsymbol{\Psi}^{\mathrm{T}} \, {}_0^2\mathbf{f} \, d^0V + \int_{0S} \boldsymbol{\Psi}^{\mathrm{T}} {}_0^2\mathbf{t} \, d^0S$$

Note that the stiffness matrix $\mathbf{K} = \mathbf{K}_1 + \mathbf{K}_2$ is a $2n \times 2n$ matrix, where n is the number of nodes in the 2-D element. It is symmetric on account of the symmetry of the matrices $\mathbf{K}_1$ and $\mathbf{K}_2$, and whose symmetry is due to the symmetry of ${}_0^1\mathbf{S}$ and ${}_0\mathbf{C}$. Also, the total and updated Lagrangian formulations are incremental formulations, where $\bar{\boldsymbol{\Delta}}$ is the incremental displacement vector and $\mathbf{K}$ is the stiffness matrix. The direct stiffness matrix is implicit in the vector $\substack{1\\0}\mathbf{F}$. For a linear analysis, $\bar{\boldsymbol{\Delta}}$ becomes the total displacement vector, $\substack{1\\0}\mathbf{F} = \mathbf{0}$, and $\mathbf{K}_2 = \mathbf{0}$.

For two-dimensional problems, the explicit forms of matrices $\mathbf{B}_1$ and $\mathbf{B}_2$ are given by

$$\mathbf{B}_1 = \mathbf{B}_1^0 + \mathbf{B}_1^u + \mathbf{B}_1^v \qquad (9.4.23)$$

$$\mathbf{B}_1^0 = \begin{bmatrix} \frac{\partial\psi_1}{\partial x} & 0 & \frac{\partial\psi_2}{\partial x} & 0 & \cdots & \frac{\partial\psi_n}{\partial x} & 0 \\ 0 & \frac{\partial\psi_1}{\partial y} & 0 & \frac{\partial\psi_2}{\partial y} & \cdots & 0 & \frac{\partial\psi_n}{\partial y} \\ \frac{\partial\psi_1}{\partial y} & \frac{\partial\psi_1}{\partial x} & \frac{\partial\psi_2}{\partial y} & \frac{\partial\psi_2}{\partial x} & & \frac{\partial\psi_n}{\partial y} & \frac{\partial\psi_n}{\partial x} \end{bmatrix} \qquad (9.4.24)$$

$$\mathbf{B}_1^u = \begin{bmatrix} \frac{\partial u}{\partial x}\frac{\partial\psi_1}{\partial x} & 0 & \frac{\partial u}{\partial x}\frac{\partial\psi_2}{\partial x} & & \cdots & & \frac{\partial u}{\partial x}\frac{\partial\psi_n}{\partial x} & 0 \\ \frac{\partial u}{\partial y}\frac{\partial\psi_1}{\partial y} & 0 & \frac{\partial u}{\partial y}\frac{\partial\psi_2}{\partial y} & & \cdots & & \frac{\partial u}{\partial y}\frac{\partial\psi_n}{\partial y} & 0 \\ \frac{\partial u}{\partial x}\frac{\partial\psi_1}{\partial y} + \frac{\partial u}{\partial y}\frac{\partial\psi_1}{\partial x} & 0 & \frac{\partial u}{\partial x}\frac{\partial\psi_2}{\partial y} + \frac{\partial u}{\partial y}\frac{\partial\psi_2}{\partial x} & \cdots & \frac{\partial u}{\partial x}\frac{\partial\psi_n}{\partial y} + \frac{\partial u}{\partial y}\frac{\partial\psi_n}{\partial x} & 0 \end{bmatrix} \qquad (9.4.25)$$

$$\mathbf{B}_1^v = \begin{bmatrix} 0 & \frac{\partial v}{\partial x}\frac{\partial\psi_1}{\partial x} & 0 & \frac{\partial v}{\partial x}\frac{\partial\psi_2}{\partial x} & & \cdots & & \frac{\partial v}{\partial x}\frac{\partial\psi_n}{\partial x} \\ 0 & \frac{\partial v}{\partial y}\frac{\partial\psi_1}{\partial y} & 0 & \frac{\partial v}{\partial y}\frac{\partial\psi_2}{\partial y} & & \cdots & & \frac{\partial v}{\partial y}\frac{\partial\psi_n}{\partial y} \\ 0 & \frac{\partial v}{\partial x}\frac{\partial\psi_1}{\partial y} + \frac{\partial v}{\partial y}\frac{\partial\psi_1}{\partial x} & 0 & \frac{\partial v}{\partial x}\frac{\partial\psi_2}{\partial y} + \frac{\partial v}{\partial y}\frac{\partial\psi_2}{\partial x} & \cdots & \frac{\partial v}{\partial x}\frac{\partial\psi_n}{\partial y} + \frac{\partial v}{\partial y}\frac{\partial\psi_n}{\partial x} \end{bmatrix} \qquad (9.4.26)$$

$$\mathbf{B}_2 = \begin{bmatrix} \frac{\partial\psi_1}{\partial x} & 0 & \frac{\partial\psi_2}{\partial x} & 0 & \cdots & \frac{\partial\psi_n}{\partial x} & 0 \\ \frac{\partial\psi_1}{\partial y} & 0 & \frac{\partial\psi_2}{\partial y} & 0 & \cdots & \frac{\partial\psi_n}{\partial y} & 0 \\ 0 & \frac{\partial\psi_1}{\partial x} & 0 & \frac{\partial\psi_2}{\partial x} & \cdots & 0 & \frac{\partial\psi_n}{\partial x} \\ 0 & \frac{\partial\psi_1}{\partial y} & 0 & \frac{\partial\psi_2}{\partial y} & \cdots & 0 & \frac{\partial\psi_n}{\partial y} \end{bmatrix} \qquad (9.4.27)$$

The finite element equations, Eq. (9.4.21), can be written in explicit form as

$$\begin{bmatrix} \mathbf{K}_1^{11} + \mathbf{K}_2^{11} & \mathbf{K}_1^{12} \\ \mathbf{K}_1^{21} & \mathbf{K}_1^{22} + \mathbf{K}_2^{22} \end{bmatrix} \left\{ \begin{matrix} \bar{\mathbf{u}} \\ \bar{\mathbf{v}} \end{matrix} \right\} = \left\{ \begin{matrix} {}_0^2\mathbf{F}^1 - {}_0^1\mathbf{F}^1 \\ {}_0^2\mathbf{F}^2 - {}_0^1\mathbf{F}^2 \end{matrix} \right\} \qquad (9.4.28)$$

where

$$K_{ij(1)}^{11} = h_e \int_{\Omega^e} \left\{ {}_0C_{11} \left(1 + \frac{\partial u}{\partial x}\right)^2 \frac{\partial \psi_i}{\partial x}\frac{\partial \psi_j}{\partial x} + {}_0C_{22} \left(\frac{\partial u}{\partial y}\right)^2 \frac{\partial \psi_i}{\partial y}\frac{\partial \psi_j}{\partial y} \right.$$

$$+ {}_0C_{12}\left(1 + \frac{\partial u}{\partial x}\right)\frac{\partial u}{\partial y}\left(\frac{\partial \psi_i}{\partial x}\frac{\partial \psi_j}{\partial y} + \frac{\partial \psi_i}{\partial y}\frac{\partial \psi_j}{\partial x}\right)$$

$$+ {}_0C_{66}\left[\left(1 + \frac{\partial u}{\partial x}\right)\frac{\partial \psi_i}{\partial y} + \frac{\partial u}{\partial y}\frac{\partial \psi_i}{\partial x}\right]$$

$$\left. \times \left[\left(1 + \frac{\partial u}{\partial x}\right)\frac{\partial \psi_j}{\partial y} + \frac{\partial u}{\partial y}\frac{\partial \psi_j}{\partial x}\right] \right\} dx\,dy$$

$$K_{ij(1)}^{12} = h_e \int_{\Omega^e} \left\{ {}_0C_{11}\left(1 + \frac{\partial u}{\partial x}\right)\frac{\partial v}{\partial x}\frac{\partial \psi_i}{\partial x}\frac{\partial \psi_j}{\partial x} + {}_0C_{22}\left(1 + \frac{\partial v}{\partial y}\right)\frac{\partial u}{\partial y}\frac{\partial \psi_i}{\partial y}\frac{\partial \psi_j}{\partial y} \right.$$

$$+ {}_0C_{12}\left[\left(1 + \frac{\partial u}{\partial x}\right)\left(1 + \frac{\partial v}{\partial y}\right)\frac{\partial \psi_i}{\partial x}\frac{\partial \psi_j}{\partial y} + \frac{\partial u}{\partial y}\frac{\partial v}{\partial x}\frac{\partial \psi_i}{\partial y}\frac{\partial \psi_j}{\partial x}\right]$$

$$+ {}_0C_{66}\left[\left(1 + \frac{\partial u}{\partial x}\right)\frac{\partial \psi_i}{\partial y} + \frac{\partial u}{\partial y}\frac{\partial \psi_i}{\partial x}\right]$$

$$\left. \times \left[\left(1 + \frac{\partial v}{\partial y}\right)\frac{\partial \psi_j}{\partial x} + \frac{\partial v}{\partial x}\frac{\partial \psi_j}{\partial y}\right] \right\} dx\,dy = K_{ji(1)}^{21}$$

$$K_{ij(1)}^{22} = h_e \int_{\Omega^e} \left\{ {}_0C_{11}\left(\frac{\partial v}{\partial x}\right)^2 \frac{\partial \psi_i}{\partial x}\frac{\partial \psi_j}{\partial x} + {}_0C_{22}\left(1 + \frac{\partial v}{\partial y}\right)^2 \frac{\partial \psi_i}{\partial y}\frac{\partial \psi_j}{\partial y} \right.$$

$$+ {}_0C_{12}\left(1 + \frac{\partial v}{\partial y}\right)\frac{\partial v}{\partial x}\left(\frac{\partial \psi_i}{\partial x}\frac{\partial \psi_j}{\partial y} + \frac{\partial \psi_i}{\partial y}\frac{\partial \psi_j}{\partial x}\right)$$

$$+ {}_0C_{66}\left[\left(1 + \frac{\partial v}{\partial y}\right)\frac{\partial \psi_i}{\partial x} + \frac{\partial v}{\partial x}\frac{\partial \psi_i}{\partial y}\right]$$

$$\left. \times \left[\left(1 + \frac{\partial v}{\partial y}\right)\frac{\partial \psi_j}{\partial x} + \frac{\partial v}{\partial x}\frac{\partial \psi_j}{\partial y}\right] \right\} dx\,dy$$

$$K_{ij(2)}^{11} = h_e \int_{\Omega^e} \left[{}_0^1 S_{xx}\frac{\partial \psi_i}{\partial x}\frac{\partial \psi_j}{\partial x} + {}_0^1 S_{xy}\left(\frac{\partial \psi_i}{\partial y}\frac{\partial \psi_j}{\partial x} + \frac{\partial \psi_i}{\partial x}\frac{\partial \psi_j}{\partial y}\right) \right.$$

$$\left. + {}_0^1 S_{yy}\frac{\partial \psi_i}{\partial y}\frac{\partial \psi_j}{\partial y} \right] dx\,dy = K_{ij(2)}^{22} \tag{9.4.29}$$

$${}_0^2 F_i^1 = h_e \int_{\Omega^e} {}_0^2 f_x \psi_i\, dx\,dy + h_e \oint_{\Gamma^e} {}_0^2 t_x \psi_i\, ds$$

$${}_0^2 F_i^2 = h_e \int_{\Omega^e} {}_0^2 f_y \psi_i\, dx\,dy + h_e \oint_{\Gamma^e} {}_0^2 t_y \psi_i\, ds$$

$${}_0^1 F_i^1 = h_e \int_{\Omega^e} \left\{ \left(1 + \frac{\partial u}{\partial x}\right)\frac{\partial \psi_i}{\partial x}\,{}_0^1 S_{xx} + \frac{\partial u}{\partial y}\frac{\partial \psi_i}{\partial y}\,{}_0^1 S_{yy} \right.$$

$$\left. + \left[\left(1 + \frac{\partial u}{\partial x}\right)\frac{\partial \psi_i}{\partial y} + \frac{\partial u}{\partial y}\frac{\partial \psi_i}{\partial x}\right]\,{}_0^1 S_{xy} \right\} dx\,dy$$

$$
{}_0^1 F_i^2 = h_e \int_{\Omega^e} \left\{ \frac{\partial v}{\partial x} \frac{\partial \psi_i}{\partial x} \, {}_0^1 S_{xx} + \left(1 + \frac{\partial v}{\partial y} \right) \frac{\partial \psi_i}{\partial y} \, {}_0^1 S_{yy} \right.
$$

$$
\left. + \left[\left(1 + \frac{\partial v}{\partial y} \right) \frac{\partial \psi_i}{\partial x} + \frac{\partial v}{\partial x} \frac{\partial \psi_i}{\partial y} \right] {}_0^1 S_{xy} \right\} dx \, dy \tag{9.4.30}
$$

where h_e is the thickness of the element, $({}_0^1 S_{xx}, {}_0^1 S_{yy}, {}_0^1 S_{xy})$ are defined in Eq. (9.4.19), and the coefficients ${}_0 C_{ij}$ are defined in Eqs. (9.2.39) and (9.2.40) for plane strain and plane stress cases, respectively.

9.4.3   Updated Lagrangian Formulation

In view of the detailed discussion of the finite element model development for the total Lagrangian formulation and the similarity between Eqs. (9.3.17) and (9.3.30), the finite element model based on the updated Lagrangian formulation can be simply written as

$$
(\mathbf{K}_1 + \mathbf{K}_2)\, \bar{\boldsymbol{\Delta}} = {}_1^2 \mathbf{F} - {}_1^1 \mathbf{F} \tag{9.4.31}
$$

where

$$
\mathbf{K}_1 = \int_{{}^1 V} (\mathbf{B}_1^0)^{\mathrm{T}} \, {}_1 \mathbf{C} \, \mathbf{B}_1^0 \, d \, {}^1 V, \quad \mathbf{K}_2 = \int_{{}^1 V} (\mathbf{B}_2)^{\mathrm{T}} \, {}^1 \boldsymbol{\sigma} \, \mathbf{B}_2 \, d \, {}^1 V
$$

$$
{}_1^1 \mathbf{F} = \int_{{}^1 V} (\mathbf{B}_1^0)^{\mathrm{T}} \, {}^1 \boldsymbol{\sigma}_v \, d \, {}^1 V, \quad {}_1^2 \mathbf{f} = \left\{ {}_1^2 f_x \atop {}_1^2 f_y \right\}, \quad {}_1^2 \mathbf{t} = \left\{ {}_1^2 t_x \atop {}_1^2 t_y \right\} \tag{9.4.32}
$$

$$
{}_1^2 \mathbf{F} = \int_{{}^1 V} \boldsymbol{\Psi}^{\mathrm{T}} \, {}_1^2 \mathbf{f} \, d \, {}^1 V + \int_{{}^1 S} \boldsymbol{\Psi}^{\mathrm{T}} \, {}_1^2 \mathbf{t} \, d \, {}^1 S
$$

where $\mathbf{B}_1^0$ and $\mathbf{B}_2$ are defined by Eqs. (9.4.23)–(9.4.27), and

$$
{}^1 \boldsymbol{\sigma} = \begin{bmatrix} {}^1 \sigma_{xx} & {}^1 \sigma_{xy} & 0 & 0 \\ {}^1 \sigma_{xy} & {}^1 \sigma_{yy} & 0 & 0 \\ 0 & 0 & {}^1 \sigma_{xx} & {}^1 \sigma_{xy} \\ 0 & 0 & {}^1 \sigma_{xy} & {}^1 \sigma_{yy} \end{bmatrix} \tag{9.4.33}
$$

$$
{}^1 \boldsymbol{\sigma}_v = \left\{ {}^1 \sigma_{xx} \atop {}^1 \sigma_{yy} \atop {}^1 \sigma_{xy} \right\} = \begin{bmatrix} {}_1 C_{11} & {}_1 C_{12} & 0 \\ {}_1 C_{12} & {}_1 C_{22} & 0 \\ 0 & 0 & {}_1 C_{66} \end{bmatrix} \left\{ {}_1^1 \varepsilon_{xx} \atop {}_1^1 \varepsilon_{yy} \atop 2\,{}_1^1 \varepsilon_{xy} \right\} \tag{9.4.34}
$$

$$
{}_1^1 \boldsymbol{\varepsilon} = \left\{ {}_1^1 \varepsilon_{xx} \atop {}_1^1 \varepsilon_{yy} \atop 2\,{}_1^1 \varepsilon_{xy} \right\} = \left\{ \begin{array}{l} \frac{\partial u}{\partial x} - \frac{1}{2}\left[\left(\frac{\partial u}{\partial x}\right)^2 + \left(\frac{\partial v}{\partial x}\right)^2 \right] \\[2mm] \frac{\partial v}{\partial y} - \frac{1}{2}\left[\left(\frac{\partial u}{\partial y}\right)^2 + \left(\frac{\partial v}{\partial y}\right)^2 \right] \\[2mm] \frac{\partial u}{\partial y} + \frac{\partial v}{\partial x} - \left(\frac{\partial u}{\partial x}\frac{\partial u}{\partial y} + \frac{\partial v}{\partial x}\frac{\partial v}{\partial y} \right) \end{array} \right\} \tag{9.4.35}
$$

and the coefficients ${}_1 C_{ij}$ are defined in Eqs. (9.2.39) and (9.2.40) for plane strain and plane stress cases, respectively.

The finite element equations (9.4.31) can be expressed as

$$\begin{bmatrix} \mathbf{K}_1^{11} + \mathbf{K}_2^{11} & \mathbf{K}_1^{12} \\ \mathbf{K}_1^{21} & \mathbf{K}_1^{22} + \mathbf{K}_2^{22} \end{bmatrix} \begin{Bmatrix} \bar{\mathbf{u}} \\ \bar{\mathbf{v}} \end{Bmatrix} = \begin{Bmatrix} {}_1^2\mathbf{F}^1 - {}_1^1\mathbf{F}^1 \\ {}_1^2\mathbf{F}^2 - {}_1^1\mathbf{F}^2 \end{Bmatrix} \tag{9.4.36}$$

where

$$K_{ij(1)}^{11} = h_e \int_{\Omega^e} \left({}_1C_{11} \frac{\partial \psi_i}{\partial x} \frac{\partial \psi_j}{\partial x} + {}_1C_{66} \frac{\partial \psi_i}{\partial y} \frac{\partial \psi_j}{\partial y} \right) dx\,dy$$

$$K_{ij(1)}^{12} = h_e \int_{\Omega^e} \left({}_1C_{12} \frac{\partial \psi_i}{\partial x} \frac{\partial \psi_j}{\partial y} + {}_1C_{66} \frac{\partial \psi_i}{\partial y} \frac{\partial \psi_j}{\partial x} \right) dx\,dy = K_{ji(1)}^{21}$$

$$K_{ij(1)}^{22} = h_e \int_{\Omega^e} \left({}_1C_{66} \frac{\partial \psi_i}{\partial x} \frac{\partial \psi_j}{\partial x} + {}_1C_{22} \frac{\partial \psi_i}{\partial y} \frac{\partial \psi_j}{\partial y} \right) dx\,dy \tag{9.4.37}$$

$$K_{ij(2)}^{11} = h_e \int_{\Omega^e} \left[{}^1\sigma_{xx} \frac{\partial \psi_i}{\partial x} \frac{\partial \psi_j}{\partial x} + {}^1\sigma_{xy} \left(\frac{\partial \psi_i}{\partial y} \frac{\partial \psi_j}{\partial x} + \frac{\partial \psi_i}{\partial x} \frac{\partial \psi_j}{\partial y} \right) \right.$$
$$\left. + {}^1\sigma_{yy} \frac{\partial \psi_i}{\partial y} \frac{\partial \psi_j}{\partial y} \right] dx\,dy = K_{ij(2)}^{22}$$

and

$$\begin{aligned}
{}_1^2F_i^1 &= h_e \int_{\Omega^e} {}_1^2f_x \psi_i \, dx\,dy + h_e \oint_{\Gamma^e} {}_1^2t_x \psi_i \, ds \\
{}_1^2F_i^2 &= h_e \int_{\Omega^e} {}_1^2f_y \psi_i \, dx\,dy + h_e \oint_{\Gamma^e} {}_1^2t_y \psi_i \, ds \\
{}_1^1F_i^1 &= h_e \int_{\Omega^e} \left(\frac{\partial \psi_i}{\partial x} {}^1\sigma_{xx} + \frac{\partial \psi_i}{\partial y} {}^1\sigma_{xy} \right) dx\,dy \\
{}_1^1F_i^2 &= h_e \int_{\Omega^e} \left(\frac{\partial \psi_i}{\partial x} {}^1\sigma_{xy} + \frac{\partial \psi_i}{\partial y} {}^1\sigma_{yy} \right) dx\,dy
\end{aligned} \tag{9.4.38}$$

where h_e is the thickness of the element ${}^1x_1 = x$ and ${}^1x_2 = y$. Note that in two-dimensional problems the applied boundary forces may be measured per unit length of the boundary and body forces per unit area of the domain (i.e. thickness h_e is already accounted for).

The formulations presented in this section are easily extendable to three-dimensional problems. For the 3-D case, the constitutive matrix ${}_0\mathbf{C}$ is 6×6 [see Eq. (9.2.35) for the values of the stiffness coefficients], ${}_0^1\mathbf{S}$ is 6×6, ${}_0^1\mathbf{S}_v$ is 6×1, $\mathbf{D}$ and $\mathbf{D_u}$ are 6×3, $\bar{\mathbf{D}}$ is 6×3, $\boldsymbol{\Psi}$ is $3 \times 3n$, $\bar{\boldsymbol{\Delta}}$ is $3n \times 1$, $\mathbf{B}_1$ is $6 \times 3n$, $\mathbf{B}_2$ is $6 \times 3n$, and the stiffness matrix $\mathbf{K}$ is of the order $3n \times 3n$, where n is the number of nodes in the 3-D element.

9.4.4 Computer Implementation

The computer implementation of the two formulations discussed in this chapter follows along the same lines as discussed for the Newton iterative procedure used for plate bending. Box 9.4.1 contains the portion of the main program where element information is passed on to the element subroutine and error check, while Box 9.4.2 contains the main parts of the element subroutine.

In the following some of the key variables are described [see Eqs. (9.4.12), (9.4.13), and (9.4.30)]:

{GPU}−Displacement vector between current and next configuration, $\bar{\boldsymbol{\Delta}}$

{GLS}−Total displacement vector, $\boldsymbol{\Delta}$

{GLF}−Imbalance force vector going into the solver BNDSYMSOLV,

$${}_{0}^{2}\mathbf{F} - {}_{0}^{1}\mathbf{F} \text{ for TLF, and } {}_{1}^{2}\mathbf{F} - {}_{1}^{1}\mathbf{F} \text{ for ULF}$$

latest displacement vector $\bar{\boldsymbol{\Delta}}$ coming out of the solver

{IRES}−Flag for solving the equations (IRES=1) or just

inverting the coefficient matrix (IRES=0)

LFORM−Type of formulation

LFORM=1: Total Lagrangian Formulation (TLF),

LFORM > 1: Updated Lagrangian Formulation (ULF)

Box 9.4.1: Fortran statements showing the transfer of element information and error check.

```
           DO 500 LOAD=1, NLS
           P=P+DP(LOAD)
           DO 200 I=1,NEQ
200    GPU(I)=0.0
210    ITER=0
           IF(NEWTON.EQ.0) GO TO 230
220    ITER=ITER+1
230    DO 240 I=1, NEQ
           GLF(I)=0.0
           DO 240 J=1, NHBW
240    GLK(I,J)=0.0
C
250    DO 340 N=1, NEM
           DO 260 I=1, NPE
           NI=NOD(N,I)
           ELXY(I,1)=GLXY(NI,1)
           ELXY(I,2)=GLXY(NI,2)
           L=NI*NDF-1
           K=I*NDF-1
           ELS(K)  =GLS(L)
           ELS(K+1)=GLS(L+1)
260    CONTINUE
270    CALL ELMATRCS2D(NGP,NN,NPE,P,ITER,NEWTON,LFORM,THKNS)
           . . .
340    CONTINUE
           . . .
           CALL BNDSYMSOLV(NRMAX,NCMAX,NEQ,NHBW,GLK,GLF,IRES)
```

Box 9.4.1: Fortran statements (continued).

```
C
C      UPDATE THE TOTAL SOLUTION VECTOR AND THE NODAL COORDINATES
C
        DO 420 I=1, NNM
        L=(I-1)*NDF+1
        GPU(L)   = GPU(L)+GLF(L)
        GPU(L+1)= GPU(L+1)+GLF(L+1)
        GLS(L)   = GLS(L)+GLF(L)
        GLS(L+1)= GLS(L+1)+GLF(L+1)
         IF(LFORM.GT.1)THEN
             GLXY(I,1) = GLXY(I,1)+GLF(L)
             GLXY(I,2) = GLXY(I,2)+GLF(L+1)
         ENDIF
420     CONTINUE
         TOL=0.0
         IF(NEWTON.GT.0) THEN
             SNORM=0.0
             ENORM=0.0
             DO 430 I=1,NEQ
             SNORM=SNORM+GLS(I)*GLS(I)
430       ENORM=ENORM+GLF(I)*GLF(I)
           TOL=DSQRT(ENORM/SNORM)
             IF(TOL.GT.EPS) THEN
                 IF(ITER.GT.ITMAX) THEN
                     WRITE (6,860)
                     STOP
                 ELSE
                     . . .
                     GO TO 220
                 ENDIF
             ENDIF
         ENDIF
         . . .
500     CONTINUE
```

Box 9.4.2: Fortran statements showing the element matrix calculations for the total Lagrangian (TL) and updated Lagrangian (UL) formulations.

```
        SUBROUTINE ELMATRCS2D(NGP,NN,NPE,F0,ITER,NEWTON,LFORM,THKNS)
        IMPLICIT REAL*8(A-H,O-Z)
        COMMON/STF/ELK(18,18),ELXY(9,2),C(4,4),ELF(18),ELS(18),ELU(18)
        DIMENSION  SF(9),GDSF(2,9),GAUSSPT(4,4),GAUSSWT(4,4)
        . . .
        NDF=NN/NPE
C    Initialize Element Force Vector and Stiffness Matrix
        DO 10 I=1,NN
        ELF(I)=0.0
        DO 10 J=1,NN
        ELK(I,J)=0.0
10      CONTINUE
```

Box 9.4.2: Fortran statements for the TL and UL formulations (continued).

```
      DO 60 NI=1,NGP
      DO 60 NJ=1,NGP
      XI =GAUSSPT(NI,NGP)
      ETA=GAUSSPT(NJ,NGP)
      CALL INTERPLN2D(NPE,XI,ETA,SF,GDSF,DET,ELXY)
      CONST=DET*GAUSSWT(NI,NGP)*GAUSSWT(NJ,NGP)*THKNS
C
C     Define the gradients of displacements and strains
      X=0.0
      U=0.0
      UX=0.0
      UY=0.0
      VX=0.0
      VY=0.0
      DO 20 I=1,NPE
      L=(I-1)*NDF+1
      X=X+SF(I)*ELXY(I,1)
      U=U+SF(I)*ELS(L)
      UX=UX+GDSF(1,I)*ELS(L)
      UY=UY+GDSF(2,I)*ELS(L)
      VX=VX+GDSF(1,I)*ELS(L+1)
20    VY=VY+GDSF(2,I)*ELS(L+1)
      UX2=UX*UX
      UY2=UY*UY
      VX2=VX*VX
      VY2=VY*VY
      IF(LFORM.EQ.1)THEN
           UXP1=1.0 + UX
           UXP2=UXP1*UXP1
           VYP1=1.0 + VY
           VYP2=VYP1*VYP1
C
C     Define the Green strain and Second Piola-Kirchhoff stress components
           EXX=UX+0.5*(UX2+VX2)
           EYY=VY+0.5*(UY2+VY2)
           EXY=UY+VX+UX*UY+VX*VY
           S11=C(1,1)*EXX+C(1,2)*EYY
           S22=C(1,2)*EXX+C(2,2)*EYY
           S12=C(3,3)*EXY
      ELSE
C
C     Define the Euler-Almansi strain and Cauchy stress tensor components
           EXX=UX-0.5*(UX2+VX2)
           EYY=VY-0.5*(UY2+VY2)
           EXY=UY+VX-UX*UY-VX*VY
           S11=C(1,1)*EXX+C(1,2)*EYY
           S22=C(1,2)*EXX+C(2,2)*EYY
           S12=C(3,3)*EXY
      ENDIF
      II=1
      DO 50 I=1,NPE
```

Box 9.4.2: Fortran statements for the TL and UL formulations (continued).

```
C
C     Imbalance force coefficients for the TOTAL Lagrangian formulation
C
      IF(LFORM.EQ.1)THEN
      ELF(II)  =ELF(II) +(FO*SF(I)-UXP1*GDSF(1,I)*S11-UY*GDSF(2,I)*S22
     *                   -(UXP1*GDSF(2,I)+UY*GDSF(1,I))*S12)*CONST
      ELF(II+1)=ELF(II+1)+(FO*SF(I)-VX*GDSF(1,I)*S11-VYP1*GDSF(2,I)*S22
     *                   -(VYP1*GDSF(1,I)+VX*GDSF(2,I))*S12)*CONST
C
C     Imbalance force coefficients for the UPDATED Lagrangian formulation
C
      ELSE
      ELF(II)  =ELF(II)  +(FO*SF(I)-GDSF(1,I)*S11-GDSF(2,I)*S12)*CONST
      ELF(II+1)=ELF(II+1)+(FO*SF(I)-GDSF(1,I)*S12-GDSF(2,I)*S22)*CONST
      ENDIF
C
      JJ=1
      DO 40 J=1,NPE
      SIG=S11*GDSF(1,I)*GDSF(1,J)+S22*GDSF(2,I)*GDSF(2,J)
     *   +S12*(GDSF(1,I)*GDSF(2,J)+GDSF(2,I)*GDSF(1,J))
C
      IF(LFORM.EQ.1)THEN
C
C     Stiffness coefficients for the TOTAL Lagrangian formulation
C
      ELK(II,JJ)=ELK(II,JJ)+(C(1,1)*UXP2*GDSF(1,I)*GDSF(1,J)
     *         +C(2,2)*UY2*GDSF(2,I)*GDSF(2,J)
     *         +C(1,2)*UXP1*UY*(GDSF(1,I)*GDSF(2,J)+GDSF(2,I)*GDSF(1,J))
     *         +C(3,3)*(UXP1*GDSF(2,I)+UY*GDSF(1,I))*
     *              (UXP1*GDSF(2,J)+UY*GDSF(1,J))+SIG)*CONST
      ELK(II+1,JJ+1)=ELK(II+1,JJ+1)+(C(1,1)*VX2*GDSF(1,I)
     *         *GDSF(1,J)+C(2,2)*VYP2*GDSF(2,I)*GDSF(2,J)
     *         +C(1,2)*VYP1*VX*(GDSF(1,I)*GDSF(2,J)+GDSF(2,I)*GDSF(1,J))
     *         +C(3,3)*(VYP1*GDSF(1,I)+VX*GDSF(2,I))*
     *              (VYP1*GDSF(1,J)+VX*GDSF(2,J))+SIG)*CONST
      ELK(II,JJ+1)=ELK(II,JJ+1)+(C(1,1)*UXP1*VX*GDSF(1,I)
     *         *GDSF(1,J)+C(2,2)*VYP1*UY*GDSF(2,I)*GDSF(2,J)
     *         +C(1,2)*(UXP1*VYP1*GDSF(1,I)*GDSF(2,J)
     *         +UY*VX*GDSF(2,I)*GDSF(1,J))
     *         +C(3,3)*(UXP1*GDSF(2,I)+UY*GDSF(1,I))*
     *              (VYP1*GDSF(1,J)+VX*GDSF(2,J)))*CONST
      ELK(II+1,JJ)=ELK(II+1,JJ)+(C(1,1)*UXP1*VX*GDSF(1,J)
     *         *GDSF(1,I)+C(2,2)*VYP1*UY*GDSF(2,J)*GDSF(2,I)
     *         +C(1,2)*(UXP1*VYP1*GDSF(1,J)*GDSF(2,I)
     *         +UY*VX*GDSF(2,J)*GDSF(1,I))
     *         +C(3,3)*(UXP1*GDSF(2,J)+UY*GDSF(1,J))*
     *              (VYP1*GDSF(1,I)+VX*GDSF(2,I)))*CONST
```

Box 9.4.2: Fortran statements for the TL and UL formulations (continued).

```
C       Stiffness coefficients for the UPDATED Lagrangian formulation
C
        ELSE
          ELK(II,JJ)=ELK(II,JJ)+(C(1,1)*GDSF(1,I)*GDSF(1,J)
     *                        +C(3,3)*GDSF(2,I)*GDSF(2,J)+SIG)*CONST
          ELK(II+1,JJ+1)=ELK(II+1,JJ+1)+(C(3,3)*GDSF(1,I)
     *                        *GDSF(1,J)+C(2,2)*GDSF(2,I)*GDSF(2,J)+SIG)*CONST
          ELK(II,JJ+1)=ELK(II,JJ+1)+(C(1,2)*GDSF(1,I)*GDSF(2,J)
     *                        +C(3,3)*GDSF(2,I)*GDSF(1,J))*CONST
          ELK(II+1,JJ)=ELK(II+1,JJ)+(C(1,2)*GDSF(2,I)*GDSF(1,J)
     *                        +C(3,3)*GDSF(1,I)*GDSF(2,J))*CONST
        ENDIF
        IF(ITEM.GT.0) THEN
          GRAM=RHO*SF(I)*SF(J)*CONST
          ELMASS(II,JJ)    =ELMASS(II,JJ)    +GRAM
          ELMASS(II+1,JJ+1)=ELMASS(II+1,JJ+1)+GRAM
        ENDIF
40      JJ=NDF*J+1
50      II=NDF*I+1
60      CONTINUE
        RETURN
        END
```

9.4.5   A Numerical Example

Here we present numerical results obtained for a cantilevered rectangular plate of length $a = 10$ in, height $b = 1$ in, and thickness $h = 0.1$ in, and subjected to uniformly distributed in-plane load $t_y = -q_0$ psi, acting downward, along its length, as shown in Fig. 9.4.1(a). The material of the plate has a modulus of $E = 1.2 \times 10^7$ psi and Poisson's ratio of $\nu = 0.3$.

First, we use a mesh of five eight-node quadratic elements (5Q8) with a load increment of $\Delta q_0 = -500$ psi [as shown in Figs. 9.4.1(b) and (c)] in both total and updated Lagrangian formulations to analyze the beam for displacements and stresses for a range of load increments. The beam is assumed to be in the *plane state of stress* and no distinction is made between $_1C_{ij}$ and $_0C_{ij}$ in the two formulations, and they are assumed to remain constant during the deformation. The applied nodal forces are assumed to remain vertical during the deformation. The 16×16 stiffness matrix coefficients (computed using 3×3 Gauss rule) of element 1 at the beginning of the first iteration of the first load step in the updated Lagrangian formulation, for example, are listed here.

```
    0.9143E+06   0.4048E+06   0.3795E+06   0.5495E+04   0.4044E+06   0.1667E+06
    0.4117E+06  -0.5495E+04  -0.5245E+06  -0.2125E+06  -0.4542E+06  -0.9524E+05
   -0.3546E+06  -0.9524E+05  -0.7766E+06  -0.1685E+06   ** Row 1 **
    0.4048E+06   0.1657E+07  -0.5495E+04   0.5700E+06   0.1667E+06   0.7330E+06
    0.5495E+04   0.8641E+06  -0.1685E+06  -0.2930E+05  -0.9524E+05  -0.1188E+07
   -0.9524E+05  -0.2784E+06  -0.2125E+06  -0.2329E+07   ** Row 2 **
```

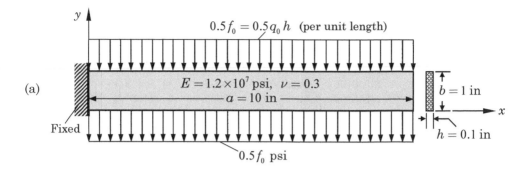

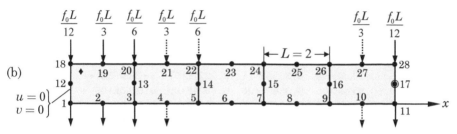

♦ Gauss point (x_0, y_0) at which the stresses are evaluated (2 × 2 Gauss rule)
◉ Node at which the displacements are tabulated/plotted

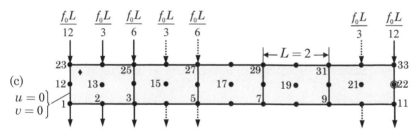

♦ Gauss point (x_0, y_0) at which the stresses are evaluated (2 × 2 Gauss rule)
◉ Node at which the displacements are tabulated/plotted

Fig. 9.4.1: (a) Geometry, (b) mesh of eight-node elements, and (c) mesh of nine-node elements used for bending of a cantilever plate under uniform load of intensity q_0 (psi).

```
 0.3795E+06 -0.5495E+04  0.9143E+06 -0.4048E+06  0.4117E+06  0.5495E+04
 0.4044E+06 -0.1667E+06 -0.5245E+06  0.2125E+06 -0.7766E+06  0.1685E+06
-0.3546E+06  0.9524E+05 -0.4542E+06  0.9524E+05  ** Row 3 **
 0.5495E+04  0.5700E+06 -0.4048E+06  0.1657E+07 -0.5495E+04  0.8641E+06
-0.1667E+06  0.7330E+06  0.1685E+06 -0.2930E+05  0.2125E+06 -0.2329E+07
 0.9524E+05 -0.2784E+06  0.9524E+05 -0.1188E+07  ** Row 4 **
 0.4044E+06  0.1667E+06  0.4117E+06 -0.5495E+04  0.9143E+06  0.4048E+06
 0.3795E+06  0.5495E+04 -0.3546E+06 -0.9524E+05 -0.7766E+06 -0.1685E+06
-0.5245E+06 -0.2125E+06 -0.4542E+06 -0.9524E+05  ** Row 5 **
 0.1667E+06  0.7330E+06  0.5495E+04  0.8641E+06  0.4048E+06  0.1657E+07
-0.5495E+04  0.5700E+06 -0.9524E+05 -0.2784E+06 -0.2125E+06 -0.2329E+07
-0.1685E+06 -0.2930E+05 -0.9524E+05 -0.1188E+07  ** Row 6 **
```

```
 0.4117E+06   0.5495E+04   0.4044E+06  -0.1667E+06   0.3795E+06  -0.5495E+04
 0.9143E+06  -0.4048E+06  -0.3546E+06   0.9524E+05  -0.4542E+06   0.9524E+05
-0.5245E+06   0.2125E+06  -0.7766E+06   0.1685E+06   ** Row 7 **
-0.5495E+04   0.8641E+06  -0.1667E+06   0.7330E+06   0.5495E+04   0.5700E+06
-0.4048E+06   0.1657E+07   0.9524E+05  -0.2784E+06   0.9524E+05  -0.1188E+07
 0.1685E+06  -0.2930E+05   0.2125E+06  -0.2329E+07   ** Row 8 **
-0.5245E+06  -0.1685E+06  -0.5245E+06   0.1685E+06  -0.3546E+06  -0.9524E+05
-0.3546E+06   0.9524E+05   0.1664E+07  -0.4002E-10   0.4366E-10  -0.3810E+06
 0.9377E+05   0.0000E+00   0.5821E-10   0.3810E+06   ** Row 9 **
-0.2125E+06  -0.2930E+05   0.2125E+06  -0.2930E+05  -0.9524E+05  -0.2784E+06
 0.9524E+05  -0.2784E+06  -0.8367E-10   0.1817E+07  -0.3810E+06   0.1746E-09
-0.1819E-10  -0.1201E+07   0.3810E+06   0.1783E-09   ** Row 10 **
-0.4542E+06  -0.9524E+05  -0.7766E+06   0.2125E+06  -0.7766E+06  -0.2125E+06
-0.4542E+06   0.9524E+05   0.4366E-10  -0.3810E+06   0.1993E+07   0.4366E-10
-0.5821E-10   0.3810E+06   0.4689E+06   0.0000E+00   ** Row 11 **
-0.9524E+05  -0.1188E+07   0.1685E+06  -0.2329E+07  -0.1685E+06  -0.2329E+07
 0.9524E+05  -0.1188E+07  -0.3810E+06   0.1746E-09   0.2910E-10   0.4812E+07
 0.3810E+06  -0.2619E-09   0.3638E-11   0.2221E+07   ** Row 12 **
-0.3546E+06  -0.9524E+05  -0.3546E+06   0.9524E+05  -0.5245E+06  -0.1685E+06
-0.5245E+06   0.1685E+06   0.9377E+05  -0.1819E-10  -0.5821E-10   0.3810E+06
 0.1664E+07  -0.5821E-10  -0.8004E-10  -0.3810E+06   ** Row 13 **
-0.9524E+05  -0.2784E+06   0.9524E+05  -0.2784E+06  -0.2125E+06  -0.2930E+05
 0.2125E+06  -0.2930E+05   0.0000E+00  -0.1201E+07   0.3810E+06  -0.2619E-09
-0.5821E-10   0.1817E+07  -0.3810E+06  -0.1692E-09   ** Row 14 **
-0.7766E+06  -0.2125E+06  -0.4542E+06   0.9524E+05  -0.4542E+06  -0.9524E+05
-0.7766E+06   0.2125E+06   0.6185E-10   0.3810E+06   0.4689E+06   0.3638E-11
-0.8004E-10  -0.3810E+06   0.1993E+07   0.4547E-10   ** Row 15 **
-0.1685E+06  -0.2329E+07   0.9524E+05  -0.1188E+07  -0.9524E+05  -0.1188E+07
 0.1685E+06  -0.2329E+07   0.3810E+06   0.1746E-09   0.7276E-11   0.2221E+07
-0.3810E+06  -0.1037E-09   0.4547E-10   0.4812E+07   ** Row 16 **
```

Table 9.4.1 contains the numerical results for displacements at the free end obtained with the updated Lagrangian formulation using the 3×3 and the 2×2 Gauss rules in the evaluation of the element coefficients (whereas the stresses are calculated, in both cases, using the 2×2 Gauss rule). Figure 9.4.2 contains plots of the end displacements versus load f_0. The 3×3 Gauss rule produces slightly lower deflections than those obtained using the 2×2 Gauss rule. Surprisingly, the solution without iteration is very good, except for initial oscillations. It took only two or three iterations to converge for each load step.

Table 9.4.2 contains the Cauchy and second Piola–Kirchhoff stresses (the values shown should be multiplied with 10^5 to obtain the actual values) at the fixed end of the beam (at the Gauss point located nearest to the top left end). All stress components are computed using the 2×2 Gauss rule. The second Piola–Kirchhoff stress components are evaluated at the point $(X, Y) = (0.422650, 0.788675)$. Figure 9.4.3 contains plots of the stresses as a function of load f_0. Most commercial codes give only the Cauchy stresses using the reduced Gauss point locations (i.e. 2×2 for a quadratic element).

Table 9.4.1: Total displacements* of node 17 (at the free end) in a cantilevered plate under uniform load; obtained with the *updated Lagrangian formulation.*

$f_0 = q_0 h$	3 × 3 Gauss rule for **K**				2 × 2 Gauss rule for **K**			
	x	$-y$	$-u$	$-v$	x	$-y$	$-u$	$-v$
50	9.9787	0.1145	0.0213	0.6145	9.9782	0.1229	0.0218	0.6229
	10.0000	0.1164	0.0000	0.6164	10.0000	0.1181	0.0000	0.6181
100	9.9159	0.7181	0.0841	1.2181	9.9138	0.7352	0.0862	1.2352
	9.9690	0.2686	0.0310	0.7686	9.9689	0.2707	0.0311	0.7707
150	9.8152	1.3010	0.1848	1.8010	9.8103	1.3274	0.1897	1.8274
	9.8901	1.1644	0.1099	1.6644	9.8896	1.1698	0.1104	1.6698
200	9.6816	1.8554	0.3184	2.3554	9.6726	1.8919	0.3274	2.3919
	9.8272	1.2658	0.1728	1.7658	9.8264	1.2719	0.1736	1.7719
250	9.5212	2.3758	0.4788	2.8758	9.5066	2.4230	0.4934	2.9230
	9.6208	2.2693	0.3792	2.7693	9.6183	2.2818	0.3817	2.7818
300	9.3402	2.8593	0.6598	3.3593	9.3187	2.9174	0.6813	3.4174
	9.5421	2.3359	0.4579	2.8359	9.5387	2.3493	0.4613	2.8493
350	9.1444	3.3046	0.8556	3.8046	9.1147	3.3738	0.8853	3.8738
	9.2429	3.2102	0.7571	3.7102	9.2347	3.2350	0.7653	3.7350
400	8.9391	3.7125	1.0609	4.2125	8.9001	3.7923	1.0999	4.2923
	9.1659	3.2796	0.8341	3.7796	9.1561	3.3051	0.8439	3.8051
450	8.7284	4.0846	1.2716	4.5846	8.6795	4.1743	1.3205	4.6743
	8.8026	4.0388	1.1974	4.5388	8.7828	4.0812	1.2172	4.5812
500	8.5160	4.4229	1.4840	4.9229	8.4566	4.5219	1.5434	5.0219
	8.7266	4.1054	1.2734	4.6054	8.7046	4.1480	1.2954	4.6480
550	8.3046	4.7302	1.6954	5.2302	8.2343	4.8377	1.7657	5.3377
	8.3545	4.7083	1.6455	5.2083	8.3179	4.7692	1.6821	5.2692
600	8.0962	5.0093	1.9038	5.5093	8.0149	5.1242	1.9851	5.6242
	8.2768	4.7839	1.7232	5.2839	8.2380	4.8448	1.7620	5.3448
650	7.8922	5.2628	2.1078	5.7628	7.8014	5.3821	2.1986	5.8821
	7.9257	5.2508	2.0743	5.7508	7.8691	5.3291	2.1309	5.8291
700	7.6937	5.4933	2.3063	5.9933	7.5928	5.6177	2.4072	6.1177
	7.8407	5.3368	2.1593	5.8368	7.7820	5.4146	2.2180	5.9146
750	7.5013	5.7031	2.4987	6.2031	7.3906	5.8320	2.6094	6.3320
	7.5280	5.6913	2.4720	6.1913	7.4510	5.7835	2.5490	6.2835
800	7.3171	5.8922	2.6829	6.3922	7.1953	6.0272	2.8047	6.5272
	7.4303	5.7881	2.5697	6.2881	7.3514	5.8796	2.6486	6.3796
850	7.1384	6.0668	2.8616	6.5668	7.0072	6.2052	2.9928	6.7052
	7.1618	6.0550	2.8382	6.5550	7.0657	6.1576	2.9343	6.657
900	6.9664	6.2268	3.0336	6.7268	6.8263	6.3679	3.1737	6.8679
	7.0488	6.1600	2.9512	6.6600	6.9505	6.2621	3.0495	6.7621

*The first row corresponds to solution with iteration and the second row corresponds to the solution with no iteration.

Table 9.4.2: Stresses* ($\times 10^{-5}$) evaluated at the left-most Gauss point nearest to the top of element 1 in a cantilevered plate under uniform load; obtained with the *updated Lagrangian formulation.*

			Cauchy stress components			Piola–Kirchhoff stress components		
$f_0 = q_0 h$	x	$-y$	$\bar{\sigma}_{xx}$	$\bar{\sigma}_{yy}$	$-\bar{\sigma}_{xy}$	$\bar{S}_{XX}$	$\bar{S}_{YY}$	$\bar{S}_{XY}$
50	0.4253	0.7861	0.7776	0.1819	0.0539	0.7885	0.1840	0.0542
	0.4254	0.7860	0.7929	0.1840	0.0545	0.8042	0.1862	0.0548
100	0.4280	0.7835	1.5457	0.3578	0.1198	1.5894	0.3661	0.1214
	0.4282	0.7833	1.5779	0.3584	0.1220	1.6236	0.3670	0.1236
150	0.4308	0.7810	2.2962	0.5268	0.1975	2.3950	0.5448	0.2014
	0.4310	0.7806	2.3476	0.5225	0.2022	2.4514	0.5416	0.2063
200	0.4335	0.7783	3.0226	0.6880	0.2861	3.1978	0.7191	0.2938
	0.4339	0.7779	3.0961	0.6761	0.2943	3.2812	0.7093	0.3024
250	0.4362	0.7757	3.7200	0.8411	0.3845	3.9918	0.8880	0.3978
	0.4367	0.7752	3.8185	0.8193	0.3970	4.1075	0.8700	0.4110
300	0.4389	0.7731	4.3854	0.9863	0.4915	4.7721	1.0509	0.5120
	0.4396	0.7724	4.5117	0.9527	0.5092	4.9255	1.0232	0.5310
350	0.4415	0.7706	5.0172	1.1237	0.6057	5.5354	1.2076	0.6354
	0.4424	0.7697	5.1731	1.0766	0.6293	5.7310	1.1693	0.6611
400	0.4441	0.7680	5.6151	1.2537	0.7258	6.2795	1.3579	0.7666
	0.4452	0.7670	5.8027	1.1921	0.7562	6.5221	1.3086	0.8001
450	0.4466	0.7655	6.1793	1.3767	0.8507	7.0025	1.5018	0.9045
	0.4478	0.7643	6.400	1.3000	0.8886	7.2971	1.4414	0.9468
500	0.4491	0.7630	6.7118	1.4933	0.9793	7.7050	1.6396	1.0479
	0.4505	0.7617	6.9668	1.4004	1.0254	8.0548	1.5682	1.1001
550	0.4514	0.7605	7.2141	1.6040	1.1108	8.3870	1.7716	1.1961
	0.4531	0.7590	7.5032	1.4948	1.1658	8.7948	1.6894	1.2590
600	0.4536	0.7580	7.6879	1.7093	1.2444	9.0489	1.8981	1.3480
	0.4556	0.7564	8.0111	1.5836	1.3088	9.5172	1.8055	1.4225
650	0.4558	0.7556	8.1352	1.8096	1.3795	9.6916	2.0193	1.5032
	0.4580	0.7540	8.4794	1.6631	1.4502	10.2038	1.9118	1.5857
700	0.4579	0.7532	8.5579	1.9055	1.5157	10.3161	2.1356	1.6609
	0.4603	0.7514	8.9337	1.7422	1.5961	10.8897	2.0184	1.7560
750	0.4600	0.7508	8.9577	1.9973	1.6525	10.9233	2.2474	1.8208
	0.4626	0.7489	9.3650	1.8176	1.7431	11.5602	2.1212	1.9291
800	0.4619	0.7486	9.3272	2.0817	1.7862	11.4997	2.3511	1.9782
	0.4649	0.7464	9.7742	1.8897	1.8907	12.2153	2.2205	2.1045
850	0.4638	0.7463	9.6857	2.1660	1.9230	12.0742	2.4544	2.1407
	0.4670	0.7440	10.1628	1.9588	2.0385	12.8557	2.3166	2.2818
900	0.4657	0.7440	10.0265	2.2473	2.0598	12.6351	2.5542	2.3045
	0.4692	0.7416	10.5321	2.0253	2.1864	13.4821	2.4097	2.4606

*The first row corresponds to the 3×3 Gauss rule for the evaluation of **K** and the second row corresponds to the 2×2 Gauss rule for the evaluation of **K**.

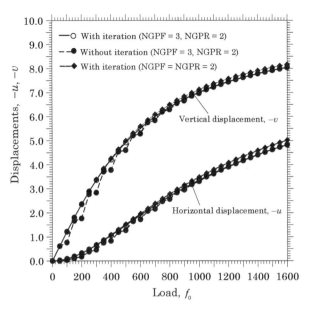

Fig. 9.4.2: Node 17 displacements $-v$ and $-u$ versus load $f_0 = q_0 h$ (obtained with the UL formulation).

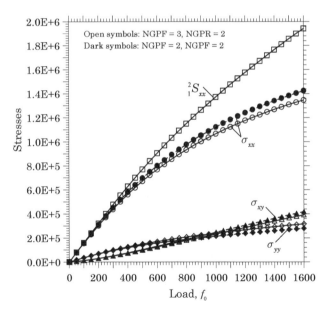

Fig. 9.4.3: Stresses versus load $f_0 = q_0 h$ (obtained with the UL formulation).

The mesh of five nine-node quadratic elements also gives almost identical results for displacements (u, v) and stresses $(\sigma_{xx}, \sigma_{xy})$, as can be seen from Table 9.4.3. The stiffness coefficients were computed using the 3×3 Gauss rule while stresses were calculated at the 2×2 Gauss point locations.

Table 9.4.3: Total displacements* of node 22 and stresses in a cantilevered plate under uniform load; obtained with the *updated Lagrangian formulation* and the mesh of 5Q9 elements.

$f_0 = q_0 h$	x	$-y$	$-u$	$-v$	x	$-y$	σ_{xx}	$-\sigma_{xy}$
50	9.9782	0.1218	0.0218	0.6218	0.4254	0.7860	0.7759	0.0549
					0.4226	0.7887	0.7877	0.0552
100	9.9142	0.7323	0.0858	1.2323	0.4282	0.7833	1.5429	0.1239
					0.4226	0.7887	1.5904	0.1253
150	9.8115	1.3214	0.1885	1.8214	0.4311	0.7806	2.2924	0.2065
					0.4226	0.7887	2.3997	0.2102
200	9.6754	1.8810	0.3246	2.3810	0.4340	0.7780	3.0175	0.3018
					0.4226	0.7887	3.2081	0.3093
250	9.5123	2.4057	0.4877	2.9057	0.4368	0.7753	3.7131	0.4084
					0.4226	0.7887	4.0087	0.4214
300	9.3286	2.8923	0.6714	3.3923	0.4396	0.7726	4.3758	0.5249
					0.4226	0.7887	4.7966	0.5452
350	9.1301	3.3400	0.8699	3.8400	0.4424	0.7699	5.0042	0.6497
					0.4226	0.7887	5.5681	0.6793
400	8.9224	3.7494	1.0776	4.2494	0.4450	0.7673	5.5980	0.7814
					0.4226	0.7887	6.3207	0.8222
450	8.7095	4.1224	1.2905	4.6224	0.4476	0.7646	6.1572	0.9185
					0.4226	0.7887	7.0524	0.9725
500	8.4953	4.4611	1.5047	4.9611	0.4501	0.7621	6.6840	1.0600
					0.4226	0.7887	7.7638	1.1291
550	8.2823	4.7684	1.7177	5.2684	0.4525	0.7595	7.1800	1.2047
					0.4226	0.7887	8.4548	1.2909
600	8.0725	5.0472	1.9275	5.5472	0.4549	0.7570	7.6470	1.3521
					0.4226	0.7887	9.1257	1.4570
650	7.8675	5.3002	2.1325	5.8002	0.4572	0.7545	8.0870	1.5012
					0.4226	0.7887	9.7774	1.6266
700	7.6681	5.5301	2.3319	6.0301	0.4594	0.7520	8.5019	1.6517
					0.4226	0.7887	10.4110	1.7992
750	7.4750	5.7392	2.5250	6.2392	0.4615	0.7496	8.8936	1.8030
					0.4226	0.7887	11.0270	1.9742
800	7.2886	5.9299	2.7114	6.4299	0.4635	0.7472	9.2638	1.9547
					0.4226	0.7887	11.6270	2.1511
850	7.1114	6.1013	2.8886	6.6013	0.4655	0.7449	9.6050	2.1020
					0.4226	0.7887	12.1960	2.3241
900	6.9392	6.2604	3.0608	6.7604	0.4674	0.7426	9.9366	2.2534
					0.4226	0.7887	12.7650	2.5033
950	6.7738	6.4064	3.2262	6.9064	0.4693	0.7403	1.0251	2.4048
					0.4226	0.7887	13.3220	2.6837
1,000	6.6150	6.5407	3.3850	7.0407	0.4711	0.7380	10.5510	2.5558
					0.4226	0.7887	13.8660	2.8651
1,050	6.4627	6.6644	3.5373	7.1644	0.4728	0.7357	10.8350	2.7064
					0.4226	0.7887	14.3990	3.0471
1,100	6.3167	6.7787	3.6833	7.2787	0.4745	0.7334	11.1060	2.8564
					0.4226	0.7887	14.9210	3.2297

*The second row corresponds to the second Piola–Kirchhoff stress components.

Table 9.4.4 contains the displacements and stresses obtained with the total Lagrangian formulation and the two different meshes. The results are in close agreement with those obtained with the updated Lagrangian formulation.

Table 9.4.4: Total displacements* and stresses in a cantilevered plate under uniform load; obtained with the *total Lagrangian formulation*.

$f_0 = q_0 h$	$-u$	$-v$	x	$-y$	σ_{xx}	$-\sigma_{xy}$	S_{xx}	$-S_{xy}$
50	0.0216	0.6144	0.4253	0.7861	0.7738	0.0508	0.7736	0.0484
	0.0221	0.6218	0.4254	0.7860	0.7713	0.0512	0.7711	0.0482
100	0.0853	1.2177	0.4278	0.7836	1.5311	0.1073	1.5299	0.0979
	0.0871	1.2320	0.4280	0.7834	1.5255	0.1090	1.5243	0.0971
150	0.1874	1.7997	0.4303	0.7810	2.2656	0.1692	2.2619	0.1487
	0.1912	1.8203	0.4306	0.7807	2.2559	1.7300	2.2521	0.1470
200	0.3225	2.3524	0.4327	0.7784	2.9724	0.2361	2.9642	0.2009
	0.3290	2.3786	0.4331	0.7780	2.9577	0.2428	2.9493	0.1980
250	0.4846	2.8706	0.4350	0.7759	3.6481	0.3074	3.6333	0.2545
	0.4940	2.9013	0.4355	0.7754	3.6276	0.3177	3.6121	0.2502
300	0.6671	3.3510	0.4372	0.7734	4.2912	0.3825	4.2672	0.3094
	0.6795	3.3854	0.4380	0.7727	4.2641	0.3970	4.2387	0.3036
350	0.8641	3.7927	0.4393	0.7709	4.9013	0.4605	4.8656	0.3656
	0.8795	3.8300	0.4400	0.7702	4.8669	0.4800	4.8287	0.3581
400	1.0702	4.1965	0.4413	0.7684	5.4792	0.5411	5.4290	0.4228
	1.0886	4.2358	0.4420	0.7676	5.4368	0.5657	5.3830	0.4136
450	1.2812	4.5640	0.4432	0.7660	6.0261	0.6235	5.9588	0.4809
	1.3023	4.6048	0.4439	0.7651	5.9753	0.6539	5.9032	0.4701
500	1.4936	4.8977	0.4450	0.7636	6.5438	0.7073	6.4570	0.5398
	1.5171	4.9394	0.4457	0.7626	6.4843	0.7438	6.3911	0.5274
550	1.7045	5.2003	0.4466	0.7613	7.0343	0.7922	6.9256	0.5994
	1.7302	5.2426	0.4474	0.7602	6.9658	0.8351	6.8491	0.5853
600	1.9122	5.4746	0.4482	0.7590	7.4997	0.8778	7.3668	0.6594
	1.9398	5.5171	0.4490	0.7578	7.4219	0.9274	7.2794	0.6438
650	2.1150	5.7234	0.4497	0.7567	7.9418	0.9638	7.7827	0.7198
	2.1444	5.7659	0.4505	0.7555	7.8546	1.0203	7.6842	0.7027
700	2.3121	5.9493	0.4511	0.7545	8.3625	1.0501	8.1754	0.7805
	2.343	5.9916	0.4520	0.7532	8.2660	1.1138	8.0656	0.7620
750	2.5028	6.1547	0.4524	0.7523	8.7638	1.1364	8.5467	0.8414
	2.5350	6.1967	0.4533	0.7509	8.6578	1.2074	8.4257	0.8216
800	2.6868	6.3418	0.4536	0.7501	9.1471	1.2227	8.8985	0.9025
	2.7201	6.3833	0.4546	0.7487	9.0316	1.3012	8.7662	0.8814
850	2.8654	6.5138	0.4548	0.7480	9.5153	1.3090	9.2337	0.9637
	2.8997	6.5548	0.4557	0.7465	9.3904	1.3951	9.0900	0.9415
900	3.0353	6.6697	0.4559	0.7459	9.8669	1.3949	9.5508	1.0248
	3.0704	6.7101	0.4569	0.7444	9.7325	1.4887	9.3960	1.0016

*The first row corresponds to the mesh of five 5Q8 elements and the second row corresponds to the mesh of 5Q9 elements.

This completes the development of finite element models of nonlinear continua. The total and updated Lagrangian formulations developed in Sections 9.3 and 9.4 can be readily extended to axisymmetric problems, three-dimensional problems, problems with assumed displacements and/or strains [4], and composite plates and shells [55]. The extension of the formulations to transient problems is straightforward in view of the discussion presented in Chapter 6. In the next section, we extend the formulation to shells by invoking the kinematic assumptions of a shell theory in the approximation of the geometry and the displacement field.

9.5　Conventional Continuum Shell Finite Element

9.5.1　Introduction

In previous chapters, beam, plate, and shell finite elements were developed using beam, plate, and shell theories, respectively, that were derived from an assumed displacement field. Such theories are often limited to geometrically linear analysis and nonlinear analysis with small strains and moderate rotations. In these elements, the geometry is not updated explicitly during the nonlinear equilibrium iterations.

The finite element model to be developed in this section is based on three-dimensional continuum equations and is known as the *continuum shell element*, which was first introduced by Ahmad, Irons, and Zienkiewicz [122]. The three-dimensional geometry of the continuum is approximated by imposing appropriate kinematic assumptions on the displacement field; although no shell theory is explicitly invoked, the resulting formulation may be identified as a shell element with qualities consistent with a first-order or third-order shear deformation shell model. Such formulations especially applicable to material and geometrical nonlinear analysis of shell-type structures in which large displacements and rotations are experienced (see [55, 123–137]). Although the shell theory-based and continuum-based shell finite elements appear quite different from each other because of the way they are constructed, the formulations are quite similar when based on the same kinematic assumptions, and they "differ only in the kind of discretization" (see Büchter and Ramm [138]).

It is well known that low-order finite element implementations for shells suffer from various forms of locking whenever a purely displacement-based formulation is adopted. The locking phenomena occurs on account of inconsistencies that arise in the discrete finite element representation of the membrane and transverse shear energies. In recent years, the issue of locking has been addressed through the use of low-order finite technology using Hu–Washizu type mixed variational principles. Among the low-order implementations are the assumed strain formulation of Dvorkin and Bathe [133] and Hinton and Huang [139], and the enhanced strain formulation of Simo and Rifai [107] are notable.

High-order finite element implementations have also been advocated in recent years as a means of eliminating the locking phenomena completely. Whenever a sufficient degree of p-refinement is employed, highly reliable locking free numerical solutions may be obtained in displacement-based models. Among these works are high-order interpolations by Pitkäranta and his colleagues [140–142] for linear analysis of isotropic shells, the least-squares finite element formulations of Pontaza and Reddy [143, 144] for the linear analysis of plates and shells, and the shell theory-based weak form Galerkin finite element models of Arciniega [145] and Arciniega and Reddy [146, 147] for the analysis of isotropic, laminated composite, and functionally graded shells.

In this section, we develop continuum shell element based on first-order shear deformation theory kinematics with five generalized displacements per node (in a general setting, a sixth degree of freedom, namely the "drilling" degree of freedom in the form of rotation about the transverse normal is introduced). The generalized displacements are introduced through the kinematics of the surface displacements while the transverse displacement is assumed to preserve the inextensibility of a transverse normal line. Consequently, the plane-stress assumption is invoked in writing the stress–strain relations.

9.5.2 Incremental Equations of Motion

Assuming that the body may experience large displacements and rotations, we wish to determine deformed configurations of the body for different times/loads. We assume that all configurations from time $t = 0$ to the current time t, both inclusive, have been determined, and the configuration Ω_2 for time $t = t + \Delta t$ is being sought next. The total Lagrangian description with the principle of virtual displacements, Eq. (9.3.17), is used to express the equilibrium of the body in configuration Ω_2.

For dynamic analysis, the statement of the principle of virtual displacements based on the total Lagrangian formulation, Eq. (9.3.17), must be modified to include inertial terms. Consider the motion in a fixed Cartesian coordinate system. In this case we have

$$\int_{2V} {}^2\rho \, {}^2\ddot{u}_i \, \delta \, {}^2u_i \, d\,{}^2V = \int_{0V} {}^0\rho \, {}^2\ddot{u}_i \, \delta \, {}^2u_i \, d\,{}^0V \tag{9.5.1}$$

and hence the mass matrix can be evaluated using the initial configuration of the body. Using Hamilton's principle we obtain the equations of motion of the moving body at time $t + \Delta t$ in the variational form as

$$\int_{0V} {}^0\rho \, {}^2\ddot{u}_i \, \delta({}^2u_i) \, d\,{}^0V + \int_{0V} {}_0C_{ijrs} \, {}_0e_{rs} \, \delta \, {}_0e_{ij} \, d\,{}^0V + \int_{0V} {}_0^1S_{ij} \, \delta \, {}_0\eta_{ij} \, d\,{}^0V$$

$$= \delta({}_0^2R) - \int_{0V} {}_0^1S_{ij} \, \delta \, {}_0e_{ij} \, d\,{}^0V \tag{9.5.2}$$

where $\delta({}^2u_i) = \delta u_i$.

9.5.3 Finite Element Model of a Continuum

Equation (9.5.2) can be used to develop the nonlinear displacement finite element model for any continuum. The basic step in deriving the finite element equations for a shell element is the selection of proper interpolation functions for the displacement field and geometry. In the case of beam and shell elements, the approximation for the geometry is chosen such that the beam or shell kinematic hypotheses are realized. First, we derive the finite element model of a continuum and then specialize it to shells. The discussion presented here is based on the papers by Liao and Reddy [136, 137].

It is important that the coordinates and displacements are interpolated using the same interpolation functions (isoparametric formulation) so that the displacement compatibility across element boundaries can be preserved in all configurations. Let

$$
{}^0x_i = \sum_{k=1}^{n} \psi_k \, {}^0x_i^k, \quad {}^1x_i = \sum_{k=1}^{n} \psi_k \, {}^1x_i^k, \quad {}^2x_i = \sum_{k=1}^{n} \psi_k \, {}^2x_i^k \tag{9.5.3}
$$

$$
{}^1u_i = \sum_{k=1}^{n} \psi_k \, {}^1u_i^k, \quad u_i = \sum_{k=1}^{n} \psi_k \, u_i^k \quad (i = 1, 2, 3) \tag{9.5.4}
$$

where k denotes the node number, ψ_k is the interpolation function corresponding to node k, i is the coordinate number, and n is the number of nodes in the element.

Substitution of Eqs. (9.5.3) and (9.5.4) in Eq. (9.5.2) yields the finite element model

$$
{}^1_0\mathbf{M}\ddot{\boldsymbol{\Delta}} + ({}^1_0\mathbf{K}_1 + {}^1_0\mathbf{K}_2)\boldsymbol{\Delta} = {}^2\mathbf{R} - {}^1_0\mathbf{F} \tag{9.5.5}
$$

where $\boldsymbol{\Delta}$ is the vector of nodal incremental displacements from time t to time $t + \Delta t$ in an element, and ${}^1_0\mathbf{M}\ddot{\boldsymbol{\Delta}}$, ${}^1_0\mathbf{K}_1\boldsymbol{\Delta}$, ${}^1_0\mathbf{K}_2\boldsymbol{\Delta}$, and ${}^1_0\mathbf{F}$ are obtained by evaluating, respectively, the following integrals:

$$
\int_{{}^0V} {}^0\rho \, {}^2\ddot{u}_i \, \delta \, {}^2u_i \, d\,{}^0V, \quad \int_{{}^0V} {}_0C_{ijrs} \, {}_0e_{rs} \, \delta \, {}_0e_{ij} \, d\,{}^0V
$$

$$
\int_{{}^0V} {}^1_0S_{ij} \, \delta \, {}_0\eta_{ij} \, d\,{}^0V, \quad \int_{{}^0V} {}^1_0S_{ij} \, \delta \, {}_0e_{ij} \, d\,{}^0V
$$

Various matrices are defined by (${}^2\mathbf{R}$ is determined using the body force and surface traction vectors, if there are any)

$$
{}^1_0\mathbf{K}_1 = \int_{{}^0A} ({}^1_0\mathbf{B}_1)^{\mathrm{T}} \, {}_0\mathbf{C} \, {}^1_0\mathbf{B}_1 \, d\,{}^0V, \quad {}^1_0\mathbf{K}_2 = \int_{{}^0V} ({}^1_0\mathbf{B}_2)^{\mathrm{T}} \, {}_0\mathbf{S} \, {}^1_0\mathbf{B}_2 \, d\,{}^0V
$$

$$
{}^1_0\mathbf{M} = \int_{{}^0V} {}^0\rho \, {}^1\mathbf{H}^{\mathrm{T}} \, {}^1\mathbf{H} \, d\,{}^0V, \quad {}^1_0\mathbf{F} = \int_{{}^0V} ({}^1_0\mathbf{B}_1)^{\mathrm{T}} \, {}^1_0\mathbf{S}_v \, d\,{}^0V \tag{9.5.6}
$$

In the above equations, $_0^1\mathbf{B}_1$ and $_0^1\mathbf{B}_2$ are the linear and nonlinear strain–displacement transformation matrices, $_0\mathbf{C}$ is the incremental stress–strain material property matrix, $_0^1\mathbf{S}$ is a matrix of second Piola–Kirchhoff stress components, $_0^1\mathbf{S}_v$ is a vector of these stresses, and $^1\mathbf{H}$ is the incremental displacement interpolation matrix [see Eq. (9.5.23)]. All matrix elements correspond to the configuration at time t and are defined with respect to the configuration at time $t = 0$. It is important to note that Eq. (9.5.5) is only an approximation to the actual solution to be determined in each time step [see Eq. (9.3.15)]. Therefore, it may be necessary to iterate in each time step until Eq. (9.3.15), with inertia terms, is satisfied to a required tolerance.

The finite element equations, Eq. (9.5.5), are second-order differential equations in time. In order to obtain numerical solutions at each time step, Eq. (9.5.5) needs to be converted to algebraic equations using a time approximation scheme, as explained in Chapter 6. In the present study the Newmark scheme is used (see Section 6.7.4 for details).

In the Newmark time integration scheme, the displacements and velocities are approximated by [see Eq. (6.7.29)]

$$
\begin{aligned}
{}^{t+\Delta t}\mathbf{\Delta} &= {}^t\mathbf{\Delta} + \Delta t \, {}^t\dot{\mathbf{\Delta}} + \frac{1}{2}(\Delta t)^2[(1-\gamma)\,{}^t\ddot{\mathbf{\Delta}} + \gamma\,{}^{t+\Delta t}\ddot{\mathbf{\Delta}}] \\
{}^{t+\Delta t}\dot{\mathbf{\Delta}} &= {}^t\dot{\mathbf{\Delta}} + \Delta t[(1-\alpha)\,{}^t\ddot{\mathbf{\Delta}} + \alpha\,{}^{t+\Delta t}\ddot{\mathbf{\Delta}}]
\end{aligned}
\tag{9.5.7}
$$

where $\alpha = \frac{1}{2}$ and $\gamma = \frac{1}{2}$ for the constant-average acceleration method, and Δt is the time step. Rearranging Eqs. (9.5.5) and (9.5.7), we obtain $[_0^1\mathbf{M} \approx _0^2\mathbf{M};$ see Eq. (6.7.33)]

$$
_0^1\hat{\mathbf{K}}\mathbf{\Delta} = {}^2\hat{\mathbf{R}}
\tag{9.5.8}
$$

where $^t\mathbf{\Delta}$ is the vector of nodal incremental displacements at time t, $\mathbf{\Delta} = {}^{t+\Delta t}\mathbf{\Delta} - {}^t\mathbf{\Delta}$, and

$$
\begin{aligned}
_0^1\hat{\mathbf{K}} &= a_3\,_0^1\mathbf{M} + _0^1\mathbf{K}_1 + _0^1\mathbf{K}_2 \\
{}^2\hat{\mathbf{R}} &= {}^2\mathbf{R} - _0^1\mathbf{F} + _0^1\mathbf{M}\left(a_3\,{}^t\mathbf{\Delta} + a_4\,{}^t\dot{\mathbf{\Delta}} + a_5\,{}^t\ddot{\mathbf{\Delta}}\right)
\end{aligned}
\tag{9.5.9}
$$

$$
a_3 = \frac{1}{\beta(\Delta t)^2}, \quad a_4 = a_3\Delta t, \quad a_5 = \frac{1}{2\beta} - 1
\tag{9.5.10}
$$

Once Eq. (9.5.8) is solved for $\mathbf{\Delta}$ at time $t + \Delta t$, the acceleration and velocity vectors are obtained using $[a_1 = \alpha\Delta t$ and $a_2 = (1-\alpha)\Delta t;$ see Eq. (6.7.36)]

$$
\begin{aligned}
{}^{t+\Delta t}\ddot{\mathbf{\Delta}} &= a_3({}^{t+\Delta t}\mathbf{\Delta} - {}^t\mathbf{\Delta}) - a_4\,{}^t\dot{\mathbf{\Delta}} - a_5\,{}^t\ddot{\mathbf{\Delta}} \\
{}^{t+\Delta t}\dot{\mathbf{\Delta}} &= {}^t\dot{\mathbf{\Delta}} + a_1\,{}^{t+\Delta t}\ddot{\mathbf{\Delta}} + a_2\,{}^t\ddot{\mathbf{\Delta}}
\end{aligned}
\tag{9.5.11}
$$

The finite element equations, Eq. (9.5.8), are solved iteratively at each time step (after assembly and imposition of boundary conditions) until Eq. (9.5.8) is satisfied within a required tolerance. The Newton–Raphson method with Wempner–Riks algorithm (see Wempner [148], Riks [149, 150], Crisfield [151], and Kweon and Hong [152]; also, see Appendix 1) is used in the present study.

9.5.4   Shell Finite Element

The shell element is deduced from the three-dimensional continuum element by imposing two kinematic constraints: (1) straight line normal to the mid-surface of the shell before deformation remains straight but not normal after deformation and (2) the transverse normal components of strain and stress are ignored in the development. However, the shell element admits arbitrarily large displacements and rotations but small strains since the shell thickness is assumed not to change and the normal is not allowed to distort.

Consider the solid three-dimensional element shown in Fig. 9.5.1. Let (ξ, η) be the curvilinear coordinates in the middle surface of the shell and ζ be the coordinate in the thickness direction. The coordinates (ξ, η, ζ) are normalized such that they vary between -1 and $+1$. The coordinates of a typical point in the element can be written as

$$x_i = \sum_{k=1}^{n} \psi_k(\xi, \eta) \left[\frac{1+\zeta}{2} (x_i^k)_{\text{top}} + \frac{1-\zeta}{2} (x_i^k)_{\text{bottom}} \right] \tag{9.5.12}$$

where n is the number of nodes in the element, and $\psi_k(\xi, \eta)$ is the finite element interpolation function associated with node k. If $\psi_k(\xi, \eta)$ are derived as interpolation functions of a parent element, square or triangular in-plane, then compatibility is achieved at the interfaces of curved space shell elements. Define

$$V_{3i}^k = (x_i^k)_{\text{top}} - (x_i^k)_{\text{bottom}}, \quad \hat{\mathbf{e}}_3^k = \mathbf{V}_3^k / |\mathbf{V}_3^k| \tag{9.5.13}$$

where $\mathbf{V}_3^k$ is the vector connecting the upper and lower points of the normal at node k. Equation (9.5.12) can be rewritten as

$$x_i = \sum_{k=1}^{n} \psi_k(\xi, \eta) \left[(x_i^k)_{\text{mid}} + \frac{\zeta}{2} V_{3i}^k \right] = \sum_{k=1}^{n} \psi_k(\xi, \eta) \left[(x_i^k)_{\text{mid}} + \frac{\zeta}{2} h_k e_{3i}^k \right] \tag{9.5.14}$$

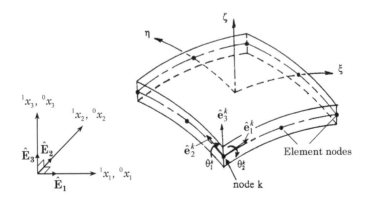

Fig. 9.5.1: Geometry and coordinate system of a curved shell element.

where $h_k = |\mathbf{V}_3^k|$ is the thickness of the shell element at node k. Hence, the coordinates of any point in the element at time t are interpolated by the expression

$$^1x_i = \sum_{k=1}^{n} \psi_k(\xi,\eta) \left[^1x_i^k + \frac{\zeta}{2} h_k\, ^1e_{3i}^k \right] \tag{9.5.15}$$

and the displacements and incremental displacements by

$$^1u_i = {}^1x_i - {}^0x_1 = \sum_{k=1}^{n} \psi_k(\xi,\eta) \left[^1u_i^k + \frac{\zeta}{2} h_k \left(^1e_{3i}^k - {}^0e_{3i}^k \right) \right] \tag{9.5.16}$$

$$u_i = {}^2u_i - {}^1u_i = \sum_{k=1}^{n} \psi_k(\xi,\eta) \left[u_i^k + \frac{\zeta}{2} h_k \left(^2e_{3i}^k - {}^1e_{3i}^k \right) \right] \tag{9.5.17}$$

Here $^1u_i^k$ and u_i^k denote, respectively, the displacement and incremental displacement components in the x_i-direction at the kth node and time t. For small rotation $d\mathbf{\Omega}$ at each node, we have

$$d\mathbf{\Omega} = \theta_2^k\, ^1\hat{\mathbf{e}}_1^k + \theta_1^k\, ^1\hat{\mathbf{e}}_2^k + \theta_3^k\, ^1\hat{\mathbf{e}}_3^k \tag{9.5.18}$$

The increment of vector $^1\hat{\mathbf{e}}_3^k$ can be written as

$$\Delta^1\hat{\mathbf{e}}_3^k = {}^2\hat{\mathbf{e}}_3^k - {}^1\hat{\mathbf{e}}_3^k = d\mathbf{\Omega} \times {}^1\hat{\mathbf{e}}_3^k = \theta_1^k\, ^1\hat{\mathbf{e}}_1^k - \theta_2^k\, ^1\hat{\mathbf{e}}_2^k \tag{9.5.19}$$

Then Eq. (9.5.17) becomes

$$u_i = \sum_{k=1}^{n} \psi_k(\xi,\eta) \left[u_i^k + \frac{\zeta}{2} h_k \left(\theta_1^k\, ^1e_{1i}^k - \theta_2^k\, ^1e_{2i}^k \right) \right] \quad (i = 1,2,3) \tag{9.5.20}$$

The unit vectors $^1\hat{\mathbf{e}}_1^k$ and $^1\hat{\mathbf{e}}_2^k$ at node k can be obtained from the relations

$$^1\hat{\mathbf{e}}_1^k = \frac{\hat{\mathbf{E}}_2 \times {}^1\hat{\mathbf{e}}_3^k}{|\hat{\mathbf{E}}_2 \times {}^1\hat{\mathbf{e}}_3^k|}, \quad {}^1\hat{\mathbf{e}}_2^k = {}^1\hat{\mathbf{e}}_3^k \times {}^1\hat{\mathbf{e}}_1^k \tag{9.5.21}$$

where $\hat{\mathbf{E}}_i$ are the unit vectors of the stationary global coordinate system $(^0x_1, {}^0x_2, {}^0x_3)$. Equation (9.5.20) can be written in matrix form as

$$\mathbf{u}^{\mathrm{T}} = \begin{Bmatrix} u_1 \\ u_2 \\ u_3 \end{Bmatrix} = {}^1\mathbf{H}\mathbf{\Delta}^e \tag{9.5.22}$$

where $\mathbf{\Delta}^e = \{u_1^k\ u_2^k\ u_3^k\ \theta_1^k\ \theta_2^k\}^{\mathrm{T}}$, $(k = 1,2,\ldots,n$, and n is the number of nodes) is the vector of nodal incremental displacements (five per node), and $^1\mathbf{H}$ is the incremental displacement interpolation matrix

$$^1\mathbf{H}_{3\times 5n} = \begin{bmatrix} \ldots\ \psi_k & 0 & 0 & \frac{1}{2}\psi_k\zeta h_k\, ^1e_{11}^k & -\frac{1}{2}\psi_k\zeta h_k\, ^1e_{21}^k & \ldots \\ \ldots\ 0 & \psi_k & 0 & \frac{1}{2}\psi_k\zeta h_k\, ^1e_{12}^k & -\frac{1}{2}\psi_k\zeta h_k\, ^1e_{22}^k & \ldots \\ \ldots\ 0 & 0 & \psi_k & \frac{1}{2}\psi_k\zeta h_k\, ^1e_{13}^k & -\frac{1}{2}\psi_k\zeta h_k\, ^1e_{23}^k & \ldots \end{bmatrix} \tag{9.5.23}$$

For each time step or iteration step one can find three unit vectors at each node from Eqs. (9.5.19) and (9.5.21).

From Eq. (9.2.14) the linear strain increments $_0\mathbf{e} = \{_0e_{11} \; _0e_{22} \; _0e_{33} \; 2_0e_{12} \; 2_0e_{13} \; 2_0e_{23}\}^{\mathrm{T}}$ can be expressed as

$$_0\mathbf{e} = {}^1\mathbf{A}_0\mathbf{u} \tag{9.5.24}$$

where $_0\mathbf{u}$ is the vector of derivatives of increment displacements

$$_0\mathbf{u} = \{_0u_{1,1} \; _0u_{1,2} \; _0u_{1,3} \; _0u_{2,1} \; _0u_{2,2} \; _0u_{2,3} \; _0u_{3,1} \; _0u_{3,2} \; _0u_{3,3}\}^{\mathrm{T}}$$

$$
{}^1\mathbf{A}_{6\times9} = \frac{1}{2}
\begin{bmatrix}
2 + {}_0^1u_{1,1} & 0 & 0 & \cdots & {}_0^1u_{3,1} & 0 & 0 \\
0 & {}_0^1u_{1,2} & 0 & \cdots & 0 & {}_0^1u_{3,2} & 0 \\
0 & 0 & {}_0^1u_{1,3} & \cdots & 0 & 0 & 2 + {}_0^1u_{3,3} \\
{}_0^1u_{1,2} & 2 + {}_0^1u_{1,1} & 0 & \cdots & {}_0^1u_{3,2} & {}_0^1u_{3,1} & 0 \\
{}_0^1u_{1,3} & 0 & 2 + {}_0^1u_{1,1} & \cdots & 2 + {}_0^1u_{3,3} & 0 & {}_0^1u_{3,1} \\
0 & {}_0^1u_{1,3} & {}_0^1u_{1,2} & \cdots & 0 & 2 + {}_0^1u_{3,3} & {}_0^1u_{3,2}
\end{bmatrix}
\tag{9.5.25}
$$

The dots in the above matrix correspond to the following entries (not displayed due to the page width):

$$
\begin{bmatrix}
{}_0^1u_{2,1} & 0 & 0 \\
0 & 2 + {}_0^1u_{2,2} & 0 \\
0 & 0 & {}_0^1u_{2,3} \\
2 + {}_0^1u_{2,2} & {}_0^1u_{2,1} & 0 \\
{}_0^1u_{2,3} & 0 & {}_0^1u_{2,1} \\
0 & {}_0^1u_{2,3} & 2 + {}_0^1u_{2,2}
\end{bmatrix}
$$

and $_0u_{i,j} = \partial u_i / \partial\, {}^0x_j$. The vectors $_0\mathbf{u}$ and $_0\mathbf{e}$ are related to the displacement increments at nodes by

$$_0\mathbf{u} = \mathbf{N}\,\mathbf{u} = \mathbf{N}\,{}^1\mathbf{H}\boldsymbol{\Delta}; \quad _0\mathbf{e} = {}^1\mathbf{A}_0\mathbf{u} = {}^1\mathbf{A}\mathbf{N}\,{}^1\mathbf{H}\boldsymbol{\Delta} \equiv {}_0^1\mathbf{B}_1\boldsymbol{\Delta}; \quad {}_0^1\mathbf{B}_1 = {}^1\mathbf{A}\mathbf{N}\,{}^1\mathbf{H} \tag{9.5.26}$$

where $\mathbf{N}^{\mathrm{T}}$ is the operator of differentials

$$
\mathbf{N}^{\mathrm{T}} =
\begin{bmatrix}
\dfrac{\partial}{\partial\, {}^0x_1} & \dfrac{\partial}{\partial\, {}^0x_2} & \dfrac{\partial}{\partial\, {}^0x_3} & 0 & 0 & 0 & 0 & 0 & 0 \\[2mm]
0 & 0 & 0 & \dfrac{\partial}{\partial\, {}^0x_1} & \dfrac{\partial}{\partial\, {}^0x_2} & \dfrac{\partial}{\partial\, {}^0x_3} & 0 & 0 & 0 \\[2mm]
0 & 0 & 0 & 0 & 0 & 0 & \dfrac{\partial}{\partial\, {}^0x_1} & \dfrac{\partial}{\partial\, {}^0x_2} & \dfrac{\partial}{\partial\, {}^0x_3}
\end{bmatrix}
\tag{9.5.27}
$$

The components of ${}^1\mathbf{A}$ include ${}_0^1u_{i,j}$. From Eq. (9.5.16) the global displacements are related to the natural curvilinear coordinates (ξ, η) and the linear coordinate ζ. Hence the derivatives of these displacements ${}_0^1u_{i,j}$ with respect to the global coordinates ${}^0x_1$, ${}^0x_2$, and ${}^0x_3$ are obtained through the relation

$$
[{}_0^1u_{i,j}] =
\begin{bmatrix}
\dfrac{\partial\, {}^1u_1}{\partial\, {}^0x_1} & \dfrac{\partial\, {}^1u_2}{\partial\, {}^0x_1} & \dfrac{\partial\, {}^1u_3}{\partial\, {}^0x_1} \\[2mm]
\dfrac{\partial\, {}^1u_1}{\partial\, {}^0x_2} & \dfrac{\partial\, {}^1u_2}{\partial\, {}^0x_2} & \dfrac{\partial\, {}^1u_3}{\partial\, {}^0x_2} \\[2mm]
\dfrac{\partial\, {}^1u_1}{\partial\, {}^0x_3} & \dfrac{\partial\, {}^1u_2}{\partial\, {}^0x_3} & \dfrac{\partial\, {}^1u_3}{\partial\, {}^0x_3}
\end{bmatrix}
= {}^0\mathbf{J}^{-1}
\begin{bmatrix}
\dfrac{\partial\, {}^1u_1}{\partial\xi} & \dfrac{\partial\, {}^1u_2}{\partial\xi} & \dfrac{\partial\, {}^1u_3}{\partial\xi} \\[2mm]
\dfrac{\partial\, {}^1u_1}{\partial\eta} & \dfrac{\partial\, {}^1u_2}{\partial\eta} & \dfrac{\partial\, {}^1u_3}{\partial\eta} \\[2mm]
\dfrac{\partial\, {}^1u_1}{\partial\zeta} & \dfrac{\partial\, {}^1u_2}{\partial\zeta} & \dfrac{\partial\, {}^1u_3}{\partial\zeta}
\end{bmatrix}
\tag{9.5.28}
$$

The Jacobian matrix $^0\mathbf{J}$ is defined as

$$
^0\mathbf{J} =
\begin{bmatrix}
\frac{\partial ^0 x_1}{\partial \xi} & \frac{\partial ^0 x_2}{\partial \xi} & \frac{\partial ^0 x_3}{\partial \xi} \\
\frac{\partial ^0 x_1}{\partial \eta} & \frac{\partial ^0 x_2}{\partial \eta} & \frac{\partial ^0 x_3}{\partial \eta} \\
\frac{\partial ^0 x_1}{\partial \zeta} & \frac{\partial ^0 x_2}{\partial \zeta} & \frac{\partial ^0 x_3}{\partial \zeta}
\end{bmatrix}
\tag{9.5.29}
$$

and is computed from the coordinate definition of Eq. (9.5.15). The derivatives of displacements $^1 u_i$ with respect to the coordinates $\xi, \eta,$ and ζ can be computed from Eq. (9.5.16). In the evaluations of element matrices in Eq. (9.5.6), the integrands of $_0^1\mathbf{B}_1, {_0}\mathbf{C}, {_0^1}\mathbf{B}_2, {_0^1}\mathbf{S}, {^1}\mathbf{H},$ and $_0^1\mathbf{S}_v$ should be expressed in the same coordinate system, namely the global coordinate system $(^0 x_1, {^0} x_2, {^0} x_3)$ or the local curvilinear system (x_1', x_2', x_3').

The number of stress and strain components are reduced to five since we neglect the transverse normal components of stress and strain. Hence, the global derivatives of displacements, $[_0^1 u_{i,j}]$ which are obtained in Eq. (9.5.26), are transformed to the local derivatives of the local displacements along the orthogonal coordinates by the following relation

$$
\begin{bmatrix}
\frac{\partial ^1 u_1'}{\partial x_1'} & \frac{\partial ^1 u_2'}{\partial x_1'} & \frac{\partial ^1 u_3'}{\partial x_1'} \\
\frac{\partial ^1 u_1'}{\partial x_2'} & \frac{\partial ^1 u_2'}{\partial x_2'} & \frac{\partial ^1 u_3'}{\partial x_2'} \\
\frac{\partial ^1 u_1'}{\partial x_3'} & \frac{\partial ^1 u_2'}{\partial x_3'} & \frac{\partial ^1 u_3'}{\partial x_3'}
\end{bmatrix}
= \boldsymbol{\Theta}^{\mathrm{T}} [_0^1 u_{i,j}] \, \boldsymbol{\Theta}
\tag{9.5.30}
$$

where $\boldsymbol{\Theta}^{\mathrm{T}}$ is the 3×3 transformation matrix between the local coordinate system (x_1', x_2', x_3') and the global coordinate system $(^0 x_1, {^0} x_2, {^0} x_3)$ at the integration point. The transformation matrix $[\theta]$ is obtained by interpolating the three orthogonal unit vectors $(^1 \hat{\mathbf{e}}_1, {^1} \hat{\mathbf{e}}_2, {^1} \hat{\mathbf{e}}_3)$ at each node:

$$
\boldsymbol{\Theta} =
\begin{bmatrix}
\sum_{k=1}^n \psi_k {^1} e_{11}^k & \sum_{k=1}^n \psi_k {^1} e_{21}^k & \sum_{k=1}^n \psi_k {^1} e_{31}^k \\
\sum_{k=1}^n \psi_k {^1} e_{12}^k & \sum_{k=1}^n \psi_k {^1} e_{22}^k & \sum_{k=1}^n \psi_k {^1} e_{32}^k \\
\sum_{k=1}^n \psi_k {^1} e_{13}^k & \sum_{k=1}^n \psi_k {^1} e_{23}^k & \sum_{k=1}^n \psi_k {^1} e_{33}^k
\end{bmatrix}
\tag{9.5.31}
$$

Since the element matrices are evaluated using numerical integration, the transformation must be performed at each integration point during the numerical integration.

In order to obtain $_0^1\mathbf{B}_1$, the vector of derivatives of incremental displacements $_0\mathbf{u}$ needs to be evaluated. Equations (9.5.28) and (9.5.30) can be used again except that $^1 u_i$ are replaced by u_i and the interpolation equation for u_i, Eq. (9.5.17), is applied.

The development of the matrix of material stiffness, $_0\mathbf{C}'$, is discussed next. Here we wish to present it for a shell element composed of orthotropic material layers with the principal material coordinates (x_1, x_2, x_3) oriented arbitrarily with respect to the local coordinate system (x_1', x_2', x_3') (with $x_3 = x_3'$). For a kth lamina of a laminated shell, the material stiffness matrix is

$$
{}_0\mathbf{C}'_{(k)} = \begin{bmatrix}
C_{11}' & C_{12}' & C_{16}' & 0 & 0 \\
C_{12}' & C_{22}' & C_{26}' & 0 & 0 \\
C_{16}' & C_{26}' & C_{66}' & 0 & 0 \\
0 & 0 & 0 & C_{44}' & C_{45}' \\
0 & 0 & 0 & C_{45}' & C_{55}'
\end{bmatrix}
\tag{9.5.32}
$$

$$
\begin{aligned}
C_{11}' &= m^4 Q_{11} + 2m^2 n^2 (Q_{12} + 2Q_{66}) + n^4 Q_{22} \\
C_{12}' &= m^2 n^2 (Q_{11} + Q_{22} - 4Q_{66}) + (m^4 + n^4) Q_{12} \\
C_{16}' &= mn[m^2 Q_{11} - n^2 Q_{22} - (m^2 - n^2)(Q_{12} + 2Q_{66})] \\
C_{22}' &= n^4 Q_{11} + 2m^2 n^2 (Q_{12} + 2Q_{66}) + m^4 Q_{22} \\
C_{26}' &= mn[n^2 Q_{11} - m^2 Q_{22} + (m^2 - n^2)(Q_{12} + 2Q_{66})] \\
C_{66}' &= m^2 n^2 (Q_{11} + Q_{22} - 2Q_{12}) + (m^2 - n^2)^2 Q_{66} \\
C_{44}' &= m^2 Q_{44} + n^2 Q_{55}, \quad C_{45}' = mn(Q_{55} - Q_{44}) \\
C_{55}' &= m^2 Q_{55} + n^2 Q_{44}, \quad m = \cos\theta_k, \quad n = \sin\theta_k
\end{aligned}
\tag{9.5.33}
$$

where Q_{ij} are the plane stress-reduced stiffness coefficients of the kth orthotropic lamina in the material coordinate system and θ_k is the angle between the x_1-axis (i.e. fiber direction) of the kth lamina and the laminate x-axis (measured in the counterclockwise direction). The Q_{ij} can be expressed in terms of engineering constants of a lamina as

$$
Q_{11} = \frac{E_1}{1 - \nu_{12}\nu_{21}}, \quad Q_{12} = \frac{\nu_{12} E_2}{1 - \nu_{12}\nu_{21}}, \quad Q_{22} = \frac{E_2}{1 - \nu_{12}\nu_{21}}
\tag{9.5.34}
$$
$$
Q_{44} = G_{23}, \quad Q_{55} = G_{13}, \quad Q_{66} = G_{12}
$$

where E_i is the modulus in the x_i-direction, G_{ij} $(i \neq j)$ are the shear moduli in the x_i-x_j plane, and ν_{ij} are the associated Poisson's ratios (see Reddy [1, 55]).

To evaluate element matrices in Eq. (9.5.6), we employ the Gauss quadrature. Since we are dealing with laminated composite structures, integration through the thickness involves individual lamina. One way is to use Gauss quadrature through the thickness direction. Since the constitutive relation ${}_0[C]$ is different from layer to layer and is not a continuous function in the thickness direction, the integration should be performed separately for each layer. This increases the computational time as the number of layers is increased. An alternative way is to perform explicit integration through the thickness and reduce the problem to a two-dimensional one. The Jacobian matrix, in general, is a function of (ξ, η, ζ). The terms in ζ may be neglected provided the thickness to curvature ratios are small. Thus the Jacobian matrix ${}^0[J]$ becomes independent of ζ and explicit integration can be employed. If ζ terms are retained in ${}^0[J]$, Gauss points through the thickness should be added.

Since the explicit integration is performed through the thickness, the expression for

$$\left[\frac{\partial^1 u_i'}{\partial x_j'}\right], \quad {}_0^1\mathbf{A}', \quad {}_0\mathbf{u}', \quad {}^1\mathbf{H}, \quad {}_0^1\mathbf{B}', \quad \{{}_0^1\varepsilon_{ij}'\}$$

are now expressed in an explicit form in terms of ζ. Hence, we can use exact integration through the thickness and use the Gauss quadrature to perform numerical integration on the mid-surface of the shell element.

For thin shell structures, in order to avoid "locking" we use reduced integration scheme to evaluate the stiffness coefficients associated with the transverse shear deformation. Hence we split the constitutive matrix ${}_0\mathbf{C}'$ into two parts, one without transverse shear moduli ${}_0\mathbf{C}'_B$, and the other with only transverse shear moduli ${}_0\mathbf{C}'_S$. Full integration is used to evaluate the stiffness coefficients containing ${}_0\mathbf{C}'_B$, and reduced integration is used for those containing ${}_0\mathbf{C}'_S$.

If a shell element is subjected to a distributed load (such as the weight or pressure), the corresponding load vector ${}^2\mathbf{R}$ from Eq. (9.5.5) is given by

$$^2\mathbf{R}_{5n\times1} = \int_{{}^0A} {}^1\mathbf{H}^{\mathrm{T}} \left\{\begin{array}{c} {}^2P_1 \\ {}^2P_2 \\ {}^2P_3 \end{array}\right\} d\,{}^0A \tag{9.5.35}$$

where ${}^2P_i$ is the component of distributed load in the ${}^0x_i$-direction at time $t + \Delta t$, ${}^0A$ is the area of upper or middle or bottom surface of the shell element depending on the position of the load and the loading is assumed deformation-independent.

Substituting ${}^1\mathbf{H}$ into Eq. (9.5.35) yields (a $5n \times 1$ vector)

$$^2\mathbf{R} = \int_{{}^0A} \begin{bmatrix} \cdots & \cdots & \cdots \\ \psi_k & 0 & 0 \\ 0 & \psi_k & 0 \\ 0 & 0 & \psi_k \\ \frac{1}{2}\zeta\psi_k h_k\,{}^1e_{11}^k & \frac{1}{2}\zeta\psi_k h_k\,{}^1e_{12}^k & -\frac{1}{2}\zeta\psi_k h_k\,{}^1e_{13}^k \\ -\frac{1}{2}\zeta\psi_k h_k\,{}^1e_{21}^k & \frac{1}{2}\zeta\psi_k h_k\,{}^1e_{22}^k & -\frac{1}{2}\zeta\psi_k h_k\,{}^1e_{23}^k \\ \cdots & \cdots & \cdots \end{bmatrix} \left\{\begin{array}{c} {}^2P_1 \\ {}^2P_2 \\ {}^2P_3 \end{array}\right\} d\,{}^0A$$

$$= \sum_{r=1}^{NGP}\sum_{s=1}^{NGP} \left\{\begin{array}{c} \cdots \\ \psi_k\,{}^2P_1 \\ \psi_k\,{}^2P_2 \\ \psi_k\,{}^2P_3 \\ \frac{1}{2}\zeta\psi_k h_k \sum_{i=1}^3 {}^2P_i\,{}^1e_{1i}^k \\ -\frac{1}{2}\zeta\psi_k h_k \sum_{i=1}^3 {}^2P_i\,{}^1e_{2i}^k \\ \cdots \end{array}\right\} W_{\xi_r} W_{\eta_s} |{}^0J|_{(\xi_r,\eta_s)} \frac{2}{h} \tag{9.5.36}$$

where $h = \sum_{k=1}^{NPE} \psi_k(\xi, \eta) h_k$ is the shell thickness at each Gauss point, and W is the weight at each Gauss point, and $|{}^0J|$ is the determinant of the Jacobian matrix in Eq. (9.5.29) at each Gauss point. Here the ζ terms are retained in Jacobian matrix and let ζ equal to 1, -1, or 0, respectively, when the distributed load is at the top, bottom, or middle surface.

9.5.5   Numerical Examples

A number of numerical examples of static bending of isotropic and orthotropic plates and shells are presented. The Riks–Wempner method is employed for tracing the nonlinear load-deflection path (see Appendix 2). For most of the problems the reduced/selective integration scheme is used to evaluate the element stiffness coefficients (see Liao and Reddy [136, 137]).

9.5.5.1   Simply-supported orthotropic plate under uniform load

Figure 9.5.2(a) shows the 2×2Q9 mesh in a quadrant of the plate with dimensions and material properties used. The boundary and symmetry conditions used are the same as those listed in Fig. 7.9.1. The results are shown in Fig. 9.5.2(b) along with the experimental results of Zaghloul and Kennedy [153]. For this simply-supported plate, the finite element results are in good agreement with the experimental results of Zaghloul and Kennedy [153].

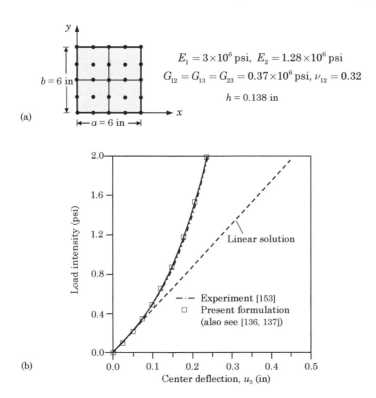

Fig. 9.5.2: Geometrically nonlinear response of an orthotropic plate. (a) Geometry and mesh. (b) Load-deflection curves.

9.5.5.2 Four-layer $(0°/90°/90°/0°)$ clamped plate under uniform load

Figure 9.5.3(a) shows a clamped, symmetrically laminated, square plate under uniform load. The material properties of a typical layer and finite element mesh are also shown in the figure. A quarter of the plate is modeled using four nine-node elements. The present results along with the experimental results of Zaghloul and Kennedy [153] are shown in Fig. 9.5.3(b). The two results are not in good agreement in this case. The difference between the theoretical and experimental results is attributed to possible difference in the support conditions used in the finite element analysis and those used in the experiment (i.e. the exact nature of clamped boundary conditions used in the test may not be the ones used in the finite element analysis), because the present results are verified against independent finite element study by Putcha and Reddy [154].

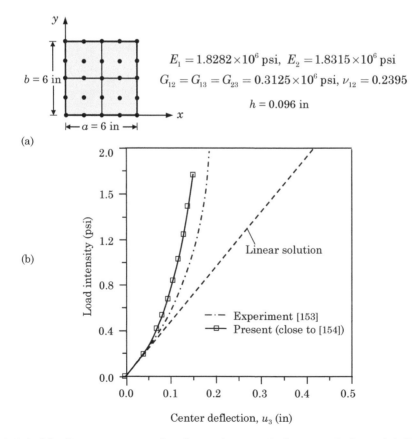

Fig. 9.5.3: Nonlinear response of a clamped cross-ply laminated plate. (a) Geometry and mesh. (b) Load-deflection curves.

9.5.5.3 Cylindrical shell roof under self-weight

Here we consider the linear analysis of a cylindrical panel under self-weight [see Example 8.3.4 and Fig. 9.5.4(a)]. This is often used as a benchmark problem (see Scordelis and Lo [117]) in the literature to assess the performance of shell finite elements. Zienkiewicz, Taylor, and Too [155] showed that the quadratic and cubic shell elements with reduced integration for all terms exhibited a more rapid convergence and better accuracy than with reduced integration of transverse shear terms only. The linear results obtained in the present study with the nine-node quadratic element and reduced integration are shown in Figs. 9.5.4(b) and 9.5.4(c). Even with one element, the results show good agreement with the exact solution [117] and for further mesh refinements the results are close to the exact one.

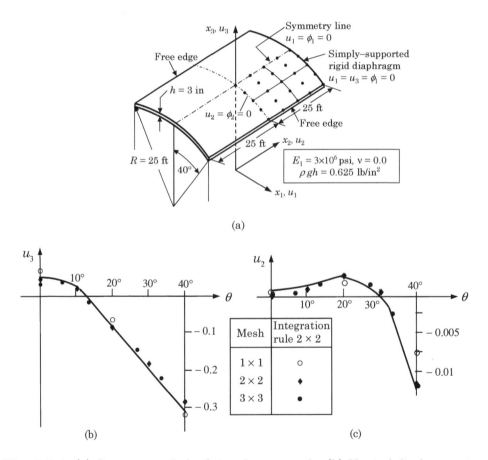

Fig. 9.5.4: (a) Geometry and the finite element mesh. (b) Vertical displacement on the mid-section. (c) Axial displacement at the support. The key shown applies to both (b) and (c).

9.5.5.4 Simply-supported spherical shell under point load

A simply-supported isotropic spherical shell panel under central point load [see Fig. 9.5.5(a)] is analyzed for its large displacement response using a meshes of sixteen four-node elements and four nine-node elements in a quarter of the shell. Figure 9.5.5(b) shows the responses calculated, including the postbuckling range, with the modified Riks–Wempner method. The figure also includes the results of Bathe and Ho [128], who have used a 5 × 5 mesh of a three-node flat triangular shell element in a quadrant. The element has six degrees of freedom per node, and its stiffness matrix is obtained by superimposing membrane and bending stiffness matrices. They have used the updated Lagrangian formulation in their analysis.

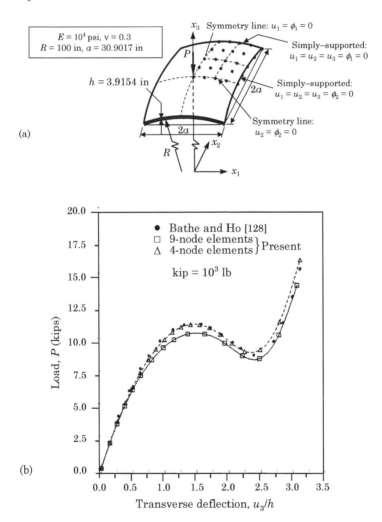

Fig. 9.5.5: A simply-supported spherical shell panel. (a) Geometry and finite element mesh. (b) Load–deflection curves.

9.5.5.5   Shallow cylindrical shell under point load

An isotropic shallow cylindrical shell hinged along the longitudinal edges and free at the curved boundaries and subjected to a point load, as shown in Fig. 9.5.6(a), is analyzed. One-quarter of the shell is modeled with four nine-node shell elements. The structure exhibits snap-through as well as snap-back phenomena, as shown in Fig. 9.5.6(b). The solution obtained by Crisfield [151] is also shown in Fig. 9.5.6(b) (also see Sabir and Lock [121]).

This completes the discussion of the use of continuum element for the large displacement analysis of a number of benchmark problems. The next section is devoted to the discussion of a refined continuum shell element.

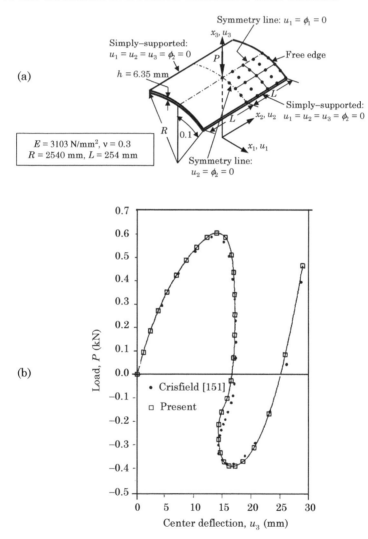

Fig. 9.5.6: Geometrically nonlinear response of a shallow cylindrical shell. (a) Geometry and finite element mesh. (b) Load–deflection curves.

9.6 A Refined Continuum Shell Finite Element

9.6.1 Background

In the majority of continuum and shell theory-based displacement finite element models, straight lines normal to the shell surface are assumed to be inextensible (i.e. thickness changes are neglected) and, as a result, the plane-stress condition must be invoked in writing the stress–strain relations. In addition a rotation tensor is typically introduced in finite rotation implementations to exactly enforce the inextensibility condition [103, 156]. The rotation tensor may be parameterized by means of rotational degrees of freedom; however, depending on the adopted parametrization, singularities and other rank-deficiencies can arise (see, for example, Betsch, Menzel, and Stein [157]). The six-parameter formulations, on the other hand, may be employed in conjunction with fully three-dimensional constitutive equations; however, such implementations are unfortunately hindered by an erroneous state of constant normal strain through the thickness, a phenomena known as the *Poisson locking* (see Bischoff and Ramm [158, 159]). We note that this form of locking is an artifact of the mathematical model and not the discrete finite element model.

In recent years there has been significant attention devoted to shell finite element formulations that may be employed with unmodified fully three-dimensional constitutive equations. Motivation for these models stems from the desire to circumvent many of the problems associated with the incorporation of the plane stress assumption. Such formulations account for thickness stretching and provide reasonable representations of all components of the through-thickness stress states of thin and thick shell structures. These models are usually called seven-parameter formulations, as they involve seven independent parameters in the kinematical description. In a seven-parameter model, the transverse displacement is expanded up to a quadratic term, which essentially mitigates Poisson locking when three-dimensional constitutive equations are adopted. Some of the notable works on seven-parameter shell formulations include Sansour [160] and Bischoff and Ramm [158, 159].

In this section, the improved first-order shear deformation continuum shell finite element developed by Payette [161] and Payette and Reddy [162] for use in the analysis of the fully geometrically nonlinear mechanical response of thin and thick isotropic, composite, and functionally graded elastic shell structures is presented (also see Arciniega [145] and Arciniega and Reddy [146, 147] for certain elements of the formulation)[2]. A seven-parameter formulation, which circumvents the need for a rotation tensor in the kinematical description and allows us to use fully three-dimensional constitutive equations in the numerical implementation, is adopted. A high-order spectral/*hp* type quadrilateral

---

[2]The theoretical development and numerical examples presented in this section come from the works of Payette [161] and Payette and Reddy [162].

finite element technology in a purely displacement-based finite element setting is adopted. This allows us to obtain: (1) highly accurate approximations of arbitrary shell geometries and (2) reliable numerical results that are completely locking-free. In the computer implementation, the Schur complement method is adopted at the element level to statically condense out all degrees of freedom interior to each element in a given finite element discretization. This procedure vastly improves computer memory requirements in the numerical implementation of the resulting shell element and allows for significant parallelization of the global solver (for more details see [161, 162]). This constitutes an important departure from the tensor based shell finite element formulation proposed previously in the work of Arciniega and Reddy [146, 147], where a chart was employed to ensure exact parametrization of the shell mid-surface.

The shell finite element framework presented in this study is applicable to the fully geometrically nonlinear analysis of elastic shell structures based on the St. Venant–Kirchhoff material model. The formulation requires the prescription of the three-dimensional coordinates of the shell mid-surface as well as two sets of directors (one set normal and the other tangent to the mid-surface) at each node in the shell finite element model as input. Each of these quantities is approximated discretely using the standard spectral/hp finite element interpolation functions within a given shell element. The prescribed tangent vector is particularly useful, as it allows for the simple construction of the local bases associated with the principal orthotropic material directions of each lamina in a given composite. This allows us to use skewed and/or arbitrarily curved quadrilateral shell elements in actual finite element simulations. Through the numerical simulation of a number of non-trivial benchmark problems, it is shown that the developed shell element is insensitive to all forms of numerical locking and severe geometric distortions.

9.6.2 Representation of Shell Mid-Surface

In theories of shells, the undeformed mid-surface Ω (where Ω, an open bounded set) of a shell is used as the reference along with the thickness h to construct the three-dimensional domain of the shell. The mid-surface is characterized using either a single or a set of two-dimensional charts from $\mathbb{R}^2$ into $\mathbb{R}^3$. In this study we dispense with this exact parametrization of Ω and instead introduce an appropriate finite element approximation of the mid-surface. To this end, we assume that the closure of Ω, denoted by $\bar{\Omega} = \Omega \bigcup \Gamma$, is approximated by a set of N high-order spectral/hp quadrilateral finite elements. We denote the resulting finite element approximation of $\bar{\Omega}$ by $\bar{\Omega}^{hp} = \bigcup_{e=1}^{N} \bar{\Omega}^e$. This description of the mid-surface leads to the following standard finite element approximation of the geometry within a given master element $\hat{\Omega}^e$:

$$\underline{\mathbf{X}} = \sum_{k=1}^{n} \psi_k(\xi^1, \xi^2)\, \underline{\mathbf{X}}^k, \quad (\xi^1, \xi^2) \in \hat{\Omega}^e \tag{9.6.1}$$

where $\mathbf{r} = \underline{\mathbf{X}}$ represents the position vector of a point on the approximate mid-surface, as shown in Fig. 9.6.1, ψ_k are the two-dimensional spectral/hp basis functions (see Section 9.6.6 for a discussion of the spectral/hp basis functions), $n = (p+1)^2$ is the number of nodes on the element, and p is the polynomial degree in each surface coordinates (ξ^1, ξ^2). We note that in this section (ξ^1, ξ^2, ξ^3) are used in place of (ξ, η, ζ) used in Section 9.5.4. In the above expression $\underline{\mathbf{X}}^k$ are the locations in $\mathbb{R}^3$ of the mid-surface nodes of the eth element (note that all finite element nodes reside on $\bar{\Omega}^{hp}$). The element nodal coordinates $\underline{\mathbf{X}}^k$, as well as all other subsequent nodal quantities, are referenced with respect to a fixed orthonormal Cartesian coordinate system with basis vectors: $\{\hat{\mathbf{E}}_1, \hat{\mathbf{E}}_2, \hat{\mathbf{E}}_3\}$; as a result $\underline{\mathbf{X}}^k = \underline{X}_i^k \hat{\mathbf{E}}_i$ (where summation on repeated indices is implied). The master element $\hat{\Omega}^e$ used in the isoparametric characterization of the approximate element mid-surface $\bar{\Omega}^e$ (i.e. $\phi^e : \hat{\Omega}^e \to \bar{\Omega}^e \subset \bar{\Omega}^{hp}$) is taken as the standard unit square $\hat{\Omega}^e = [-1, 1]^2$. The p-refinement (i.e. increasing the polynomial-order p of the spectral/hp basis functions) allows us to reduce the error in approximating $\bar{\Omega}$ by $\bar{\Omega}^{hp}$.

At each point of the mid-surface $\bar{\Omega}^e$ of a given element, we define the vectors

$$\mathbf{a}_\alpha = \frac{\partial \mathbf{r}}{\partial \xi^\alpha} = \frac{\partial \underline{\mathbf{X}}}{\partial \xi^\alpha} \equiv \underline{\mathbf{X}}_{,\alpha} \tag{9.6.2}$$

which are linearly independent and hence form a local basis of the tangent plane. We use the usual convention that Greek indices range over 1 and 2 and

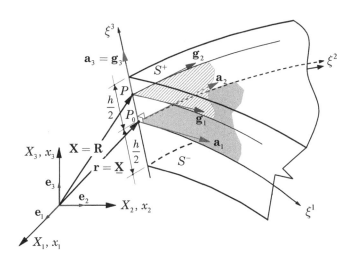

Fig. 9.6.1: Shell geometry and coordinate system.

Latin indices over 1, 2, and 3. The unit normal vector may be expressed as

$$\mathbf{a}_3 = \frac{\mathbf{a}_1 \times \mathbf{a}_2}{||\mathbf{a}_1 \times \mathbf{a}_2||} \tag{9.6.3}$$

For each $(\xi^1, \xi^2) \in \hat{\Omega}^e$, the vectors $\mathbf{a}_i$ define a basis for $\mathbb{R}^3$. The finite element approximation of the unit normal within a given element is

$$\hat{\mathbf{n}} = \sum_{k=1}^{n} \psi_k(\xi^1, \xi^2)\, \hat{\mathbf{n}}^k \tag{9.6.4}$$

Thus, we input the mid-surface locations $\underline{\mathbf{X}}$ and the unit normals $\hat{\mathbf{n}}^k$, both evaluated at the element nodes.

We now can represent the three-dimensional geometry of the undeformed configuration of a typical shell element $\bar{\mathcal{B}}_0^e$ and as a consequence $\bar{\mathcal{B}}_0^{hp}$ (i.e. the finite element approximation of the three-dimensional undeformed shell configuration $\bar{\mathcal{B}}_0$). Assuming a constant thickness h, we define the position vector $\mathbf{R} = \mathbf{X}$ in the shell element as

$$\mathbf{X} = \sum_{k=1}^{n} \psi_k(\xi^1, \xi^2) \left(\underline{\mathbf{X}}^k + \xi^3 \frac{h}{2}\, \hat{\mathbf{n}}^k \right) \tag{9.6.5}$$

where $\xi^3 \in [-1, 1]$. The process of parameterizing $\bar{\mathcal{B}}_0^e$ is summarized in Fig. 9.6.2.

At each point of the shell element $\bar{\mathcal{B}}_0^e$ (not necessary on the mid-surface $\bar{\Omega}^e$) we define a set of covariant basis vectors

$$\mathbf{g}_i = \frac{\partial \mathbf{X}}{\partial \xi^i} \equiv \mathbf{X}_{,i} \tag{9.6.6}$$

Using Eq. (9.6.5) allows us to express the shell basis vectors as

$$\mathbf{g}_\alpha = \mathbf{a}_\alpha + \xi^3 \frac{h}{2}\, \hat{\mathbf{n}}_{,\alpha}, \qquad \mathbf{g}_3 = \frac{h}{2}\, \hat{\mathbf{n}} \tag{9.6.7}$$

Figure 9.6.3 shows an illustration of the vectors $\mathbf{a}_\alpha$ and $\mathbf{g}_\alpha$ at points A and B, respectively, in a typical shell element $\bar{\mathcal{B}}_0^e$; point A resides on the mid-surface $\bar{\Omega}^e$, while B lies directly above A in the direction of the unit normal $\hat{\mathbf{n}}$.

The covariant basis vectors $\mathbf{g}_i$ allow us to write a differential line element in $\bar{\mathcal{B}}_0^e$ in terms of the curvilinear coordinates (ξ^1, ξ^2, ξ^3) as

$$d\mathbf{X} = d\mathbf{X}_1 + d\mathbf{X}_2 + d\mathbf{X}_3 = \mathbf{g}_1 d\xi^1 + \mathbf{g}_2 d\xi^2 + \mathbf{g}_3 d\xi^3 \tag{9.6.8}$$

which can be expressed in matrix form as

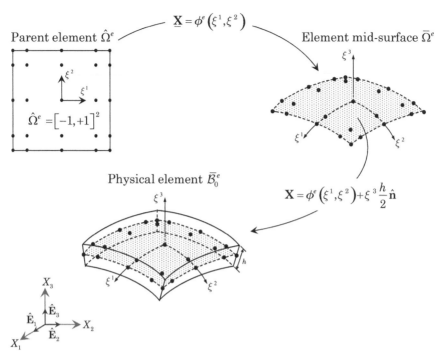

Fig. 9.6.2: Approximation of the three-dimensional geometry of a shell element in the reference configuration based on an isoparametric map from the parent element to the finite element approximation of the mid-surface, followed by an additional map to account for the shell thickness.

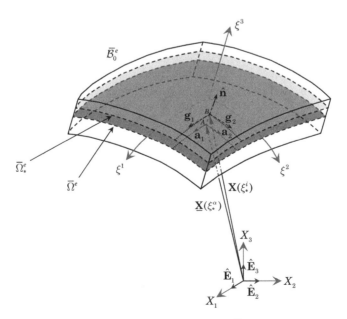

Fig. 9.6.3: Geometry of a typical shell finite element $\bar{\mathcal{B}}_0^e$ in the reference configuration. The basis vectors $\mathbf{a}_\alpha$ and $\mathbf{g}_\alpha$ and the unit normal $\hat{\mathbf{n}}$ are shown.

$$\begin{Bmatrix} dX_1 \\ dX_2 \\ dX_3 \end{Bmatrix} = \{d\xi\}^{\mathrm{T}}[J] = \begin{Bmatrix} d\xi^1 \\ d\xi^2 \\ d\xi^3 \end{Bmatrix}^{\mathrm{T}} \begin{bmatrix} \frac{\partial X_1}{\partial \xi^1} & \frac{\partial X_2}{\partial \xi^1} & \frac{\partial X_3}{\partial \xi^1} \\ \frac{\partial X_1}{\partial \xi^2} & \frac{\partial X_2}{\partial \xi^2} & \frac{\partial X_3}{\partial \xi^2} \\ \frac{\partial X_1}{\partial \xi^3} & \frac{\partial X_2}{\partial \xi^3} & \frac{\partial X_3}{\partial \xi^3} \end{bmatrix} \tag{9.6.9}$$

Here $[J]$ denotes the Jacobian matrix (or the matrix of deformation gradient), and its determinant is denoted as J. The inverse (assumed to exist) of $[J]$ is denoted here as $[J^\star]$. A differential volume element in $\bar{\mathcal{B}}_0^e$ is given as

$$d\bar{\mathcal{B}}_0^e = d\mathbf{X}_1 \cdot (d\mathbf{X}_2 \times d\mathbf{X}_3) = J d\xi^1 d\xi^2 d\xi^3 \tag{9.6.10}$$

We associate with the covariant basis, a dual or contravariant basis defined by

$$\mathbf{g}^i \cdot \mathbf{g}_j = \delta_j^i \tag{9.6.11}$$

where δ_j^i is the Kronecker delta. The contravariant basis vectors may also be determined from the following formulae

$$\mathbf{g}^1 = (\mathbf{g}_2 \times \mathbf{g}_3)/J, \qquad \mathbf{g}^2 = (\mathbf{g}_3 \times \mathbf{g}_1)/J, \qquad \mathbf{g}^3 = (\mathbf{g}_1 \times \mathbf{g}_2)/J \tag{9.6.12}$$

The covariant and contravariant basis vectors may be alternatively defined in terms of the components of Jacobian matrix and its inverse

$$\mathbf{g}_i = J_{ij}\hat{\mathbf{E}}_j, \qquad \mathbf{g}^i = J_{ji}^\star \hat{\mathbf{E}}_j \tag{9.6.13}$$

9.6.3   Displacement and Strain Fields

We now consider the motion $\boldsymbol{\chi}(\mathbf{X}, t)$ of the shell from the reference finite element configuration $\bar{\mathcal{B}}_0^{hp}$ to the current configuration $\bar{\mathcal{B}}_t^{hp}$. The displacement of a material point from the reference configuration to the current configuration is

$$\mathbf{u}(\mathbf{X}, t) = \boldsymbol{\chi}(\mathbf{X}, t) - \mathbf{X} = \mathbf{x}(\mathbf{X}, t) - \mathbf{X} \tag{9.6.14}$$

Next, we assume that at any point within a typical shell element $\bar{\mathcal{B}}_0^e$, the displacement vector may expanded in Taylor's series expansion with respect to the thickness coordinate ξ^3

$$\mathbf{u}(\mathbf{X}(\xi^i), t) \cong \mathbf{u}^{(0)}(\xi^\alpha, t) + \xi^3 \mathbf{u}^{(1)}(\xi^\alpha, t) + \frac{(\xi^3)^2}{2}\mathbf{u}^{(2)}(\xi^\alpha, t) + \cdots \tag{9.6.15}$$

where $\mathbf{u}^{(j)}(\xi^\alpha, t) = \partial^j \mathbf{u}(\xi^i, t)/\partial(\xi^3)^j|_{\xi^3=0}$.

The Taylor series for $\mathbf{u}$ is truncated such that the resulting shell model is asymptotically consistent with three-dimensional elasticity (see Chapple and

Bathe [156]), thereby allowing for the use of fully three-dimensional constitutive equations in the mathematical model and subsequent numerical implementation. Therefore, the displacement field is restricted to the following seven-parameter expansion:

$$\mathbf{u}(\xi^i) = \underline{\mathbf{u}}(\xi^\alpha) + \xi^3 \frac{h}{2} \boldsymbol{\varphi}(\xi^\alpha) + (\xi^3)^2 \frac{h}{2} \boldsymbol{\psi}(\xi^\alpha) \tag{9.6.16}$$

where each $\mathbf{u}^{(j)}(\xi^\alpha, t)$ ($j = 0, 1$ and 2) has been renamed, and for $j = 1$ and 2, scaled by some factor of h. For the sake of brevity, the time parameter t is omitted from the above expressions and the subsequent discussion. The generalized displacements $\mathbf{u}$, $\boldsymbol{\varphi}$ and $\boldsymbol{\psi}$ may be expressed as

$$\underline{\mathbf{u}}(\xi^\alpha) = \underline{u}_i(\xi^\alpha)\hat{\mathbf{E}}_i, \qquad \boldsymbol{\varphi}(\xi^\alpha) = \varphi_i(\xi^\alpha)\hat{\mathbf{E}}_i, \qquad \boldsymbol{\psi}(\xi^\alpha) = \Psi(\xi^\alpha)\hat{\mathbf{n}}(\xi^\alpha) \tag{9.6.17}$$

The quantity $\underline{\mathbf{u}}$ represents the mid-plane displacement and $\boldsymbol{\varphi}$ is the so-called difference vector (which gives the change in the mid-surface director). The seventh parameter Ψ is included to circumvent spurious stresses in the thickness direction, caused in the six-parameter formulation by an artificial constant normal strain (a phenomena referred to as *Poisson locking* [158]).

The position occupied by a material point belonging to $\bar{\mathcal{B}}_0^e$ at the current time t may be evaluated by substituting the assumed displacement field into Eq. (9.6.14) which upon rearrangement yields

$$\mathbf{x} = \mathbf{X} + \mathbf{u} = \underline{\mathbf{x}} + \xi^3 \frac{h}{2} \hat{\hat{\mathbf{n}}} + (\xi^3)^2 \frac{h}{2} \Psi \hat{\mathbf{n}} \tag{9.6.18}$$

where $\underline{\mathbf{x}} = \underline{\mathbf{X}} + \underline{\mathbf{u}}$ (a point on the deformed mid-surface) and $\hat{\hat{\mathbf{n}}} = \hat{\mathbf{n}} + \boldsymbol{\varphi}$ (a *pseudo*-director associated with the deformed mid-surface). We note that, unlike $\hat{\mathbf{n}}$, the director $\hat{\hat{\mathbf{n}}}$ is in general neither a unit vector nor it is normal to the deformed mid-surface.

The finite element approximation of the displacement field given in Eq. (9.6.16) is represented as (i.e. isoparametric formulation is used)

$$\mathbf{u}(\xi^i) = \sum_{k=1}^{n} \psi_k(\xi^1, \xi^2) \left(\underline{\mathbf{u}}^k + \xi^3 \frac{h}{2} \boldsymbol{\varphi}^k + (\xi^3)^2 \frac{h}{2} \Psi^k \hat{\mathbf{n}}(\xi^\alpha) \right) \tag{9.6.19}$$

where $\hat{\mathbf{n}}(\xi^\alpha)$ is given by Eq. (9.6.4). Note that we interpolate Ψ and $\hat{\mathbf{n}}$ separately in the finite element approximation of $\boldsymbol{\psi}$, as opposed to interpolating the product $\Psi \hat{\mathbf{n}}$ as a single entity. The derivative of the displacement field with respect to the curvilinear coordinates of the element may be expressed as

$$\mathbf{u}_{,\alpha} = \sum_{k=1}^{n} \frac{\partial \psi_k}{\partial \xi^\alpha} \left[\underline{\mathbf{u}}^k + \xi^3 \frac{h}{2} \boldsymbol{\varphi}^k + (\xi^3)^2 \frac{h}{2} \left(\Psi^k \hat{\mathbf{n}}(\xi^\beta) + \hat{\mathbf{n}}^k \Psi(\xi^\beta) \right) \right] \tag{9.6.20a}$$

$$\mathbf{u}_{,3} = h \sum_{k=1}^{n} \psi_k(\xi^1, \xi^2) \left(\frac{1}{2} \boldsymbol{\varphi}^k + \xi^3 \boldsymbol{\Psi}^k \hat{\mathbf{n}}(\xi^\beta) \right) \tag{9.6.20b}$$

The deformation gradient $\mathbf{F}$ is given by [see Eq. (9.1.3)]

$$\mathbf{F} = (\boldsymbol{\nabla}_0 \mathbf{x}(\mathbf{X}, t))^{\mathrm{T}} = \mathbf{x}_{,i} \mathbf{g}^i = \bar{\mathbf{g}}_j \mathbf{g}^j \tag{9.6.21}$$

where $\bar{\mathbf{g}}_j = \mathbf{g}_{,j} + \mathbf{u}_{,j}$ are the covariant basis vectors associated with the deformed finite element configuration of the three-dimensional shell $\bar{\mathcal{B}}_t^{hp}$. The symbol $\boldsymbol{\nabla}_0$ denotes the material gradient operator. The Green–Lagrange strain tensor $\mathbf{E}$ is

$$\mathbf{E} = \frac{1}{2} \left(\mathbf{F}^{\mathrm{T}} \mathbf{F} - \mathbf{I} \right) = \frac{1}{2} \left(\mathbf{u}_{,i} \cdot \mathbf{g}_j + \mathbf{g}_i \cdot \mathbf{u}_{,j} + \mathbf{u}_{,i} \cdot \mathbf{u}_{,j} \right) \mathbf{g}^i \mathbf{g}^j \tag{9.6.22}$$

The covariant components of the Green–Lagrange strain tensor [i.e. the coefficients of the second-order tensor contravariant bases $\mathbf{g}^i \mathbf{g}^j$ appearing in Eq. (9.6.22)] may be expanded in terms of the thickness coordinate ξ^3 as

$$E_{ij}(\xi^m) = \varepsilon_{ij}^{(0)} + \xi^3 \varepsilon_{ij}^{(1)} + (\xi^3)^2 \varepsilon_{ij}^{(2)} + (\xi^3)^3 \varepsilon_{ij}^{(3)} + (\xi^3)^4 \varepsilon_{ij}^{(4)} \tag{9.6.23}$$

where $\varepsilon_{ij}^{(n)} = \varepsilon_{ij}^{(n)}(\xi^\alpha)$. In the present formulation all covariant components of $\mathbf{E}$ that are of higher order than linear in ξ^3 [i.e. the underlined terms in Eq. (9.6.23)] are neglected. The covariant components of the retained strain terms may be determined from

$$\varepsilon_{\alpha\beta}^{(0)} = \frac{1}{2} \left(\underline{\mathbf{u}}_{,\alpha} \cdot \mathbf{a}_\beta + \mathbf{a}_\alpha \cdot \underline{\mathbf{u}}_{,\beta} + \underline{\mathbf{u}}_{,\alpha} \cdot \underline{\mathbf{u}}_{,\beta} \right) \tag{9.6.24a}$$

$$\varepsilon_{\alpha\beta}^{(1)} = \frac{h}{4} \left[\underline{\mathbf{u}}_{,\alpha} \cdot \left(\hat{\mathbf{n}}_{,\beta} + \boldsymbol{\varphi}_{,\beta} \right) + \left(\hat{\mathbf{n}}_{,\alpha} + \boldsymbol{\varphi}_{,\alpha} \right) \cdot \underline{\mathbf{u}}_{,\beta} + \boldsymbol{\varphi}_{,\alpha} \cdot \mathbf{a}_\beta + \mathbf{a}_\alpha \cdot \boldsymbol{\varphi}_{,\beta} \right] \tag{9.6.24b}$$

$$\varepsilon_{\alpha 3}^{(0)} = \frac{h}{4} \left[\underline{\mathbf{u}}_{,\alpha} \cdot \left(\hat{\mathbf{n}} + \boldsymbol{\varphi} \right) + \mathbf{a}_\alpha \cdot \boldsymbol{\varphi} \right] \tag{9.6.24c}$$

$$\varepsilon_{\alpha 3}^{(1)} = \frac{h}{2} \left\{ \frac{h}{4} \left[\boldsymbol{\varphi}_{,\alpha} \cdot \hat{\mathbf{n}} + \left(\hat{\mathbf{n}}_{,\alpha} + \boldsymbol{\varphi}_{,\alpha} \right) \cdot \boldsymbol{\varphi} \right] + \left(\mathbf{a}_\alpha + \underline{\mathbf{u}}_{,\alpha} \right) \cdot \boldsymbol{\psi} \right\} \tag{9.6.24d}$$

$$\varepsilon_{33}^{(0)} = \frac{h^2}{8} \left(2\hat{\mathbf{n}} + \boldsymbol{\varphi} \right) \cdot \boldsymbol{\varphi} \tag{9.6.24e}$$

$$\varepsilon_{33}^{(1)} = \frac{h^2}{2} \left(\hat{\mathbf{n}} + \boldsymbol{\varphi} \right) \cdot \boldsymbol{\psi} \tag{9.6.24f}$$

The six-parameter formulation is obtained as a special case by setting $\boldsymbol{\psi} = \mathbf{0}$; that is, the strain component $\varepsilon_{33}^{(1)}$ is identically zero.

9.6.4 Constitutive Relations

The underlying kinematic assumptions of the adopted shell finite element formulation can be applied in the context of a multitude of material models (e.g.

Cauchy elastic, hyperelastic, viscoelastic, elasto–plastic, etc.). In this work we assume that the material response remains in the elastic regime. Furthermore, we assume that the second Piola–Kirchhoff stress tensor $\mathbf{S}$ is related to the Green–Lagrange strain tensor $\mathbf{E}$ by the following relation

$$\mathbf{S} = \mathbf{C} : \mathbf{E} \tag{9.6.25}$$

where $\mathbf{C} = C^{ijkl}\mathbf{g}_i\,\mathbf{g}_j\,\mathbf{g}_k\,\mathbf{g}_l$ is the fourth-order elasticity tensor [with symmetries given in Eq. (2.8.15)]. We assume the elasticity tensor to be independent of the shell deformation. However, we do allow $\mathbf{C}$ to be non-homogeneous (i.e. a function of $\mathbf{X}$). In the numerical implementation, we rely on the following component representation of the set of constitutive equations

$$S^{ij} = C^{ijkl}\,E_{kl} \tag{9.6.26}$$

The adopted material model may also be expressed in matrix form as

$$
\begin{Bmatrix} S^{11} \\ S^{22} \\ S^{33} \\ S^{23} \\ S^{13} \\ S^{12} \end{Bmatrix}
=
\begin{bmatrix}
C^{1111} & C^{1122} & C^{1133} & C^{1123} & C^{1113} & C^{1112} \\
C^{1122} & C^{2222} & C^{2233} & C^{2223} & C^{2213} & C^{2212} \\
C^{1133} & C^{2233} & C^{3333} & C^{3323} & C^{3313} & C^{3312} \\
C^{1123} & C^{2223} & C^{3323} & C^{2323} & C^{2313} & C^{2312} \\
C^{1113} & C^{2213} & C^{3313} & C^{2313} & C^{1313} & C^{1312} \\
C^{1112} & C^{2212} & C^{3312} & C^{2312} & C^{1312} & C^{1212}
\end{bmatrix}
\begin{Bmatrix} E_{11} \\ E_{22} \\ E_{33} \\ 2E_{23} \\ 2E_{13} \\ 2E_{12} \end{Bmatrix}
\tag{9.6.27}
$$

where the coefficient matrix $[C^{ijkl}]$ appearing in the above expression is the matrix form of the contravariant components of the elasticity tensor $\mathbf{C}$. It should be evident that there are in general 21 unique contravariant components of $\mathbf{C}$.

For completeness, we recall that the second Piola–Kirchhoff stress tensor $\mathbf{S}$ is related to the Cauchy stress tensor $\boldsymbol{\sigma}$ by [see Eq. (9.1.10)]

$$\mathbf{S} = J_{\mathbf{F}}\,\mathbf{F}^{-1}\,\boldsymbol{\sigma}\,\mathbf{F}^{-T} \tag{9.6.28}$$

where $J_{\mathbf{F}} = \det(\mathbf{F})$. The symmetry of $\mathbf{S}$ follows from the symmetry of $\boldsymbol{\sigma}$.

9.6.4.1 Isotropic and functionally graded shells

We now specialize the assumed constitutive model for use in the context of isotropic shells. We consider the homogeneous case and also the scenario where the material is functionally graded through the thickness of the shell. Homogeneous shells are abundant and can be found in piping, pressure vessels, ship hulls, large roofs, and the bodies of automobiles. Functionally graded shells on the other hand have been advocated for use in high-temperature environments with applications in reactor vessels, turbines, and other machine parts (see Reddy [55]). These materials are typically composed of metals and ceramics to maximize the strength and toughness properties of the former and the thermal and corrosion resistance attributes of the latter.

For isotropic materials, the fourth-order elasticity tensor $\mathbf{C}$ may be expressed as

$$\mathbf{C} = \lambda \mathbf{I}\mathbf{I} + 2\mu \mathbf{\Pi} \tag{9.6.29}$$

Here $\mathbf{I} = \delta_{ij}\hat{\mathbf{E}}_i\,\hat{\mathbf{E}}_j$ is the second-order unit tensor and $\mathbf{\Pi} = \frac{1}{2}(\delta_{ik}\,\delta_{jl}+\delta_{il}\,\delta_{jk})\hat{\mathbf{E}}_i\,\hat{\mathbf{E}}_j\,\hat{\mathbf{E}}_k\,\hat{\mathbf{E}}_l$ is the fourth-order unit tensor. These tensors may also be expressed with respect to the covariant basis vectors $\mathbf{g}_i$ as

$$\mathbf{I} = \mathbf{G} = g^{ij}\mathbf{g}_i\,\mathbf{g}_j, \quad \mathbf{\Pi} = \frac{1}{2}(g^{ik}g^{jl} + g^{il}g^{jk})\mathbf{g}_i\,\mathbf{g}_j\,\mathbf{g}_k\,\mathbf{g}_l \tag{9.6.30}$$

where $g^{ij} = \mathbf{g}^i \cdot \mathbf{g}^j$ are the contravariant components of the *Riemannian metric* tensor $\mathbf{G}$ in the reference configuration. We can therefore express the contravariant components of $\mathbf{C}$ as

$$C^{ijkl} = \lambda g^{ij}g^{kl} + \mu(g^{ik}g^{jl} + g^{il}g^{jk}) \tag{9.6.31}$$

Although $\mathbf{C}$ depends on only the Lamé parameters, the 21 unique contravariant components associated with the matrix $[C^{ijkl}]$ are, in general, distinct from one another. The Lamé parameters λ and μ are related to Young's modulus E and Poisson's ratio ν by the following expressions:

$$\lambda = \frac{\nu E}{(1+\nu)(1-2\nu)}, \quad \mu = \frac{E}{2(1+\nu)} \tag{9.6.32}$$

For the homogeneous case, Young's modulus E and Poisson's ratio ν are constant throughout the shell structure. For functionally graded structures, we assume that the shell is composed of two isotropic constituents that are combined in a prescribed fashion through the shell thickness. In such cases, we allow Young's modulus to vary, while Poisson's ratio is constant, with respect to the shell thickness coordinate ξ^3 as follows (see Reddy [55]):

$$E(\xi^3) = (E^+ - E^-)f^+(\xi^3) + E^- \tag{9.6.33}$$

where

$$f^+(\xi^3) = \left(\frac{\xi^3 + 1}{2}\right)^n \tag{9.6.34}$$

The quantities E^- and E^+ constitute Young's moduli at the bottom ($\xi^3 = -1$) and top ($\xi^3 = +1$) surfaces of the shell, respectively. Equation (9.6.33) constitutes a power-law variation of E through the shell thickness (where the non-negative constant n is the power-law parameter). Note that E^- and E^+ are recovered throughout the thickness in the limits where $n \to \infty$ and $n \to 0$, respectively. As in the homogeneous case, functionally graded shells may also be described using Eq. (9.6.31) if the Lamé parameters are taken as functions of ξ^3.

9.6.4.2 Laminated composite shells

In this work we are also concerned with the numerical simulation of laminated composite shell structures. A composite laminae is a thin sheet (plate or shell like) of material, typically composed of two distinct constituents, which together possess desirable mechanical properties that cannot be exhibited by the individual materials acting in bulk alone. A laminated composite shell is a collection of stacked laminae (where the stacking sequence is typically prescribed in a manner which maximizes the desired stiffness of the composite). In our analysis, we treat each laminae as an orthotropic layer of material. We further assume that for a given structure, perfect bonding exists between each layer and that the continuum hypothesis holds.

To simplify the discussion, we initially restrict our attention to a shell composed of a single orthotopic layer (i.e. one lamina). Next, we define at each node in $\bar{\Omega}^{hp}$ a unit vector $\hat{\mathbf{t}}$ that is tangent to the finite element approximation of the mid-plane. The discrete tangents are utilized to define a continuous tangent vector field in $\bar{\Omega}^{hp}$. Within each element, the tangent field is represented as

$$\hat{\mathbf{t}} = \sum_{k=1}^{n} \psi_k(\xi^1, \xi^2)\, \hat{\mathbf{t}}^k \tag{9.6.35}$$

The tangent vector $\hat{\mathbf{t}}$ is prescribed in a manner that allows us to easily construct a local orthogonal Cartesian basis $\{\hat{\mathbf{e}}_1, \hat{\mathbf{e}}_2, \hat{\mathbf{e}}_3\}$ associated with the principal directions of the orthotropic laminae. Figure 9.6.4 contains the geometry, nodes, unit normals $\hat{\mathbf{n}}^k$, and unit tangents $\hat{\mathbf{t}}^k$ for a typical high-order spectral/hp shell finite element. Note that the direction of $\hat{\mathbf{t}}$ need not coincide with the direction of either $\mathbf{a}_1$ or $\mathbf{a}_2$.

The elasticity tensor $\mathbf{C}$ can be expressed with respect to the local basis $\{\hat{\mathbf{e}}_1, \hat{\mathbf{e}}_2, \hat{\mathbf{e}}_3\}$ (which we will soon define) as

$$\mathbf{C} = \bar{C}_{ijkl}\, \hat{\mathbf{e}}_i\, \hat{\mathbf{e}}_j\, \hat{\mathbf{e}}_k\, \hat{\mathbf{e}}_l \tag{9.6.36}$$

Assuming an orthotropic material model, we can express the coefficients $\bar{C}_{ijkl}$ in matrix form as follows:

$$[\bar{C}] = \begin{bmatrix} \bar{C}_{1111} & \bar{C}_{1122} & \bar{C}_{1133} & 0 & 0 & 0 \\ \bar{C}_{1122} & \bar{C}_{2222} & \bar{C}_{2233} & 0 & 0 & 0 \\ \bar{C}_{1133} & \bar{C}_{2233} & \bar{C}_{3333} & 0 & 0 & 0 \\ 0 & 0 & 0 & \bar{C}_{2323} & 0 & 0 \\ 0 & 0 & 0 & 0 & \bar{C}_{1313} & 0 \\ 0 & 0 & 0 & 0 & 0 & \bar{C}_{1212} \end{bmatrix} \tag{9.6.37}$$

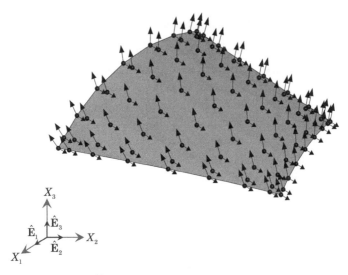

Fig. 9.6.4: The mid-surface $\bar{\Omega}^e$ of a typical high-order spectral/hp shell finite element (element shown is for $p = 8$).

The components of the coefficient matrix $[\bar{C}]$ may be determined in terms of the nine engineering material parameters: E_1, E_2, E_3, ν_{12}, ν_{13}, ν_{23}, G_{12}, G_{13}, and G_{23} as

$$[\bar{C}] = \begin{bmatrix} \frac{1}{E_1} & -\frac{\nu_{12}}{E_1} & -\frac{\nu_{13}}{E_1} & 0 & 0 & 0 \\ -\frac{\nu_{12}}{E_1} & \frac{1}{E_2} & -\frac{\nu_{23}}{E_2} & 0 & 0 & 0 \\ -\frac{\nu_{13}}{E_1} & -\frac{\nu_{23}}{E_2} & \frac{1}{E_3} & 0 & 0 & 0 \\ 0 & 0 & 0 & \frac{1}{G_{23}} & 0 & 0 \\ 0 & 0 & 0 & 0 & \frac{1}{G_{13}} & 0 \\ 0 & 0 & 0 & 0 & 0 & \frac{1}{G_{12}} \end{bmatrix}^{-1} \quad (9.6.38)$$

We next address construction of the local basis $\{\hat{\mathbf{e}}_1, \hat{\mathbf{e}}_2, \hat{\mathbf{e}}_3\}$ for a typical shell element. Without loss of generality we take $\hat{\mathbf{e}}_3 = \hat{\mathbf{n}}$. Next we assume that $\hat{\mathbf{e}}_1$ may be obtained locally in terms of a proper finite rotation of the tangent $\hat{\mathbf{t}}$ about the unit normal $\hat{\mathbf{n}}$, where the angle of rotation is θ. In this work we always define $\hat{\mathbf{t}}$ such that it is sufficient to take θ as constant in $\bar{\Omega}^e$ and throughout $\bar{\Omega}^{hp}$. Given the preceding assumptions, we can show from geometry that the local basis vectors for the principal directions of the material are given as

$$\hat{\mathbf{e}}_1 = \hat{\mathbf{t}}\cos\theta + \hat{\mathbf{n}} \times \hat{\mathbf{t}}\sin\theta, \quad \hat{\mathbf{e}}_2 = -\hat{\mathbf{t}}\sin\theta + \hat{\mathbf{n}} \times \hat{\mathbf{t}}\cos\theta, \quad \hat{\mathbf{e}}_3 = \hat{\mathbf{n}} \quad (9.6.39)$$

When $\theta = 0$, the in-plane basis vectors of the principal material directions reduce to $\hat{\mathbf{e}}_1 = \hat{\mathbf{t}}$ and $\hat{\mathbf{e}}_2 = \hat{\mathbf{n}} \times \hat{\mathbf{t}}$. We note that the in-plane *material* basis vectors $\hat{\mathbf{e}}_\alpha$ are constructed independently from the in-plane *natural* basis vectors $\mathbf{a}_\alpha$. As a result, we may freely employ unstructured skewed and/or curved quadrilateral finite elements in the numerical discretization of complex shell structures. Key to the success of the present formulation is an appropriate prescription of the discrete tangent vector $\hat{\mathbf{t}}$ and angle of rotation θ. In Fig. 9.6.5 we show the unit normal $\hat{\mathbf{n}}$, unit tangent $\hat{\mathbf{t}}$, rotation angle θ, and local material basis vectors $\hat{\mathbf{e}}_i$ at a point on the mid-surface of a typical shell element.

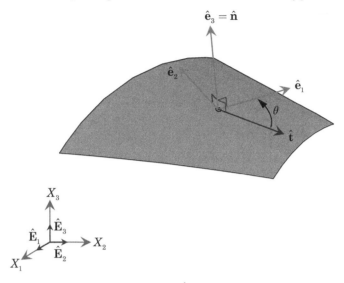

Fig. 9.6.5: The unit normal $\hat{\mathbf{n}}$, unit tangent $\hat{\mathbf{t}}$, rotation angle θ, and local basis vectors $\{\hat{\mathbf{e}}_1, \hat{\mathbf{e}}_2, \hat{\mathbf{e}}_3\}$ at a point on the mid-surface $\bar{\Omega}^e$ of a typical shell finite element.

In the numerical implementation, the contravariant components C^{ijkl} of the elasticity tensor $\mathbf{C}$ are needed. These may be obtained by contracting Eq. (9.6.36) with $\mathbf{g}^i \, \mathbf{g}^j \, \mathbf{g}^k \, \mathbf{g}^l$, which yields

$$C^{ijkl} = T_{im} T_{jn} T_{kp} T_{lq} \bar{C}_{mnpq} \tag{9.6.40}$$

where the components of T_{ij} are defined as

$$T_{ij} = \mathbf{g}^i \cdot \hat{\mathbf{e}}_j = J_{ki}^\star \bar{E}_{jk} \tag{9.6.41}$$

and $\bar{E}_{jk} = \hat{\mathbf{e}}_j \cdot \hat{\mathbf{E}}_k$. We see that evaluation of C^{ijkl} requires five matrix multiplications. In the actual numerical implementation, however, the C++ statements are generated for evaluating the 21 independent coefficients in C^{ijkl} using the symbolic algebra software Maple©. The expressions are quite involved and are hence not provided here (see Payette [161]).

The above discussion has been limited to the analysis of composite shells composed of a single orthotropic layer. For multi-layered composites, a unit tangent vector $\hat{\mathbf{t}}$ along with a set of orientation angles $\boldsymbol{\theta} = (\theta_1, \ldots, \theta_{\mathrm{NL}})$ associated with each ply (where NL is the total number of layers) is defined. Each layer in the laminated composite shell is numbered from the bottom ply to the top laminae in the positive ξ^3-direction. Within a given layer (say, layer k) we obtain the local material basis vectors $\{\hat{\mathbf{e}}_1^k, \hat{\mathbf{e}}_2^k, \hat{\mathbf{e}}_3^k\}$ using θ_k in place of θ in Eq. (9.6.39). Once the local basis vectors are known, one may determine the components C^{ijkl} throughout the kth ply using Eq. (9.6.40).

9.6.5 The Principle of Virtual Displacements and its Discretization

The finite element model is formulated using the principle of virtual displacements. The formulation presented here is restricted to static analysis by omitting the inertial terms (see Section 9.5.3 for transient analysis). The principle of virtual displacements may be stated for the continuous problem at hand as follows: find $\Phi \in \mathcal{V}$ such that for all $\delta\Phi \in \mathcal{W}$ the following weak statement holds:

$$\mathcal{G}(\delta\Phi, \Phi) = \delta W_I(\delta\Phi, \Phi) + \delta W_E(\delta\Phi, \Phi) \equiv 0 \qquad (9.6.42)$$

where $\Phi = (\mathbf{u}, \varphi, \Psi)$ and δW_I and δW_E are the internal and external virtual works, respectively, due to actual forces in moving through virtual displacements. These quantities may be defined with respect to the undeformed configuration as

$$\delta W_I = \int_{\mathcal{B}_0} \delta\mathbf{E} : \mathbf{S}\, d\mathcal{B}_0 \qquad (9.6.43a)$$

$$\delta W_E = -\int_{\mathcal{B}_0} \rho_0\, \delta\mathbf{u} \cdot \mathbf{f}_0\, d\mathcal{B}_0 - \int_{\Gamma_\sigma} \delta\mathbf{u} \cdot \mathbf{t}_0\, ds_0 \qquad (9.6.43b)$$

where ρ_0 is the density, $\mathbf{f}_0$ is the body force, and $\mathbf{t}_0$ is the traction vector, which are expressed with respect to the reference configuration. The space of admissible configurations $\mathcal{V}$ and the vector space of admissible variations $\mathcal{W}$ are defined for the continuous problem as

$$\mathcal{V} := \left\{\ \Phi = (\underline{\mathbf{u}}, \varphi, \Psi) : \qquad \Phi \in \mathbf{H}^1(\omega) \times H^1(\omega) \times L_2(\omega), \quad \Phi = \Phi^{\mathrm{P}} \text{ on } \Gamma^{\mathrm{D}}\right\}$$
$$(9.6.44a)$$

$$\mathcal{W} := \left\{\delta\Phi = (\delta\underline{\mathbf{u}}, \delta\varphi, \delta\Psi) : \quad \delta\Phi \in \mathbf{H}^1(\omega) \times H^1(\omega) \times L_2(\omega), \quad \delta\Phi = \mathbf{0} \quad \text{on } \Gamma^{\mathrm{D}}\right\}$$
$$(9.6.44b)$$

where $\mathbf{H}^1(\Omega) = [H^1(\Omega)]^2$, and $H^1(\Omega)$ and $L_2(\Omega)$ are standard Sobolev and Lebesgue spaces on Ω, respectively (see Section 1.7). Furthermore, Γ^{D} is the part of the boundary on which Φ is specified.

In the numerical implementation, we restrict Φ and $\delta\Phi$ to their appropriate spectral/hp finite element subspaces: $\Phi_{hp} \in \mathcal{V}^{hp}$ and $\delta\Phi_{hp} \in \mathcal{W}^{hp}$. This results in the following discrete variation problem: find $\Phi_{hp} \in \mathcal{V}^{hp}$ such that

$$\mathcal{G}(\delta\Phi_{hp}, \Phi_{hp}) = \delta W_I(\delta\Phi_{hp}, \Phi_{hp}) + \delta W_E(\delta\Phi_{hp}, \Phi_{hp}) \equiv 0, \quad \forall \delta\Phi_{hp} \in \mathcal{W}^{hp}$$
$$(9.6.45)$$

Evaluation of the internal virtual work statement for the eth element of the discrete problem yields

$$\delta W_I^e = \int_{\mathcal{B}_0^e} \left(\delta\boldsymbol{\varepsilon}^{(0)} + \xi^3 \delta\boldsymbol{\varepsilon}^{(1)}\right) : \mathbf{C} : \left(\boldsymbol{\varepsilon}^{(0)} + \xi^3 \boldsymbol{\varepsilon}^{(1)}\right) d\mathcal{B}_0^e$$

$$= \int_{\hat{\Omega}^e} \int_{-1}^{+1} \left(\delta\varepsilon_{ij}^{(0)} + \xi^3 \delta\varepsilon_{ij}^{(1)}\right) C^{ijkl} \left(\varepsilon_{kl}^{(0)} + \xi^3 \varepsilon_{kl}^{(1)}\right) J \, d\xi^3 d\hat{\Omega}^e$$

$$= \int_{\hat{\Omega}^e} \left[A^{ijkl}\delta\varepsilon_{ij}^{(0)}\varepsilon_{kl}^{(0)} + B^{ijkl}\left(\delta\varepsilon_{ij}^{(0)}\varepsilon_{kl}^{(1)} + \delta\varepsilon_{ij}^{(1)}\varepsilon_{kl}^{(0)}\right) + D^{ijkl}\delta\varepsilon_{ij}^{(1)}\varepsilon_{kl}^{(1)} \right] d\hat{\Omega}^e$$

$$(9.6.46)$$

where $\int_{\hat{\Omega}^e} (\cdot) \, d\hat{\Omega}^e = \int_{-1}^{+1}\int_{-1}^{+1} (\cdot) \, d\xi^1 d\xi^2$. The quantities A^{ijkl}, B^{ijkl}, and D^{ijkl} are the contravariant components of the effective *extensional*, *extensional-bending coupling*, and *bending* fourth-order stiffness tensors, respectively, defined as

$$\left\{A^{ijkl}, B^{ijkl}, D^{ijkl}\right\} = \int_{-1}^{+1} \left\{1, \xi^3, (\xi^3)^2\right\} C^{ijkl} J \, d\xi^3 \qquad (9.6.47)$$

We note that the stiffness components have been defined such that they include the Jacobian determinant J. In the computer implementation, we perform the above integration numerically using the Gauss–Legendre quadrature rule (with 50 quadrature points taken along the thickness direction of each laminae). Therefore, no thin- or shallow-shell type approximating assumptions are imposed on either J or C^{ijkl} in the finite element formulation.

The derivation of virtual internal energy in Eq. (9.6.46) is not a trivial task. Even for isotropic materials these expressions have an extremely complex form when displacements and rotations are large. Table 9.6.1, taken from Arciniega [145], gives the number of terms of the virtual strain energy for the shell model presented here and for different other problems.

The external virtual work consists of body forces and tractions. For each element, we decompose the boundary of the shell into top $\Gamma_{\sigma,+}^e$, bottom $\Gamma_{\sigma,-}^e$, and lateral $\Gamma_{\sigma,S}^e$ surfaces. As a result, the external virtual work for a typical shell element may be expressed as

$$\delta W_E^e = -\int_{\mathcal{B}_0^e} \rho_0 \delta\mathbf{u} \cdot \mathbf{f}_0 \, d\mathcal{B}_0^e - \int_{\Gamma_{\sigma,+}^e} \delta\mathbf{u} \cdot \mathbf{t}_0^+ \, ds^+ - \int_{\Gamma_{\sigma,-}^e} \delta\mathbf{u} \cdot \mathbf{t}_0^- \, ds^-$$

$$- \int_{\Gamma_{\sigma,s}^e} \delta\mathbf{u} \cdot \mathbf{t}_0^s \, ds^s \qquad (9.6.48)$$

The traction boundary conditions along the top and bottom of the shell element may be expressed as

Table 9.6.1: Number of terms of the virtual internal energy for different physical problems.

| | | | Virtual strain energy | | |
| | | Kinematic | | | |
Theory	Application	variables	Linear	Nonlin.	Total
Beam (EBT)*	Moderate rotations	2	2	3	5
2-D Navier–Stokes equations**	––	3	10	4	14
Beam (Present)*	Finite deformations	5	13	109	122
Cylindrical shells (Sanders)	Moderate rotations	5	106	193	299
Rectangular plates (Present)	Finite deformations	7	136	2,245	2,381
Circular plates (Present)	Finite deformations	7	232	5,197	5,429
Spherical shells (Present)	Finite deformations	7	666	19,630	20,296
Hyperboloidal shells (Present)	Finite deformations	7	699	19,424	20,123

* Isotropic cases. **Newtonian fluid with constant viscosity.

$$\int_{\Gamma_{\sigma,+}^e} \delta\mathbf{u} \cdot \mathbf{t}_0^+ \, ds^+ = \int_{\hat{\Omega}^e} \sum_{k=1}^{n} \psi_k(\xi^1, \xi^2)\left(\delta\underline{\mathbf{u}}^k + \frac{h}{2}\delta\varphi^k + \frac{h}{2}\delta\Psi^k\hat{\mathbf{n}}\right) \cdot \mathbf{t}_0^+ J^+ \, d\hat{\Omega}^e$$

$$\int_{\Gamma_{\sigma,-}^e} \delta\mathbf{u} \cdot \mathbf{t}_0^- \, ds^- = \int_{\hat{\Omega}^e} \sum_{k=1}^{n} \psi_k(\xi^1, \xi^2)\left(\delta\underline{\mathbf{u}}^k - \frac{h}{2}\delta\varphi^k + \frac{h}{2}\delta\Psi^k\hat{\mathbf{n}}\right) \cdot \mathbf{t}_0^- J^- \, d\hat{\Omega}^e$$

$$(9.6.49)$$

where

$$J^+ = \|\mathbf{g}_1^+ \times \mathbf{g}_2^+\|, \qquad J^- = \|\mathbf{g}_1^- \times \mathbf{g}_2^-\| \qquad (9.6.50a)$$

$$\mathbf{g}_\alpha^+ = \mathbf{g}_\alpha(\xi^1, \xi^2, +1), \qquad \mathbf{g}_\alpha^- = \mathbf{g}_\alpha(\xi^1, \xi^2, -1) \qquad (9.6.50b)$$

9.6.6 The Spectral/*hp* Basis Functions

In this work we employ a family of quadrilateral finite elements constructed using high polynomial order spectral/*hp* basis functions. The quantity h (not to be confused with the shell thickness) in the definition of the sub-spaces $\mathcal{V}^{hp}$ and $\mathcal{W}^{hp}$ represents the average size of all the elements in a given finite element discretization. Likewise, the symbol p denotes the polynomial degree of the finite element interpolation functions associated with each element in the model. As a result, the approximate shell geometry and discrete solution may be refined by either increasing the number of elements in $\bar{\Omega}^{hp}$ (h-refinement), increasing the polynomial order p within each element $\bar{\Omega}^e$ (p-refinement) or through an appropriate and systematic combination of both h-refinement and p-refinement.

The high-order two-dimensional basis functions $\psi_i(\xi^1, \xi^2)$ adopted here are constructed from tensor products of the following one-dimensional C^0 spectral nodal interpolation functions (see Karniadakis and Sherwin [163]):

$$\varphi_j(\xi) = \frac{(\xi - 1)(\xi + 1)L'_p(\xi)}{p(p + 1)L_p(\xi_j)(\xi - \xi_j)} \tag{9.6.51}$$

where $L_p(\xi)$ is the Legendre polynomial of order p and $L'_p(\xi) = \partial L_p/\partial \xi$. The quantities ξ_j are the locations of the nodes associated with the one-dimensional interpolants and are obtained as the roots of the following expression:

$$(\xi - 1)(\xi + 1)L'_p(\xi) = 0 \quad \text{in } [-1, 1] \tag{9.6.52}$$

The nodal points $\{\xi_j\}_{j=1}^{p+1}$ found in solving Eq. (9.6.52) are known as the Gauss–Lobatto–Legendre (GLL) points. Whenever $p \leq 2$, the GLL points are equally spaced within the standard interval $[-1, +1]$. When $p > 2$, the GLL points are distributed unequally with discernable bias given to the end points of the interval. The bias associated with the spacing of the GLL points increases with p. Figure 9.6.6 contains a plot of the interpolation functions $\{\varphi_j\}_{j=1}^{p+1}$ for $p = 6$. The figure contains the interpolation functions associated with both an equal as well as a GLL spacing of the nodal points in the standard bi-unit interval. The interpolation functions constructed using equal nodal spacing clearly exhibit oscillations (often termed the *Runge effect*) near the end points of the standard interval. These oscillations become more pronounced as p is increased. The spectral interpolation functions, on the other hand, are free of the *Runge effect*. Finite element coefficient matrices constructed using spectral interpolation functions, as a result, are better conditioned than matrices formulated using elements with equally spaced nodes.

The basis functions required by the shell element developed in this work are constructed using the tensor product of one-dimensional functions φ_j:

$$\psi_i(\xi^1, \xi^2) = \varphi_j(\xi^1)\,\varphi_k(\xi^2), \quad (\xi^1, \xi^1) \in \hat{\Omega}^e \tag{9.6.53}$$

where $i = j + (k - 1)(p + 1)$ and $j, k = 1, \ldots, p + 1$. Figure 9.6.7 contains two-dimensional master elements for $p = 1, 2, 4, 8$.

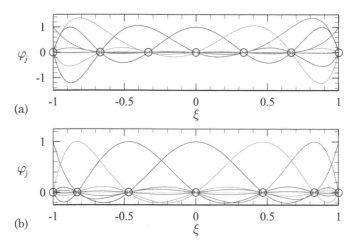

Fig. 9.6.6: High-polynomial order one-dimensional C^0 Lagrange interpolation functions. Cases shown are for $p = 6$ with: (a) equal spacing of the element nodes and (b) unequal nodal spacing associated with GLL points.

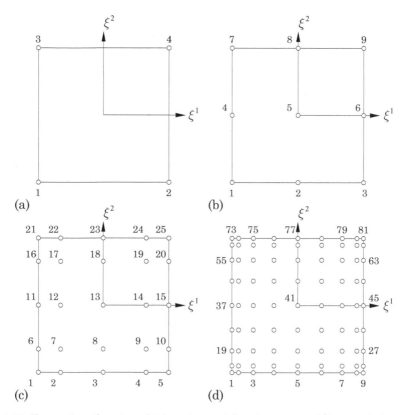

Fig. 9.6.7: Examples of various high-polynomial order spectral/hp quadrilateral master elements $\hat{\Omega}^e$: (a) a 4-node element, $p = 1$; (b) a 9-node element, $p = 2$; (c) a 25-node element, $p = 4$; and (d) an 81-node element, $p = 8$.

9.6.7   Finite Element Model and Solution of Nonlinear Equations

Here we employ Newton's method in the solution of the resulting nonlinear algebraic equations. To facilitate a numerical solution for problems involving very large deformations, we further embed the iterative Newton procedure within an incremental load stepping algorithm. For post-buckling analysis we employ a cylindrical arc-length solution procedure. In an effort to improve system memory requirements in the numerical implementation, we further adopt element-level static condensation, wherein, the interior degrees of freedom of each element are implicitly eliminated prior to global assembly of the sparse finite element coefficient matrix.

9.6.7.1   The Newton procedure

The Newton iteration proceeds as follows: given a known solution state Φ^r_{hp} we seek $\Delta\Phi^{r+1}_{hp}$ satisfying the following linearized problem:

$$\mathcal{G}(\delta\Phi_{hp}, \Phi^r_{hp}) + D\mathcal{G}(\delta\Phi_{hp}, \Phi^r_{hp})[\Delta\Phi^{r+1}_{hp}] = 0 \qquad (9.6.54)$$

where $\Delta\Phi^{r+1}_{hp} = \Phi^{r+1}_{hp} - \Phi^r_{hp}$. To simplify the present discussion, we introduce the following notation for the discrete quantities: $\hat{\Phi}^r = \Phi^r_{hp}$, $\delta\hat{\Phi} = \delta\Phi_{hp}$, and $\Delta\hat{\Phi}^{r+1} = \Delta\Phi^{r+1}_{hp}$. The discrete tangent operator $D\mathcal{G}(\delta\Phi_{hp}, \Phi^r_{hp})[\Delta\Phi^{r+1}_{hp}]$ may be evaluated within a typical element as

$$D\mathcal{G}^e(\delta\hat{\Phi}, \hat{\Phi}^r)[\Delta\hat{\Phi}^{r+1}] = D\mathcal{G}^e_G(\delta\hat{\Phi}, \hat{\Phi}^r)[\Delta\hat{\Phi}^{r+1}] + D\mathcal{G}^e_M(\delta\hat{\Phi}, \hat{\Phi}^r)[\Delta\hat{\Phi}^{r+1}] \quad (9.6.55)$$

The geometric tangent operator $D\mathcal{G}^e_G(\delta\hat{\Phi}, \hat{\Phi}^r)[\Delta\hat{\Phi}^{r+1}]$ and material tangent operator $D\mathcal{G}^e_M(\delta\hat{\Phi}, \hat{\Phi}^r)[\Delta\hat{\Phi}^{r+1}]$ are determined within the element as

$$D\mathcal{G}^e_G(\delta\hat{\Phi}, \hat{\Phi}^r)[\Delta\hat{\Phi}^{r+1}] = \int_{\mathcal{B}^e_0} \left(D\delta\varepsilon^{(0)}_{ij}[\Delta\hat{\Phi}^{r+1}] + \xi^3 D\delta\varepsilon^{(1)}_{ij}[\Delta\hat{\Phi}^{r+1}]\right) S^{ij}\, d\mathcal{B}^e_0$$

$$(9.6.56a)$$

$$= \int_{\hat{\Omega}^e} \left(D\delta\varepsilon^{(0)}_{ij}[\Delta\hat{\Phi}^{r+1}]N^{ij} + D\delta\varepsilon^{(1)}_{ij}[\Delta\hat{\Phi}^{r+1}]M^{ij}\right) d\hat{\Omega}^e$$

$$D\mathcal{G}^e_M(\delta\hat{\Phi}, \hat{\Phi}^r)[\Delta\hat{\Phi}^{r+1}] = \int_{\mathcal{B}^e_0} \left(\delta\varepsilon^{(0)}_{ij} + \xi^3 \delta\varepsilon^{(1)}_{ij}\right) C^{ijkl} \left(D\varepsilon^{(0)}_{kl}[\Delta\hat{\Phi}^{r+1}]\right. \qquad (9.6.56b)$$

$$\left. + \xi^3 D\varepsilon^{(1)}_{kl}[\Delta\hat{\Phi}^{r+1}]\right) d\mathcal{B}^e_0$$

$$= \int_{\hat{\Omega}^e} \left[\left(A^{ijkl}\delta\varepsilon^{(0)}_{ij} + B^{ijkl}\delta\varepsilon^{(1)}_{ij}\right) D\varepsilon^{(0)}_{kl}[\Delta\hat{\Phi}^{r+1}]\right.$$

$$\left. + \left(B^{ijkl}\delta\varepsilon^{(0)}_{ij} + D^{ijkl}\delta\varepsilon^{(1)}_{ij}\right) D\varepsilon^{(1)}_{kl}[\Delta\hat{\Phi}^{r+1}]\right] d\hat{\Omega}^e$$

The contravariant components of the internal stress resultants N^{ij} and M^{ij} appearing in the discrete tangent operator may be evaluated as

$$N^{ij} = \int_{-1}^{+1} S^{ij} J \, d\xi^3 = \left(A^{ijkl} \varepsilon_{kl}^{(0)} + B^{ijkl} \varepsilon_{kl}^{(1)} \right) \Big|_{\Phi_{hp} = \hat{\Phi}^r} \tag{9.6.57a}$$

$$M^{ij} = \int_{-1}^{+1} \xi^3 S^{ij} J \, d\xi^3 = \left(B^{ijkl} \varepsilon_{kl}^{(0)} + D^{ijkl} \varepsilon_{kl}^{(1)} \right) \Big|_{\Phi_{hp} = \hat{\Phi}^r} \tag{9.6.57b}$$

Upon substitution of the finite element approximations into the linearized virtual work statement, we arrive at the finite element model (i.e. a system of nonlinear algebraic equations) for the *e*th element

$$[K^e]^{(r)} \{\delta\Delta^e\}^{(r+1)} = \{F^e\}^{(r)} \tag{9.6.58}$$

where $[K^e]^{(r)}$ is the element tangent coefficient matrix at the beginning of the *r*th iteration, $\{F^e\}^{(r)}$ is the element imbalance force vector and $\{\delta\Delta^e\}^{(r+1)} = \{\Delta^e\}^{(r+1)} - \{\Delta^e\}^{(r)}$ is the incremental solution to be determined. Due to the incredible complexity of the above system of equations (there are 22,050 unique terms in the discrete tangent operator $D\mathcal{G}^e(\delta\hat{\Phi}, \hat{\Phi}^r)[\Delta\hat{\Phi}^{r+1}]$), the symbolic algebra software Maple© has been utilized in the construction of $[K^e]^{(r)}$ and $\{F^e\}^{(r)}$. We partition Eq. (9.6.58) into the following equivalent form

$$\begin{bmatrix} [^e K^{11}]^{(r)} & \cdots & [^e K^{17}]^{(r)} \\ \vdots & \ddots & \vdots \\ [^e K^{71}]^{(r)} & \cdots & [^e K^{77}]^{(r)} \end{bmatrix} \begin{Bmatrix} \{^e \delta\Delta^{(1)}\}^{(r+1)} \\ \vdots \\ \{^e \delta\Delta^{(7)}\}^{(r+1)} \end{Bmatrix} = \begin{Bmatrix} \{^e F^{(1)}\}^{(r)} \\ \vdots \\ \{^e F^{(7)}\}^{(r)} \end{Bmatrix} \tag{9.6.59}$$

where

$$^e K_{ij}^{\alpha\beta(r)} = \int_{\hat{\Omega}^e} \sum_{l=0}^{2} \sum_{m=0}^{2} \mathcal{C}_{lm}^{\alpha\beta}(\mathbf{X}(\xi^1, \xi^2), \hat{\mathbf{r}}(\xi^1, \xi^2), \Phi_{hp}^r(\xi^1, \xi^2)) \mathcal{S}_{ij}^{lm}(\xi^1, \xi^2) d\hat{\Omega}^e \tag{9.6.60a}$$

$$^e F_i^{(\alpha)(r)} = \int_{\hat{\Omega}^e} \sum_{l=0}^{2} \mathcal{F}_i^{\alpha}(\mathbf{X}(\xi^1, \xi^2), \hat{\mathbf{r}}(\xi^1, \xi^2), \Phi_{hp}^n(\xi^1, \xi^2)) \mathcal{T}_i^{l}(\xi^1, \xi^2) d\hat{\Omega}^e \tag{9.6.60b}$$

for $i, j = 1, \ldots, (p+1)^2$ and $\alpha, \beta = 1, \ldots, 7$. The functions $\mathcal{S}_{ij}^{lm}$ and $\mathcal{T}_i^{l}$ are of the form

$$\mathcal{S}_{ij}^{00} = \psi_i \psi_j, \quad \mathcal{S}_{ij}^{0m} = \psi_i \frac{\partial\psi_j}{\partial\xi^m}, \quad \mathcal{S}_{ij}^{l0} = \frac{\partial\psi_i}{\partial\xi^l}\psi_j, \quad \mathcal{S}_{ij}^{lm} = \frac{\partial\psi_i}{\partial\xi^l}\frac{\partial\psi_j}{\partial\xi^m} \tag{9.6.61a}$$

$$\mathcal{T}_i^0 = \psi_i, \quad \mathcal{T}_i^1 = \frac{\partial\psi_i}{\partial\xi^1}, \quad \mathcal{T}_i^2 = \frac{\partial\psi_i}{\partial\xi^2} \tag{9.6.61b}$$

where l and m each range from 1 to 2. The coefficients $\mathcal{C}_{lm}^{\alpha\beta}$ and $\mathcal{F}_l^{\alpha}$ (which are independent of i and j) are quite involved; in the numerical implementation, these quantities are obtained symbolically using Maple$^©$ and have then translated the resulting expressions into C^{++} code. At this point it is worth noting that interpolating Ψ and $\hat{\mathbf{n}}$ separately in the finite element approximation of ψ [refer to Eq. (9.6.19)] is crucial in ensuring that $\mathcal{C}^{\alpha\beta}_{lm}$ and $\mathcal{F}^{\alpha}_{l}$ are indeed independent of the i and j indices in $^{e}K^{\alpha\beta(r)}_{ij}$ and $^{e}F^{(\alpha)(r)}_{i}$.

The components of the element coefficient matrix and force vector are obtained numerically using the Gauss–Legendre quadrature rule, where $p + 1$ quadrature points are taken in each coordinate direction of $\hat{\Omega}^{e}$. At a given integration point $(\xi^{1}_{I}, \xi^{2}_{J}) \in \hat{\Omega}^{e}$ we evaluate numerically, based on Eq. (9.6.47), the components of A^{ijkl}, B^{ijkl}, and D^{ijkl}. Once the effective stiffnesses are known we determine $C^{\alpha\beta}_{lm}(\xi^{1}_{I}, \xi^{2}_{J})$ and $\mathcal{F}^{\alpha}_{l}(\xi^{1}_{I}, \xi^{2}_{J})$ and then apply the summation procedure of the Gauss–Legendre quadrature rule to the components of $^{e}K^{\alpha\beta(r)}_{ij}$ and $^{e}F^{(\alpha)(r)}_{i}$. Repeating this process at each quadrature point in $\hat{\Omega}^{e}$ ensures an efficient numerical implementation.

9.6.7.2 The cylindrical arc-length procedure

For a vast majority of nonlinear problems, the Newton iterative procedure is adequate to solve the assembled set of nonlinear finite element equations. However, in the numerical simulation of the post-buckling of shell structures, such a strategy may fail to trace the equilibrium path through the limit points (i.e. points where the tangent matrix is singular). For these problems an arc-length procedure is adopted, wherein a constraint equation is proposed to control the load factor associated with a given load step. For general details on the historical development of the arc-length method one may consult the works of Riks [149, 150] and Crisfield [151] (detailed explanations of the method may also be found in the textbook of Bathe [4] and Appendix 2).

In both solution schemes, we assume that the external loads are applied in increments. Then we express the discrete weak formulation, given by Eq. (9.6.42), at the current load step as

$$\{R\}^{(r+1)} = \{F^{\text{int}}\}^{(r+1)} - \lambda^{(r+1)}\{F^{\text{ext}}\} \equiv 0 \qquad (9.6.62)$$

where $\{F^{\text{int}}\}^{(r+1)}$ is a column vector obtained from the internal virtual work and $\{F^{\text{ext}}\}$ is a constant vector (independent of the load step) constructed from the externally applied virtual work. The quantity $\lambda^{(r+1)}$ is the load factor associated with the current load step. Linearizing the above expression using Newton's method yields

$$[K]^{(r)}\{\delta\Delta\}^{(r+1)} = -\{R\}^{(r)} + \delta\lambda^{(r+1)}\{F^{\text{ext}}\} \qquad (9.6.63)$$

where $\{\delta\Delta\}^{(r+1)}$ and $\delta\lambda^{(r+1)}$ are defined as

$$\{\delta\Delta\}^{(r+1)} = \{\Delta\}^{(r+1)} - \{\Delta\}^{(r)} \qquad (9.6.64a)$$

$$\delta\lambda^{(r+1)} = \lambda^{(r+1)} - \lambda^{(r)} \tag{9.6.64b}$$

In the Newton solution procedure, $\lambda^{(r+1)}$ is prescribed by the user, and hence $\delta\lambda^{(r+1)} = 0$. In the arc-length procedure, we solve for a sequence of shell configurations associated with the prescribed load parameters $\{^k\lambda\}_{k=1}^{N}$. In solving for the next configuration, the coefficient matrix $[K]^{(0)}$ and residual $\{R\}^{(0)}$ are constructed using the converged solution from the previous load step.

In the arc-length solution procedure, the following additive decomposition of the incremental solution is introduced:

$$\{\delta\Delta\}^{(r+1)} = \{\delta\bar{\Delta}\}^{(r+1)} + \delta\lambda^{(r+1)}\{\delta\tilde{\Delta}\}^{(r+1)} \tag{9.6.65}$$

Using the above expression along with Eq. (9.6.63) allows us to obtain the following two sets of linearized equations for $\{\delta\bar{\Delta}\}^{(r+1)}$ and $\{\delta\tilde{\Delta}\}^{(r+1)}$:

$$[K]^{(r)}\{\delta\bar{\Delta}\}^{(r+1)} = -\{R\}^{(r)} \tag{9.6.66a}$$

$$[K]^{(r)}\{\delta\tilde{\Delta}\}^{(r+1)} = \{F^{\text{ext}}\} \tag{9.6.66b}$$

Once the above equations have been solved (and assuming, of course, that $\delta\lambda^{(r+1)}$ is known), we may obtain $\{\delta\Delta\}^{(r+1)}$ using Eq. (9.6.65). Next we define the solution increments $\{\hat{\Delta}\}^{(r+1)}$ and $\hat{\lambda}^{(r+1)}$, between the current configurations and the configuration being sought, as

$$\{\hat{\Delta}\}^{(r+1)} = \{\Delta\}^{(r+1)} - \{\Delta\}$$

$$= \{\hat{\Delta}\}^{(r)} + \{\delta\bar{\Delta}\}^{(r+1)} + \delta\lambda^{(r+1)}\{\delta\tilde{\Delta}\}^{(r+1)} \tag{9.6.67a}$$

$$\hat{\lambda}^{(r+1)} = \hat{\lambda}^{(r)} + \delta\lambda^{(r+1)} \tag{9.6.67b}$$

With the aid of the above formulae, we can define the standard spherical arc-length constraint equation for $\delta\lambda^{(r+1)}$ as

$$\begin{aligned}\mathcal{K}^{(r+1)} &= ||\{\hat{\Delta}\}^{(r+1)}||^2 + \beta(\hat{\lambda}^{(r+1)})^2||\{F^{\text{ext}}\}||^2 - (\Delta L)^2 \\ &= a_1(\delta\lambda^{(r+1)})^2 + a_2\delta\lambda^{(r+1)} + a_3 = 0\end{aligned} \tag{9.6.68}$$

where ΔL is the arc-length, β is a scaling parameter, and $||\cdot||$ denotes the Euclidean norm. The constraint $\mathcal{K}^{(r+1)}$ is a quadratic equation in $\delta\lambda^{(r+1)}$ with coefficients a_1, a_2, and a_3 defined as

$$a_1 = ||\{\delta\tilde{\Delta}\}^{(r+1)}||^2 + \beta||\{F^{\text{ext}}\}||^2 \tag{9.6.69a}$$

$$a_2 = 2\left[\left(\{\hat{\Delta}\}^{(r)} + \{\delta\bar{\Delta}\}^{(r+1)}\right)^{\text{T}}\{\delta\tilde{\Delta}\}^{(r+1)} + \beta\hat{\lambda}^{(r)}||\{F^{\text{ext}}\}||^2\right] \tag{9.6.69b}$$

$$a_3 = ||\{\hat{\Delta}\}^{(r)} + \{\delta\bar{\Delta}\}^{(r+1)}||^2 + \beta(\hat{\lambda}^{(r)})^2||\{F^{\text{ext}}\}||^2 - (\Delta L)^2 \tag{9.6.69c}$$

The two possible solutions for the constraint equation may be expressed as

$$\delta\lambda_1^{(r+1)} = \frac{-a_2 + \sqrt{a_2^2 - 4a_1a_3}}{2a_1}, \qquad \delta\lambda_2^{(r+1)} = \frac{-a_2 - \sqrt{a_2^2 - 4a_1a_3}}{2a_1} \tag{9.6.70}$$

We select $\delta\lambda_i^{(r+1)}$ such that the inner product of $\{\hat{\Delta}\}^{(r+1)}$ with $\{\hat{\Delta}\}^{(r)}$ is positive. This ensures that we do not march backwards along the previously computed solution path. In the event that both $\delta\lambda_1^{(r+1)}$ and $\delta\lambda_2^{(r+1)}$ yield positive inner products of $^{t+\Delta t}\{\hat{\Delta}\}^{(n+1)}$ with $^{t+\Delta t}\{\hat{\Delta}\}^{(r)}$, we select $\{\delta\Delta\}^{(r+1)}$ such that $^{t+\Delta t}\{\hat{\Delta}\}^{(r+1)}$ is closest to $^{t+\Delta t}\{\hat{\Delta}\}^{(r)}$ in the Euclidean metric. For the first iteration of a given load step, we select $\{\hat{\Delta}\}^{(1)}$ such that the inner product of $\{\hat{\Delta}\}^{(1)}$ with $\{\hat{\Delta}\}$ is positive (where $\{\hat{\Delta}\}$ is the converged incremental solution from the previous load step). In the numerical implementation, we take $\beta = 0$, which results in the well-known cylindrical arc-length procedure (see Crisfield [151] for a discussion on the importance of β).

To initialize the arc-length solution method (at the first load step and the beginning of the first iteration: $r = 0$) we take $\lambda^{(0)} = 0$, *prescribe* an appropriate value for $\delta\lambda^{(1)}$ (typically we define $\{F^{\text{ext}}\}$ such that it is sufficient to take $\delta\lambda^{(1)} = 1$) and then solve Eq. (9.6.66b) for $\{\delta\tilde{\Delta}\}^{(1)}$. We then take $\{\delta\Delta\}^{(1)} = \delta\lambda^{(1)}\{\delta\tilde{\Delta}\}^{(1)}$ and define the arc-length ΔL for the subsequent nonlinear iterations as

$$^1\Delta L = \delta\lambda^{(1)}||\delta\tilde{\Delta}^{(1)}|| \tag{9.6.71}$$

To improve the efficiency of the arc-length method, the arc-length $^2\Delta L$ is adjusted for the next load step, depending on how many iterations were required to achieve nonlinear solution convergence at the immediate previous load step. Here the following formula from the literature (see Kweon and Hong [152]) is adopted:

$$^2\Delta L = {}^1\Delta L\sqrt{^2I/^1I} \tag{9.6.72}$$

where ^1I is the *actual* number of iterations required for convergence at the immediate previous load step and ^2I is the *desired* number of iterations required to satisfy the convergence criterion at the current load step. Typically we may take $4 \leq {}^2I \leq 6$, which naturally reduces the arc-length in the vicinity of limit points and increases the arc-length whenever nonlinear convergence is quickly achieved.

9.6.7.3   Element-level static condensation and assembly of elements

A major disadvantage of high-order finite elements is that the number of interior nodes increases with the polynomial order p. As a result, the coefficient matrix of the resulting global system of finite element equations will in general be more *dense* than a system of equations obtained using a low-order discretization (assuming of course that both discretizations possess the same number of degrees of freedom).

To improve the competitiveness of the shell finite elements based on higher-order spectral interpolation functions, an element-level static condensation is adopted, wherein the element-level equations are rearranged for element e [given by Eq. (9.6.59)] with respect to the element boundary $\{{}^e\delta\Delta^{(\mathcal{B})}\}^{(r+1)}$ and element interior $\{{}^e\delta\Delta^{(\mathcal{I})}\}^{(r+1)}$ solution increments as

$$
\begin{bmatrix} [{}^eK^{\mathcal{BB}}]^{(r)} & [{}^eK^{\mathcal{BI}}]^{(r)} \\ [{}^eK^{\mathcal{IB}}]^{(r)} & [{}^eK^{\mathcal{II}}]^{(r)} \end{bmatrix} \begin{Bmatrix} \{{}^e\delta\Delta^{(\mathcal{B})}\}^{(r+1)} \\ \{{}^e\delta\Delta^{(\mathcal{I})}\}^{(r+1)} \end{Bmatrix} = \begin{Bmatrix} \{{}^eF^{(\mathcal{B})}\}^{(r)} \\ \{{}^eF^{(\mathcal{I})}\}^{(r)} \end{Bmatrix}
\tag{9.6.73}
$$

The interior nodal degrees of freedom for element e do not have connectivity to other elements in the mesh and therefore they may be implicitly removed from the above system of equations to yield (note that this is possible because the forces corresponding to internal nodal degrees of freedom are known)

$$
[\bar{K}^e]^{(r)}\{{}^e\delta\Delta^{(\mathcal{B})}\}^{(r+1)} = \{\bar{F}^e\}^{(r)}
\tag{9.6.74}
$$

where the effective element coefficient matrix $[\bar{K}^e]^{(r)}$ and force vector $\{\bar{F}^e\}^{(r)}$ are of the form

$$
[\bar{K}^e]^{(r)} = [{}^eK^{\mathcal{BB}}]^{(r)} - [{}^eK^{\mathcal{BI}}]^{(r)}[{}^eK^{\mathcal{II}}]^{(r)\,-1}[{}^eK^{\mathcal{IB}}]^{(r)}
\tag{9.6.75a}
$$

$$
\{\bar{F}^e\}^{(r)} = \{{}^eF^{(\mathcal{B})}\}^{(r)} - [{}^eK^{\mathcal{BI}}]^{(r)}[{}^eK^{\mathcal{II}}]^{(r)\,-1}\{{}^eF^{(\mathcal{I})}\}^{(r)}
\tag{9.6.75b}
$$

The effective element coefficient matrix and force vector may be constructed using the dense matrix routines available in LAPACK.

The statically-condensed element-level equations are assembled into the following global system of linearized algebraic equations:

$$
[\bar{K}]^{(r)}\{\delta\Delta^{(\mathcal{B})}\}^{(r+1)} = \{\bar{F}\}^{(r)}
\tag{9.6.76}
$$

whose solution yields the degree of freedom increments for the collection of nodes residing on the element boundaries. The interior degree of freedom increments may be recovered by post computing (for each element) the solution using the following set of equations

$$
[{}^eK^{\mathcal{II}}]^{(r)}\{{}^e\delta\Delta^{(\mathcal{I})}\}^{(r+1)} = \{{}^eF^{(\mathcal{I})}\}^{(r)} - [{}^eK^{\mathcal{IB}}]^{(r)}\{{}^e\delta\Delta^{(\mathcal{B})}\}^{(r+1)}
\tag{9.6.77}
$$

We note that the present high-order spectral/hp finite element formulation is very much comparable with standard low-order shell finite element implementations in terms of system memory requirements associated with constructing and storing the sparse form of $[\bar{K}]^{(r)}$. Figure 9.6.8 shows an example of a mesh of p-level four elements before and after static condensation. In addition, the present formulation admits straightforward parallelization in the construction of the effective element matrices and also the post-computation of the interior solution increments. In addition, unlike many low-order discretizations, the present shell finite element formulation is completely displacement-based.

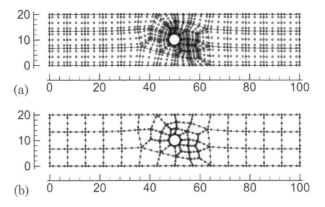

Fig. 9.6.8: A high-order spectral/hp finite element discretization ($p = 4$) of a 2-D region: (a) finite element mesh showing elements and nodes and (b) a statically condensed version of the same mesh showing the elements and nodes.

9.6.8 Numerical Examples

In this section we present numerical results for various standard shell benchmark problems. The problems have been solved using a C^{++} OpenMP-based parallel implementation of the high-order spectral/hp shell finite element formulation described in this section (for details, see the dissertation of Payette [161]). For the global solver, we utilize the UMFPACK library (a set of routines for solving sparse unsymmetric linear systems directly using the multifrontal method; see Davis and Duff [164, 165] and Davis [166, 167]). The problems discussed here have been selected from the literature to demonstrate the capabilities of the shell finite element presented in this section; they are some of the challenging finite deformation problems for elastic shells found in the literature. Several of the problems presented here were also analyzed by Arciniega and Reddy [146, 147], who used a tensor-based shell finite element model in which a given shell geometry was prescribed exactly at the quadrature points while high-order Lagrange type basis functions (with *equal* spacing of the element nodes) were utilized for the numerical solution.

We construct the finite element approximation of the undeformed midsurface geometry for each example problem by mapping the nodal positions of a conforming finite element discretization of $\bar{\omega} \subset \mathbb{R}^2$ (a closed and bounded region) into the nodal locations associated with $\bar{\Omega}^{hp} \subset \mathbb{R}^3$. The coordinates of $\mathbb{R}^2$ are denoted as (ω^1, ω^2) and unless otherwise stated we take $\bar{\omega} = [0, 1] \times [0, 1]$. The discrete mapping used to characterize the nodal coordinates of $\bar{\Omega}^{hp}$ is also employed to prescribe the nodal values for $\hat{\mathbf{n}}$ and $\hat{\mathbf{t}}$. In most examples, no units for geometrical and material parameters are specified, and the reader may interpret the results in terms of an appropriate system of units. A convergence criterion of 10^{-6} is adopted in all numerical examples.

9.6.8.1   A cantilevered plate strip under an end transverse load

As a first example, we consider the mechanical response of a cantilevered plate strip subjected to a distributed end shear load q as shown in Fig. 9.6.9, where $L = 10$, $b = 1$, and $h = 0.1$. We consider an isotropic plate and also a multi-layered composite laminate with material properties given as

$$\text{Isotropic:} \quad E = 1.2 \times 10^6, \quad \nu = 0.0 \tag{9.6.78a}$$

$$\text{Orthotropic:} \begin{cases} E_1 = 1.0 \times 10^6, & E_2 = E_3 = 0.3 \times 10^6 \\ G_{23} = 0.12 \times 10^6, & G_{13} = G_{12} = 0.15 \times 10^6 \\ \nu_{23} = 0.25, & \nu_{13} = \nu_{12} = 0.25 \end{cases} \tag{9.6.78b}$$

The isotropic case has been considered by many authors (see, for example, Parisch [168], Simo, Rifai, and Fox [106], Saleeb *et al.* [169], and Sze and his colleagues [170–172]), while a laminated composite and FGM version of the problem has been considered by Arciniega [145] and Arciniega and Reddy [146, 147].

We employ a regular finite element mesh consisting of four elements, with $p = 4$. The unit normal and unit tangent vectors are prescribed as $\hat{\mathbf{n}} = \hat{\mathbf{E}}_3$ and $\hat{\mathbf{t}} = \hat{\mathbf{E}}_1$, respectively. Figure 9.6.10 contains the computed axial and vertical deflections of the plate tip for the isotropic case. The calculated deflections are in excellent agreement with the numerical results reported by Sze, Liu, and Lo [172]. Figure 9.6.11 depicts the undeformed and various deformed mid-surface configurations of the isotropic plate strip.

In Fig. 9.6.12 plots of the transverse tip deflections versus the applied load q for four different lamination schemes are shown. We see that the stacking sequence $(90°/0°/90°)$ yields the most flexible response while the $(0°/90°/0°)$ laminate exhibits the greatest stiffness. As expected, the non-symmetric stacking sequence $(-45°/45°/-45°/45°)$ also leads to lateral deflection of the plate in the direction of the X_2 coordinate. The composite plate results compare nicely with the results reported by Arciniega [145].

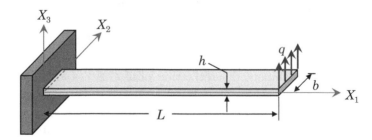

Fig. 9.6.9: A cantilevered plate strip subjected to a vertical load at the free end.

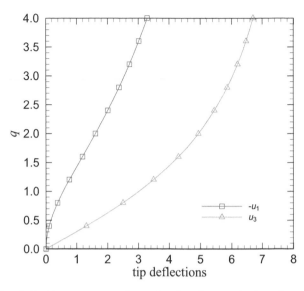

Fig. 9.6.10: Tip deflections versus shear load q for an isotropic cantilevered plate strip.

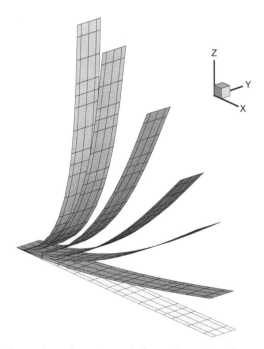

Fig. 9.6.11: Undeformed and various deformed mid-surface configurations of an isotropic cantilevered plate strip subjected at its end to a vertical load ($q = 0.4, 1.2, 2, 4, 10,$ and 20).

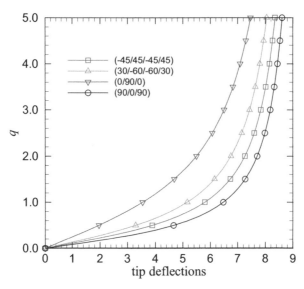

Fig. 9.6.12: Vertical tip deflections u_3 versus transverse line load q for laminated composite cantilevered plate strips under end load.

9.6.8.2   Post-buckling of a plate strip under axial compressive load

In this example we wish to determine the post-buckling behavior of an isotropic plate strip subjected to an end compressive load q as shown in Fig. 9.6.13. The material properties for the problem are the same as those employed by Massin and Al Mikdad [173], given as

$$E = 2.0 \times 10^{11}, \quad \nu = 0.3 \tag{9.6.79}$$

In addition, the geometric parameters are prescribed as $L = 0.5$, $b = 0.075$, and $h = 0.0045$. The analytical solution, first obtained by Leonhard Euler, may be found in the textbook on the linearized theory of elasticity by Timoshenko and Goodier [174].

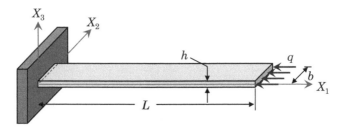

Fig. 9.6.13: A cantilevered plate strip subjected to a compressive axial load at the free end.

To instigate post-buckling behavior of the plate beyond the limit point, we introduce a perturbation technique, wherein the load is prescribed slightly out-of-plane at an angle of 1/1000 radians (see Massin and Al Mikdad [173] for a similar approach). In Fig. 9.6.14 we trace the axial and transverse deflections of the plate tip versus applied load P, where P is the net resultant force associated with the distributed axial load q. We also show in this figure the critical buckling load based on the Euler–Bernoulli beam theory, $P_{\text{cr}} = EI(\pi/2L)^2 = 1124.21$, where $I = bh^3/12$, is the second moment of area about the X_2 axis. We see that post-buckling occurs in the numerical simulation in the immediate vicinity of the critical load P_{cr}. We find that our computed tip deflections are in excellent agreement with the numerical results reported by Arciniega and Reddy [146]. In Fig. 9.6.15 we further show the undeformed and various post-buckled mid-surface configurations of the plate strip. Although the cylindrical arc-length method may be employed for this problem, the reported results have been obtained using the incremental/iterative Newton procedure. This is admissible since the applied load is non-decreasing when traversing the limit point.

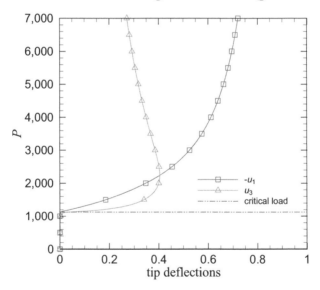

Fig. 9.6.14: Tip deflections versus compressive load P for the cantilevered plate strip (the critical Euler buckling load P_{cr} is also shown).

9.6.8.3   An annular plate with a slit under an end transverse load

We now examine a plate whose geometry cannot be exactly characterized in terms of the isoparametric map given in Eq. (9.6.1). The problem consists of a slit cantilevered annular plate as shown in Fig. 9.6.16 that is subjected to a line shear load q at its free end. We take $R_i = 6$, $R_o = 10$, and $h = 0.03$. We consider an isotropic plate and also a multi-layered composite laminate, where

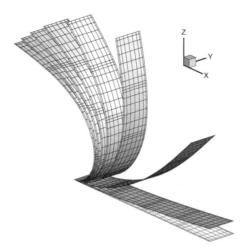

Fig. 9.6.15: Undeformed and various post-buckled deformed mid-surface configurations of the axially loaded cantilevered plate strip ($P = 1125$, 1250, 2000, 3000, 4000, 5000, 6000, and 7000).

the material properties are taken as

Isotropic: $E = 21.0 \times 10^6$, $\nu = 0.0$ (9.6.80a)

Orthotropic: $\begin{cases} E_1 = 20.0 \times 10^6,\ E_2 = E_3 = 6.0 \times 10^6 \\ G_{23} = 2.4 \times 10^6,\ G_{13} = G_{12} = 3.0 \times 10^6 \\ \nu_{23} = 0.25, \qquad \nu_{13} = \nu_{12} = 0.3 \end{cases}$ (9.6.80b)

Numerical solutions for the isotropic case may be found in the works of Büchter and Ramm [138], Brank, Korelc, and Ibrahimbegović [175], Barut, Madenci, and Tessler [176], Mohan and Kapania [177], Sansour and Kollmann [178], and Sze, Liu, and Lo [172], among others, while a laminated composite version of the problem has been solved by Arciniega and Reddy [146].

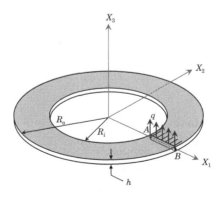

Fig. 9.6.16: An annular plate with a slit, fixed at one end and subjected to a vertical load at the other end.

We employ a finite element mesh consisting of four elements with the p-level taken as 8. The nodal coordinates of the mid-surface $\bar{\Omega}^{hp}$ are obtained using the following formula:

$$\underline{\mathbf{X}} = [R_i + (R_o - R_i)\omega^1][\cos(2\pi\omega^2)\hat{\mathbf{E}}_1 + \sin(2\pi\omega^2)\hat{\mathbf{E}}_2] \qquad (9.6.81)$$

The unit normal vector is given as $\hat{\mathbf{n}} = \hat{\mathbf{E}}_3$ and the unit tangent vector (used for the laminated composite problem) is defined at the nodes as

$$\hat{\mathbf{t}} = \cos(2\pi\omega^2)\hat{\mathbf{E}}_1 + \sin(2\pi\omega^2)\hat{\mathbf{E}}_2 \qquad (9.6.82)$$

Each numerical simulation is conducted using the incremental/iterative Newton procedure with 80 load steps.

The transverse tip deflections versus the net applied force $P = (R_o - R_i)q$ at points A and B are shown for the isotropic case in Fig. 9.6.17. The computed deflections agree very well with the tabulated displacement values reported by Sze, Liu, and Lo [172]. In Fig. 9.6.18 we trace the tip deflections at point B versus the applied load P for four distinct lamination schemes. Our computed results are found to be in excellent agreement with the displacements reported by Arciniega and Reddy [146] for each set of stacking sequences. In Fig. 9.6.19 we show the undeformed and various deformed mid-surface configurations of the isotropic plate and the $(-45^\circ/45^\circ/-45^\circ/45^\circ)$ laminated composite structure. Clearly, both structures undergo very large deformations which are qualitatively quite similar.

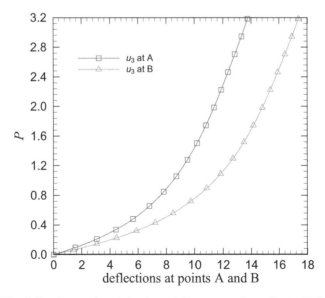

Fig. 9.6.17: Tip deflections at points A and B versus shear force P for the isotropic slit annular plate.

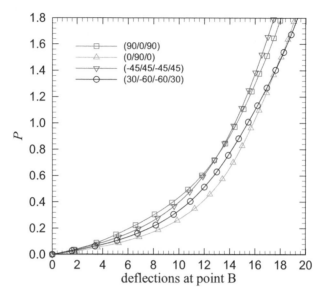

Fig. 9.6.18: Vertical tip deflections u_3 at point B versus shear force P for various laminated composite slit annular plates.

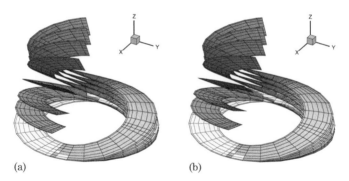

Fig. 9.6.19: Undeformed and various deformed mid-surface configurations of two annular plates: (a) an isotropic plate, where $P = 0.16$, 0.32, 0.64, 1.28, 1.92, 2.56, and 3.20 and (b) a laminated composite plate with $(-45°/45°/-45°/45°)$ stacking sequence, where $P = 0.09$, 0.18, 0.36, 0.72, 1.08, 1.44, and 1.80.

9.6.8.4   A cylindrical panel subjected to a point load

We next examine the mechanical response of various thin cylindrical roof-like panels, each subjected to a point force P as shown in Fig. 9.6.20 (see Example 8.3.7). Variants of this problem are found throughout the literature (see, for example, Parisch [168], Brank, Korelc, and Ibrahimbegović [175], Sze and Zheng [170], and Arciniega and Reddy [146, 147], among others), and they are especially popular on account of the snap-through behavior exhibited in these problems. In the present example we take $\alpha = 0.1$ rad, $a = 508$ mm, and

$R = 2,540$ mm, where R is the radius of the undeformed mid-surface. We perform a parametric study by considering the following three cases for the shell
thickness: $h = 25.4$, 12.7, and 6.35 mm. We investigate isotropic, laminated
composite and functionally graded shell configurations with material properties
given as

Isotropic: $\qquad\qquad E = 3,102.75$ N/mm^2 (or MPa), $\nu = 0.3$ (9.6.83a)

Orthotropic: $\quad\begin{cases} E_1 = 3,300 \text{ N/mm}^2, \ E_2 = E_3 = 1,100 \text{ N/mm}^2 \\ G_{23} = 440 \text{ N/mm}^2, \quad G_{13} = G_{12} = 660 \text{ N/mm}^2 \\ \nu_{23} = 0.25, \qquad\qquad \nu_{13} = \nu_{12} = 0.25 \end{cases}$

$$(9.6.83b)$$

Functionally graded: $\begin{cases} E^- = 70 \text{ GPa}, \ E^+ = 151 \text{ GPa} \\ \nu^- = 0.3, \qquad \nu^+ = 0.3 \end{cases}$ (9.6.83c)

For the laminated composite shell problems, we consider the following lamination schemes:
$(90°/0°/90°)$; $(0°/90°/0°)$; $(-45°/45°/-45°/45°)$; and $(30°/-60°/-60°/30°)$.

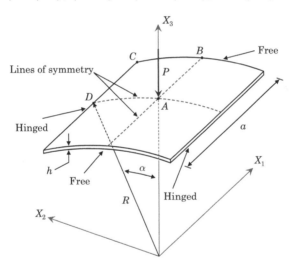

Fig. 9.6.20: A shallow cylindrical panel subjected at its center to a vertical point
load.

The finite element nodal values for the mid-surface coordinates and the unit
normal vector are obtained using the following formulae:

$$\underline{\mathbf{X}} = \frac{a}{2}\omega^1\hat{\mathbf{E}}_1 + R[\sin(\alpha\omega^2)\hat{\mathbf{E}}_2 + \cos(\alpha\omega^2)\hat{\mathbf{E}}_3] \qquad (9.6.84a)$$

$$\hat{\mathbf{n}} = \sin(\alpha\omega^2)\hat{\mathbf{E}}_2 + \cos(\alpha\omega^2)\hat{\mathbf{E}}_3 \qquad (9.6.84b)$$

where the full physical domain may be parameterized by taking $\bar{\boldsymbol{\omega}} = [-1, 1]^2$.
The unit tangent vector is prescribed as $\hat{\mathbf{t}} = \hat{\mathbf{E}}_1$. With the exception of the

angled-ply laminates $(-45°/45°/-45°/45°)$ and $(30°/-60°/-60°/30°)$, all numerical simulations are conducted using one quarter of the physical domain by taking $\bar{\omega} = [0,1]^2$ and invoking appropriate symmetry boundary conditions. We employ a uniform 2×2 mesh for the quarter model and a 4×4 discretization for the full domain using a p-level of four. Along the hinged edges, we take the nodal translations and X_1 component of the difference vector as zero.

In Fig. 9.6.21 we show the deflection of the isotropic shell at point A versus the applied load P for the case where $h = 25.4$ mm. The results, which agree strongly with those reported by Arciniega [145], have been obtained using the incremental/iterative Newton procedure. In Figs. 9.6.22 through 9.6.25 we trace the center deflections versus P for the isotropic and laminated composite panels for the cases where $h = 12.7$ and 6.35 mm. Each numerical simulation has been conducted using the cylindrical arc-length method. The results are in excellent agreement with the tabulated values given by Sze, Liu, and Lo [172]; and for the angled-ply laminates, the solutions presented by Arciniega and Reddy [146]. It is evident that decreasing the shell thickness greatly increases the complexity of the equilibrium path associated with the arc-length based numerical solution. For example, we observe from Figs. 9.6.24 and 9.6.25 that laminates $(0°/90°/0°)$, $(-45°/45°/-45°/45°)$, and $(30°/-60°/-60°/30°)$ exhibit highly nonlinear equilibrium paths when $h = 6.35$ mm.

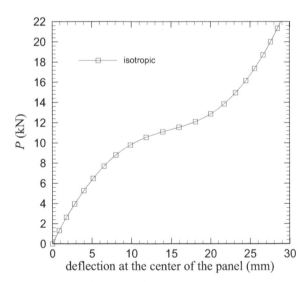

Fig. 9.6.21: Vertical deflection at the center of a shallow isotropic cylindrical panel under point load (case shown is for $h = 25.4$ mm).

Numerical results for metal-ceramic functionally graded panels, for the cases where $h = 12.7$ and 6.35 mm, are shown in Figs. 9.6.26 and 9.6.27. The metal (aluminum) is taken as the bottom material and the ceramic (zirconia) as the top constituent, with the elastic properties given in Eq. (9.6.83c). As in the isotropic and laminated composite cases, the complexity of the equilibrium paths of the functionally graded panels increases as the shell thickness h is reduced. We adopt the cylindrical arc-length procedure and vary the power-law parameter n to obtain the numerical solutions. The results shown in Figs. 9.6.26 and 9.6.27 are visually in agreement with the deflection curves provided by Arciniega and Reddy [146, 147].

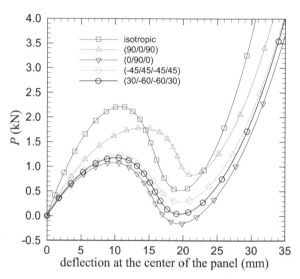

Fig. 9.6.22: Vertical deflection at the center of an isotropic and laminated composite shallow cylindrical panels under point load (cases shown are for $h = 12.7$ mm).

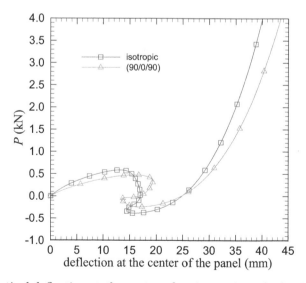

Fig. 9.6.23: Vertical deflection at the center of an isotropic and a laminated composite shallow cylindrical panel under point load (cases shown are for $h = 6.35$ mm).

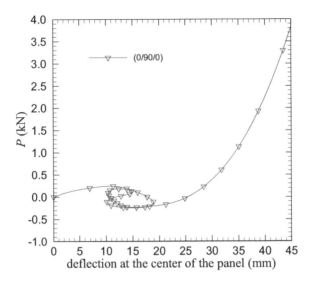

Fig. 9.6.24: Vertical deflection at the center of a laminated composite shallow cylindrical panel under point load (case shown is for $h = 6.35$ mm).

An artifact of the snap-through phenomena is the mathematical existence of multiple solution configurations for certain loading scenarios. For example, the 6.35 mm thick laminate panel with stacking sequence $(0°/90°/0°)$ possesses five equilibrium configurations for the case where $P = 0$ kN. These configurations (including the undeformed configuration) are shown in Fig. 9.6.28, from left to right and top to bottom, in the order in which they occur in traveling along the equilibrium path (shown in Fig. 9.6.24).

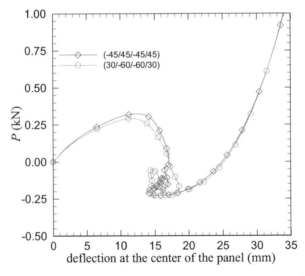

Fig. 9.6.25: Vertical deflection at the center of laminated composite shallow cylindrical panels under point load (cases shown are for $h = 6.35$ mm).

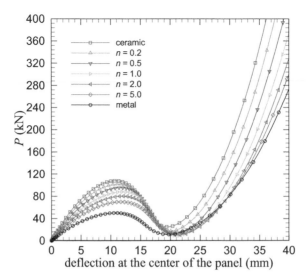

Fig. 9.6.26: Vertical deflection at the center of functionally graded metal-ceramic shallow cylindrical panels under point load (cases shown are for $h = 12.7$ mm).

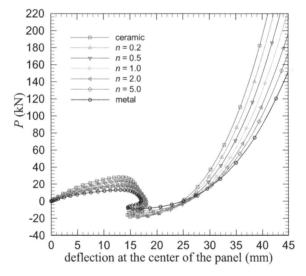

Fig. 9.6.27: Vertical deflection at the center of functionally graded metal-ceramic shallow cylindrical panels under point load (cases shown are for $h = 6.35$ mm).

9.6.8.5   Pull-out of an open-ended cylindrical shell

In this example we consider the mechanical deformation of an open-ended cylinder subjected to two pull-out point forces P, as shown in Fig. 9.6.29. Unlike the previous example, in this problem we apply the loads such that the shell undergoes very large displacements and rotations. As a result, this problem

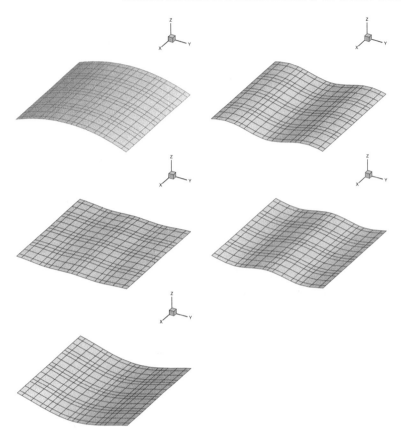

Fig. 9.6.28: Undeformed and various deformed mid-surface configurations of the $(0°/90°/0°)$ stacking sequence laminated composite shallow cylindrical panel (cases shown are for $h = 6.35$ mm and $P = 0$ kN). The vertical component of each mid-surface configuration has been scaled by a factor of 4.

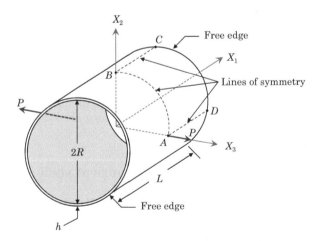

Fig. 9.6.29: An open-ended cylindrical shell subjected to two point loads.

constitutes a severe test of shell finite element formulations and has been addressed by Brank, Korelc, and Ibrahimbegović [175], Mohan and Kapania [177], Sansour and Kollmann [178], Sze, Chan, and Pian [171], Sze, Liu, and Lo [172], and Arciniega and Reddy [146], among others. The isotropic material properties are taken as

$$E = 10.5 \times 10^6, \quad \nu = 0.3125 \tag{9.6.85}$$

The geometric parameters are taken as: $L = 10.35$, $h = 0.094$, and $R = 4.953$; here we have taken R as the radius of the *undeformed* mid-surface as opposed to the radius of the *inner* surface of the shell.

Symmetry in the geometry, material properties, and loading allow us to model only an octant of the actual open-ended cylinder. For the numerical model we employ a regular 2×2 mesh with $p = 8$ in the shell octant containing points A, B, C, and D. The incremental/iterative Newton procedure is adopted using a total of 80 load steps.

The radial deflections versus the net applied pulling force P are shown in Fig. 9.6.30 for points A, B, and C. The computed deflections are in excellent agreement with results of Sze, Liu, and Lo [172] and Arciniega and Reddy [146]. The mechanical response of the shell is interesting in that the deformation is initially bending dominated; however, membrane forces clearly play an increasingly significant role as the load is intensified, resulting in a pronounced overall stiffening of the structure. Figure 9.6.31 contains the undeformed and various deformed mid-surface configurations for the problem. The overall deflections and rotations are clearly quite large, especially for the final shell configuration (i.e. the case where $P = 40,000$).

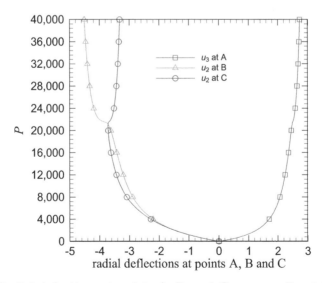

Fig. 9.6.30: Radial deflections at points A, B, and C versus pull-out force P for the open-ended cylindrical shell.

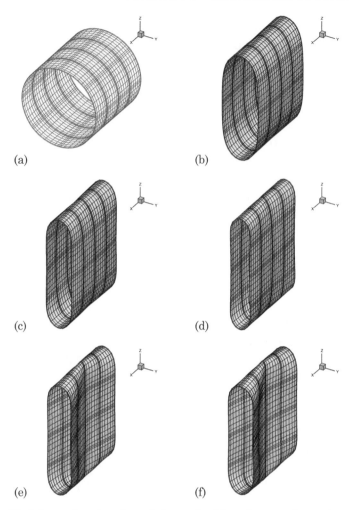

Fig. 9.6.31: Undeformed and various deformed mid-surface configurations of the open-ended cylindrical shell: (a) undeformed configuration, (b) $P = 5{,}000$, (c) $P = 10{,}000$, (d) $P = 20{,}000$, (e) $P = 30{,}000$, and (f) $P = 40{,}000$.

9.6.8.6   A pinched half-cylindrical shell

In this example we consider a half-cylindrical shell subjected to a single point force P as shown in Fig. 9.6.32. Numerical solutions for this problem may be found in the works of Brank, Korelc, and Ibrahimbegović [175], Mohan and Kapania [177], Sansour and Kollmann [178], Sze and his colleagues [170–172], and Arciniega and Reddy [146], among others. We employ the following material properties for the isotropic and laminated composite versions of the problem:

$$\text{Isotropic:} \quad E = 2.0685 \times 10^{7}, \quad \nu = 0.3 \qquad (9.6.86a)$$

$$\text{Orthotropic:} \begin{cases} E_1 = 2,068.5, \qquad E_2 = E_3 = 517.125 \\ G_{23} = 198.8942, G_{13} = G_{12} = 795.6 \\ \nu_{23} = 0.3, \qquad\quad \nu_{13} = \nu_{12} = 0.3 \end{cases} \qquad (9.6.86b)$$

The geometric parameters are taken as $L = 3.048$, $R = 1.016$, and $h = 0.03$ for the isotropic shell and $L = 304.8$, $R = 101.6$, and $h = 3.0$ for the orthotropic shell (where R is the radius of the mid-surface).

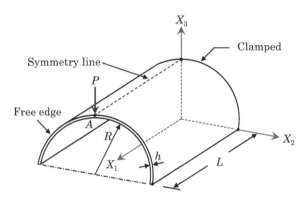

Fig. 9.6.32: A half-cylindrical shell subjected to a single point load.

As in the previous example, we exploit symmetry of the problem by performing the finite element simulations using half of the physical domain of the shell (see the line of symmetry shown in Fig. 9.6.32); a regular 4×4 mesh is adopted for each simulation with $p = 8$. For the support boundary conditions along the bottom longitudinal edges, we take the vertical deflection and X_3 component of the difference vector as zero. For the laminated composite simulations, with stacking sequences given as: $(90°/0°/90°)$ and $(0°/90°/0°)$, we prescribe the unit tangent vector as $\hat{\mathbf{t}} = \hat{\mathbf{E}}_1$.

In Figs. 9.6.33 and 9.6.34 we trace the vertical displacements at point A of the isotropic and laminated composite cylinders. The cylindrical arc-length procedure has been used in each numerical simulation to smoothly traverse the limit points. We find that the computed displacements agree well with the results reported by Arciniega and Reddy [146]. Finally, in Fig. 9.6.35 we show the undeformed and various deformed mid-surface configurations of the pinched isotropic half-cylinder.

9.6.8.7   A pinched cylinder with rigid diaphragms

Consider the isotropic, linearly elastic, pinched circular cylinder problem of Example 8.3.5 (see Kreja, Schmidt, and Reddy [102] and Cho and Roh [120]). The cylinder is restrained radially by diaphragms at its ends and loaded by two equal opposing forces (where the applied loads act on the shells mid-surface),

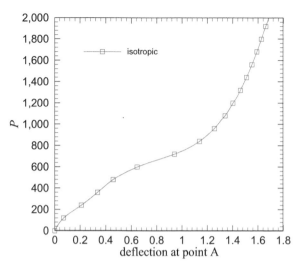

Fig. 9.6.33: Vertical deflection at point A of an isotropic half-cylindrical shell under point load.

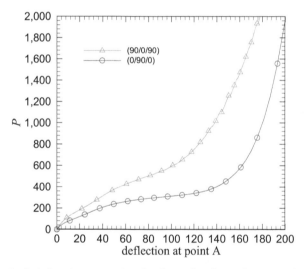

Fig. 9.6.34: Vertical deflection at point A of two laminated composite half-cylindrical shells under point load.

as shown in Fig. 9.6.36(a). The geometric and material data used are the same as given in Eq. (8.3.17). The computational domain (i.e. one octant of the cylinder) is shown in Fig. 9.6.36(b) and the deformed configuration for $P = 1.0$ ($P/4$ is used in the quadrant) is shown in Fig. 9.6.36(c). The numerical results obtained using the developed shell element are in excellent agreement with the most accurate numerical results found in the literature. In fact, the shell element is able to accurately predict the displacements along sections AB and

DC of the pinched cylinder, as shown in Fig. 9.6.36. This is by far the most challenging convergence study for this problem, a result that is rarely reported in the literature.

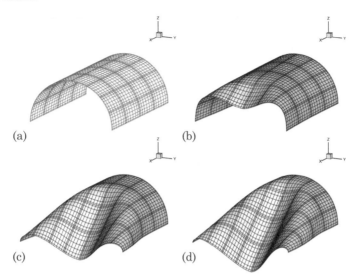

Fig. 9.6.35: Undeformed and various deformed mid-surface configurations of the isotropic pinched half-cylindrical shell: (a) undeformed configuration, (b) $P = 600$, (c) $P = 1,200$ and (d) $P = 2,000$.

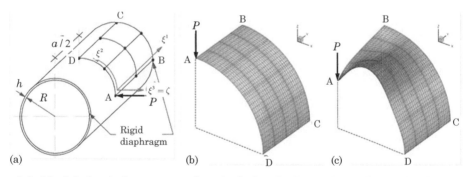

Fig. 9.6.36: Mechanical response of a pinched cylinder with rigid end diaphragms. (a) Geometry of pinched cylinder. (b) Finite element discretization of one octant of the undeformed mid-surface of the cylinder. (c) Finite element solution of deformed mid-surface of the cylinder ($P = 1.0$; magnified by a factor of 5×10^6).

9.6.8.8 A pinched hemisphere with an $18°$ hole

Consider a pinched isotropic hemisphere with an $18°$ circular cutout. This problem is widely recognized as one of the most severe shell benchmark problems involving finite deformations and has been addressed by many researchers

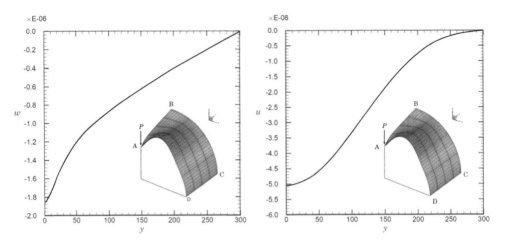

Fig. 9.6.37: Mechanical response of a pinched cylinder with rigid end diaphragms. Left figure shows the vertical displacement w along section AB of the cylinder and the right figure shows the radial displacement profile u along section DC of the cylinder.

(see, for example, Simo, Rifai, and Fox [106], Büchter and Ramm [138], Mohan and Kapania [177], Sansour and Kollmann [178], Jiang and Chernuka [179], and Sze and his colleagues [170–172], among others). The computational domain (i.e. one quarter of the hemisphere) is shown in Fig. 9.6.38. The external loads for the problem consist of four alternating radial point forces $0.5P$, prescribed along the equator at $90°$ intervals. The mid-surface radius and shell thickness are taken as $R = 10.0$ and $h = 0.04$, respectively. The material properties are prescribed as

$$E = 6.825 \times 10^7, \quad \nu = 0.3 \tag{9.6.87}$$

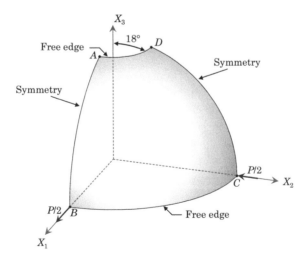

Fig. 9.6.38: A pinched hemisphere with an $18°$ hole (the computational domain shown above is one quarter of the physical domain of the shell).

For the finite element mesh we employ a regular 8×8 discretization with $p = 4$. The finite element nodes on the mid-surface $\bar{\Omega}^{hp}$ are obtained using the following formula

$$
\underline{\mathbf{X}} = R \Big\{ \sin[\alpha + (\pi/2 - \alpha)\omega^1][\cos(\pi\omega^2/2)\hat{\mathbf{E}}_1 + \sin(\pi\omega^2/2)\hat{\mathbf{E}}_2]
$$
$$
+ \cos[\alpha + (\pi/2 - \alpha)\omega^1]\hat{\mathbf{E}}_3 \Big\} \tag{9.6.88}
$$

where $\alpha = 18° = \pi/10$ rad. The incremental/iterative Newton method is used in the solution procedure with 80 load steps and $P_{\max}$ is taken as 400. In addition to the symmetry boundary conditions, we also require the X_3 component of the displacement of the node located at point B to be zero.

Figure 9.6.39 shows the radial deflections at points B and C versus the applied pinching force P. Our reported deflections compare quite well with the numerical results tabulated by Sze, Liu, and Lo [172]. In Fig. 9.6.40 we show the undeformed and three deformed mid-surface configurations of the pinched hemisphere.

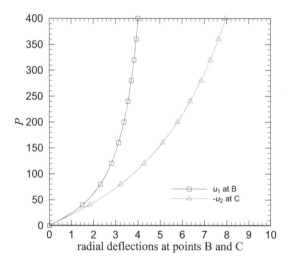

Fig. 9.6.39: Radial deflections at points B and C of the pinched hemisphere.

9.6.8.9   A pinched composite hyperboloidal shell

As a final numerical example, we consider the finite deformation of a laminated composite hyperboloidal shell that is loaded by four alternating radial point forces P. The computational domain (i.e. one octant of the actual hyperboloid) is shown in Fig. 9.6.41. This challenging benchmark, originally proposed by

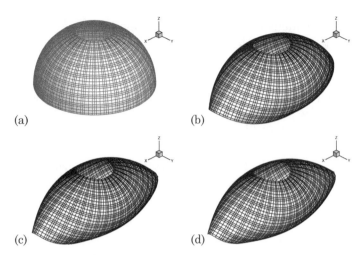

Fig. 9.6.40: Undeformed and various deformed mid-surface configurations of the pinched hemispherical shell: (a) undeformed configuration, (b) $P = 150$, (c) $P = 300$, and (d) $P = 400$.

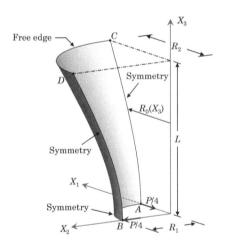

Fig. 9.6.41: A pinched laminated composite hyperboloidal shell (the computational domain shown above is one octant of the physical domain of the shell).

Başar, Ding, and Schultz [113], was designed to test the capabilities of shell elements in handling geometrically complex shell structures undergoing very large displacements and rotations. The problem has been considered by Wagner and Gruttmann [180], Balah and Al-Ghamedy [114], and by Arciniega and Reddy [146]. The orthotropic material properties for each lamina are taken as

$$
\begin{aligned}
E_1 &= 40.0 \times 10^6, & E_2 = E_3 &= 1.0 \times 10^6 \\
G_{23} &= 0.6 \times 10^6, & G_{13} = G_{12} &= 0.6 \times 10^6 \\
\nu_{23} &= 0.25, & \nu_{13} = \nu_{12} &= 0.25
\end{aligned}
\tag{9.6.89}
$$

We employ three finite element discretizations of the computational domain (see Fig. 9.6.42) including: a structured 4×4 mesh, an unstructured 4×4 mesh and a structured 5×5 mesh; where in all cases p is taken as eight. The unstructured mesh is utilized to showcase the ability of the proposed shell element to accurately solve nontrivial laminated composite shell problems using arbitrarily curved elements. Each mesh is generated by mapping the nodal coordinates of an appropriate conforming discretization of $\bar{\omega} = [0,1]^2$ onto the finite element approximation of the mid-surface $\bar{\Omega}^{hp}$ of the composite hyperboloid using the following formula

$$\underline{\mathbf{X}} = R_0(\omega^2) \left[\cos(\pi\omega^1/2)\hat{\mathbf{E}}_1 + \sin(\pi\omega^1/2)\hat{\mathbf{E}}_2 \right] + L\omega^2\hat{\mathbf{E}}_3 \qquad (9.6.90)$$

where $R_0(\omega^2) = R_1\sqrt{1 + (L\omega^2/C)^2}$. The geometric parameters are taken as $R_1 = 7.5$, $C = 20\sqrt{3}$, $L = 20.0$, and $h = 0.04$. The unit normal and tangent vectors are defined at the finite element nodes using the following expressions:

$$\hat{\mathbf{n}} = \frac{\partial\underline{\mathbf{X}}/\partial\omega^1 \times \partial\underline{\mathbf{X}}/\partial\omega^2}{||\partial\underline{\mathbf{X}}/\partial\omega^1 \times \partial\underline{\mathbf{X}}/\partial\omega^2||} \qquad (9.6.91\text{a})$$

$$\hat{\mathbf{t}} = -\sin(\pi\omega^1/2)\hat{\mathbf{E}}_1 + \cos(\pi\omega^1/2)\hat{\mathbf{E}}_2 \qquad (9.6.91\text{b})$$

In Figs. 9.6.43 and 9.6.44 we show various displacement components versus the applied load P at points A, B, C, and D of the hyperboloidal shell for the composite lamination schemes: $(0°/90°/0°)$ and $(90°/0°/90°)$. The computed displacements for the stacking sequence $(0°/90°/0°)$, obtained using both structured and unstructured meshes, are in excellent agreement with the results of Başar, Ding, and Schultz [113] and Arciniega and Reddy [146]. The displacements calculated for the laminate $(90°/0°/90°)$ [obtained using the regular 5×5

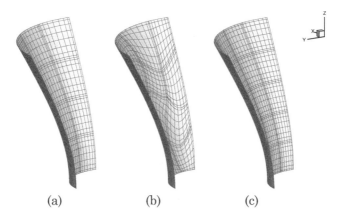

(a) (b) (c)

Fig. 9.6.42: Finite element discretizations of the composite hyperboloid, where the p-level is 8: (a) a 4×4 structured discretization, (b) a 4×4 unstructured discretization, and (c) a 5×5 structured discretization.

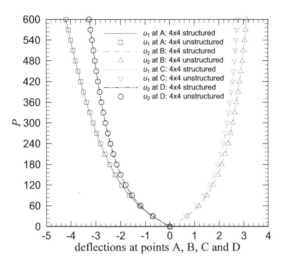

Fig. 9.6.43: Deflections at points A, B, C, and D of the pinched laminated $(0°/90°/0°)$ composite hyperboloidal shell.

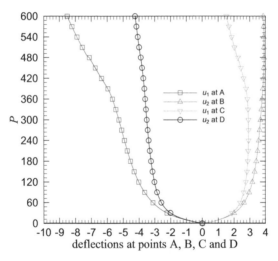

Fig. 9.6.44: Deflections at points A, B, C, and D of the pinched laminated $(90°/0°/90°)$ composite hyperboloidal shell.

discretization shown in Fig. 9.6.43(c)], however, are greater than the values reported by Başar, Ding, and Schultz [113] but are also somewhat less than the results obtained by Arciniega and Reddy [146]. In Fig. 9.6.45 we show various mid-surface configurations of the hyperboloid for each composite laminate. All numerical results have been obtained via the incremental/iterative Newton procedure using 120 load steps.

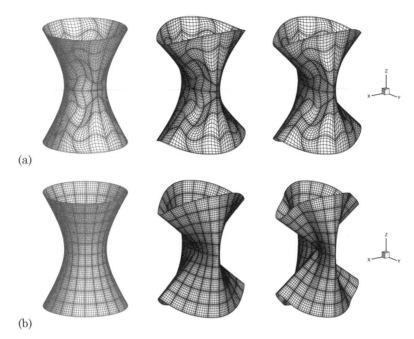

Fig. 9.6.45: Undeformed and various deformed mid-surface configurations of two pinched laminated composite hyperboloidal shells. (a) $(0°/90°/0°)$: undeformed configuration, $P = 250$, and $P = 500$ (from left to right). (b) $(90°/0°/90°)$: undeformed configuration, $P = 250$, and $P = 500$ (from left to right).

9.7 Summary

Nonlinear continuum finite element formulation of elastic solids is discussed, and updated and total Lagrangian finite element formulations are presented using the principle of virtual displacements. As special cases, finite element models of plane elastic solids and shells are presented. Several examples of application of the continuum plane elasticity and shell elements are presented.

Lastly, the refined seven-parameter nonlinear shell element developed in [161, 162] is discussed. The element: (a) accounts for thickness stretch, (b) uses three-dimensional constitutive relations, and (c) employs high-order spectral/hp approximations. The element uses a modified first-order shell theory kinematics that involves a seven-parameter expansion of the displacement field. The seventh parameter was included to allow for the use of fully three-dimensional constitutive equations in the numerical implementation. The high-order spectral/hp continuum shell finite element can be used for the numerical simulation of the *fully* finite deformation elastic response of isotropic, laminated composite, and functionally graded elastic shell structures. Through the numerical simulation of several benchmark problems, it is shown that the shell element is insensitive to all forms of numerical locking and severe geometric distortions. Additional examples, especially applications to composite and functionally graded plates and shells, can be found in [145–147, 161, 162]. Extensions of the element to include inelastic material behavior and damage prediction are awaiting attention.

Problems

9.1 Investigate the behavior of the 4-node 2-D continuum element based on the updated Lagrangian formulation by analyzing the cantilevered plate problem in Fig. 9.4.1 with 10×2 and 20×4 meshes.

9.2 Use the updated Lagrangian formulation to analyze the cantilevered plate problem in Fig. 9.4.1, except that the load is replaced by an end shear load $-\tau_0 h = -q_0$ (lb/in), as shown in Fig. P9.2. Use 5×1Q9 mesh, an increment of $\Delta q_0 = 500$ lb/in (or $f_0 = 50$ lb/in.), and 32 load steps with convergence tolerance of $\epsilon = 10^{-3}$. Plot the load versus vertical deflection at node 22 and load versus Cauchy stresses σ_{xx} and σ_{xy} at the Gauss point nearest to the top left end.

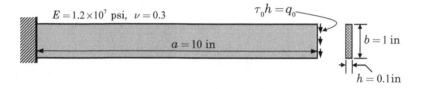

Fig. P9.2

9.3 Repeat Problem 9.2 using the total Lagrangian formulation.

9.4 Investigate the behavior of the 4-node 2-D continuum element based on the updated Lagrangian formulation by analyzing Problem 9.2 with 10×2 and 20×4 meshes.

9.5 Write a computer subroutine to calculate stresses in the plane elastic continuum element by updated and total Lagrangian formulations.

Weak-Form Finite Element Models of Flows of Viscous Incompressible Fluids

I have steadily endeavoured to keep my mind free so as to give up any hypothesis, however much beloved (and I cannot resist forming one on every subject), as soon as the facts are shown to be opposed to it. ———— Charles Darwin

10.1 Introduction

Flows of viscous incompressible fluids can be classified into two major types: a smooth, orderly motion is called *laminar flow*, and a random fluctuating motion is called *turbulent flow*. Flows of viscous fluids can be characterized by a non-dimensional parameter known as the *Reynolds number*, $Re = \rho V L/\mu$, which is defined as the ratio of inertial forces ρV^2 to viscous forces $\mu V/L$. Here ρ denotes the density of the fluid, μ the fluid viscosity, V the characteristic flow velocity, and L is a characteristic dimension of the flow region. High viscosity fluids and/or small velocities produce relatively small Reynolds numbers and a laminar flow. The case of $Re << 1$ corresponds to the flow (called *Stokes flow*) in which the inertial effects are small compared to the viscous effects and are therefore neglected. The flow of less viscous fluids and/or higher velocities lead to higher Reynolds numbers and a turbulent flow.

Tremendous progress has been achieved over the last few decades in the field of computational fluid dynamics. The advent of the digital computer and, in particular, parallel processing has made it possible to numerically simulate complex flow patterns that could only have been investigated using experimental procedures and dimensional analyses. Much of the success and breakthroughs in the numerical simulation of the Navier–Stokes equations for incompressible fluids have come in the context of low-order finite difference and finite volume technologies. Although the finite element method has become the dominant method of choice in the numerical analysis of solids, it has yet to receive such widespread acceptance when applied to fluid flow problems. It is well known, however, that the finite element method offers many advantages over the finite difference and finite volume methods. In particular, the finite element method can naturally deal with complex regions and complicated boundary conditions, and it possesses a rich mathematical foundation. As a result, there has been a renewed interest in recent years in developing efficient and accurate finite element models of the Navier–Stokes equations governing flows of viscous incompressible fluids.

Generally, whenever any weak formulation is equivalent to the problem of minimizing an unconstrained convex quadratic functional, the finite element model inherits the following desirable mathematical and computational properties:

1. The numerical solution becomes an orthogonal projection of the exact solution onto the trial space of a given conforming finite element approximation. As a result, the numerical solution represents the "best approximation" of the exact solution in the trial space (as measured by a well defined *energy* norm).

2. No restrictive compatibility requirements, such as the discrete inf-sup condition, ever arise that must be additionally satisfied by the discrete conforming function spaces of the various dependent variables.

3. The resulting linear algebraic system of global finite element equations are always symmetric and positive-definite – a property that may be exploited by both direct as well as iterative solvers.

The vast majority of finite element models for fluids are based on weak-form Galerkin formulations. However, weak-form Galerkin formulations of the Navier–Stokes equations are not equivalent to minimization of unconstrained convex quadratic functionals. Therefore, the approximations used for velocity field and pressure must satisfy ceratin restrictive conditions, as will be discussed in the forthcoming sections.

In the present chapter we develop weak-form Galerkin finite element models using basic physical variables, namely, velocities and pressure. Both Newtonian and non-Newtonian (power-law) fluids are considered here. Computer implementation of the finite element models developed herein is also discussed, and some representative examples of applications of the finite element models are presented.

10.2   Governing Equations

10.2.1   Introduction

Here we review the governing equations of a continuous medium based on the Eulerian description, as presented in Chapter 2 (also see Reddy [1]). The notation used here is the same as that used in Chapter 2. The vector form as well as the rectangular Cartesian form of the equations are presented. The summation convention (of summing repeated indices over their ranges) is used.

10.2.2 Equation of Mass Continuity

Application of the principle of conservation of mass to an element of the region (called control volume) results in the following equation, known as the *continuity equation*:

$$\frac{\partial \rho}{\partial t} + \nabla \cdot (\rho \mathbf{v}) = 0 \tag{10.2.1}$$

where ρ is the density (kg/m^3) of the medium, $\mathbf{v}$ is the velocity vector (m/s), and ∇ is the vector differential operator. Introducing the *material derivative* or *Eulerian derivative* operator d/dt

$$\frac{d}{dt} = \frac{\partial}{\partial t} + \mathbf{v} \cdot \nabla, \qquad \frac{d}{dt} = \frac{\partial}{\partial t} + v_j \frac{\partial}{\partial x_j} \tag{10.2.2}$$

Eq. (10.2.1) can be expressed in the alternative form

$$\frac{d\rho}{dt} + \rho \nabla \cdot \mathbf{v} = 0, \qquad \frac{d\rho}{dt} + \rho \frac{\partial v_i}{\partial x_i} = 0 \tag{10.2.3}$$

For steady-state conditions, the continuity equation becomes

$$\nabla \cdot (\rho \mathbf{v}) = 0, \qquad \frac{\partial}{\partial x_i} (\rho v_i) = 0 \tag{10.2.4}$$

When the density changes following a fluid particle are negligible, the continuum is termed *incompressible* and we have $d\rho/dt = 0$. The continuity equation, Eq. (10.2.3), then becomes

$$\nabla \cdot \mathbf{v} = 0, \qquad \frac{\partial v_i}{\partial x_i} = 0 \tag{10.2.5}$$

which is often referred to as the *incompressibility condition*. The condition in Eq. (10.2.5) expresses the fact that the volume change is zero for an incompressible fluid during its deformation.

10.2.3 Equations of Motion

The principle of conservation of linear momentum (or Newton's Second Law of Motion) gives rise to the equations of motion

$$\rho \frac{d\mathbf{v}}{dt} = \nabla \cdot \boldsymbol{\sigma}^{\mathrm{T}} + \rho \mathbf{f}, \qquad \rho \frac{dv_i}{dt} = \frac{\partial \sigma_{ji}}{\partial x_j} + \rho f_i \tag{10.2.6}$$

where $\boldsymbol{\sigma}$ is the Cauchy stress tensor (N/m^2) and $\mathbf{f}$ is the body force vector, measured per unit mass. Equation (10.2.6) is known as the Navier–Stokes equations.

The principle of conservation of angular momentum, in the absence of distributed couples, leads to the symmetry of the stress tensor:

$$\boldsymbol{\sigma} = (\boldsymbol{\sigma})^{\mathrm{T}}, \qquad \sigma_{ij} = \sigma_{ji} \tag{10.2.7}$$

where the superscript T denotes the transpose of the enclosed quantity.

10.2.4   Energy Equation

The principle of conservation of energy (or the first law of thermodynamics) as applied to an incompressible fluid results in the equation

$$\rho c_v \frac{dT}{dt} = -\boldsymbol{\nabla} \cdot \mathbf{q} + Q + \Phi, \quad \rho c_v \frac{dT}{dt} = -\frac{\partial q_i}{\partial x_i} + Q + \Phi \qquad (10.2.8)$$

where T denotes the temperature (°C), $\mathbf{q}$ is the heat flux vector (W/m^2), Q is the internal heat generation (W/m^3), Φ is the viscous dissipation function,

$$\Phi = \boldsymbol{\tau} : \mathbf{d} = \tau_{ij} \, d_{ij} \qquad (10.2.9)$$

and c_v is the specific heat [J/(kg ·°C)] at constant volume. In Eq. (10.2.9), $\boldsymbol{\tau}$ denotes the viscous part of the Cauchy stress tensor $\boldsymbol{\sigma}$ and $\mathbf{d}$ is the symmetric part of the velocity gradient tensor as defined in Eq. (2.4.29). Other types of internal heat generation may arise from other physical processes such as chemical reactions and Joule heating.

10.2.5   Constitutive Equations

For most of this study we assume that the fluid under consideration to be Newtonian (i.e. the constitutive relations are linear). Non-Newtonian fluids will be considered in Section 10.9. Further, the fluids are assumed to be incompressible, and the flow is laminar. For flows involving buoyancy forces, an extended form of the Boussinesq approximation may be invoked, which allows the density ρ to vary with temperature T according to the relation

$$\rho = \rho_0[1 - \beta(T - T_0)] \qquad (10.2.10)$$

where β is the coefficient of thermal expansion (m/m/°C) and the subscript zero indicates a reference condition. The variation of density as given in Eq. (10.2.10) is permitted only in the description of the body force; the density in all other situations is assumed to be that of the reference state, $\rho = \rho_0$.

For viscous incompressible fluids the total stress $\boldsymbol{\sigma}$ can be decomposed into hydrostatic and viscous parts:

$$\boldsymbol{\sigma} = \boldsymbol{\tau} + (-P)\mathbf{I}, \quad \sigma_{ij} = \tau_{ij} - P \, \delta_{ij} \qquad (10.2.11)$$

where P is the hydrostatic pressure, $\mathbf{I}$ is the unit tensor, and $\boldsymbol{\tau}$ is the viscous stress tensor. For Newtonian fluids, the viscous stress tensor is related to the symmetric part of the velocity gradient tensor, $\mathbf{d}$, by

$$\boldsymbol{\tau} = \mathbf{C} : \mathbf{d}, \quad \tau_{ij} = C_{ijk\ell} \, d_{k\ell} \qquad (10.2.12)$$

where $\mathbf{C}$ is the fourth-order tensor of fluid properties. The strain rate tensor $\mathbf{d}$ is defined by

$$\mathbf{d} = \frac{1}{2}\left[(\boldsymbol{\nabla}\mathbf{v}) + (\boldsymbol{\nabla}\mathbf{v})^{\mathrm{T}}\right], \quad d_{ij} = \frac{1}{2}\left(\frac{\partial v_i}{\partial x_j} + \frac{\partial v_j}{\partial x_i}\right); \quad d_{ii} = \frac{\partial v_i}{\partial x_i} \qquad (10.2.13)$$

For an isotropic fluid (i.e. whose material properties are independent of direction), the fourth-order tensor $\mathbf{C}$ can be expressed in terms of two constants, λ and μ, called the Lamé constants, and Eq. (10.2.12) takes the form

$$\boldsymbol{\tau} = \lambda(\operatorname{tr}\mathbf{d})\mathbf{I} + 2\mu\mathbf{d}, \quad \tau_{ij} = \lambda d_{kk}\delta_{ij} + 2\mu\, d_{ij} \tag{10.2.14}$$

where $(\operatorname{tr}\mathbf{d})$ denotes the *trace* (or sum of the diagonal elements) of the matrix representing the second-order tensor $\mathbf{d}$. For an incompressible fluid, we have $\operatorname{tr}\mathbf{d} = d_{ii} = 0$ and Eq. (10.2.14) becomes

$$\boldsymbol{\tau} = 2\mu\mathbf{d}, \quad \tau_{ij} = 2\mu\, d_{ij} \tag{10.2.15}$$

The Fourier heat conduction law states that the heat flux is proportional to the gradient of temperature

$$\mathbf{q} = -\mathbf{k}\cdot\boldsymbol{\nabla}T, \quad q_i = -k_{ij}\frac{\partial T}{\partial x_j} \tag{10.2.16}$$

where $\mathbf{k}$ denotes the conductivity tensor of order two. The negative sign in Eq. (10.2.16) indicates that heat flows from high-temperature regions to low-temperature regions. For an isotropic medium, $\mathbf{k}$ is of the form

$$\mathbf{k} = k\mathbf{I} = k\,\delta_{ij}\,\hat{\mathbf{e}}_i\,\hat{\mathbf{e}}_j \tag{10.2.17}$$

where k denotes the thermal conductivity [W/(m·°C)] of the medium.

The material coefficients, μ, c_v, β, and $\mathbf{k}$ are generally functions of position and temperature. Conductivity tensor is a symmetric, second-order tensor (i.e. $\mathbf{k}^{\mathrm{T}} = \mathbf{k}$). The volumetric heat source for the fluid and/or solid may be a function of temperature, time, and spatial location. In developing the finite element models, the dependence of the material properties on the spatial location is assumed. The dependence of the viscosity and conductivity on temperature and strain rates is discussed in [1, 10].

10.2.6   Boundary Conditions

The boundary conditions for the flow problem involve specifying one element of the pair $(\mathbf{v}, \mathbf{t})$ at every point on the boundary:

$$\mathbf{v} = \hat{\mathbf{v}} \quad (v_i = \hat{v}_i) \quad \text{on } \Gamma_v \tag{10.2.18a}$$
$$\mathbf{t} \equiv \hat{\mathbf{n}}\cdot\boldsymbol{\sigma} = \hat{\mathbf{t}} \quad (t_i = n_j\sigma_{ji} = \hat{t}_i) \quad \text{on } \Gamma_\sigma \tag{10.2.18b}$$

where $\hat{\mathbf{n}}$ is the unit normal to the boundary and Γ_v and Γ_σ are the boundary portions on which the velocity $\mathbf{v}$ and tractions (stress vector) $\mathbf{t}$ are specified, respectively (see Fig. 10.2.1).

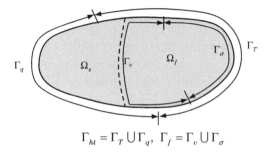

$$\Gamma_{ht} = \Gamma_T \cup \Gamma_q, \quad \Gamma_f = \Gamma_v \cup \Gamma_\sigma$$

Fig. 10.2.1: A schematic of various boundary portions.

The boundary conditions for the heat transfer problem (i.e. energy equation) are

$$T = \hat{T} \quad \text{on} \quad \Gamma_T \tag{10.2.19a}$$

$$q_n \equiv \hat{\mathbf{n}} \cdot \mathbf{q} = \hat{q}_n \quad \text{on} \quad \Gamma_q \tag{10.2.19b}$$

where Γ_T and Γ_q are the boundary portions on which the temperature and heat flux are specified, respectively. A more general boundary condition for the heat transfer problem is given by

$$q_{\text{cond}} + q_{\text{conv}} + q_{\text{rad}} = \hat{q}_n \quad \text{on} \quad \Gamma_q \tag{10.2.20}$$

where q_{cond}, q_{conv}, and q_{rad} are the conductive, convective, and radiative parts of heat flux, respectively,

$$q_{\text{cond}} = -\mathbf{k} \cdot \boldsymbol{\nabla} T, \quad q_{\text{conv}} = h_c(T - T_c), \quad q_{\text{rad}} = h_r(T - T_r) \tag{10.2.21}$$

Here h_c and h_r denote the heat transfer coefficients associated with convective and radiative heat transfer, respectively. In general, they are functions of position, time, and temperature.

The various portions of the boundary Γ must satisfy the requirements

$$\Gamma_{\mathbf{v}} \cup \Gamma_\sigma = \Gamma, \quad \Gamma_{\mathbf{v}} \cap \Gamma_\sigma = 0 \tag{10.2.22a}$$

$$\Gamma_T \cup \Gamma_q = \Gamma, \quad \Gamma_T \cap \Gamma_q = 0 \tag{10.2.22b}$$

where $\{0\}$ denotes the empty set. Equations (10.2.22a,b) imply that the total boundary is composed of two disjoint portions, $\Gamma_{\mathbf{v}}$ and Γ_σ, or Γ_T and Γ_q, whose union is the total boundary. Of course, in the analysis of practical problems one may encounter situations where, for example, both velocity and traction are known at a point. This is a mathematical singularity, and one must pick one of the two conditions but not both. Often the primary variables, that is $\mathbf{v}$ and T, are picked over the secondary variables $\mathbf{t}$ and q_n.

10.3 Summary of Governing Equations

10.3.1 Vector Form

Equations (10.2.3), (10.2.6), and (10.2.8) can be expressed in terms of the primitive variables $(\mathbf{v}, P, T)$ through Eqs. (10.2.12)–(10.2.16). The results are summarized below for isotropic, Newtonian, viscous, incompressible fluids in the presence of buoyancy forces:

$$\nabla \cdot \mathbf{v} = 0 \tag{10.3.1}$$

$$\rho \left(\frac{\partial \mathbf{v}}{\partial t} + \mathbf{v} \cdot \nabla \mathbf{v} \right) = -\nabla P + \mu \nabla \cdot \left[(\nabla \mathbf{v}) + (\nabla \mathbf{v})^{\mathsf{T}} \right]$$
$$+ \rho \mathbf{f} + \rho \mathbf{g} \, \beta (T - T_0) \tag{10.3.2}$$

$$\rho c_v \left(\frac{\partial T}{\partial t} + \mathbf{v} \cdot \nabla T \right) = \nabla \cdot (k \nabla T) + g + \Phi \tag{10.3.3}$$

where $\mathbf{v}$ is the velocity vector, ρ is the density, $\mathbf{g}$ is the gravity force vector per unit mass, T is the temperature, c_v is the specific heat of the fluid at constant volume, and g is the rate of internal heat generation per unit volume.

The above equations are valid for the fluid region Ω_f. In the solid region Ω_s, the fluid velocity is zero, $\mathbf{v} = \mathbf{0}$, and the only relevant equation is (10.3.3). The energy equation, Eq. (10.3.3), for the solid region is given by

$$\rho_s c_s \frac{\partial T}{\partial t} = \nabla \cdot (k_s \nabla T) + g_s \tag{10.3.4}$$

In writing Eq. (10.3.4) it is assumed that the solid is stationary with respect to the coordinate frame such that the convective transport of energy [i.e. the nonlinear part of Eq. (10.3.3)] need not be considered. The finite element model of Eq. (10.3.4) was discussed in Chapter 6 when the conductivity k_s is a function, in general, of x, y, and T.

10.3.2 Cartesian Component Form

The vector forms of Eqs. (10.3.1)–(10.3.3) allow us to express them in any coordinate system by expressing the vector operator ∇ and all other vectors in that coordinate system. In the Cartesian coordinate system (x_1, x_2, x_3), the kinematic and constitutive relations take the form

$$d_{ij} = \frac{1}{2} \left(\frac{\partial v_i}{\partial x_j} + \frac{\partial v_j}{\partial x_i} \right) \tag{10.3.5}$$

$$\sigma_{ij} = -P\delta_{ij} + \tau_{ij} \; ; \quad \tau_{ij} = 2\mu d_{ij} \tag{10.3.6}$$

The conservation equations can be expressed as

$$\frac{\partial v_i}{\partial x_i} = 0 \tag{10.3.7}$$

$$\rho\left(\frac{\partial v_i}{\partial t} + v_j\frac{\partial v_i}{\partial x_j}\right) = \frac{\partial}{\partial x_j}\left[-P\delta_{ij} + \mu\left(\frac{\partial v_i}{\partial x_j} + \frac{\partial v_j}{\partial x_i}\right)\right]$$
$$+ \rho f_i - \rho g_i\beta(T - T_0) \tag{10.3.8}$$

$$\rho c_v\left(\frac{\partial T}{\partial t} + v_j\frac{\partial T}{\partial x_j}\right) = \frac{\partial}{\partial x_i}\left(k\frac{\partial T}{\partial x_i}\right) + g + 2\mu d_{ij}d_{ij} \tag{10.3.9}$$

for the fluid region Ω_f and

$$\rho_s c_s\frac{\partial T}{\partial t} = \frac{\partial}{\partial x_i}\left(k_s\frac{\partial T}{\partial x_i}\right) + g_s \tag{10.3.10}$$

for the solid region Ω_s. Equations (10.3.5)–(10.3.10) are written for a Cartesian geometry in an Eulerian reference frame, with the indices $i, j = 1, 2, 3$ (or $i, j = 1, 2$ for two-dimensional problems); the Einstein summation convention on repeated indices is used.

In the present study, we shall consider two different finite element models of Eqs. (10.3.7) and (10.3.8). The first one is a natural formulation in which the weak forms of Eqs. (10.3.7) and (10.3.8) are used to construct the finite element model. The resulting finite element model is termed the *velocity–pressure model* or *mixed model*. The phrase "mixed" is used because velocity variables are mixed with the force-like variable, pressure, and both types of variables are retained in a single formulation. The second model is based on the interpretation that the continuity equation, Eq. (10.3.7), is an additional relation among the velocity components (i.e. a constraint among the v_i), and the constraint is satisfied in a least-squares (i.e. approximate) sense. This particular method of including the constraint in the formulation is known as the *penalty function method*, and the model is termed the *penalty–finite element model*. In this case, the pressure variable is effectively eliminated from the formulation. It is informative to note that the velocity–pressure (or mixed) formulation is the same as the Lagrange multiplier formulation, wherein the constraint is included by means of the Lagrange multiplier method. The Lagrange multiplier turns out to be the negative of the pressure.

There exist in computational fluid dynamics literature hundreds of papers on finite element models of incompressible flows and their applications. Interested readers may consult the books by Reddy and Gartling [10] and Gresho and Sani [181] for many of these references. Some of these works will be cited at appropriate places of this book.

10.4   Velocity–Pressure Finite Element Model

10.4.1   Weak Forms

The starting point for the development of the finite element models of Eqs. (10.3.7) and (10.3.8) is their weak forms. Here we consider the two-dimensional case, and the three-dimensional case follows in a straightforward manner. First, we write Eqs. (10.3.7) and (10.3.8) for the two-dimensional case using the notation $v_1 = v_x, v_2 = v_y, x_1 = x$, and $x_2 = y$. Keeping in mind that the sequence of variables is (v_x, v_y, P), we write the momentum equations followed by the continuity equation. We have

$$0 = \rho \left(\frac{\partial v_x}{\partial t} + v_x \frac{\partial v_x}{\partial x} + v_y \frac{\partial v_x}{\partial y} \right) - \frac{\partial}{\partial x} \left(2\mu \frac{\partial v_x}{\partial x} - P \right)$$
$$- \frac{\partial}{\partial y} \left[\mu \left(\frac{\partial v_x}{\partial y} + \frac{\partial v_y}{\partial x} \right) \right] - \rho f_x \tag{10.4.1}$$

$$0 = \rho \left(\frac{\partial v_y}{\partial t} + v_x \frac{\partial v_y}{\partial x} + v_y \frac{\partial v_y}{\partial y} \right) - \frac{\partial}{\partial y} \left(2\mu \frac{\partial v_y}{\partial y} - P \right)$$
$$- \frac{\partial}{\partial x} \left[\mu \left(\frac{\partial v_x}{\partial y} + \frac{\partial v_y}{\partial x} \right) \right] - \rho f_y \tag{10.4.2}$$

$$0 = \frac{\partial v_x}{\partial x} + \frac{\partial v_y}{\partial y} \tag{10.4.3}$$

where (f_x, f_y) are the components of the body force vector $\mathbf{f}$.

The weighted-integral statements of these equations over a typical element Ω^e are given by

$$0 = \int_{\Omega^e} w_x \left\{ \rho \left(\frac{\partial v_x}{\partial t} + v_x \frac{\partial v_x}{\partial x} + v_y \frac{\partial v_x}{\partial y} \right) - \frac{\partial}{\partial x} \left(2\mu \frac{\partial v_x}{\partial x} - P \right) \right.$$
$$\left. - \frac{\partial}{\partial y} \left[\mu \left(\frac{\partial v_x}{\partial y} + \frac{\partial v_y}{\partial x} \right) \right] - \rho f_x \right\} dx\, dy \tag{10.4.4}$$

$$0 = \int_{\Omega^e} w_y \left\{ \rho \left(\frac{\partial v_y}{\partial t} + v_x \frac{\partial v_y}{\partial x} + v_y \frac{\partial v_y}{\partial y} \right) - \frac{\partial}{\partial y} \left(2\mu \frac{\partial v_y}{\partial y} - P \right) \right.$$
$$\left. - \frac{\partial}{\partial x} \left[\mu \left(\frac{\partial v_x}{\partial y} + \frac{\partial v_y}{\partial x} \right) \right] - \rho f_y \right\} dx\, dy \tag{10.4.5}$$

$$0 = \int_{\Omega^e} Q \left(\frac{\partial v_x}{\partial x} + \frac{\partial v_y}{\partial y} \right) dx\, dy \tag{10.4.6}$$

where (w_x, w_y, Q) are weight functions, which will be equated, in the Ritz or weak-form Galerkin finite element models, to the interpolation functions used for (v_x, v_y, P), respectively.

Equation (10.4.6) will remain unaffected, as integration-by-parts does not help reduce differentiability on (v_x, v_y). In addition, the boundary terms obtained will be in conflict with the physical boundary conditions. The same comment applies to the time-derivative and nonlinear terms in Eqs. (10.4.4) and (10.4.5). Note that trading of differentiability between the weight functions and problem variables is subjected to the restriction that the resulting boundary expressions are physically meaningful. Otherwise, the secondary variables of the formulation may not be the quantities the physical problem admits as boundary quantities. An examination of the boundary stress components t_α in Eq. (10.2.18b) shows that the pressure term appears as a part of the expression [also see Eq. (10.3.6)]. Therefore, the pressure term must also be integrated-by-parts to keep the boundary stresses intact. The integration-by-parts also allows the pressure variable to have lower-order approximation.

The final expressions for the weak forms are given by

$$
0 = \int_{\Omega^e} \left[w_x \, \rho \left(\frac{\partial v_x}{\partial t} + v_x \frac{\partial v_x}{\partial x} + v_y \frac{\partial v_x}{\partial y} \right) + 2\mu \frac{\partial w_x}{\partial x} \frac{\partial v_x}{\partial x} - \frac{\partial w_x}{\partial x} P \right.
$$
$$
\left. + \mu \frac{\partial w_x}{\partial y} \left(\frac{\partial v_x}{\partial y} + \frac{\partial v_y}{\partial x} \right) - \rho w_x f_x \right] dx\, dy - \oint_{\Gamma^e} w_x t_x \, ds \quad (10.4.7)
$$

$$
0 = \int_{\Omega^e} \left[w_y \, \rho \left(\frac{\partial v_y}{\partial t} + v_x \frac{\partial v_y}{\partial x} + v_y \frac{\partial v_y}{\partial y} \right) + 2\mu \frac{\partial w_y}{\partial y} \frac{\partial v_y}{\partial y} - \frac{\partial w_y}{\partial y} P \right.
$$
$$
\left. + \mu \frac{\partial w_y}{\partial x} \left(\frac{\partial v_x}{\partial y} + \frac{\partial v_y}{\partial x} \right) - \rho w_y f_y \right] dx\, dy - \oint_{\Gamma^e} w_y t_y \, ds \quad (10.4.8)
$$

$$
0 = \int_{\Omega^e} Q \left(\frac{\partial v_x}{\partial x} + \frac{\partial v_y}{\partial y} \right) dx\, dy \quad (10.4.9)
$$

where (t_x, t_y) are the boundary stress (or traction) components

$$
\begin{aligned}
t_x &= \left(2\mu \frac{\partial v_x}{\partial x} - P \right) n_x + \mu \left(\frac{\partial v_x}{\partial y} + \frac{\partial v_y}{\partial x} \right) n_y \\
t_y &= \mu \left(\frac{\partial v_x}{\partial y} + \frac{\partial v_y}{\partial x} \right) n_x + \left(2\mu \frac{\partial v_y}{\partial y} - P \right) n_y
\end{aligned}
\quad (10.4.10)
$$

and (n_x, n_y) are the direction cosines of the unit normal vector $\hat{\mathbf{n}}$ on the boundary Γ^e.

10.4.2   Semidiscrete Finite Element Model

Since we are developing the weak-form Galerkin finite element models, the choice of the weight functions is restricted to the spaces of approximation functions used for the pressure and velocity fields. Suppose that the dependent variables (v_x, v_y, P) are approximated by expansions of the form

$$
v_x(x, y, t) \approx v_x^h = \sum_{j=1}^m v_x^j(t)\, \psi_j(x, y)
$$
$$
v_y(x, y, t) \approx v_y^h = \sum_{j=1}^m v_y^j(t)\, \psi_j(x, y) \quad (10.4.11)
$$
$$
P(x, y, t) \approx P_h = \sum_{j=1}^n P_j(t)\, \phi_j(x, y)
$$

where ψ_j and ϕ_j are interpolation (or shape) functions, and (v_x^j, v_y^j, P_j) are nodal values of (v_x, v_y, P), respectively. The weight functions (w_x, w_y, Q) have the following correspondence:

$$w_x \sim v_x, \quad w_y \sim v_y, \quad Q \sim P \tag{10.4.12}$$

Substitution of Eq. (10.4.11) into Eqs. (10.4.7)–(10.4.9) results in the following finite element equations:

$$
\begin{bmatrix} \mathbf{M} & 0 & 0 \\ 0 & \mathbf{M} & 0 \\ 0 & 0 & 0 \end{bmatrix}
\begin{Bmatrix} \dot{\mathbf{v}}_x \\ \dot{\mathbf{v}}_y \\ \dot{\mathbf{P}} \end{Bmatrix}
+
\begin{bmatrix} \mathbf{C(v)} & 0 & 0 \\ 0 & \mathbf{C(v)} & 0 \\ 0 & 0 & 0 \end{bmatrix}
\begin{Bmatrix} \mathbf{v}_x \\ \mathbf{v}_y \\ \mathbf{P} \end{Bmatrix}
$$

$$
+
\begin{bmatrix} 2\mathbf{S}^{11}+\mathbf{S}^{22} & (\mathbf{S}^{12})^{\mathrm{T}} & -\tilde{\mathbf{S}}^{10} \\ \mathbf{S}^{12} & \mathbf{S}^{11}+2\mathbf{S}^{22} & -\tilde{\mathbf{S}}^{20} \\ -(\tilde{\mathbf{S}}^{10})^{\mathrm{T}} & -(\tilde{\mathbf{S}}^{20})^{\mathrm{T}} & 0 \end{bmatrix}
\begin{Bmatrix} \mathbf{v}_x \\ \mathbf{v}_y \\ \mathbf{P} \end{Bmatrix}
=
\begin{Bmatrix} \mathbf{F}^1 \\ \mathbf{F}^2 \\ 0 \end{Bmatrix}
\tag{10.4.13}
$$

The coefficient matrices shown in Eq. (10.4.13) are defined by

$$
\begin{aligned}
M_{ij} &= \int_{\Omega^e} \rho \psi_i \psi_j \, dx\, dy, & \tilde{S}_{ij}^{10} &= \int_{\Omega^e} \frac{\partial \psi_i}{\partial x} \phi_j \, dx\, dy \\
\tilde{S}_{ij}^{20} &= \int_{\Omega^e} \frac{\partial \psi_i}{\partial y} \phi_j \, dx\, dy, & S_{ij}^{12} &= \int_{\Omega^e} \mu \frac{\partial \psi_i}{\partial x} \frac{\partial \psi_j}{\partial y} \, dx\, dy \\
S_{ij}^{11} &= \int_{\Omega^e} \mu \frac{\partial \psi_i}{\partial x} \frac{\partial \psi_j}{\partial x} \, dx\, dy, & S_{ij}^{22} &= \int_{\Omega^e} \mu \frac{\partial \psi_i}{\partial y} \frac{\partial \psi_j}{\partial y} \, dx\, dy \\
C_{ij} &= \int_{\Omega^e} \rho \psi_i \left(v_x \frac{\partial \psi_j}{\partial x} + v_y \frac{\partial \psi_j}{\partial y} \right) dx\, dy \\
F_i^1 &= \int_{\Omega^e} \rho \psi_i f_x \, dx\, dy + \oint_{\Gamma^e} \psi_i t_x \, ds \\
F_i^2 &= \int_{\Omega^e} \rho \psi_i f_y \, dx\, dy + \oint_{\Gamma^e} \psi_i t_y \, ds
\end{aligned}
\tag{10.4.14}
$$

The two sets of interpolation functions used in Eq. (10.4.11) should be of the Lagrange type, that is, derived by interpolating only the values of the functions and not their derivatives. There are two different finite elements associated with the two sets of field variables, (v_x, v_y) and P, and hence there are two different finite element meshes corresponding to the two variables over the same domain, Ω. If one of the meshes contains the other mesh as a subset, then we choose to display the first mesh and indicate the nodal degrees of freedom associated with the nodes of a typical element of the mesh.

The interpolation used for the pressure variable should be different from that used for the velocities, because the weak forms in Eqs. (10.4.7)–(10.4.9) contain only the first derivatives of the velocities v_x and v_y and no derivatives of the pressure P. Also, note that the governing equations, Eqs. (10.4.1) and

(10.4.2), contain second-order derivatives of v_x and v_y and first-order derivatives of P. Thus, for consistency, v_x and v_y must be interpolated using functions that are one order higher than those used for P, unless very high-order functions are used for all variables (in which case the consistency will be satisfied by adjusting degrees of polynomial used for velocities and pressure). In addition, the essential boundary conditions of the formulation do not include specification of the pressure; it enters the boundary conditions as a part of the natural boundary conditions (i.e. in specifying t_x and t_y). This implies that the pressure variable need not be carried as a variable that is continuous across inter-element boundaries. These observations lead to the conclusion that the pressure variable should be interpolated with functions that are one order less than those used for the velocity field and that the approximation may be discontinuous (i.e. not continuous from one element to other). Thus, quadratic interpolation of v_i and discontinuous linear interpolation of P are admissible. Models that use equal interpolation of the velocities and pressure with this formulation are known to give inaccurate results (see [182–189]).

10.4.3 Fully Discretized Finite Element Model

The semidiscrete finite element model developed in the previous section can be converted to the fully discretized model using the time approximations discussed in Chapter 6. Equation (10.4.13) can be written in a compact form as

$$\mathbf{M}\dot{\mathbf{\Delta}} + (\mathbf{C} + \mathbf{K})\mathbf{\Delta} = \mathbf{F} \tag{10.4.15}$$

where $\mathbf{\Delta} = (\mathbf{v}, P)$. In view of Eqs. (6.7.17) and (6.7.18), we obtain

$$\hat{\mathbf{K}}(\mathbf{v}_{s+1})\mathbf{\Delta}_{s+1} = \tilde{\mathbf{K}}(\mathbf{v}_s)\mathbf{\Delta}_s + \hat{\mathbf{F}}_{s,s+1} \tag{10.4.16}$$

where

$$\begin{aligned}
\hat{\mathbf{K}}(\mathbf{v}_{s+1}) &= \mathbf{M} + a_1\left[\mathbf{C}(\mathbf{v}_{s+1}) + \mathbf{K}(\mathbf{v}_{s+1})\right] \\
\tilde{\mathbf{K}}(\mathbf{v}_s) &= \mathbf{M} - a_2\left[\mathbf{C}(\mathbf{v}_s) + \mathbf{K}(\mathbf{v}_s)\right] \\
\hat{\mathbf{F}}_{s,s+1} &= a_1\mathbf{F}_{s+1} + a_2\mathbf{F}_s, \quad a_1 = \alpha\Delta t, \quad a_2 = (1-\alpha)\Delta t
\end{aligned} \tag{10.4.17}$$

where α is the parameter that defines the time approximation scheme as defined in Eq. (6.7.16), and Δt is the time step. Dependence of $\mathbf{C}$ on the nodal values $\mathbf{v}$ is due to the convective terms, and dependence of $\mathbf{K}$ on $\mathbf{v}$ is due to viscosity being a function of the strain rates (e.g. power-law fluids). If the viscosity is constant, that is Newtonian fluids, then $\mathbf{K}$ is independent of the nodal values $\mathbf{v}$ of the velocity field. For direct iteration solution, Eq. (10.4.16) takes the form

$$\hat{\mathbf{K}}(\mathbf{v}_{s+1}^r)\mathbf{\Delta}_{s+1}^{r+1} = \tilde{\mathbf{K}}(\mathbf{v}_s)\mathbf{\Delta}_s + \hat{\mathbf{F}}_{s,s+1} \tag{10.4.18}$$

10.5 Penalty Finite Element Models

10.5.1 Introduction

The penalty function method, like the Lagrange multiplier method, allows us to reformulate a problem with constraints as one without constraints (see [2, 10, 181–187]). In order to use the penalty function method for the flow of a viscous incompressible fluid, first it is reformulated as a variational problem subjected to a constraint. For the purpose of describing the penalty function method, we consider the steady-state Stokes flow problem (i.e. without time-dependent and nonlinear terms) in two dimensions. Then the penalty method is applied to the variational problem with a constraint. The development will then be extended to unsteady Navier–Stokes equations.

Consider the weak forms in Eqs. (10.4.7)–(10.4.9), and omit the time derivative and nonlinear terms. These can be expressed in the form

$$B((\mathbf{w}, Q), (\mathbf{v}, P)) = \ell(\mathbf{w}, Q) \tag{10.5.1}$$

where $(\mathbf{w}, Q)$ are the weight functions used for the momentum and continuity equations, respectively, $B(\cdot, \cdot)$ is a *bilinear form* [i.e. an expression that is linear in $(\mathbf{w}, Q)$ as well as $(\mathbf{v}, P)$] and $\ell(\cdot)$ is a *linear form*, defined by

$$B((\mathbf{w}, Q), (\mathbf{v}, P))$$

$$= \int_{\Omega^e} \mu \left[2 \left(\frac{\partial w_x}{\partial x} \frac{\partial v_x}{\partial x} + \frac{\partial w_y}{\partial y} \frac{\partial v_y}{\partial y} \right) + \left(\frac{\partial w_x}{\partial y} + \frac{\partial w_y}{\partial x} \right) \left(\frac{\partial v_x}{\partial y} + \frac{\partial v_y}{\partial x} \right) \right] dx\, dy$$

$$- \int_{\Omega^e} \left[\left(\frac{\partial w_x}{\partial x} + \frac{\partial w_y}{\partial y} \right) P + \left(\frac{\partial v_x}{\partial x} + \frac{\partial v_y}{\partial y} \right) Q \right] dx\, dy \tag{10.5.2a}$$

$$\ell(\mathbf{w}, Q) = \int_{\Omega^e} \rho \left(w_x f_x + w_y f_y \right) dx\, dy + \oint_{\Gamma^e} \left(w_x t_x + w_y t_y \right) ds \tag{10.5.2b}$$

and (t_x, t_y) are the boundary stress components defined in Eq. (10.4.10). The statement in Eq. (10.5.1) is known as the *variational problem* associated with steady-state Stokes problem.

The finite element model based on the variational problem, Eq. (10.5.1), is a special case of the mixed finite element model in Eq. (10.4.13). Equation (10.4.13) is more general than the problem at hand in that Eq. (10.4.13) is valid for time-dependent Navier–Stokes equations. To make it simple to understand the penalty function method, only the steady-state Stokes problem is considered. The inertial (i.e. time-dependent) and convective (i.e. nonlinear) terms may be added to the equations of the penalty formulation as they are not connected to the divergence-free (i.e. incompressibility) condition, which is central to the penalty formulation.

10.5.2   Penalty Function Method

Suppose that the velocity field $(\mathbf{v})$ is such that the continuity equation, Eq. (10.4.3), is satisfied identically. Then the weight function $(\mathbf{w})$, being a virtual variation of the velocity vector, also satisfies the continuity equation

$$\nabla \cdot \mathbf{w} = \frac{\partial w_x}{\partial x} + \frac{\partial w_y}{\partial y} = 0 \qquad (10.5.3)$$

As a result, the second integral expression in Eq. (10.5.2a) drops out, and the pressure, and hence the weight function Q, does not appear explicitly in the variational problem, Eq. (10.5.1). The resulting variational problem now can be stated as follows: among all vectors $\mathbf{v} = v_x \hat{\mathbf{e}}_x + v_y \hat{\mathbf{e}}_y$ that satisfies the continuity equation, Eq. (10.4.3), find the one that satisfies the variational problem

$$B_0((w_x, w_y), (v_x, v_y)) = \ell_0(w_x, w_y) \qquad (10.5.4)$$

for all admissible weight functions $\mathbf{w} = w_x \hat{\mathbf{e}}_x + w_y \hat{\mathbf{e}}_y$ [i.e. which satisfies the condition in Eq. (10.5.3)]. The bilinear form and linear form in Eq. (10.5.4) are defined by

$$B_0(\mathbf{w}, \mathbf{v}) =$$
$$\int_{\Omega^e} \mu \left[2 \left(\frac{\partial w_x}{\partial x} \frac{\partial v_x}{\partial x} + \frac{\partial w_y}{\partial y} \frac{\partial v_y}{\partial y} \right) + \left(\frac{\partial w_x}{\partial y} + \frac{\partial w_y}{\partial x} \right) \left(\frac{\partial v_x}{\partial y} + \frac{\partial v_y}{\partial x} \right) \right] dx\, dy$$
$$(10.5.5a)$$

$$\ell_0(\mathbf{w}) = \int_{\Omega^e} \rho \left(w_x f_x + w_y f_y \right) dx\, dy + \oint_{\Gamma^e} \left(w_x t_x + w_y t_y \right) ds \qquad (10.5.5b)$$

The variational problem in Eq. (10.5.4) is a constrained variational problem, because the solution vector $\mathbf{v}$ is constrained to satisfy the continuity equation.

We recall from Eq. (1.7.27) that $B_0(\mathbf{w}, \mathbf{v})$ is symmetric in $\mathbf{w}$ and $\mathbf{v}$ if and only if

$$B_0(\mathbf{w}, \mathbf{v}) = B_0(\mathbf{v}, \mathbf{w}) \qquad (10.5.6)$$

holds, and it is linear in $\mathbf{w}$ as well as $\mathbf{v}$ only if the following relations hold for any vectors $\mathbf{w}_1, \mathbf{w}_2, \mathbf{v}_1$, and $\mathbf{v}_2$ and arbitrary constants α and β:

$$\begin{aligned} B_0(\alpha \mathbf{w}_1 + \beta \mathbf{w}_2, \mathbf{v}) &= \alpha B_0(\mathbf{w}_1, \mathbf{v}) + \beta B_0(\mathbf{w}_2, \mathbf{v}) \\ B_0(\mathbf{w}, \alpha \mathbf{v}_1 + \beta \mathbf{v}_2) &= \alpha B_0(\mathbf{w}, \mathbf{v}_1) + \beta B_0(\mathbf{w}, \mathbf{v}_2) \end{aligned} \qquad (10.5.7)$$

Thus, $B_0(\cdot, \cdot)$ is called *bilinear* if and if only it satisfies conditions in Eq. (10.5.7). Similarly $\ell_0(\cdot)$ is called linear in $\mathbf{w}$ if and only if it satisfies the condition

$$\ell_0(\alpha \mathbf{w}_1 + \beta \mathbf{w}_2) = \alpha \ell_0(\mathbf{w}_1) + \beta \ell_0(\mathbf{w}_2) \qquad (10.5.8)$$

Whenever the bilinear form of a variational problem is symmetric in its arguments, it is possible to construct a quadratic functional such that the minimum of the quadratic functional is equivalent to the variational problem (see [3, 13, 15, 16]). The quadratic functional is given by the expression

$$I_0(\mathbf{v}) = \frac{1}{2}B_0(\mathbf{v}, \mathbf{v}) - \ell_0(\mathbf{v}) \tag{10.5.9}$$

Now we can state that Eqs. (10.4.1)–(10.4.3) governing the steady flow of viscous incompressible fluids are equivalent to minimizing the quadratic functional $I_0(\mathbf{v})$ [$\mathbf{v} = (v_x, v_y)$] subjected to the constraint

$$G(\mathbf{v}) \equiv \frac{\partial v_x}{\partial x} + \frac{\partial v_y}{\partial y} = 0 \tag{10.5.10}$$

At this point it should be remembered that the discussion presented in this section thus far is to reformulate the problem as one of a constrained problem so that the penalty function method can be used. The advantage of the constrained problem is that the pressure variable P does not appear in the formulation.

In the penalty function method, the constrained problem is reformulated as an unconstrained problem as follows: minimize the modified functional

$$I_p(\mathbf{v}) \equiv I_0(\mathbf{v}) + \frac{\gamma_e}{2}\int_{\Omega^e} [G(\mathbf{v})]^2 d\mathbf{x} \tag{10.5.11}$$

where γ_e is called the *penalty parameter*. Note that the constraint is included in a least-squares sense into the functional. Seeking the minimum of the modified functional $I_p(\mathbf{v})$ is equivalent to seeking the minimum of both $I_0(\mathbf{v})$ and $G(\mathbf{v})$, the latter with respect to the weight γ_e. The larger the value of γ_e, the more exactly the constraint is satisfied. The necessary condition for the minimum of I_p is

$$\delta I_p = 0 \tag{10.5.12}$$

We have

$$0 = \int_{\Omega^e}\left[2\mu\frac{\partial \delta v_x}{\partial x}\frac{\partial v_x}{\partial x} + \mu\frac{\partial \delta v_x}{\partial y}\left(\frac{\partial v_x}{\partial y} + \frac{\partial v_y}{\partial x}\right)\right]dx\,dy - \int_{\Omega^e}\rho f_x\,\delta v_x\,dx\,dy$$
$$- \oint_{\Gamma^e}\delta v_x\,t_x\,ds + \int_{\Omega^e}\gamma_e\frac{\partial \delta v_x}{\partial x}\left(\frac{\partial v_x}{\partial x} + \frac{\partial v_y}{\partial y}\right)dx\,dy \tag{10.5.13a}$$

$$0 = \int_{\Omega^e}\left[2\mu\frac{\partial \delta v_y}{\partial y}\frac{\partial v_y}{\partial y} + \mu\frac{\partial \delta v_y}{\partial x}\left(\frac{\partial v_x}{\partial y} + \frac{\partial v_y}{\partial x}\right)\right]dx\,dy - \int_{\Omega^e}\rho f_y\,\delta v_y\,dx\,dy$$
$$- \oint_{\Gamma^e}\delta v_y\,t_y\,ds + \int_{\Omega^e}\gamma_e\frac{\partial \delta v_y}{\partial y}\left(\frac{\partial v_x}{\partial x} + \frac{\partial v_y}{\partial y}\right)dx\,dy \tag{10.5.13b}$$

These two statements provide the weak forms for the penalty finite element model with $\delta v_x = w_x$ and $\delta v_y = w_y$. We note that the pressure does not appear explicitly in the weak forms, Eqs. (10.5.13a,b), although it is a part of the

538

boundary stresses. An approximation for the pressure can be post-computed (at the reduced-order Gauss points) from the relation

$$P = -\gamma_e \left(\frac{\partial v_x}{\partial x} + \frac{\partial v_y}{\partial y} \right) \tag{10.5.14}$$

where $\mathbf{v} = \mathbf{v}(\gamma_e)$ is the velocity field known from the penalty finite element model.

The time derivative terms and nonlinear terms can be added to Eqs. (10.5.13a,b) without affecting the above discussion. We obtain

$$\begin{aligned}
0 = \int_{\Omega^e} & \left[\delta v_x \, \rho \left(\frac{\partial v_x}{\partial t} + v_x \frac{\partial v_x}{\partial x} + v_y \frac{\partial v_x}{\partial y} \right) + 2\mu \frac{\partial \delta v_x}{\partial x} \frac{\partial v_x}{\partial x} \right. \\
& \left. + \mu \frac{\partial \delta v_x}{\partial y} \left(\frac{\partial v_x}{\partial y} + \frac{\partial v_y}{\partial x} \right) + \gamma_e \frac{\partial \delta v_x}{\partial x} \left(\frac{\partial v_x}{\partial x} + \frac{\partial v_y}{\partial y} \right) \right] dx\, dy \\
& - \int_{\Omega^e} \rho f_x \, \delta v_x \, dx\, dy - \oint_{\Gamma^e} \delta v_x t_x \, ds
\end{aligned} \tag{10.5.15a}$$

$$\begin{aligned}
0 = \int_{\Omega^e} & \left[\delta v_y \, \rho \left(\frac{\partial v_y}{\partial t} + v_x \frac{\partial v_y}{\partial x} + v_y \frac{\partial v_y}{\partial y} \right) + 2\mu \frac{\partial \delta v_y}{\partial y} \frac{\partial v_y}{\partial y} \right. \\
& \left. + \mu \frac{\partial \delta v_y}{\partial x} \left(\frac{\partial v_x}{\partial y} + \frac{\partial v_y}{\partial x} \right) + \gamma_e \frac{\partial \delta v_y}{\partial y} \left(\frac{\partial v_x}{\partial x} + \frac{\partial v_y}{\partial y} \right) \right] dx\, dy \\
& - \int_{\Omega^e} \rho f_y \, \delta v_y \, dx\, dy - \oint_{\Gamma^e} \delta v_y t_y \, ds
\end{aligned} \tag{10.5.15b}$$

10.5.3 Reduced Integration Penalty Model

The penalty finite element model is obtained from Eqs. (10.5.15a,b) by substituting the finite element interpolation from Eq. (10.4.11) for the velocity field, and $\delta v_x = \psi_i^e$ and $\delta v_y = \psi_i^e$:

$$\begin{aligned}
& \left(\begin{bmatrix} \mathbf{C(v)} & \mathbf{0} \\ \mathbf{0} & \mathbf{C(v)} \end{bmatrix} + \begin{bmatrix} 2\mathbf{S}^{11} + \mathbf{S}^{22} & (\mathbf{S}^{12})^{\mathrm{T}} \\ \mathbf{S}^{12} & \mathbf{S}^{11} + 2\mathbf{S}^{22} \end{bmatrix} + \begin{bmatrix} \bar{\mathbf{S}}^{11} & \bar{\mathbf{S}}^{12} \\ (\bar{\mathbf{S}}^{12})^{\mathrm{T}} & \bar{\mathbf{S}}^{22} \end{bmatrix} \right) \begin{Bmatrix} \mathbf{v}_x \\ \mathbf{v}_y \end{Bmatrix} \\
& \qquad\qquad + \begin{bmatrix} \mathbf{M} & \mathbf{0} \\ \mathbf{0} & \mathbf{M} \end{bmatrix} \begin{Bmatrix} \dot{\mathbf{v}}_x \\ \dot{\mathbf{v}}_y \end{Bmatrix} = \begin{Bmatrix} \mathbf{F}^1 \\ \mathbf{F}^2 \end{Bmatrix}
\end{aligned} \tag{10.5.16}$$

where $\mathbf{C(v)}$, $\mathbf{S}^{\alpha\beta}$ $(\alpha, \beta = 1, 2)$, $\mathbf{M}$, $\mathbf{F}^1$, and $\mathbf{F}^2$ are the same as those defined in Eq. (10.4.14), and

$$\bar{S}_{ij}^{\alpha\beta} = \int_{\Omega^e} \gamma_e \frac{\partial \psi_i^e}{\partial x_\alpha} \frac{\partial \psi_j^e}{\partial x_\beta} \, dx\, dy \quad (\alpha, \beta = 1, 2; \; x_1 = x, x_2 = y) \tag{10.5.17}$$

Equation (10.5.16) can be expressed symbolically as $(\mathbf{v} = \{\mathbf{v}_x, \mathbf{v}_y\}^{\mathrm{T}})$

$$[\mathbf{C}(\rho, \mathbf{v}) + \mathbf{K}_\mu + \mathbf{K}_\gamma] \, \mathbf{v} + \mathbf{M}\dot{\mathbf{v}} = \mathbf{F} \tag{10.5.18}$$

where $\mathbf{K}_\mu$ and $\mathbf{K}_\gamma$ are matrices that depend only on μ and γ, respectively; whereas $\mathbf{C}$ depends on ρ and $\mathbf{v}$. The numerical construction of the penalty terms presented in Eq. (10.5.18) requires special consideration, the details of which are given in Section 10.6.3.

10.5.4   Consistent Penalty Model

An alternative formulation of the penalty finite element model is based directly on the use of Eq. (10.5.14). In this formulation, pressure P in the momentum equations, Eqs. (10.4.1) and (10.4.2), is replaced with Eq. (10.5.14), and Eq. (10.4.3) is replaced with Eq. (10.5.14). The weak form of Eq. (10.5.14) is

$$0 = \int_{\Omega^e} \delta P \left[P + \gamma_e \left(\frac{\partial v_x}{\partial x} + \frac{\partial v_y}{\partial y} \right) \right] dx \, dy \tag{10.5.19}$$

The finite element model of Eq. (10.5.19), with P interpolated as in Eq. (10.4.11), is given by

$$\mathbf{M}^p\mathbf{P} + \gamma_e \{ (\tilde{\mathbf{S}}^{10})^{\mathrm{T}} \;\; (\tilde{\mathbf{S}}^{20})^{\mathrm{T}} \} \begin{Bmatrix} \mathbf{v}_x \\ \mathbf{v}_y \end{Bmatrix} = \mathbf{0} \tag{10.5.20}$$

where

$$M_{ij}^p = \int_{\Omega^e} \phi_i \phi_j \, dx \, dy, \quad \tilde{S}_{ij}^{10} = \int_{\Omega^e} \frac{\partial \psi_i}{\partial x} \phi_j \, dx \, dy, \quad \tilde{S}_{ij}^{20} = \int_{\Omega^e} \frac{\partial \psi_i}{\partial y} \phi_j \, dx \, dy$$

Since $\mathbf{M}^p$ is invertible, we can write

$$\mathbf{P} = -\gamma_e (\mathbf{M}^p)^{-1} \{ (\tilde{\mathbf{S}}^{10})^{\mathrm{T}} \;\; (\tilde{\mathbf{S}}^{20})^{\mathrm{T}} \} \begin{Bmatrix} \mathbf{v}_x \\ \mathbf{v}_y \end{Bmatrix} \tag{10.5.21}$$

Next, we write Eq. (10.4.13) in the alternative form

$$\begin{bmatrix} \mathbf{M} & \mathbf{0} \\ \mathbf{0} & \mathbf{M} \end{bmatrix} \begin{Bmatrix} \dot{\mathbf{v}}_x \\ \dot{\mathbf{v}}_y \end{Bmatrix} + \begin{bmatrix} \mathbf{C}(\mathbf{v}) & \mathbf{0} \\ \mathbf{0} & \mathbf{C}(\mathbf{v}) \end{bmatrix} \begin{Bmatrix} \mathbf{v}_x \\ \mathbf{v}_y \end{Bmatrix} - \begin{Bmatrix} \tilde{\mathbf{S}}^{10} \\ \tilde{\mathbf{S}}^{20} \end{Bmatrix} \mathbf{P}$$

$$+ \begin{bmatrix} 2\mathbf{S}^{11} + \mathbf{S}^{22} & (\mathbf{S}^{12})^{\mathrm{T}} \\ \mathbf{S}^{12} & \mathbf{S}^{11} + 2\mathbf{S}^{22} \end{bmatrix} \begin{Bmatrix} \mathbf{v}_x \\ \mathbf{v}_y \end{Bmatrix} = \begin{Bmatrix} \mathbf{F}^1 \\ \mathbf{F}^2 \end{Bmatrix} \tag{10.5.22}$$

When Eq. (10.5.21) is substituted for the pressure into Eq. (10.5.22), we obtain

$$\begin{bmatrix} \mathbf{M} & \mathbf{0} \\ \mathbf{0} & \mathbf{M} \end{bmatrix} \begin{Bmatrix} \dot{\mathbf{v}}_x \\ \dot{\mathbf{v}}_y \end{Bmatrix} + \begin{bmatrix} \mathbf{C}(\mathbf{v}) & \mathbf{0} \\ \mathbf{0} & \mathbf{C}(\mathbf{v}) \end{bmatrix} \begin{Bmatrix} \mathbf{v}_x \\ \mathbf{v}_y \end{Bmatrix}$$

$$+ \gamma^e \begin{bmatrix} \tilde{\mathbf{S}}^{10}(\mathbf{M}^p)^{-1}(\tilde{\mathbf{S}}^{10})^{\mathrm{T}} & \tilde{\mathbf{S}}^{10}(\mathbf{M}^p)^{-1}(\tilde{\mathbf{S}}^{20})^{\mathrm{T}} \\ \tilde{\mathbf{S}}^{20}(\mathbf{M}^p)^{-1}(\tilde{\mathbf{S}}^{10})^{\mathrm{T}} & \tilde{\mathbf{S}}^{20}(\mathbf{M}^p)^{-1}(\tilde{\mathbf{S}}^{20})^{\mathrm{T}} \end{bmatrix} \begin{Bmatrix} \mathbf{v}_x \\ \mathbf{v}_y \end{Bmatrix}$$

$$+ \begin{bmatrix} 2\mathbf{S}^{11} + \mathbf{S}^{22} & (\mathbf{S}^{12})^{\mathrm{T}} \\ \mathbf{S}^{12} & \mathbf{S}^{11} + 2\mathbf{S}^{22} \end{bmatrix} \begin{Bmatrix} \mathbf{v}_x \\ \mathbf{v}_y \end{Bmatrix} = \begin{Bmatrix} \mathbf{F}^1 \\ \mathbf{F}^2 \end{Bmatrix} \tag{10.5.23}$$

which can be written in compact form as

$$\mathbf{M}\dot{\mathbf{v}} + [\mathbf{C}(\mathbf{v}) + \mathbf{K}_\mu + \mathbf{K}_{p\gamma}]\mathbf{v} = \mathbf{F} \tag{10.5.24}$$

where [see Eq. (10.4.14) for the definition of M_{ij} and S_{ij}^{mn}]

$$\mathbf{M} = \begin{bmatrix} \mathbf{M} & 0 \\ 0 & \mathbf{M} \end{bmatrix}, \quad \mathbf{C} = \begin{bmatrix} \mathbf{C}(\mathbf{v}) & 0 \\ 0 & \mathbf{C}(\mathbf{v}) \end{bmatrix}, \quad \mathbf{v} = \begin{Bmatrix} \mathbf{v}_x \\ \mathbf{v}_y \end{Bmatrix}$$

$$\mathbf{K}_{p\gamma} = \gamma^e \begin{bmatrix} \tilde{\mathbf{S}}^{10}(\mathbf{M}^p)^{-1}(\tilde{\mathbf{S}}^{10})^{\mathrm{T}} & \tilde{\mathbf{S}}^{10}(\mathbf{M}^p)^{-1}(\tilde{\mathbf{S}}^{20})^{\mathrm{T}} \\ \tilde{\mathbf{S}}^{20}(\mathbf{M}^p)^{-1}(\tilde{\mathbf{S}}^{10})^{\mathrm{T}} & \tilde{\mathbf{S}}^{20}(\mathbf{M}^p)^{-1}(\tilde{\mathbf{S}}^{20})^{\mathrm{T}} \end{bmatrix} \tag{10.5.25}$$

$$\mathbf{K}_\mu = \begin{bmatrix} 2\mathbf{S}^{11} + \mathbf{S}^{22} & (\mathbf{S}^{12})^{\mathrm{T}} \\ \mathbf{S}^{12} & \mathbf{S}^{11} + 2\mathbf{S}^{22} \end{bmatrix}$$

The penalty finite element model described here is termed a *consistent penalty model* because it is derived from the discretized form of Eq. (10.5.14). This is in contrast to the reduced penalty model described earlier, which falls into the category known as the *reduced integration penalty (RIP)* model (see Oden [188]). Equation (10.5.24) has the same general form as Eq. (10.5.18). However, while $\mathbf{K}_\mu$ defined in Eqs. (10.5.25) and (10.5.16) are the same, $\mathbf{K}_{p\gamma}$ defined in Eq. (10.5.25) and $\mathbf{K}_\gamma$ defined in Eq. (10.5.16) are not the same. The overall size of the penalty finite element model in Eqs. (10.5.18) and (10.5.24) is reduced in comparison to the mixed finite element model in Eq. (10.4.13) because of the elimination of the pressure variable. To recover the pressure in the consistent penalty finite element model, the inverted form of Eq. (10.5.21) is used with the velocity field known from Eq. (10.5.24). The numerical implementation of the consistent penalty method relies on the ability to invert $\mathbf{M}^p$ at the element level. This restricts the choice of the basis functions used to represent the pressure.

10.6 Computational Aspects

10.6.1 Properties of the Finite Element Equations

Some of the properties of the matrix equations in Eqs. (10.4.13), (10.5.18), and (10.5.24) are listed below because they greatly influence the choice of a solution procedure for the various types of problems [10].

1. The matrix equations in Eqs. (10.4.13), (10.5.18), and (10.5.24) represent discrete analogs of the basic conservation equations with each term representing a particular physical process. For example, $\mathbf{M}$ represents the mass matrix, $\mathbf{C}$ represents the velocity dependent convective transport term,

$\mathbf{K}_\mu$ represents the viscous terms, and $\mathbf{K}_\gamma$ represents the divergence-free condition. The right-hand side $\mathbf{F}$ contains body forces and surface forces.

2. An inspection of the structure of the individual matrices in Eq. (10.4.14) shows that $\mathbf{M}$ and $\mathbf{K}_\mu$ are symmetric, while $\mathbf{C}$ is unsymmetric. This makes the coefficient matrices of the vector $\mathbf{v}$ in Eqs. (10.4.13), (10.5.18), and (10.5.24) unsymmetric, and the solution procedure must deal with an unsymmetric system. When material properties are constant and flow velocities are sufficiently small, the convective terms are negligible and the equations are linear and symmetric.

3. An additional difficulty of the mixed finite element model is the presence of zeroes on the matrix diagonals corresponding to the pressure variables [see Eq. (10.4.13)]. Direct equation solving methods must use some type of pivoting strategy, while the use of iterative solvers is severely handicapped by poor convergence behavior attributable mainly to the form of the constraint equation.

4. Equations (10.4.13), (10.5.18), and (10.5.24) represent a set of ordinary differential equations in time. The fact that the pressure does not appear explicitly in the continuity equation [see Eq. (10.4.13)] makes the system time-singular in the pressure and precludes the use of purely explicit time-integration methods.

5. The choice of the penalty parameter is largely dictated by the ratio of the magnitude of penalty terms to the viscous and convective terms (or compared to the Reynolds number, Re), the mesh, and the precision of the computer. The following range of γ is used in computations

$$\gamma = 10^4 Re \text{ to } \gamma = 10^{12} Re \qquad (10.6.1)$$

10.6.2 Choice of Elements

As is clear from the weak statements, the finite element models of conductive heat transfer as well as viscous incompressible flows require only the C^0-continuous functions to approximate the field variables (i.e. temperature, velocities, and pressure). Thus, any of the Lagrange and serendipity family of interpolation functions are admissible for the interpolation of the velocity field in mixed and penalty finite element models.

The choice of interpolation functions used for the pressure variable in the mixed finite element model is further constrained by the special role the pressure plays in incompressible flows. Recall that the pressure can be interpreted as a Lagrange multiplier that serves to enforce the incompressibility constraint on the velocity field. From Eq. (10.4.11) it is seen that the approximation functions ϕ_i used for pressure is the weighting function for the continuity equation.

In order to prevent an over-constrained system of discrete equations, the interpolation used for pressure must be at least one order lower than that used for the velocity field (i.e. unequal order interpolation). Further, pressure need not be made continuous across elements because the pressure variable does not constitute a primary variable of the weak form presented in Eqs. (10.4.7)–(10.4.9). Note that the unequal order interpolation criteria can be relaxed for certain "stabilized" formulations [189], which will not be discussed here.

Convergent finite element approximations of problems with constraints are governed by the ellipticity requirement and the *Ladyzhenskaya–Babuska–Brezzi (LBB) condition* (see Reddy [15, pp. 454–461], Oden [188], and others [190–194]). It is by no means a simple task to rigorously prove whether every new element developed for the viscous incompressible flows satisfies the LBB condition. The discussion of the LBB condition is beyond the scope of the present study and will not be discussed here.

Commonly used elements for two-dimensional flows of viscous incompressible fluids are shown in Fig. 10.6.1. In the case of linear elements, pressure is treated as discontinuous between elements; otherwise, the whole domain will have the same pressure. Two different pressure approximations have been used when the velocities are approximated by quadratic Lagrange functions. The first is a continuous linear approximation, in which the pressure is defined at the corner nodes of the element and is made continuous across element boundaries. The second pressure approximation involves a discontinuous (between elements) linear variation defined on the element by

$$
\mathbf{\Phi} = \left\{ \begin{array}{c} 1 \\ x \\ y \end{array} \right\} \tag{10.6.2}
$$

Here the unknowns are not nodal point values of the pressure but correspond to the coefficients in $P = a \cdot 1 + b \cdot x + c \cdot y$. In Eq. (10.6.2) the interpolation functions are written in terms of the global coordinates (x, y) for the problem.

When the eight-node quadratic element is used to represent the velocity field, a continuous-linear pressure approximation may be selected. When a discontinuous pressure variation is used with this element, the constant pressure representation over each element must be used. The quadratic quadrilateral elements shown in Fig. 10.6.1 are known to satisfy the LBB condition and thus give reliable solutions for velocity and pressure fields. Other elements may yield acceptable solutions for the velocity field but the pressure field is often in error (e.g. pressure values may appear as in the so-called checker-board pattern). In such cases, pressure may be determined from the Poisson equation for the pressure (see Reddy and Gartling [10]).

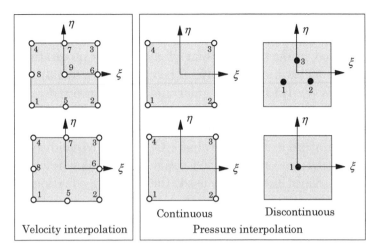

Fig. 10.6.1: The quadrilateral elements used for the mixed and penalty finite element models.

10.6.3 Evaluation of Element Matrices in Penalty Models

The numerical evaluation of the coefficient matrices appearing in Eq. (10.5.18) requires special consideration. This aspect is discussed here for the steady-state case. For the steady-state flows with constant material properties, Eq. (10.5.18) is of the form

$$(\rho\hat{\mathbf{C}}(\mathbf{v}) + \mu\hat{\mathbf{K}} + \gamma\bar{\mathbf{K}})\mathbf{v} = \mathbf{F} \tag{10.6.3}$$

where $\hat{\mathbf{C}}$ is the contribution due to the convective terms, $\hat{\mathbf{K}}$ is the contribution from the viscous terms, and $\bar{\mathbf{K}}$ is from the penalty terms, which come from the incompressibility constraint. In theory, as we increase the value of γ, the conservation of mass is satisfied more exactly. However, in practice, for some large value of γ, the contribution from the viscous terms would be negligibly small compared to the penalty terms in a computer. Thus, if $\bar{\mathbf{K}}$ is a non-singular (i.e. invertible) matrix, the solution of Eq. (10.6.3) for a large value of γ is trivial, $\{\mathbf{v}\} = \{\mathbf{0}\}$. While the solution satisfies the continuity equation, it does not satisfy the momentum equations. In this case the discrete problem, Eq. (10.6.3), is said to be over-constrained or "locked". If $\bar{\mathbf{K}}$ is singular, then the sum $(\rho\hat{\mathbf{C}} + \mu\hat{\mathbf{K}} + \gamma\bar{\mathbf{K}})$ is non-singular (because $\rho\hat{\mathbf{C}} + \mu\hat{\mathbf{K}}$ is non-singular), and a non-trivial solution to the problem is obtained.

The numerical problem described above is eliminated by proper evaluation of the integrals in $\hat{\mathbf{C}}$, $\hat{\mathbf{K}}$, and $\bar{\mathbf{K}}$. It is found that if the coefficients of $\bar{\mathbf{K}}$ (i.e. penalty matrix coefficients) are evaluated using a numerical integration rule of an order less than that required to integrate them exactly, the finite element equation, Eq. (10.6.3), give acceptable solutions for the velocity field. This technique of under-integrating the penalty terms is known in the literature as *reduced integration*. For example, if a linear quadrilateral element is used to approximate the velocity field, the matrix coefficients $\hat{\mathbf{C}}$ and $\hat{\mathbf{K}}$ (as well as $\mathbf{M}$ for unsteady problems) are evaluated using the 2×2 Gauss quadrature, and $\bar{\mathbf{K}}$ is evaluated using the one-point (1×1) Gauss quadrature. The one-point quadrature yields a singular $\bar{\mathbf{K}}$. Therefore, Eq. (10.6.3) can be solved because $(\rho\hat{\mathbf{C}} + \mu\hat{\mathbf{K}} + \gamma\bar{\mathbf{K}})$ is non-singular and can be inverted (after assembly and imposition of boundary conditions) to obtain a good finite element solution of the original problem. When a quadratic quadrilateral element is used, the 3×3 Gauss quadrature is used to evaluate $\hat{\mathbf{C}}$, $\hat{\mathbf{K}}$, and $\mathbf{M}$, and the 2×2 Gauss quadrature is used to evaluate $\bar{\mathbf{K}}$. Of course, as the degree of interpolation goes up, or very refined meshes are used, the resulting equations become less sensitive to locking.

Concerning the post-computation of pressure in the penalty model, the pressure should be computed by evaluating Eq. (10.5.14) at integration points corresponding to the reduced Gauss rule. This is equivalent to using an interpolation for pressure that is one order less than the one used for the velocity field. The pressure computed using Eq. (10.5.14) at the reduced integration points is not always reliable and accurate. The pressures predicted using the linear elements, especially for coarse meshes, are seldom acceptable. Quadratic elements are known to yield more reliable results. In general, triangular elements do not yield stable solutions for pressures. Various techniques have been proposed in the literature to obtain accurate pressure fields (see [185, 195–197]).

10.6.4 Post-Computation of Pressure and Stresses

The analysis of a flow problem generally includes calculation of not only the velocity field and pressure but also the computation of the stress field. For a plane two-dimensional flow, the stress components $(\sigma_{xx}, \sigma_{yy}, \sigma_{xy})$ are given by

$$
\sigma_{xx} = 2\mu\frac{\partial v_x}{\partial x} - P, \quad \sigma_{yy} = 2\mu\frac{\partial v_y}{\partial y} - P, \quad \sigma_{xy} = \mu\left(\frac{\partial v_x}{\partial y} + \frac{\partial v_y}{\partial x}\right) \tag{10.6.4}
$$

where μ is the viscosity of the fluid.

First, pressure P is calculated in mixed finite element model from

$$
P(x, y) = \sum_{j=1}^{n} \phi_j(x, y)P_j \tag{10.6.5}
$$

and in reduced-integration penalty finite element model from

$$
P_\gamma(x, y) = -\gamma \sum_{j=1}^{m}\left(\frac{\partial\psi_j}{\partial x}v_x^j + \frac{\partial\psi_j}{\partial y}v_y^j\right) \tag{10.6.6}
$$

In the latter, the expression is evaluated using reduced integration rule. Substitution of the finite element approximations, Eq. (10.4.11) for the velocity field and pressure from Eq. (10.6.5) or Eq. (10.6.6) into Eq. (10.6.4) yields

$$
\sigma_{xx} = 2\mu \sum_{j=1}^{m} \frac{\partial \psi_j}{\partial x} v_x^j - P, \quad \sigma_{yy} = 2\mu \sum_{j=1}^{m} \frac{\partial \psi_j}{\partial y} v_y^j - P
$$

$$
\sigma_{xy} = \mu \sum_{j=1}^{m} \left(\frac{\partial \psi_j}{\partial y} v_x^j + \frac{\partial \psi_j}{\partial x} v_y^j \right)
$$

(10.6.7)

The spatial derivatives of the interpolation functions in Eq. (10.6.7) must be evaluated using the reduced Gauss point rule. Thus, the stresses (as well as the pressure) are computed using the one-point Gauss rule for linear elements and the 2×2 Gauss rule for quadratic elements. The stresses computed at interior integration points can be extrapolated to the nodes by a simple linear extrapolation procedure, and they may be appropriately averaged between adjacent elements to produce a continuous stress field. In the consistent penalty finite element model, pressure is calculated at the nodes using Eq. (10.5.21).

10.7 Computer Implementation

10.7.1 Mixed Model

The computer implementation of the mixed model is some what complicated by the fact that the element contains variable degrees of freedom and the coefficient matrix is not positive-definite due to the appearance of zeros on the diagonal. Here, we discuss computer implementation of mixed model with quadratic approximation of the velocity field and continuous linear approximation of the pressure. A eight- or nine-node element depicting all nodal values of the formulation will have three degrees of freedom (v_x, v_y, P) at the corner nodes and two degrees of freedom (v_x, v_y) at the mid-side and interior nodes, as shown in Fig. 10.7.1. This complicates the calculation of element matrices as well the assembly of element equations to form the global system of equations.

To facilitate the assembly, we create a companion array NFD to the connectivity array NOD. Recall that NOD(I, J) denotes the global node number corresponding to the Jth node of element I in the mesh. Array NFD is similar to array NOD but it connects the degrees of freedom rather than the global node numbers associated with the element nodes:

NFD$(I, J) =$ The *last* global degree of freedom number associated with the Jth node of the Ith element.

The word 'last' refers to the third degree of freedom at the corner nodes and the second degree of freedom at the mid-side and interior nodes. To see the

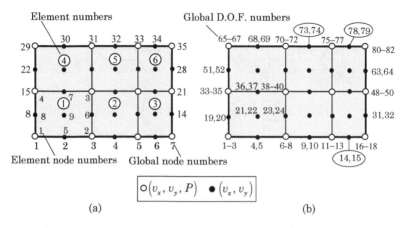

Fig. 10.7.1: (a) Element and global node numbers and (b) global degrees of freedom of a mesh of nine-node elements used for the mixed formulation.

meaning of array NFD, consider the mesh of six nine-node elements shown in Fig. 10.7.1(a). First note that the connectivity array NOD for the mesh shown in Fig. 10.7.1(a) is given by

$$
[\text{NOD}] =
\begin{array}{c}
 1 \;\; 2 \;\; 3 \;\; 4 \;\;\; 5 \;\; 6 \;\; 7 \;\;\; 8 \;\; 9 \\
\left[
\begin{array}{ccccccccc}
1 & 3 & 17 & 15 & 2 & 10 & 16 & 8 & 9 \\
3 & 5 & 19 & 17 & 4 & 12 & 18 & 10 & 11 \\
5 & 7 & 21 & 19 & 6 & 14 & 20 & 12 & 13 \\
15 & 17 & 31 & 29 & 16 & 24 & 30 & 22 & 23 \\
17 & 19 & 33 & 31 & 18 & 26 & 32 & 24 & 25 \\
19 & 21 & 35 & 33 & 20 & 28 & 34 & 26 & 27
\end{array}
\right]
\begin{array}{c}
1 \\ 2 \\ 3 \\ 4 \\ 5 \\ 6
\end{array}
\end{array}
$$

The NFD array is [see Fig. 10.7.1(b)] as follows:

$$
[\text{NFD}] =
\begin{array}{c}
 1 \;\; 2 \;\;\; 3 \;\; 4 \;\;\; 5 \;\; 6 \;\; 7 \;\;\; 8 \;\; 9 \\
\left[
\begin{array}{ccccccccc}
3 & 8 & 40 & 35 & 5 & 24 & 37 & 20 & 22 \\
8 & 13 & 45 & 40 & 10 & 28 & 42 & 24 & 26 \\
13 & 18 & 50 & 45 & 15 & 32 & 47 & 28 & 30 \\
35 & 40 & 72 & 67 & 37 & 56 & 69 & 52 & 54 \\
40 & 45 & 77 & 72 & 42 & 60 & 74 & 56 & 58 \\
45 & 50 & 82 & 77 & 47 & 64 & 79 & 60 & 62
\end{array}
\right]
\begin{array}{c}
1 \\ 2 \\ 3 \\ 4 \\ 5 \\ 6
\end{array}
\end{array}
$$

In addition, we define an array of the total degrees of freedom of nodes 1 through 9 of the element

$$\{\text{NFR}\}=\{\; 3 \;\; 3 \;\; 3 \;\; 3 \;\; 2 \;\; 2 \;\; 2 \;\; 2 \;\; 2 \;\}$$

With the help of these arrays, we can assemble the element matrices. Fortran statements of the assembly are given in Box 10.7.1.

Box 10.7.1: Fortran statements of the assembly procedure for the mixed model.

```
C
C      Assembly of element coefficient matrix and right-hand column vector
C      for a typical element in mixed formulation.
C
          DO 90 I=1,NPE
          NDFI=NFR(I)
          NR=NFD(N,I)-NDFI
          DO 90 II=1,NDFI
          NR=NR+1
          IF(I.LE.5)THEN
              L=(I-1)*3+II
          ELSE
              L=12+(I-5)*2+II
          ENDIF
          GLF(NR) = GLF(NR)+ELF(L)
C
          DO 90 J=1,NPE
          NDFJ=NFR(J)
          NCL=NFD(N,J)-NDFJ
          DO 90 JJ=1,NDFJ
          NCL=NCL+1
          IF(J.LE.5)THEN
              M=(J-1)*3+JJ
          ELSE
              M=12+(J-5)*2+JJ
          ENDIF
          GLK(NR,NCL)=GLK(NR,NCL)+ELK(L,M)
   90     CONTINUE
```

Fortran statements of the element calculations for the mixed finite element model are given in Box 10.7.2. Here the array ELV denotes the element nodal velocity vector, AMU is the viscosity μ, and RHO is the density ρ. The meaning of other variables remain the same as defined in the earlier discussions:

$$\text{NGP} = \text{number of Gauss points}$$
$$\text{GAUSPT(I,J)} = \text{array of Gauss points}$$
$$\text{GAUSWT(I,J)} = \text{array of Gauss weights}$$
$$\text{DET} = \text{determinant of the Jacobian matrix}$$
$$\text{NPE} = \text{number of nodes per element, 8 or 9}$$
$$\text{SF(I)} = \psi_i, \text{ the Lagrange interpolation functions of the quadratic element}$$
$$\text{SFL(I)} = \text{Lagrange interpolation functions of the linear element}$$
$$\text{GDSF}(j,i) = \frac{\partial \psi_i}{\partial x_j}, \ x_1 = x, \ x_2 = y, \text{ etc.}$$

Box 10.7.2: Fortran statements for the calculation of element matrices of the mixed finite element model.

```
          DO 100 NI=1,NGPF
          DO 100 NJ=1,NGPF
          XI=GAUSPT(NI,NGPF)
          ETA=GAUSPT(NJ,NGPF)
          CALL SHP2DPV(NPE,XI,ETA,ELXY,DET)
          CONST=DET*GAUSWT(NI,NGPF)*GAUSWT(NJ,NGPF)
          VX=0.0
          VY=0.0
          DO 60 I=1,NPE
          L=2*I-1
          VX=VX+SF(I)*ELV(L)
          VY=VY+SF(I)*ELV(L+1)
   60     CONTINUE
          II=1
          KI = 3
          DO 80 I = 1,NPE
          JJ=1
          KJ = 3
          DO 70 J = 1,NPE
          CONV = SF(I)*(VX*GDSF(1,J)+VY*GDSF(2,J))*CONST
          S11 = GDSF(1,I)*GDSF(1,J)*CONST
          S22 = GDSF(2,I)*GDSF(2,J)*CONST
          S12 = GDSF(1,I)*GDSF(2,J)*CONST
          S21 = GDSF(2,I)*GDSF(1,J)*CONST
          IF(J.LE.4)THEN
              S10 = GDSF(1,I)*SFL(J)*CONST
              S20 = GDSF(2,I)*SFL(J)*CONST
          ENDIF
          IF(I.LE.4)THEN
              S01 = GDSF(1,J)*SFL(I)*CONST
              S02 = GDSF(2,J)*SFL(I)*CONST
          ENDIF
          ELK(II,JJ)     = ELK(II,JJ)+AMU*(2.0*S11+S22)+RHO*CONV
          ELK(II,JJ+1)   = ELK(II,JJ+1)+AMU*S21
          ELK(II+1,JJ+1) = ELK(II+1,JJ+1)+AMU*(S11+2.0*S22)+RHO*CONV
          ELK(II+1,JJ)   = ELK(II+1,JJ)  +AMU*S12
          IF(J .LE. 4)THEN
             ELK(II,JJ+2)   = ELK(II,JJ+2)   - S10
             ELK(II+1,JJ+2) = ELK(II+1,JJ+2) - S20
          ENDIF
          IF(I .LE. 4)THEN
             ELK(II+2,JJ)   = ELK(II+2,JJ)   - S01
             ELK(II+2,JJ+1) = ELK(II+2,JJ+1) - S02
          ENDIF
          IF(J .GT. 4)KJ=2
   70     JJ = JJ + KJ
          IF(I .GT. 4)KI=2
   80     II = II + KI
  100     CONTINUE
```

10.7.2   Penalty Model

Computer implementation of the penalty finite element model is quite straightforward and is the same as any multi-degree of freedom systems (see Chapters 5 and 7 and Box 10.7.3), and hence not discussed further. Only note should be made of the fact that the element calculations involve two Gauss loops: a full integration loop for the evaluations of all terms except for the penalty terms and the other one is a reduced integration loop for the evaluation of the penalty terms of the coefficient matrix $\bar{\mathbf{K}}$ [see Eq. (10.6.3)]. Box 10.7.3 contains a Fortran listing of the subroutine to evaluate the coefficient matrix in the penalty finite element model of the Navier–Stokes equations (for the direct iteration only). Box 10.7.4 contains Fortran statements for the computation of stresses and pressure in the penalty finite element model [based on Eqs. (10.6.4) and (10.6.6)].

Box 10.7.3: Fortran statements for the calculation of element matrices of the penalty finite element model.

```
      SUBROUTINE PNLTY2D(NGPF,NPE,NN,NONLIN,GAMMA,RHO,AMU)
C
C     Calculates the coefficient matrix for penalty model
C
      IMPLICIT REAL*8(A-H,O-Z)
      COMMON/STF1/STIF(18,18),ELF(18),ELXY(9,2),VEL(18)
      COMMON/SHP/SFL(9),GDSFL(2,9)
      DIMENSION GAUSPT(4,4),GAUSWT(4,4)
      DATA GAUSPT/4*.0D0,-.57735027D0,.57735027D0,2*.0D0,-.77459667D0,
     *    0.0D0,.77459667D0,.0D0,-.86113631D0,-.33998104D0,.33998104D0,
     *    0.86113631D0/
      DATA GAUSWT/2.0D0,3*0.D0,2*1.0D0,2*.0D0,.55555555D0,.88888888D0,
     *    0.55555555D0,0.0D0,0.34785485D0,2*0.65214515D0,0.34785485D0/
C
      NDF=NN/NPE
      NGPR=NGPF-1
      DO 20 I=1,NN
      ELF(I)=0.0
      DO 20 J=1,NN
   20 STIF(I,J)=0.0
      DO 100 NI=1,NGPF
      DO 100 NJ=1,NGPF
      XI =GAUSPT(NI,NGPF)
      ETA=GAUSPT(NJ,NGPF)
      CALL INTERPLN2D(NPE,XI,ETA,DET,ELXY)
      X=0.0D0
      DO 30 I=1,NPE
   30 X=X+ELXY(I,1)*SFL(I)
      CONST=DET*GAUSWT(NI,NGPF)*GAUSWT(NJ,NGPF)
      IF(NONLIN.GT.0)THEN
         U=0.0D0
         V=0.0D0
```

Box 10.7.3: Fortran statements for the penalty model (continued).

```
          DO 40 I=1,NPE
          L = (I-1)*NDF+1
          U=VEL(L)*SFL(I)+U
   40     V=VEL(L+1)*SFL(I)+V
     ENDIF
     II=1
     DO 90 I=1,NPE
     JJ=1
     DO 80 J=1,NPE
     S11=GDSFL(1,I)*GDSFL(1,J)*CONST
     S22=GDSFL(2,I)*GDSFL(2,J)*CONST
     S12=GDSFL(1,I)*GDSFL(2,J)*CONST
     S21=GDSFL(2,I)*GDSFL(1,J)*CONST
     STIF(II,JJ)    =STIF(II,JJ)    +AMU*(2.0*S11+S22)
     STIF(II+1,JJ+1)=STIF(II+1,JJ+1)+AMU*(S11+2.0*S22)
     STIF(II,JJ+1)  =STIF(II,JJ+1)  +AMU*S21
     STIF(II+1,JJ)  =STIF(II+1,JJ)  +AMU*S12
     IF(NONLIN.GT.0)THEN
        CNV=SFL(I)*(U*GDSFL(1,J)+V*GDSFL(2,J))*CONST
        STIF(II,JJ)    =STIF(II,JJ)    + RHO*CNV
        STIF(II+1,JJ+1)=STIF(II+1,JJ+1)+ RHO*CNV
     ENDIF
  80 JJ=NDF*J+1
  90 II=NDF*I+1
 100 CONTINUE
C    Evaluate the penalty terms using reduced integration
     DO 150 NI=1,NGPR
     DO 150 NJ=1,NGPR
     XI =GAUSPT(NI,NGPR)
     ETA=GAUSPT(NJ,NGPR)
     CALL INTERPLN2D(NPE,XI,ETA,DET,ELXY)
     DO 110 I=1,NPE
 110 X=X+ELXY(I,1)*SFL(I)
     CONST=DET*GAUSWT(NI,NGPR)*GAUSWT(NJ,NGPR)
     II=1
     DO 140 I=1,NPE
     JJ=1
     DO 130 J=1,NPE
     S11=GDSFL(1,I)*GDSFL(1,J)*CONST
     S22=GDSFL(2,I)*GDSFL(2,J)*CONST
     S12=GDSFL(1,I)*GDSFL(2,J)*CONST
     S21=GDSFL(2,I)*GDSFL(1,J)*CONST
     STIF(II,JJ)    = STIF(II,JJ)    + GAMMA*S11
     STIF(II+1,JJ+1) = STIF(II+1,JJ+1) + GAMMA*S22
     STIF(II,JJ+1)  = STIF(II,JJ+1)  + GAMMA*S12
     STIF(II+1,JJ)  = STIF(II+1,JJ)  + GAMMA*S21
 130 JJ=NDF*J+1
 140 II=NDF*I+1
 150 CONTINUE
     RETURN
     END
```

Box 10.7.4: Fortran statements for stress and pressure calculation in the penalty finite element model.

```
      SUBROUTINE STRS2D(GAMMA,AMU,NGPR,NPE,ELXY,VEL,NONLIN)
C
C     Computes viscous stresses, pressure, and the divergence of velocity
C           at the integration points of the reduced integration rule
C
      IMPLICIT REAL*8 (A-H,O-Z)
      COMMON/SHP/SFL(9),GDSFL(2,9)
      DIMENSION ELXY(9,2),VEL(18),GAUSPT(4,4)
      DATA GAUSPT/4*0.0D0,-.57735027D0,.57735027D0,2*0.0D0,-.77459667D0,
     *0.0D0,0.77459667D0,0.0D0,-0.86113631D0,-0.33998104D0,.33998104D0,
     *0.86113631D0/
C
C     Calculation of strain-rates and stresses in the element
C
      IT=6
      NDF=2
      DO 50 NI=1, NGPR
      DO 50 NJ=1, NGPR
      XI  = GAUSPT(NI,NGPR)
      ETA = GAUSPT(NJ,NGPR)
      CALL INTERPLN2D(NPE,XI,ETA,DET,ELXY)
      X = 0.0
      Y = 0.0
      U =0.0
      UX=0.0
      UY=0.0
      VX=0.0
      VY=0.0
      DO 30 I=1,NPE
      X = X + ELXY(I,1)*SFL(I)
      Y = Y + ELXY(I,2)*SFL(I)
      L=(I-1)*NDF+1
      UX=UX+VEL(L)*GDSFL(1,I)
      UY=UY+VEL(L)*GDSFL(2,I)
      VX=VX+VEL(L+1)*GDSFL(1,I)
      VY=VY+VEL(L+1)*GDSFL(2,I)
   30 CONTINUE
      DIV=UX+VY
      PRS=-GAMMA*DIV
      SX =2.0*AMU*UX-PRS
      SY =2.0*AMU*VY-PRS
      SXY=AMU*(UY+VX)
   50 WRITE (IT,60) X,Y,SX,SY,SXY,PRS
   60 FORMAT(6(2X,E11.4))
      RETURN
      END
```

10.7.3   Transient Analysis

Computer implementation of nonlinear time-dependent problems is complicated by the fact that one must keep track of the solution vectors at different Reynolds numbers, times, and iterations. Often, for a fixed value of the Reynolds number, one obtains the transient solution. Therefore, the outer loop is on the number of Reynolds number increments, followed by a loop on the number of time steps, and the inner most loop being on nonlinear iterations.

10.8   Numerical Examples

10.8.1   Preliminary Comments

In this section, a small number of flow problems solved using the finite element models developed herein are presented. Additional examples may be found in the book by Reddy and Gartling [10]. The examples presented herein were solved using the reduced integration penalty finite element model (RIP) and mixed finite element model. The objective of the first several examples, which deal with linear solutions, is to evaluate the accuracy of the penalty and mixed finite element models by comparing with the available analytical or numerical results and to illustrate the effect of the penalty parameter on the accuracy of the solutions. The remaining examples are for Reynolds numbers greater than unity (i.e. convective terms are included), and the results were obtained using the reduced integration penalty finite element model.

For high Reynolds number flows, the total value of Re is divided into suitable increments, ΔRe, much the same way the total load was divided in a structural problem. The converged nonlinear solution of the preceding Reynolds number is used as the initial guess in the first iteration of the next Reynolds number. In general, for very high Reynolds numbers under-relaxation may be necessary to achieve convergence. Recall that under-relaxation involves use of a weighted average of velocities from the last two iterations (for ELV)

$$\bar{\mathbf{v}}^r = \beta\,\mathbf{v}^{(r)} + (1 - \beta)\,\mathbf{v}^{(r-1)} \tag{10.8.1}$$

to compute the element matrix. Here β is the acceleration parameter, whose value is problem dependent. One may begin with $\beta = 0.5$ and numerically experiment to determine one that gives convergent solution.

10.8.2   Linear Problems

Here we consider three examples that have either analytical (not exact) or numerical solutions available in the literature. In all three cases, the theoretical background and computational aspects are provided for verification purpose.

Example 10.8.1

Fluid Squeezed Between Parallel Plates. Consider the slow flow of a viscous incompressible material squeezed between two long parallel plates, as shown in Fig. 10.8.1(a). Let V_0 be the velocity with which the two plates are moving toward each other (i.e. squeezing out the fluid), and let $2b$ and $2a$ denote, respectively, the distance between and the length of the plates at a fixed instant of time. Assuming that the length of the plates (into the plane of the paper) is very large compared to both the width of and the distance between the plates, the problem can be treated as one of a plane flow. Use the biaxial symmetry (prove it to yourself), use a 5×3 nonuniform mesh of nine-node quadratic elements (Q9) in the mixed model, a 10×6 mesh of the four-node linear elements (L4), and 5×3 mesh of nine-node quadratic elements in the penalty model, as shown in Fig. 10.8.1(b), to analyze the problem for velocity and pressure fields.

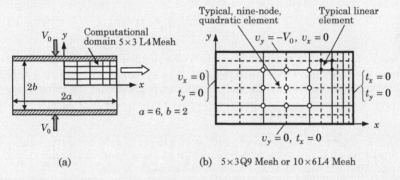

Fig. 10.8.1: (a) Geometry and the computational domain and (b) the finite element mesh used for the analysis of slow flow of viscous incompressible fluid between parallel plates.

Solution: First, we identify the boundary conditions for the computational domain, $\Omega : 0 < x < a = 6, 0 < y < b = 2$. At every boundary point, we must know one element of each of the two pairs: (v_x, t_x) and (v_y, t_y). We have [see Fig. 10.8.1(b)]

$$\text{At } x = 0: \quad v_x = 0, \quad t_y = -\sigma_{xy} = 0; \quad \text{At } y = 0: \quad v_y = 0, \quad t_x = -\sigma_{xy} = 0$$
$$\text{At } x = a: \quad t_x = \sigma_{xx} = 0, \quad t_y = \sigma_{xy} = 0; \quad \text{At } y = b: \quad v_x = 0, \quad v_y = -V_0$$

We note that the computational domain has singularities in the boundary conditions at all four corner points, namely, at $(x, y) = (0, 0)$, $(x, y) = (0, b)$, $(x, y) = (a, 0)$, and $(x, y) = (a, b)$. At $(x, y) = (0, 0), (0, b)$, and $(a, 0)$, the singularity is due to the fact that both v_y and t_y are specified (to be zero); at $(x, y) = (a, b)$ we have double singularity because both elements of the two pairs are specified. In the finite element method, we always give priority to the specified primary variables (in this case, velocities) when conflict between the primary and secondary variables arise. However, we cannot totally avoid the singularity at $(x, y) = (a, b)$, because the stress-free condition on $x = a$ line (outflow boundary) must be imposed as we do not have any conditions on the velocity components there. By refining the mesh in the vicinity of these points, we can localize the effect (but not eliminate). If we do not impose any boundary conditions there, it amounts, in the finite element formulation, to requiring $t_x = t_y = 0$ in the integral sense. In the mixed finite element model, it is necessary to specify the pressure at least at one node. In the present case, the node at $(x, y) = (a, 0)$ is specified to have zero pressure.

The presence of singularities requires us to use nonuniform meshes, with smaller elements near the free surface (i.e. at $x = a = 6$); the following element lengths are used to generate the mesh:

$$DX(I) = \{1.0 \ 1.0 \ 1.0 \ 1.0 \ 0.5 \ 0.5 \ 0.25 \ 0.25 \ 0.25 \ 0.25\}$$

$$DY(I) = \{0.25 \ 0.25 \ 0.5 \ 0.5 \ 0.25 \ 0.25\}$$

The meshes used for both models have exactly the same number of nodes, although the total number of degrees of freedom are not the same due to the pressure variable in the mixed model.

The velocities $v_x(x, 0)$ obtained with the two finite element models are compared in Table 10.8.1 with the approximate analytical solution of Nadai [198], which is given by

$$v_x(x, y) = \frac{3V_0 x}{2b}\left(1 - \frac{y^2}{b^2}\right), \quad v_y(x, y) = -\frac{3V_0 y}{2b}\left(3 - \frac{y^2}{b^2}\right)$$

$$P(x, y) = \frac{3\mu V_0}{2b^3}(a^2 + y^2 - x^2)$$

(1)

The problem cannot have an exact solution due to the singularities discussed above. The nine-node element gives very good results for both the penalty and mixed models. The influence of the penalty parameter on the accuracy of the solution is clear from the results. Whether the element is linear or quadratic, it is necessary to use a large value of the penalty parameter.

Table 10.8.1: Comparison of finite element solution $v_x(x, 0)$ with the analytical solution for fluid squeezed between plates; 5×3Q9 mesh is used in the mixed model, and 5×3Q9 and 10×6 L4 meshes are used in the penalty model*.

| | $\gamma = 1.0$ | | $\gamma = 100$ | | $\gamma = 10^8$ | | Mixed (nine-node) | Nadai's series solution |
x	Four-node	Nine-node	Four-node	Nine-node	Four-node	Nine-node		
1.00	0.0303	0.0310	0.6563	0.6513	0.7576	0.7505	0.7497	0.7500
2.00	0.0677	0.0691	1.3165	1.3062	1.5135	1.4992	1.5031	1.5000
3.00	0.1213	0.1233	1.9911	1.9769	2.2756	2.2557	2.2561	2.2500
4.00	0.2040	0.2061	2.6960	2.6730	3.0541	3.0238	3.0203	3.0000
4.50	0.2611	0.2631	3.0718	3.0463	3.4648	3.4307	3.4292	3.3750
5.00	0.3297	0.3310	3.4347	3.3956	3.8517	3.8029	3.8165	3.7500
5.25	0.3674	0.3684	3.6120	3.5732	4.0441	3.9944	3.9893	3.9375
5.50	0.4060	0.4064	3.7388	3.6874	4.1712	4.1085	4.1204	4.1250
5.75	0.4438	0.4443	3.8316	3.7924	4.2654	4.2160	4.2058	4.3125
6.00	0.4793	0.4797	3.8362	3.7862	4.2549	4.1937	4.2364	4.5000

*The 2×2 Gauss rule for non-penalty terms and one-point Gauss rule for penalty terms are used for linear elements (L4), whereas the 3×3 Gauss rule for non-penalty terms and the 2×2 Gauss rule for penalty terms are used for quadratic elements (Q9).

Figure 10.8.2(a) contains plots of the velocity field $v_x(x, y)$ versus y at $x = 4$ and $x = a = 6$, obtained with two different meshes, 5×3Q9 and 10×8Q9, and the penalty finite element model, and Fig. 10.8.2(b) contains the same plots obtained with the mixed finite element model. Both models predict essentially the same results (see also Fig. 10.8.3). One should not conclude that the finite element solutions are in error when compared to Nadai's approximate solution; rather one may say that Nadai's solution [198] is not a bad approximation of the true solution.

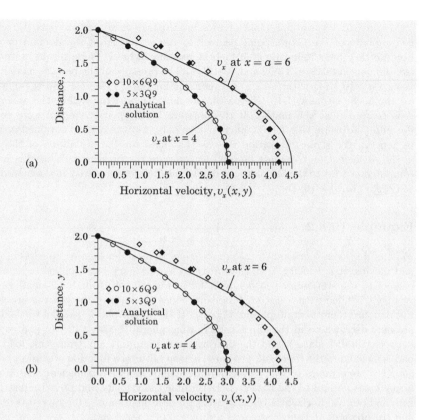

Fig. 10.8.2: Velocity fields $v(x_0, y)$ at $x_0 = 4$ and $x_0 = 6$ for fluid squeezed between parallel plates. (a) Penalty model. (b) Mixed model.

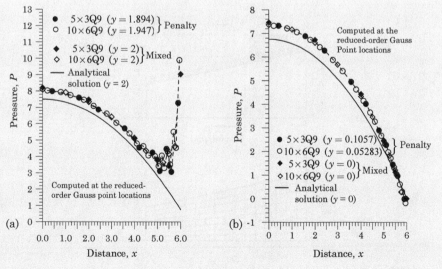

Fig. 10.8.3: Pressure $P(x, y_0)$ versus x for fluid squeezed between parallel plates. (a) Near or at the top plate. (b) Near or on the centerline of the domain

Figure 10.8.3 contains plots of pressure $P(x,y)$ versus x, for $y = y_0$, where y_0 is the y-coordinate of the Gauss point nearest to the centerline of the domain or the top plate, because they are the locations of highest pressure values. These results were obtained with two different meshes: 5×3Q9 and 10×6Q9. The pressure in the penalty model was computed using Eq. (10.5.14) with the 2×2 Gauss rule for the Q9 element and the one-point formula for the linear element (L4), whereas in the mixed model (as well as the analytical solution) it is computed at the nodes. If the pressure in the penalty model were computed using the full quadrature rule for rectangular elements, we would have obtained erroneous values. In general, the same integration rule as that used for the evaluation of the penalty terms in the coefficient matrix must be used to compute the pressure. The oscillations in pressure computed nearest to the top plate are due to the double-singularity in the boundary conditions at $(x,y) = (a,b) = (6,2)$.

Example 10.8.2

Flow of a viscous lubricant in a slider bearing. The slider bearing consists of a short sliding pad moving at a velocity $v_x = V_0$ relative to a stationary pad inclined at a small angle with respect to the stationary pad, as depicted in Fig. 10.8.4(a), and the small gap between the two pads is filled with a viscous lubricant. Since the ends of the bearing are generally open, the pressure there is atmospheric, $P = P_0$. If the upper pad is parallel to the base plate, the pressure everywhere in the gap must be atmospheric (because dP/dx is a constant for flow between parallel plates), and the bearing cannot support any transverse load. If the upper pad is inclined to the base pad, pressure (in general, P is a function of x and y) develops in the gap. For large values of V_0, the pressure developed can be of sufficient magnitude to support heavy loads normal to the base pad. Determine the velocity and pressure distributions in the gap, and plot (a) horizontal velocity $v_x(x_0, y)$ as a function of y for $x_0 = 0, 0.18$ and $x_0 = 0.36$ and (b) pressure and shear stress as a function of x for $y = 0$.

Solution: In the finite element analysis, we solve the Stokes equations [i.e. neglect the convective terms in Eqs. (10.4.1) and (10.4.2)] using a mesh of 16×8 linear quadrilateral elements to analyze the problem. Figure 10.8.4(b) shows the mesh (graded uniformly) and the boundary conditions.

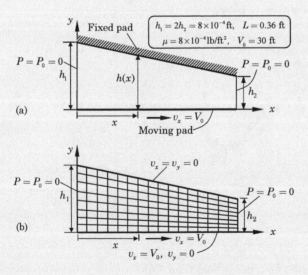

Fig. 10.8.4: Schematic and the finite element mesh for slider bearing.

Schlichting [199] developed an analytical solution of the problem by assuming that the width of the gap and the angle of inclination are small so that $v_y = 0$ and the pressure is not a function of y. Assuming a two-dimensional state of flow and a small angle of inclination, and neglecting the normal stress gradient (in comparison with the shear stress gradient), the equations governing the flow of the lubricant between the pads can be reduced to

$$\mu\frac{\partial^2 v_x}{\partial y^2} = \frac{dP}{dx}, \quad 0 < x < L \tag{1}$$

where the pressure gradient is given by

$$\frac{dP}{dx} = \frac{6\mu V_0}{h^2}\left(1 - \frac{H}{h}\right), \quad H = \frac{2h_1 h_2}{h_1 + h_2} \tag{2}$$

The boundary conditions are

$$v_x(x,0) = V_0, \quad v_x(x,h) = 0, \quad h(x) = h_1 + \frac{h_2 - h_1}{L}x \tag{3}$$

The solution of Eq. (1) subject to boundary conditions in Eq. (3) is

$$v_x(x,y) = \left(V_0 - \frac{h^2}{2\mu}\frac{dP}{dx}\frac{y}{h}\right)\left(1 - \frac{y}{h}\right) \tag{4}$$

$$P(x) = \frac{6\mu V_0 L(h_1 - h)(h - h_2)}{h^2(h_1^2 - h_2^2)} \tag{5}$$

$$\sigma_{xy}(x,y) = \mu\frac{\partial v_x}{\partial y} = \frac{dP}{dx}\left(y - \frac{h}{2}\right) - \mu\frac{V_0}{h} \tag{6}$$

Table 10.8.2 contains a comparison of the finite element solutions and analytical solutions for the velocity, pressure, and shear stress. The penalty parameter is chosen to be $\gamma = \mu \times 10^8$. Figure 10.8.5 contains plots of the horizontal velocity v_x versus $y = h(x)$ at $x = 0$ ft, $x = 0.18$ ft, and $x = 0.36$ ft. Figure 10.8.6 contains plots of pressure $P \times 10^{-2}$ and shear stress $-\sigma_{xy}$ as a function of x. The analytical solutions for (v_x, P, σ_{xy}) are evaluated for $y = 0$. Although the finite element solutions for P and σ_{xy} are essentially independent of y, they were computed at the center of the first row of elements along the moving block. The results for the velocities are in excellent agreement with the analytical solutions (4)–(6), validating the assumptions made in the development of the analytical solution. The pressure and shear stress are also in good agreement.

Example 10.8.3

Lid-driven cavity flow. Consider the laminar flow of a viscous, incompressible fluid in a square cavity bounded by three stationary walls and a lid moving at a constant velocity in its own plane, as shown in Fig. 10.8.7. Singularities exist at each corner where the moving lid meets a fixed wall. This example is one that has been extensively studied by analytical, numerical, and experimental methods (see [200–202], for example), and it is often used as a benchmark problem to verify a computational scheme. Assuming a unit square ($a = 1$), the velocity of the top wall to be unity ($v_0 = 1$), and using uniform meshes of (a) 8×8 linear elements and (b) 4×4 nine-node quadratic elements, determine the velocity and pressure fields. At the singular points, namely at the top corners of the lid, assume that $v_x(x,1) = v_0 = 1.0$.

Table 10.8.2: Comparison of finite element solutions velocities with the analytical solutions for viscous fluid in a slider bearing.

$\bar{y}$	$v_x(0, y)$ FEM	$v_x(0, y)$ Analyt.	$\bar{y}$	$v_x(0.18, y)$ FEM	$v_x(0.18, y)$ Analyt.	$\bar{y}$	$v_x(0.36, y)$ FEM	$v_x(0.36, y)$ Analyt.
0.0	30.000	30.000	0.00	30.000	30.000	0.00	30.000	30.000
1.0	22.923	22.969	0.75	25.139	25.156	0.50	29.564	29.531
2.0	16.799	16.875	1.50	20.596	20.625	1.00	28.182	28.125
3.0	11.626	11.719	2.25	16.372	16.406	1.50	25.853	25.781
4.0	7.403	7.500	3.00	12.465	12.500	2.00	22.577	22.500
5.0	4.130	4.219	3.75	8.874	8.906	2.50	18.354	18.281
6.0	1.805	1.875	4.50	5.600	5.625	3.00	13.184	13.125
7.0	0.429	0.469	5.25	2.642	2.656	3.50	7.066	7.031
8.0	0.000	0.000	6.00	0.000	0.000	4.00	0.000	0.000

	Analytical			FEM		
x	$\bar{P}(x,0)$	$-\sigma_{xy}(x,0)$	$\bar{x}$	$\bar{y}$	$\bar{P}$	$-\sigma_{xy}$
0.01	7.50	59.99	0.1125	0.4922	8.46	56.61
0.03	22.46	59.89	0.3375	0.4766	25.46	56.60
0.05	37.29	59.67	0.5625	0.4609	42.31	56.47
0.07	51.89	59.30	0.7875	0.4453	58.76	56.17
0.09	66.12	58.77	1.0125	0.4297	74.69	55.69
0.27	129.60	38.40	2.5875	0.3203	134.40	41.77
0.29	118.57	32.71	2.8125	0.3047	125.60	36.93
0.31	99.58	25.70	3.0375	0.2891	107.60	30.76
0.33	70.30	17.04	3.2625	0.2734	77.39	22.89
0.35	27.61	6.31	3.4875	0.2578	30.80	12.82

$\bar{x} = 10x, \quad \bar{y} = y \times 10^4, \quad \bar{P} = P \times 10^{-2}.$

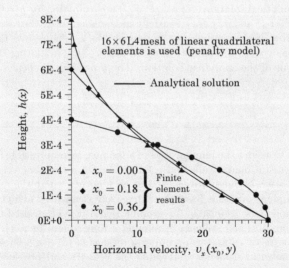

Fig. 10.8.5: Velocity distributions for the slider bearing problem.

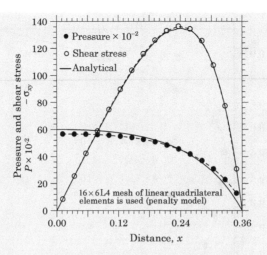

Fig. 10.8.6: Pressure and shear stress distributions for the slider bearing problem.

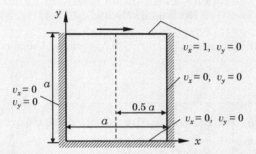

Fig. 10.8.7: Geometry and boundary conditions for the wall-driven cavity problem.

Solution: The linear solution for the horizontal velocity along the vertical centerline obtained with the two meshes, 8×8L4 and 4×4Q9, and two different values of the penalty parameter, $\gamma = 10^2$ and $\gamma = 10^8$, are presented in Table 10.8.3 (see also Fig. 10.8.8), and the variation of pressure along the top wall (computed at the reduced Gauss points) is shown in Fig. 10.8.9. It is clear that the value of the penalty parameter between $\gamma = 10^2$ and $\gamma = 10^8$ has small effect on the solution.

Table 10.8.3: Velocity $v_x(0.5, y)$ obtained with various values of the penalty parameter γ (linear solution).

	Mesh: 8×8L4		Mesh: 4×4Q9	
y	$\gamma = 10^2$	$\gamma = 10^8$	$\gamma = 10^2$	$\gamma = 10^8$
0.125	-0.0557	-0.0579	-0.0589	-0.0615
0.250	-0.0938	-0.0988	-0.0984	-0.1039
0.375	-0.1250	-0.1317	-0.1320	-0.1394
0.500	-0.1354	-0.1471	-0.1442	-0.1563
0.625	-0.0818	-0.0950	-0.0983	-0.1118
0.750	0.0958	0.0805	0.0641	0.0481
0.875	0.4601	0.4501	0.4295	0.4186

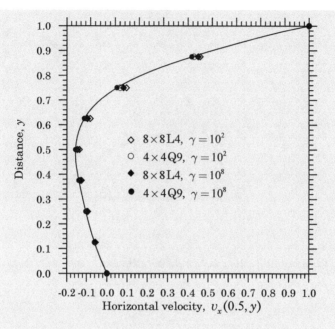

Fig. 10.8.8: Plots of horizontal velocity $v_x(0.5, y)$ along the vertical centerline of the cavity.

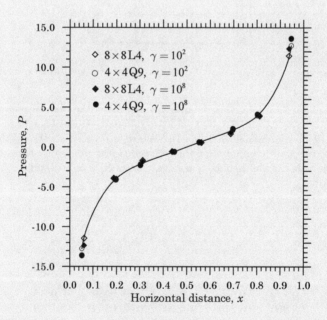

Fig. 10.8.9: Plots of pressure $P(x, y_0)$ along the top wall of the cavity.

10.8.3 Nonlinear Problems

Here we consider a couple of examples involving the solution of the Navier–Stokes equations. First we revisit the lid-driven cavity flow.

Example 10.8.4

Lid-driven cavity flow. Consider the lid-driven cavity problem of Example 10.8.3 for nonlinear analysis (i.e. solve Navier–Stokes equations for $Re > 0$). For the problem at hand, the Reynolds number ($Re = \rho v_0 a/\mu$) can be varied by varying the density ρ while keeping the viscosity μ and lid velocity v_0 fixed. Thus, take (in addition to the choice of the characteristic length $a = 1$) $\mu = 1$ and $v_0 = 1$ so that $Re = \rho$. Solve the problem for $Re = 250, 500$, and 750, using uniform 8×8 mesh of linear elements as well as 4×4 mesh of nine-node quadratic elements, penalty parameter $\gamma = 10^8$, and convergence tolerance of $\epsilon = 10^{-3}$.

Solution: In solving this problem, the mesh used should be such that the boundary layer thickness, which is of the order of $Re^{-\frac{1}{2}}$, near the walls is resolved. The numerical results obtained with the two meshes and a relaxation parameter $\beta = 0.25$ [see Eq. (10.8.1)], after some numerical tries, are presented in Table 10.8.4 for $Re = 250, 500$, and 750. The convergence may not be reached for $Re = 750$ if $\beta = 0$.

Figure 10.8.10 shows plots of the horizontal velocity along the cavity centerline obtained with 8×10 mesh of nine-node quadratic elements and 16×20 mesh of linear elements for $Re = 0, 500$, and 1,000 ($\beta = 0.5$). The mesh grading used for the Q9 elements is as follows:

$$\{DX\} = \{0.125 \ 0.125 \ 0.125 \ 0.125 \ 0.125 \ 0.125 \ 0.125 \ 0.125\}$$

$$\{DY\} = \{0.125 \ 0.125 \ 0.125 \ 0.125 \ 0.125 \ 0.125 \ 0.0625 \ 0.0625 \ 0.0625 \ 0.0625\}$$

For the linear elements the above lengths are subdivided. The results show excellent convergence with p refinement. The pressure and shear stress near to the moving lid obtained with the two meshes are shown in Figs. 10.8.11 and 10.8.12, respectively. Clearly, both pressure and shear stress exhibit oscillations.

Table 10.8.4: Velocity $v_x(0.5, y)$ obtained with linear and quadratic elements and for various values of the Reynolds number (values in parentheses are linear solutions).

y		Mesh: 8×8L4			Mesh: 4×4Q9	
$Re \rightarrow$	250(9)*	500(9)	750(10)	250(10)	500(12)	750(20)
0.125	−0.0367	−0.0239	−0.0128	−0.0412	−0.0131	0.0146
	(−0.0579)			(−0.0615)		
0.250	−0.0688	−0.0502	−0.0320	−0.0851	−0.0520	0.0017
	(−0.0988)			(−0.1039)		
0.375	−0.0944	−0.0733	−0.0533	−0.1283	−0.1133	−0.0481
	(−0.1317)			(−0.1393)		
0.500	−0.0911	−0.0696	−0.0569	−0.1305	−0.1284	−0.1086
	(−0.1471)			(−0.1563)		
0.625	−0.0176	0.0043	0.0020	−0.0437	−0.0494	−0.0901
	(−0.0950)			(−0.1118)		
0.750	0.0469	0.0414	0.0323	0.0753	0.1042	0.0546
	(0.0805)			(0.0481)		
0.875	0.2616	0.1712	0.1207	0.2833	0.2137	0.1495
	(0.4500)			(0.4186)		

* Number in parentheses denotes the number of iterations taken for convergence.

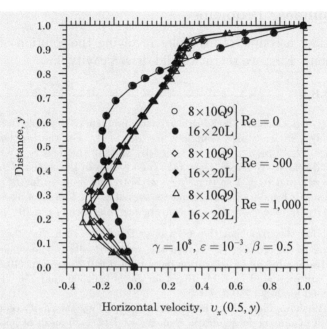

Fig. 10.8.10: Velocity $v_x(0.5, y)$ versus y for Reynolds numbers $Re = 0, 500$, and $1,000$ (obtained with $8 \times 10Q9$ and $16 \times 20L4$ meshes).

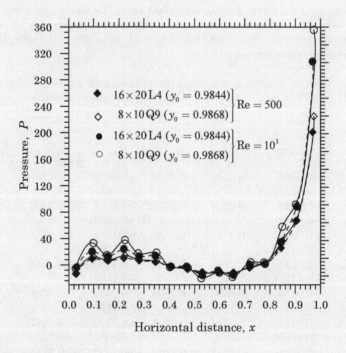

Fig. 10.8.11: Plots of pressure $P(x, y_0)$ along the top wall of the cavity ($8 \times 10Q9$ and $16 \times 20L4$ meshes).

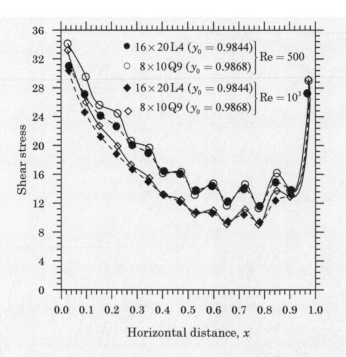

Fig. 10.8.12: Plots of shear stress $\sigma_{xy}(x, y_0)$ along the top wall of the cavity (8×10Q9 and 16×20L4 meshes).

Finally, a refined nonuniform mesh of linear elements was used (see Fig. 10.8.13), where a penalty parameter of $\gamma = 10^8$ and convergence tolerance of $\epsilon = 10^{-3}$ were adopted. Convergence was achieved with three iterations at each Reynolds number step $\Delta Re = 100$. Figures 10.8.14 and 10.8.15 show the results for $Re = 400$ and $Re = 10^3$.

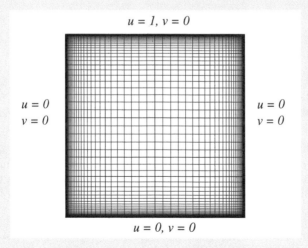

Fig. 10.8.13: Mesh used for the wall-driven cavity problems.

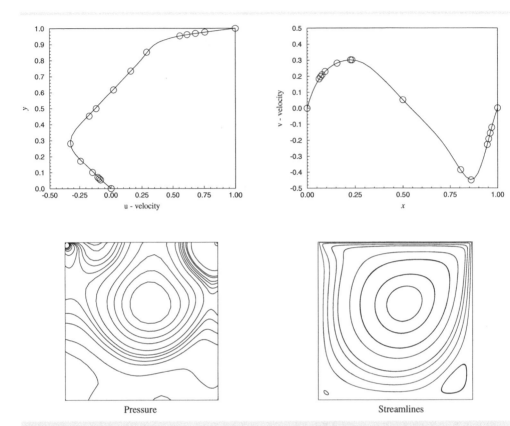

Fig. 10.8.14: Horizontal and vertical velocity profiles along the vertical and horizontal mid-sections of the cavity (— computed, ○ Ghia et al. [202]) and pressure and streamline contours ($Re = 400$).

Example 10.8.5

Backward-Facing Step. Analyze the well-known backward-facing problem (see Gartling [203] and Reddy and Gartling [10]) using the penalty finite element model. The geometry and boundary conditions of the computational domain are shown in Fig. 10.8.16. A penalty parameter of $\gamma = 10^8 Re$ and convergence tolerance of $\varepsilon \leq 10^{-2}$ are to be used.

Solution: Figure 10.8.17 shows contour plots of the streamlines and pressure, while Fig. 10.8.18 contains pressure profiles along the upper and lower walls for Reynolds number $Re = 800$ (convergence is achieved with five iterations). The results compare well with those of Gartling [203]. This problem will be revisited in the next chapter in the context of the least-squares finite element analysis.

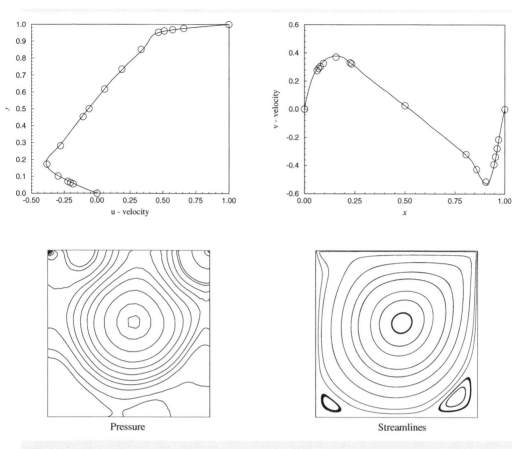

Fig. 10.8.15: Horizontal and vertical velocity profiles along the vertical and horizontal mid-sections of the cavity for $Re = 10^3$ (— computed, ○ Ghia et al. [202]). Pressure and streamline contours.

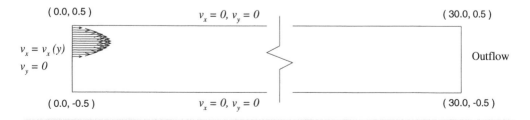

Fig. 10.8.16: Geometry and boundary conditions for flow over a backward-facing step.

Streamlines

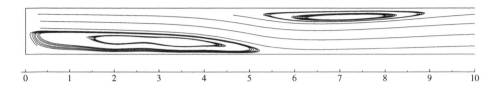

Pressure

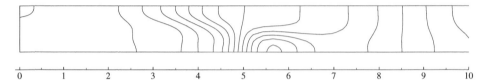

Fig. 10.8.17: Streamlines and pressure contours for flow over a backward-facing step ($Re = 800$).

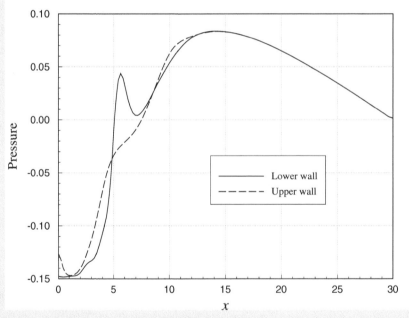

Fig. 10.8.18: Plots of pressure along the upper and lower walls of the channel for flow over a backward-facing step ($Re = 800$).

This completes the steady-state analysis of flows of incompressible fluids. In Section 10.8.4, we study the unsteady (or transient) analysis of the flows of viscous incompressible fluids using the weak-form Galerkin finite element formulation.

10.8.4 Transient Analysis

The weak-form penalty finite element models developed in Section 10.5 already include time-dependent terms [see Eq. (10.5.16) or (10.5.18)]. The fully discretized equations in Eq. (10.5.18) are discussed here. We have

$$\hat{\mathbf{K}}(\mathbf{v}_{s+1})\mathbf{v}_{s+1} = \tilde{\mathbf{K}}(\mathbf{v}_s)\mathbf{v}_s + \hat{\mathbf{F}}_{s,s+1} \tag{10.8.2}$$

where

$$\begin{aligned}
\hat{\mathbf{K}}(\mathbf{v}_{s+1}) &= \mathbf{M} + a_1\left[\mathbf{C}(\mathbf{v}_{s+1}) + \mathbf{K}_\mu(\mathbf{v}_{s+1}) + \mathbf{K}_\gamma\right] \\
\tilde{\mathbf{K}}(\mathbf{v}_s) &= \mathbf{M} - a_2\left[\mathbf{C}(\mathbf{v}_s) + \mathbf{K}_\mu(\mathbf{v}_s) + \mathbf{K}_\gamma\right] \\
\hat{\mathbf{F}}_{s,s+1} &= a_1\mathbf{F}_{s+1} + a_2\mathbf{F}_s
\end{aligned} \tag{10.8.3}$$

where $\mathbf{M}$, $\mathbf{C}$, $\mathbf{K}_\mu$, and $\mathbf{K}_\gamma$ are defined in Eqs. (10.5.16) and (10.5.18).

Dependence of $\mathbf{C}$ on the nodal values $\mathbf{v}$ is due to the convective terms and dependence of $\mathbf{K}_\mu$ on $\mathbf{v}$ is due to viscosity being a function of the strain rates (e.g. power-law fluids). If viscosity is constant, that is Newtonian fluids, then $\mathbf{K}_\mu$ is independent of the nodal values $\mathbf{v}$ of the velocity field. For direct iteration solution, Eq. (10.8.2) takes the form

$$\hat{\mathbf{K}}(\mathbf{v}_{s+1}^r)\mathbf{v}_{s+1}^{r+1} = \tilde{\mathbf{K}}(\mathbf{v}_s)\mathbf{v}_s + \hat{\mathbf{F}}_{s,s+1} \tag{10.8.4}$$

Example 10.8.6 ──

Transient analysis of the lid-driven cavity problem. Carry out the transient analysis, for $Re = 0$, $Re = 1{,}000$, and $Re = 2{,}500$, of a viscous fluid inside a wall-driven cavity (see Example 10.8.3). Use $16 \times 20\text{Q}4$ nonuniform mesh of four-node rectangular elements in the domain. Take the mesh size in each coordinate direction to be

```
{DX}={0.0625   0.0625   0.0625   0.0625   0.0625   0.0625   0.0625   0.0625
      0.0625   0.0625   0.0625   0.0625   0.0625   0.0625   0.0625   0.0625}
{DY}={0.0625   0.0625   0.0625   0.0625   0.0625   0.0625   0.0625   0.0625
      0.0625   0.0625   0.0625   0.0625   0.03125  0.03125  0.03125  0.03125
      0.03125  0.03125  0.03125  0.03125}
```

Use the Crank–Nicolson method ($\alpha = 0.5$) with two different time steps $\Delta t = 0.01$ and $\Delta t = 0.001$ are used.

Solution: Table 10.8.5 contains the velocity field $v_x(0.5, y, t) \times 10$ for times $t = 0.01$, 0.05, and 0.1 for $Re = 0$. The solution reaches the steady state ($\epsilon = 10^{-2}$) at time $t = 0.1$ when $\Delta t = 0.01$ is used. Figure 10.8.19 shows the evolution of $v_x(0.5, y, t)$.

Next, we consider nonlinear transient analysis. The same mesh, namely, the mesh of 16×20 four-node quadrilateral elements, is used with $\Delta t = 0.001$. The transient response is calculated for Reynolds numbers $Re = 1{,}000$ and $Re = 2{,}500$ separately. Figures 10.8.20(a) and (b) show plots of the nonlinear steady-state and transient center horizontal velocity $v_x(0.5, y, t)$ versus y for various times.

Table 10.8.5: The horizontal velocity field $v_x(0.5, y, t) \times 10$ versus time t for the wall-driven cavity problem ($16 \times 20Q4$ mesh).

y	$t = 0.01$ $\Delta t = 0.01$	$t = 0.01$ $\Delta t = 0.001$	$t = 0.05$ $\Delta t = 0.01$	$t = 0.05$ $\Delta t = 0.001$	$t = 0.10$ $\Delta t = 0.01$	Steady-state
0.0625	−0.1342	−0.1953	−0.3103	−0.3247	−0.3655	−0.3688
0.1250	−0.1936	−0.3140	−0.5624	−0.5841	−0.6558	−0.6631
0.1875	−0.2314	−0.3940	−0.7888	−0.8163	−0.9108	−0.9198
0.2500	−0.2691	−0.4651	−1.0122	−1.0435	−1.1499	−1.1593
0.3125	−0.3157	−0.5475	−1.2346	−1.2746	−1.3802	−1.3886
0.3750	−0.3759	−0.6536	−1.4790	−1.5053	−1.5967	−1.6028
0.4375	−0.4516	−0.7902	−1.6964	−1.7151	−1.7793	−1.7820
0.5000	−0.5435	−0.9605	−1.8536	−1.8643	−1.8906	−1.8895
0.5625	−0.6465	−1.1577	−1.8846	−1.8878	−1.8700	−1.8652
0.6250	−0.7474	−1.3479	−1.7011	−1.6946	−1.6336	−1.6250
0.6875	−0.8097	−1.4428	−1.1889	−1.1653	−1.0700	−1.0572
0.7500	−0.7536	−1.1523	−0.2093	−0.1693	−0.0520	−0.0382
0.7813	−0.6325	−0.7744	0.5100	0.5471	0.6713	0.6820
0.8125	−0.4077	−0.1695	1.4014	1.4197	1.5467	1.5526
0.8438	−0.0054	0.7336	2.4885	2.4716	2.5918	2.5965
0.8750	0.6329	1.9318	3.7259	3.6716	3.7646	3.7824
0.9063	1.7000	3.5232	5.1185	5.0707	5.1198	5.1616
0.9375	3.3334	5.3837	6.5139	6.5756	6.6082	6.6410
0.9688	5.9470	7.5970	7.9975	8.2488	8.3805	8.2838

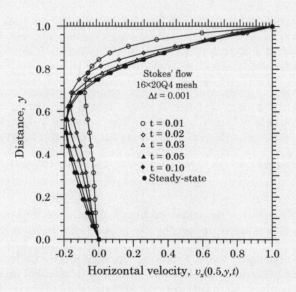

Fig. 10.8.19: Evolution of the horizontal velocity $v_x(0.5, y, t)$ inside a wall-driven cavity.

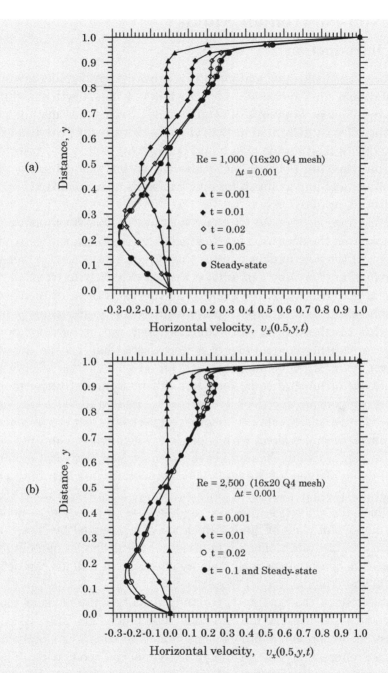

Fig. 10.8.20: Evolution of the horizontal velocity field $v_x(0.5, y, t)$ for the wall-driven cavity problem (nonlinear analysis). (a) Re=1,000 and (b) Re=2,500.

This completes the transient analysis of flows of viscous incompressible fluids. Additional examples of transient analysis will be presented in the next chapter.

10.9   Non-Newtonian Fluids

10.9.1   Introduction

Some of the most challenging and important areas currently under investigation in computational fluid mechanics concern flows of non-Newtonian fluids (i.e. stresses are nonlinear functions of strain rates). Such fluids may or may not have memory of past deformation (i.e. viscoelastic or not). Practical examples of such fluids are multi-grade oils, liquid detergents, paints, and printing inks. Polymer solutions and polymer melts also fall within this category. All such flows are extremely important in forming processes of various kinds applied to metals, plastics, or glass.

Numerical approaches used for analyzing flows of non-Newtonian fluids differ very little from the ones used for Newtonian fluids. When the shear viscosity is a function of the rate of deformation tensor (i.e. for so-called power-law fluids), the equation of motion can still be written explicitly in terms of velocity components. For such fluids, the same formulations as those of the Newtonian fluids can be used. For example, the pressure–velocity and penalty finite element models discussed in the preceding sections can be used for power-law fluids. The constitutive equations of a viscoelastic fluid can be described in terms of the *extra stress* components and are given in terms of either differential equations or integral equations. In a differential constitutive model, the extra stress components and their derivatives are related to the velocity components and their derivatives and, therefore, require us to treat stress components as independent variables along with the velocity components and pressure.

Most existing works in the finite element analysis of viscoelastic fluids have considered different types of Maxwell/Oldroyd models, in which both viscosity and relaxation time are constants. The analysis was based on mixed formulations in which the velocities, pressure, and stresses are treated as nodal variables. An important class of fluids, which are characterized by shear viscosity and elasticity require additional study from numerical simulation point of view. The viscoelastic effects along with memory effects of such fluids can be studied using the White–Metzner model (see [204–207]).

The objective of this section is to study penalty finite element models of power-law and viscoelastic White–Metzner type fluids [204–212]. As a part of this discussion, we present finite element models of equations governing flows of viscous incompressible fluids through axisymmetric geometries. Therefore, first we rewrite the conservation of mass and momentum in the cylindrical coordinate system.

10.9.2   Governing Equations in Cylindrical Coordinates

Equations (10.3.1)–(10.3.3) can be expressed in a cylindrical coordinate system, (r, θ, z), by writing all vectors and tensors, including the del operator, in terms of components in a cylindrical coordinate system, as shown in Table 1.6.1 (also see Problems 1.8–1.10 for details). For example, the del operator and the material time derivative operators in the cylindrical coordinate system are given by

$$\nabla = \hat{\mathbf{e}}_r \frac{\partial}{\partial r} + \hat{\mathbf{e}}_\theta \frac{1}{r} \frac{\partial}{\partial \theta} + \hat{\mathbf{e}}_z \frac{\partial}{\partial z} \tag{10.9.1}$$

$$\frac{d}{dt} = \frac{\partial}{\partial t} + v_r \frac{\partial}{\partial r} + \frac{v_\theta}{r} \frac{\partial}{\partial \theta} + v_z \frac{\partial}{\partial z} \tag{10.9.2}$$

where $(\hat{\mathbf{e}}_r, \hat{\mathbf{e}}_\theta, \hat{\mathbf{e}}_z)$ are the unit basis vectors and (v_r, v_θ, v_z) are the velocity components in the r, θ, and z directions, respectively. Recall from Table 1.6.1 that the derivatives of the basis vectors $\hat{\mathbf{e}}_r$ and $\hat{\mathbf{e}}_\theta$ with respect to the coordinates r and z are zero while the derivatives with respect to θ are given by

$$\frac{\partial \hat{\mathbf{e}}_r}{\partial \theta} = \hat{\mathbf{e}}_\theta; \quad \frac{\partial \hat{\mathbf{e}}_\theta}{\partial \theta} = -\hat{\mathbf{e}}_r \tag{10.9.3}$$

The components of the symmetric part of the velocity gradient tensor **d** in Eq. (10.2.13) and constitutive relations, Eqs. (10.2.11) and (10.2.15), take the form

$$
\begin{aligned}
d_{rr} &= \frac{\partial v_r}{\partial r}; & d_{\theta\theta} &= \frac{1}{r}\frac{\partial v_\theta}{\partial \theta} + \frac{v_r}{r} \\
d_{zz} &= \frac{\partial v_z}{\partial z}; & 2d_{r\theta} &= \frac{\partial v_\theta}{\partial r} - \frac{v_\theta}{r} + \frac{1}{r}\frac{\partial v_r}{\partial \theta} \\
2d_{\theta z} &= \frac{1}{r}\frac{\partial v_z}{\partial \theta} + \frac{\partial v_\theta}{\partial z}; & 2d_{zr} &= \frac{\partial v_r}{\partial z} + \frac{\partial v_z}{\partial r}
\end{aligned}
\tag{10.9.4}
$$

$$
\begin{aligned}
\sigma_{rr} &= -P + 2\mu d_{rr}; & \sigma_{\theta\theta} &= -P + 2\mu d_{\theta\theta}; & \sigma_{zz} &= -P + 2\mu d_{zz} \\
\sigma_{r\theta} &= 2\mu d_{r\theta}; & \sigma_{\theta z} &= 2\mu d_{\theta z}; & \sigma_{zr} &= 2\mu d_{zr}
\end{aligned}
\tag{10.9.5}
$$

The governing equations, Eqs. (10.3.1)–(10.3.3), can be expressed in the cylindrical coordinate system as

$$\frac{1}{r}\frac{\partial}{\partial r}(r v_r) + \frac{1}{r}\frac{\partial v_\theta}{\partial \theta} + \frac{\partial v_z}{\partial z} = 0 \tag{10.9.6}$$

$$\rho\left(\frac{dv_r}{dt} - \frac{v_\theta^2}{r}\right) = \rho f_r + \frac{1}{r}\left[\frac{\partial(r\sigma_{rr})}{\partial r} + \frac{\partial \sigma_{r\theta}}{\partial \theta} + \frac{\partial(r\sigma_{zr})}{\partial z}\right] - \frac{\sigma_{\theta\theta}}{r} \tag{10.9.7}$$

$$\rho\left(\frac{dv_\theta}{dt} + \frac{v_r v_\theta}{r}\right) = \rho f_\theta + \frac{1}{r}\left[\frac{\partial(r\sigma_{r\theta})}{\partial r} + \frac{\partial \sigma_{\theta\theta}}{\partial \theta} + \frac{\partial(r\sigma_{\theta z})}{\partial z}\right] + \frac{\sigma_{r\theta}}{r} \tag{10.9.8}$$

$$\rho\left(\frac{dv_z}{dt}\right) = \rho f_z + \frac{1}{r}\left[\frac{\partial(r\sigma_{zr})}{\partial r} + \frac{\partial \sigma_{z\theta}}{\partial \theta} + \frac{\partial(r\sigma_{zz})}{\partial z}\right] \tag{10.9.9}$$

$$\rho c_v \frac{dT}{dt} = \frac{1}{r}\frac{\partial}{\partial r}\left(r k_{rr}\frac{\partial T}{\partial r}\right) + \frac{1}{r^2}\frac{\partial}{\partial \theta}\left(k_{\theta\theta}\frac{\partial T}{\partial \theta}\right) + \frac{\partial}{\partial z}\left(k_{zz}\frac{\partial T}{\partial z}\right) + \Phi + g \tag{10.9.10}$$

where the stress components are known in terms of the velocity components via Eqs. (10.9.4) and (10.9.5). The viscous dissipation Φ is given by

$$\Phi = 2\mu \left(d_{rr}^2 + d_{\theta\theta}^2 + d_{zz}^2 \right) + 4\mu \left(d_{r\theta}^2 + d_{rz}^2 + d_{z\theta}^2 \right) \tag{10.9.11}$$

Note that the time derivative of a vector (or tensor) in a rotating reference frame is given by (see Reddy and Rasmussen [13], pp. 69–74)

$$\left[\frac{d(\cdot)}{dt} \right]_{nonrot} = \left[\frac{d(\cdot)}{dt} \right]_{rot} + \omega \times (\cdot)$$

where $\omega = \frac{d\theta}{dt}\hat{\mathbf{e}}_z = \frac{v_\theta}{r}\hat{\mathbf{e}}_z$ is the angular velocity vector of the rotating frame of reference. Therefore, we have

$$\left(\frac{d\mathbf{v}}{dt} \right)_{nonrot} = \left(\frac{d\mathbf{v}}{dt} \right)_{rot} + \left(\frac{v_\theta}{r}\hat{\mathbf{e}}_z \right) \times \mathbf{v}$$

$$= \hat{\mathbf{e}}_r \left(\frac{dv_r}{dt} - v_\theta \frac{v_\theta}{r} \right) + \hat{\mathbf{e}}_\theta \left(\frac{dv_\theta}{dt} + v_r \frac{v_\theta}{r} \right) + \hat{\mathbf{e}}_z \frac{dv_z}{dt} \tag{10.9.12}$$

Equation (10.9.12) could be used in conjunction with Eqs. (10.9.6)–(10.9.11) to provide a description of fluid motion in a rotating cylindrical coordinate system.

For axisymmetric conditions (i.e. all variables are independent of θ and $f_\theta = v_\theta = 0$), the governing equations are simplified to

$$\frac{1}{r}\frac{\partial}{\partial r}(rv_r) + \frac{\partial v_z}{\partial z} = 0 \tag{10.9.13}$$

$$\rho \left(\frac{\partial v_r}{\partial t} + v_r \frac{\partial v_r}{\partial r} + v_z \frac{\partial v_r}{\partial z} \right) = \rho f_r + \frac{1}{r}\frac{\partial(r\sigma_{rr})}{\partial r} + \frac{\partial \sigma_{zr}}{\partial z} - \frac{\sigma_{\theta\theta}}{r} \tag{10.9.14}$$

$$\rho \left(\frac{\partial v_z}{\partial t} + v_r \frac{\partial v_z}{\partial r} + v_z \frac{\partial v_z}{\partial z} \right) = \rho f_z + \frac{1}{r}\frac{\partial(r\sigma_{zr})}{\partial r} + \frac{\partial \sigma_{zz}}{\partial z} \tag{10.9.15}$$

$$\rho c_v \left(\frac{\partial T}{\partial t} + v_r \frac{\partial T}{\partial r} + v_z \frac{\partial T}{\partial z} \right) = \frac{1}{r}\frac{\partial}{\partial r}\left(rk_{rr}\frac{\partial T}{\partial r} \right) + \frac{\partial}{\partial z}\left(k_{zz}\frac{\partial T}{\partial z} \right) + \Phi + g \tag{10.9.16}$$

10.9.3   Power-Law Fluids

Many viscous fluids used in industrial applications are characterized by so-called power-law constitutive behavior. Power-law fluids exhibit nonlinear material behavior according to the relation (sum on i and j is implied)

$$\boldsymbol{\sigma} = \boldsymbol{\tau} - P\mathbf{I}, \quad \boldsymbol{\tau} = 2\mu(I_2)\mathbf{D}, \quad \mu = \mu_0 \left(I_2 \right)^{\frac{(n-1)}{2}}$$

$$I_2 = \frac{1}{2} \left(d_{rr}^2 + d_{\theta\theta}^2 + d_{zz}^2 \right) + d_{r\theta}d_{r\theta} + d_{rz}d_{rz} + d_{\theta z}d_{\theta z} \tag{10.9.17}$$

where $\boldsymbol{\tau}$ is the viscous part of the stress tensor $\boldsymbol{\sigma}$, $\mathbf{I}$ is the unit tensor, I_2 is the second invariant of $\mathbf{d}$, and parameters μ_0 and n characterize the fluid (determined experimentally). For many non-Newtonian fluids, the viscosity μ decreases with increasing shear rate. These are called *shear thinning fluids* and have the power law index $n < 1$. Fluids with power law index $n > 1$ are called *shear-thickening fluids*. For such fluids, μ increases with increasing shear rate. For $n = 1$, Eq. (10.9.17) gives the Newtonian relation ($\mu = \mu_0$, constant). The power-law constitutive relation in Eq. (10.9.17) makes the problem nonlinear for $n \neq 1$, even for the case where the convective term $\mathbf{v} \cdot \nabla \mathbf{v}$ is negligible.

To illustrate how the power-law constitutive equation affects the finite element equations, we consider steady, isothermal, axisymmetric flows of viscous incompressible fluids. Assuming negligible viscous dissipation and body forces, Eqs. (10.9.13)–(10.9.16) can be expressed in terms of (v_r, v_z) as

$$\frac{1}{r}\frac{\partial}{\partial r}(rv_r) + \frac{\partial v_z}{\partial z} = 0 \tag{10.9.18}$$

$$\rho\left(v_r\frac{\partial v_r}{\partial r} + v_z\frac{\partial v_r}{\partial z}\right) = \frac{1}{r}\frac{\partial}{\partial r}\left(2\mu r\frac{\partial v_r}{\partial r}\right) + \frac{\partial}{\partial z}\left[\mu\left(\frac{\partial v_r}{\partial z} + \frac{\partial v_z}{\partial r}\right)\right]$$
$$- 2\mu\frac{v_r}{r^2} - \frac{\partial P}{\partial r} \tag{10.9.19}$$

$$\rho\left(v_r\frac{\partial v_z}{\partial r} + v_z\frac{\partial v_z}{\partial z}\right) = \frac{1}{r}\frac{\partial}{\partial r}\left[\mu r\left(\frac{\partial v_r}{\partial z} + \frac{\partial v_z}{\partial r}\right)\right] + \frac{\partial}{\partial z}\left(2\mu\frac{\partial v_z}{\partial z}\right) - \frac{\partial P}{\partial z}$$
$$\tag{10.9.20}$$

The penalty-finite element model of Eqs. (10.9.18)–(10.9.20) can be obtained as before. After replacing P in Eqs. (10.9.19) and (10.9.20) with

$$P = -\gamma_p\left(\frac{\partial v_r}{\partial r} + \frac{v_r}{r} + \frac{\partial v_z}{\partial z}\right) \tag{10.9.21}$$

one may construct their weak forms and go on to develop the finite element model [note that Eq. (10.9.18) is no longer used]. The value of the penalty parameter γ_p that is most suitable for this class problems must be determined by conducting numerical experiments with some benchmark problems. The finite element model has the form [see Eq. (10.5.18) or (10.5.24)]

$$[\mathbf{C}(\mathbf{v}) + \mathbf{K}_\mu + \mathbf{K}_\gamma]\mathbf{v} = \mathbf{F} \quad \text{or} \quad \hat{\mathbf{K}}(\mathbf{v})\mathbf{v} = \mathbf{F} \tag{10.9.22}$$

where $\mathbf{C}$, $\mathbf{K}_\mu$, and $\mathbf{K}_\gamma$ are the convective, diffusive, and penalty contributions, respectively, to the coefficient matrix. Because of the dependence of μ on $\mathbf{v}$, both $\mathbf{C}$ and $\mathbf{K}_\mu$ are functions of the unknown velocity field $\mathbf{v}$, as can be seen from Eqs. (10.5.16) and (10.4.13).

The direct iteration scheme for this case is given by

$$\hat{\mathbf{K}}(\mathbf{v}^{(r)})\mathbf{v}^{r+1} = \mathbf{F} \tag{10.9.23}$$

where $\hat{\mathbf{K}}$ is evaluated using the viscosity $\mu = \mu(D_{ij})$ computed according to Eq. (10.9.17) at the rth iteration. It would be more appropriate to compute μ at the

Gauss points of the element and use it in the evaluation of $\hat{K}_{ij}$, than to assume μ is an element-wise constant. For example, consider the diffusion/viscous portion of the element coefficients K_{ij}^{11}:

$$K_{ij}^{11} = \int_{\Omega^e} \mu \left(2 \frac{\partial \psi_i}{\partial r} \frac{\partial \psi_j}{\partial r} + \frac{\partial \psi_i}{\partial z} \frac{\partial \psi_j}{\partial z} \right) r\,dr dz \qquad (10.9.24)$$

For numerical evaluation of K_{ij}^{11} we use the Gauss quadrature rule

$$K_{ij}^{11} = \sum_{I,J=1}^{NGP} AMU(I,J) * [2.0 * GDSF(1,i) * GDSF(1,j)$$
$$+ GDSF(2,i) * GDSF(2,j)\,] * CONST \qquad (10.9.25)$$

where $AMU(I,J)$ is the value of μ at the (I,J)th Gauss point. Obviously, $AMU(I,J)$ must be evaluated using Eq. (10.9.17) prior to using it in Eq. (10.9.25).

10.9.4   White–Metzner Fluids

The general constitutive equation for fluids dominated by shear viscosity is given by (see [204–206])

$$\boldsymbol{\tau} + \lambda \left\{ \frac{\partial \boldsymbol{\tau}}{\partial t} + \mathbf{v} \cdot \nabla \boldsymbol{\tau} - [(\nabla \mathbf{v})^{\mathrm{T}} \cdot \boldsymbol{\tau} + \boldsymbol{\tau} \cdot (\nabla \mathbf{v})] \right\} = 2\eta \mathbf{d} \qquad (10.9.26)$$

where λ is the relaxation time, η is the shear viscosity of the fluid, and $\mathbf{d}$ is the symmetric part of the velocity gradient tensor.

To characterize the White–Metzner fluid, it is necessary to know the viscosity curve as a function of shear rate and the first normal stress difference. The extra stress tensor $\boldsymbol{\tau}$ is separated into purely viscous part $\boldsymbol{\tau}^2$ and viscoelastic part $\boldsymbol{\tau}^1$:

$$\boldsymbol{\tau} = \boldsymbol{\tau}^1 + \boldsymbol{\tau}^2 \qquad (10.9.27)$$

$$\boldsymbol{\tau}^1 + \lambda \left\{ \frac{\partial \boldsymbol{\tau}^1}{\partial t} + \mathbf{v} \cdot \nabla \boldsymbol{\tau}^1 - [(\nabla \mathbf{v})^{\mathrm{T}} \cdot \boldsymbol{\tau}^1 + \boldsymbol{\tau}^1 \cdot (\nabla \mathbf{v})] \right\} = 2\eta_1 \mathbf{d} \qquad (10.9.28)$$

$$\boldsymbol{\tau}^2 = 2\eta_2 \mathbf{d} \qquad (10.9.29)$$

and η_1 and η_2 can be defined as the fractions of the shear viscosity η. The fluid which obeys Eq. (10.9.27) is characterized by η_1, η_2, and λ, which are the functions of the symmetric part of the velocity gradient tensor, $\mathbf{d}$.

Here, the relaxation time λ is assumed to depend on the shear rate γ according to

$$\lambda(\gamma) = a + b(\log \gamma) + c(\log \gamma)^2, \quad \gamma = \sqrt{4I_2} \qquad (10.9.30)$$

All constants in Eq. (10.9.30) are evaluated by curve fitting data of polymeric melts (see [208]). Also, η is assumed to be of the particular form

$$\eta = \eta(I_2), \quad I_2 = \frac{1}{2}\left(d_{rr}^2 + d_{\theta\theta}^2 + d_{zz}^2\right) + d_{r\theta}d_{r\theta} + d_{rz}d_{rz} + d_{\theta z}d_{\theta z} \quad (10.9.31)$$

Although Eq. (10.9.31) gives the general functional form for the viscosity function, experimental observation and a theoretical basis must be used to provide a specific model for non-Newtonian viscosities. A variety of models can be used to calculate viscosity. Here, the power law model in Eq. (10.9.17) is used (replace μ with η).

In view of Eq. (10.9.27), the equation of motion (in the absence of body forces) for the White–Metzner viscoelastic fluids can be written as

$$\boldsymbol{\nabla}\cdot(\boldsymbol{\tau}^1 + 2\eta_2\mathbf{d}) - \boldsymbol{\nabla}P = \rho\left(\frac{\partial\mathbf{v}}{\partial t} + \mathbf{v}\cdot\boldsymbol{\nabla}\mathbf{v}\right) \quad (10.9.32)$$

For simplicity, we drop the superscript 1 from $\boldsymbol{\tau}^1$ in Eq. (10.9.32) as well as in Eq. (10.9.28). Equations (10.9.32) and (10.9.28), together with the continuity equation $\boldsymbol{\nabla}\cdot\mathbf{v} = 0$, represent the system of governing equations for the White–Metzner fluids. The viscosity η_1 is calculated using Eq. (10.9.28) and λ from Eq. (10.9.30). Viscosity η_2 is often taken as a function of η_1. Note that the power-law constitutive equation can be obtained as a special case from Eq. (10.9.28).

For axisymmetric flows of White–Metzner fluids, the governing equations are given by (with $u = v_r$ and $w = v_z$)

$$\frac{\partial u}{\partial r} + \frac{u}{r} + \frac{\partial w}{\partial z} = 0 \quad (10.9.33)$$

$$\frac{\partial}{\partial r}\left(\tau_{rr} + 2\eta_2\frac{\partial u}{\partial r}\right) + \frac{1}{r}\left(\tau_{rr} + 2\eta_2\frac{\partial u}{\partial r} - \tau_{\theta\theta} - 2\eta_2\frac{u}{r}\right) +$$
$$\frac{\partial}{\partial z}\left[\tau_{rz} + \eta_2\left(\frac{\partial u}{\partial z} + \frac{\partial w}{\partial r}\right)\right] - \frac{\partial P}{\partial r} = \rho\left(\frac{\partial u}{\partial t} + u\frac{\partial u}{\partial r} + w\frac{\partial u}{\partial z}\right) \quad (10.9.34)$$

$$\frac{\partial}{\partial r}\left[\tau_{rz} + \eta_2\left(\frac{\partial u}{\partial z} + \frac{\partial w}{\partial r}\right)\right] + \frac{1}{r}\left[\tau_{rz} + \eta_2\left(\frac{\partial u}{\partial z} + \frac{\partial w}{\partial r}\right)\right] +$$
$$\frac{\partial}{\partial z}\left(\tau_{zz} + 2\eta_2\frac{\partial w}{\partial z}\right) - \frac{\partial P}{\partial z} = \rho\left(\frac{\partial w}{\partial t} + u\frac{\partial w}{\partial r} + w\frac{\partial w}{\partial z}\right) \quad (10.9.35)$$

$$\tau_{rr} + \lambda\left[\frac{\partial\tau_{rr}}{\partial t} + u\frac{\partial\tau_{rr}}{\partial r} + w\frac{\partial\tau_{rr}}{\partial z} - 2\left(\frac{\partial u}{\partial r}\tau_{rr} + \frac{\partial u}{\partial z}\tau_{rz}\right)\right] = 2\eta_1\frac{\partial u}{\partial r} \quad (10.9.36)$$

$$\tau_{zz} + \lambda\left[\frac{\partial\tau_{zz}}{\partial t} + u\frac{\partial\tau_{zz}}{\partial r} + w\frac{\partial\tau_{zz}}{\partial z} - 2\left(\frac{\partial w}{\partial z}\tau_{zz} + \frac{\partial w}{\partial r}\tau_{rz}\right)\right] = 2\eta_1\frac{\partial w}{\partial z} \quad (10.9.37)$$

$$\tau_{rz} + \lambda \left[\frac{\partial \tau_{rz}}{\partial t} + u \frac{\partial \tau_{rz}}{\partial r} + w \frac{\partial \tau_{rz}}{\partial z} - 2 \left(\frac{\partial u}{\partial r} \tau_{rz} + \frac{\partial w}{\partial z} \tau_{rz} + \frac{\partial u}{\partial z} \tau_{zz} + \frac{\partial w}{\partial r} \tau_{rr} \right) \right]$$

$$= \eta_1 \left(\frac{\partial w}{\partial r} + \frac{\partial u}{\partial z} \right) \tag{10.9.38}$$

$$\tau_{\theta\theta} + \lambda \left[\frac{\partial \tau_{\theta\theta}}{\partial t} + u \frac{\partial \tau_{\theta\theta}}{\partial r} + w \frac{\partial \tau_{\theta\theta}}{\partial z} - 2 \frac{u}{r} \tau_{\theta\theta} \right] = 2\eta_1 \frac{u}{r} \tag{10.9.39}$$

where u and w denote the velocity components in the radial and axial directions, respectively.

The (mixed) penalty finite element model of the above equations can be developed by constructing the weak forms of Eqs. (10.9.34)–(10.9.39) [after replacing the pressure with Eq. (10.9.21)] and interpolating velocities (u, w) and extra stress components $(\tau_{rr}, \tau_{\theta\theta}, \tau_{zz}, \tau_{rz})$ as follows:

$$u(r, z, t) = \sum_{j=1}^{n} u_j(t)\psi_j(r, z), \qquad w(r, z, t) = \sum_{j=1}^{n} w_j(t)\psi_j(r, z)$$

$$\tau_{rr}(r, z, t) = \sum_{j=1}^{L} \tau_{rr}^j(t)\psi_j(r, z), \qquad \tau_{zz}(r, z, t) = \sum_{j=1}^{L} \tau_{zz}^j(t)\psi_j(r, z) \tag{10.9.40}$$

$$\tau_{rz}(r, z, t) = \sum_{j=1}^{L} \tau_{rz}^j(t)\psi_j(r, z), \qquad \tau_{\theta\theta}(r, z, t) = \sum_{j=1}^{L} \tau_{\theta\theta}^j(t)\psi_j(r, z)$$

where, (u_j, w_j) are the nodal velocities, $(\tau_{rr}^j, \tau_{zz}^j, \tau_{rz}^j, \tau_{\theta\theta}^j)$ are the nodal extra stress components, and ψ_j are the Lagrange interpolation functions. The finite element model has the general form

$$\mathbf{M}^e \dot{\boldsymbol{\Delta}}^e + \mathbf{K}^e \boldsymbol{\Delta}^e = \mathbf{F}^e \tag{10.9.41}$$

where

$$(\boldsymbol{\Delta}^e)^{\mathrm{T}} = \{\mathbf{u} \ \mathbf{w} \ \boldsymbol{\tau}_{rr} \ \boldsymbol{\tau}_{zz} \ \boldsymbol{\tau}_{rz} \ \boldsymbol{\tau}_{\theta\theta}\} \tag{10.9.42}$$

The nonzero coefficients of the matrices are listed below.

$$K_{ij}^{11} = \int_{\Omega^e} \left[\rho\psi_i \left(\bar{u} \frac{\partial \psi_j}{\partial r} + \bar{w} \frac{\partial \psi_j}{\partial z} \right) + \eta_2 \left(2 \frac{\partial \psi_i}{\partial r} \frac{\partial \psi_j}{\partial r} + \frac{\partial \psi_i}{\partial z} \frac{\partial \psi_j}{\partial z} + 2 \frac{\psi_i \psi_j}{r^2} \right) \right] r \, dr \, dz$$

$$+ \gamma_p \int_{\Omega^e} \frac{1}{r} \frac{\partial(r\psi_i)}{\partial r} \frac{\partial(r\psi_j)}{\partial r} \, dr \, dz$$

$$K_{ij}^{12} = \int_{\Omega^e} \eta_2 \frac{\partial \psi_i}{\partial z} \frac{\partial \psi_j}{\partial r} r \, dr \, dz + \gamma_p \int_{\Omega^e} \frac{\partial(r\psi_i)}{\partial r} \frac{\partial \psi_j}{\partial z} \, dr \, dz$$

$$K_{ij}^{13} = \int_{\Omega^e} \frac{\partial(r\psi_i)}{\partial r} \psi_j \, dr \, dz, \quad K_{ij}^{15} = \int_{\Omega^e} \frac{\partial \psi_i}{\partial z} \psi_j \, r \, dr \, dz, \quad K_{ij}^{16} = \int_{\Omega^e} \psi_i \psi_j \, dr \, dz$$

$$K_{ij}^{21} = \int_{\Omega^e} \eta_2 \left(\frac{\partial \psi_i}{\partial r} \frac{\partial \psi_j}{\partial z} - \frac{1}{r} \psi_i \frac{\partial \psi_j}{\partial z} \right) r \, dr \, dz + \gamma_p \int_{\Omega^e} \frac{\partial \psi_i}{\partial z} \frac{\partial(r\psi_j)}{\partial r} \, dr \, dz$$

$$K_{ij}^{22} = \int_{\Omega^e} \left[\rho\psi_i \left(\bar{u}\frac{\partial\psi_j}{\partial r} + \bar{w}\frac{\partial\psi_j}{\partial z} \right) + \eta_2 \left(\frac{\partial\psi_i}{\partial r}\frac{\partial\psi_j}{\partial r} + 2\frac{\partial\psi_i}{\partial z}\frac{\partial\psi_j}{\partial z} \right) \right.$$
$$\left. + \eta_2 \psi_i \frac{\partial\psi_j}{\partial r} \right] r\, dr\, dz + \gamma_p \int_{\Omega^e} \frac{\partial\psi_i}{\partial z}\frac{\partial\psi_j}{\partial z} r\, dr\, dz$$

$$K_{ij}^{24} = \int_{\Omega^e} \frac{\partial\psi_i}{\partial z}\psi_j \, r\, dr\, dz, \quad K_{ij}^{25} = \int_{\Omega^e} \frac{\partial(r\psi_i)}{\partial r}\psi_j \, dr\, dz$$

$$K_{ij}^{31} = -2\int_{\Omega^e} \eta_1\psi_i\frac{\partial\psi_j}{\partial r} \, r\, dr\, dz, \quad K_{ij}^{35} = -2\int_{\Omega^e} \lambda\frac{\partial\bar{u}}{\partial z}\psi_i\psi_j \, r\, dr\, dz$$

$$K_{ij}^{33} = \int_{\Omega^e} \left\{ \psi_i\psi_j + \lambda \left[\psi_i \left(\bar{u}\frac{\partial\psi_j}{\partial r} + \bar{w}\frac{\partial\psi_j}{\partial z} \right) - 2\left(\frac{\partial\bar{u}}{\partial r}\right)\psi_i\psi_j \right] \right\} r\, dr\, dz$$

$$K_{ij}^{42} = -2\int_{\Omega^e} \eta_1\psi_i\frac{\partial\psi_j}{\partial z} \, r\, dr\, dz$$

$$K_{ij}^{44} = \int_{\Omega^e} \left\{ \psi_i\psi_j + \lambda \left[\psi_i \left(\bar{u}\frac{\partial\psi_j}{\partial r} + \bar{w}\frac{\partial\psi_j}{\partial z} \right) - 2\frac{\partial\bar{w}}{\partial z}\psi_i\psi_j \right] \right\} r\, dr\, dz$$

$$K_{ij}^{45} = -2\int_{\Omega^e} \lambda\frac{\partial\bar{w}}{\partial r}\psi_i\psi_j \, r\, dr\, dz, \quad K_{ij}^{51} = -2\int_{\Omega^e} \eta_1\psi_i\frac{\partial\psi_i}{\partial z} \, r\, dr\, dz$$

$$K_{ij}^{52} = -2\int_{\Omega^e} \eta_1\psi_i\frac{\partial\psi_j}{\partial r} \, r\, dr\, dz, \quad K_{ij}^{53} = -2\int_{\Omega^e} \lambda\frac{\partial\bar{w}}{\partial r}\psi_i\psi_j \, r\, dr\, dz$$

$$K_{ij}^{54} = -2\int_{\Omega^e} \lambda\frac{\partial\bar{u}}{\partial z}\psi_i\psi_j \, r\, dr\, dz, \quad K_{ij}^{61} = -2\int_{\Omega^e} \eta_1\psi_i\psi_j \, dr\, dz$$

$$K_{ij}^{55} = \int_{\Omega^e} \left\{ \psi_i\psi_j + \lambda \left[\psi_i \left(\bar{u}\frac{\partial\psi_j}{\partial r} + \bar{w}\frac{\partial\psi_j}{\partial z} \right) - 2\left(\frac{\partial\bar{u}}{\partial r} + \frac{\partial\bar{w}}{\partial z}\right)\psi_i\psi_j \right] \right\} r\, dr\, dz$$

$$K_{ij}^{66} = \int_{\Omega^e} \left\{ \psi_i\psi_j + \lambda \left[\psi_i \left(\bar{u}\frac{\partial\psi_j}{\partial r} + \bar{w}\frac{\partial\psi_j}{\partial z} \right) - \frac{2}{r}\bar{u}\psi_i\psi_j \right] \right\} r\, dr\, dz$$

$$M_{ij}^{11} = \int_{\Omega^e} \rho\psi_i\psi_j \, r\, dr\, dz, \quad M_{ij}^{22} = \int_{\Omega^e} \rho\psi_i\psi_j \, r\, dr\, dz$$

$$M_{ij}^{33} = \int_{\Omega^e} \lambda\psi_i\psi_j \, r\, dr\, dz, \quad M_{ij}^{44} = \int_{\Omega^e} \lambda\psi_i\psi_j \, r\, dr\, dz$$

$$M_{ij}^{55} = \int_{\Omega^e} \lambda\psi_i\psi_j \, r\, dr\, dz, \quad M_{ij}^{66} = \int_{\Omega^e} \lambda\psi_i\psi_j \, r\, dr\, dz \tag{10.9.43}$$

A bar over the velocities u and w indicate that they are calculated using the nodal values from the previous iteration.

For viscoelastic fluids, specification of velocities is insufficient on account of fluid memory. If the boundary of the domain contains an entry region, then fully developed flow conditions may be assumed. All extra stress components must be specified as essential boundary conditions along the entry region. Failing to do so may lead to the propagation of errors throughout the flow domain when relaxation time λ becomes large.

10.9.5   Numerical Examples

Consider the steady flow of a power-law fluid in a uniform pipe ($R = 1$, $L = 6$, $w_0 = 10$, $n = 0.2$, $\mu_0 = 1.0$, and $\gamma_p = 10^8$). The geometry and boundary conditions of the computational domain are shown in Fig. 10.9.1. Three different meshes shown in Fig. 10.9.2 are used [208]. The finite element results obtained using the velocity finite element model in Eq. (10.9.22) along with the analytical results for the axial velocities as a function of the radial distance are presented in Table 10.9.1, where the analytical solution for the axial velocity is

$$w(r, L) = w_0 \left(\frac{3n + 1}{n + 1} \right) \left[1 - \left(\frac{r}{R} \right)^{1 + \frac{1}{n}} \right]$$

Next, the same problem (with $R = 1$ and $L = 5$) is studied with the mixed finite element model in Eq. (10.9.38). Both power-law ($n = 0.25$, $\eta_0 = 10^4$, $\eta_2 = 0$) and White–Metzner ($n = 0.25$, $\eta_0 = 10^4$, $\eta_2 = 0$, $a = 0.435$, $b = -0.453$, $c = 0.1388$) fluids are analyzed. A uniform mesh of 10×6, 4-node, quadrilateral elements is used. In addition to the velocity boundary conditions shown in Fig. 10.9.1, the stresses are specified to be zero at the entrance. The penalty parameter is taken to be $\gamma_p = 10^8$.

Figures 10.9.3 and 10.9.4 contain plots of the axial velocity profiles at $z = 0$, $z = 2.0$, and $z = 5.0$, for power-law and White–Metzner fluids, respectively. We note that the velocity profiles obtained with the velocity model (VM) in Eq. (10.9.22) and mixed model (MM) in Eq. (10.9.38) are essentially the same when stress boundary conditions are not imposed. The stress boundary conditions (which can be imposed point-wise only in the mixed model) do have an effect: specification of the stresses at the inlet increases the centerline velocity for both power-law fluids as well as White–Metzner fluids. If the stress boundary conditions are not imposed, the mixed model for the White–Metzner fluid does not yield converged solution. Figure 10.9.5 shows the effect of the penalty parameter on the velocity field.

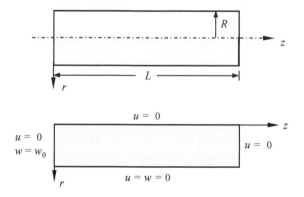

Fig. 10.9.1: Geometry and boundary conditions for flow through a pipe.

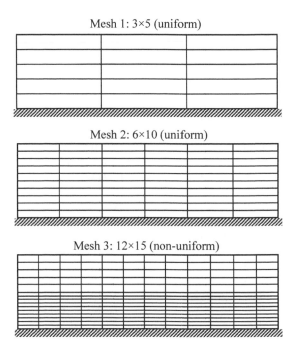

Fig. 10.9.2: Three different meshes used for the pipe flow.

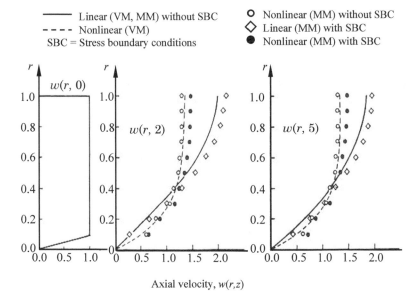

Fig. 10.9.3: Axial velocity profiles for the flow of power-law fluid through a pipe.

Table 10.9.1: A comparison of the velocity field $w(r, 6)$ for the flow of a power-law fluid in a pipe ($R = 1$, $L = 6$, $w_0 = 10$, $n = 0.2$, $\mu_0 = 1.0$, and $\gamma_p = 10^8$).

r	Mesh 1	Mesh2	Mesh 3	Analytical
0.00	11.894	12.833	13.402	13.333
0.05	–	–	–	13.333
0.10	–	12.832	13.401	13.333
0.15	–	–	–	13.333
0.20	11.865	12.826	13.395	13.332
0.25	–	–	–	13.330
0.30	–	12.795	13.360	13.324
0.35	–	–	–	13.309
0.40	11.721	12.686	13.243	13.279
0.45	–	–	–	13.223
0.50	–	12.401	12.940	13.125
0.55	–	–	12.679	12.964
0.60	10.852	11.783	12.305	12.711
0.65	–	–	11.787	12.328
0.70	–	10.610	11.086	11.765
0.75	–	–	10.156	10.960
0.80	7.916	8.563	8.946	9.838
0.85	–	–	7.396	8.305
0.90	–	5.133	5.451	6.248
0.95	–	–	2.993	3.532
1.00	0.000	0.000	0.000	0.000

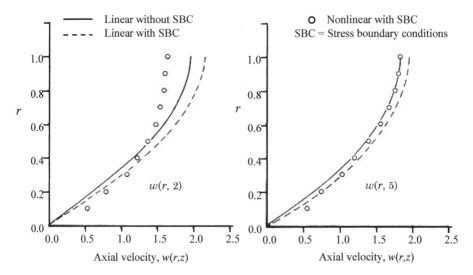

Fig. 10.9.4: Axial velocity profiles for the flow of White–Metzner fluid through a pipe.

Table 10.9.2: Effect of the convective terms* on the velocity field $w(r, 5)$ for the flow of a power-law fluid in a pipe ($R = 1$, $L = 5$, $w_0 = 10$, $n = 0.2$, and $10 \times 6Q4$ mesh).

r	Newtonian	Non-Newtonian		
		$Re = 0$	$Re = 10^2$	$Re = 10^3$
0.0	1.8358	1.3357	1.1312	1.0267
0.1	1.8001	1.3355	1.1302	1.0265
0.2	1.7398	1.3338	1.1289	1.0266
0.3	1.6461	1.3271	1.1270	1.0268
0.4	1.5186	1.3081	1.1241	1.0271
0.5	1.3578	1.2657	1.1197	1.0275
0.6	1.1651	1.1849	1.1097	1.0274
0.7	0.9397	1.0468	1.0799	1.0265
0.8	0.6712	0.8273	0.9809	1.0209
0.9	0.3079	0.4795	0.6709	0.8967
1.0	0.0000	0.0000	0.0000	0.0000

$$* \; Re = \frac{\rho w_0 R}{\eta_0}$$

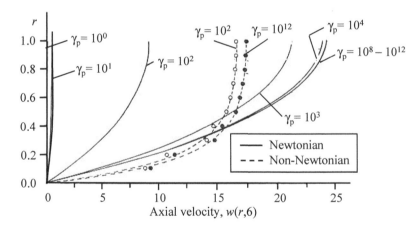

Fig. 10.9.5: The effect of the penalty parameter on the flow of a power-law fluid.

The effect of the convective terms on the velocity field is also investigated for the power-law fluid ($n = 0.25$, $\eta_0 = 10^4$, $\eta_2 = 0$, $R = 1$, $L = 5$, and $w_0 = 1$) using the velocity model in Eq. (10.9.22). The results are presented in Table 10.9.2. The effect is to flatten the velocity profile from a parabolic one. Figure 10.9.6 shows the axial velocity profile for flow through a plane channel ($a = 6$, $b = 1$, $n = 0.2$, $\eta_0 = \eta_1 = 1$). The results are in good agreement with those reported in [208].

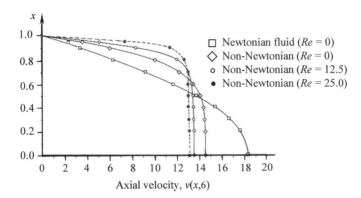

Fig. 10.9.6: Axial velocity profiles for flow of power-law fluid through a plane channel.

10.10   Coupled Fluid Flow and Heat Transfer

10.10.1   Finite Element Models

There exist many engineering systems where the fluid flow is affected by the heat transfer to or from the system and vice versa. Convective cooling of a heated body (like in an internal combustion engine) is a generic example of the coupling. In such cases, we solve the governing equations of fluid flow (i.e. the Navier–Stokes equations and continuity equation) as well as the heat transfer (i.e. the energy equation), as discussed in Section 10.3. The energy equation is given by [see Eq. (10.3.3)]

$$
\rho c_v \left[\left(u \frac{\partial T}{\partial x} + v \frac{\partial T}{\partial y} \right) + \frac{\partial T}{\partial t} \right] = \frac{\partial}{\partial x} \left(k_{11} \frac{\partial T}{\partial x} \right) + \frac{\partial}{\partial y} \left(k_{22} \frac{\partial T}{\partial y} \right) + \Phi + g \quad (10.10.1)
$$

where c_v is specific heat at constant volume, Φ is the dissipation energy

$$
\Phi = \sigma_{ij} \varepsilon_{ij} \quad (10.10.2)
$$

and g is the rate of internal heat generation.

The finite element model of the energy equation is given by

$$
\mathbf{C}^e \dot{\mathbf{T}}^e + \mathbf{H}^e \mathbf{T}^e = \mathbf{Q}^e \quad (10.10.3)
$$

where

$$
H_{ij}^e = \rho c_v \int_{\Omega^e} \psi_i^{(2)} \left(\bar{u} \frac{\partial \psi_j^{(2)}}{\partial x} + \bar{v} \frac{\partial \psi_j^{(2)}}{\partial y} \right) dx dy
$$

$$
+ \int_{\Omega^e} \left(k_{11} \frac{\partial \psi_i^{(2)}}{\partial x} \frac{\partial \psi_j^{(2)}}{\partial x} + k_{22} \frac{\partial \psi_i^{(2)}}{\partial y} \frac{\partial \psi_j^{(2)}}{\partial y} \right) dx dy
$$

$$C_{ij}^e = \rho c_v \int_{\Omega^e} \psi_i^{(2)} \psi_j^{(2)} \, dx dy$$

$$Q_i^e = \int_{\Omega^e} \psi_i^{(2)} (g + \Phi) \, dx dy + \oint_{\Gamma^e} q_n \psi_i^{(2)} \, ds \qquad (10.10.4)$$

u and v being the velocity components (that couple to the fluid flow problem), and $\psi_i^{(2)}$ the Lagrange interpolation functions used to interpolate the temperature field.

Equation (10.10.3) must be solved along with the flow equations

$$\mathbf{M}^e \dot{\boldsymbol{\Delta}}^e + \mathbf{K}^e \boldsymbol{\Delta}^e = \mathbf{F}^e \qquad (10.10.5)$$

developed earlier. Both equations, Eqs. (10.10.3) and (10.10.5), must be solved iteratively, using the latest velocity field and temperature field to compute the coefficient matrices. It should be noted that fluid flow equations are coupled to the heat transfer equation through the body force terms f_x and f_y in the Navier–Stokes equations. For buoyancy-driven flows, the body force in the direction of the gravity is a function of the temperature, that is, $f_y = -\rho a_g \beta (T - T_0)$ if the y-axis is taken vertically up, where a_g is acceleration due to gravity and β is thermal expansion coefficient. Thus, the momentum equations are fully coupled with the energy equation.

The following strategy is found suitable for the convective heat transfer problems. Solve the energy equation, Eq. (10.10.3), with an assumed velocity field (say, zero). Then use the assumed velocity and temperature fields in Eq. (10.10.5) and solve for the new velocity field. The initial guess of the velocity field can be either the linear (Stokes) solution or the solution of a problem at lower Rayleigh number when solving for high Rayleigh number flows. The Rayleigh number Ra and Prandtl number Pr (characteristic numbers used in convective heat transfer) are defined by

$$Ra = \frac{\beta \, a_g \, L^3 \, \Delta T}{\kappa \, \nu}, \quad \nu = \frac{\mu}{\rho}, \quad \kappa = \frac{k}{\rho c_p}, \quad Pr = \frac{\nu}{\kappa} \qquad (10.10.6)$$

where L is the characteristic dimension of the flow region, ΔT is the temperature difference between hot and cold walls, ρ is the density, μ is the viscosity, k is the conductivity, and c_p is the specific heat at constant pressure.

10.10.2 Numerical Examples

Here we include a couple of sample problems, taken from Reddy and Gartling [10], to illustrate the ideas presented above.

10.10.2.1 Heated cavity

Consider a closed square cavity filled with a viscous incompressible fluid. The top and bottom faces of the cavity are assumed to be insulated while the vertical faces subjected to different temperatures, as shown in Fig. 10.10.1(a). A typical 16×16 mesh of eight-node quadratic elements is shown in Fig. 10.10.1(b).

This problem was solved using the earliest versions of NACHOS code [213], which is based on velocity–pressure (mixed) formulation and used the direct iteration (Picard) scheme. More recent solutions have been obtained using Newton's method for the combined equation set equivalent to Eqs. (10.10.3) and (10.10.5). The streamline and isotherm plots are shown in Figs. 10.10.2 and 10.10.3 for Rayleigh numbers of $Ra = 10^4$ and 10^6, respectively ($Pr = \nu/\kappa = 0.71$). For the lower Rayleigh number, the flow is relatively weak and the thermal field is only slightly perturbed from a conduction solution. At the higher Rayleigh number, the flow field develops a considerable structure while the thermal field becomes vertically stratified in the core of the cavity with high heat flux regions along the vertical boundaries (see [213]).

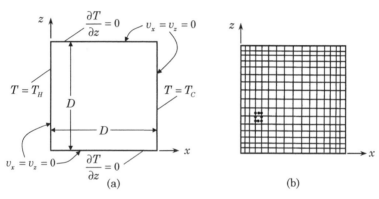

Fig. 10.10.1: (a) Geometry of a heated square cavity. (b) A typical finite element mesh.

10.10.2.2   Solar receiver

Figure 10.10.4 shows a cross-section of an annular solar receiver tube surrounded by an eccentrically located glass envelope. The inner tube carries a heat transfer fluid that is heated by a flux that varies with position around the tube. The incident flux is due to solar energy being concentrated on the tube by a parabolic trough collector. The glass envelope provides a shield to reduce the forced convection (wind) heat loss from the collector tube [10, 214].

Figures 10.10.5–10.10.7 contain streamline and isotherm plots for an air-filled annulus for various temperature and geometric configurations. The flow pattern and heat flux distribution are quite sensitive to variations in these parameters even though the Rayleigh number is the same for all cases [3, 214].

Additional details on the formulation as well as applications of coupled heat transfer and fluid flow can be found in [213–219] (in particular, see [219] and Chapter 5 of the book by Reddy and Gartling [10] and references therein).

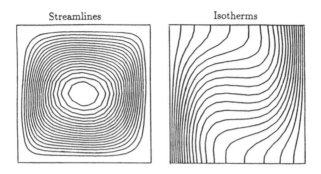

Fig. 10.10.2: Streamlines and isotherms for natural convection in a square cavity filled with viscous fluid ($Ra = 10^4$, $Pr = 0.71$).

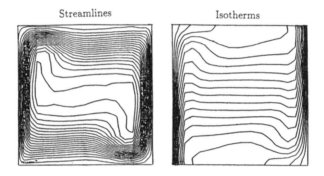

Fig. 10.10.3: Streamlines and isotherms for natural convection in a square cavity filled with viscous fluid ($Ra = 10^6$, $Pr = 0.71$).

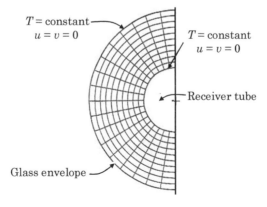

Fig. 10.10.4: Mesh and boundary conditions for the annular solar receiver.

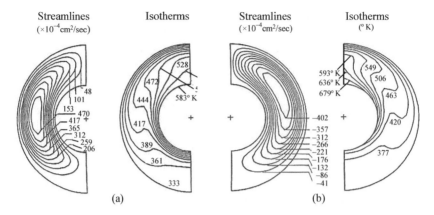

Fig. 10.10.5: Plots of streamlines and isotherms for the solar receiver: (a) uniform wall temperature, $Ra = 1.2 \times 10^4$ and (b) asymmetric wall temperature, *hot on top* $(Ra = 1.2 \times 10^4)$.

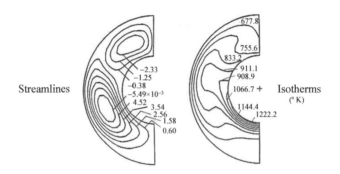

Fig. 10.10.6: Plots of streamlines and isotherms for the solar receiver; nonuniform wall temperature, *hot on bottom* $(Ra = 1.2 \times 10^4)$.

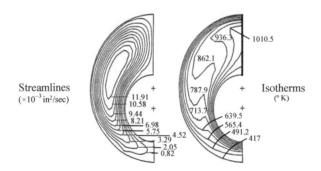

Fig. 10.10.7: Plots of streamlines and isotherms for the solar receiver; uniform wall temperature, *eccentric geometry* $(Ra = 1.2 \times 10^4)$.

10.11 Summary

In this chapter, finite element models of flows of viscous incompressible fluids are presented for Newtonian and non-Newtonian (power-law) fluids. Two different finite element models are formulated, namely, the mixed finite element model involving velocities and pressure and the penalty finite element model that involves only velocity degrees of freedom. A number of examples are presented. Formulation of the equations of motion for transient analysis is also discussed. Finally, finite element models of coupled fluid flow and heat transfer, using the penalty method, accounting for buoyancy effects is presented and a number of numerical examples are included.

Problems

10.1 Consider the vector equations, Eqs. (10.3.1) and (10.3.2). Develop the weak statements of the equations in vector form. Under what conditions can these equations be obtained from the minimum of a quadratic functional? See Oden and Reddy [220] for details.

10.2 Consider Eqs. (10.4.1)–(10.4.3) in cylindrical coordinates (r, θ, z). For axisymmetric viscous incompressible flows (i.e. the flow field is independent of the θ coordinate), and when the convective (nonlinear) terms are neglected, we have

$$\rho \frac{dv_r}{dt} = \frac{1}{r}\frac{\partial}{\partial r}(r\sigma_{rr}) - \frac{\sigma_{\theta\theta}}{r} + \frac{\partial \sigma_{rz}}{\partial z} + f_r \tag{1}$$

$$\rho \frac{dv_z}{dt} = \frac{1}{r}\frac{\partial}{\partial r}(r\sigma_{rz}) + \frac{\partial \sigma_{zz}}{\partial z} + f_z \tag{2}$$

$$0 = \frac{1}{r}\frac{\partial}{\partial r}(rv_r) + \frac{\partial v_z}{\partial z} \tag{3}$$

where

$$\sigma_{rr} = -P + 2\mu\frac{\partial v_r}{\partial r}, \quad \sigma_{\theta\theta} = -P + 2\mu\frac{v_r}{r}$$

$$\sigma_z = -P + 2\mu\frac{\partial v_z}{\partial z}, \quad \sigma_{rz} = \mu\left(\frac{\partial v_r}{\partial z} + \frac{\partial v_z}{\partial r}\right) \tag{4}$$

and (v_r, v_z) are the components of the velocity vector $\mathbf{v} = \hat{\mathbf{e}}_r v_r + \hat{\mathbf{e}}_z v_z$. Develop the semidiscrete mixed finite element model of the equations.

10.3 Develop the semidiscrete penalty finite element model of the equations in Problem 10.2.

10.4 The equations governing unsteady slow flow (i.e. Stokes flow) of viscous incompressible fluids in the x-y plane can be expressed in terms of vorticity ω and stream function ψ:

$$\rho\frac{\partial \omega}{\partial t} - \mu\nabla^2\omega = 0 \tag{1}$$

$$-\omega - \nabla^2\psi = 0 \tag{2}$$

Develop the semidiscrete finite element model of the equations. Discuss the meaning of the secondary variables. Use the α-family of approximation to reduce the ordinary differential equations to algebraic equations.

10.5 Compute the tangent coefficient matrix for the penalty finite element model in Eq. (10.5.16).

10.6 Verify Eqs. (10.9.7)–(10.9.9).

10.7 Carry out steady-state analysis of the *lid-driven cavity* problem (see Example 10.8.4) for Re=5,000 using the 16×20 mesh of bilinear elements. Use $DX(I)$ and $DY(I)$ as given in Example 10.8.6. You may experiment with the acceleration parameter β (for convergence) and use a convergence tolerance of 10^{-2}. Plot the horizontal velocity v_x at the vertical centerline of the cavity versus the coordinate y for the linear and nonlinear cases. Note that you must pick ρ and μ to obtain the desired Reynolds number.

10.8 Carry out the transient analysis of the lid-driven cavity of Problem 10.7. Use the data of Example 10.8.6.

11

Least-Squares Finite Element Models of Flows of Viscous Incompressible Fluids

Let the mind be enlarged, according to its capacity, to the grandeur of the mysteries, and not the mysteries contracted to the narrowness of the mind. —— Francis Bacon

11.1 Introduction

Finite element models based on least-squares method often offer an appealing alternative to the more popular weak-form Galerkin approach used for flows of viscous incompressible fluids. Although not as popular as weak-form Galerkin formulations, least-squares models of partial differential equations have been an active field of research since at least the early 1970's [221–223]. In 1976, Eason [49] compiled an extensive review containing well over 200 references to least-squares methods as applied to the solution of partial differential equations. Since the publication of this review article, least-squares finite element models have continued to receive substantial attention and discussion in the literature.

Least-squares variational formulations allow us to define an unconstrained convex least-squares functional $\mathcal{J}(\mathbf{u})$ in terms of the sum of the squares of the norms of the partial differential equation residuals [38], where standard Sobolev norms are employed (e.g. norms associated with $L_2(\Omega)$ or $H^k(\Omega)$, where $k \in \mathbb{N}$). If the governing equations (augmented by the appropriate boundary conditions) are well posed, it can be readily shown that the exact solution coincides with the minimizer of the least-squares functional. As a result, in the least-squares method the variational problem is obtained via direct minimization of $\mathcal{J}(\mathbf{u})$. In the case of linear differential equations, it is always possible to associate with the least-squares functional a well-defined energy norm $\|\mathbf{u}\|_E$. If it can be shown that the energy norm induced by the least-squares functional is equivalent to an appropriate standard norm, such as the $\mathbf{H}^1(\Omega)$ norm, optimal convergence rates can be established for the least-squares finite element model. Under such conditions the least-squares finite element formulation constitutes an ideal *variational setting*, regardless of whether such a setting is achieved by the associated weak-form Galerkin finite element formulation [39].

J.N. Reddy, *An Introduction to Nonlinear Finite Element Analysis*, Second Edition. ©J.N. Reddy 2015. Published in 2015 by Oxford University Press.

To maintain practicality in the numerical implementation, it becomes computationally advantageous to construct the least-squares functional in terms of the sum of the squares of the $L_2(\Omega)$ norms of the first-order form of the partial differential equation residuals. Regrettably, it is not always possible to establish *a priori* norm equivalence (or $H^1(\Omega)$-coercivity) of the resulting least-squares formulation. As identified by Bochev [224] using Agmon, Douglis, and Nirenberg (ADN) elliptic theory [225], it is typically possible to construct a least-squares functional that is $H^1(\Omega)$-coercive [39, 222, 226, 227]. Unfortunately, the optimal choice of norms can: (a) depend on the nature of the boundary conditions of a given problem and (b) result in an unattractive computational implementation. It is important to note that departure from the ideal variational setting (i.e. using a least-squares functional that is non-$H^1(\Omega)$-coercive) does not typically result in disastrous consequences for least-squares finite element models. Even when a given formulation is non-$H^1(\Omega)$-coercive, the least-squares finite element model always: (a) possesses the best approximation property with respect to a well-defined norm (i.e. the energy norm $\|\mathbf{u}\|_E$) and (b) avoids restrictive compatibility requirements on the finite element function spaces (i.e. the discrete inf-sup condition never arises). That the least-squares method is always based on a minimization principle ensures a robust setting that is often lacking in Galerkin based weak formulations. It is well known, however, that non-$H^1(\Omega)$-coercive low-order finite element implementations are often prone to locking whenever full numerical integration techniques are employed in evaluating the coefficient matrices. In the context of the Navier–Stokes equations, it has been shown that such issues may be largely avoided through the use of collocation or selective reduced integration strategies (see Jiang and his colleagues [38, 179, 228–230]). On the other hand, the combination of high-order finite element technology with least-squares variational principles has also shown great promise in recent years. In particular, building off of the earlier work of Jiang and Sonnad [229], Bell and Surana [231, 232], Gerrtisma and coworkers [233–237], Pontaza and Reddy [43, 44, 238], Prabhakar and Reddy [45, 239, 240], Prabhakar, Pontaza, and Reddy [241], and Payette and Reddy [50] have demonstrated numerically that hp-least-squares finite element models are capable of yielding highly accurate results even when the least-squares functional cannot be shown to be $H^1(\Omega)$-coercive *a priori*.

Least-squares finite element models offer several additional attractive features as compared with weak-form Galerkin formulations. In the case of linear analysis, the least-squares formulation always admits a symmetric positive-definite (SPD) coefficient matrix, regardless of whether or not such symmetry is manifest in the governing partial differential equations. As a result, extremely robust direct as well as iterative solution algorithms (such as the preconditioned conjugate gradient method) can be employed in the solution process and only half of the global coefficient matrix need be stored in memory. This is not the case when the weak-form Galerkin scheme is applied to non-self-adjoint systems of equations. As mentioned previously, the least-squares formulation does not suffer from the restrictive inf-sup condition [242, 243]. This is highly desirable in the numerical discretization of fluid mechanics problems, as it allows the velocity and pressure to be approximated using the same bases of interpolation. Finally, least-squares formulations are also free from the need for numerical dissipation through the use of upwind techniques. As a result, *ad hoc* stabilization is not needed in the analysis of convection dominated problems [38].

Least-squares formulations are certainly not without their own deficiencies. Most problems in physics possess at the very minimum second-order spatial differential operators. Since no weakening of these operators is typically possible through the employment of Green's identities (as can be readily accomplished in weak form Galerkin formulations), least-squares models typically require higher regularity of the approximate solution within each element. Higher regularity requirements negatively affect the condition number of the coefficient matrix and also the continuity requirement of the solution across element boundaries. High regularity requirements may be avoided by constructing the least-squares finite element model in terms of an equivalent lower-order system by the introduction of additional independent auxiliary variables [38]. The resulting *mixed* formulation permits the use of standard Lagrange interpolation functions and also improves the conditioning of the global coefficient matrix [244, 245]. However, such benefits are gained at the expense of an increase in size of the global system of equations. It can be argued that such a formulation is at least somewhat useful, however, as the auxiliary variables often represent important physical quantities of interest (e.g. the heat flux, vorticity, stress, etc.).

Other drawbacks to least-squares formulations, in the context of fluid mechanics, include lack of local mass conservation and poor coupling between the velocity and pressure in transient problems [246]. This is especially true whenever low-order elements are employed in a given finite element discretization. The discrete violation of the requirement that the velocity be a solenoidal vector is often attributed to the fact that, in least-squares formulations, local satisfaction of the governing PDEs is sacrificed in favor of global minimization of the governing equation residuals [247]. In transient flow problems, lack of velocity–pressure coupling has also been identified as a source for poor local mass conservation [248]. We must emphasize that the violation *is not* merely numerical noise and, depending on the nature of the domain and boundary conditions, can actually be quite substantial [247].

Several techniques have been proposed to improve local mass conservation in least-squares finite element models. For example, Deang and Gunzburger [246] advocated weighting the continuity equation residual in the definition of the least-squares functional, where the chosen weight may be either uniform across the whole problem domain or distinct for each element. Chang and Nelson [247], on the other hand, combined the least-squares method with Lagrange multipliers to exactly enforce element-level mass conservation. In this approach the continuity equation is treated as an additional constraint for each element that is enforced in the discrete setting through a set of NE Lagrange multipliers. Although successful, this approach comes at the expense of increasing the system size of the finite element equations and compromising the unconstrained minimization setting that is so attractive for least-squares finite element models. Heys *et al.* [249] demonstrated improved mass balance using a least-squares functional based on a novel first-order reformulation of the incompressible Navier–Stokes equations. It is worth noting that increasing the *p*-level also tends to improves mass conservation, as demonstrated by Pontaza and Reddy [238].

For non-stationary flows, lack of strong velocity–pressure coupling can also compromise local mass conservation and further lead to total instability in space-time decoupled finite element simulations [238]. Pontaza [250] showed that the employment of a regularized form of the continuity equation in least-squares formulations can greatly enhance velocity–pressure coupling and as a direct consequence local mass conservation. Similar approaches have also been advocated in the iterative penalty formulations of Prabhakar and Reddy [239, 240]. For a more mathematical analysis of such optimization methods as applied to the Stokes problem, one may consult the book of Bochev and Gunzburger [39].

In this chapter[1] we consider the least-squares finite element formulation of the Navier–Stokes equations governing viscous incompressible fluids. An L_2-norm based least-squares functional is constructed as a (possibly, a weighted) sum of the norms of the residuals in the governing equations due to the approximation of the dependent unknowns. Weak forms are obtained by minimizing the least-squares functional with respect to each of the dependent variable. When the governing equations are nonlinear (i.e. the least-squares functional is cubic or higher-order function of the dependent unknowns), the weak forms are obtained via direct minimization of the least-squares functional with the aid of the Gâteaux derivative. Although, in theory, the least-squares principle requires that minimization ought to be performed prior to linearization, such an approach is often impractical and not necessary. We illustrate the differences between the various linearization schemes adopted in the abstract formulation, by numerically solving several nonlinear two-dimensional verification benchmark boundary-value problems using least-squares finite element models.

---

[1]This chapter is based on the dissertation of the author's graduate student, Dr. Gregory Payette, *Spectral/hp Finite Element Models for Fluids and Structures*, Texas A&M University, College Station, May 2012; also, see [161].

11.2 Least-Squares Finite Element Formulation

11.2.1 The Navier–Stokes Equations of Incompressible Fluids

We next turn our attention to the incompressible form of the stationary Navier–Stokes equations, which constitutes a very popular application for least-squares variational principles. The classical problem for the non-dimensional form of the incompressible Navier–Stokes equations can be stated as follows: find the velocity vector $\mathbf{v}(\mathbf{x})$ and pressure $P(\mathbf{x})$ such that

$$\mathbf{v} \cdot \nabla\mathbf{v} = -\nabla P + \frac{1}{\text{Re}}\nabla \cdot (\nabla\mathbf{v} + \nabla\mathbf{v}^{\text{T}}) + \mathbf{b} \qquad \text{in } \Omega \qquad (11.2.1\text{a})$$

$$\nabla \cdot \mathbf{v} = 0 \qquad \text{in } \Omega \qquad (11.2.1\text{b})$$

$$\mathbf{v} = \mathbf{v}^{\text{P}} \qquad \text{on } \Gamma^{\text{D}} \qquad (11.2.1\text{c})$$

$$\hat{\mathbf{n}} \cdot \boldsymbol{\sigma} = \mathbf{t}^{\text{P}} \qquad \text{on } \Gamma^{\text{N}} \qquad (11.2.1\text{d})$$

where Re is the Reynolds number, $\mathbf{b}$ is the body force, $\boldsymbol{\sigma}$ is the Cauchy stress tensor, and $\hat{\mathbf{n}}$ is the outward unit normal. The Cauchy stress is given in terms of the following constitutive equation

$$\boldsymbol{\sigma} = -P\mathbf{I} + \frac{1}{\text{Re}}(\nabla\mathbf{v} + \nabla\mathbf{v}^{\text{T}}) \qquad (11.2.2)$$

There are many first-order formulations of the Navier–Stokes equations that have been presented in the literature that can be used to construct finite element models of least-squares type. One of the most popular schemes is the velocity–pressure–vorticity $(\mathbf{v}, P, \boldsymbol{\omega})$ formulation. In this formulation the vorticity vector $\boldsymbol{\omega} = \nabla \times \mathbf{v}$ is introduced, along with the vector identity

$$\nabla \times (\nabla \times \mathbf{v}) = -\nabla^2\mathbf{v} + \nabla(\nabla \cdot \mathbf{v}) \qquad (11.2.3)$$

As a result, the original problem can be restated as one of finding the velocity $\mathbf{v}(\mathbf{x})$, pressure $P(\mathbf{x})$, and vorticity $\boldsymbol{\omega}(\mathbf{x})$ such that

$$\mathbf{v} \cdot \nabla\mathbf{v} + \nabla P + \frac{1}{\text{Re}}\nabla \times \boldsymbol{\omega} = \mathbf{b} \qquad \text{in } \Omega \qquad (11.2.4\text{a})$$

$$\boldsymbol{\omega} - \nabla \times \mathbf{v} = \mathbf{0} \qquad \text{in } \Omega \qquad (11.2.4\text{b})$$

$$\nabla \cdot \mathbf{v} = 0 \qquad \text{in } \Omega \qquad (11.2.4\text{c})$$

$$\mathbf{v} = \mathbf{v}^{\text{P}} \qquad \text{on } \Gamma_{\mathbf{v}} \qquad (11.2.4\text{d})$$

$$\boldsymbol{\omega} = \boldsymbol{\omega}^{\text{P}} \qquad \text{on } \Gamma_{\omega} \qquad (11.2.4\text{e})$$

$$\hat{\mathbf{n}} \cdot \tilde{\boldsymbol{\sigma}} = \tilde{\mathbf{t}}^{\text{P}} \qquad \text{on } \Gamma^{\text{N}} \qquad (11.2.4\text{f})$$

where Γ^{D} has been partitioned such that $\Gamma^{\text{D}} = \Gamma_{\mathbf{v}} \bigcup \Gamma_{\omega}$ and $\Gamma_{\mathbf{v}} \bigcap \Gamma_{\omega} = \varnothing$. We note that the incompressibility constraint has been imposed in the construction of the momentum equation. The pseudo-traction boundary condition $\tilde{\mathbf{t}}^{\mathrm{P}}$ is given in terms of the pseudo-stress tensor $\tilde{\boldsymbol{\sigma}}$ defined as

$$\tilde{\boldsymbol{\sigma}} = -P\mathbf{I} + \frac{1}{\mathrm{Re}}\nabla\mathbf{v} \tag{11.2.5}$$

For three-dimensional analysis it is helpful to augment the above equations by the compatibility condition $\nabla \cdot \boldsymbol{\omega} = 0$.

We associate with the stationary first-order form of the incompressible Navier–Stokes equations the following *true* least-squares functional

$$\mathcal{J}(\mathbf{v}, P, \boldsymbol{\omega}; \mathbf{b}, \tilde{\mathbf{t}}^{\mathrm{P}}) = \frac{1}{2}\left(\|\mathbf{v} \cdot \nabla\mathbf{v} + \nabla P + \frac{1}{\mathrm{Re}}\nabla \times \boldsymbol{\omega} - \mathbf{b}\|_{\Omega,0}^2 + \|\nabla \cdot \mathbf{v}\|_{\Omega,0}^2\right.$$

$$\left. + \|\boldsymbol{\omega} - \nabla \times \mathbf{v}\|_{\Omega,0}^2\right) \tag{11.2.6}$$

The outflow boundary condition may also be directly accounted for in the definition of the least-squares functional. Linearized versions of the least-squares functional may be obtained by replacing the nonlinear convective term with following Picard or Newton approximations:

$$\mathbf{v} \cdot \nabla\mathbf{v}|_{\mathrm{Pic}} = \mathbf{v}_0 \cdot \nabla\mathbf{v} \tag{11.2.7a}$$

$$\mathbf{v} \cdot \nabla\mathbf{v}|_{\mathrm{New}} = \mathbf{v} \cdot \nabla\mathbf{v}_0 + \mathbf{v}_0 \cdot \nabla\mathbf{v} - \mathbf{v}_0 \cdot \nabla\mathbf{v}_0 \tag{11.2.7b}$$

The linearized least-squares based weak form resulting from invoking the minimization principle may be stated as follows: find $\mathbf{u} \in \mathcal{V}$ such that for all $\delta\mathbf{u} \in \mathcal{W}$ the following expression holds:

$$\mathcal{B}(\delta\mathbf{u}, \mathbf{u}) = \mathcal{F}(\delta\mathbf{u}) \tag{11.2.8}$$

where $\mathbf{u} = (\mathbf{v}, P, \boldsymbol{\omega})$, $\delta\mathbf{u} = (\delta\mathbf{v}, \delta P, \delta\boldsymbol{\omega})$, and $\mathcal{V}$ and $\mathcal{W}$ are appropriate function spaces. The forms $\mathcal{B}(\delta\mathbf{u}, \mathbf{u})$ and $\mathcal{F}(\delta\mathbf{u})$ are

$$\mathcal{B}(\delta\mathbf{u}, \mathbf{u}) = \int_\Omega \left[\left(\mathbf{v}_0 \cdot \nabla\delta\mathbf{v} + \nabla\delta P + \frac{1}{\mathrm{Re}}\nabla \times \delta\boldsymbol{\omega}\right) \cdot \left(\mathbf{v}_0 \cdot \nabla\mathbf{v} + \nabla P\right.\right.$$

$$\left. + \frac{1}{\mathrm{Re}}\nabla \times \boldsymbol{\omega}\right) + (\nabla \cdot \delta\mathbf{v})(\nabla \cdot \mathbf{v}) + (\delta\boldsymbol{\omega} - \nabla \times \delta\mathbf{v}) \cdot (\boldsymbol{\omega} - \nabla \times \mathbf{v})$$

$$\left. + \underline{(\delta\mathbf{v} \cdot \nabla\mathbf{v}_0) \cdot (\mathbf{v} \cdot \nabla\mathbf{v}_0)}\right] d\Omega \tag{11.2.9a}$$

$$\mathcal{F}(\delta\mathbf{u}) = \int_\Omega \left[\left(\mathbf{v}_0 \cdot \nabla\delta\mathbf{v} + \nabla\delta p + \frac{1}{\mathrm{Re}}\nabla \times \delta\boldsymbol{\omega}\right) \cdot \mathbf{b} - \underline{(\delta\mathbf{v} \cdot \nabla\mathbf{v}_0)}\cdot\right.$$

$$\left. \underline{\left(\nabla p_0 + \frac{1}{\mathrm{Re}}\nabla \times \boldsymbol{\omega}_0 - \mathbf{b}\right)}\right] d\Omega \tag{11.2.9b}$$

when the Picard linearization scheme is employed, and

$$\mathcal{B}(\delta\mathbf{u}, \mathbf{u}) = \int_\Omega \left[\left(\delta\mathbf{v} \cdot \nabla\mathbf{v}_0 + \mathbf{v}_0 \cdot \nabla\delta\mathbf{v} + \nabla\delta P + \frac{1}{\mathrm{Re}}\nabla \times \delta\boldsymbol{\omega}\right) \cdot\right.$$

$$
\left(\mathbf{v} \cdot \boldsymbol{\nabla} \mathbf{v}_0 + \mathbf{v}_0 \cdot \boldsymbol{\nabla} \mathbf{v} + \boldsymbol{\nabla} P + \frac{1}{\mathrm{Re}} \boldsymbol{\nabla} \times \boldsymbol{\omega} \right) + (\boldsymbol{\nabla} \cdot \delta \mathbf{v})(\boldsymbol{\nabla} \cdot \mathbf{v})
$$
$$
+ (\delta \boldsymbol{\omega} - \boldsymbol{\nabla} \times \delta \mathbf{v}) \cdot (\boldsymbol{\omega} - \boldsymbol{\nabla} \times \mathbf{v}) + \underline{(\delta \mathbf{v} \cdot \boldsymbol{\nabla} \mathbf{v} + \mathbf{v} \cdot \boldsymbol{\nabla} \delta \mathbf{v}) \cdot}
$$
$$
\underline{\left(\mathbf{v}_0 \cdot \boldsymbol{\nabla} \mathbf{v}_0 + \boldsymbol{\nabla} p_0 + \frac{1}{\mathrm{Re}} \boldsymbol{\nabla} \times \boldsymbol{\omega}_0 - \mathbf{b} \right)} \Big] d\Omega \qquad (11.2.10\mathrm{a})
$$

$$
\mathcal{F}(\delta \mathbf{u}) = \int_\Omega \Big[\left(\delta \mathbf{v} \cdot \boldsymbol{\nabla} \mathbf{v}_0 + \mathbf{v}_0 \cdot \boldsymbol{\nabla} \delta \mathbf{v} + \boldsymbol{\nabla} \delta P + \frac{1}{\mathrm{Re}} \boldsymbol{\nabla} \times \tilde{\boldsymbol{\omega}} \right) \cdot (\mathbf{b} + \mathbf{v}_0 \cdot \boldsymbol{\nabla} \mathbf{v}_0)
$$
$$
+ \underline{(\delta \mathbf{v} \cdot \boldsymbol{\nabla} \mathbf{v}_0 + \mathbf{v}_0 \cdot \boldsymbol{\nabla} \delta \mathbf{v}) \cdot \left(\mathbf{v}_0 \cdot \boldsymbol{\nabla} \mathbf{v}_0 + \boldsymbol{\nabla} p_0 + \frac{1}{\mathrm{Re}} \boldsymbol{\nabla} \times \boldsymbol{\omega}_0 - \mathbf{b} \right)} \Big] d\Omega
$$

$$(11.2.10\mathrm{b})$$

when Newton's method is applied. The underlined terms appear when minimization is performed prior to linearization.

11.2.2 Numerical Examples

11.2.2.1 Low Reynolds number flow past a circular cylinder

In this example we consider flow past a circular cylinder, a problem that has been studied extensively by way of experiment [251–253] and is a standard benchmark for numerical computation [254–256]. It is well-known from both experimentation [253] and numerical modeling [257] that for moderately low Reynolds numbers ($5 < \mathrm{Re} < 46.1$) the flow is spatially stationary and characterized by two symmetric regions of circulation directly downwind of the cylinder. The size of the standing vortices in the wake region is proportional to the Reynolds number.

Ideally we would like to model the flow in a manner such that the effects due to the truncation of the problem's infinite domain to a finite computational domain do not corrupt the numerical solution. To this end we take $\bar{\Omega}$ to be the set difference between the rectangular region $[-25, 25] \times [-15, 15]$ and an open circular region with unit diameter centered about the origin. The computational domain $\bar{\Omega}^{hp} \cong \bar{\Omega}$ consists of 240 nonuniform finite elements, with 15 element layers in the radial direction and 16 along the circumference of the cylinder, as shown in Fig. 11.2.1. The smallest elements are placed in the vicinity of the cylinder to ensure adequate numerical resolution in the anticipated wake region. The mesh geometry is characterized using an isoparametric formulation which, when combined with high-order finite element technology, allows for a highly accurate approximation of the cylinder surface. As in the Poisson benchmark problem, we refine the mesh by systematically increasing the p-level of the finite element approximation functions. We consider the cases where $p = 2, 4, 6$, and 8; which amounts to 3,968, 15,616, 34,944, and 61,952 total degrees of freedom for each corresponding finite element discretization. The outflow boundary condition is enforced weakly through the least-squares functional with $\mathbf{t}^{\mathrm{P}}$ taken as zero along the right hand side of $\bar{\Omega}$. Nonlinear convergence is declared for a given numerical simulation once the relative error in the solution is less than 10^{-6}.

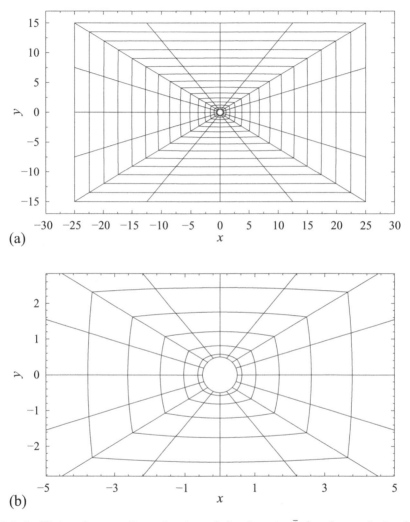

Fig. 11.2.1: Finite element discretization of the domain $\bar{\Omega}$ for the analysis of steady flow past a circular cylinder: (a) complete mesh and (b) mesh near the cylinder.

Since this problem does not admit an analytic solution, we obtain a reasonable *a posteriori* estimate for the error via a numerical evaluation of $\mathcal{J}$ in the post-computation. Exponential decay of the least-squares functional, shown in Fig. 11.2.2, is clearly seen as the polynomial order of the numerical solution is increased. As expected, each least-squares model produces identical converged results (for a given p) with the exception of the Picard scheme (applied prior to minimization). The value of $\mathcal{J}$ for this scheme is only slightly greater than the values predicted by the other three models.

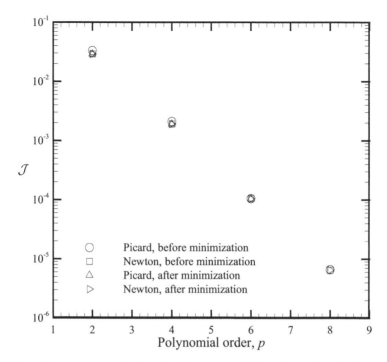

Fig. 11.2.2: Convergence of the least-squares finite element solutions under p-refinement as measured in terms of the least-squares functional $\mathcal{J}$ for steady flow past a circular cylinder at Re = 40.

Figure 11.2.3 shows the computed pressure coefficients along the surface of the cylinder. When the mesh is coarse ($p = 2$), linearization before minimization using Picard's method yields substantially different results than does Newton's method (where by Newton's method we mean either Newton schemes, as they both yield identical results). Neither solution at this p-level, however, constitutes an appropriate converged solution for the problem. We see that as the p is increased to four, the Picard (before minimization) and Newton schemes begin to yield the same results. Finally, at $p = 6$ we observe virtually no difference between the results of both schemes. The computed values for $p = 6$ and higher were found to be in excellent agreement with the empirical work conducted by Grove *et al.* [252]. Figure 11.2.4 contains plots of the pressure field and velocity component v_y in the vicinity of the cylinder for the Newton solution at $p = 8$. Streamlines are also shown highlighting the size of the circulation regions. The numerical simulation predicts the wake region to extend 4.5 cylinder radii downstream of the cylinder, which is in excellent agreement with the numerical results reported by Kawaguti and Jain [254].

A word on the nonlinear iterative convergence behavior of each finite element scheme is also in order. In each formulation we solved the equations without the employment of load steps. A summary of the total number of iterations needed to achieve the desired termination criteria is summarized in

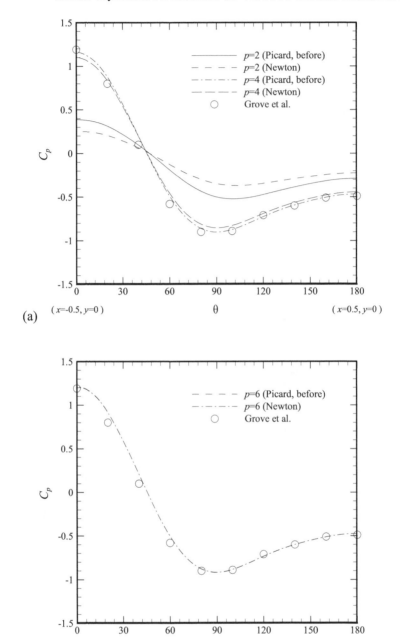

Fig. 11.2.3: A comparison of the numerically computed pressure coefficient C_p along the surface of the cylinder at Re $= 40$ with the experimental data obtained by Grove et al. [252]: (a) non-converged numerical solutions and (b) fully-converged numerical solutions.

Table 11.2.1. When linearization was performed prior to minimization, solution convergence was possible without the need for relaxation. For the case of the Picard linearization after minimization, convergence was extremely slow and could not be achieved without relaxation. Due to such poor rates of convergence, a solution at $p = 8$ was not attempted. Using the Newton linearization scheme after minimization produced divergent results, with or without the use of a relaxation parameter. Therefore, a mixed Newton scheme was considered, where three Newton (before minimization) iterations were used prior to subsequent Newton (after minimization) iterations. This method produced very good results in terms of minimizing the total number of iterations required for convergence. For low p values this approach slightly outperformed the Newton (before minimization) scheme. Overall, the Newton (before minimization and mixed) schemes exhibited much better convergence rates than the Picard formulation.

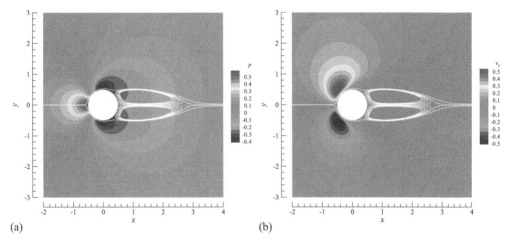

Fig. 11.2.4: Steady flow past a circular cylinder at Re $= 40$: (a) pressure field and streamlines and (b) velocity component v_y and streamlines.

Table 11.2.1: Steady flow past a circular cylinder: number of iterations required to satisfy the nonlinear solution convergence criterion for various least-squares finite element implementations (convergence tolerance $\epsilon = 10^{-6}$).

		Number of nonlinear iterations			
Least-squares formulation	ω_0	$p = 2$	$p = 4$	$p = 6$	$p = 8$
Picard (before)	1.0	11	15	15	15
Newton (before)	1.0	10	9	7	6
Picard (after)	0.5	39	161	179	–
Newton (mixed)	1.0	7	7	7	7

11.2.2.2   Steady flow over a backward facing step

In this example the flow of a viscous incompressible fluid over a backward facing step is considered. This problem was studied by way of experiment and also numerical simulation by Armaly *et al.* [258]. Laminar, transition, and turbulent flows were empirically assessed for $70 < \text{Re} < 8{,}000$, and numerically simulated for steady-state cases up to a Reynolds number of 1,250. In the present study, the stationary solution of the two-dimensional problem is evaluated at $\text{Re} = 800$, using the simplified step configuration proposed in the benchmark solution of Garting [203].

The computational domain for the problem is taken as $\bar{\Omega} = [0, 30] \times [-0.5, 0.5]$ as shown in Fig. 11.2.5. The fluid enters the domain on the left hand side of $\bar{\Omega}$ on $0 \leq y \leq 0.5$. The velocity vector at the inlet is assumed to be horizontal with the x-component given by the parabolic expression $\bar{v}_x = 24y(0.5 - y)$ for $0 < y < 0.5$ and $\bar{v}_x = 0$ for $-0.5 < y < 0$. The components of the velocity are taken to be zero along all solid surfaces in accordance with the non-slip condition. The outflow boundary condition is enforced weakly by taking $\mathbf{t}^P = \mathbf{0}$ in the least-squares functional.

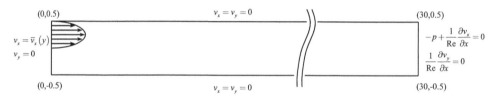

Fig. 11.2.5: Geometry and boundary conditions for steady flow of an incompressible viscous fluid over a backward facing step at $\text{Re} = 800$.

We discretize the computational domain into a set of 40 rectangular finite elements, with 20 elements along the channel length and 2 along the channel height as shown in Fig. 11.2.6. The majority of the elements are positioned within 15 units of the channel inlet to ensure proper resolution of all variables within the flow separation regions anticipated downstream of the step. We once again refine the discrete solution by systematically increasing the number of nodes in each finite element. We arrive at the numerical solution at $\text{Re} = 800$, by solving a series of problems at intermediate Reynolds numbers. We begin by solving the problem at $\text{Re} = 100$ followed by $\text{Re} = 200$ and so on until we reach $\text{Re} = 800$. For each intermediate problem, we utilize the converged solution from the previous problem in the series as the initial guess. As an initial guess for the problem where $\text{Re} = 100$, we assume all variables to be zero. Nonlinear convergence is declared for each problem when the Euclidean norm of the difference between the nonlinear solution increments is less than 10^{-4}.

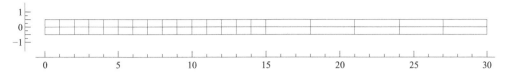

Fig. 11.2.6: Finite element mesh for analysis of stationary incompressible viscous flow over a backward facing step.

In Fig. 11.2.7 we show the least-squares finite element solution of the problem as determined using a polynomial of order 10 within each element. The velocity vectors are depicted along with contour plots of the velocity components and pressure field. The flow is characterized by a large recirculation zone directly behind the step on the low side of the channel that extends roughly 6.1 units beyond the step. A second region of flow separation and recirculation is also present on the top side of the channel that develops around 4.9 units downstream of the step and extends to approximately $x = 10.5$.

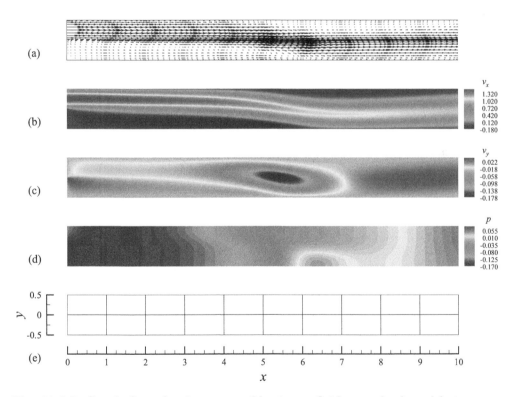

Fig. 11.2.7: Steady flow of an incompressible viscous fluid over a backward facing step at Re $= 800$: (a) velocity vector field at finite element nodes, (b) velocity component v_x, (c) velocity component v_y, (d) pressure field, and (e) finite element mesh directly behind step.

In Fig. 11.2.8 we compare the numerical solutions for the components of the velocity vector along $x = 7$ and $x = 15$ with the results reported by Gartling

[203], where a weak-form Galerkin finite element model was used. The converged results for the Picard (linearization before minimization) and Newton schemes are in excellent agreement with the published data. As expected, the Picard and Newton formulations yield different results when the finite element mesh is too coarse to allow for convergence. It is interesting to note, however, that the Picard scheme offers a somewhat better approximation of the velocity components on the coarse mesh than does Newton's method at $x = 7$ and $x = 15$. The reason for this phenomenon is unclear; however, it is expected that both schemes converge to the same solution under proper mesh refinement.

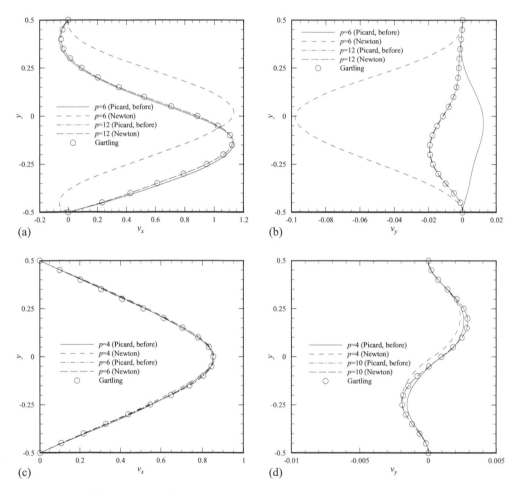

Fig. 11.2.8: Comparison of numerically computed velocity components for the steady flow of a viscous fluid over a backward facing step at Re = 800 with the published results of Gartling [203] (also see Reddy and Gartling [10]): (a) horizontal velocity profile at $x = 7$, (b) vertical velocity profile at $x = 7$, (c) horizontal velocity profile at $x = 15$, and (d) vertical velocity profile at $x = 15$.

In Fig. 11.2.9 the value of the least-squares functional for the Picard (before minimization) and Newton (before minimization) schemes as a function of p is shown. Although $\mathcal{J}$ is always greater for the Picard scheme as compared with Newton's method, the actual numerical values are nearly identical. A summary of the total number of iterations required to reach convergence at each Reynolds number is provided in Table 11.2.2 for $p = 6$. The Picard and Newton solutions (where linearization was performed prior to minimization) were obtained without the use of relaxation. For the Picard (after minimization) formulation, a relaxation parameter of $\omega_0 = 0.5$ was utilized. Even with the aid of relaxation, however, the scheme still suffered from a poor rate of convergence. A converged solution for the Newton (after minimization) formulation, with or without relaxation, could not be obtained. As in the previous example, once again a mixed Newton formulation was introduced in an attempt to recover a convergent solution. In the mixed approach, the Newton (before minimization) formulation was utilized in the iterative solution scheme until ϵ was less than 0.05, at which point a switch was made to the Newton (after minimization) formulation. Use of relaxation was found to be unnecessary in the mixed approach. On average the mixed formulation performed comparably to the Newton (before minimization) scheme and in some cases superior at this p-level. However, at higher p-levels we find little difference between the convergence behaviors of these Newton formulations (especially at Re = 700, 800). As expected, the Newton schemes require far fewer iterations than their Picard counterparts.

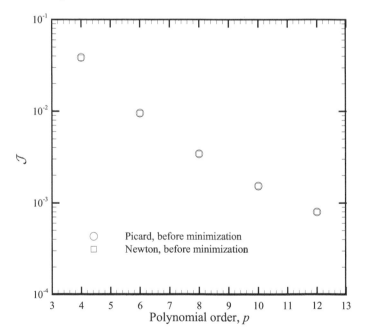

Fig. 11.2.9: Convergence of the least-squares finite element solutions under p-refinement as measured in terms of the least-squares functional $\mathcal{J}$ for steady flow over a backward facing step at Re = 800.

Table 11.2.2: Steady flow over backward facing step: number of iterations required to satisfy the nonlinear solution convergence criterion for various least-squares finite element implementations, where $p = 6$ (convergence tolerance $\epsilon = 10^{-4}$).

	Number of nonlinear iterations			
Reynolds number	Picard (before)	Newton (before)	Picard (after)	Newton (mixed)
100	13	6	81	6
200	20	4	313	4
300	28	5	616	5
400	39	6	774	6
500	53	7	641	8
600	63	8	501	7
700	69	11	405	8
800	74	15	380	10

11.2.2.3   Lid-driven cavity flow

As a final verification benchmark, we consider the classical two-dimensional lid-driven cavity flow problem. The computational domain is taken as the unit square given as $\bar{\Omega} = [0, 1] \times [0, 1]$. On the bottom and left and right sides of the cavity the components of the velocity are specified to be zero. Along the top surface a horizontal velocity profile is specified using the following expression

$$v_x(x) = \begin{cases} \tanh(50x) & 0 \leqslant x \leqslant 0.5 \\ -\tanh[50(x - 1)] & 0.5 < x \leqslant 1.0 \end{cases} \tag{11.2.11}$$

These boundary condition essentially prescribes v_x as unity along the majority of the top surface of the cavity with a smooth and abrupt transition to $v_x = 0$ at the corners. The boundary condition is applied in this way to avoid *singularities* in the solution in the vicinity of the upper corners [241]. High-order methods are sensitive to such singularities, and the above boundary condition ensures, in this sense, a well-posed problem. The pressure is taken to be zero at the single point $(x, y) = (0.5, 0)$. We seek the steady-state solution of the problem at a Reynolds number of 3,200, and compare our numerical solutions with the tabulated finite difference results reported by Ghia, Ghia, and Shin [202], who used $v_x = 1$ at all points of the lid except at $x = 0$ and $x = 1$, where they used $v_x = 0$.

The domain is discretized into a nonuniform set of 144 rectangular finite elements as shown in Fig. 11.2.10. The mesh is graded such that smaller elements are placed near the boundaries to ensure proper resolution of the numerical solution in the regions of the boundary layers and anticipated vortices. As in previous examples, the mesh is refined by increasing the p-level of the solution within each finite element. We utilize polynomials of orders $p = 4$, 5, 6, 7, 8, and 9 in our analysis which correspond with 9,604, 14,884, 21,316, 28,900, 37,636, and 47,524 total degrees of freedom. As in the previous example, the desired solution at Re = 3,200 is obtained by solving a series of problems at intermediate Reynolds numbers. In this case we solve a series of seven problems beginning with the first problem posed with Re = 457.1428 and culminating with the final desired solution at Re = 3,200. Nonlinear convergence is considered to be achieved in each problem once the Euclidean norm of the difference between the nonlinear iterative solution increments is less than 10^{-4}. Due to poor convergence properties, a numerical solution using the Picard scheme (after minimization) was not attempted.

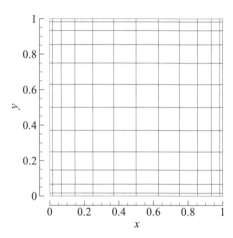

Fig. 11.2.10: Finite element mesh for the lid-driven cavity flow problem.

The velocity components, pressure field, and streamlines are shown in Fig. 11.2.11 for the numerical solution obtained using Newton's method (before minimization) at $p = 9$. The flow is characterized by a large region of rotation that is just off-set from the geometric center of the cavity. Secondary vortices are also present in the regions near the bottom (left and right) and top left corners of the domain. The streamline patterns match well with the published results of Ghia, Ghia, and Shin [202].

In Fig. 11.2.12 we compare our least-squares finite element solutions along the vertical and horizontal mid-planes of the cavity with the tabulated results of Ghia, Ghia, and Shin [202]. We once again find that when the mesh is too coarse to yield a convergent solution, the numerical results differ for the Picard (linearization before minimization) and Newton schemes. However, as expected both schemes yield identical results when the mesh is properly refined. As in the previous example, we are surprised to find that the Picard scheme offers a slightly better approximation of the velocity components on the coarse mesh along the mid-lines of the cavity.

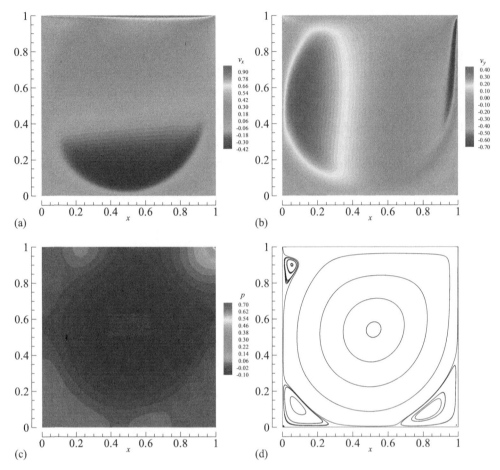

Fig. 11.2.11: Two-dimensional lid-driven cavity flow at Re = 3,200: (a) velocity component v_x, (b) velocity component v_y, (c) pressure field, and (d) streamline patterns in cavity highlighting standing vortices.

The value of the least-squares functional as a function of the p-level exhibits characteristics similar to those discussed in previous examples. In particular, the value of $\mathcal{J}$ in both Newton schemes is always slightly less than (although nearly identical to) the value as determined using the Picard (before minimization) method. We were once again able to obtain convergent solutions using the Picard and Newton schemes (with linearization performed before minimization) without the need for solution relaxation. Solution convergence could not be achieved for the Newton (after minimization) scheme with or without relaxation. As a result, we again utilized a mixed Newton scheme, where Newton (before minimization) iterations where performed until the relative error ϵ was less than 0.01 at which point we switched to the Newton (after minimization) scheme. The total number of iterations required for solution convergence at a p-level of 6 is summarized for each scheme in Table 11.2.3.

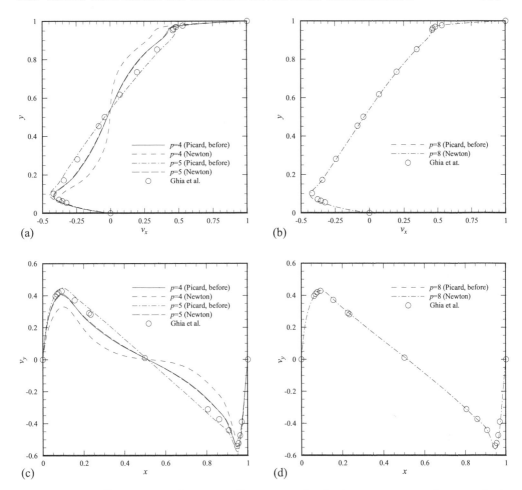

Fig. 11.2.12: Comparison of numerically computed velocity components along vertical and horizontal mid-lines of lid-driven cavity with published results of Ghia, Ghia, and Shin [202] at Re = 3,200: (a) non-converged numerical solutions for horizontal velocity v_x profiles along vertical centerline, (b) converged numerical solutions for horizontal velocity v_x profiles along vertical centerline, (c) non-converged numerical solutions for vertical velocity v_y profiles along horizontal centerline, and (d) converged numerical solutions for vertical velocity v_y profiles along horizontal centerline.

11.3 A Least-Squares Finite Element Model with Enhanced Element-Level Mass Conservation

11.3.1 Introduction

In this section a novel least-squares finite element model for both the steady and unsteady solutions of viscous incompressible fluids based on the standard velocity–pressure–vorticity first-order system, but with enhanced element-level mass conservation, is presented. The formulation may be viewed as a direct extension of the works of Chang and Nelson [247] and Pontaza [250]. For the

Table 11.2.3: Two-dimensional lid-driven cavity problem: number of iterations required to satisfy the nonlinear solution convergence criterion for various least-squares finite element implementations, where $p = 6$ (convergence tolerance $\epsilon = 10^{-4}$).

	Number of nonlinear iterations		
Reynolds number	Picard (before)	Newton (before)	Newton (mixed)
457.14	12	7	7
914.29	12	5	5
1371.43	12	4	4
1828.57	14	4	4
2285.71	17	4	4
2742.86	20	4	4
3200.00	23	3	3

steady flow case, the constrained minimization problem of Chang and Nelson [247] is recast into an unconstrained minimization problem through the use of the penalty method [2, 10]. This approach is quite natural, as the traditional least-squares method is itself, in a sense, a multi-equation penalty formulation (where penalization is applied *to all* of the residuals). For unsteady flows, a penalty formulation is presented that enhances local mass conservation while improving velocity–pressure coupling and overall numerical stability.

11.3.2   Unsteady Flows

The Navier–Stokes equations governing the unsteady (i.e. non-stationary) flow of a viscous incompressible fluid may be stated in non-dimensional form as follows: find the velocity $\mathbf{v}(\mathbf{x}, t)$ and pressure $p(\mathbf{x}, t)$ such that

$$\frac{\partial \mathbf{v}}{\partial t} + \mathbf{v} \cdot \nabla \mathbf{v} + \nabla p - \frac{1}{\mathrm{Re}} \nabla \cdot [\nabla \mathbf{v} + (\nabla \mathbf{v})^{\mathrm{T}}] = \mathbf{b} \qquad \text{in } \Omega \times (0, \tau] \qquad (11.3.1\mathrm{a})$$

$$\nabla \cdot \mathbf{v} = 0 \qquad \text{in } \Omega \times (0, \tau] \qquad (11.3.1\mathrm{b})$$

$$\mathbf{v}(\mathbf{x}, 0) = \bar{\mathbf{v}}(\mathbf{x}) \qquad \text{in } \Omega \qquad (11.3.1\mathrm{c})$$

$$\mathbf{v} = \mathbf{v}^{\mathrm{P}}(\mathbf{x}, t) \qquad \text{on } \Gamma^{\mathrm{D}} \times (0, \tau] \quad (11.3.1\mathrm{d})$$

$$\hat{\mathbf{n}} \cdot \boldsymbol{\sigma} = \mathbf{t}^{\mathrm{P}}(\mathbf{x}, t) \qquad \text{on } \Gamma^{\mathrm{N}} \times (0, \tau] \quad (11.3.1\mathrm{e})$$

where $\tau \in \mathbb{R}^{+}$ is the time parameter, Re is the Reynolds number, $\mathbf{b}$ is the body force, $\boldsymbol{\sigma}$ is the Cauchy stress tensor, and $\hat{\mathbf{n}}$ is the outward unit normal vector to the boundary. In addition, $\bar{\mathbf{v}}(\mathbf{x})$ is the initial velocity profile in Ω, $\mathbf{v}^{\mathrm{P}}(\mathbf{x}, t)$ is the prescribed velocity on Γ^{D} and $\mathbf{t}^{\mathrm{P}}(\mathbf{x}, t)$ is the traction specified on Γ^{N}. We assume that $\nabla \cdot \bar{\mathbf{v}} = 0$ in Ω and that the problem is well posed. Whenever $\Gamma^{\mathrm{N}} = \varnothing$, we further prescribe the pressure at a single point in $\bar{\Omega}$.

11.3.2.1 The velocity–pressure–vorticity first-order system

As discussed in Section 11.2, the Navier–Stokes equations as expressed in terms of the primitive variables $\mathbf{v}(\mathbf{x}, t)$ and $p(\mathbf{x}, t)$ are ill-suited for direct implementation in a least-squares finite element formulation. To allow for the use of C^0 basis functions in the numerical implementation we introduce the vorticity vector $\omega = \nabla \times \mathbf{v}$ to allow us to recast the Navier–Stokes equations in terms of an equivalent first-order system, which amounts to finding the velocity $\mathbf{v}(\mathbf{x}, t)$, pressure $p(\mathbf{x}, t)$, and vorticity $\omega(\mathbf{x}, t)$ such that

$$\frac{\partial \mathbf{v}}{\partial t} + \mathbf{v} \cdot \nabla \mathbf{v} + \nabla p + \frac{1}{\mathrm{Re}} \nabla \times \omega = \mathbf{b} \qquad \text{in } \Omega \times (0, \tau] \qquad (11.3.2\mathrm{a})$$

$$\nabla \cdot \mathbf{v} = 0 \qquad \text{in } \Omega \times (0, \tau] \qquad (11.3.2\mathrm{b})$$

$$\omega - \nabla \times \mathbf{v} = 0 \qquad \text{in } \Omega \times (0, \tau] \qquad (11.3.2\mathrm{c})$$

$$\mathbf{v}(\mathbf{x}, 0) = \bar{\mathbf{v}}(\mathbf{x}) \qquad \text{in } \Omega \qquad (11.3.2\mathrm{d})$$

$$\mathbf{v} = \mathbf{v}^{\mathrm{P}}(\mathbf{x}, t) \qquad \text{on } \Gamma_{\mathbf{v}} \times (0, \tau] \qquad (11.3.2\mathrm{e})$$

$$\omega = \omega^{\mathrm{P}}(\mathbf{x}, t) \qquad \text{on } \Gamma_{\omega} \times (0, \tau] \qquad (11.3.2\mathrm{f})$$

$$\hat{\mathbf{n}} \cdot \tilde{\sigma} = \tilde{\mathbf{t}}^{\mathrm{P}}(\mathbf{x}, t) \qquad \text{on } \Gamma^{\mathrm{N}} \times (0, \tau] \qquad (11.3.2\mathrm{g})$$

In the above expressions $\omega^{\mathrm{P}}(\mathbf{x}, t)$ is the prescribed vorticity on Γ_{ω}, $\tilde{\sigma} = -P\mathbf{I} + 1/\mathrm{Re}\nabla\mathbf{v}$ is the pseudo stress tensor and $\tilde{\mathbf{t}}^{\mathrm{P}}(\mathbf{x}, t)$ is the pseudo traction vector specified on Γ^{N}. The Dirichlet part of the boundary has been partitioned such that $\Gamma^{\mathrm{D}} = \Gamma_{\mathbf{v}} \bigcup \Gamma_{\omega}$ and $\Gamma_{\mathbf{v}} \bigcap \Gamma_{\omega} = \varnothing$. The expression $\nabla(\nabla \cdot \mathbf{v})$ has been eliminated from the momentum equation on account of the solenoidal nature of the velocity field; as a result, the outflow condition given in Eq. (11.3.2g) is preferred over Eq. (11.3.1e) (see Sani and Gresho [259]). For three-dimensional problems it is helpful to augment the first-order system with the seemingly redundant compatibility condition $\nabla \cdot \omega = 0$ in $\Omega \times (0, \tau]$ [38].

11.3.2.2 Temporal discretization

In this work we employ a space-time decoupled finite element approximation of the dependent variables. At each time step we approximate the time derivative of the velocity field using the backwards difference formula of order n (or BDFn)

$$\frac{\partial \mathbf{v}^{s+1}}{\partial t} \cong \frac{1}{\Delta t_{s+1}} \left(\gamma_0^n \mathbf{v}^{s+1} - \sum_{q=0}^{n-1} \beta_q^n \mathbf{v}^{s-q} \right) \qquad (11.3.3)$$

where $\Delta t_{s+1} = t_{s+1} - t_s$ is the time increment and γ_0^n and β_q^n are the temporal integration parameters. It is well-known that the backward difference formulae are particularly useful in the numerical solutions of stiff partial differential equations and differential-algebraic equations (DAEs). The backward difference formulae are especially valuable in achieving numerical stability and typically provide sufficient numerical dissipation of spurious high-frequency modes [260].

In this chapter we adopt the BDF1 and BDF2 formulae. Since the BDF2 time integrator is non-self-starting, we employ the BDF1 formula in the first few time steps.

11.3.2.3   The standard L_2-norm based least-squares model

The standard least-squares functional associated with the first-order vorticity form of the Navier–Stokes equations is constructed in terms of the sum of the squares of the L_2 norms of the partial differential equation residuals. In the space-time decoupled formulation, we define the least-squares functional associated with the current time step $t = t_{s+1}$ as

$$
\mathcal{J}_{\Delta t}(\mathbf{v}, P, \boldsymbol{\omega}; \tilde{\mathbf{b}}, \tilde{\mathbf{t}}^{\mathrm{P}}) = \frac{1}{2}\Big(\alpha \| \lambda_0^n \mathbf{v} + \mathbf{v} \cdot \boldsymbol{\nabla}\mathbf{v}_0 + \mathbf{v}_0 \cdot \boldsymbol{\nabla}\mathbf{v} + \boldsymbol{\nabla}P + \frac{1}{\mathrm{Re}}\boldsymbol{\nabla} \times \boldsymbol{\omega} - \tilde{\mathbf{b}} \|_{\Omega,0}^2
$$

$$
+ \| \boldsymbol{\nabla} \cdot \mathbf{v} \|_{\Omega,0}^2 + \| \boldsymbol{\omega} - \boldsymbol{\nabla} \times \mathbf{v} \|_{\Omega,0}^2 + \| \mathbf{n} \cdot \tilde{\boldsymbol{\sigma}} - \tilde{\mathbf{t}}^{\mathrm{P}} \|_{\Gamma^{\mathrm{N}},0}^2 \Big)
$$

$$(11.3.4)$$

where the quantities λ_0^n and $\tilde{\mathbf{b}}$ are defined as

$$
\lambda_0^n = \gamma_0^n/\Delta t, \qquad \tilde{\mathbf{b}} = \mathbf{b} + \mathbf{v}_0 \cdot \boldsymbol{\nabla}\mathbf{v}_0 + \frac{1}{\Delta t}\sum_{q=0}^{n-1}\beta_q^n \mathbf{v}^{s-q} \qquad (11.3.5)
$$

All quantities appearing in the definition of $\mathcal{J}_{\Delta t}$ are evaluated at the current time step $t = t_{s+1}$ unless explicitly noted otherwise. Newton's method has been employed in linearizing the momentum equation prior to minimization (see Section 11.2 and Payette and Reddy [50]). The weighting parameter α is taken as $\alpha = (\Delta t)^2$ to ensure the discrete minimization problem is not extraneously dominated by the momentum equation residual in the limit as $\Delta t \to 0$.

The least-squares based weak formulation resulting from minimization of $\mathcal{J}_{\Delta t}$ may be stated as follows: find $\mathbf{u} \in \mathcal{V}$ such that

$$
\mathcal{B}_{\Delta t}(\delta\mathbf{u}, \mathbf{u}) = \mathcal{F}_{\Delta t}(\delta\mathbf{u}) \quad \text{for all } \delta\mathbf{u} \in \mathcal{W} \qquad (11.3.6)
$$

where $\mathbf{u} = (\mathbf{v}, P, \boldsymbol{\omega})$, $\delta\mathbf{u} = (\delta\mathbf{v}, \delta P, \delta\boldsymbol{\omega})$, and $\mathcal{V}$ and $\mathcal{W}$ are appropriate function spaces. The form $\mathcal{B}_{\Delta t}(\delta\mathbf{u}, \mathbf{u})$ and linear functional $\mathcal{F}_{\Delta t}(\delta\mathbf{u})$ are given as

$$
\mathcal{B}_{\Delta t}(\delta\mathbf{u}, \mathbf{u}) = \int_\Omega \Big[\alpha\Big(\lambda_0^n\delta\mathbf{v} + \delta\mathbf{v} \cdot \boldsymbol{\nabla}\mathbf{v}_0 + \mathbf{v}_0 \cdot \boldsymbol{\nabla}\delta\mathbf{v} + \boldsymbol{\nabla}\delta P + \frac{1}{\mathrm{Re}}\boldsymbol{\nabla} \times \delta\boldsymbol{\omega} \Big) \cdot
$$

$$
\Big(\lambda_0^n\mathbf{v} + \mathbf{v} \cdot \boldsymbol{\nabla}\mathbf{v}_0 + \mathbf{v}_0 \cdot \boldsymbol{\nabla}\mathbf{v} + \boldsymbol{\nabla}P + \frac{1}{\mathrm{Re}}\boldsymbol{\nabla} \times \boldsymbol{\omega} \Big)
$$

$$
+ (\boldsymbol{\nabla} \cdot \delta\mathbf{v})(\boldsymbol{\nabla} \cdot \mathbf{v}) + (\delta\boldsymbol{\omega} - \boldsymbol{\nabla} \times \delta\mathbf{v}) \cdot (\boldsymbol{\omega} - \boldsymbol{\nabla} \times \mathbf{v}) \Big] d\Omega
$$

$$
+ \int_{\Gamma^{\mathrm{N}}} \Big(-\delta P\hat{\mathbf{n}} + \frac{1}{\mathrm{Re}}\hat{\mathbf{n}} \cdot \boldsymbol{\nabla}\delta\mathbf{v} \Big) \cdot \Big(-P\hat{\mathbf{n}} + \frac{1}{\mathrm{Re}}\hat{\mathbf{n}} \cdot \boldsymbol{\nabla}\mathbf{v} \Big) d\Gamma^{\mathrm{N}}
$$

$$(11.3.7a)$$

$$\mathcal{F}_{\Delta t}(\delta\mathbf{u}) = \int_\Omega \alpha \left(\lambda_0^n \delta\mathbf{v} + \delta\mathbf{v} \cdot \nabla\mathbf{v}_0 + \mathbf{v}_0 \cdot \nabla\delta\mathbf{v} + \nabla\delta P + \frac{1}{\mathrm{Re}} \boldsymbol{\nabla} \times \delta\boldsymbol{\omega} \right) \cdot \tilde{\mathbf{b}} \, d\Omega$$

$$+ \int_{\Gamma^{\mathrm{N}}} \left(-\delta P\hat{\mathbf{n}} + \frac{1}{\mathrm{Re}} \hat{\mathbf{n}} \cdot \boldsymbol{\nabla}\delta\mathbf{v} \right) \cdot \tilde{\mathbf{t}}^{\mathrm{P}} \, d\Gamma^{\mathrm{N}} \tag{11.3.7b}$$

The least-squares finite element model associated with the above standard $(\mathbf{v}, P, \omega)$-space-time decoupled least-squares functional $\mathcal{J}_{\Delta t}$ often suffers from poor *local* mass conservation and can lead to an ill-behaved response (most notability in the pressure) as we march in time; this is especially true when Δt is small and α is taken as unity.

11.3.2.4 A modified L_2-norm based least-squares model with improved element-level mass conservation

The purpose of this section is to present a modified least-squares finite element model that both enhances local mass conservation and improves velocity–pressure coupling. To this end we first recall that in the traditional weak-form Galerkin model, the pressure may be identified as the Lagrange multiplier whose role is to enforce the divergence free constraint on the velocity field. In least-squares finite element models, however, the pressure no longer plays this well-defined role. In an effort to improve the function of the pressure in enforcing the continuity equation, Pontaza [250] proposed a penalty least-squares finite element model (also see Prabhakar and his colleagues [240, 241]) based on the following *regularized* form of the divergence free condition for the velocity:

$$\boldsymbol{\nabla} \cdot \mathbf{v} = -\epsilon_p \Delta P \qquad \text{in } \Omega \times (0, \tau] \tag{11.3.8}$$

where ϵ_p is a small parameter, $\Delta P = P_{k+1} - P_k$, and the index $k \in \mathbb{N}$ pertains to the iterative penalization of the divergence free constraint. The incompressibility constraint is recovered in either the limit as $\epsilon_p \to 0$ or $k \to \infty$ (assuming that the sequence $\{\Delta P\}_{k=0}^\infty$ is a Cauchy sequence). In practice the regularization may be adopted in conjunction with the iterative Newton solution procedure. Pontaza [250] demonstrated numerically that using Eq. (11.3.8) in place of Eq. (11.3.2b) in the construction of $\mathcal{J}_{\Delta t}$ results in a significant improvement in the evolution of P for unsteady flows.

With the regularized continuity equation in mind, we propose a novel *unconstrained* least-squares formulation that both enhances element-level mass conservation for steady flows and improves the temporal evolution of the pressure for non-stationary flows. The basic idea is to add directly to $\mathcal{J}_{\Delta t}$ a penalized sum of the squares of the appropriately *normalized* element-level integrals of Eq. (11.3.8). To make the concept clear, we consider the integral of Eq. (11.3.8) over an *arbitrary*, possibly time dependent, region $\mathcal{P}(t)$

$$\hat{Q}^{\mathcal{P}}(t) = \oint_{\partial\mathcal{P}(t)} \hat{\mathbf{n}} \cdot \mathbf{v} \, d\Gamma^{\mathcal{P}(t)} + \int_{\mathcal{P}(t)} \epsilon_p \Delta P \, d\mathcal{P}(t) \tag{11.3.9}$$

When the second term on the right-hand side is neglected, the quantity $\hat{Q}^{\mathcal{P}}(t)$ may be identified in the discrete setting as the volumetric flow rate imbalance associated with region $\mathcal{P}(t)$. Replacing $\mathcal{P}(t)$ with Ω^e in the above equation allows us to obtain the following expression for the eth element of the finite element discretization:

$$\hat{Q}^e(t) = \oint_{\Gamma^e} \hat{\mathbf{n}} \cdot \mathbf{v} \, d\Gamma^e + \int_{\Omega^e} \epsilon_p \Delta P \, d\Omega^e \tag{11.3.10}$$

We find it useful to normalize the above expression as $Q^e(t) = \hat{Q}^e(t)/\mu(\Omega^e)$ where $\mu(\Omega^e)$ denotes the Lebesgue measure or nd-dimensional volume of Ω^e. As a result, $Q^e(t)$ represents (when $\epsilon_p = 0$) the volumetric flow rate imbalance per nd-dimensional volume of Ω^e.

We now can define the following *modified* space-time decoupled least-squares functional for the first-order vorticity form of the Navier–Stokes equations:

$$\mathcal{J}_{\Delta t}^{\star}(\mathbf{v}, P, \omega; \tilde{\mathbf{b}}, \tilde{\mathbf{t}}^{\mathrm{P}}) = \mathcal{J}_{\Delta t}(\mathbf{v}, P, \omega; \tilde{\mathbf{b}}, \tilde{\mathbf{t}}^{\mathrm{P}}) + \frac{\gamma}{2} \sum_{e=1}^{\mathrm{NE}} (Q^e)^2 \tag{11.3.11}$$

where γ is a global penalty parameter. The *modified* least-squares problem resulting from minimization of $\mathcal{J}_{\Delta t}^{\star}$ may be stated as follows: find $\mathbf{u} \in \mathcal{V}$ such that

$$\mathcal{B}_{\Delta t}^{\star}(\delta \mathbf{u}, \mathbf{u}) = \mathcal{F}_{\Delta t}^{\star}(\delta \mathbf{u}) \quad \text{for all } \delta \mathbf{u} \in \mathcal{W} \tag{11.3.12}$$

where $\mathcal{B}_{\Delta t}^{\star}(\delta \mathbf{u}, \mathbf{u})$ and $\mathcal{F}_{\Delta t}^{\star}(\delta \mathbf{u})$ are

$$\mathcal{B}_{\Delta t}^{\star}(\delta \mathbf{u}, \mathbf{u}) = \mathcal{B}_{\Delta t}(\delta \mathbf{u}, \mathbf{u}) + \gamma \sum_{e=1}^{\mathrm{NE}} \left(\oint_{\Gamma^e} \hat{\mathbf{n}} \cdot \delta \mathbf{v} \, d\Gamma^e + \int_{\Omega^e} \epsilon_p \delta P \, d\Omega^e \right) \times$$

$$\left(\oint_{\Gamma^e} \hat{\mathbf{n}} \cdot \mathbf{v} \, d\Gamma^e + \int_{\Omega^e} \epsilon_p P \, d\Omega^e \right) / \mu(\Omega^e)^2 \tag{11.3.13a}$$

$$\mathcal{F}_{\Delta t}^{\star}(\delta \mathbf{u}) = \mathcal{F}_{\Delta t}(\delta \mathbf{u}) + \gamma \sum_{e=1}^{\mathrm{NE}} \left(\oint_{\Gamma^e} \hat{\mathbf{n}} \cdot \delta \mathbf{v} \, d\Gamma^e + \int_{\Omega^e} \epsilon_p \, \delta P \, d\Omega^e \right) \times$$

$$\int_{\Omega^e} \epsilon_p P_0 \, d\Omega^e / \mu(\Omega^e)^2 \tag{11.3.13b}$$

Unlike $\mathcal{J}_{\Delta t}$, the modified least-squares functional $\mathcal{J}_{\Delta t}^{\star}$ clearly includes both element-level mass conservation as well as velocity–pressure coupling. Working in terms of $\mathcal{J}_{\Delta t}^{\star}$ leads to an unconstrained minimization problem that may be viewed as an attractive alternative to the Lagrange multiplier based least-squares model of the Stokes equations proposed by Chang and Nelson [247]. For steady flows, we find that it is sufficient to take $\epsilon_p = 0$.

11.3.3 Numerical Examples: Verification Problems

In this section we present numerical results obtained using the modified least-squares formulation. The problems have been selected to assess the capabilities of the formulation to: (a) generally improve element-level mass conservation and (b) enhance velocity–pressure coupling and overall numerical stability in non-stationary flows.

First we test the performance of the proposed least-squares formulation to improve mass conservation for problems involving steady fluid flows. For steady flows, we utilize the stationary least-squares functionals $\mathcal{J}$ and $\mathcal{J}^{\star}$, obtained by setting $\alpha = 1$ and $\gamma_0^n = \beta_q^n = 0$ (where $q = 0, \ldots, n-1$) in the definitions of $\mathcal{J}_{\Delta t}$ and $\mathcal{J}_{\Delta t}^{\star}$, respectively.

11.3.3.1 Steady Kovasznay flow

In this first example, we numerically examine a well-known incompressible fluid flow problem possessing an analytic solution. The solution is due to Kovasznay [261] and is posed on a two-dimensional square region defined as $\bar{\Omega} = [-0.5, 1.5] \times [-0.5, 1.5]$. The proposed solution is of the form

$$v_x = 1 - e^{\lambda x} \cos(2\pi y), \qquad v_x = \frac{\lambda}{2\pi} e^{\lambda x} \sin(2\pi y), \qquad P = P_{\text{ref}} - \frac{1}{2} e^{2\lambda x}$$
$$(11.3.14)$$

where the parameter λ is given as $\lambda = \text{Re}/2 - [(\text{Re}/2)^2 + (2\pi)^2]^{1/2}$ and P_{ref} is a reference pressure (which in the current study is taken to be zero).

We discretize the domain $\bar{\Omega}$ into 8 nonuniform rectangular finite elements as depicted in Fig. 11.3.1(a). Figure 11.3.1(b) shows the numerically computed horizontal velocity component v_x. The boundary conditions for the problem are applied by specifying the exact solution for the velocity vector $\mathbf{v}$ along the entire boundary through an employment of Eq. (11.3.14). We specify no boundary conditions for the vorticity and only prescribe the pressure at the single point $\mathbf{x} = (-0.5, 0)$. In this study the mesh is refined by systematically increasing the p-level of the finite element approximation within each element. Nonlinear convergence for a given numerical simulation is declared once the relative Euclidean norm of the solution residuals, $\|\Delta^k - \Delta^{k-1}\|/\|\Delta^k\|$, is less than 10^{-6}. All reported numerical results are obtained for a Reynolds number of 40.

In Table 11.3.1 we present the decay of the $L_2(\Omega)$-norm error measures for the velocity, pressure, and vorticity fields under p-refinement, where the penalty parameter γ is varied from 0 to 100. We also show the decay of the *unmodified* least-squares functional $\mathcal{J}$. We observe exponential decay in the error measures for all variables as the p-level is increased. This observation is true for all values of γ considered. Figure 11.3.2 shows the evolution of the error measures under p-refinement for the case where $\gamma = 100$. Clearly, the inclusion of element-level

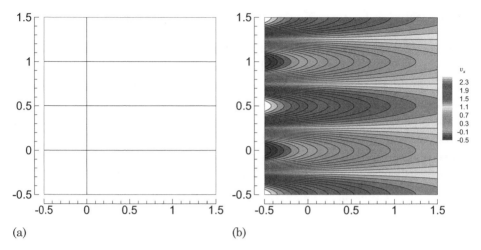

(a)   (b)

Fig. 11.3.1: Kovasznay flow: (a) spectral/hp finite element discretization of domain $\bar{\Omega}$ and (b) numerical solution of horizontal velocity field v_x for Re = 40.

mass conservation in the definition of $\mathcal{J}^\star$ does not pollute the integrity of the finite element solution.

Table 11.3.1: Kovasznay flow: decay of the unmodified least-squares functional $\mathcal{J}$ and convergence of the numerically computed velocity components, pressure, and vorticity in the $L_2(\Omega)$-norm under p-refinement for various values of the penalty parameter γ.

γ	p	$\mathcal{J}^{1/2}$	$\|v_x - v_x^{hp}\|_0$	$\|v_y - v_y^{hp}\|_0$	$\|P - P_{hp}\|_0$	$\|\omega - \omega_{hp}\|_0$
0	3	1.5061 E-01	1.7124 E-02	6.2962 E-03	7.3897 E-03	1.9290 E-01
	5	4.5738 E-03	2.6574 E-04	1.1853 E-04	1.4078 E-04	3.9751 E-03
	7	6.7688 E-05	2.1663 E-06	1.1941 E-06	6.9539 E-07	4.4836 E-05
	9	5.8185 E-07	1.2070 E-08	8.0882 E-09	5.4114 E-09	3.2600 E-07
1	3	1.5061 E-01	1.7089 E-02	6.2757 E-03	7.4704 E-03	1.9274 E-01
	5	4.5738 E-03	2.6434 E-04	1.1868 E-04	1.3214 E-04	3.9711 E-03
	7	6.7688 E-05	2.1659 E-06	1.1941 E-06	6.9467 E-07	4.4836 E-05
	9	5.8185 E-07	1.2067 E-08	8.0881 E-09	5.3552 E-09	3.2600 E-07
10	3	1.5061 E-01	1.7080 E-02	6.2675 E-03	7.4982 E-03	1.9267 E-01
	5	4.5739 E-03	2.6425 E-04	1.1873 E-04	1.2996 E-04	3.9703 E-03
	7	6.7688 E-05	2.1658 E-06	1.1941 E-06	6.9442 E-07	4.4836 E-05
	9	5.8185 E-07	1.2067 E-08	8.0881 E-09	5.3407 E-09	3.2600 E-07
100	3	1.5061 E-01	1.7079 E-02	6.2662 E-03	7.5023 E-03	1.9266 E-01
	5	4.5739 E-03	2.6424 E-04	1.1874 E-04	1.2968 E-04	3.9702 E-03
	7	6.7688 E-05	2.1657 E-06	1.1941 E-06	6.9438 E-07	4.4836 E-05
	9	5.8185 E-07	1.2067 E-08	8.0881 E-09	5.3381 E-09	3.2600 E-07

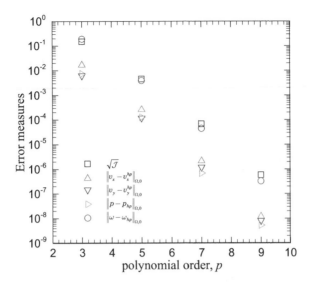

Fig. 11.3.2: Convergence of numerically computed velocity, pressure and vorticity under p-refinement to the analytic solution of Kovasznay for $\gamma = 100$. The decay of the square root of the unmodified least-squares functional $\mathcal{J}$ is also shown.

Figure 11.3.3 shows the decay of the normalized volumetric flow rate imbalance Q^e for element 1 under p-refinement for $\gamma = 0, 1, 10$, and 100, where $\bar{\Omega}^1 = [-0.5, 0] \times [-0.5, 0]$. The normalized volumetric flow rate imbalance Q^e associated with each element in $\bar{\Omega}^{hp}$ for p-levels 3 and 7 is also provided in Fig. 11.3.4. Although p-refinement clearly improves local mass conservation, significant additional enhancement may be obtained through constructing the least-squares finite element model using the modified least-squares functional $\mathcal{J}^\star$ as opposed to the standard least-squares functional $\mathcal{J}$.

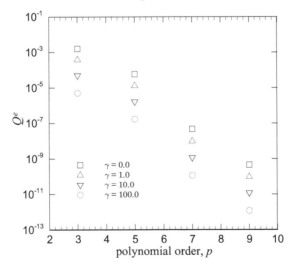

Fig. 11.3.3: Decay of normalized volumetric flow rate imbalance Q^e under p-refinement for various values of γ for Kovasznay flow. Results are for element 1, where $\bar{\Omega}^1 = [-0.5, 0] \times [-0.5, 0]$.

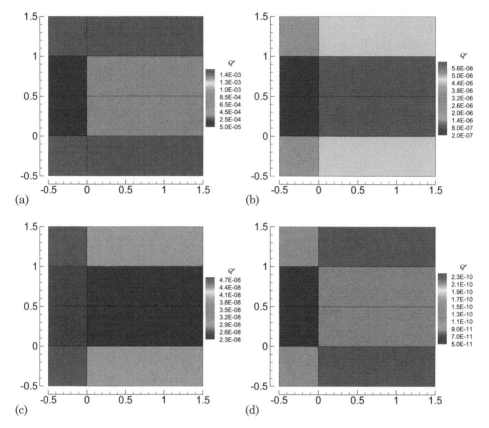

Fig. 11.3.4: Normalized volumetric flow rate imbalance Q^e for each finite element in $\bar{\Omega}^{hp}$ for Kovasznay flow. Results are for various polynomial orders and values of γ: (a) p-level = 3 and $\gamma = 0$, (b) p-level = 3 and $\gamma = 100$, (c) p-level = 7 and $\gamma = 0$, and (d) p-level = 7 and $\gamma = 100$.

11.3.3.2   Steady flow in a 1×2 rectangular cavity

In this example, we test the modified least-squares finite element model with a two-dimensional lid-driven cavity flow problem consisting of a rectangular domain, $\bar{\Omega} = [0, 1] \times [0, 2]$, with an aspect ratio of 2. Along the bottom and left and right sides of the cavity, the velocity is taken to be zero in accordance with the no slip condition. It is a common practice in the literature to prescribe a unit value for the horizontal velocity component v_x along the entire top surface of the cavity. In the context of high-order finite elements, however, such a boundary condition produces undesirable singularities in the vicinity of the corners of the upper cavity. In an effort to avoid an ill-posed *discrete* problem, the horizontal velocity profile is prescribed in terms of the following expression:

$$v_x^{\mathrm{p}}(x) = \begin{cases} \tanh(50x) & 0 \leqslant x \leqslant 0.5 \\ -\tanh[50(x - 1)] & 0.5 < x \leqslant 1.0 \end{cases} \qquad (11.3.15)$$

which allows for a smooth transition from unity to zero in the neighborhoods of the corners, as can be seen in Fig. 11.3.5. The pressure is taken to be zero at the point $\mathbf{x} = (0.5, 0)$. The Reynolds number for the problem is 1,500.

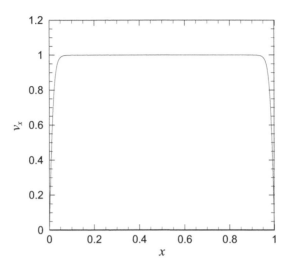

Fig. 11.3.5: Specified horizontal velocity profile v_x along the top surface of the 2-D lid-driven cavity flow problem with aspect ratio of 2.

A 12×24 (288 elements) finite element discretization with p-level equal to nine, as shown in Fig. 11.3.6(a), is used. The mesh is graded so as to adequately resolve the anticipated boundary layers and regions of circulation near the cavity walls. Convergence of the iterative solution procedure is declared once the relative norm of the solution residual, $\|\Delta^k - \Delta^{k-1}\| / \|\Delta^k\|$, is less than 10^{-6}. We employ a continuation approach, wherein the solution at Re $= 1,500$ is arrived by solving a series of problems posed at intermediate Reynolds numbers. We begin by solving the problem at Re $= 300$, followed by Re $= 600$, and so on until we reach Re $= 1,500$. For each problem, the converged solution taken from the immediate previous problem in the series is used as the initial guess. The problem is solved using the modified least-squares finite element model, taking γ equal to 0, 0.1, 1.0, and 10.0.

Figure 11.3.6(b) shows the vorticity field and streamlines for the problem. The flow is characterized by two large regions of circulation, with smaller vortex regions also present in the vicinity of the bottom as well as the upper left hand corners of the domain. Figure 11.3.6(c) shows the horizontal velocity component v_x along the vertical centerline of the domain as determined using a p-level of 9 and $\gamma = 10$. The streamlines and horizontal velocity component v_x along the vertical centerline are visually in excellent agreement with the results reported by Gupta and Kalita [262].

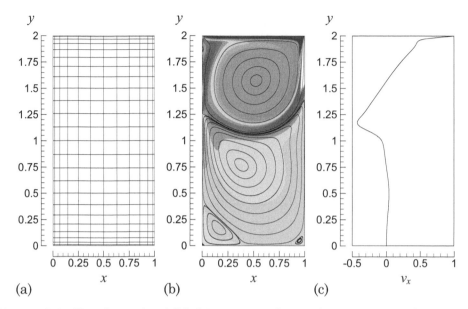

(a)   (b)   (c)

Fig. 11.3.6: Two-dimensional lid-driven cavity flow with aspect ratio of 2 at Re = 1,500. Numerical results obtained for $p = 9$: (a) finite element mesh, (b) vorticity field and streamlines and (c) horizontal velocity component v_x along vertical cavity centerline.

Figure 11.3.7 shows the decay of the normalized volumetric flow rate imbalance Q^e for element 107, where the geometric centroid of the element is located at $\mathbf{x} = (0.3103, 0.8053)$. In this figure, both the p and the penalty parameter γ are varied. We also provide in Fig. 11.3.8 an illustration of the normalized volumetric flow rate imbalance Q^e for all elements in the discretization at $p = 5$ and

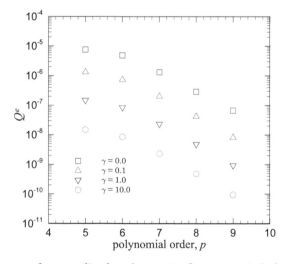

Fig. 11.3.7: Decay of normalized volumetric flow rate imbalance Q^e under p-refinement for various values of γ for 2-D lid-driven flow in a rectangular cavity. Results shown are for element 107 with geometric centroid located at $\mathbf{x} = (0.3103, 0.8053)$.

$p = 7$ for various values for the penalty parameter γ. Clearly, both the polynomial order as well as the value of γ are significant factors in improving element level mass conservation for this problem. It is interesting to note that substantial improvement in element level mass conservation is obtained even for the case where $\gamma = 1.0$.

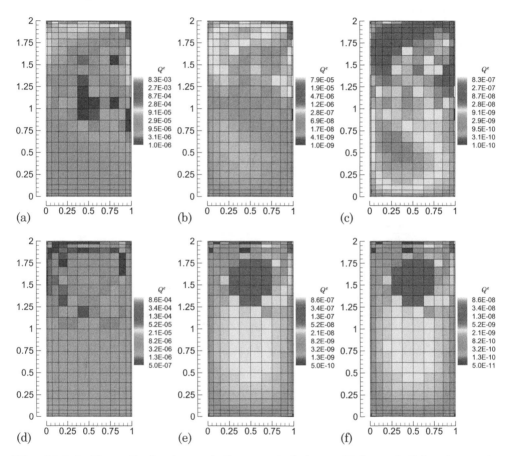

Fig. 11.3.8: Normalized volumetric flow rate imbalance Q^e for each finite element in $\bar{\Omega}^{hp}$ for 2-D lid-driven flow in a rectangular cavity. (a) $p = 5$ and $\gamma = 0$, (b) $p = 5$ and $\gamma = 1$, (c) $p = 5$ and $\gamma = 10$, (d) $p = 7$ and $\gamma = 0$, (e) $p = 7$ and $\gamma = 1$, and (f) $p = 7$ and $\gamma = 10$.

11.3.3.3 Steady flow past a large cylinder in a narrow channel

As a final steady flow example, we consider a problem constituting a much more rigorous test for mass conservation, namely, flow past a circular cylinder in a narrow channel (which is similar to the problem considered by Chang and Nelson [247]). The domain $\bar{\Omega}$ is the set difference between the closed rectangular region $[-5, 10] \times [-1, 1]$ and an open circular region with unit diameter centered about the origin. Along the inflow part of the boundary we prescribe a parabolic horizontal velocity profile $v_x^{\mathrm{p}}(y) = \frac{3}{2}(1 - y^2)$, which is consistent with Poiseuille flow. The pseudo-traction is taken to be zero on the right hand side of the domain and a no-slip condition is utilized on all other parts of the boundary. The Reynolds number, based on the diameter of the cylinder and the average horizontal velocity at the inlet is 40.

The finite element mesh utilized in the study is shown in Fig. 11.3.9(a). The problem is solved by varying p incrementally from 2 to 7. At each p-level we further investigate the influence of the penalty parameter on improving mass conservation by solving the problem for $\gamma = 0$, 1, 10, and 100. We adopt the same nonlinear convergence criteria for the iterative solution procedure that was used in the two previous steady-state benchmark problems. The horizontal velocity

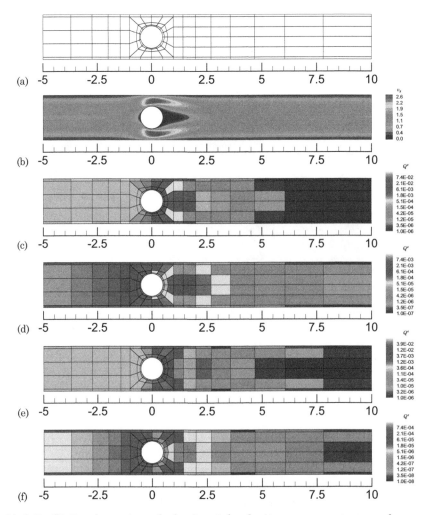

Fig. 11.3.9: Finite element mesh, horizontal velocity component v_x, and normalized volumetric flow rate imbalance Q^e for each finite element in $\bar{\Omega}^{hp}$ for steady flow past a large circular cylinder in a narrow channel at Re = 40: (a) finite element mesh, (b) horizontal velocity component v_x, (c) Q^e for $p = 2$ and $\gamma = 0$, (d) Q^e for $p = 2$ and $\gamma = 100$, (e) Q^e for $p = 3$ and $\gamma = 0$, and (f) Q^e for $p = 3$ and $\gamma = 100$.

profile in the domain is shown in Fig. 11.3.9(b). The element-level normalized volumetric flow rate imbalance Q^e for each element in the discretization are shown in Fig. 11.3.9(c)–(f) for p-levels 2 and 3, where γ is taken as either 0 or 100. In Fig. 11.3.10 we show the horizontal velocity profile along the gap between the cylinder and the top of the channel at $x = 0$. The "exact" solution in this figure is the finite element solution obtained using $p = 7$ and $\gamma = 0$.

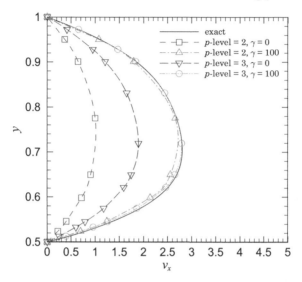

Fig. 11.3.10: Horizontal velocity $v_x(0, y)$ profiles along the gap between the top of the circular cylinder and the channel wall at $x = 0$ for flow past a large cylinder in a narrow channel.

Figure 11.3.11 shows the observed volumetric flow rate imbalance for element 115 of the finite element model, where both the p and penalty parameter γ are varied. The normalized volumetric flow rate past the crown of the cylinder for the various finite element discretizations is summarized in Table 11.3.2. Clearly, mass conservation is improved by constructing the finite element model in terms of the modified least-squares functional $\mathcal{J}^\star$.

Next, we assess the performance of the least-squares formulation with improved mass conservation, velocity–pressure coupling, and overall numerical stability in the numerical simulation of unsteady flows. Unless otherwise stated, we take $\alpha = (\Delta t)^2$ in the definition of $\mathcal{J}^\star_{\Delta t}$ for each numerical simulation.

11.3.3.4 Unsteady flow past a circular cylinder

As the first unsteady example we consider the problem of flow past a circular cylinder, where the Reynolds number is taken as 100. For the computational domain Ω, we take the set difference between the open square region $(-15.5, 25.5) \times (-20.5, 20.5)$ and a closed unit-diameter circle that is centered about the origin. The spatial discretizations $\overline{\Omega}^{hp}$ that are employed in the finite

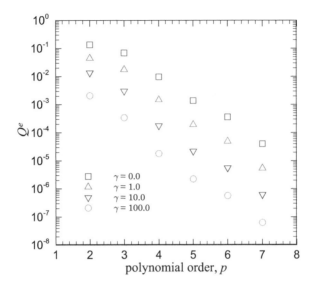

Fig. 11.3.11: Decay of normalized volumetric flow rate imbalance Q^e under p-refinement for various values of γ for flow past a large cylinder in a narrow channel. Results shown are for element 115 with geometric centroid located at $\mathbf{x} = (-0.8365, 0)$.

Table 11.3.2: Normalized volumetric flow rate past the crown $(x = 0, y)$ of the large circular cylinder.

	Normalized volumetric flow rate			
p	$\gamma = 0$	$\gamma = 1$	$\gamma = 10$	$\gamma = 100$
2	0.34870	0.53887	0.81899	0.97124
3	0.65973	0.81925	0.95983	0.99538
4	0.95250	0.98485	0.99767	0.99975
5	0.99335	0.99951	0.99971	0.99997
6	0.99828	0.99951	0.99993	0.99999

element simulations are shown in Fig. 11.3.12. The top mesh contains 2,004 quadratic elements (i.e. $p = 2$). Likewise, the bottom mesh contains 501 elements with $p = 4$. Each discretization contains 8,216 nodes and 32,864 total degrees of freedom.

All flow fields are initially taken to be zero. The horizontal velocity component v_x is then gradually increased in time along the left, top, and bottom sides of $\bar{\Omega}^{hp}$ in accordance with the formula $v_x^p(t) = v_\infty \tanh(t)$; the free-stream velocity v_∞ is taken to be 1.0. A no-slip boundary condition is used along the circular cylinder and the outflow boundary condition (along the right hand side of the domain) is enforced weakly by taking the pseudo-traction $\tilde{\mathbf{t}}^p$ as zero in the definition of $\mathcal{J}_{\Delta t}^\star$. We employ the BDF2 time integrator with a uniform time step size of $\Delta t = 0.1$ s. Since the BDF2 integration formula is non-self-starting,

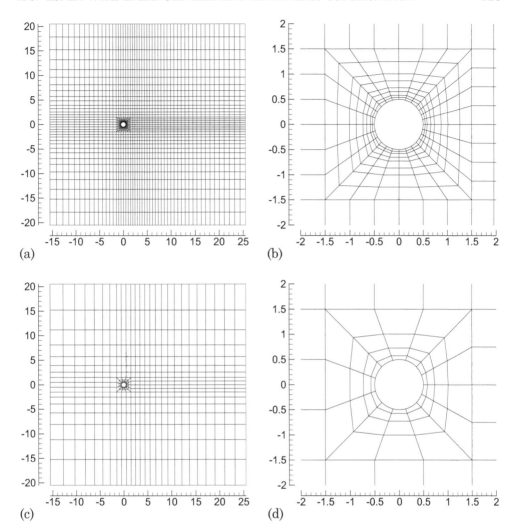

Fig. 11.3.12: Spectral/hp finite element discretizations used to numerically simulate the unsteady viscous flow of an incompressible fluid past a circular cylinder: (a) computational domain $\bar{\Omega}^{hp}$ for a p-level of 2, (b) close-up view of $\bar{\Omega}^{hp}$ in the vicinity of the cylinder for a p-level of 2, (c) computational domain $\bar{\Omega}^{hp}$ for $p = 4$, and (d) close-up view of $\bar{\Omega}^{hp}$ in the vicinity of the cylinder for $p = 4$.

we utilize the BDF1 formula for the first ten time steps. A total of 3,000 time steps are employed in each transient finite element simulation.

A error tolerance for the nonlinear convergence of $\epsilon = 10^{-6}$ (the error is defined in terms of the relative Euclidean norm of the residuals in the nodal velocities between two successive iterations, is adopted at each time step). This typically requires only two or three nonlinear iterations. The linearized algebraic equations are constructed and solved using the solution procedures for sparse system of equations, as outlined by Payette [161]. The UMFPACK direct solver library [164–167] is utilized in the numerical solution of the global sparse set of finite element equations. We solve for the temporal evolution of the fluid using the modified least-squares functional $\mathcal{J}_{\Delta t}^{\star}$ for the cases where γ is either 0 or 100. Although all results reported below have been obtained by taking ϵ_p as zero, we note in passing that we have also obtained reliable solutions using $\epsilon_p = 0.005$ and 0.01. The additional velocity–pressure coupling associated with a non-zero value for ϵ_p, however, typically demands a greater number of nonlinear iterations to meet the nonlinear convergence tolerance of $\epsilon = 10^{-6}$.

In Fig. 11.3.13 we show the time history of the velocity components, vorticity, and pressure at the spatial point $(x, y) = (1, 0)$ as computed using the

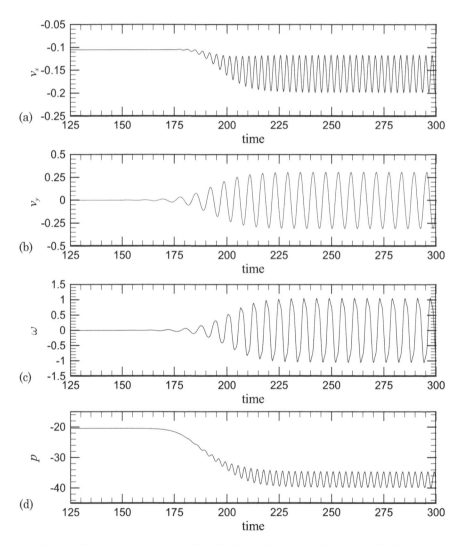

Fig. 11.3.13: Time history of the flow fields behind the circular cylinder at $(x, y) = (1, 0)$ as determined using a p-level of 4 in the spatial discretization: (a) horizontal velocity component v_x, (b) vertical velocity component v_y, (c) vorticity ω, and (d) pressure field p.

finite element mesh shown in Fig. 11.3.12(c) with γ taken as 100. The dimensional pressure field shown in Fig. 11.3.13(d) has been obtained by scaling the non-dimensional pressure field by a factor of 100. The *virtually* stationary flow pattern that forms during the early stages of the simulation becomes noticeably unstable between 150 and 175 s. The instability eventually results in a well-defined periodic swirling of vortices that are shed in the wake region immediately downwind of the cylinder. This oscillatory behavior is commonly referred to as the von Kármán vortex street. We measure the non-dimensional period to be $\mathcal{T} = 6.035$; this translates into a non-dimensional shedding frequency (or Strouhal number) of St $= 0.1657$. This is in very close agreement with St $= 0.1653$ reported by Pontaza and Reddy [238] using a space-time coupled spectral/hp least-squares finite element simulation.

Instantaneous contours of the velocity components v_x and v_y along with the pressure field P in the wake region are depicted in Fig. 11.3.14 at $t = 280$ s. We also provide in Fig. 11.3.15 snapshots of the vorticity field ω during the course of a single shedding cycle. Both figures have been generated using the

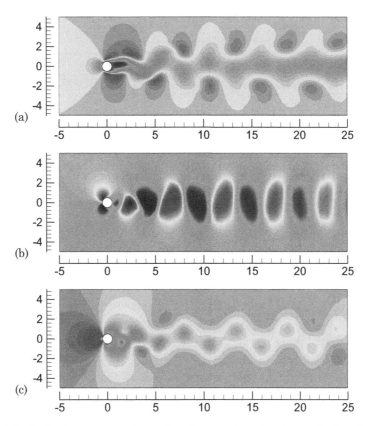

Fig. 11.3.14: Instantaneous contours for flow past a circular cylinder at $t = 280$ s, where the finite element mesh associated with $p = 4$ has been used: (a) horizontal velocity component v_x, (b) vertical velocity component v_y, and (c) pressure field P.

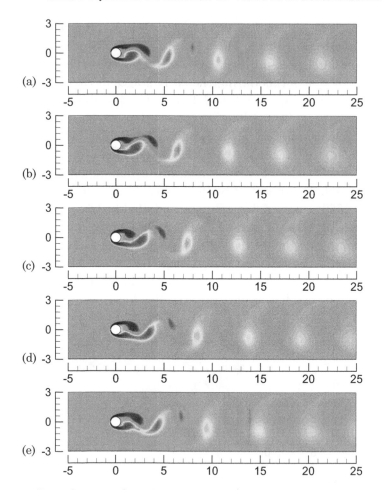

Fig. 11.3.15: Time history of vorticity contours behind the circular cylinder at five successive discrete instances in time. The finite element mesh associated with a p-level of 4 has been employed: (a) $t = 280.0$ s, (b) $t = 281.2$ s, (c) $t = 282.4$ s, (d) $t = 283.6$ s, and (e) $t = 284.8$ s.

numerical results obtained using a p-level of 4 and $\gamma = 100$. We see that within a given period $\mathcal{T}$ two eddies are shed from the cylinder into the wake region, one originating from the top and the other from the bottom of the cylinder. The former eddy spins clockwise while the latter rotates in the counterclockwise direction. The outflow boundary condition, imposed weakly through the least-squares functional $\mathcal{J}_{\Delta t}^{\star}$, clearly allows the fluid to leave the computational domain in a physically reasonable manner.

In an effort to showcase the performance of the modified least-squares finite element model in improving local mass conservation, we present in Fig. 11.3.16 the normalized volumetric flow rate imbalance Q^e for the finite elements in a neighborhood of the wake region behind the circular cylinder. The reported results are for the numerical solution obtained at $t = 260$ s using the spatial discretization shown in Fig. 11.3.12(c). It is clear that element-level mass conservation has clearly improved by taking γ as 100 as opposed to 0 in the modified least-squares formulation. The improvement in the solution is particularly noticeable for the smaller elements in $\bar{\Omega}^{hp}$ that are closest to the cylinder. To assess general mass conservation for the fluid flowing past the circular cylinder, we post-compute the absolute value of the volumetric flow rate

$$Q(t) = \left| \oint_{\Gamma^s} \hat{\mathbf{n}} \cdot \mathbf{v}_{hp}(t) d\Gamma^s \right| \tag{11.3.16}$$

across the closed surface $\Gamma^s = \partial\Omega^s$, where $\Omega^s = (-1.5, 1.5)^2$. At each time step we evaluate Q using the Gauss–Legendre quadrature rule. In Fig. 11.3.17 the time history of Q is traced for both spatial discretizations, where γ is again taken as either 0 or 100. For both spatial discretizations we observe significant improvement in mass conservation across Γ^s when γ is taken as 100.

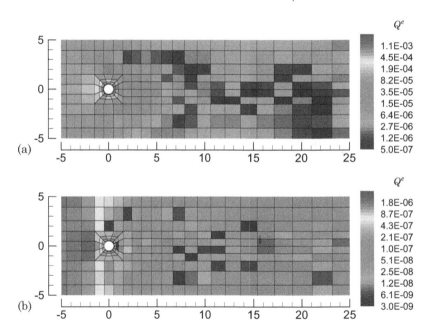

Fig. 11.3.16: Normalized volumetric flow rate imbalance Q^e for the finite elements in the vicinity of the wake region behind the circular cylinder. The results shown are a snapshot taken at $t = 260$ s using a p-level of 4 in the spatial discretization: (a) $\gamma = 0$ and (b) $\gamma = 100$.

11.3.3.5 Unsteady flow past a large cylinder in a narrow channel

In this next example we revisit the obstructed channel flow problem introduced previously as a steady flow benchmark. To obtain a non-stationary problem, we raise the Reynolds number from 40 to 100 and vertically translate the circular cylinder 0.01 spatial units upward. The channel is again taken to be 15 units in length and 2 in height, with the center of the cylinder placed 5 units from the

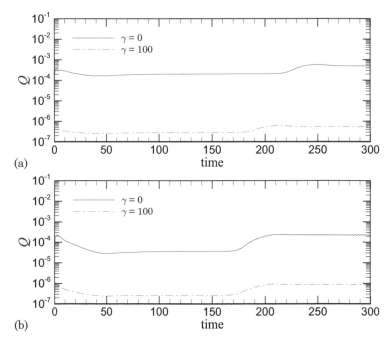

(a)

(b)

Fig. 11.3.17: Time history of the volumetric flow rate Q past the closed surface Γ^s, where $\Gamma^s = \partial\Omega^s$ is the boundary associated with the region $\Omega^s = (-1.5, 1.5)^2$. The reported results are obtained using the spectral/hp spatial discretizations shown in Fig. 11.3.12 where: (a) the p-level is 2 and (b) the p-level is 4.

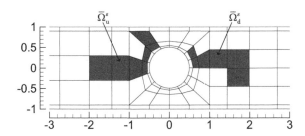

Fig. 11.3.18: Close-up view of the finite element mesh used to simulate unsteady flow through a channel with a circular obstruction. The shaded regions $\bar{\Omega}_u^s$ and $\bar{\Omega}_d^s$ are control volumes used in the post-processing stage to assess the severity of mass conservation violation for a given finite element simulation.

inlet side of the domain. The finite element mesh is nearly identical to the one employed in the stationary flow problem [see Fig. 11.3.9(a)]. A close-up view of the mesh (in the vicinity of the cylinder) used in the current study is shown in Fig. 11.3.18.

All flow fields are taken initially to be zero. For the inflow boundary condition, taken along the left-hand side of the computational domain, we specify a time-dependent parabolic horizontal velocity profile $v_x^p(y, t) = \frac{3}{2}(1 - y^2)\tanh(t)$. The prescribed outflow and no slip boundary conditions are the same as those described for the steady-state version of the problem. As in the previous non-stationary example, we again employ the BDF2 time integrator (where the BDF1 integrator is utilized for the first ten time steps). We solve the problem over a total time interval of 100 s. The nonlinear convergence criteria, defined in terms of the relative Euclidean norm of the residuals in the nodal velocities between two successive iterations, is taken as $\epsilon = 10^{-4}$ at each time step. The UMFPACK direct solver library is again utilized in the solution of the sparse global system of finite element equations. The problem is solved using the modified least-squares functional $\mathcal{J}^{\star}_{\Delta t}$ for all possible combinations of the following parameters: $\gamma = 0$ and 100, $\epsilon_p = 0$ and 0.005, $p = 4$ and 6, and $\Delta t = 0.05$. To verify that the numerical solutions are indeed sufficiently resolved in time, we also solved the problem using a time increment of $\Delta t = 0.02$ for the case where $p = 6$, $\gamma = 100$, and $\epsilon_p = 0.005$.

The time history of the vertical velocity component v_y at the spatial point $(x, y) = (2, 0)$ is shown in Fig. 11.3.19. The non-symmetric domain allows the instability in the flow to propagate quickly such that a well-defined periodic response is reached at around 50 s into the simulation. From Fig. 11.3.19 we measure the non-dimensional period for a typical vortex shedding cycle to be $\mathcal{T} = 1.93$ which corresponds with a non-dimensional shedding frequency of $\mathrm{St} = 0.5181$. The presence of the channel clearly results in a much shorter shedding cycle than what was observed in the previously presented *external* flow past a cylinder problem.

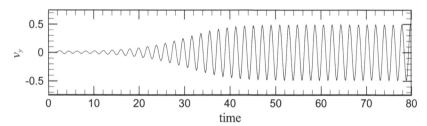

Fig. 11.3.19: Time history of the vertical velocity component v_y downstream from the circular cylinder at $(x, y) = (2, 0)$ as determined using $p = 6$ in the spatial discretization. The time step is $\Delta t = 0.02$, $\gamma = 100$, and $\epsilon_p = 0.005$.

Contours of the instantaneous pressure P, velocity component v_y and vorticity ω in the wake region are shown in Fig. 11.3.20 at $t = 75.5$ s for $p = 6$. The swirling of vortices that develops in the wake region clearly becomes suppressed (due to the channel walls) as the fluid travels further downstream past the cylinder.

In Fig. 11.3.21 we present a snapshot of the normalized volumetric flow rate imbalance Q^e for all elements in the computational discretization $\bar{\Omega}^{hp}$ at $t = 75.5$ s, using $p = 4$. We take $\epsilon_p = 0$ in the post-processing of Q^e for each element. In Fig. 11.3.22 we trace the time histories of the volumetric flow rate Q [obtained using Eq. (11.3.16)] through the closed boundaries of the upstream and downstream control regions $\bar{\Omega}_{\mathrm{u}}^s$ and $\bar{\Omega}_{\mathrm{d}}^s$ shown in Fig. 11.3.18, again using a spatial discretization with $p = 4$. Noticeable improvement in mass conservation is observed for each element $\bar{\Omega}^e$ in $\bar{\Omega}^{hp}$ and also for the control regions for the case where $\gamma = 100$.

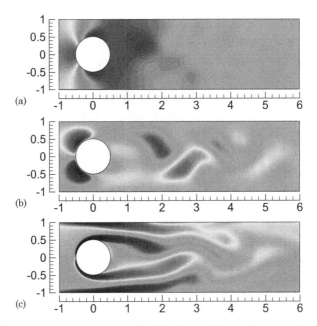

(a)

(b)

(c)

Fig. 11.3.20: Instantaneous contours for flow in a channel past a circular cylinder at $t = 75.5$ s, where the finite element mesh associated with $p = 6$ is used: (a) pressure field P, (b) vertical velocity component v_y, and (c) vorticity ω. The results shown are for the case where $\Delta t = 0.05$ s, $\gamma = 100$, and $\epsilon_p = 0.005$.

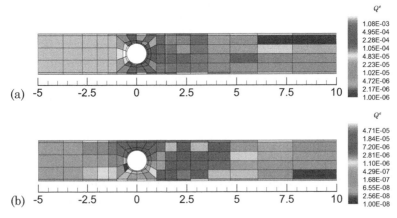

(a)

(b)

Fig. 11.3.21: Normalized volumetric flow rate imbalance Q^e for each finite element in $\bar{\Omega}^{hp}$ for flow in a channel past a circular cylinder at $t = 75.5$ s. The results shown have been obtained using a time increment of $\Delta t = 0.05$ s: (a) $p = 4$ and $\gamma = 0$, and $\epsilon_p = 0.005$. (b) $p = 4$, $\gamma = 100$, and $\epsilon_p = 0.005$.

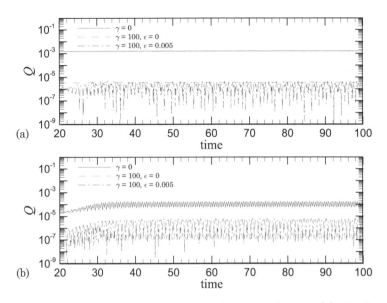

Fig. 11.3.22: Time histories of the volumetric flow rate Q past: (a) the closed surface Γ_u^s and (b) the closed surface Γ_d^s. The surfaces $\Gamma_u^s = \partial\Omega_u^s$ and $\Gamma_d^s = \partial\Omega_d^s$ are the boundaries of the upstream and downstream control volumes shown in Fig. 11.3.18. The results are for the case where $p = 4$ and $\Delta t = 0.05$.

Time histories of the normalized volumetric flow rate Q past the crown of the cylinder are plotted in Fig. 11.3.23 for different values of γ and ϵ_p at $p = 4$ and 6. The results shown have been normalized by the long term prescribed inlet volumetric flow rate. Similar results were observed at the domain exit and also at $x = 1.0$. In the upper two plots (where the p-level is 4 and 6, respectively) we observe excellent mass conservation when $\gamma = 100$ for both values chosen for ϵ_p.

In an effort to demonstrate that the proposed formulation improves velocity–pressure coupling and overall numerical stability in the simulation of transient flows, we show in the lower two plots of Fig. 11.3.23 the normalized volumetric flow rates past the crown of the cylinder using a non-scaled version of $\mathcal{J}_{\Delta t}^\star$ (i.e. taking $\alpha = 1$) for a p-level of 6. For the standard least-squares formulation, obtained by setting $\gamma = 0$, we observe spurious oscillations in all fields, which eventually leads to total instability of the finite element solution procedure; this simulation was, therefore, manually terminated at $t = 20$ s. The cases where $\gamma = 100$ yield reliable results for all fields; furthermore, excellent mass conservation is observed despite the fact that no scaling of the momentum equation residual in the definition of the least-squares functional is used. When $\gamma = 100$ and $\epsilon_p = 0.005$, we observe "exact" mass conservation up to three decimal places for all times. For the current example problem it is clear that the modified least-squares formulation has the ability to (a) improve mass conservation and (b) enhance numerical stability in the simulation of unsteady flows.

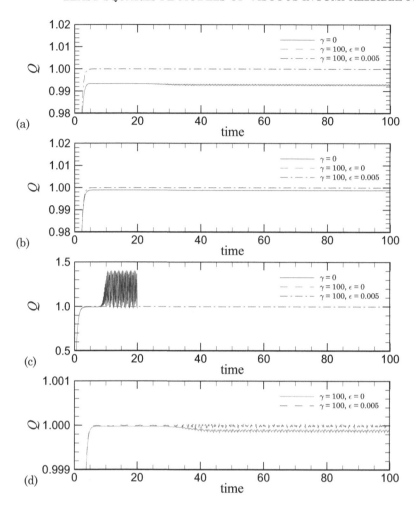

Fig. 11.3.23: Time histories of the normalized volumetric flow rate Q past the crown $(x = 0, y)$ of the large circular cylinder. All results have been obtained using a time increment of $\Delta t = 0.05$ s: (a) $p = 4$ and $\alpha = (\Delta t)^2$, (b) $p = 6$ and $\alpha = (\Delta t)^2$, (c) $p = 4$ and $\alpha = 1.0$, and (d) $p = 6$ and $\alpha = 1.0$.

11.4   Summary and Future Direction

In this chapter, finite element formulations for flows of viscous incompressible fluids using high-order spectral/hp finite element technology and least-squares finite element formulation have been presented. The primary objective has been to present novel mathematical models and innovative discretization procedures in the numerical simulation of fluid mechanics problems, wherein the additional benefits of employing high-order spectral/hp finite element technology and least-squares formulations are pronounced. As a result, *ad hoc* tricks (e.g. reduced integration and/or mixed interpolations) required to stabilize low-order finite element formulations are unnecessary.

Following the works of Payette [161] and Payette and Reddy [50], a novel least-squares finite element model of the steady-state and non-stationary incompressible Navier–Stokes equations with enhanced local mass conservation is also presented. The presented model is a modification of the standard L_2-norm least-squares formulation of the Navier–Stokes equations based on the equivalent velocity–pressure–vorticity first-order system. In the new model, we modified the standard least-squares functional to also include an appropriately penalized sum of the squares of the element-level integrals of a regularized form of the continuity equation. As a consequence, the resulting finite element model directly inherited terms in the bilinear form and linear functional (which could be adjusted based on the penalty parameter) that tend to improve element-level mass conservation. A notable quality of the formulation was that improved mass conservation could be attained without introducing additional variables or compromising the unconstrained minimization setting for the numerical solution. Numerical simulations confirmed that the proposed formulation could significantly improve mass conservation for both steady and non-stationary flows. For transient flows, the formulation was further shown to enhance velocity–pressure coupling and overall numerical stability (most notably for cases where the momentum equation residual, appearing in the least-squares functional, was not weighted by the square of the time step).

In the way of an extension of the works by Payette [161] and Payette and Reddy [50], problems of fluid–structure interaction is the most promising. Here we briefly describe an example problem that is solved by Payette [161] using a least-squares finite element model discussed in this chapter that can handle moving fluid boundaries (a preliminary step towards the development of a fluid–structure interaction computational methodology).

We consider the flow of a viscous incompressible fluid inside a square cavity, where the cavity under consideration is a unit square centered at the origin. A 0.28 units diameter solid circular cylinder is positioned at the origin of the cavity at $t = 0$ s. Immediately following $t = 0$ s, the cylinder begins to translate with an instantaneous unit velocity in the x direction. We impose no-slip type boundary conditions along all solid surfaces, including the cylinder Γ_{cyl} and cavity walls Γ_{walls}, where $\Gamma = \Gamma_{\text{cyl}} \cup \Gamma_{\text{walls}}$. This amounts to specifying $\mathbf{v} = \mathbf{0}$ on Γ_{walls} and $v_x = 1$ on Γ_{cyl}. We prescribe the pressure to be zero at the single node located at $(x, y) = (-1, 0)$. The Reynolds number for the flow is taken as $\mathrm{Re} = 100$ by specifying $\rho = 1$, $\mu = 1/100$ and a characteristic unit length. The initial boundary-value problem is posed on the time interval $t = (0, 0.7]$ s.

The computational domain is discretized into 480 nonuniform finite elements, where we place 40 element layers along the circumference of the cylinder and 12 in the radial direction. A depiction of the finite element mesh at $t = 0$ and $t = 0.70$ is shown in Fig. 11.4.1. Mesh refinement is employed near the cylinder to ensure acceptable numerical resolution of all variables in the wake region downstream of the cylinder. We solve the problem using a p-level of 4 in each finite element, which amounts to 31,360 total degrees of freedom in the numerical model. We employ Newton's method to linearize the finite element equations and adopt a time step size of $\Delta t = 0.005$. The α-family of time approximation is utilized in the temporal discretization (with α taken as 0.5); the first-order backward difference scheme is employed in the first few iterations.

The evolution of the deforming fluid mesh is determined at each time step using a standard pseudo-elasticity formulation (see Belytschko, Liu, and Moran [5]) that is implemented in conjunction with the arbitrary Lagrangian–Eulerian (ALE) formulation. In this approach, we solve a linear elasticity boundary-value problem with Dirichlet boundary conditions at each time step on the fluid domain. The position of the cylinder at the current time step is used directly as a boundary condition in the mesh motion scheme to determine the new locations and velocities of the nodes of the fluid mesh. A weak-form Galerkin finite element model of the pseudo-elasticity equations is employed. To prevent excessive distortion of elements in the model we specify Young's modulus for the eth finite element as $E_e = E_0 \mu(\Omega^e)^{-0.5}$ where $\mu(\Omega^e)$ is the area of the element; the non-negative quantity E_0 is arbitrary. In Fig. 11.4.2 we present snapshots of the numerical results for the velocity components and pressure at $t = 0.25$, 0.50, and 0.70. We see that at this Reynolds number the flow field is symmetric about the y-axis. Our numerical results agree well with the high-order weak form spectral element solution presented by Bodard, Bouffanais, and Deville [263].

Based on these preliminary results, the possibility of extending the work of Payette [161] to the numerical simulation of fluid–structure interaction problems using high-order spectral/hp finite element procedures appears promising. Use of a weak-form Galerkin finite element model for solids and a suitable least-squares model for viscous incompressible fluids seem to be the most appropriate computational approach for the analysis of fluid–structure interaction problems.

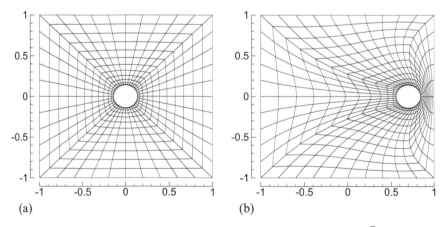

Fig. 11.4.1: Finite element discretization of computational domain $\bar{\Omega}$ for the analysis of transient incompressible flow inside a square cavity induced by the motion of a circular cylinder: (a) fluid mesh at $t = 0$ s and (b) fluid mesh at $t = 0.70$ s.

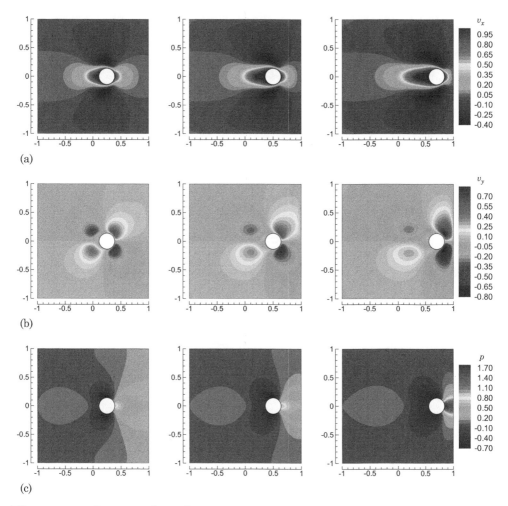

Fig. 11.4.2: Transient flow of an incompressible viscous fluid inside a square cavity induced by a moving cylinder at $t = 0.25, 0.50$, and 0.70 (from the left to the right): (a) velocity component v_x, (b) velocity component v_y, and (c) pressure field P.

Problems

11.1 Develop the least-squares finite element model with the velocity field and pressure as variables of the Navier–Stokes equations governing axisymmetric flows (see Problem 10.2).

11.2 Develop the least-squares finite element model with velocity field, pressure, and vorticity as variables of the Stokes equations governing axisymmetric flows.

11.3 Establish the expressions in Eqs. (11.2.9a) and (11.2.9b) from the least-squares functional in Eq. (11.2.6) and the Picard linearization in Eq. (11.2.7a).

11.4 Establish the expressions in Eqs. (11.2.10a) and (11.2.10b) from the least-squares functional in Eq. (11.2.6) and the Newton linearization in Eq. (11.2.7b).

1

Appendix: Solution Procedures for Linear Equations

A1.1 Introduction

All finite element equations, after assembly and imposition of boundary (and initial) conditions, can be expressed as a set of linear algebraic equations that must be solved. In general, the finite element equations are of the form

$$[A]\{X\} = \{B\} \qquad (A1.1.1)$$

where $[A]$ is the coefficient matrix resulting from the assembly of element matrices, $\{X\}$ is the column of unknowns (typically nodal values), and $\{B\}$ is the known column vector. In the finite element method, $[A]$ is a banded sparse matrix that may be either symmetric or unsymmetric depending on the characteristics of the governing differential equation(s) describing the physical problem, and possibly the nonlinear solution method used (see Appendix 2).

A banded matrix is one in which all elements beyond a diagonal parallel to the main diagonal are zero. The maximum number of non-zero diagonals above or below the main diagonal plus one (to account for the main diagonal) is called *half bandwidth* (NHBW) of the matrix [see Fig. A1.1.1(a)]. When the matrix is symmetric, it is sufficient to store elements in upper or lower half bandwidth of the matrix $[A]$ [see Fig. A1.1.1(b)] and write solvers to take note of the fact that $a_{ij} = a_{ji}$ for all rows i and columns j of the matrix. When the matrix is not symmetric, one must store all elements in the full bandwidth (2*NHBW-1) of the matrix[1]. Here we include the Fortran subroutines of a symmetric banded equation solver (SYMSOLVR) and unsymmetric banded equation solver (UNSYMSLV).

---

[1]In the computer program UNSYMSLV, NBW=2*NHBW is used for full bandwidth. This gives one extra column in the coefficient matrix, which is used to store the global source vector.

J.N. Reddy, *An Introduction to Nonlinear Finite Element Analysis*, Second Edition. ©J.N. Reddy 2015. Published in 2015 by Oxford University Press.

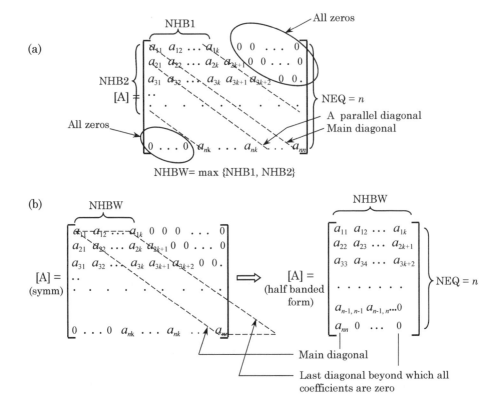

Fig. A1.1.1: Banded symmetric and unsymmetric matrices.

A set of linear algebraic equations can be solved by either a direct or an iterative method. Direct methods, like the Gauss elimination method, provide the solution after a fixed number of steps and are less sensitive to the conditioning of the matrix (the condition number of a matrix is the ratio of the largest to smallest eigenvalue associated with the matrix). Apart from the round-off errors introduced in the computer, direct methods yield the "exact" solution to the equations. However, the main deficiency of the direct solvers is that they require the coefficient matrix be stored in an ordered format to carry out the algebraic operations on the coefficients of the augmented matrix $[A|B]$ and obtain the solution. In recent years, the direct solvers have been refined to reduce this deficiency through innovative data management techniques (e.g. frontal solvers, skyline solvers, and others; see Carey and Oden [264] and Hood [265]). These improvements enable users to solve moderately large systems of equations efficiently. However, they have been found to be unsuitable for solving very large systems of equations because they demand out-of-core storage of the equations and hence require large data transfers. In addition, direct methods are difficult to organize on multiprocessor, parallel computers.

The limitations on CPU time and storage requirements preclude the use of direct solvers for complex problems with more than a quarter million equations, and iterative methods are found to be more efficient in that they require less storage and CPU time. Iterative methods are approximate and only converged solutions will be close to the true solutions. In iterative methods, the global matrix formation may be avoided. The major operation in iterative methods is the matrix-vector multiplication as compared to the matrix reduction (by elementary operations) in direct methods. A significant advantage of iterative methods is that a given set of equations can be divided into as many subsets of equations as there are processors and calculations can be performed in parallel on the array of processors. However, convergence characteristics of iterative methods depend on the condition number of the system of linear equations, and a suitable preconditioner is a must to achieve convergence. For additional details, the reader may consult [265–268].

A1.2   Direct Methods

A1.2.1   Preliminary Comments

Direct methods are those in which simultaneous linear algebraic equations are solved "exactly" (within the computational round-off error) by successive elimination of variables and back substitution. The Gauss elimination method is a fixed-step technique [266–268], and frontal [265] and skyline [4] solution methods are examples of direct solution methods that use the Gauss elimination technique efficiently. The direct methods are the most commonly used techniques when the number of equations involved is not too large. The number of elementary operations for Gauss elimination is of the order $n^3/3 + \mathrm{O}(n^2)$, where n denotes the number of unknowns.

The *frontal* solution procedure [265] is faster than most direct solvers; it requires less core space as long as active variables can be kept in the core, and it allows for partial pivoting. An additional advantage is that no specific node numbering scheme is needed, though a judicious element numbering helps to minimize the front width.

Due to the fact that the approximation functions are defined only within an element, the coefficient matrix in the finite element method is banded; that is, $a_{ij} = 0$ for $j > i + n_b$, where n_b is the half bandwidth of the matrix $[A]$. This greatly reduces the number of operations in solving the equations, if we make note of the fact that elements outside the bandwidth are zero. Of course, the bandwidth size depends on the global node numbering. The skyline technique is one in which bandedness of the finite element equations is exploited by storing the row number m_j of the first nonzero element in column j. The variables m_i, $i = 1, 2, \ldots, n$, define the skyline of the matrix (see Bathe [4] for details).

A1.2.2 Symmetric Solver

In the case of a symmetric solver, the global coefficient matrix $[A] \equiv [GLK]$ = [BAND] is stored in symmetric banded form (NEQ× NHBW), where NEQ is the number of global equations and NHBW is the half bandwidth. Fortran statements for the assembly of element coefficient matrices [ELK] and element source vector {ELF} for the symmetric case are included in Box A1.2.1. Note that BETA is the acceleration parameter, β [see Eq. (4.4.6)], and the array {GP2} denotes the solution vector from the $(r-2)$ iteration while {GP1} denotes the solution from the $(r-1)$ iteration. Note that the global source vector is stored in array {GLF}={RHS}. When IRES > 0, right-hand elimination is skipped. After Gauss elimination is completed (by setting IRES = 0) in SYMSOLVR, the solution is returned in array {GLF} (see Box A1.2.2 and Fortran programs FEM1D and FEM2D discussed in the textbook by Reddy [2]).

Box A1.2.1: Fortran statements for the calculation of the half bandwidth and assembly of element coefficient matrices.

```
C
C    {ELU} - Element solution vector
C    {ELF} - Element source vector
C    {GLF} - Assembled source vector (before entering solver)
C    [ELK] - Element coefficient matrix
C    [GLK] - Assembled coefficient matrix
C    [NOD] - Connectivity array
C
         DO 140 N=1,NEM
              DO 110 I=1,NPE
              NI=NOD(N,I)
              ELU(I)=BETA*GP2(NI)+(1.0-BETA)*GP1(NI)
110           ELX(I)=GLX(NI)
              CALL ELMATRCS(IEL,NPE,NONLIN,F0)
              DO 130 I=1,NPE
              NR=(NOD(N,I)-1)*NDF
              DO 130 II=1,NDF
                NR=NR+1
                L=(I-1)*NDF+II
                GLF(NR)=GLF(NR)+ELF(L)
                DO 120 J=1,NPE
                    NCL=(NOD(N,J)-1)*NDF
                    DO 120 JJ=1,NDF
                    M=(J-1)*NDF+JJ
                    NC=NCL-NR+JJ+1
                    IF(NC.GT.0)THEN
                        GLK(NR,NC)=GLK(NR,NC)+ELK(L,M)
                    ENDIF
120             CONTINUE
130         CONTINUE
140  CONTINUE
```

Box A1.2.2: Fortran subroutine for solution of banded *symmetric* equations.

```
      SUBROUTINE SYMSOLVR(NRM,NCM,NEQNS,NBW,BAND,RHS,IRES)
C
C     Solution of a banded, symmetric, system of algebraic equations
C     [BAND]{U} = {RHS} using the Gauss elimination method. The actual
C     dimensions of the matrix [BAND] in the calling program, are
C     NRM by NCM.
C
      IMPLICIT REAL*8(A-H,O-Z)
      DIMENSION BAND(NRM,NCM),RHS(NRM)
      MEQNS=NEQNS-1
      IF(IRES.LE.0) THEN
          DO 30 NPIV=1,MEQNS
          NPIVOT=NPIV+1
          LSTSUB=NPIV+NBW-1
          IF(LSTSUB.GT.NEQNS) THEN
             LSTSUB=NEQNS
          ENDIF
          DO 20 NROW=NPIVOT,LSTSUB
          NCOL=NROW-NPIV+1
          FACTOR=BAND(NPIV,NCOL)/BAND(NPIV,1)
          DO 10 NCOL=NROW,LSTSUB
          ICOL=NCOL-NROW+1
          JCOL=NCOL-NPIV+1
10        BAND(NROW,ICOL)=BAND(NROW,ICOL)-FACTOR*BAND(NPIV,JCOL)
20        RHS(NROW)=RHS(NROW)-FACTOR*RHS(NPIV)
30        CONTINUE
      ELSE
40        DO 60 NPIV=1,MEQNS
          NPIVOT=NPIV+1
          LSTSUB=NPIV+NBW-1
          IF(LSTSUB.GT.NEQNS) THEN
             LSTSUB=NEQNS
          ENDIF
          DO 50 NROW=NPIVOT,LSTSUB
          NCOL=NROW-NPIV+1
          FACTOR=BAND(NPIV,NCOL)/BAND(NPIV,1)
50        RHS(NROW)=RHS(NROW)-FACTOR*RHS(NPIV)
60        CONTINUE
      ENDIF
C
C   Back substitution
C
      DO 90 IJK=2,NEQNS
      NPIV=NEQNS-IJK+2
      RHS(NPIV)=RHS(NPIV)/BAND(NPIV,1)
      LSTSUB=NPIV-NBW+1
      IF(LSTSUB.LT.1) THEN
         LSTSUB=1
      ENDIF
```

Box A1.2.2: Listing of subroutine **SYMSOLVR** (continued).

```
         NPIVOT=NPIV-1
         DO 80 JKI=LSTSUB,NPIVOT
         NROW=NPIVOT-JKI+LSTSUB
         NCOL=NPIV-NROW+1
         FACTOR=BAND(NROW,NCOL)
   80  RHS(NROW)=RHS(NROW)-FACTOR*RHS(NPIV)
   90  CONTINUE
         RHS(1)=RHS(1)/BAND(1,1)
         RETURN
         END
```

A1.2.3 Unsymmetric Solver

In the case of an unsymmetric solver the matrix [GLK] is stored in full band-width form (NEQ × NBW), where NEQ is the number of global equations and NBW=2*NHBW is twice the half bandwidth. Since there is an additional column in NBW, we use it to store the global source vector; that is GLF(I) → GLK(I,NBW). Fortran statements for the assembly of element matrices for the unsymmetric case are included in Box A1.2.3. After Gauss elimination is completed in UNSYMSLV (see Box A1.2.4), the solution is returned in the last column of the coefficient matrix {GLK(I,NBW)}.

Box A1.2.3: Fortran statements for assembly of the element matrices and source vector into a banded *unsymmetric* global matrix.

```
         DO 140 N=1,NEM
             DO 110 I=1,NPE
             NI=NOD(N,I)
             ELU(I)=BETA*GP2(NI)+(1.0-BETA)*GP1(NI)
  110        ELX(I)=GLX(NI)
             CALL ELEKMF(IEL,NPE,NONLIN,FO)
             DO 130 I=1,NPE
             NR=(NOD(N,I)-1)*NDF
             DO 130 II=1,NDF
                 NR=NR+1
                 L=(I-1)*NDF+II
                 GLK(NR,NBW)=GLK(NR,NBW)+ELF(L)
                 DO 120 J=1,NPE
                 NCL=(NOD(N,J)-1)*NDF
                 DO 120 JJ=1,NDF
                     M=(J-1)*NDF+JJ
                     NC=NCL-NR+JJ+NHBW
                     IF(NC.GT.0)THEN
                         GLK(NR,NC)=GLK(NR,NC)+ELK(L,M)
                     ENDIF
  120            CONTINUE
  130        CONTINUE
  140    CONTINUE
```

Box A1.2.4: Fortran subroutine for solution of banded *unsymmetric* equations.

```
        SUBROUTINE UNSYMSLV(A,NRMAX,NCMAX,N,ITERM)
C
C     Solver for BANDED UNSYMMETRIC system of algebraic equations
C
        IMPLICIT REAL*8 (A-H,O-Z)
        DIMENSION A(NRMAX,NCMAX)
        CERO=1.0D-15
        PARE=CERO**2
        NBND=2*ITERM
        NBM=NBND-1
C
C     Begin elimination of the lower left
C
        DO 80 I=1,N
        IF (DABS(A(I,ITERM)).LT.CERO) GO TO 10
        GO TO 20
10      IF (DABS(A(I,ITERM)).LT.PARE) GO TO 110
20      JLAST=MINO(I+ITERM-1,N)
        L=ITERM+1
        DO 40 J=I,JLAST
        L=L-1
        IF (DABS(A(J,L)).LT.PARE) GO TO 40
        B=A(J,L)
        DO 30 K=L,NBND
30      A(J,K)=A(J,K)/B
        IF (I.EQ.N) GO TO 90
40      CONTINUE
        L=0
        JFIRST=I+1
        IF (JLAST.LE.I) GO TO 80
        DO 70 J=JFIRST,JLAST
        L=L+1
        IF (DABS(A(J,ITERM-L)).LT.PARE) GO TO 70
        DO 50 K=ITERM,NBM
50      A(J,K-L)=A(J-L,K)-A(J,K-L)
        A(J,NBND)=A(J-L,NBND)-A(J,NBND)
        IF (I.GE.N-ITERM+1) GO TO 70
        DO 60 K=1,L
60      A(J,NBND-K)=-A(J,NBND-K)
70      CONTINUE
80      CONTINUE
90      L=ITERM-1
        DO 100 I=2,N
        DO 100 J=1,L
        IF (N+1-I+J.GT.N) GO TO 100
        A(N+1-I,NBND)=A(N+1-I,NBND)-A(N+1-I+J,NBND)*A(N+1-I,ITERM+J)
100     CONTINUE
        RETURN
```

Box A1.2.4: Listing of subroutine **UNSYMSLV** (continued).

```
110 WRITE (6,140) I,A(I,ITERM)
    STOP
140 FORMAT (/,2X,'Computation stopped in UNSYMSLV because zero appears
    * on the main diagonal *** Eqn no. and value:',I5,E12.4)
    END
```

A1.3   Iterative Methods

Among the various iterative methods that are available in the literature, the conjugate gradient (CG) method [269] is most widely used because it is a finite step method (i.e. apart from round-off errors, the solution is achieved in a fixed number of iterations) and it can be used to determine the inverse. However, the number of iterations required depends on the condition number of the coefficient matrix. The convergence of the conjugate gradient method, and iterative methods in general, can be improved by preconditioning and/or scaling the equations [270–272].

The limitations on storage can be overcome by solving the equations at the element level, that is, use the Gauss–Seidel iteration idea for the set of variables associated with the element. This approach avoids assembly of element matrices to form the global coefficient matrix. This idea of using the element-by-element data structure of the coefficient matrix was first pointed out by Fox and Stanton [273] and Fried [274]. The phrase *element-by-element* refers to a particular data structure for finite element techniques wherein information is stored and maintained at the element level rather than assembled into a global data structure. In this method the matrix–vector multiplications are carried out at the element level and the assembly is carried out on the resultant vector. This idea proves to be very attractive when solving large problems, because the matrix–vector multiplication can be done in parallel on a series of processors. Another advantage of this method is that the resultant savings in storage, compared to direct solvers, allows solution of large problems on small computers.

The advantages of the element-by-element data structure over assembling the global coefficient matrix are: (1) the need for formation and storage of a global matrix is eliminated, and therefore the total storage and computational costs are low, (2) the amount of storage is independent of the node numbering and mesh topology and depends on the number and type of elements in the mesh, and (3) the element-by-element solution algorithms can be vectorized for efficient use on supercomputers. The major disadvantage of the element-by-element data structure is the limited number of preconditioners that can be formulated from the unassembled matrices. A review of the literature on element-by-element algorithms can be found in [275], and the methods have been investigated by numerous investigators [197, 275–278].

2

Appendix:
Solution Procedures
for Nonlinear Equations

A2.1 Introduction

Finite element formulations of nonlinear differential equations lead to nonlinear algebraic equations for each element of the finite element mesh. The element equation is of the form,

$$[K^e(\{u^e\})]\{u^e\} = \{F^e\} \qquad (A2.1.1)$$

where

$[K^e]$ — element coefficient matrix (or "stiffness" matrix), which
depends on the solution vector $\{u^e\}$,

$\{u^e\}$ — column vector of element nodal values, and $\qquad$ (A2.1.2)

$\{F^e\}$ — column vector of element nodal "forces".

When $[K^e]$ is independent of $\{u^e\}$, the matrix coefficients can be evaluated for all elements and the assembled equations can be solved for the global nodal values $\{u\}$ after imposing boundary conditions. When $[K^e]$ depends on the unknown solution vector $\{u^e\}$, the matrix coefficients cannot be evaluated. If we can find an approximation to $\{u^e\}$, say $\{u^e\}^1$, then $[K^e(\{u^e\}^1)]$ can be evaluated and assembled. This amounts to linearizing the nonlinear equations, Eq. (A2.1.1). Then a next approximation to the solution can be obtained by solving the assembled equations,

$$\{u\}^2 = [K(\{u\}^1)]^{-1}\{F\} \qquad (A2.1.3)$$

This procedure can be repeated until the approximate solution comes close to the actual solution in some measure. Such a procedure is called an iterative procedure.

J.N. Reddy, *An Introduction to Nonlinear Finite Element Analysis*, Second Edition. ©J.N. Reddy 2015. Published in 2015 by Oxford University Press.

Here we discuss the following commonly used iterative procedures:

1. The Picard Iteration (or Direct Iteration) Method

2. The Newton Iteration Method[1]

3. The Riks Method

The details of these methods are discussed next, with the aid of a single nonlinear equation. For a more complete presentation of these methods, the reader may consult [148–150] and [279–284].

Consider the nonlinear equation,

$$K(u)\,u = F \quad \text{or} \quad R(u) = 0 \tag{A2.1.4}$$

where u is the solution to be determined, $K(u)$ is a known function of u, F is the known 'force', and R is the residual

$$R(u) = K(u)\,u - F \tag{A2.1.5}$$

A plot of the equilibrium path, $R(u, F) = 0$, is shown in Fig. A2.1.1. For any value u_1, $K(u_1)$ denotes the secant of the curve at $u = u_1$, and $\left(\frac{\partial R}{\partial u}\right)\big|_{u_1}$ denotes the tangent of the curve at $u = u_1$.

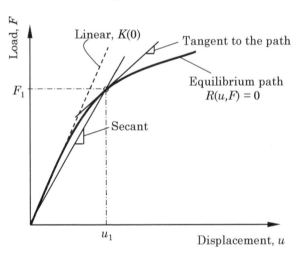

Fig. A2.1.1: Typical force–displacement curve.

A2.2   The Picard Iteration Method

In the Picard iteration, also known as the *direct iteration* method, we begin with an initial guess for u, say $u^{(0)}$, ($u^{(0)} = 0$ in Fig. A2.2.1) and determine a first approximation of u by solving the equation

---

[1]Newton's method is often referred to in the literature as the Newton–Raphson method.

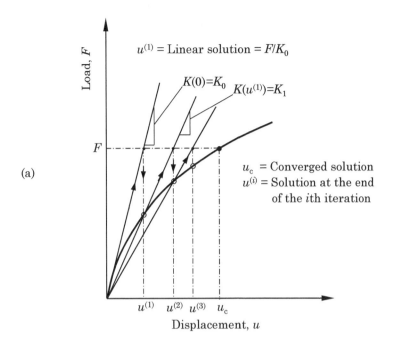

(a)

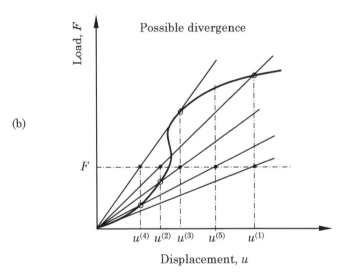

(b)

Fig. A2.2.1: Direct iteration scheme. (a) Case of convergence. (b) Case of divergence.

$$u^{(1)} = \left(K(u^{(0)}) \right)^{-1} F \qquad (A2.2.1)$$

$u^{(1)} \neq u$, and a second approximation for u is sought by using the last approximation to evaluate K

$$u^{(2)} = \left(K(u^{(1)}) \right)^{-1} F \qquad (A2.2.2)$$

This procedure is continued until the difference between two consecutive approximations of u differ by a preselected value. Thus, the algorithm and criterion for convergence may be written as

$$\text{Algorithm:} \qquad u^{(r)} = \left(K(u^{(r-1)}) \right)^{-1} F \qquad (A2.2.3)$$

$$\text{Convergence criterion:} \qquad \sqrt{\frac{(u^{(r)} - u^{(r-1)})^2}{(u^{(r)})^2}} < \epsilon \qquad (A2.2.4)$$

where ϵ denotes the *convergence tolerance* and r denotes the iteration number.

A geometric interpretation of the procedure described above is illustrated in Fig. A2.2.1(a) for an initial guess of $u^{(0)} = 0$. At the beginning of iteration r, the secant of the curve $R(u) = 0$ is found at the point $u = u^{(r-1)}$ and the solution $u^{(r)}$ is computed using Eq. (A2.2.3). Figure A2.2.1(a) shows the convergence to the true solution u_c, whereas Fig. A2.2.1(b) shows a possible divergence of the algorithm. Thus, the success of the algorithm depends on the nature of the nonlinear curve $R(u) = 0$, the initial guess, and the load increment.

In the direct iteration method discussed above, the secant is evaluated at each iteration and inverted to obtain the next approximate solution. This can be computationally very expensive when the number of algebraic equations to be solved is large, that is, when K is a matrix and $[K]^{-1}$ is its inverse. When K has a linear portion, and in most problems of interest to us it does, an *alternative direct iteration* algorithm can be formulated. Let

$$K(u) = K_L + K_N(u) \qquad (A2.2.5)$$

where K_L and $K_N(u)$ are the linear and nonlinear parts of K. Note that K_L is the slope at $u = 0$ of the curve $R(u) = 0$. Then we can write

$$u^{(r)} = (K_L)^{-1}[F - K_N(u^{(r-1)}) \, u^{(r-1)}] \qquad (A2.2.6)$$

This scheme involves evaluating the nonlinear part K_N at each iteration, which is computationally less expensive when compared to evaluating $[K(u^{r-1})]$ and inverting it. The inversion of K_L is required only once, and it should be saved for subsequent use. The criterion in Eq. (A2.2.4) can be used to check for convergence.

Geometrically, the alternative direct iteration algorithm uses the initial slope, that is, slope at the origin of the curve for *all* iterations, while updating the *effective* force

$$\hat{F} \equiv F - K_N(u^{(r-1)}) \, u^{(r-1)}$$

at each iteration (see Fig. A2.2.2). The rate of convergence of this algorithm, if at all it converges, is slower than that in Eq. (A2.2.3).

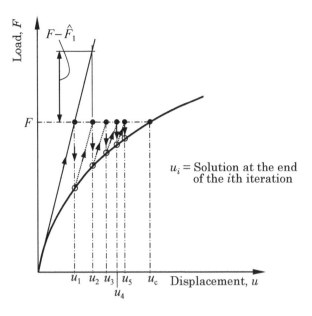

Fig. A2.2.2: Modified direct iteration scheme.

As applied to the assembly of finite element equations in Eq. (A2.1.1), the two algorithms take the following forms:

$$\text{Algorithm 1:} \quad \mathbf{U}^r = (\mathbf{K}(\mathbf{U}^{(r-1)}))^{-1}\mathbf{F} \qquad (A2.2.7)$$

$$\text{Algorithm 2:} \quad \mathbf{U}^r = (\mathbf{K}_L)^{-1}\left(\mathbf{F} - \mathbf{K}_N(\mathbf{U}^{(r-1)})\mathbf{U}^{(r-1)}\right) \qquad (A2.2.8)$$

$$\text{Convergence Criterion:} \quad \sqrt{\frac{\sum_{I=1}^{N}(U_I^{(r)} - U_I^{(r-1)})^2}{\sum_{I=1}^{N}(U_I^{(r)})^2}} < \epsilon \qquad (A2.2.9)$$

where $\mathbf{U}$ denotes the global solution vector. The global matrices $\mathbf{K}$, $\mathbf{K}_L$, and $\mathbf{K}_N$ are assembled from the corresponding element matrices ($\mathbf{u}$ denotes the solution vector at the element level):

$$\begin{aligned}
\mathbf{K} \text{ is assembled using} \quad & \mathbf{K}^e(\mathbf{u}^{(r-1)}) \\
\mathbf{K}_L \text{ is assembled using} \quad & \mathbf{K}_L^e \\
\mathbf{K}_N \text{ is assembled using} \quad & \mathbf{K}_N^e(\mathbf{u}^{(r-1)})
\end{aligned} \qquad (A2.2.10)$$

where

$$\mathbf{K}^e = \mathbf{K}_L^e + \mathbf{K}_N^e \qquad (A2.2.11)$$

The rate of convergence of the iterative procedure can be accelerated, in certain type of nonlinear behavior, by a *relaxation* procedure in which a weighted average of the last two solutions is used to evaluate $\mathbf{K}^e$ (or $\mathbf{K}_N^e$): $\mathbf{K}^e(\bar{\mathbf{u}}^e)$, where

$\bar{\mathbf{u}}^e = \beta \mathbf{u}^{(r-2)} + (1-\beta)\mathbf{u}^{(r-1)}$, and $0 \le \beta \le 1$ is called the relaxation or acceleration parameter. In this case Eqs. (A2.2.7) and (A2.2.8) take the form,

$$\text{Algorithm 1} \qquad \mathbf{U}^r = (\mathbf{K}(\bar{\mathbf{U}}))^{-1}\mathbf{F} \qquad\qquad\qquad \text{(A2.2.12)}$$

$$\text{Algorithm 2} \qquad \mathbf{U}^r = (\mathbf{K}_L)^{-1}(\mathbf{F} - \mathbf{K}_N(\bar{\mathbf{U}})\bar{\mathbf{U}}) \qquad \text{(A2.2.13)}$$

$$\bar{\mathbf{U}} = \beta\mathbf{U}^{(r-2)} + (1-\beta)\mathbf{U}^{(r-1)} \qquad \text{(A2.2.14)}$$

The actual value of β varies from problem to problem.

The computational algorithm of the direct iteration is summarized below. At each load level follow the steps:

(1) Compute element matrices $\mathbf{K}^e$ and $\mathbf{F}^e$ (for transient problems, $\mathbf{K}^e$ and $\mathbf{F}^e$ are to be replaced by $\hat{\mathbf{K}}^e$ and $\hat{\mathbf{F}}^e$) using the solution $\mathbf{U}^{(r-1)}$ from the previous iteration (of current load and/or time). For the first iteration of the subsequent load steps, use $\mathbf{U}_c$, the converged solution of the last load step. For the first load step use $\mathbf{U}_c = \mathbf{0}$, provided $\mathbf{K}$ is not singular, to compute the linear solution.

(2) Assemble the element matrices $\mathbf{K}^e$ and $\mathbf{F}^e$ (or $\hat{\mathbf{K}}^e$ and $\hat{\mathbf{F}}^e$).

(3) Apply the boundary conditions on the assembled set of equations.

(4) Solve the assembled equations.

(5) Check for convergence using Eq. (A2.2.9).

(6a) If the convergence criterion is satisfied, increase the load to next level, initialize the counter on iterations, and repeat Steps 1–5.

(6b) If the convergence criterion is not satisfied, check if the maximum number of allowable iterations is exceeded. If yes, terminate the computation printing a message to that effect. If the number of iterations did not exceed the maximum allowed, update $\mathbf{U}^{(r-1)}$ and $\mathbf{U}^{(r)}$ and repeat Steps 1–5.

A2.3   The Newton Iteration Method

Suppose that we know solution $u^{(r-1)}$ of Eq. (A2.1.1) at $(r-1)$st iteration and interested in seeking solution at the rth iteration. We expand $R(u)$ about the known solution $u^{(r-1)}$ in Taylor's series,

$$R(u) = R(u^{(r-1)}) + \left(\frac{\partial R}{\partial u}\right)\bigg|_{u^{(r-1)}} \delta u + \frac{1}{2}\left(\frac{\partial^2 R}{\partial u^2}\right)\bigg|_{u^{(r-1)}} (\delta u)^2 + \cdots = 0 \quad \text{(A2.3.1)}$$

where δu is the increment,

$$\delta u^{(r)} = u^{(r)} - u^{(r-1)} \qquad\qquad\qquad \text{(A2.3.2)}$$

Assuming that the second- and higher-order terms in δu are negligible, we can write Eq. (A2.3.1) as

$$\delta u^{(r)} = - \left(K_T(u^{(r-1)})\right)^{-1} R(u^{(r-1)})$$

$$= \left(K_T(u^{(r-1)}) \right)^{-1} \left(F - K(u^{(r-1)}) u^{(r-1)} \right) \qquad \text{(A2.3.3)}$$

where K_T is the slope (tangent) of the curve $R(u)$ at $u^{(r-1)}$ (see Fig. A2.3.1):

$$K_T = \left. \frac{\partial R}{\partial u} \right|_{u^{(r-1)}} \qquad \text{(A2.3.4)}$$

The residual or *imbalance force*, $R(u^{(r-1)})$ is gradually reduced to zero if the procedure converges. Equation (A2.3.3) gives the increment of u at the rth iteration so that the total solution is

$$u^{(r)} = u^{(r-1)} + \delta u^{(r)} \qquad \text{(A2.3.5)}$$

(a)

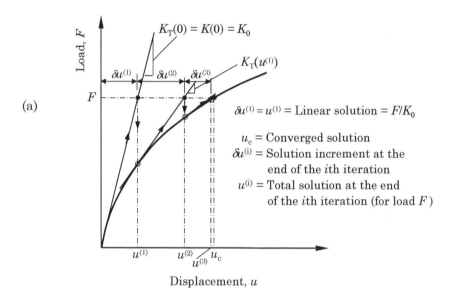

(b)

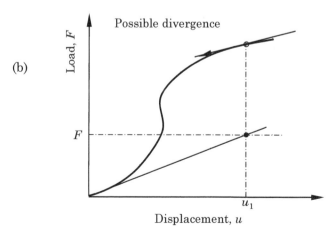

Fig. A2.3.1: The Newton scheme. (a) A case of convergence. (b) A case of divergence.

The iteration is continued until a convergence criterion, say Eq. (A2.2.4), is satisfied. Other convergence criteria include checking the magnitude of the imbalance force. A geometrical interpretation of the Newton procedure is shown in Fig. A2.3.1(a). For most problems, the method has faster convergence characteristics. Figure A2.3.1(b) illustrates possible divergence of the iterative procedure for certain problems.

The Newton method requires that the tangent K_T be computed at each iteration. This can be very expensive when many degrees of freedom are involved. A *modified Newton's technique* involves, for a fixed load step, either keeping K_T fixed while updating the imbalance force at each iteration (see Fig. A2.3.2) or updating K_T only at each preselected number of iterations while updating the imbalance force at each iteration. The latter is known as the Newton–Raphson method. There are several other modifications of the procedure.

The Newton and modified Newton procedures take the following forms when applied to the assembly of element equations (A2.3.3):

Newton's Procedure

$$\delta \mathbf{U} = -(\mathbf{K}_T)^{-1}\mathbf{R} \tag{A2.3.6}$$

Modified Newton's Procedure

$$\delta \mathbf{U} = -(\bar{\mathbf{K}}_T)^{-1}\bar{\mathbf{R}} \tag{A2.3.7}$$

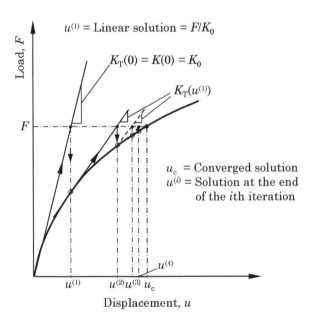

Fig. A2.3.2: Modified Newton's scheme.

where $\mathbf{K}_T$ is the tangent matrix (is calculated at the element level first and then assembled)

$$\mathbf{K}_T = \left.\frac{\partial \mathbf{R}}{\partial \mathbf{U}}\right|_{\mathbf{U}^{r-1}} \quad \text{or} \quad (K_T)_{ij} = \frac{\partial R_i}{\partial U_j} = K_{ij} + \sum_{k=1}^{N} \frac{\partial K_{ik}}{\partial U_j} U_k \qquad \text{(A2.3.8a)}$$

where N is the total number of equations in the mesh, and

$$\mathbf{R} = \mathbf{K}(\mathbf{U}^{(r-1)})\mathbf{U}^{(r-1)} - \mathbf{F} \qquad \text{(A2.3.8b)}$$

$$\bar{\mathbf{K}}_T = \mathbf{K}_T(\bar{\mathbf{U}}), \quad \bar{\mathbf{K}} = \mathbf{K}(\bar{\mathbf{U}}) \qquad \text{(A2.3.8c)}$$

and $\bar{\mathbf{U}}$ is the solution at the beginning of the current load step, and

$$\mathbf{U}^{(r)} = \mathbf{U}^{(r-1)} + \delta \mathbf{U}^{(r)} \qquad \text{(A2.3.9)}$$

The relaxation procedure described in Eq. (A2.2.14) can be used to accelerate convergence for certain nonlinear problems.

Since only the increment of the solution is computed in each iteration of the Newton iteration, the incremental equations, Eqs. (A2.3.6) and (A2.3.7), are subject to the homogeneous form of the specified essential boundary conditions of the problem. Thus after the first iteration of the first load step, any specified non-zero values of $\mathbf{U}$ should be set to zero so that in the subsequent iterations and loads, $\delta \mathbf{U}$ is subjected to homogeneous boundary conditions.

For each load step, the following computations are required for the Newton or modified Newton procedure:

(1) Evaluate element matrices $\mathbf{K}^e$ and $\mathbf{F}^e$ (or $\hat{\mathbf{K}}^e$ and $\hat{\mathbf{F}}^e$ for the transient problems), and compute $\mathbf{K}_T^e$ and $\mathbf{R}^e$ using Eqs. (A2.3.8a) and (A2.3.8b) for an element.

(2) Assemble element matrices $\mathbf{K}_T^e$ and $\mathbf{R}^e$; for the modified Newton iteration procedure, save either assembled $\mathbf{K}_T$ or its inverse for use in subsequent iterations.

(3) Apply the boundary conditions on the assembled set of equations. *Note:* Set the specified boundary conditions on $\mathbf{U}$ to zero after Step 3 in the first iteration of the first load step.

(4) Solve the assembled equations.

(5) Update the solution vector using Eq. (A2.3.9).

(6) Check for convergence.

(7a) If the convergence criterion is satisfied, increase the load, initialize the iteration counter, and repeat Steps 1–6. For the modified Newton iteration, compute $\mathbf{F}^e - \bar{\mathbf{K}}^e \mathbf{U}^{(r-1)}$ in Step 1 and go to Step 2.

(7b) If the convergence criterion is not satisfied, check if the maximum number of iterations allowed is exceeded. If it is, terminate the computation by printing a message. If the maximum allowable number of iterations is not exceeded, go to Step 1.

In order to reduce the number of operations per iteration, in the modified Newton method the same system matrices are used for several iterations. These matrices are updated only at the beginning of each load step or only when the convergence rate becomes poor. The modified Newton method may require more iterations to reach a new equilibrium point.

A2.4   The Riks and Modified Riks Methods

The Newton method and its modifications are often used to trace nonlinear solution paths. However, the Newton method fails to trace the nonlinear equilibrium path through the limit point (see Fig. A2.4.1), because in the vicinity of a limit point the tangent matrix $[K_T]$ becomes singular and the iteration procedure diverges. Wempner [148] and Riks [149] suggested a procedure to predict the nonlinear equilibrium path through limit points. The method provides the Newton method and its modifications with a technique to control progress along the equilibrium path. The theoretical development of this method and its modification can be found in [279–284]. In the modified Riks method (see [279, 280]) the load increment for each load step is considered to be an unknown and solved as a part of the solution.

The basic idea of the Riks technique can be described for a single nonlinear equation as follows (see Fig. A2.4.2): The length Δs of the tangent to the current equilibrium point is prescribed, and the new point is found as the intersection of the *normal to the tangent* with the equilibrium path [see Fig. A2.4.2(a)]. Then iteration is performed along the normal toward the new equilibrium point, as illustrated in Fig. A2.4.2(a). Crisfield [151] suggested using a circular arc

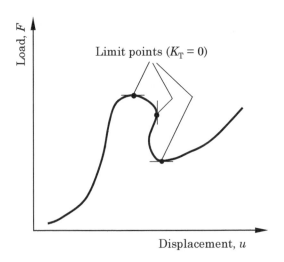

Fig. A2.4.1: Load–deflection curve with limit points.

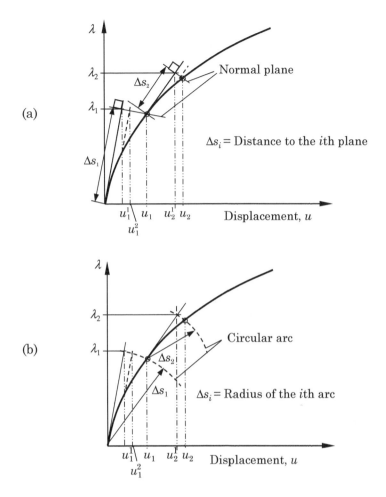

Fig. A2.4.2: The Riks method. (a) Normal plane scheme. (b) Circular arc scheme.

in place of the normal [see Fig. A2.4.2(b)]. The center of the circle is at the current equilibrium point and Δs is its radius. For multidimensional problems the normal and circular arcs become a plane and sphere, respectively, Crisfield [151] updated the tangent stiffness matrix only at the beginning of each load increment (i.e. modified Newton's method). In the present study we describe the modified Riks method due to Crisfield [151].

We wish to solve Eq. (A2.1.4) for u as a function of the source term F. If F is independent of the geometry, we can write it as

$$F = \lambda \bar{F} \tag{A2.4.1}$$

where λ is a scalar, called *load parameter*, which is considered as an unknown parameter. Equation (A2.1.5) becomes

$$R = K(u)\,u - \lambda \bar{F} \tag{A2.4.2}$$

Now suppose that the solution $(u_n^{(r-1)}, \lambda_n^{(r-1)})$ at $(r-1)$st iteration of the nth load step is known and we wish to determine the solution $(u_n^{(r)}, \lambda_n^{(r)})$ at the rth iteration. Expanding R, which is now a function of λ and u, in Taylor's series about the known solution, we have,

$$R(u_n^{(r)}, \lambda_n^{(r)}) = R(u_n^{(r-1)}, \lambda_n^{(r-1)}) + \left(\frac{\partial R}{\partial \lambda}\right)^{(r-1)} \delta\lambda_n^{(r)} + \left(\frac{\partial R}{\partial u}\right)^{(r-1)} \delta u_n^{(r)} + \cdots$$

$$= 0$$

Omitting the higher-order terms involving the increments $\delta\lambda_n^{(r)}$ and $\delta u_n^{(r)}$, we obtain

$$0 = R_n^{(r-1)} - \bar{F}\,\delta\lambda_n^{(r)} + (K_T)^{(r-1)}\,\delta u_n^{(r)} \qquad (A2.4.3)$$

where $K_T = \partial R/\partial u$ is the tangent matrix [see Eq. (A2.3.4)]. The incremental solution at the current iteration of the nth load step is given by

$$\delta u_n^{(r)} = -K_T^{-1}(R_n^{(r-1)} - \bar{F}\,\delta\lambda_n^{(r)})$$
$$\equiv \delta\bar{u}_n^{(r)} + \delta\lambda_n^{(r)}\,\delta\hat{u}_n \qquad (A2.4.4a)$$

where $\delta u_n^{(r)}$ is the usual increment in displacement due to known out-of-balance force vector $R_n^{(r-1)}$ with known $\lambda_n^{(r-1)}$ and K_T is the tangent at the beginning of the current load increment (i.e. modified Newton's method is used)

$$\delta\bar{u}_n^{(r)} = -K_T^{-1}R_n^{(r-1)} \qquad (A2.4.4b)$$

and $\delta\hat{u}_n$ is the tangential solution (see Fig. A2.4.3)

$$\delta\hat{u}_n = K_T^{-1}\bar{F} \qquad (A2.4.4c)$$

Note that K_T is evaluated using the converged solution u_{n-1} of the last load step,

$$K_T = \left(\frac{\partial R}{\partial u}\right)\bigg|_{u=u_{n-1}} = K(u_{n-1}) + \left(\frac{\partial K}{\partial u}\right)\bigg|_{u=u_{n-1}} u_{n-1} \qquad (A2.4.5)$$

and $\delta\hat{u}_n$ is computed at the beginning of each load step.

The solution at the rth iteration of the current load step is given by

$$u_n = u_{n-1} + \Delta u_n^{(r)} \qquad (A2.4.6a)$$
$$\Delta u_n^{(r)} = \Delta u_n^{(r-1)} + \delta u_n^{(r)}, \quad \lambda_n^{(r)} = \lambda_n^{(r-1)} + \delta\lambda_n^{(r)} \qquad (A2.4.6b)$$

For the very first iteration of the first load step, we assume $u = u_0$ and a value for the incremental load parameter $\delta\lambda_1^{(0)}$ and solve the equation

$$\delta\hat{u}_1 = (K_T)^{-1}\bar{F} \qquad (A2.4.7)$$

Then we compute $\delta u_1^{(0)} = \delta\lambda_1^{(0)}\,\delta\hat{u}_1$.

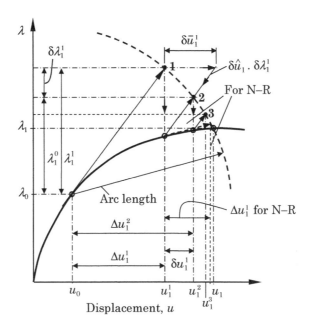

Fig. A2.4.3: Modified Riks scheme.

Select the arc length Δs to be the length of the vector

$$(\Delta s)^2 = \Delta u_n^{(r)} \, \Delta u_n^{(r)} \tag{A2.4.8}$$

Substituting for $\Delta u_n^{(r)}$ from Eq. (A2.4.6b) [and $\delta u_n^{(r)}$ from Eq. (A2.4.4a)] we obtain the following quadratic equation for the increment in the load parameter, $\delta \lambda_n^{(r)}$:

$$a_1[\delta \lambda_n^{(r)}]^2 + 2a_2[\delta \lambda_n^{(r)}] + a_3 = 0 \tag{A2.4.9a}$$

$$
\begin{aligned}
a_1 &= \delta \hat{u}_n \, \delta \hat{u}_n \\
a_2 &= (\Delta u_n^{(r-1)} + \delta \bar{u}_n^{(r)}) \, \delta \hat{u}_n \\
a_3 &= (\Delta u_n^{(r-1)} + \delta \bar{u}_n^{(r)}) \, (\Delta u_n^{(r-1)} + \delta \bar{u}_n^{(r)}) - (\Delta s)^2
\end{aligned} \tag{A2.4.9b}
$$

Let us denote the roots of this quadratic equation as $\delta \lambda_{n1}^{(r)}$ and $\delta \lambda_{n2}^{(r)}$. To avoid "tracing back" the equilibrium path (i.e. going back on the known equilibrium path), we require the angle between the incremental solution vectors at two consecutive iterations, $\Delta u_n^{(r-1)}$ and $\Delta u_n^{(r)}$, be positive. Corresponding to the two roots, $\delta \lambda_{n1}^{(r)}$ and $\delta \lambda_{n2}^{(r)}$, there correspond two values of $\Delta u_n^{(r)}$, denoted $\Delta u_{n1}^{(r)}$ and $\Delta u_{n2}^{(r)}$. The root that gives the positive angle is the one we select from $(\delta \lambda_{n1}^{(r)}, \delta \lambda_{n2}^{(r)})$. The "angle" is defined to be product of the vector $\Delta u_n^{(r-1)}$ and $\Delta u_n^{(r)}$. Then we check to see which one of the following two products is positive:

$$\Delta u_n^{(r-1)} \, \Delta u_{n1}^{(r)} \quad \text{and} \quad \Delta u_n^{(r-1)} \, \Delta u_{n2}^{(r)} \tag{A2.4.10a}$$

If both roots are positive, then we select the ones closest to the linear solution,

$$\delta\lambda_n^{(r)} = -\frac{a_3}{2a_2} \qquad (A2.4.10b)$$

The first arc length Δs is computed using Eqs. (A2.4.7) and (A2.4.8):

$$\Delta s = \delta\lambda_1^0 \sqrt{\delta\hat{u}_1 \, \delta\hat{u}_1} \qquad (A2.4.11)$$

To control the number of iterations taken to converge in the subsequent load increments, Δs can be scaled,

$$\Delta s_n = \Delta s_{n-1} \frac{I_d}{I_0} \qquad (A2.4.12)$$

where Δs_{n-1} is the arc length used in the last iteration of the $(n-1)$st load step, I_d is the number of desired iterations (usually < 5) and I_0 is the number of iterations required for convergence in the previous step. Equation (A2.4.12) will automatically give small arc lengths in the areas of the most severe nonlinearity and longer lengths when the response is linear or nearly linear. To avoid convergence of the solution at a higher equilibrium path, maximum arc lengths should be specified.

For all load steps after the first, the initial incremental load parameter $\delta\lambda_n^{(0)}$ is calculated from

$$\delta\lambda_n^{(0)} = \pm\Delta s_n \left[\delta\hat{u}_n \, \delta\hat{u}_n\right]^{-\frac{1}{2}} \qquad (A2.4.13)$$

The plus sign is for continuing the load increment in the same direction as the previous load step and the negative sign is to reverse the load step. The sign follows that of the previous increment unless the value of determinant of the tangent matrix has changed in sign.

The modified Riks procedure described above for a single equation can be extended to the finite element equations in Eq. (A2.1.1). We introduce a load parameter λ as an additional dependent variable,

$$\mathbf{F} = \lambda\bar{\mathbf{F}} \qquad (A2.4.14)$$

In writing Eq. (A2.4.14), loads are assumed to be independent of the deformation. The assembled equations associated with Eq. (A2.1.1) become

$$\mathbf{R}(\mathbf{U}, \lambda) = \mathbf{KU} - \lambda\bar{\mathbf{F}} = \mathbf{0} \qquad (A2.4.15)$$

The residual vector $\mathbf{R}$ is now considered as a function of both $\mathbf{U}$ and λ.

Now suppose that the solution $(\mathbf{U}_n^{(r-1)}, \lambda_n^{(r-1)})$ at the $(r-1)$st iteration of the nth load step is known. Expanding $\mathbf{R}$ in Taylor's series about $(\mathbf{U}_n^{(r-1)}, \lambda_n^{(r-1)})$, we have

$$R_i = R_i(\mathbf{U}_n^{(r-1)}, \lambda_n^{(r-1)}) + \left(\frac{\partial R_i}{\partial\lambda}\right)_{\mathbf{U}^{r-1}, \lambda_n^{r-1}} \delta\lambda^{(r)} - \left(\frac{\partial R_i}{\partial U_j}\right)_{\mathbf{U}_j^{(r-1)}, \lambda_n^{r-1}} \delta U_j^{(r)} + \cdots$$

where the subscript 'n' is omitted for brevity. Omitting second- and higher-order terms involving $\delta\lambda^{(r)}$ and $\delta U_j^{(r)}$, we can write

$$0 = \mathbf{R}^{(r-1)} - \delta\lambda_n^{(r)}\bar{\mathbf{F}} + \mathbf{K}_T\delta\mathbf{U}^{(r)}$$

or

$$\delta\mathbf{U}_n^{(r)} = -(\mathbf{K}_T)^{-1}\mathbf{R}^{(r-1)} + \delta\lambda_n^{(r)}(\mathbf{K}_T)^{-1}\bar{\mathbf{F}} \qquad (A2.4.16a)$$
$$\equiv \delta\bar{\mathbf{U}}_n^{(r)} + \delta\lambda_n^{(r)}\delta\hat{\mathbf{U}}_n \qquad (A2.4.16b)$$

where $\mathbf{F}$ is the load vector, $\mathbf{R}_n^{(r-1)}$ the unbalanced force vector at iteration $(r-1)$,

$$\delta\bar{\mathbf{U}}_n^{(r)} = -(\mathbf{K}_T)^{-1}\mathbf{R}_n^{(r-1)} \qquad (A2.4.17a)$$

and $\delta\lambda_n^{(r)}$ the load increment [given by Eq. (A2.4.10b)], and

$$\delta\hat{\mathbf{U}}_n = (\mathbf{K}_T)^{-1}\bar{\mathbf{F}} \qquad (A2.4.17b)$$

For the first iteration of any load step, $\delta\lambda_n^{(0)}$ is computed from [see Eq. (A2.4.13)]:

$$\delta\lambda_n^{(0)} = \pm\Delta s_n(\delta\hat{\mathbf{U}}_n^{\mathrm{T}}\delta\hat{\mathbf{U}}_n)^{-1/2} \qquad (A2.4.18)$$

where [see Eq. (A2.4.12)]

$$\Delta s_n = \Delta s_{n-1}\frac{I_d}{I_0} \qquad (A2.4.19a)$$

and Δs_{n-1} is the arc length computed using the relation

$$\Delta s_{n-1} = \sqrt{\Delta\mathbf{U}_{n-1}^{\mathrm{T}}\Delta\mathbf{U}_{n-1}} \qquad (A2.4.19b)$$

$\Delta\mathbf{U}_{n-1}$ being the converged solution increment of the previous load step.

For the first iteration of the first load step, we use:

$$\Delta s = \delta\lambda_1^0\sqrt{\delta\hat{\mathbf{U}}_1^{\mathrm{T}}\delta\hat{\mathbf{U}}_1} \qquad (A2.4.20a)$$

$$\delta\hat{\mathbf{U}}_1 = (\mathbf{K}_T)^{-1}\bar{\mathbf{F}}, \quad \mathbf{K}_T = \mathbf{K}_T(\mathbf{U}_0) \qquad (A2.4.20b)$$

where $\delta\lambda_1^0$ is an assumed load increment and $\mathbf{U}_0$ is an assumed solution vector (often we assume $\delta\lambda_1^0 = 1$ and $\mathbf{U}_0 = \mathbf{0}$).

The solution increment is updated using

$$\Delta\mathbf{U}_n^r = \Delta\mathbf{U}_n^{(r-1)} + \delta\mathbf{U}_n^{(r)} \qquad (A2.4.21)$$

and the total solution at the current load step is given by

$$\mathbf{U}_n = \mathbf{U}_{n-1} + \Delta\mathbf{U}_n^{(r)} \qquad (A2.4.22)$$

The constants in Eq. (A2.4.9b) are computed using

$$
\begin{aligned}
a_1 &= \delta\hat{\mathbf{U}}_n^{\mathrm{T}}\delta\hat{\mathbf{U}}_n \\
a_2 &= (\Delta\mathbf{U}_n^{(r-1)} + \delta\bar{\mathbf{U}}_n^{(r)})^{\mathrm{T}}\delta\hat{\mathbf{U}}_n \\
a_3 &= (\Delta\mathbf{U}_n^{(r-1)} + \delta\bar{\mathbf{U}}_n^{(r)})^{\mathrm{T}}(\Delta\mathbf{U}_n^{(r-1)} + \delta\bar{\mathbf{U}}_n^{(r)}) - (\Delta s)_n^2
\end{aligned}
\tag{A2.4.23}
$$

The computational algorithm of the modified Riks method is summarized next.

First iteration of first load step

(1) Choose a load increment $\delta\lambda_1^0$ (say, $\delta\lambda_1^0 = 1$) and solution vector $\mathbf{U}_0$ (say, $\mathbf{U}_0 = \mathbf{0}$).
(2) Form element matrices $\mathbf{K}_T^e$ and

$$
\mathbf{R}^e = \mathbf{K}^e\mathbf{u}_0^e - \mathbf{F}^e
$$

(3) Assemble element matrices.
(4) Apply the boundary conditions.
(5) Solve for $\delta\hat{\mathbf{U}}_1$ and $\delta\bar{\mathbf{U}}_1^{(1)}$ using Eqs. (A2.4.17a) and (A2.4.17b).
(6) Compute the solution increment [see Eqs. (A2.4.16) and (A2.4.21)] and update the solution

$$
\delta\mathbf{U}_1^{(1)} = \delta\bar{\mathbf{U}}_1^{(1)} + \delta\lambda_1^0\delta\hat{\mathbf{U}}_1; \quad \Delta\mathbf{U}_1^{(1)} = \delta\mathbf{U}_1^{(1)}; \quad \mathbf{U}_1 = \mathbf{U}_0 + \delta\mathbf{U}_1^{(1)}
$$

(7) Update the load increment $\lambda_1^{(1)} = \delta\lambda_1^0$.
(8) Compute the arc length [see Eq. (A2.4.20a)]

$$
\Delta s = \delta\lambda_0\sqrt{\delta\hat{\mathbf{U}}_1^{\mathrm{T}}\delta\hat{\mathbf{U}}_1}
$$

(9) Go to Step 9 of the rth iteration of any load step [see Eq. (A2.4.26)].

First iteration of any load step except the first

(1) Calculate the system matrices $\mathbf{K}^e$, $\mathbf{K}_T^e$, and $\mathbf{F}^e$.
(2) Assemble the element matrices.
(3) Apply the boundary conditions.
(4) Compute the tangential solution

$$
\delta\hat{\mathbf{U}}_{n-1} = \mathbf{K}_T^{-1}\bar{\mathbf{F}}
$$

(5) Compute the initial incremental load parameter $\delta\lambda_n^{(0)}$ by Eq. (A2.4.18):

$$
\delta\lambda_n^{(0)} = \pm(\Delta s)_n[\delta\hat{\mathbf{U}}_n^{\mathrm{T}}\,\delta\hat{\mathbf{U}}_n]^{-1/2}
\tag{A2.4.24}
$$

(6) Compute the incremental solution using Eq. (A2.4.17a):

$$\bar{\mathbf{U}}_n^{(1)} = -(\mathbf{K}_T)^{-1}\mathbf{R}_n^{(0)}$$

(7) Update the total solution vector and load parameter:

$$\delta\mathbf{U}_n^{(1)} = \delta\bar{\mathbf{U}}_n^{(1)} + \delta\lambda_n^{(0)}\delta\hat{\mathbf{U}}_n^{(1)}$$
$$\mathbf{U}_n = \mathbf{U}_{n-1} + \delta\mathbf{U}_n^{(1)}$$
$$\lambda_n^{(1)} = \lambda_n^{(0)} + \delta\lambda_n^{(0)}, \quad \Delta\mathbf{U}_n^{(1)} = \delta\mathbf{U}_n^{(1)} \qquad (\text{A2.4.25})$$

(8) Check for convergence [see Eq. (A2.4.28)]. If convergence is achieved, go to step 15 below. If not, continue with the 2nd iteration by going to step 9.

The rth iteration of any load step $(r = 2, 3, \ldots)$

(9) Update the external load vector

$$\mathbf{F}^{(r-1)} = \lambda_n^{(r-1)}\bar{\mathbf{F}} \qquad (\text{A2.4.26})$$

(10) Update the system matrices (skip forming of $\mathbf{K}_T$ for modified Newton iteration).

(11) Solve for $\delta\bar{\mathbf{U}}_n^{(r)}$ and $\delta\hat{\mathbf{U}}_n$ from the two sets of equations in Eqs. (A2.4.17a) and (A2.4.17b); for the modified Newton method Eq. (A2.4.17b) need not be resolved.

(12) Compute the incremental load parameter $\delta\lambda[= \delta\lambda_n^{(r)}]$ from the following quadratic equation:

$$a_1(\delta\lambda)^2 + 2a_2\,\delta\lambda + a_3 = 0$$

where

$$a_1 = \delta\hat{\mathbf{U}}_n^{\mathrm{T}}\delta\hat{\mathbf{U}}_n$$
$$a_2 = (\delta\bar{\mathbf{U}}_n^{(r)} + \Delta\mathbf{U}_n^{(r-1)})^{\mathrm{T}}\delta\hat{\mathbf{U}}_n$$
$$a_3 = (\delta\bar{\mathbf{U}}_n^{(r)} + \Delta\mathbf{U}_n^{(r-1)})^{\mathrm{T}}(\delta\bar{\mathbf{U}}^{(r)} + \Delta\mathbf{U}_n^{(r-1)}) - (\Delta s)_n^2$$

and Δs is the arc length of the current load step. Two solutions $\delta\lambda_1$ and $\delta\lambda_2$ of this quadratic equation are used to compute two corresponding vectors $\Delta\mathbf{U}_{n1}^{(r)}$ and $\Delta\mathbf{U}_{n2}^{(r)}$. The $\delta\lambda$ that gives positive value to the product $\Delta\mathbf{U}_n^{(r-1)}\Delta\mathbf{U}_n^{(r)}$ is selected. If both $\delta\lambda_1$ and $\delta\lambda_2$ give positive values of the product, we use the one giving the smallest value of $(-a_3/a_2)$.

(13) Compute the correction to the solution vector

$$\delta U_n^{(r)} = \delta \bar{U}_n^{(r)} + \delta \lambda \, \delta \hat{U}_n$$

and update the incremental solution vector, the total solution vector, and the load parameter:

$$\begin{aligned}
\Delta U_n^{(r)} &= \Delta U_n^{(r-1)} + \delta U_n^{(r)} \\
U_n &= U_{n-1} + \Delta U_n^{(r)} \\
\lambda_n^{(r)} &= \lambda_n^{(r-1)} + \delta \lambda_n^{(r)}
\end{aligned} \qquad \text{(A2.4.27)}$$

(14) Repeat steps 9–13 until the following convergence criterion is satisfied:

$$\left[\frac{(U_n^{(r)} - U_n^{(r-1)})^{\mathrm{T}}(U_n^{(r)} - U_n^{(r-1)})}{(U_n^{(r)})^{\mathrm{T}} U_n^{(r)}} \right]^{\frac{1}{2}} < \epsilon \qquad \text{(A2.4.28)}$$

(15) Adjust the arc length for the subsequent load steps by [see Eq. (A2.4.12)]
$\Delta s_n = \Delta s_{n-1}(I_d/I_0)$.

(16) Start a new load step by returning to Step 1.

References

The references cited in the book are listed here by chapter. Some references are cited in multiple chapters but they are not listed again under those chapters.

CHAPTER 1

1. J. N. Reddy, *An Introduction to Continuum Mechanics*, 2nd ed., Cambridge University Press, New York (2013).
2. J. N. Reddy, *An Introduction to the Finite Element Method*, 3rd ed., McGraw–Hill, New York (2006).
3. J. N. Reddy, *Energy Principles and Variational Methods in Applied Mechanics*, 2nd ed., John Wiley & Sons, New York (2002).
4. K. J. Bathe, *Finite Element Procedures*, Prentice–Hall, Englewood Cliffs, NJ (2006).
5. T. Belytschko, W. K. Liu, B. Moran, and K. I. Elkhodary, *Nonlinear Finite Elements for Continua and Structures*, 2nd ed., John Wiley, Chichester, UK (2014).
6. C. A. Crisfield, *Non-Linear Finite Element Analysis of Solids and Structures, Vol. 1: Essentials*, John Wiley, Chichester, UK (1991).
7. C. A. Crisfield, *Non-Linear Finite Element Analysis of Solids and Structures, Vol. 2: Advanced Topics*, John Wiley, Chichester, UK (1997).
8. E. Hinton (Ed.), *NAFEMS Introduction to Nonlinear Finite Element Analysis*, NAFEMS, Glasgow, UK (1992).
9. D. R. J. Owen and E. Hinton, *Finite Elements in Plasticity: Theory and Practice*, Pineridge Press, Swansea, UK (1991).
10. J. N. Reddy and D. K. Gartling, *The Finite Element Method in Heat Transfer and Fluid Dynamics*, 3rd ed., CRC Press, Boca Raton, FL (2010).
11. P. Wriggers, *Nonlinear Finite Element Methods*, Springer–Verlag, Berlin (2008).
12. O. C. Zienkiewicz and R. L. Taylor, *The Finite Element Method*, 4th ed., *Vol. 1: Basic Formulation and Linear Problems*, *Vol. 2: Solid and Fluid Mechanics, Dynamics and Non-linearity*, McGraw–Hill, London, UK (1989); other editions and versions have appeared since 1989.
13. J. N. Reddy and M. L. Rasmussen, *Advanced Engineering Analysis*, John Wiley & Sons, New York (1982); reprinted by Krieger, Malabar, FL (1991).
14. A. W. Naylor and G. R. Sell, *Linear Operator Theory in Engineering and Science*, Holt, Rinehart, and Winston, New York (1971).
15. J. N. Reddy, *Applied Functional Analysis and Variational Methods in Engineering*, McGraw–Hill, New York (1984); reprinted by Krieger Publishing, Malabar, Florida (1991).
16. J. N. Reddy, *Lecture Notes on Applied Mathematical Analysis Methods for Engineers*, Department of Mechanical Engineering, Texas A&M University, College Station, Texas (2008).
17. R. G. Sargent, "An overview of verification and validation of simulation models," *Proceedings of the 1987 Winter Simulation Conference*, Society of Computer Simulation (1987).
18. I. Babuška and T. Strouboulis, *The Finite Element Method and its Reliability*, Oxford University Press, Oxford, UK (2001).

19. I. Babuška and J. T. Oden, "Verification and validation in computational engineering and science: basic concepts," *Computer Methods in Applied Mechanics and Engineering,* **193**, 4057–4066 (2004).

20. *Guide for Verification and Validation in Computational Solid Mechanics,* ASME V&V 10-2006, American Society of Mechanical Engineers, New York (2006).

21. B. Szabó and I. Babuška, *Introduction to Finite Element Analysis: Formulation, Verification and Validation,* John Wiley & Sons, New York (2011).

CHAPTER 2

22. G. A. Holzapfel, *Nonlinear Solid Mechanics,* John Wiley & Sons, New York (2000).

23. J. Bonet and R. D. Wood, *Nonlinear Continuum Mechanics for Finite Element Analysis*, 2nd ed., Cambridge University Press, New York (2008).

24. L. E. Malvern, *Introduction to the Mechanics of a Continuous Medium,* Prentice–Hall, Englewood Cliffs, NJ (1969).

CHAPTER 3

25. J. T. Oden and G. F. Carey, *Finite Elements, Mathematical Aspects*, Vol. IV, Prentice–Hall, Englewood Cliffs, NJ (1983).

26. F. Brezzi and M. Fortin, *Mixed and Hybrid Finite Element Methods,* Springer–Verlag, Berlin (1991).

27. F. Brezzi and K. J. Bathe, "The inf-sup condition, equivalent forms and applications," in *Reliability of Methods for Engineering Analysis*, K. J. Bathe and D. R. J. Owen (Eds.), Pineridge Press, Swansea, U.K. (1986).

28. T. Belytschko, Y. Krongauz, D. Organ, M. Fleming, and P. Krysl, "Meshless Methods: An overview and recent developments," *Computer Methods in Applied Mechanics and Engineering,* **139**, 3–47 (1996).

29. C. A. Duarte and J. T. Oden, "*H-p* clouds – an *h-p* meshless method," *Numerical Methods for Partial Differential Equations,* **12**, 673–705 (1996).

30. I. Babuska and M. Melenk, "The partition of unity method," *International Journal for Numerical Methods in Engineering,* **40**, 727–758 (1997).

31. S. N. Atluri and S. Shen, *The Meshless Local Petrov–Galerkin (MLPG) Method,* Tech Science Press, Encino, CA (2002).

32. S. N. Atluri, *The Meshless Method (MLPG) for Domain & BIE Discretization,* Tech Science Press, Encino, CA (2004).

33. T. Belytschko and J. S. Chen, *Meshfree and Particle Methods,* John Wiley & Sons, New York (2007).

34. A. Huerta, T. Belytschko, S. Fernández-Méndez, and T. Rabczuk, "Meshfree methods", *Encyclopedia of Computational Mechanics,* **1**(10), 279–309 (2004).

35. S. Li and W. K. Liu, *Meshfree Particle Methods,* Springer–Verlag, Berlin (2004).

36. G. R. Liu and Y. T. Gu, *An Introduction to Meshfree Methods and Their Programming,* Springer–Verlag, AA Dordrecht, The Netherlands (2005).

37. G. R. Liu, *Meshfree Methods. Moving Beyond the Finite Element Method,* 2nd ed., CRC Press, Boca Raton, FL (2010).

38. B.-N. Jiang, *The Least-Squares Finite Element Method. Theory and Applications in Computational Fluid Dynamics and Electromagnetics,* Springer–Verlag, New York (1998).

39. P. B. Bochev and M. D. Gunzburger, *Least-Squares Finite Element Methods*, Springer–Verlag, New York (2009).

40. K. S. Surana, A. R. Ahmadi, and J. N. Reddy, "The k-version of finite element method for non-self-adjoint operators in bvp," *International Journal of Computational Engineering Science*, 4(4), 737–812 (2003).

41. K. S. Surana, A. R. Ahmadi, and J. N. Reddy, "The k-version of finite element method for nonlinear operators in BVP," *International Journal of Computational Engineering Science,* 5(1), 133–207 (2004).

42. K. S. Surana, S. Allu, P. W. TenPas, and J. N. Reddy, "k-Version finite element method in gas dynamics: higher order global differentiability numerical solutions," *International Journal for Numerical Methods in Engineering,* 69(6), 1109–1157 (2007).

43. J. P. Pontaza and J. N. Reddy, "Spectral/hp least-squares finite element formulation for the Navier–Stokes equations," *Journal of Computational Physics,* 190(2), 523–549 (2003).

44. J. P. Pontaza and J. N. Reddy, "Least-squares finite element formulations for viscous compressible and incompressible fluid flows," *Computer Methods in Applied Mechanics and Engineering,* 195, 2454–2494 (2006).

45. V. Prabhakar and J. N. Reddy, "Spectral/hp penalty least-squares finite element formulation for the steady incompressible Navier–Stokes equations," *Journal of Computational Physics,* 215(1), 274–297 (2006).

46. T. J. R. Hughes and A. N. Brooks, "A multidimensional upwind scheme with no cross-wind diffusion," in *Finite Element Methods for Convection Dominated Flows,* T. J. R. Hughes (Ed.), ASME, AMD 34, NY, 19–35 (1979).

47. A. N. Brooks and T. J. R. Hughes, "Streamline-upwind/Petrov–Galerkin formulations for convection dominated flows with particular emphasis on incompressible Navier–Stokes equation," *Computer Methods in Applied Mechanics and Engineering,* 32, 199–259 (1982).

48. T. J. R. Hughes, "Multiscale phenomena: Green's functions, the Dirichlet-to-Neumann formulation, subgrid scale models, bubbles and the origin of stabilized methods," *Computer Methods in Applied Mechanics and Engineering,* 127, 387–401 (1995).

49. E. D. Eason, "A Review of Least-Squares Methods for Solving Partial Differential Equations", *International Journal for Numerical Methods in Engineering,* 10(5), 1021–1046 (1976).

50. G. S. Payette and J. N. Reddy, "On the roles of minimization and linearization in least squares finite element models of nonlinear boundary-value problems," *Journal of Computational Physics,* 230(9), 3589–3613 (2011).

CHAPTER 4

51. P. L. Sachdev, *A Compendium on Nonlinear Ordinary Differential Equations,* John Wiley, New York (1997).

52. M. Kubicek and V. Hlavacek, *Numerical Solution of Nonlinear Boundary Value Problems with Applications,* Prentice–Hall, Englewood Cliffs, NJ (1983).

CHAPTER 5

53. J. N. Reddy, "Microstructure-dependent couple stress theories of functionally graded beams," *Journal of the Mechanics and Physics of Solids,* 59, 2382–2399 (2011).

CHAPTER 6

54. M. N. Özisik, *Finite Difference Methods in Heat Transfer,* CRC Press, Boca Raton, FL (1994).

CHAPTER 7

55. J. N. Reddy, *Mechanics of Laminated Plates and Shells, Theory and Analysis*, 2nd ed., CRC Press, Boca Raton, FL (2004).

56. J. N. Reddy, *Theory and Analysis of Elastic Plates and Shells*, 2nd ed., CRC Press, Boca Raton, FL (2007).

57. R. J. Melosh, "Basis of derivation of matrices for the direct stiffness method," *AIAA Journal*, **1**, 1631–1637 (1963).

58. O. C. Zienkiewicz and Y. K. Cheung, "The finite element method for analysis of elastic isotropic and orthotropic slabs," *Proceedings of the Institute of Civil Engineers*, London, **28**, 471–488 (1964).

59. F. K. Bogner, R. L. Fox, and L. A. Schmidt, Jr., "The generation of interelement-compatible stiffness and mass matrices by the use of interpolation formulas," *Proceedings of the Conference on Matrix Methods in Structural Mechanics*, Air Force Institute of Technology, Wright–Patterson Air Force Base, OH, 397–443 (1965).

60. J. M. Gere and W. Weaver, Jr., *Analysis of Framed Structures*, D. von Nostrand, New York (1965).

61. B. Fraeijis de Veubeke, "A conforming finite element for plate bending," *International Journal of Solids and Structures*, **4**(1), 95–108 (1968).

62. J. Barlow, "Optimal stress location in finite element models," *International Journal for Numerical Methods in Engineering*, **10**, 243–251 (1976).

63. J. Barlow, "More on optimal stress points – reduced integration element distortions and error estimation," *International Journal for Numerical Methods in Engineering*, **28**, 1486–1504 (1989).

64. S. Lévy, "Bending of rectangular plates with large deflections," Report No. 737, NACA (1942).

65. C. T. Wang, "Bending of rectangular plates with large deflections," Report No. 1462, NACA (1948).

66. T. Kawai and N. Yoshimura, "Analysis of large deflection of plates by the finite element method," *International Journal for Numerical Methods in Engineering*, **1**, 123–133 (1969).

67. C. W. Bert, S. K. Jang, and A. G. Striz, "Nonlinear bending analysis of orthotropic rectangular plates by the method of differential quadrature," *Computational Mechanics*, **5**, 217–226 (1989).

68. S. Lévy, "Square Plate with Clamped Edges Under Pressure Producing Large Deflections," Tech. Note 847, NACA (1942).

69. A. Pica, R. D. Wood, and E. Hinton, "Finite element analysis of geometrically nonlinear plate behaviour using a Mindlin formulation," *Computers and Structures*, **11**, 203–215 (1980).

70. J. N. Reddy, "Analysis of layered composite plates accounting for large deflections and transverse shear strains," in *Recent Advances in Non-Linear Computational Mechanics*, E. Hinton, D. R. J. Owen, and C. Taylor (Eds.), Ch. 6, pp. 155–202, Pineridge Press, Swansea, UK (1982).

71. J. N. Reddy, "Simple finite elements with relaxed continuity for non-linear analysis of plates," *Proceedings of the Third International Conference in Australia on Finite Element Methods*, A. P. Kabaila and V. A. Pulmano (Eds.), University of New South Wales, Kensington, New South Wales, Australia (1979).

72. J. N. Reddy, "A penalty plate-bending element for the analysis of laminated anisotropic composite plates," *International Journal for Numerical Methods in Engineering*, **15**(8), 1187–1206 (1980).

73. H. C. Huang and E. Hinton, "A nine node Lagrangian Mindlin plate element with enhanced shear interpolation," *Engineering Computations*, **1**(4), 369–379 (1984).

74. R. C. Averill and J. N. Reddy, "Behavior of plate elements based on the first-order shear deformation theory," *Engineering Computations*, **7**, 57–74 (1990).

75. O. C. Zienkiewicz, J. J. M. Too, and R. L. Taylor, "Reduced integration technique in general analysis of plates and shells," *International Journal for Numerical Methods in Engineering*, **3**, 275–290 (1971).

76. T. J. R. Hughes, M. Cohen, and M. Haroun, "Reduced and selective integration techniques in the finite element analysis of plates," *Nuclear Engineering and Design*, **46**, 203–222 (1981).

77. T. Belytschko, C. S. Tsay, and W. K. Liu, "Stabilization matrix for the bilinear Mindlin plate element," *Computer Methods in Applied Mechanics and Engineering*, **29**, 313–327 (1981).

78. K. J. Bathe and E. N. Dvorkin, "A four-node plate bending element based on Mindlin/ Reissner plate theory and mixed interpolation," *International Journal for Numerical Methods in Engineering*, **21**, 367–383 (1985).

79. J. N. Reddy and W. C. Chao, "A comparison of closed-form and finite element solutions of thick laminated anisotropic rectangular plates," *Nuclear Engineering and Design*, **64**, 153–167 (1981).

80. J. N. Reddy, "On mixed finite-element formulations of a higher-order theory of composite laminates," *Finite Element Methods for Plate and Shell Structures*, T. J. R. Hughes and E. Hinton (Eds.), pp. 31–57, Pineridge Press, Swnasea, UK (1986).

CHAPTER 8

81. S. A. Ambartsumyan, "Calculation of laminated anisotropic shells," *Izvestiia Akademiia Nauk Armenskoi SSR, Ser. Fiz. Mat. Est. Tekh. Nauk.*, **6**(3), p. 15 (1953).

82. S. A. Ambartsumyan, *Theory of Anisotropic Shells*, NASA Report TT F-118 (1964).

83. S. A. Ambartsumyan, *Theory of Anisotropic Shells*, Moscow, 1961; English translation, NASA-TT-F-118 (1964).

84. W. Flügge, *Stresses in Shells,* Springer, Berlin (1960).

85. H. Kraus, *Thin Elastic Shells,* John Wiley, New York (1967).

86. S. Timoshenko and S. Woinowsky-Krieger, *Theory of Plates and Shells*, McGraw–Hill, New York (1959).

87. A. W. Leissa, *Vibration of Shells*, NASA SP–288, Washington, D.C. (1973).

88. C. L. Dym, *Introduction to the Theory of Shells,* Pergamon, New York (1974).

89. A. C. Ugural, *Stresses in Beams, Plates, and Shells*, 3rd ed., CRC Press, Boca Raton, FL (2009).

90. M. Farshad, *Design and Analysis of Shell Structures*, Kluwer Academic Publishers, Dordrecht, The Netherlands (1992).

91. J. L. Sanders, Jr., "Nonlinear theories for thin shells," *Quarterly Journal of Applied Mathematics*, **21**(1), 21–36 (1963).

92. B. Budiansky and J. L. Sanders, "On the 'best' first order linear shell theory," *Progress in Applied Mechanics, The Prager Anniversary Volume,* Macmillan, New York, pp. 129–140 (1963).

93. A. N. Palazotto and S. T. Dennis, *Nonlinear Analysis of Shell Structures*, AIAA Education Series, Washington, DC (1992).

94. V. Z. Vlasov, *General Theory of Shells and Its Applications in Engineering*, (Translation of *Obshchaya teoriya obolocheck i yeye prilozheniya v tekhnike*), NASA TT F-99, National Aeronautics and Space Administration, Washington, DC (1964).

95. W. T. Koiter, "Foundations and basic equations of shell theory. A survey of recent progress," *Theory of Shells,* F. I. Niordson (Ed.), IUTAM Symposium, Copenhagen, 93–105 (1967).

96. P. M. Naghdi, "Foundations of elastic shell theory," in *Progress in Solid Mechanics,* volume IV, I. N. Sneddon and R. Hill (Eds.), North–Holland, Amsterdam, Netherlands (1963).

97. P. M. Naghdi, "The theory of shells and plates," in *Principles of Classical Mechanics and Field Theory,* volume VI, S. Flügge (Ed.), *The Encyclopedia of Physics,* Part A2, pp. 425–640, Springer–Verlag, Berlin (1972).

98. L. Librescu, *Elastostatics and Kinetics of Anisotropic and Heterogeneous Shell-Type Structures,* Noordhoff, Leyden, The Netherlands (1975).

99. R. Schmidt and J. N. Reddy, "A refined small strain and moderate rotation theory of elastic anisotropic shells," *Journal of Applied Mechanics,* **55**, 611–617 (1988).

100. A. F. Palmerio, and J. N. Reddy, and R. Schmidt, "On a moderate rotation theory of laminated anisotropic shells, Part 1: Theory," *International Journal of Non-Linear Mechanics,* **25**, 687–700 (1990).

101. A. F. Palmerio, J. N. Reddy, and R. Schmidt, "On a moderate rotation theory of laminated anisotropic shells, Part 2: Finite element analysis," *International Journal of Non-Linear Mechanics,* **25**, 701–714 (1990).

102. I. Kreja, R. Schmidt, and J. N. Reddy, "Finite elements based on a first-order shear deformation moderate rotation theory with applications to the analysis of composite structures," *International Journal of Non-Linear Mechanics,* **32**(6), 1123–1142 (1997).

103. J. C. Simo and D. D. Fox, "On a stress resultant geometrically exact shell model. Part I: Formulation and optimal parametrization," *Computer Methods in Applied Mechanics and Engineering,* **72**, 267–304 (1989).

104. J. C. Simo, D. D. Fox, and M. S. Rifai, "On a stress resultant geometrically exact shell model. Part II: the linear theory," *Computer Methods in Applied Mechanics and Engineering,* **73**, 53–92 (1989).

105. J. C. Simo, D. D. Fox, and M. S. Rifai, "On a stress resultant geometrically exact shell model. Part III: Computational aspects of the nonlinear theory," *Computer Methods in Applied Mechanics and Engineering,* **79**, 21–70 (1990).

106. J. C. Simo, M. S. Rifai, and D. D. Fox, "On a stress resultant geometrically exact shell model. Part IV: Variable thickness shells with through-the-thickness stretching," *Computer Methods in Applied Mechanics and Engineering,* **81**, 53–91 (1990).

107. J. C. Simo and M. S. Rifai, "A class of mixed assumed strain methods and the method of incompatible modes," *International Journal for Numerical Methods in Engineering,* **29**, 1595–1638 (1990).

108. W. Pietraszkiewicz, *Finite Rotations and Lagrangian Description in the Nonlinear Theory of Shells,* Polish Scientific Publishers, Warsaw, Poland (1979).

109. W. Pietraszkiewicz and J. Badur, "Finite rotations in the description of continuum deformation," *International Journal of Engineering Science,* **21**, 1097–1115 (1983).

110. W. Pietraszkiewicz, "Lagrangian description and incremental formulation in the nonlinear theory of thin shells," *International Journal of Non-Linear Mechanics,* **19**, 115–140 (1984).

111. Y. Başar, "A consistent theory of geometrically nonlinear shells with an independent rotation vector," *International Journal of Solids and Structures,* **23**, 1401–1415 (1987).

112. Y. Başar, "Finite-rotation theories for composite laminates," *Acta Mechanica,* **98**, 159–176 (1983).

113. Y. Başar, Y. Ding, and R. Schultz, "Refined shear-deformation models for composite laminates with finite rotations," *International Journal of Solids and Structures,* **30** 2611–2638 (1993).

114. M. Balah and H. N. Al-Ghamedy, "Finite element formulation of a third-order laminated finite rotation shell element," *Computers and Structures,* **80**, 1975–1990 (2002).

115. K. P. Rao, "A rectangular laminated anisotropic shallow thin shell finite element," *Computer Methods in Applied Mechanics and Engineering,* **15**, 13–33 (1978).

116. C. Brebbia and J. Connor, "Geometrically nonlinear finite element analysis," *Journal of Engineering Mechanics,* 463–483 (1969).

117. A. C. Scordelis and K. S. Lo, "Computer analysis of cylindrical shells," *Journal of the American Concrete Institute,* **61**(5), 539–562 (1964).

118. G. Cantin and R. W. Clough, "A curved, cylindrical-shell, finite element," *AIAA Journal,* **6**(6), 1057–1062 (1968).

119. O. C. Zienkiewicz, *The Finite Element Method,* McGraw–Hill, New York (1977).

120. M. Cho and H. Y. Roh, "Development of geometrically exact new elements based on general curvilinear coordinates," *International Journal for Numerical Methods in Engineering,* **56**, 81–115 (2003).

121. A. B. Sabir and A. C. Lock, "The applications of finite elements to large deflection geometrically nonlinear behaviour of cylindrical shells," *Variational Methods in Engineering, Vol. II,* pp. 7/66–7/75, Southampton University Press, Southampton, UK (1972).

CHAPTER 9

122. S. Ahmad, B. M. Irons, and O. C. Zienkiewicz, "Analysis of thick and thin shell structures by curved finite elements," *International Journal for Numerical Methods in Engineering,* **2**(3), 419–451 (1970).

123. G. Horrigmoe and P. G. Bergan, "Incremental variational principle and finite element models for nonlinear problems," *Computer Methods in Applied Mechanics and Engineering,* **7**, 201–217 (1976).

124. P. G. Bergan and T. Soreide, "Solution of large displacement and instability problems using the current stiffness parameter," in *Finite Elements in Non-Linear Mechanics,* P. G. Bergan *et al.* (Eds.), pp. 647–669, Tapir Publishers, Trondheim, Norway (1978).

125. W. Wunderlich, "Incremental formulations for geometrically nonlinear problems," in *Formulations and Algorithm in Finite Element Analysis,* K. J. Bathe, J. T. Oden, and W. Wunderlich (Eds.), pp. 193–239, MIT Press, Boston, MA (1977).

126. E. Ramm, "A plate/shell element for large deflections and rotations," in *Formulations and Computational Algorithms in Finite Element Analysis,* K. J. Bathe, J. T. Oden, and W. Wunderlich (Eds.), MIT Press, Boston, MA (1977).

127. K. J. Bathe and S. Bolourchi, "A geometric and material nonlinear plate and shell element," *Computers and Structures,* **11**, 23–48 (1980).

128. K. J. Bathe and L. W. Ho, "A simple and effective element for analysis of general shell structures," *Computers and Structures,* **13**, 673–681 (1981).

129. T. Y. Chang and K. Sawamiphakdi, "Large deformation analysis of laminated shells by finite element method," *Computers and Structures,* **13**, 331–340 (1981).

130. W. Kanok-Nukulchai, R. L. Taylor, and T. J. R. Hughes, "A large deformation formulation for shell analysis by the finite element method," *Computers and Structures,* **13**, 19–27 (1981).

131. K. S. Surana, "Geometrically nonlinear formulation for the three dimensional solid-shell transition finite elements," *Computers and Structures,* **15**, 549–566 (1982).

132. W. C. Chao and J. N. Reddy, "Analysis of laminated composite shells using a degenerated 3-D element," *International Journal for Numerical Methods in Engineering,* **20**, 1991–2007 (1984).

133. E. Dvorkin and K. J. Bathe, "A continuum mechanics based four-node shell element for general nonlinear analysis," *Engineering Computations*, **1**, 77–88 (1984).

134. G. M. Stanley and C. A. Felippa, "Computational procedures for postbuckling for composite shells," in *Finite Element Methods for Nonlinear Problems*, P. G. Bergan, K. J. Bathe, and W. Wunderlich (Eds.), pp. 359–385, Springer–Verlag, Berlin (1986).

135. C. L. Liao, J. N. Reddy, and S. P. Engelstad, "A solid-shell transition element for geometrically nonlinear analysis of laminated composite structures," *International Journal for Numerical Methods in Engineering*, **26**, 1843–1854 (1988).

136. C. L. Liao and J. N. Reddy, "A continuum-based stiffened composite shell element for geometrically nonlinear analysis," *AIAA Journal*, **27**(1), 95–101 (1989).

137. C. L. Liao and J. N. Reddy, "Analysis of anisotropic, stiffened composite laminates using a continuum shell element," *Computers and Structures*, **34**(6), 805–815 (1990).

138. N. Büchter and E. Ramm, "Shell theory versus degeneration – a comparison in large rotation finite element analysis," *International Journal for Numerical Methods in Engineering*, **34**(1), 39–59 (1992).

139. E. Hinton and H. C. Huang, "A family of quadrilateral Mindlin plate elements with substitute shear strain fields," *Computers and Structures*, **23**(3), 409–431 (1986).

140. L. Leino and J. Pitkäranta, "On the membrane locking of h-p finite elements in a cylindrical shell problem," *International Journal for Numerical Methods in Engineering*, **37**, 1053–1070 (1994).

141. J. Pitkäranta, Y. Leino, O. Ovaskainen, and J. Piila, "Shell deformation states and the finite element method: A benchmark study of cylindrical shells," *Computer Methods in Applied Mechanics and Engineering*, **128**, 81–121 (1995).

142. H. Hakula, Y. Leino, and J. Pitkäranta, "Scale resolution, locking, and high-order finite element modelling shells," *Computer Methods in Applied Mechanics and Engineering*, **133**, 157–182 (1996).

143. J. P. Pontaza and J. N. Reddy, "Mixed plate bending elements based on least-squares formulation," *International Journal for Numerical Methods in Engineering*, **60**(5), 891–922 (2004).

144. J. P. Pontaza and J. N. Reddy, "Least-square finite element formulation for shear deformable shells," *Computer Methods in Applied Mechanics and Engineering*, **194**, 2464–2493 (2005).

145. R. A. Arciniega, *On a tensor-based finite element model for the analysis of shell structures*, Ph.D. Dissertation, Department of Mechanical Engineering, Texas A&M University, College Station, Texas, USA, December 2005.

146. R. A. Arciniega and J. N. Reddy, "Tensor-based finite element formulation for geometrically nonlinear analysis of shell structures," *Computer Methods in Applied Mechanics and Engineering*, **196**(4-6), 1048–1073 (2007).

147. R. A. Arciniega and J. N. Reddy, "Large deformation analysis of functionally graded shells," *International Journal of Solids and Structures*, **44**(6), 2036–2052 (2007).

148. G. A. Wempner, "Discrete approximations related to nonlinear theories of solids," *International Journal of Solids and Structures*, **7**, 1581–1599 (1971).

149. E. Riks, "The application of Newton's method to the problem of elastic stability," *Journal of Applied Mechanics*, **39**, 1060–1066 (1972).

150. E. Riks, "An incremental approach to the solution of snapping and buckling problems," *International Journal for Numerical Methods in Engineering*, **15**, 524–551 (1979).

151. M. A. Crisfield, "A fast incremental/iterative solution procedure that handles 'snap-through', " *Computers and Structures*, **13**(1–3), 55–62 (1981).

152. J. H. Kweon and C. S. Hong, "An improved arc-length method for postbuckling analysis of composite cylindrical panels," *Computers and Structures,* **53**(3), 541–549 (1994).

153. S. A. Zaghloul and J. B. Kennedy, "Nonlinear behavior of symmetrically laminated plates," *Journal of Applied Mechanics,* **42**(1), 234–236 (1975).

154. N. S. Putcha and J. N. Reddy, "A Refined Mixed Shear Flexible Finite Element for the Nonlinear Analysis of Laminated Plates," *Computers and Structures,* **22**, 529–538 (1986).

155. O. C. Zienkiewicz, R. L. Taylor, and J. M. Too, "Reduced integration techniques in general analysis of plates and shells," *International Journal for Numerical Methods in Engineering,* **3**, 275–290 (1971).

156. D. Chapelle and K. J. Bathe, *The Finite Element Analysis of Shells – Fundamentals,* Springer–Verlag, New York (2003).

157. P. Betsch, A. Menzel, and E. Stein, "On the parametrization of finite rotations in computational mechanics: A classification of concepts with application to smooth shells," *Computational Methods in Applied Mechanics and Engineering,* **155**, 273–305 (1998).

158. M. Bischoff and E. Ramm, "Shear deformable shell elements for large strains and rotations," *International Journal for Numerical Methods in Engineering,* **40**, 4427–4449 (1997).

159. M. Bischoff and E. Ramm, "On the physical significance of higher order kinematic and static variables in a three-dimensional shell formulation," *International Journal of Solids and Structures,* **37**, 6933–6960 (2000).

160. C. Sansour, "A theory and finite element formulation of shells at finite deformations involving thickness change: Circumventing the use of a rotation tensor," *Archives of Applied Mechanics,* **65**, 194–216 (1995).

161. G. S. Payette, "Spectral/hp finite element models for fluids and structures," Ph.D. Dissertation, Department of Mechanical Engineering, Texas A&M University, College Station, Texas, USA, May 2012.

162. G. S. Payette and J. N. Reddy, "A seven-parameter spectral/hp finite element formulation for isotropic, laminated composite and functionally graded shell structures," *Computer Methods in Applied Mechanics and Engineering,* **278**, 664–704 (2014).

163. G. E. Karniadakis and S. J. Sherwin, *Spectral/hp Element Methods for CFD,* Oxford University Press, New York (1999).

164. T. A. Davis and I. S. Duff, "An unsymmetric-pattern multifrontal method for sparse LU factorization," *SIAM Journal of Matrix Analysis and Applications,* **18**(1), 140–158 (1997).

165. T. A. Davis and I. S. Duff, "A combined unifrontal/multifrontal method for unsymmetric sparse matrices," *ACM Transactions on Mathematical Software,* **25**(1), 1–20 (1999).

166. T. A. Davis, "A column pre-ordering strategy for the unsymmetric-pattern multifrontal method," *ACM Transactions on Mathematical Software,* **30**(2), 165–195 (2004).

167. T. A. Davis, "Algorithm 832: UMFPACK V4.3 – an unsymmetric-pattern multifrontal method," *ACM Transactions on Mathematical Software,* **30**(2), 196–199 (2004).

168. H. Parisch, "Large displacements of shells including material nonlinearities," *Computer Methods in Applied Mechanics and Engineering,* **27**(2), 183–214 (1981).

169. A. F. Saleeb, T. Y. Chang, W. Graf, and S. Yingyeunyong, "Hybrid/mixed model for non-linear shell analysis and its applications to large-rotation problems," *International Journal for Numerical Methods in Engineering,* **29**(2), 407–446 (1990).

170. K. Y. Sze and S.-J. Zheng, "A stabilized hybrid-stress solid element for geometrically nonlinear homogeneous and laminated shell analyses," *Computer Methods in Applied Mechanics and Engineering,* **191**, 1945–1966 (2002).

171. K. Y. Sze, W. K. Chan, and T. H. H. Pian, "An eight-node hybrid-stress solid-shell

element for geometric non-linear analysis of elastic shells," *International Journal for Numerical Methods in Engineering*, **55**(7), 853–878 (2002).

172. K. Y. Sze, X. H. Liu, and S. H. Lo, "Popular benchmark problems for geometric nonlinear analysis of shells," *Finite Elements in Analysis in Design*, **40**(11), 1551–1569 (2004).

173. P. Massin and M. Al Mikdad, "Nine node and seven node thick shell elements with large displacements and rotations," *Computers and Structures*, **80**(9-10), 835–847 (2002).

174. S. Timoshenko and J. N. Goodier, *Theory of Elasticity*, 3rd ed., McGraw–Hill, New York (1970, 1987).

175. B. Brank, J. Korelc, and A. Ibrahimbegović, "Nonlinear shell problem formulation accounting for through-the-thickness stretching and its finite element implementation," *Computers and Structures*, **80**, 699–717 (2002).

176. A. Barut, E. Madenci, and A. Tessler, "Nonlinear analysis of laminates through a Mindlin-type shear deformable shallow shell element," *Computer Methods in Applied Mechanics and Engineering*, **143**(1-2), 155–173 (1997).

177. P. Mohan and R. K. Kapania, "Updated Lagrangian formulation of a flat triangular element for thin laminated shells," *AIAA Journal*, **36**(2), 273–281 (1998).

178. C. Sansour and F. G. Kollmann, "Families of 4-node and 9-node finite elements for a finite deformation shell theory. An assessment of hybrid stress, hybrid strain and enhanced strain elements," *Computational Mechanics*, **24**, 435–447 (2000).

179. L. Jiang and M. W. Chernuka, "A simple four-noded corotational shell element for arbitrarily large rotations," *Computers and Structures*, **53**(5), 1123–1132 (1994).

180. W. Wagner and F. Gruttmann, "A simple finite rotation formulation for composite shell elements," *Engineering Computations*, **11**(2), 145–176 (1994).

CHAPTER 10

181. P. M. Gresho and R. L. Sani, *Incompressible Flow and the Finite Element Method. Advection-Diffusion and Isothermal Laminar Flow*, John Wiley & Sons, Chichester, UK (1999).

182. J. N. Reddy, "On the accuracy and existence of solutions to primitive variable models of viscous incompressible fluids," *International Journal of Engineering Science*, **16**, 921–929 (1978).

183. J. N. Reddy, "On the finite element method with penalty for incompressible fluid flow problems," in *The Mathematics of Finite Elements and Applications III*, J. R. Whiteman (Ed.), Academic Press, New York (1979).

184. J. N. Reddy, "On penalty function methods in the finite element analysis of flow problems," *International Journal of Numerical Methods in Fluids*, **2**, 151–171 (1982).

185. M. P. Reddy and J. N. Reddy, "Finite-element analysis of flows of non-Newtonian fluids in three-dimensional enclosures," *International Journal of Non-Linear Mechanics*, **27**, 9–26 (1992).

186. R. S. Marshall, J. C. Heinrich, and O. C. Zienkiewicz, "Natural convection in a square enclosure by a finite element, penalty function method, using primitive fluid variables," *Numerical Heat Transfer*, **1**, 315–330 (1978).

187. T. J. R. Hughes, W. K. Liu, and A. Brooks, "Review of finite element analysis of incompressible viscous flows by penalty function formulation," *Journal of Computational Physics*, **30**, 1–60 (1979).

188. J. T. Oden, "RIP methods for Stokesian flows," in R. H. Gallagher, O. C. Zienkiewicz, J. T. Oden, and D. Norrie (Eds.), *Finite Element Method in Flow Problems*, Vol. IV, John Wiley, London, UK (1982).

189. T. J. R. Hughes, L. P. Franca, and M. Balestra, "A new finite element formulation for computational fluid dynamics. V. Circumventing the Babuska–Brezzi condition: A stable Petrov–Galerkin formulation for the Stokes problem accommodating equal-order interpolations," *Computer Methods in Applied Mechanics and Engineering*, **59**, 85–99 (1986).

190. F. Brezzi and M. Fortin, *Mixed and Hybrid Finite Element Methods*, Springer–Verlag, Berlin, Germany (1991).

191. P. Le Tallac and V. Ruas, "On the convergence of the bilinear-velocity constant-pressure finite element method in viscous flow," *Computer Methods in Applied Mechanics and Engineering*, **54**, 235–243 (1986).

192. R. Temam, *Theory and Numerical Analysis of the Navier–Stokes Equations*, North–Holland, Amsterdam, The Netherlands (1977).

193. M. Bercovier and M. Engelman, "A finite element for the numerical solution of viscous incompressible flows," *Journal of Computational Physics*, **30**, 181–201 (1979).

194. M. Engelman, R. L. Sani, P. M. Gresho, and M. Bercovier, "Consistent vs. reduced integration penalty methods for incompressible media using several old and new elements," *International Journal of Numerical Methods in Fluids*, **2**, 25–42 (1982).

195. E. M. Salonen and J. Aalto, "A pressure determination scheme," *Proceedings of the Fifth International Conference on Numerical Methods in Laminar and Turbulent Flow*, C. Taylor, M. D. Olson, P. M. Gresho, and W. G. Habashi (Eds.), Pineridge Press, Swansea, UK (1985).

196. T. Shiojima and Y. Shimazaki, "A pressure-smoothing scheme for incompressible flow problems," *International Journal of Numerical Methods in Fluids*, **9**, 557–567 (1989).

197. M. P. Reddy, J. N. Reddy, and H. U. Akay, "Penalty finite element analysis of incompressible flows using element by element solution algorithms," *Computer Methods in Applied Mechanics and Engineering*, **100**, 169–205 (1992).

198. A. Nadai, *Theory of Flow and Fracture of Solids*, McGraw–Hill, New York (1963).

199. H. Schlichting, *Boundary-Layer Theory* (translated by J. Kestin), 7th ed., McGraw–Hill, New York (1979).

200. O. R. Burggraf, "Analytical and numerical studies of the structure of steady separated flows," *Journal of Fluid Mechanics*, **24**(1), 113–151 (1966).

201. F. Pan and A. Acrivos, "Steady flow in rectangular cavities," *Journal of Fluid Mechanics*, **28**(4), 643–655 (1967).

202. U. Ghia, K. N. Ghia, and C. T. Shin, "High-resolution for incompressible flow using the Navier–Stokes equations and the multigrid method," *Journal of Computational Physics*, **48**, 387–411 (1982).

203. D. K. Gartling, "A test problem for outflow boundary conditions – flow over a backward facing step," *International Journal for Numerical Methods in Fluids*, **11**, 953–967 (1990).

204. J. L. White and A. B. Metzner, "Development of constitutive equations for polymeric melts and solutions," *Journal of Applied Polymer Science*, **7**, 1867–1889 (1963).

205. M. J. Crochet and K. Walters, "Numerical methods in non-Newtonian fluid mechanics," *Annual Review of Fluid Mechanics*, **15**, 241–260 (1983).

206. M. J. Crochet, A. R. Davies, and K. Walters, *Numerical Simulation of Non-Newtonian Flow*, Elsevier, Amsterdam, The Netherlands (1984).

207. D. K. Gartling and N. Phan Thien, "A numerical simulation of a plastic fluid in a parallel-plate plastometer," *Journal of Non-Newtonian Fluid Mechanics*, **14**, 347–360 (1984).

208. J. N. Reddy and V. A. Padhye, "A penalty finite element model for axisymmetric flows of non-Newtonian fluids," *Numerical Methods for Partial Differential Equations*, **4**, 33–56 (1988).

209. M. Iga and J. N. Reddy, "Penalty finite element analysis of free surface flows of power-law fluids," *International Journal of Non-Linear Mechanics*, **24**(5), 383–399 (1989).

210. M. P. Reddy and J. N. Reddy, "Finite element analysis of flows of non-Newtonian fluids in three-dimensional enclosures," *International Journal of Non-Linear Mechanics*, **27**(1), 9–26 (1992).

211. M. P. Reddy and J. N. Reddy, "Numerical simulation of forming processes using a coupled fluid flow and heat transfer model," *International Journal for Numerical Methods in Engineering*, **35**, 807–833 (1992).

212. C. J. Coleman, "A finite element routine for analyzing non-Newtonian fluids, Part 1," *Journal of Non-Newtonian Fluid Mechanics*, **7**, 289–301 (1980).

213. D. K. Gartling, "Convective heat transfer analysis by the finite element method," *Computer Methods in Applied Mechanics and Engineering*, **12**, 365–382 (1977).

214. R. S. Marshall, J. C. Heinrich, and O. C. Zienkiewicz, "Natural convection in a square enclosure by a finite element, penalty function method, using primitive fluid variables," *Numerical Heat Transfer*, **1**, 315–330 (1978).

215. J. N. Reddy and A. Satake, "A comparison of various finite element models of natural convection in enclosures," *Journal of Heat Transfer*, **102**, 659–666 (1980).

216. J. N. Reddy, "Penalty-finite-element analysis of 3-D Navier–Stokes equations," *Computer Methods in Applied Mechanics and Engineering*, **35**, 87–106 (1982).

217. G. de Vahl Davis and I. P. Jones, "Natural convection in a square cavity: a comparison exercise," *International Journal for Numerical Methods in Fluids*, **3**, 227–248 (1983).

218. M. Dhaubhadel, D. Telionis, and J. N. Reddy, "Finite element analysis of fluid flow and heat transfer for staggered bundles of cylinders in cross flow," *International Journal for Numerical Methods in Fluids*, **7**, 1325–1342 (1987).

219. D. H. Pelletier, J. N. Reddy, and J. A. Schetz, "Some recent developments and trends in finite element computational natural convection," in *Annual Review of Numerical Fluid Mechanics and Heat Transfer*, Vol. 2, C. L. Tien and T. C. Chawla (Eds.), Hemisphere, New York, pp. 39–85 (1987).

220. J. T. Oden and J. N. Reddy, *Variational Methods in Theoretical Mechanics*, 2nd ed., Springer–Verlag, New York (1983).

CHAPTER 11

221. J. H. Bramble and A. H. Schatz, "Least squares methods for 2*m*th order elliptic boundary-value problems," *Mathematics of Computation*, **25**(113), 1–32 (1971).

222. J. H. Bramble and J. E. Pasciak, "Least-squares methods for Stokes equations based on a discrete minus one inner product," *Journal of Computational and Applied Mathematics*, **74**(1,2), 155–173 (1996).

223. J. H. Bramble, R. D. Lazarov, and J. E. Pasciak, "Least-squares for second-order elliptic problems," *Computer Methods in Applied Mechanics and Engineering*, **152**(1,2), 195–210 (1998).

224. P. B. Bochev, "Analysis of least-squares finite element methods for the Navier–Stokes equations," *SIAM Journal of Numerical Analysis*, **34**, 1817–1844 (1997).

225. S. Agmon, A. Douglis, and L. Nirenberg, "Estimates near the boundary for solutions of elliptic partial differential equations satisfying general boundary conditions II," *Communications in Pure and Applied Mathematuics*, **17**(1), 35–92 (1964).

226. P. B. Bochev, Z. Cai, T. A. Manteuffel, and S. F. McCormick, "Analysis of velocity-flux first-order system least-squares principles for the Navier–Stokes equations, Part I," *SIAM Journal of Numerical Analysis*, **35**, 990–1009 (1998).

227. P. B. Bochev, Z. Cai, T. A. Manteuffel, and S. F. McCormick, "Analysis of velocity-flux first-order system least-squares principles for the Navier–Stokes equations, Part II," *SIAM Journal of Numerical Analysis,* **36**(4), 1125–1144 (1999).

228. B.-N. Jiang, "A least-squares finite element method for incompressible Navier–Stokes problems," *International Journal for Numerical Methods in Fluids,* **14**, 843–859 (1992).

229. B.-N. Jiang and V. Sonnad, "Least-squares solution of incompressible Navier–Stokes equations with the p-version of finite elements," *Computational Mechanics* **15**, 129–136 (1994).

230. B.-N. Jiang, T. L. Lin, and L. A. Povinelli, "Large-scale computation of incompressible viscous flow by the least-squares finite element method," *Computer Methods in Applied Mechanics and Engineering,* **114**, 213–231 (1994).

231. B. C. Bell and K. S. Surana, "A space-time coupled p-version least-squares finite element formulation for unsteady fluid dynamics problems," *International Journal for Numerical Methods in Engineering,* **37**, 3545–3569 (1994).

232. B. C. Bell and K. S. Surana, "A space-time coupled p-version least-squares finite element formulation for unsteady two-dimensional Navier–Stokes equations," *International Journal for Numerical Methods in Engineering,* **39**, 2593–2618 (1996).

233. M. M. J. Proot and M. I. Gerritsma, "A least-squares spectral element formulation for the Stokes problem," *Journal of Scientific Computing,* **17**(1-4), 285–296 (2002).

234. M. I. Gerritsma and M. M. J. Proot, "Analysis of a discontinuous least squares spectral element method," *Journal of Scientific Computing,* **17**(1-4), 297–306 (2002).

235. M. M. J. Proot and M. I. Gerritsma, "Least-squares spectral elements applied to the Stokes problem," *Journal of Computational Physics,* **181**, 454–477 (2002).

236. M. I. Gerritsma, "Direct minimization of the discontinuous least-squares spectral element method for viscoelastic fluids," *Journal of Scientific Computing,* **27** (1-3), 245–256 (2006).

237. B. De Maerschalck and M. I. Gerritsma, "Least-squares spectral element method for non-linear hyperbolic differential equations," *Journal of Computational and Applied Mathematics,* **215**(2), 357–367 (2008).

238. J. P. Pontaza and J. N. Reddy, "Space-time coupled spectral/hp least-squares finite element formulation for the incompressible Navier–Stokes equations," *Journal of Computational Physics,* **197**(2), 418–459 (2004).

239. V. Prabhakar and J. N. Reddy, "A stress-based least-squares finite-element model for incompressible Navier–Stokes equations," *International Journal for Numerical Methods in Fluids,* **54**(11), 1369–1385 (2007).

240. V. Prabhakar and J. N. Reddy, "Spectral/hp penalty least-squares finite element formulation for unsteady incompressible flows," *International Journal of Numerical Methods for Fluids,* **58**(3), 287–306 (2008).

241. V. Prabhakar, J. P. Pontaza, and J. N. Reddy, "A collocation penalty least-squares finite element formulation for incompressible flows," *Computer Methods in Applied Mechanics and Engineering,* **197**, 449–463 (2008).

242. I. Babuska, "The finite element method with Lagrange multipliers," *Numerische Mathematik,* **20**, 179–192 (1973).

243. F. Brezzi, "On the existence, uniqueness and approximation saddle-point problems arising from Lagrangian multipliers," *Revue Francaise d'Automatique Informatique Recherche Operationelle, Analyse Numérique,* **8**, 129–151 (1974).

244. G. F. Carey and B.-N. Jiang, "Least-squares finite element method and preconditioned conjugate gradient solution," *International Journal for Numerical Methods in Fluids,* **24**(7), 1283–1296 (1987).

245. G. F. Carey and B.-N. Jiang, "Nonlinear preconditioned conjugate gradient and least-squares finite elements," *Computer Methods in Applied Mechanics and Engineering,* **62**(2), 145–154 (1987).

246. J. M. Deang and M. D. Gunzburger, "Issues related to least-squares finite element methods for the Stokes equations," *SIAM Journal of Scientific Computing,* **20**(3), 878–906 (1998).

247. C. L. Chang and J. J. Nelson, "Least-squares finite element method for the Stokes problem with zero residual of mass conservation," *SIAM Journal of Numerical Analysis,* **34**(2), 480–489 (1997).

248. J. P. Pontaza, "Least-squares variational principles and the finite element method: theory, formulations, and models for solid and fluid mechanics," *Finite Elements in Analysis and Design,* **41**(7-8), 703–728 (2005).

249. J. J. Heys, E. Lee, T. A. Manteuffel, and S. F. McCormick, "An alternative least-squares formulation of the Navier–Stokes equations with improved mass conservation," *Journal of Computational Physics,* **226**(1), 994–1006 (2007).

250. J. P. Pontaza, "A least-squares finite element formulation for unsteady incompressible flows with improved velocity-pressure coupling," *Journal of Computational Physics,* **217**(2), 563–588 (2006).

251. D. J. Tritton, "Experiments on the flow past a circular cylinder at low Reynolds numbers," *Journal of Fluid Mechanics,* **6**, 547–567 (1959).

252. A. S. Grove, F. H. Shair, E. E. Petersen, and A. Acrivos, "An experimental investigation of the steady separated flow past a circular cylinder," *Journal of Fluid Mechanics,* **19**, 60–80 (1964).

253. M. Provansal, C. Mathis, and L. Boyer, "Bénard-von Kármán instability: Transient and forced regimes," *Journal of Fluid Mechanics,* **182**, 1–22 (1987).

254. M. Kawaguti and P. Jain, "Numerical study of a viscous fluid past a circular cylinder," *Journal of Physical Society of Japan,* **21**, 2055–2062 (1966).

255. S. C. R. Dennis and G. Z. Chang, "Numerical solutions for steady flow past a circular cylinder at Reynolds numbers up to 100," *Journal of Fluid Mechanics,* **42**, 471–489 (1970).

256. B. Fornberg, "A numerical study of steady viscous flow past a circular cylinder," *Journal of Fluid Mechanics,* **98**(4), 819–855 (1980).

257. C. P. Jackson, "A finite-element study of the onset of vortex shedding in flow past variously shaped bodies," *Journal of Fluid Mechanics,* **182**, 23–45 (1987).

258. B. F. Armaly, F. Durst, J. C. F. Pereira, and B. Schonung, "Experimental and theoretical investigation of backward-facing step flow," *Journal of Fluid Mechanics,* **127**, 473–496 (1983).

259. R. L. Sani and P. M. Gresho, "Résumé and remarks on the open boundary condition minisymposium," *International Journal for Numerical Methods in Fluids,* **18**(10), 983–1008 (1994).

260. R. W. Johnson, *The Handbook of Fluid Dynamics,* CRC Press, Boca Raton, FL (1988).

261. L. I. G. Kovasznay, "Laminar flow behind a two-dimensional grid," *Proceedings of the Cambridge Philosophical Society,* **44**, 58–62 (1948).

262. M. M. Gupta and J. C. Kalita, "A new paradigm for solving Navier–Stokes equations: streamfunction–velocity formulation," *Journal of Computational Physics,* **207**(1), 52–68 (2005).

263. N. Bodard, R. Bouffanais, and M. O. Deville, "Solution of moving-boundary problems by the spectral element method," *Applied Numerical Methods,* **58**, 968–984 (2008).

APPENDIX 1

264. G. F. Carey and J. T. Oden, *Finite Elements: Computational Aspects,* Prentice–Hall, Englewood Cliffs, NJ (1984).
265. P. Hood, "Frontal solution program for unsymmetric matrices," *International Journal for Numerical Methods in Engineering,* **10**, 379–399 (1976); also see **10**, 1055 (1976) for a correction.
266. K. E. Atkinson, *An Introduction to Numerical Analysis,* John Wiley, New York (1978).
267. B. Carnahan, H. A. Luther, and J. O. Wilkes, *Applied Numerical Methods,* John Wiley, New York (1969).
268. D. K. Faddeev, V. N. Faddeeva, and C. Robert, *Computational Methods of Linear Algebra,* W. H. Freeman & Co., San Francisco, CA (1963).
269. M. R. Hestenes and E. L. Stiefel, "Methods of conjugate gradients for solving linear systems," *National Bureau of Standards Journal of Research,* **49**, 409–436 (1952).
270. G. H. Golub and C. F. Van Loan, *Matrix Computations,* 2nd ed., The Johns Hopkins University Press, Baltimore, MD (1989).
271. K. C. Jea and D. M. Young, "On the simplification of generalized conjugate–gradient methods for nonsymmetrizable linear systems," *Linear Algebra Applications,* **52**, 399–417 (1983).
272. D. J. Evans, "Use of preconditioning in iterative methods for solving linear equations with symmetric positive definite matrices," *Computer Journal,* **4**, 73–78 (1961).
273. R. L. Fox and E. L. Stanton, "Developments in structural analysis by direct energy minimization," *AIAA Journal,* **6**, 1036–1042 (1968).
274. I. Fried, "A gradient computational procedure for the solution of large problems arising from the finite element discretization method," *International Journal for Numerical Methods in Engineering,* **2**, 477–494 (1970).
275. A. J. Wathen, "An analysis of some element-by-element techniques," *Computer Methods in Applied Mechanics and Engineering,* **74**, 271–287 (1989).
276. J. L. Hayes and P. Devloo, "An element-by-element block iterative method for large non-linear problems," in *Proceedings of the International Conference on Innovative Methods for Nonlinear Problems,* W. K. Liu, T. Belytschko, and K.C. Park (Eds.), Pineridge Press, Swansea, UK, pp. 51–62 (1984).
277. Y. Saad and M. H. Schultz, "GMRES: A generalized minimal residual algorithm for solving nonsymmetric linear systems," *SIAM Journal of Scientific and Statistical Computations,* **7**(3), 856–869 (1986).
278. J. A. Mitchell and J. N. Reddy, "A multilevel hierarchical preconditioner for thin elastic solids," *International Journal for Numerical Methods in Engineering,* **43**, 1383–1400 (1998).

APPENDIX 2

279. J. L. Batoz and G. Dhatt, "Incremental displacement algorithms for non-linear problems," *International Journal for Numerical Methods in Engineering,* **14**, 1262–1267 (1979).
280. P. G. Bergan, G. Horrigmoe, B. Krakeland, and T. H. Soreide, "Solution techniques for non-linear finite element problems," *International Journal for Numerical Methods in Engineering,* **12**, 1677–1696 (1978).
281. I. Newton, "De analysis per aequationes infinitas (1690)," in *The Mathematical Papers of Isaac Newton, Vol. 11 (1667–1670),* D. T. Whiteside (Ed.), Cambridge University Press, Cambridge, UK, 207–247 (1968).

282. J. Raphson, "Analysis aequationum universalis seu ad aequationes algebraicas resolvendas methodus generalis, & expedita, ex nova infinitarum serierum methodo, deducta ac demonstrata," London (1690) (original in British Library, London); published in English by RareBooksClub.com (2012).

283. F. Cajori, "Historical note on the Newton–Raphson method of approximation," *American Mathematical Monthly*, **18**(2), 29–32 (1911).

284. N. Bićanić and K. H. Johnson, "Who was '-Raphson'?," *International Journal for Numerical Methods in Engineering*, **14**(1), 148–152 (1979).

Index

A

Acceleration parameter, 182, 228, 239, 552, 587, 640, 650

Admissible variation, 85, 142, 486

α-family of approximation, 284, 294, 294, 587, 634

Alternating symbol, 18

Alternative direct iteration algorithm, 648

Analytical solution, 2–4, 40, 136, 237, 342, 360, 500, 537, 538, 554–558, 578

Angular:
    displacement, 2–5, 8–11, 54, 218
    momentum, 2, 70, 74, 88, 525
    motion, 8
    velocity, 166, 571

Anisotropic, 267

Approximation:
    finite element, 98, 102, 104, 118
    functions, 97, 100, 104, 114, 118, 122, 129, 141, 152, 190, 247, 267, 275, 371, 440, 532, 595, 639
    linear, 104, 109
    of the boundary data, 6
    of the derivative, 6
    quadratic, 105, 109
    Taylor's series, 6
    time,

Area coordinates, 126

Assembly of:
    coefficients, 198, 547, 640
    elements, 108, 130, 194, 495
    equations, 154, 179, 194, 269, 284, 545, 652
    matrices, 200, 637, 642, 644

Associative law, 13, 28, 29

Axisymmetric:
    bending, 166, 261, 263
    deformation, 166, 369
    flows, 572–574
    geometry, 570
    problems, 136–139, 271–273

B

Backward difference method, 285, 289, 300

Backward gradient, 22, 54

Balance of:
    angular momentum, 2, 70, 74
    energy, 70, 176
    linear momentum, 2, 3, 70, 72

Banach space, 31, 36

Basis vectors, 16–18, 21, 42, 49, 57, 475–480, 484, 571

Beams:
    Euler–Bernoulli, 167, 213–242, 311, 501
    finite element models of, 222, 245
    functionally graded, 256–260
    nonlinear bending of, 213–261
    stiffness coefficients of, 222–227, 246–250
    Timoshenko, 167, 242–255

Benchmark problem, 41, 406, 470, 474, 521, 573, 595, 620

Bending deformation, 248, 258, 311

Bilinear:
    form, 37–39, 103, 142, 535, 633
    functional, 102

Body force, 63, 73, 85, 110, 165, 205, 429, 435, 445, 460, 486, 525, 531, 541, 583, 593